Lecture Notes in Computer Science

Vol. 379: A. Kreczmar, G. Mirkowska (Eds.), Mathematical Foundations of Computer Science 1989. Proceedings, 1989. VIII, 605 pages. 1989.

Vol. 380: J. Csirik, J. Demetrovics, F. Gécseg (Eds.), Fundamentals of Computation Theory. Proceedings, 1989. XI, 493 pages. 1989.

Vol. 381: J. Dassow, J. Kelemen (Eds.), Machines, Languages, and Complexity. Proceedings, 1988. VI, 244 pages. 1989.

Vol. 382: F. Dehne, J.-R. Sack, N. Santoro (Eds.), Algorithms and Data Structures. WADS '89. Proceedings, 1989. IX, 592 pages. 1989.

Vol. 383: K. Furukawa, H. Tanaka, T. Fujisaki (Eds.), Logic Programming '88. Proceedings, 1988. VII, 251 pages. 1989 (Subseries LNAI).

Vol. 384: G.A. van Zee, J.G.G. van de Vorst (Eds.), Parallel Computing 1988. Proceedings, 1988. V, 135 pages. 1989.

Vol. 385: E. Börger, H. Kleine Büning, M.M. Richter (Eds.), CSL '88. Proceedings, 1988. VI, 399 pages. 1989.

Vol. 386: J.E. Pin (Ed.), Formal Properties of Finite Automata and Applications. Proceedings, 1988. VIII, 260 pages. 1989.

Vol. 387: C. Ghezzi, J. A. McDermid (Eds.), ESEC '89. 2nd European Software Engineering Conference. Proceedings, 1989. VI, 496 pages. 1989.

Vol. 388: G. Cohen, J. Wolfmann (Eds.), Coding Theory and Applications. Proceedings, 1988. IX, 329 pages. 1989.

Vol. 389: D.H. Pitt, D.E. Rydeheard, P. Dybjer, A.M. Pitts, A. Poigné (Eds.), Category Theory and Computer Science. Proceedings, 1989. VI, 365 pages. 1989.

Vol. 390: J.P. Martins, E.M. Morgado (Eds.), EPIA 89. Proceedings, 1989. XII, 400 pages. 1989 (Subseries LNAI).

Vol. 391: J.-D. Boissonnat, J.-P. Laumond (Eds.), Geometry and Robotics. Proceedings, 1988. VI, 413 pages. 1989.

Vol. 392: J.-C. Bermond, M. Raynal (Eds.), Distributed Algorithms. Proceedings, 1989. VI, 315 pages. 1989.

Vol. 393: H. Ehrig, H. Herrlich, H.-J. Kreowski, G. Preuß (Eds.), Categorical Methods in Computer Science. VI, 350 pages. 1989.

Vol. 394: M. Wirsing, J.A. Bergstra (Eds.), Algebraic Methods: Theory, Tools and Applications. VI, 558 pages. 1989.

Vol. 395: M. Schmidt-Schauß, Computational Aspects of an Order-Sorted Logic with Term Declarations. VIII, 171 pages. 1989 (Subseries LNAI).

Vol. 396: T.A. Berson, T. Beth (Eds.), Local Area Network Security. Proceedings, 1989. IX, 152 pages. 1989.

Vol. 397: K.P. Jantke (Ed.), Analogical and Inductive Inference. Proceedings, 1989. IX, 338 pages. 1989 (Subseries LNAI).

Vol. 398: B. Banieqbal, H. Barringer, A. Pnueli (Eds.), Temporal Logic in Specification. Proceedings, 1987. VI, 448 pages. 1989.

Vol. 399: V. Cantoni, R. Creutzburg, S. Levialdi, G. Wolf (Eds.), Recent Issues in Pattern Analysis and Recognition. VII, 400 pages. 1989.

Vol. 400: R. Klein, Concrete and Abstract Voronoi Diagrams. IV, 167 pages. 1989.

Vol. 401: H. Djidjev (Ed.), Optimal Algorithms. Proceedings, 1989. VI, 308 pages. 1989.

Vol. 402: T.P. Bagchi, V.K. Chaudhri, Interactive Relational Database Design. XI, 186 pages. 1989.

Vol. 403: S. Goldwasser (Ed.), Advances in Cryptology – CRYPTO '88. Proceedings, 1988. XI, 591 pages. 1990.

Vol. 404: J. Beer, Concepts, Design, and Performance Analysis of a Parallel Prolog Machine. VI, 128 pages. 1989.

Vol. 405: C.E. Veni Madhavan (Ed.), Foundations of Software Technology and Theoretical Computer Science. Proceedings, 1989. VIII, 339 pages. 1989.

Vol. 406: C.J. Barter, M.J. Brooks (Eds.), AI '88. Proceedings, 1988. VIII, 463 pages. 1990 (Subseries LNAI).

Vol. 407: J. Sifakis (Ed.), Automatic Verification Methods for Finite State Systems. Proceedings, 1989. VII, 382 pages. 1990.

Vol. 408: M. Leeser, G. Brown (Eds.), Hardware Specification, Verification and Synthesis: Mathematical Aspects. Proceedings, 1989. VI, 402 pages. 1990.

Vol. 409: A. Buchmann, O. Günther, T. R. Smith, Y.-F. Wang (Eds.), Design and Implementation of Large Spatial Databases. Proceedings, 1989. IX, 364 pages. 1990.

Vol. 410: F. Pichler, R. Moreno-Diaz (Eds.), Computer Aided Systems Theory – EUROCAST '89. Proceedings, 1989. VII, 427 pages. 1990.

Vol. 411: M. Nagl (Ed.), Graph-Theoretic Concepts in Computer Science. Proceedings, 1989. VII, 374 pages. 1990.

Vol. 412: L. B. Almeida, C. J. Wellekens (Eds.), Neural Networks. Proceedings, 1990. IX, 276 pages. 1990.

Vol. 413: R. Lenz, Group Theoretical Methods in Image Processing. VIII, 139 pages. 1990.

Vol. 414: A. Kreczmar, A. Salwicki, M. Warpechowski, LOGLAN '88 – Report on the Programming Language. X, 133 pages. 1990.

Vol. 415: C. Choffrut, T. Lengauer (Eds.), STACS 90. Proceedings, 1990. VI, 312 pages. 1990.

Vol. 416: F. Bancilhon, C. Thanos, D. Tsichritzis (Eds.), Advances in Database Technology – EDBT '90. Proceedings, 1990. IX, 452 pages. 1990.

Vol. 417: P. Martin-Löf, G. Mints (Eds.), COLOG-88. International Conference on Computer Logic. Proceedings, 1988. VI, 338 pages. 1990.

Vol. 418: K.H. Bläsius, U. Hedtstück, C.-R. Rollinger (Eds.), Sorts and Types in Artificial Intelligence. Proceedings, 1989. VIII, 307 pages. 1990. (Subseries LNAI).

Vol. 419: K. Weichselberger, S. Pöhlmann, A Methodology for Uncertainty in Knowledge-Based Systems. VIII, 136 pages. 1990 (Subseries LNAI).

Vol. 420: Z. Michalewicz (Ed.), Statistical and Scientific Database Management, V SSDBM. Proceedings, 1990. V, 256 pages. 1990.

Vol. 421: T. Onodera, S. Kawai, A Formal Model of Visualization in Computer Graphics Systems. X, 100 pages. 1990.

Vol. 422: B. Nebel, Reasoning and Revision in Hybrid Representation Systems. XII, 270 pages. 1990 (Subseries LNAI).

Vol. 423: L. E. Deimel (Ed.), Software Engineering Education. Proceedings, 1990. VI, 164 pages. 1990.

Vol. 424: G. Rozenberg (Ed.), Advances in Petri Nets 1989. VI, 524 pages. 1990.

Vol. 425: C.H. Bergman, R.D. Maddux, D.L. Pigozzi (Eds.), Algebraic Logic and Universal Algebra in Computer Science. Proceedings, 1988. XI, 292 pages. 1990.

Vol. 426: N. Houbak, SIL – a Simulation Language. VII, 192 pages. 1990.

Vol. 427: O. Faugeras (Ed.), Computer Vision – ECCV 90. Proceedings, 1990. XII, 619 pages. 1990.

Vol. 428: D. Bjørner, C.A.R. Hoare, H. Langmaack (Eds.), VDM '90. VDM and Z – Formal Methods in Software Development. Proceedings, 1990. XVII, 580 pages. 1990.

Vol. 429: A. Miola (Ed.), Design and Implementation of Symbolic Computation Systems. Proceedings, 1990. XII, 284 pages. 1990.

Vol. 430: J.W. de Bakker, W.-P. de Roever, G. Rozenberg (Eds.), Stepwise Refinement of Distributed Systems. Models, Formalisms, Correctness. Proceedings, 1989. X, 808 pages. 1990.

Vol. 431: A. Arnold (Ed.), CAAP '90. Proceedings, 1990. VI, 285 pages. 1990.

Vol. 432: N. Jones (Ed.), ESOP '90. Proceedings, 1990. IX, 436 pages. 1990.

Vol. 433: W. Schröder-Preikschat, W. Zimmer (Eds.), Progress in Distributed Operating Systems and Distributed Systems Management. Proceedings, 1989. V, 206 pages. 1990.

Vol. 434: J.-J. Quisquater, J. Vandewalle (Eds.), Advances in Cryptology – EUROCRYPT '89. Proceedings, 1989. X, 710 pages. 1990.

Vol. 435: G. Brassard (Ed.), Advances in Cryptology – CRYPTO '89. Proceedings, 1989. XIII, 634 pages. 1990.

Vol. 436: B. Steinholtz, A. Sølvberg, L. Bergman (Eds.), Advanced Information Systems Engineering. Proceedings, 1990. X, 392 pages. 1990.

Lecture Notes in Computer Science

Edited by G. Goos and J. Hartmanis

488

Ronald V. Book (Ed.)

Rewriting Techniques and Applications

4th International Conference, RTA-91
Como, Italy, April 10–12, 1991
Proceedings

Springer-Verlag
Berlin Heidelberg New York London Paris
Tokyo Hong Kong Barcelona Budapest

Volume Editor

Ronald V. Book
Department of Mathematics, University of California
Santa Barbara, CA 93106, USA

CR Subject Classification (1987): D.3, F.3.2, F.4, I.1, I.2.2–3

ISBN 3-540-53904-2 Springer-Verlag Berlin Heidelberg New York
ISBN 0-387-53904-2 Springer-Verlag New York Berlin Heidelberg

Printed in Germany

Printing and binding: Druckhaus Beltz, Hemsbach/Bergstr.
2145/3140-543210 – Printed on acid-free paper

Preface

This volume contains the proceedings of the Fourth International Conference on Rewriting Techniques and Applications (RTA-91). The conference was held April 10-12, 1991, in Como, Italy. The three previous conferences were as follows (the proceedings were also published in Lecture Notes in Computer Science, Springer Verlag):

Dijon, May 1985 (Vol. 202);
Bordeaux, May 1987 (Vol. 256);
Chapel Hill, May 1989 (Vol. 355);

There were 87 submissions to RTA-91 that were received on time. The Program Committee met on November 9, 1990, at the Institut für Informatik, Universität Würzburg, Germany, and chose 35 papers for presentation. Each paper was evaluated by at least two members of the committee or their outside referees. Also included in this volume are short descriptions of four implemented equational reasoning systems that were demonstrated at the meeting. After the Program Committee met, the Chairman accepted a proposal to present a list of open problems in the proceedings.

The Program Committee consisted of the following:
Leo Bachmair (Stony Brook), Harald Ganzinger (Dortmund), Claude Kirchner (Nancy), Ursula Martin (London), Michael O'Donnell (Chicago), Gert Smolka (Saarbrücken), Friedrich Otto (Kassel), and Franz Winkler (Linz).

Professor Giovanni Degli Antoni and Dr. Marelva Bianchi of the University of Milan were responsible for the local arrangements. They were assisted by Ms. Manuela Troglio and Ms. Donatella Marchegiano.

An Organizing Committee was appointed to assist the Chairman in policy issues. This committee consisted of the following:
Maria Paola Bonacina (Milano), Nachum Dershowitz (Illinois), Jean Gallier (Pennsylvania), Jean Pierre Jouannaud (Orsay), Deepak Kapur (Albany), Claude Kirchner (Nancy), Pierre Lescanne (Nancy), David Musser (Rensselaer), and David Plaisted (North Carolina).

RTA-91 was sponsored by the Department of Computer Science of the University of Milan with support from the Institut für Informatik, Universität Würzburg and from INRIA - Nancy. It was held in cooperation with the European Association for Theoretical Computer Science, the IEEE Technical Committee on Mathematical Foundations of Computer Science, and the Association for Computing Machinery - SIGACT, SIGART, and SIGSAM.

The Chairman is grateful to Professor Klaus Wagner and the Institut für Informatik, Universität Würzburg for support of the Program Committee's work, to Ms. Shilo Brooks (Santa Barbara) and Mr. Michael Wolfrath (Würzburg) for their help with many tasks that had to be performed, and to the members of the Program Committee and the Organizing Committee for their efforts.

Würzburg
February 1991

Ronald V.Book
Chairman of the Program and
Organizing Committees

List of Referees

The Program Committee gratefully acknowledges the help of the following referees:

M. Adi
J. Avenhaus
F. Baader
S. Bailey
H. Bertling
A. Bockmayer
M. Bonacina
A. Boudet
H.-J. Bückert
H. Chen
Y. Chen
A. Cichon
H. Comon
E. Contejean
B. Courcelle
H. Curien
O. Danvy
N. Dershowitz
V. Diekert
E. Domenjoud
D. Dougherty
H. Ehrig
P. Enjalbert
J. Gallier
S. Garland
R. Giegerich
B. Gramlich
M. Hanus
R. Heckmann
M. Hermann
B. Hoffman
J. Hsiang
H. Hußmann
U. Hustadt
M. Jantzen
J.-P. Jouannaud
D. Kapur
H. Kirchner
F. Klay
J. Klop
D. Korff
E. Kounalis
P. Kursawe
P. Lescanne
D. Lugiez
K. Madlener
A. Megrelis
P. Narendran
T. Nipkow
W. Nutt
H. Ohlbach
V. van Oostrom
P. Padawitz
D. Perrin
D. Plaisted
D. Plump
L. Puel
I. Ramakrishnan
S. Rebelsky
J.-L. Remy
P. Rety
M. Rusinowitch
A. Sattler-Klein
R. Schott
R. Sekar
S. Sherman
K. Sieber
W. Snyder
R. Socher-Ambrosius
J. Steinbach
A. Suarez
M. Thomas
S. Tison
R. Treinen
P. Viry
J. de Vries
U. Waldmann
U. Wertz
R. Wilhelm
D. Wolfram

Table of Contents

Transfinite Reductions in Orthogonal Term Rewriting Systems

(Extended abstract)

J.R. KENNAWAY [1], J.W. KLOP [2], M.R. SLEEP [3] & F.J. DE VRIES [4]
(1) jrk@sys.uea.ac.uk, (2) jwk@cwi.nl, (3) mrs@sys.uea.ac.uk, (4) ferjan@cwi.nl

(2,4) *CWI, Centre for Mathematics and Computer Science, Amsterdam*
(1,3) *School of Information Systems, University of East Anglia, Norwich*

Abstract. Strongly convergent reduction is the fundamental notion of reduction in infinitary orthogonal term rewriting systems (OTRSs). For these we prove the Transfinite Parallel Moves Lemma and the Compressing Lemma. Strongness is necessary as shown by counterexamples. Normal forms, which we allow to be infinite, are unique, in contrast to ω-normal forms. Strongly converging fair reductions result in normal forms.

In general OTRSs the infinite Church-Rosser Property fails for strongly converging reductions. However for Böhm reduction (as in Lambda Calculus, subterms without head normal forms may be replaced by ⊥) the infinite Church-Rosser property does hold. The infinite Church-Rosser Property for non-unifiable OTRSs follows. The top-terminating OTRSs of Dershowitz c.s. are examples of non-unifiable OTRSs.

1985 Mathematics Subject Classification: 68Q50
1987 CR Categories: F4.1, F4.2
Keywords and Phrases: orthogonal term rewriting systems, infinitary rewriting, strong converging reductions, infinite Church-Rosser Properties, normal forms, Böhm Trees, head normal forms, non-unifiable term rewriting systems.

All authors were partially sponsored by SEMAGRAPH, ESPRIT Basic Research Action 3074. The first author was also partially surported by a SERC Advanced Fellowship, and by SERC grant no. GR/F 91582.

1. INTRODUCTION

The theory of Orthogonal Term Rewrite Systems (OTRS) is now well established within theoretical computer science. Comprehensive surveys have appeared recently in [Der90a, Klo91]. In this paper we consider extensions of the established theory to cover infinite terms and infinite reductions.

1.1. Motivation

At first sight, the motivation for such extensions might appear of theoretical interest only, with little practical relevance. However, it turns out that both infinite terms and infinite rewriting sequences do have practical relevance.

A practical motivation for studying infinite terms and term rewriting arises in the context of lazy functional languages such as Miranda [Tur85] and Haskell [Hud88]. In such languages it is possible to work with infinite terms, such as the list of all Fibonacci numbers or the list of all primes. This style of programming has been advocated by Turner [Tur85], Peyton-Jones [Pey87] and others. Of course the outcome of a particular computation must be finite, but it is pleasant to define such results as finite portions of an infinite term. It would be even more pleasant to know that nice properties (for example Church-Rosserness) hold for infinite as well as finite rewriting, but the standard theory does not tell us this. As we show below, Church-Rosserness is one of several standard results which does *not* hold for infinite rewriting in general, although it does hold for terms which have an infinite normal form (Theorem 4.1.3).

A second practical motivation for considering infinite reduction sequences arises from the common graph-rewrite based implementations of functional languages. The correspondence between graph rewriting and term rewriting was studied in [Bar87] for acyclic graphs. When cyclic graphs are considered, the correspondence with term rewriting immediately requires consideration of infinite terms and infinite reductions. The correspondence with graphs is the motivation for [Far89].

1.2. Overview

With these motivations in mind, we set out to identify precise foundations for transfinite rewriting. A certain amount of care is needed to establish appropriate notions and we do this in Section 2. One can take a topological approach as in [Der89a,b&90] and consider infinite reduction sequences that are converging to a limit in the metric completion of the space of finite terms. However, converging reductions fail to satisfy some natural properties for orthogonal TRSs. Instead we concentrate on strongly converging reductions as introduced by [Far89], which turn out to be better behaved.

Basic facts for infinitary orthogonal term rewrite systems

	converging reductions	*strongly converging reductions*
Transf. Parallel Moves Lemma	NO (3.1.3)	YES (3.1.2)
Inf. Church-Rosser Property	NO (4.1.1)	NO (4.1.1)
Unique ω–normal forms	NO (4.1.1)	NO (4.1.1)
Unique normal forms	YES (3.3.6)	YES (3.3.6)
Compressing Lemma	NO [Far89], (3.2.1)	YES (3.2.5) partial result in [Far89]
Fair reductions result in	ω-normal forms [Der90b], (3.4.2.i)	normal forms (3.4.2.ii)

(Table 1.1)

In Section 3 we prove the fundamental results for infinitary orthogonal rewrite systems, as summarized in Table 1.1. Then in Section 4 we show the failure of the infinite Church-Rosser Property for general OTRSs. The counterexample refutes not only the CR-property for strongly converging but also the CR-property for converging reductions studied by Dershowitz c.s. Introducing ideas from Lambda Calculus we eliminate subterms that have no head normal form by reducing them to ⊥. The new Böhm-reduction $\rightarrow_\perp$ has the infinite Church Rosser Property for strongly converging reductions. Normal forms for $\rightarrow_\perp$-reduction are so called Böhm Trees: they are unique. Finally we show that orthogonal TRSs in which there are no rule in which a left-hand side of a rule can be unified with the right-hand side have the infinite Church-Rosser Property. This class of orthogonal TRSs includes the top-terminating orthogonal TRSs of Dershowitz c.s.

The present paper is an extended abstract of a longer paper in preparation by the same authors [Ken90a]. There it will be proved that the infinite Church-Rosserproperty holds for strongly converging reductions in OTRSs that contain at most one collapsing rule, which then has to be of the form $I(x) \to x$. The full paper will further contain extensions of the theory of needed redexes to infinitary orthogonal term rewriting systems and will unravel the connections between graph rewriting and infinitary term rewriting.

We acknowledge critical reading of an earlier draft by Aart Middeldorp.

2. INFINITARY ORTHOGONAL TERM REWRITING SYSTEMS

We briefly recall the definition of a finitary term rewriting system, before we define infinitary orthogonal term rewriting systems involving both finite and infinite terms. For more details the reader is referred to [Der90a] and [Klo91].

2.1. Finitary term rewriting systems

A *finitary term rewriting system* over a signature Σ is a pair $(Ter(\Sigma),R)$ consisting of the set $Ter(\Sigma)$ of finite terms over the signature Σ and a set of rewrite rules $R \subseteq Ter(\Sigma)\times Ter(\Sigma)$.

The *signature* Σ consists of a countably infinite set Var_Σ of variables (x,y,z,...) and a non-empty set of function symbols (A,B,C,...,F,G,...) of various finite arities ≥ 0. Constants are function symbols with arity 0. The set $Ter(\Sigma)$ of *finite* terms (t,s,...) over Σ can be defined as usual: the smallest set containing the variables and closed under function application.

The set $O(t)$ of *occurrences* in t is defined by induction on the structure of t as follows: $O(t) = \{<>\}$ if t is a variable and $O(t) = \{<>\} \cup \{<i,u> \mid 1\leq i\leq n \text{ and } <u>\in O(t_i)\}$ if t is of the form $F(t_1,...,t_n)$. If $u\in O(t)$ then the subterm t/u at occurrence u is defined by induction: $t/<> = t$ and $F(t_1,...,t_n)/<i,u> = t_i/u$. The *depth* of a subterm of t at u is the length of u.

Contexts are terms in $Ter(\Sigma\cup\{\square\})$, in which the special constant $\square$, denoting an empty place, occurs exactly once. Contexts are denoted by $C[\ ]$ and the result of substituting a term t in place of $\square$ is $C[t]\in Ter(\Sigma)$. A *proper* context is a context not equal to $\square$.

Substitutions are maps $\sigma: Var_\Sigma \to Ter(\Sigma)$ satisfying $\sigma(F(t_1,...,t_n)) = F(\sigma(t_1),...,\sigma(t_n))$.

The set R of *rewrite rules* contains pairs (l,r) of terms in $Ter(\Sigma)$, written as $l \to r$, such that the left-hand side l is not a variable and the variables of the right-hand side r are contained in l. The result l^σ of the application of the substitution of σ to the term l is called an instance of l. A *redex* (reducible expression) is an instance of a left-hand side of a rewrite rule. A *reduction step* $t \to s$ is a pair of terms of the form $C[l^\sigma] \to C[r^\sigma]$, where $l \to r$ is a rewrite rule in R. Concatenating reduction steps we get a *finite reduction sequence* $t_0 \to t_1 \to \ldots \to t_n$, which we also denote by $t_0 \to_n t_n$, or an infinite reduction sequence $t_0 \to t_1 \to \ldots$.

2.2. Infinitary orthogonal term rewriting systems

An *infinitary term rewriting system* over a signature Σ is a pair $(Ter^\infty(\Sigma),R)$ consisting of the set $Ter^\infty(\Sigma)$ of finite and infinite terms over the signature Σ and a set of rewrite rules $R\subseteq Ter(\Sigma)\times Ter^\infty(\Sigma)$. We don't consider rewrite rules with infinite left-hand sides, but right-hand sides may be infinite in order to interpret various liberal forms of graph rewriting in infinitary term rewriting. In [Der90b] only finite left- and right-sides are considered.

It takes some elaboration to define the set $Ter^{\infty}(\Sigma)$ of *finite and infinite terms*. Finite terms may be represented as finite trees, well-labelled with variables and function symbols. Well-labelled means that a node with $n \geq 1$ successors is labelled with a function symbol of arity n and that a node with no successors is labelled either with a constant or a variable. Now *infinite terms* are infinite well-labelled trees with nodes at finite distance to the root. Substitutions, contexts and reduction steps generalize trivially to the set of infinitary terms $Ter^{\infty}(\Sigma)$.

To introduce the *prefix ordering* $\leq$ on terms we extend the signature Σ with a fresh symbol Ω. The prefix ordering $\leq$ on $Ter^{\infty}(\Sigma \cup \{\Omega\})$ is defined inductively: $x \leq x$ for any variable x, $\Omega \leq t$ for any term t and if $t_1 \leq s_1, ..., t_n \leq s_n$ then $F(t_1,...,t_n) \leq F(s_1,...,s_n)$.

If all function symbols of Σ occur in R we will write just R for $(Ter^{\infty}(\Sigma),R)$. The usual properties for finitary TRSs extend verbatim to infinitary TRSs:

2.2.1. DEFINITION. Let R be an infinitary TRS.

(i) R is *left-linear* if no variable occurs more than once in a left-hand side of R's rewrite rules;

(ii) (informally) R is *non-overlapping* (or non-ambiguous) if non-variable parts of different rewrite rules don't overlap and non-variable parts of the same rewrite rule overlap only entirely:

(ii') (formally) R is *non-overlapping* if for any two left-hand sides s and t, any occurrence u in t, and any substitutions σ and $\tau: Var_{\Sigma} \to Ter(\Sigma)$ it holds that if $(t/u)^{\sigma} = s^{\tau}$ then either t/u is a variable or t and s are left-hand sides of the same rewrite rule and u is the empty occurrence < >, the occurrence of the root.

(iii) R is *orthogonal* if R is both left-linear and non-overlapping.

It is well-known (cf. [Ros73], [Klo91]) that finitary orthogonal TRSs satisfy the finitary Church-Rosser property, i.e., ${}^*\!\leftarrow \cdot \to^* \subseteq \to^* \cdot {}^*\!\leftarrow$, where $\to^*$ is the transitive, reflexive closure of the relation $\to$. It is obvious that infinitary orthogonal TRSs inherit this finitary property.

In the present infinitary context it is natural to define that a term is a *normal form* if it contains no redexes, just like in the finitary context. A term t has a normal form s if there is a reduction $t \to_{\alpha} s$. Dershowitz, Kaplan and Plaisted [Der89a, Der89b and Der90b] consider a weaker, more liberal notion of normal form: the ω-*normal forms*. An ω-normal form is a term such that if this term can reduce, then it reduces in one step to itself. One sees easily that restricted to finite terms normal forms and ω-normal forms are already different concepts: in the TRS with rule $A \to A$ the term A is an ω-normal form, but not a normal form.

2.3. Converging and strongly converging transfinite reductions

Generalizing the finite situation we would like to express that there is a reduction of length $\alpha+1$ that transforms t_0 into t_{α}, where α may be any ordinal. Compare the following three reductions of length ω, the corresponding TRSs are easy to imagine:

(i) $A \to B \to A \to B \to ...$,

(ii) $C \to S(C) \to S(S(C)) \to ...$,

(iii) $D(E) \to D(S(E)) \to D(S(S(E))) \to ...$.

Clearly in the first reduction A will not be transformed in the limit to anything fixed, in contrast to C and D(E) in the second and third reduction. It is tempting to say that the limit of C will be S^{ω}, an infinite reduction of S (plus all the necessary brackets), and similar D(E) should have as limit $D(S^{\omega})$. Cauchy convergence is the natural formalism in which to express all this.

The set $\mathrm{Ter}(\Sigma)$ of finite terms for a signature Σ can be provided with an ultra-metric d: $\mathrm{Ter}(\Sigma)\times\mathrm{Ter}(\Sigma) \to [0,1]$ (cf. e.g. [Arn80]). The distance $d(t,s)$ of two terms t and s is 0 if t and s are equal, and otherwise 2^{-k}, where $k\in\mathbb{N}$ is the largest number such that the labels of all nodes of s and t at depth less than or equal to k are equally labelled. The metric completion of $\mathrm{Ter}(\Sigma)$ is isomorphic to the set of infinitary terms $\mathrm{Ter}^{\infty}(\Sigma)$ (cf. [Arn80])

In the complete metric space $\mathrm{Ter}^{\infty}(\Sigma)$ all Cauchy sequences of ordinal length α have a limit. We will now recall the transfinite converging reductions by Dershowitz, Kaplan and Plaisted [Der90b].

2.3.1. DEFINITION. A *sequence* of length α is a set of elements indexed by some ordinal $\alpha \geq 1$: notation $(t_\beta)_{\beta<\alpha}$. Instead of $(t_\beta)_{\beta<\alpha+1}$ we often write $(t_\beta)_{\beta\leq\alpha}$.

2.3.2. DEFINITION. By induction on the ordinal α we define when a sequence $(t_\beta)_{\beta\leq\alpha}$ is a *converging sequence* towards its limit t_α (notation: $t_0 \overset{c}{\to}_\alpha t_\alpha$):

(i) $t_0 \overset{c}{\to}_0 t_0$,

(ii) $t_0 \overset{c}{\to}_{\beta+1} t_{\beta+1}$ if $t_0 \overset{c}{\to}_\beta t_\beta$,

(iii) $t_0 \overset{c}{\to}_\lambda t_\lambda$ if $t_0 \overset{c}{\to}_\beta t_\beta$ for all $\beta<\lambda$ and $\forall\varepsilon>0\ \exists\beta<\lambda\ \forall\gamma\ (\beta<\gamma<\lambda \to d(t_\gamma,t_\lambda) < \varepsilon)$.

This definition of transfinite convergence is an instance of the so-called Moore-Smith convergence over nets (cf. for instance [Kel55]). Limits are unique: if the topological space is a Hausdorff space then each net in the space converges to at most one point; the spaces $\mathrm{Ter}(\Sigma)$ and $\mathrm{Ter}^{\infty}(\Sigma)$ are Hausdorff spaces.

2.3.3. DEFINITION. A *reduction of length* $\alpha \geq 1$ is a sequence $(t_\beta)_{\beta<\alpha}$ such that $t_\beta \to t_{\beta+1}$ for all β such that $\beta+1 < \alpha$. The redex contracted by the step $t_\beta \to t_{\beta+1}$ will be denoted by R_β, its depth as subterm of t_β by d_β.

We will now define strong reductions as reductions in which the depth of the reduced redexes tends to infinity. We present the definition for reductions of arbitrary transfinite length.

2.3.4. DEFINITION. By induction on the ordinal $\alpha \geq 1$ we define when a reduction $(t_\beta)_{\beta<\alpha}$ is a *strong* reduction:

(i) $(t_\beta)_{\beta<1}$ is a strong reduction;
(ii) $(t_\gamma)_{\gamma<\beta+1}$ is a strong reduction if $(t_\gamma)_{\gamma<\beta}$ is a strong reduction;
(iii) $(t_\gamma)_{\gamma<\lambda}$ is a strong reduction if for all $\beta<\lambda$ the reduction $(t_\gamma)_{\gamma<\beta}$ is strong and $\forall d>0\ \exists\beta<\lambda\ \forall\gamma\ (\beta\leq\gamma<\lambda \to d_\gamma>d)$.

2.3.5. DEFINITION. A *strongly converging* reduction is a converging sequence that is a strong reduction.

The strongly converging reductions are of importance for the theory of infinitary term rewriting. Therefore we denote a strongly converging reduction $(t_\beta)_{\beta\leq\alpha}$ by $t_0 \to_\alpha t_\alpha$ By $t \to_{\leq\alpha} s$ we denote the existence of a strong reduction of length less than or equal to α converging towards limit s. We use a similar notation $t \overset{c}{\to}_{\leq\alpha} s$ for converging reductions of length less than or equal to α.

The second example of this section is an example of a strongly converging reduction. Other examples of strongly converging reductions are found in (3.2.1.ii) and (4.1.1).

2.4. Counting steps in strongly converging reductions

Convergent transfinite reductions exist of any length. Consider for example the TRS with the single rule A → A. Reductions of the form $A \overset{c}{\rightarrow}_{\alpha} A$ are converging for any ordinal α. However these sequences are not strongly convergent. The example $A \overset{c}{\rightarrow}_{\alpha} A$ shows also that in a converging reduction any number of reduction steps may be performed below some depth. For strongly converging reductions this is different:

2.4.1. THEOREM. *If $t_0 \rightarrow_{\lambda} t_{\lambda}$ is strongly convergent, then the number of steps in $t_0 \rightarrow_{\lambda} t_{\lambda}$ reducing a redex at depth ≤ n is finite.*

PROOF. Assume $t_0 \rightarrow_{\lambda} t_{\lambda}$ is strongly convergent. As this reduction is strong there is a last step $t_{\alpha} \rightarrow t_{\alpha+1}$ at which a redex is contracted at depth ≤ n. Consider the initial segment $t_0 \rightarrow_{\alpha} t_{\alpha}$, and repeat the argument. By the well-ordering of the ordinals (no infinite descending chains of ordinals) this process stops in finitely many steps. □

2.4.2. COROLLARY. *A strongly converging transfinite reduction has countable length.*

PROOF. By the previous Theorem 2.4.1 a strongly convergent transfinite reduction can only perform finitely many reductions at any given depth $d \in \mathbf{N}$. □

3. FUNDAMENTAL FACTS OF INFINITARY TERM REWRITING

From now on we consider infinitary orthogonal term rewriting systems, except in 3.4.

3.1. The Transfinite Parallel Moves Lemma

In t → s let s be obtained by contraction of the redex S in t. Recall the notation u\S of the set descendants of a redex occurrence u of t in the contraction of S (cf. [Hue79]). Descendance can be extended to transfinite reductions:

3.1.1. DEFINITION. Let $t_0 \rightarrow_{\alpha} t_{\alpha}$ be a transfinite strongly converging reduction such that for all β<α t_{β} reduces to $t_{\beta+1}$ by contraction of the redex R_{β}. By induction on the ordinal α we define the set of descendants u\α in t_{α} that descend from the redex occurrence u in t_0:

(i) $u\backslash 0 = \{u\}$

(ii) $u\backslash(\beta+1) = \bigcup\{v\backslash R_{\beta} \mid v \in u\backslash\beta\}$

(iii) $u\backslash\lambda = \{v \mid \exists\beta<\lambda\ \forall\gamma\ (\beta \leq \gamma < \lambda \rightarrow v \in u\backslash\gamma)\}$

3.1.2. TRANSFINITE PARALLEL MOVES LEMMA.
Let $t_0 \rightarrow_{\alpha} t_{\alpha}$ be a strongly converging reduction sequence of t_0 with limit t_{α} and let $t_0 \rightarrow s_0$ be a reduction of a redex S of t_0. Then for each β ≤ α a term s_{β} can be constructed by outermost contraction of all descendants of S in t_{β} such that $s_{\beta} \rightarrow^* s_{\beta+1}$ for each β ≤ α and all these reductions together form a strongly converging reduction from s_0 to s_{α}. (See Figure 3.1)

$$t_0 \xrightarrow{} t_1 \xrightarrow{} \cdots \xrightarrow{} t_\beta \xrightarrow{R_\beta} t_{\beta+1} \xrightarrow{} \cdots \xrightarrow{} t_\alpha$$

$$S\downarrow \quad \downarrow* \quad * \quad \downarrow^{\leq\omega} \quad * \downarrow^{\leq\omega} \quad * \quad \downarrow^{\leq\omega}$$

$$s_0 \xrightarrow{*} s_1 \xrightarrow{*} \cdots \xrightarrow{} s_\beta \xrightarrow{} s_{\beta+1} \xrightarrow{} \cdots \xrightarrow{} s_\alpha$$

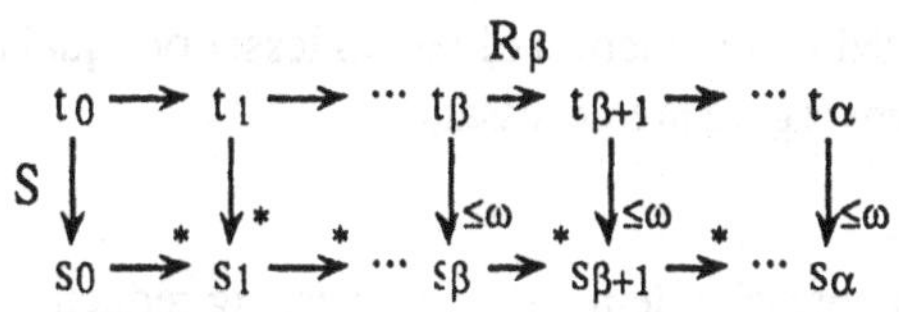

(Figure 3.1)

PROOF. First note that outermost reduction of a finite or an infinite number of disjoint redexes in some term gives a strongly converging reduction. hence all vertical reductions in Figure 3.1 are strongly converging.

We prove the lemma by induction on the ordinal α. The case with zero is easy. Next, let α be of the form $\beta+1$. This goes like the traditional proof, taking care of the possible infinite right-hand sides. Finally, let α be a limit ordinal λ. Assume as induction hypothesis that we have the Transfinite Parallel Moves Lemma for $\beta < \lambda$. There are two possibilities: there exists a $\beta < \lambda$ such that the actual length of the reduction sequence $t_\beta \to_{\leq\omega} s_\beta$ is zero, that is there are no descendants of S in t_β, or there is no such β. The first possibility is easy: we find that $t_\gamma = s_\gamma$ for all γ with $\beta \leq \gamma < \lambda$. It follows that s_0 strongly converges to s_λ.

So let us pursue the second possibility and suppose there is no such β.
Let $(v_\beta)_{\beta \leq \mu}$ be the reduction of the bottom line of Figure 3.1 obtained by refining the sequence $(s_\beta)_{\beta \leq \lambda}$ with reductions $s_\beta \to_{\leq\omega} s_{\beta+1}$ for each $\beta < \alpha$. That such a μ exists follows by an exercise on well-orderings: refining a well-ordering with well-orderings gives again a well-ordering. In order to conclude $s_0 = v_0 \to_\mu v_\mu = s_\lambda$ we have to show: (i) the reduction $(v_\beta)_{\beta \leq \mu}$ is strong, (ii) the reduction $(v_\beta)_{\beta \leq \mu}$ is converging. But this is straigthforward. □

It seems natural to ask whether a transfinite parallel moves lemma exists for the larger class of converging reductions. The following example shows that the construction embodied in the Transfinite Parallel Moves Lemma for strongly converging reductions does not generalize.

3.1.3. COUNTEREXAMPLE.
  Rules: $A(x,y) \to A(y,x),\ C \to D$
  Sequences: $A(C,C) \to\ A(C,C) \to\ A(C,C) \to\ A(C,C) \to\ \cdots \to^c_\omega\ A(C,C)$
               $\downarrow \qquad\quad \downarrow \qquad\quad \downarrow \qquad\quad \downarrow \qquad\qquad\qquad \not\!\!\lambda$
               $A(C,D) \to A(D,C) \to A(C,D) \to A(D,C) \to \cdots$ NO LIMIT

The bottom infinite reduction obtained by standard projection over the one step reduction $C \to D$ does not converge to any limit. □

Note that this example is a counterexample not to the Parallel Moves Lemma, but to a method of proving it. It might be possible that by altering the construction, perhaps by considering a more liberal notion of descendant, the parallel moves lemma holds for transfinite converging reductions. After all, every term occurring in the counterexample can reduce to $A(D,D)$.

3.2. The Compressing Lemma

In this section we will prove the Compressing Lemma for infinitary left-linear TRSs: if $t \to_\alpha s$ is strongly converging, then $t \to_{\leq\omega} s$. That is: any strongly converging reduction from t into s of length $\alpha+1$ can be compressed in a reduction of length lesser or equal than $\omega+1$. The conditions left-linearity and strongly converging are necessary:

3.2.1. COUNTEREXAMPLES.

(i) Example against a compressing lemma for converging reductions in orthogonal TRSs.

Rules: $A(x) \to A(B(x)), B(x) \to E(x)$

Sequence: $A(C) \to_{\omega} A(B(B^{\omega})) \to A(E(B^{\omega}))$.

Note: $A(C)$ cannot reduce to $A(E(B^{\omega}))$ in $\leq \omega$ steps. The reduction is converging but not strong.

(ii) Example of [Der89a] against a compressing lemma for strongly converging reductions in non-left-linear, non-overlapping TRSs.

Rules: $A \to S(A), B \to S(B), H(x,x) \to C$

Sequence: $H(A,B) \to^{*} H(S(A),S(B)) \to^{*} H(S(S(A)),S(S(B))) \to_{\omega} H(S^{\omega},S^{\omega}) \to C$

Note: The term $H(A,B)$ of Dershowitz and Kaplan (cf. [Der89a]) can reduce via the limit $H(S^{\omega},S^{\omega})$ to C. But not $H(A,B) \to_{\leq\omega} C$. The sequence is strongly converging. □

The proof of the Compressing Lemma will go in two steps whose proofs we skip.

3.2.2. COMPRESSING LEMMA for $\omega+1$. *If* $t \to_{\omega+1} s$ *is strongly converging, then* $t \to_{\leq\omega} s$.

3.2.4. COMPRESSING LEMMA for limit ordinals. *If* $t_0 \to_{\lambda} t_{\lambda}$ *is strongly convergent, then there exists a strongly convergent reduction* $t_0 \to_{\leq\omega} t_{\lambda}$.

3.2.5. GENERAL COMPRESSING LEMMA. *For any ordinal* α *if* $t \to_{\alpha} t_{\alpha}$ *is strongly convergent, then there exists a strongly convergent reduction* $t \to_{\leq\omega} t_{\alpha}$.

PROOF. Together 3.2.4 and 3.2.2 establish the Compressing Lemma. Every infinite ordinal α has the form $\lambda+n$, for a limit ordinal λ and a finite n. For any strongly convergent sequence $t \to_{\lambda+n} t_{\alpha}$, we apply Theorem 3.3.4 to the first λ steps, to obtain a sequence $t \to_{\leq\, \omega+n} t_{\alpha}$, then apply Theorem 3.2.2 n times to obtain $t \to_{\leq\omega} t_{\alpha}$. □

3.3. The unique normal form property

We will show for infinitary orthogonal TRSs that each term has at most one normal form. In contrast, Example 4.1.1 shows that the unique ω-normal form property does not hold in general. To obtain the positive result we need the notion of a stable reduction. Informally, an infinite reduction is *stable* if the sequence of stable prefixes of its terms converges to its limit: a stable prefix of a term t is a prefix of t such that no occurrence of that prefix can become an occurrence of a redex in any strongly converging reduction starting from t. Stable reductions will be strongly converging.

3.3.1. DEFINITION. (i) A prefix $s \leq t$ is *stable with respect to a reduction* if no occurrence of s becomes an occurrence of a redex during that reduction.

(ii) A prefix $s \leq t$ is *stable* if s is stable for all possible strongly converging reduction sequences from t.

The restriction in part (ii) to strong reductions is technically convenient. For terms having a normal form, it is in fact unnecesssary; the following proposition may be proved by use of the Transfinite Parallel Moves Lemma. We omit the proof.

3.3.2. PROPOSITION. *In an orthogonal TRS: If a prefix* t *of* t_0 *is stable with respect to a strong reduction from* t_0 *which converges to normal form, then it is stable.*

3.3.3. DEFINITION. Let $\Sigma(t)$ denote the maximal stable prefix of t. A converging reduction $t_0 \rightarrow_{\leq\omega} t_\omega$ is called *stable* if $\forall d \exists N \forall k \geq N \ |\Sigma(t_k)| > d$, where $|t|$ denotes the minimal distance of an occurrence of Ω in t to the root, if there is any, otherwise $|t| = \infty$.

Stability is a very strong condition. The limit of an infinite stable reduction sequence is a normal form, from which it easily follows that stable reduction is Church-Rosser. The proof of the following lemma is routine and therefore omitted.

3.3.4. LEMMA. (i) *If* $t \rightarrow s$ *then* $\Sigma(t) \leq \Sigma(s)$.
(ii) *For reductions: stable* $\Rightarrow$ *strongly convergent* $\Rightarrow$ *convergent. But not conversely.*
(iii) *The limit of a stable reduction sequence is a normal form.* □

3.3.5. THEOREM. *The following are equivalent:*
(i) $t \rightarrow_{\leq\omega} s$ *is a converging reduction to normal form;*
(ii) $t \rightarrow_{\leq\omega} s$ *is a strong converging reduction to normal form;*
(iii) $t \rightarrow_{\leq\omega} s$ *is a stable reduction*

Some comments on the proof: It is trivial to see that (iii) ⇒ (ii) ⇒ (i).The proof of (i) ⇒ (ii) is a *reductio ad absurdum*. The proof of (ii) ⇒ (iii) has become easy by Proposition 3.3.2.

3.3.6. UNIQUE NORMAL FORM PROPERTY. *Normal forms are unique in orthogonal TRSs.*

PROOF. Suppose a term t admits two converging reductions $t \rightarrow s_1 \rightarrow s_2 \rightarrow \ldots \rightarrow^{c}_{\leq\omega} s$ and $t \rightarrow r_1 \rightarrow r_2 \rightarrow \ldots \rightarrow^{c}_{\leq\omega} r$ to normal form. By Theorem 3.3.5 these reductions are stable. By the finite Church-Rosser property, for each n there exists u_n such that $s_n \rightarrow^* u_n$ and $r_n \rightarrow^* u_n$. We obtain $t \rightarrow^* u_1 \rightarrow^* u_2 \rightarrow^* \ldots$. Using Lemma 3.3.4 (i) the newly constructed reduction $(u_n)_{n\in \mathbf{N}}$ inherits its stableness from the stable reductions $(s_n)_{n\in \mathbf{N}}$ and $(r_n)_{n\in \mathbf{N}}$. Thus we see by Theorem 3.3.5 that the limit u of (u_n) is a normal form. By Lemma 3.3.4 (i) we see that $\Sigma(s_n) \leq \Sigma(u_n)$ and $\Sigma(r_n) \leq \Sigma(u_n)$. Hence $s \equiv \lim_{n\to\infty} \Sigma(s_n) \leq \lim_{n\to\infty} \Sigma(u_n) \equiv u \geq \lim_{n\to\infty} \Sigma(r_n) \equiv r$. Since normal forms are maximal in the prefix ordering (in contrast to ω-normal forms) s and r are equal. □

It is not difficult to show that any normal form that can be reached via a converging reduction, might also be reached via a strongly converging reduction.

3.4. Fair reductions

Theorem 3.3.5 implies that stable converging reductions result in normal forms. If we add a fairness condition to strongly converging reductions, then their limits will also be normal forms. The same fairness condition added to converging reductions results in converging reductions to ω-normal form [Der89b]. Fairness of a reduction will express that, whenever a redex occurs in a term during this reduction, the redex itself or a term containing the redex will be reduced within a finite number of steps.

3.4.1. DEFINITION. (i) Let r be a redex of t at occurrence u. A reduction $t \rightarrow_{\leq\omega} t'$ *preserves* r if no step of this reduction performs a contraction at an occurrence $\leq$ u.

(ii) A reduction $t \rightarrow_{\leq\omega} t'$ is *fair* if for every term t" in the reduction, and every redex r of t" some finite part of this reduction starting at t" does not preserve r.

Note that a finite sequence is fair if and only if it ends in a normal form, and fair reductions don't need to be converging. Note also that orthogonality guarantees that if the reduction $t \rightarrow_{\leq\omega} t'$ preserves a redex in t of a certain rule, then t' contains a redex of the same rule.

We skip the proof of the following theorem. The proof is straightforward.

3.4.2. THEOREM. (i) [Der89b] *The limit of a fair, converging reduction is an ω-normal form.*
(ii) *The limit of a fair, strongly converging reduction is a normal form.*

3.4.3. COROLLARY. *A reduction sequence is fair, strongly convergent if and only if it is stable.*

4. THE INFINITE CHURCH-ROSSER PROPERTY

4.1. Failure of the infinite Church-Rosser Property for orthogonal TRSs

In the standard theory of orthogonal TRSs one proves the finite Church-Rosser Property after establishing the Finite Parallel Moves Lemma. The following counterexample shows that, despite the Transfinite Parallel Moves Lemma, the infinite Church Rosser property

$$\leftarrow_{\omega} \cdot {}_{\omega}\rightarrow \subseteq {}_{\leq\omega}\rightarrow \cdot \leftarrow_{\leq\omega}$$

does not hold for strongly converging reductions.

4.1.1. COUNTEREXAMPLE.
Rules: $A(x) \rightarrow x$, $B(x) \rightarrow x$, $C \rightarrow A(B(C))$
Sequences: $C \rightarrow A(B(C)) \rightarrow A(C) \rightarrow A(A(B(C))) \rightarrow A(A(C)) \rightarrow_{\omega} A^{\omega}$
$C \rightarrow A(B(C)) \rightarrow B(C) \rightarrow B(A(B(C))) \rightarrow B(B(C)) \rightarrow_{\omega} B^{\omega}$
Hence $C \rightarrow_{\leq\omega} A^{\omega}$ as well as $C \rightarrow_{\leq\omega} B^{\omega}$. But there is no term t such that $A^{\omega} \rightarrow_{\leq\omega} t \leftarrow_{\leq\omega} B^{\omega}$ be it converging or strongly converging. □

4.2. Böhm trees

The counterexample and Theorem 4.1 suggest that terms having ω-normal forms that are not normal forms are blocking a proof of the Infinitary Church-Rosser Property for converging reductions. From Lambda Calculus (cf. [Bar84]) we will borrow the notion head normal form (hnf), for terms that cannot be reduced to a redex and the idea for a reduction relation $\rightarrow_{\perp}$ extending $\rightarrow$ with an extra rule: $t \rightarrow \perp$ if t has no hnf. ⊥ is a fresh symbol that we add to the signature of the TRS.

4.2.1. DEFINITION. A term *is a head normal form* (hnf) if the term cannot be reduced to a redex, and a term *has a hnf* if it can be reduced to a hnf.

4.2.2. DEFINITION. (i) Let us denote by $\stackrel{\perp}{\rightarrow}$ the rewrite relation {<C[t],C[⊥]> | t has no head normal form for $\rightarrow$, C[ ] is a one-place context}.
(ii) Let the rewrite relation underlying *Böhm reduction* (notation $\rightarrow_{\perp}$) be $\rightarrow \cup \stackrel{\perp}{\rightarrow}$.

(iii) A term t has a *Böhm tree* if there exists a strongly converging Böhm reduction from t to $\rightarrow_{\perp}$-normal form.
(iv) Let *strict Böhm reduction* (notation $\rightarrow_{[\perp]}$) be the subreduction of $\rightarrow_{\perp}$ in which $\overset{\perp}{\rightarrow}$-reduction has priority over $\rightarrow$-reduction.

We skip the proof of the following lemma and theorem.

4.2.3. LEMMA. (i) $\overset{\perp}{\rightarrow}$ *is finitely CR.*
(ii) $\rightarrow_{\perp}$ *is finitely CR.*
(iii) *Each finite part of a Böhm tree can be found in finitely many steps.*
(iv) *A term has at most one Böhm tree.*
(v) *Böhm reduction* $\rightarrow_{\perp}$ *and strict Böhm reduction* $\rightarrow_{\perp}$ *have the same normal forms.*
(vi) $\rightarrow_{[\perp]}$*-reductions are strongly convergent and of lenght not more than* ω
(vii) *Every term has a normal form with respect to* $\rightarrow_{[\perp]}$.

4.2.4. THEOREM. *For both strongly convergent* $\rightarrow_{\perp}$*-reduction and convergent* $\rightarrow_{\perp}$*-reduction the infinite Church-Rosser Property holds.*

4.3. Non-unifiable orthogonal TRSs have the infinite Church-Rosser Property

From the work of Dershowitz, Plaisted and Kaplan on convergent reductions it follows that any left-linear, top-terminating and semi-ω-confluent TRS satisfies the infinite Church-Rosser property:

$$ {}_{\omega}\overset{c}{\leftarrow} \cdot \overset{c}{\rightarrow}_{\omega} \subseteq \overset{c}{\rightarrow}_{\leq\omega} \cdot {}_{\leq\omega}\overset{c}{\leftarrow} $$

(cf. [Der90b]: combine Theorem 1, Proposition 2 with Theorem 9.). A TRS is *top-terminating* if there are no top-terminating reductions of length ω, that is reductions with infinitely many rewrites at the root of the initial term of the reduction. Semi-ω-confluency, that is

$$ \overset{*}{\leftarrow} \cdot \overset{c}{\rightarrow}_{\omega} \subseteq \overset{c}{\rightarrow}_{\leq\omega} \cdot {}_{\leq\omega}\overset{c}{\leftarrow} $$

holds if the Transfinite Parallel Moves Lemma holds for converging reductions. On the assumption that we are in a orthogonal TRS in which all convergent reductions are strong the infinite Church-Rosser Property holds for this TRS. Top-termination implies this assumption. Hence in top-terminating orthogonal TRSs the infinite Church-Rosser Property holds.
Using our techniques we can explain and improve this result.

4.3.1. DEFINITION. A TRS is called *unifiable* if the TRS contains a *unifiable* rule, that is a rule $l \rightarrow r$ such that for some substitution σ with finite and infinite terms for variables $l^{\sigma} = r^{\sigma}$.

Note that unifiability in the space of finite and infinite terms means unifiability "without the occurs check": the terms I(x) and x are unifiable in this setting, and their most general unifier is the infinite term I^{ω}. Collapsing rules, i.e. rules which right-hand side is a variable are unifiable.

4.3.2. LEMMA. *The following are equivalent for an orthogonal TRS:*
(i) *the TRS is non-unifiable,*
(ii) *all convergent reductions of the TRS are strong,*

(iii) *all convergent reductions are top-terminating.*

4.3.3. THEOREM. *Any non-unifiable orthogonal TRS has the infinite Church-Rosser Property for converging reductions.*

The theorem follows from the quoted results of Dershowitz, Kaplan and Plaisted. Space prevents us to explain another proof: Non-unifiable TRSs are a special instance of non-collapsing TRSs, i.e. TRSs in which there are no rules whose right-hand side is a single variable. In non-collapsing orthogonal TRSs the infinite Church-Rosser Property holds for strongly converging reductions (cf. [Ken90a,b]). In fact we may admit one single collapsing rule of the form I(x) → x and still retain the infinite Church-Rosser property (cf. [Ken90a,b]). Note, for instance, that the collapsing rule for K in Combinatory Logic disturbs the infinite Church-Rosser property. A counterexample is not difficult to construct.

5. REFERENCES

[Arn80] A. ARNOLD and M. NIVAT, The metric space of infinite trees. Algebraic and topological properties, Fundamenta Informatica, 4 (1980) 445-76.

[Bar84] H.P. BARENDREGT, *The Lambda Calculus, its Syntax and Semantics*, 2nd ed., (North-Holland, 1984).

[Bar87] H.P. BARENDREGT, M.C.J.D. VAN EEKELEN, J.R.W. GLAUERT, J.R. KENNAWAY, M.J. PLASMEIJER, and M.R. SLEEP, *Term graph rewriting*, Proc. PARLE Conference vol II, LNCS vol. 259, pp. 141–158 (Springer-Verlag, 1987).

[Der89a] N. DERSHOWITZ and S. KAPLAN, *Rewrite, rewrite, rewrite, rewrite, rewrite*, Principles of programming languages, Austin, Texas, 1989, pp. 250–259.

[Der89b] N. DERSHOWITZ, S. KAPLAN and D.A. PLAISTED, *Infinite Normal Forms (plus corrigendum)*, ICALP 1989, pp. 249–262.

[Der90a] N. DERSHOWITZ and J.P. JOUANNAUD, *Rewrite Systems*, to appear in *Handbook of Theoretical Computer Science* (ed. J. van Leeuwen) vol.B, chapter 15, North-Holland.

[Der90b] N. DERSHOWITZ, S. KAPLAN and D.A. PLAISTED, *Rewrite, rewrite, rewrite, rewrite, rewrite* to appear.

[Far89] W.M. FARMER and R.J. WATRO, *Redex capturing in term graph rewriting*, in *Computing with the Curry Chip* (eds. W.M. Farmer, J.D. Ramsdell and R.J. Watro), Report M89-59, MITRE, 1989.

[Hud88] P. HUDAK et al, *Report on the Functional Programming Language Haskell.* Draft Proposed Standard, 1988.

[Hue79] G. HUET and J.-J. LÉVY, *Call by need computations in non-ambiguous linear term rewriting systems*, Report 359, INRIA, 1979.

[Kel55] J.L. KELLEY, *General Topology*, Graduate Texts in Mathematics 27, Springer-Verlag New York, 1955, Second Printing 1985.

[Ken90a] J.R.KENNAWAY, J.W. KLOP, M.R. SLEEP and F.J. DE VRIES, *Transfinite reductions in orthogonal term rewriting systems* (Full paper), report CS-R9041, CWI, Amsterdam, 1990.

[Ken90b] J.R.KENNAWAY, J.W. KLOP, M.R. SLEEP and F.J. DE VRIES, *An infinitary Church-Rosser property for non-collapsing orthogonal term rewriting systems*, report CS-R9043, CWI, Amsterdam, 1990.

[Klo80] J.W. KLOP, *Combinatory reduction systems*, Mathematical Centre Tracts no. 127, CWI, Amsterdam, 1980.

[Klo91] J.W. KLOP, *Term rewriting systems*, to appear in *Handbook of Logic in Computer Science*, Vol I (eds. S. Abramsky, D. Gabbay and T. Maibaum), Oxford University Press, 1991.

[Pey87] S.L. PEYTON JONES, *The Implementation of Functional Programming Languages*, (Prentice-Hall, 1987).

[Ros73] B.K. ROSEN, *Tree manipulating systems and Church Rosser theorems*, JACM 20 (1973) 160-187.

[Tur85] D.A. TURNER, *Miranda: a non-strict functional language with polymorphic types*, in J.-P. Jouannaud (ed.), Proc. ACM Conf. on Functional Programming Languages and Computer Architecture, Lecture Notes in Computer Science, vol. 201, Springer-Verlag, 1985.

Redex Capturing in Term Graph Rewriting (Concise Version)*

William M. Farmer and **Ronald J. Watro**
The MITRE Corporation, Burlington Road, A156
Bedford, MA 01730 USA

Abstract

Term graphs are a natural generalization of terms in which structure sharing is allowed. Structure sharing makes term graph rewriting a time- and space-efficient method for implementing term rewrite systems. Certain structure sharing schemes can lead to a situation in which a term graph component is rewritten to another component that contains the original. This phenomenon, called *redex capturing*, introduces cycles into the term graph which is being rewritten—even when the graph and the rule themselves do not contain cycles. In some applications, redex capturing is undesirable, such as in contexts where garbage collectors require that graphs be acyclic. In other applications, for example in the use of the fixed-point combinator Y, redex capturing acts as a rewriting optimization. We show, using results about infinite rewritings of trees, that term graph rewriting with arbitrary structure sharing (including redex capturing) is sound for left-linear term rewrite systems.

1 Introduction

Term rewriting is a model of computation and a model of formula manipulation that is employed in many areas of computer science, including abstract data types, functional programming, and automated reasoning. The most direct way of implementing term rewriting is via string rewriting or tree rewriting. However, these approaches are inefficient because there is no *structure sharing*, that is, it is not possible to represent multiple occurrences of a subterm by a single substring or subtree. A better approach to implementing term rewriting is through the use of a certain kind of graph rewriting. The basic idea is to represent terms as "term graphs" in which structure sharing is possible. Terms are then rewritten by rewriting their graphical representations. A single rewriting of a term graph can correspond to several rewritings of the term represented by the term graph, due to structure sharing. Thus term graph rewriting is a time-efficient as well as a space-efficient method for implementing term rewrite systems.

Just as a term rewrite rule is an ordered pair of terms, a term graph rewrite rule is an ordered pair of term graphs. We will define term graph rewriting in a manner slightly different from previous work. Our definition is nondeterministic in the sense that a redex can be reduced in several ways, depending on how structure is shared. (Other approaches

*The full version of this paper will appear in the *International Journal of Foundations of Computer Science* [7].

(such as [1]) have assumed that the rewriting rule itself determines the structure sharing.) We use the nondeterministic approach because our goal is to prove a general soundness theorem that applies to all forms of structure sharing.

Term graphs without cycles are usually called simply *directed acyclic graphs (DAGs)*. Proofs about acyclic systems are simplified by the fact that every finite DAG corresponds to a unique term. Staples [12] and Barendregt et al. [1] have studied graph rewriting without cycles. In [1], a strong soundness theorem is given, stating that when term rewriting is implemented by acyclic graph rewriting, then each result obtainable by graph rewriting is also obtainable by term rewriting. It is also shown in [1] that, in regular term rewrite systems, term graph rewriting is *complete* with respect to term rewriting.

Term graphs with cycles are useful for representing infinite objects and for efficiently implementing certain term rewrite rules, such as the rule for the fixed-point combinator Y (see Example 4.4). For some term graph rewrite rules, certain structure sharing schemes can lead to a situation in which a term graph component is rewritten to another component that contains the original. This phenomenon, called *redex capturing*, introduces cycles into the term graph being rewritten—even when the graph and the rule themselves do not contain cycles. Redex capturing can be undesirable in some situations. This is obviously the case in contexts where term graphs must be acyclic, such as in the presence of garbage collectors which only work correctly on acyclic graphs. Also, correctness proofs are complicated by the fact that cyclic term graphs do not correspond to unique finite terms. On the other hand, redex capturing acts as a very desirable rewriting optimization for the Y combinator. An interesting example of a system which illustrates the issues of structure sharing and redex capturing in term graph rewriting is the Clio verification system [2].

The nature of graph rewriting with cycles, and with redex capturing in particular, has not been adequately addressed in the literature. When cycles are allowed in term graph rewriting, it is no longer sound with respect to (finite) term rewriting in the strong sense described above. In this paper, we give a careful description of what redex capturing is and how it occurs. We show that redex capturing results from the application of certain rewrite rules under certain structure sharing schemes. We also show that there is a natural sense in which term graph rewriting with redex capturing is a sound method for implementing left-linear term rewrite systems.

This latter result follows as a corollary from the main theorem of the paper, the Soundness Theorem, which says that, if only left-linear, left-finite, left-acyclic term graph rewrite rules are employed, a term graph rewriting of finite length with redex capturing corresponds to a certain infinite, but "convergent" tree rewriting. The Soundness Theorem makes no assumption about how structure sharing is performed. The proof of the theorem uses a lemma, called the $\omega\omega$-Lemma, which is the major technical result of the paper. From the Soundness Theorem we obtain as an immediate corollary a proof of the soundness of the cyclic rewrite rule for the fixed-point combinator Y. In an earlier paper, Farmer, Ramsdell and Watro [6] proved (without using infinite rewriting) that graph head reduction of combinator expressions using the cyclic Y-rule is a correct implementation of term head reduction. The soundness result presented here is a broad generalization of one part of the earlier correctness theorem.

The paper consists of five sections in addition to this introduction. In Sections 2 and 3, we present the basic definitions of term graphs and term graph rewriting. We emphasize that the nature of term graph rewriting is heavily dependent on the kind of structure sharing scheme that is employed. Section 4 is devoted to the phenomenon of redex capturing. We show that to avoid redex capturing there is a cost to be paid in terms of generality, implementation efficiency, or implementation simplicity. In Section 5, we discuss tree rewriting—which we view as a special case of term graph rewriting. Section 5 contains the statement of the $\omega\omega$-Lemma. The Soundness Theorem and the corollaries described above are given in Section 6.

All the proofs and several examples have been omitted from this concise version of the paper; they can be found in the full version [7].

2 Term graphs

We present in this section the definition of a term graph and several related concepts. A term graph is a natural generalization of a labeled tree. Informally, term graphs are rooted, directed, ordered graphs in which each node is labeled by a function symbol or a variable. Unlike trees, term graphs allow structure to be shared, and in particular, may contain cycles. Also, term graphs "unravel" into trees. Most of the definitions and propositions given below are minor adaptations of work of Barendregt et al. [1], Kennaway [8] and Raoult [11]. Unlike [1] and [11], we do allow term graphs to have more than one node labeled by the same variable.

Let $\mathcal{F}$ be a set of *function symbols* and let $\mathcal{V}$ be a set of *variables*. We assume that each function symbol f has a non-negative integer arity, denoted by $\text{arity}(f)$, and we define $\text{arity}(v) = 0$ for all variables v. For any set S, let S^* be the set of all finite words over S. The length of a word w is denoted by $|w|$.

A *term graph (over $\mathcal{F}$ and $\mathcal{V}$)* is a quadruple $(N, \text{lab}, \text{suc}, \rho)$ where N is a set of *nodes* with $\rho \in N$, $\text{lab} : N \to \mathcal{F} \cup \mathcal{V}$, and $\text{suc} : N \to N^*$ such that $|\text{suc}(\alpha)| = \text{arity}(\text{lab}(\alpha))$ for all $\alpha \in N$. For $\alpha \in N$, $\text{lab}(\alpha)$ is called the *label* of α and the members of $\text{suc}(\alpha)$ are called the *successors* of α. ρ is called the *root* of the term graph. Let $i\text{-suc}(\alpha) = \alpha_i$ for $\text{suc}(\alpha) = \alpha_1 \cdots \alpha_n$. In the rest of this paper term graphs are simply called *graphs* and are denoted by G, H, etc. The components of a graph G are denoted by $N_G, \text{lab}_G, \text{suc}_G$, and ρ_G. When there is no possibility for confusion, the components of a graph G_s will be denoted by $N_s, \text{lab}_s, \text{suc}_s$, and ρ_s.

A *path in* a graph G is a finite sequence $p = \langle(\alpha_1, i_1), \ldots, (\alpha_n, i_n), \alpha_{n+1}\rangle$ $(n \geq 0)$ where $\alpha_1, \ldots, \alpha_{n+1} \in N_G$, $i_1, \ldots, i_n \in \omega$ (the natural numbers), and $\alpha_{m+1} = i_m\text{-suc}(\alpha_m)$ for all m with $1 \leq m \leq n$. The path p is said to be *from* α_1 *to* α_{n+1} and to have *length* n. A graph is *root connected* if there is a path from the root to each node in the graph. The *depth of a node* α in G, written $\text{dp}_G(\alpha)$, is the least $n \geq 0$ such that there is a path from ρ_G to α with length n. If $\alpha \in N_G$ and there is no path from ρ_G to α, $\text{dp}_G(\alpha)$ is undefined.

In many applications of graph rewriting, nodes with undefined depth are irrelevant and are removed (whenever possible). The task of detecting and removing such nodes is called *garbage collection*. In this paper, little attention will be given to nodes with undefined depth and the issue of garbage collection will be treated trivially.

A *cycle* is a path from a node to itself of length ≥ 1. A graph is *cyclic* if there is a cycle in the graph. A *tree* is a graph G such that there is exactly one path from the root of G to each other node of G. A *term* is a tree T such that N_T is finite.

Let G be a graph with $\beta \in N_G$. The *subgraph of G at β*, written G/β is the graph $(N', \text{lab}', \text{suc}', \beta)$ where

$$N' = \{\alpha \in N : \text{ there is a path from } \beta \text{ to } \alpha\}$$

and lab′ and suc′ are the restrictions of lab and suc to N'. Notice that G/β is a root connected graph. An *extension* of G is a graph H such that $\rho_H = \rho_G$ and G is a subgraph of H.

A map $\varphi : N_G \to N_H$ is *homomorphic at* $\alpha \in N_G$ if

1. $\mathrm{lab}_H(\varphi(\alpha)) = \mathrm{lab}_G(\alpha)$ and
2. $\mathrm{suc}_H(\varphi(\alpha)) = \bar{\varphi}(\mathrm{suc}_G(\alpha))$

where $\bar{\varphi}(\alpha_1 \cdots \alpha_n) = \varphi(\alpha_1) \cdots \varphi(\alpha_n)$. φ is a *homomorphism from G to H* if φ is homomorphic at every $\alpha \in N_G$. Notice that in a homomorphism there is no distinction between nodes labeled with 0-ary function symbols and nodes labeled with variables.

A homomorphism φ from G to H is *rooted* if $\varphi(\rho_G) = \rho_H$. A bijective homomorphism φ from G to H is an *isomorphism* from G to H. G and H are *equivalent*, written $G \approx H$, if there is a rooted isomorphism from G to H.

Proposition 2.1 *Let G be a root connected term graph.*

1. *For every $\alpha \in N_H$, there is at most one homomorphism φ from G to H with $\varphi(\rho_G) = \alpha$.*
2. *A rooted homomorphism from a graph G' to G must be surjective.*
3. *If there is a rooted homomorphism from G to a tree T, then $G \approx T$.*

An *unraveling* of G is a tree T such that there is a rooted homomorphism from T to G/ρ_G. For each graph G, let $u(G)$ be some unraveling of G. Every graph has an unraveling, and any two unraveling of a graph are equivalent.

G and H are *tree equivalent*, written $G \approx_t H$, if $u(G) \approx u(H)$. A graph G is *consolidated* if no two distinct subgraphs of G are tree equivalent.

Proposition 2.2
1. *$G \approx H$ implies $G \approx_t H$.*
2. *If T and U are trees, then $T \approx_t U$ implies $T \approx U$.*
3. *For every graph G there is a consolidated graph G' such that $G \approx_t G'$.*
4. *If there is a rooted homomorphism from a consolidated graph G to a root connected graph H, then $G \approx H$.*

A *weak homomorphism from G to H* is a map $\varphi : N_G \to N_H$ such that:

1. φ is homomorphic at every $\alpha \in N_G$ with $\mathrm{lab}_G(\alpha) \in \mathcal{F}$.
2. For all $\alpha, \beta \in N_G$ *with* $\mathrm{lab}_G(\alpha) = \mathrm{lab}_G(\beta) \in \mathcal{V}$,

$$H/\varphi(\alpha) \;\approx_t\; H/\varphi(\beta).$$

Weak homomorphisms create bindings for domain nodes that are labeled by variables. Let φ_i be a weak homomorphism from G_i to H for $i = 1, 2$. φ_1 and φ_2 are *compatible* if, for all $\alpha_1 \in N_1$ and all $\alpha_2 \in N_2$ with $\mathrm{lab}_1(\alpha_1) = \mathrm{lab}_2(\alpha_2) \in \mathcal{V}$,

$$H/\varphi_1(\alpha_1) \;\approx_t\; H/\varphi_2(\alpha_2).$$

When presenting specific graphs, we shall use a linear notation and sometimes a pictorial notation. The linear notation is derived from the linear notation for graphs given in [1]. However, we write $f^\alpha(-,\ldots,-)$ in place of $\alpha : f(-,\ldots,-)$; the expression f^α denotes a node α with label f. Also, the leading node-label pair in a linear notation corresponds to the root of the graph being described. With the pictorial notation, the root of a graph is always the top-most node, and nodes in the graph with undefined depth are not depicted. Otherwise, the pictorial notation is basically self-explanatory.

3 Term graph rewriting

A *(graph) rewrite rule* is a pair $r = \langle G_L, G_R \rangle$ of root connected graphs. r is a *term* [*tree*] *rewrite rule* if G_L and G_R are terms [trees]. r is *left-linear* if there are no distinct nodes $\alpha_1, \alpha_2 \in N_L$ with $\mathrm{lab}_L(\alpha_1) = \mathrm{lab}_L(\alpha_2) \in \mathcal{V}$. r is *left-finite* if N_L is finite. r is *left-acyclic* if G_L is acyclic.

A *redex in* a graph G is a pair $\Delta = (r, \theta)$ where r is a rewrite rule $\langle G_L, G_R \rangle$ and θ is a weak homomorphism from G_L to G. $\theta(\rho_L)$ is called the *root* of Δ. The *depth* of Δ *in* G, written $\mathrm{dp}_G(\Delta)$, is $\mathrm{dp}_G(\theta(\rho_L))$.

Let $\Delta = (\langle G_L, G_R \rangle, \theta_L)$ be a redex in G. To *reduce* Δ in G means to construct a new graph H from G in two steps as follows:

1. *Build step.* Choose an extension H^* of G and a weak homomorphism θ_R from G_R to H^* such that θ_L and θ_R are compatible.

2. *Redirection step.* For all $\alpha \in N_{H^*}$ and integers i such that $i\text{-suc}_{H^*}(\alpha) = \theta_L(\rho_L)$, redefine $i\text{-suc}_{H^*}(\alpha)$ to be $\theta_R(\rho_R)$. Also, if $\rho_{H^*} = \theta_L(\rho_L)$, redefine ρ_{H^*} to be $\theta_R(\rho_R)$. Let H be this new graph obtained from H^*.

Notice that the build step is nondeterministic but the redirection step is deterministic (once H^* and θ_R have been chosen). Suppose there is a redex $\Delta = (r, \theta_L)$ in G with root α. Then to *apply* r *at* α *in* G (*using* H^* *and* θ_R) means to reduce the redex Δ in G (using H^* and θ_R).

In our definition we have not specified any relationship between G and the image of G_R under θ_R. Thus the image of G_R is allowed to share the structure of G. There are several schemes for controlling structure sharing between G and the image of G_R. Four structure sharing schemes are briefly described below:

(1) *No structure sharing.* The simplest scheme is to allow no structure sharing at all between G and the image of G_R. When G is a term and $\langle G_L, G_R \rangle$ is a term rewrite rule, redex reduction under this scheme is the same as redex reduction in ordinary term rewriting. Of course, any implementation of rewriting based on this scheme will be highly inefficient.

(2) *Minimal structure sharing.* A very natural way of sharing structure between G and the image of G_R is to require that the following condition holds. Suppose $\alpha \in N_R$ and there is some $\beta \in N_L$ such that $\mathrm{lab}_R(\alpha) = \mathrm{lab}_L(\beta) \in \mathcal{V}$. Then $\theta_R(\alpha) = \theta_L(\gamma)$ for some $\gamma \in N_L$ such that $\mathrm{lab}_L(\gamma) = \mathrm{lab}_R(\alpha)$. Intuitively, this means that the image of a node in G_R labeled with the variable x is equal to the image of some node in G_L labeled with x (provided such a node in G_L exists). This mode of structure sharing is incorporated in nearly all structuring sharing schemes used in implementations of term graph rewriting.

(3) *Maximal structure sharing.* Another scheme is to have maximal structure sharing between G and the image of G_R. H^* is chosen so that H is consolidated relative to

G, i.e., if $\alpha \in N_H - N_G$, then there is no $\beta \in N_G \cup N_H$ distinct from α such that $H/\beta \approx_t H/\alpha$. (Maximal structure sharing is similar to *hash consing* in some pure Lisp implementations.) Redex reduction based on this scheme can be costly to implement because it requires global information about the structure of G.

(4) *Rule-based structure sharing.* The final scheme that we shall discuss is presented in [1]. In this scheme, structure sharing is determined entirely by structure sharing between G_L and G_R. θ_R must satisfy two conditions: (1) if $\alpha \in N_L \cap N_R$, then $\theta_R(\alpha) = \theta_L(\alpha)$, and (2) if $\alpha \in N_R - N_L$, then $\theta_R(\alpha) \in N_{H^*} - N_G$. This scheme leads to less structure sharing than the former scheme, but it requires only local information about the structure of G.

Example 3.1 Suppose we would like to apply the rewrite rule

$$\langle h(x^\alpha), g(x^\alpha, x^\alpha)\rangle$$

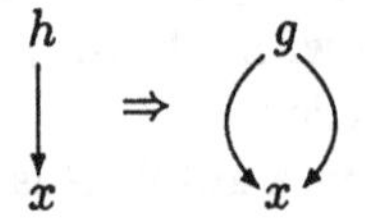

to the graph

$$G = f(g^\beta(a^\gamma, a^\gamma), h(a^\gamma)).$$

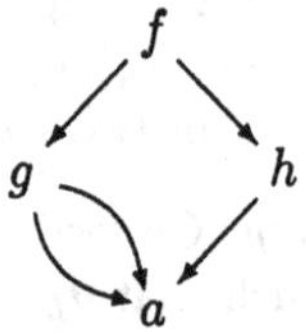

If we use the no structure sharing scheme above, the result is

$$H_1 = f(g^\beta(a^\gamma, a^\gamma), g(a^\delta, a^\delta)) + h(a^\gamma).$$

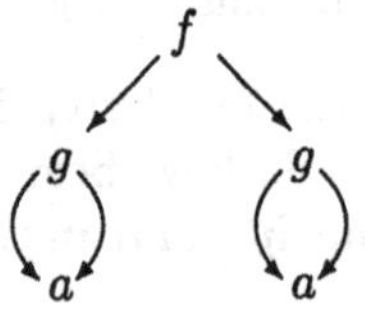

If we use the minimal or rule-based structure sharing scheme, the result is

$$H_2 = f(g^\beta(a^\gamma, a^\gamma), g(a^\gamma, a^\gamma)) + h(a^\gamma).$$

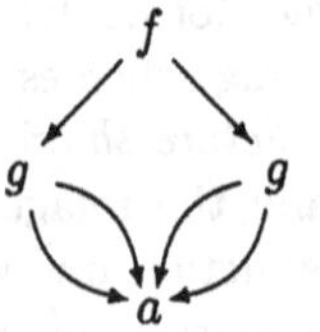

And, if we use the maximal structuring sharing scheme, the result is

$$H_3 = f(g^\beta(a^\gamma, a^\gamma), g^\beta) + h(a^\gamma).$$

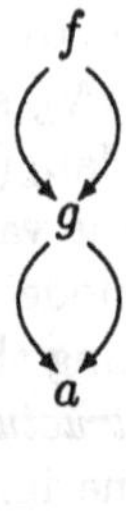

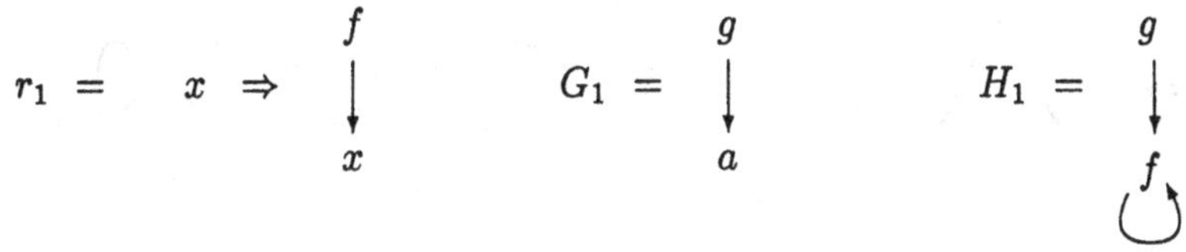

Figure 1: Example 4.1.

Our definition of redex reduction provides several opportunities for nodes with undefined depth to be introduced into H. For instance, in the example above, the node labeled with h in H_i ($i = 1,2,3$) has undefined depth (and does not appear in the pictorial notation). In general, $\theta_L(\rho_L)$ will have undefined depth in H if $\theta_L(\rho_L) \neq \theta_R(\rho_R)$. In many applications of term graph rewriting, the data structure representing the node $\theta_L(\rho_L)$ is overwritten so that it represents the node $\theta_R(\rho_R)$. This provides a very efficient way of performing the redirection step of redex reduction.

Let $\mathcal{R}$ be a set of rewrite rules. An *$\mathcal{R}$-redex* in G is a redex (r,θ) in G where $r \in \mathcal{R}$. For root connected graphs G and H, $G \rightarrow_{\mathcal{R}} H$ means that $H = H'/\rho_{H'}$ where H' is the result of reducing an $\mathcal{R}$-redex Δ in G. We write $G \rightarrow^{\Delta}_{\mathcal{R}} H$ when we want to indicate the redex Δ. Let $\rightarrow^{*}_{\mathcal{R}}$ denote the reflexive and transitive closure of the relation $\rightarrow_{\mathcal{R}}$.

We shall assume in the rest of the paper that all rewrite rules are members of a fixed set $\mathcal{R}$ of rewrite rules. Since $\mathcal{R}$ is fixed we shall use $\rightarrow$, $\rightarrow^{\Delta}$, etc. as abbreviations for $\rightarrow_{\mathcal{R}}$, $\rightarrow^{\Delta}_{\mathcal{R}}$, etc.

A *graph rewriting* of a graph G_0 via $\mathcal{R}$ is either (1) a finite sequence $\Gamma = \langle (G_0,\Delta_0),\ldots,(G_n,\Delta_n),G_{n+1}\rangle$ ($n \geq 0$) such that

$$G_0 \rightarrow^{\Delta_0} G_1 \rightarrow^{\Delta_1} G_2 \rightarrow^{\Delta_2} \cdots \rightarrow^{\Delta_n} G_{n+1}$$

or (2) an infinite sequence $\Gamma = \langle (G_0,\Delta_0),(G_1,\Delta_1),(G_2,\Delta_2),\ldots\rangle$ such that

$$G_0 \rightarrow^{\Delta_0} G_1 \rightarrow^{\Delta_1} G_2 \rightarrow^{\Delta_2} \cdots.$$

4 Redex capturing

Let $r = \langle G_L, G_R\rangle$ be a rewrite rule, $\Delta = (r,\theta_L)$ a redex in a graph G, H^* an extension of G, and θ_R a weak homomorphism from G_R to H^* such that θ_L and θ_R are compatible. Suppose that $\theta_L(\rho_L) \neq \theta_R(\rho_R)$ and that there exists a path from $\theta_R(\rho_R)$ to $\theta_L(\rho_L)$. If Δ is reduced using H^* and θ_R, the redirection step will create a cycle that begins and ends with $\theta_R(\rho_R)$. We call this phenomenon *redex capturing* (since the image of G_R "captures" the redex Δ). It is illustrated by the following two examples.

Example 4.1 [Figure 1] Let $r_1 = \langle x^\alpha, f(x^\alpha)\rangle$ and $G_1 = g(a^\beta)$. When r_1 is applied to the redex in G_1 with root β using either the minimal, maximal, or rule-based structure sharing scheme, the result is $H_1 = g(f^\gamma(f^\gamma)) + a^\beta$.

Example 4.2 [Figure 2] Let $r_2 = \langle f(x^\alpha), g(x^\alpha, x^\alpha)\rangle$ and $G_2 = f^\beta(f^\beta)$. When r_2 is applied to G_2 using either the minimal, maximal, or rule-based structure sharing scheme, the result is $H_2 = g^\gamma(g^\gamma, g^\gamma) + f^\beta(g^\gamma)$.

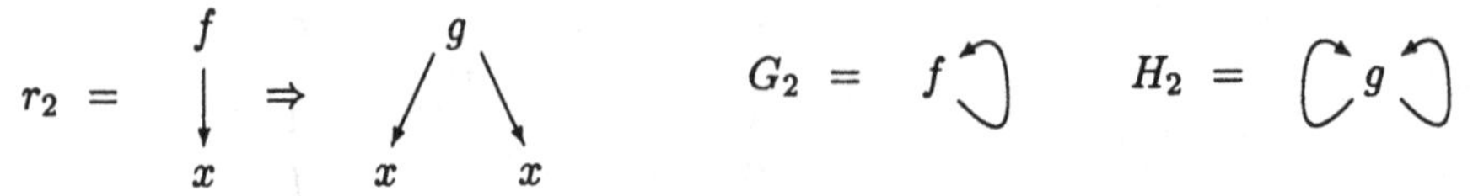

Figure 2: Example 4.2.

Examples 4.1 shows that redex capturing can create cycles even when there are no cycles in the graph being rewritten or in the rewrite rule being applied. This means that, if cyclic graphs are undesired (e.g., so that simple garbage collectors can be used), redex capturing must be avoided. Example 4.2 shows that the application of the most innocuous sort of rewrite rule can, in the right situation, involve redex capturing.

Proposition 4.3 *Let $r = \langle G_L, G_R \rangle$ be a rewrite rule such that G_R contains a node labeled by a variable. Then there is some application of r involving redex capturing.*

For many applications, redex capturing is definitely undesirable because it introduces cycles and possibly unwanted rewritings. However, for some applications, redex capturing can be used as an optimization technique. This is illustrated in the following example.

Example 4.4 [Figure 3] The rewrite rule for the fixed-point combinator Y is

$$r_Y = \langle A(Y, x), A(x, A(Y, x)) \rangle.$$

This rule is interesting because the left side of the rule is a subgraph of the right side. A tempting alternative rule for Y, in which structure sharing is embodied, is

$$r'_Y = \langle A^\alpha(Y, x^\beta), A(x^\beta, A^\alpha) \rangle.$$

However, redex capturing results when this rule is applied using the rule-based structure sharing scheme. The effect is exactly equivalent to applying the cyclic rule

$$r^c_Y = \langle A(Y, x), A^\alpha(x, A^\alpha) \rangle$$

using the minimal structure sharing scheme. In implementations of functional programming languages, r^c_Y is usually used instead of r_Y because it is computationally more efficient (for a discussion of this subject, see [10] and [13]). In particular, iterative procedures defined recursively with Y execute in constant space when r^c_Y is employed, but not when r_Y is used.

As we have shown above, redex capturing can be either desirable or undesirable, depending on the application that is involved. We have not, however, addressed the very important question of whether redex capturing ever leads to an unsound implementation of term rewriting. The rest of the paper is devoted to showing that term graph rewriting with redex capturing is in fact sound with respect to term rewriting, provided only left-linear term rewrite rules are employed (Corollary 6.2). We shall show this by proving that a finite rewriting with redex capturing corresponds to a (possibly infinite) "convergent tree rewriting". This result will be proved in Section 6, and the terminology and machinery behind "convergent tree rewritings" will be given in the next section.

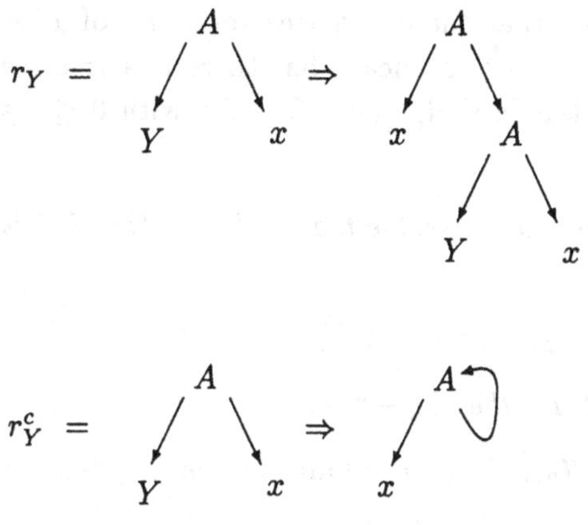

Figure 3: Example 4.4.

5 Tree rewriting

In this section, we explore term graph rewriting restricted to trees and define a notion of convergence for infinite tree rewriting sequences. This subject is also examined by Dershowitz, Kaplan and Plaisted [4] and Kennaway et al. [9].

Let T, U be trees and $d \geq 0$. Define $N_T^d = \{\alpha \in N_T : \mathrm{dp}_T(\alpha) \leq d\}$. A *rooted isomorphism from T to U to depth d* is a map $\varphi : N_T \to N_U$ such that (1) φ is homomorphic at each $\alpha \in N_T^d$ and (2) φ is a bijection from N_T^d to N_U^d. T is *equivalent to U to depth d*, written $T \approx_d U$, if there is a rooted isomorphism from T to U to depth d.

Let $\sigma = \langle T_0, T_1, T_2, \ldots \rangle$ be an infinite sequence of trees. The *limit* of σ, written $\lim(\sigma)$, is a tree T (which is unique, if it exists) such that

$$\forall d \geq 0 \; \exists m \geq 0 \; \forall n \geq m \; [T_n \approx_d T].$$

Define $u(\mathcal{R})$ to be the following set of tree rewrite rules:

$$\{\langle u(G_L), u(G_R) \rangle : \langle G_L, G_R \rangle \in \mathcal{R}\}.$$

Proposition 5.1 *If $G \to_{\mathcal{R}} H$, then $G \to_{u(\mathcal{R})} H$.*

A *tree rewriting via $\mathcal{R}$* is a graph rewriting via $u(\mathcal{R})$ composed entirely of trees. For an infinite tree rewriting $\Gamma = \langle (T_0, \Delta_0), (T_1, \Delta_1), (T_2, \Delta_2), \ldots \rangle$, define $\sigma(\Gamma) = \langle T_0, T_1, T_2, \ldots \rangle$. $\to_{\mathrm{t},\mathcal{R}}$ is the relation $\to_{u(\mathcal{R})}$ restricted to trees, and $\to_{\mathrm{t},\mathcal{R}}^*$ is the reflexive and transitive closure of $\to_{\mathrm{t},\mathcal{R}}$. As before, we shall use $\to_{\mathrm{t}}$, $\to_{\mathrm{t}}^{\Delta}$, etc. as abbreviations for $\to_{\mathrm{t},\mathcal{R}}$, $\to_{\mathrm{t},\mathcal{R}}^{\Delta}$, etc.

Let Γ be a tree rewriting. If $\Gamma = \langle (T_0, \Delta_0), \ldots, (T_n, \Delta_n), T_{n+1} \rangle$, Γ *converges to T* means $T = T_{n+1}$. If $\Gamma = \langle (T_0, \Delta_0), (T_1, \Delta_1), (T_2, \Delta_2), \ldots \rangle$, Γ *converges to T* means that

$$T = \lim(\sigma(\Gamma)) \quad \text{and} \quad \lim_{n \to \infty} \mathrm{dp}_{T_n}(\Delta_n) = \infty.$$

Proposition 5.2 *Let $\Gamma = \langle (T_0, \Delta_0), (T_1, \Delta_1), (T_2, \Delta_2), \ldots \rangle$ be an infinite tree rewriting. If $\lim_{n \to \infty} \mathrm{dp}_{T_n}(\Delta_n) = \infty$, then $\lim(\sigma(\Gamma))$ exists.*

Let $T \rightarrow_{\mathrm{t}}^{\omega} U$ mean that there is a tree rewriting of T which converges to U. For an integer $d \geq 0$, let $T \rightarrow_{\mathrm{t}}^{d,*} U$ mean that there is a tree rewriting $\Gamma = \langle (T_0, \Delta_0), \ldots, (T_n, \Delta_n), T_{n+1} \rangle$ such that $d \leq \mathrm{dp}_{T_i}(\Delta_i)$ for all i with $0 \leq i \leq n$. $T \rightarrow_{\mathrm{t}}^{d,\omega} U$ is defined similarly.

Lemma 5.3 ($\omega\omega$-Lemma) *Assume that each member of $\mathcal{R}$ is left-linear, left-finite, and left-acyclic.*

1. *If $T_0 \rightarrow_{\mathrm{t}}^{\omega} T_1 \rightarrow_{\mathrm{t}} T_2$, then $T_0 \rightarrow_{\mathrm{t}}^{\omega} T_2$.*
2. *If $T_0 \rightarrow_{\mathrm{t}}^{\omega} T_1 \rightarrow_{\mathrm{t}}^{\omega} T_2$, then $T_0 \rightarrow_{\mathrm{t}}^{\omega} T_2$.*
3. *Let $\sigma = \langle T_0, T_1, T_2, \ldots \rangle$ be an infinite sequence of trees such that*
   (a) $T_0 \rightarrow_{\mathrm{t}}^{d_0,\omega} T_1 \rightarrow_{\mathrm{t}}^{d_1,\omega} T_2 \rightarrow_{\mathrm{t}}^{d_2,\omega} \cdots$.
   (b) $\lim_{n\rightarrow\infty} d_n = \infty$.

   Then $T_0 \rightarrow_{\mathrm{t}}^{\omega} \lim(\sigma)$.

The proof of this result is given in the full paper [7]. Our definition of convergence was conceived from the following example:

Example 5.4 [Figure 4] Let $T_1 = u(g(f^{\alpha}(f^{\alpha}), b))$, $T_2 = u(g(f^{\alpha}(f^{\alpha}), c))$, and $\mathcal{R} = \{\langle g(x, b), g(f(x), b)\rangle, \langle b, c\rangle\}$. Consider the infinite tree rewriting $\Gamma = \langle (g(a, b), \Delta_0), (g(f(a), b), \Delta_1), (g(f(f(a)), b), \Delta_2), \ldots \rangle$. Clearly, the depth of each redex Δ_i in its respective tree is 0, and so Γ does not converge. However, $\lim(\sigma(\Gamma)) = T_1$, and $T_1 \rightarrow_{\mathrm{t}} T_2$, but there is no infinite tree rewriting Γ' beginning with $g(a, b)$ such that $\sigma(\Gamma') = T_2$. Hence the $\omega\omega$-Lemma would not be true if $T \rightarrow_{\mathrm{t}}^{\omega} U$ were defined without the condition on the depth of redexes.

There are also simple examples (given in the full paper [7]) which show that the hypotheses of the $\omega\omega$-Lemma cannot be eliminated. The $\omega\omega$-Lemma has been generalized to all ordinals in [4] and [9].

6 Soundness of term graph rewriting

The following theorem is proved using the $\omega\omega$-Lemma:

Theorem 6.1 (Soundness Theorem) *Assume that each member of $\mathcal{R}$ is left-linear, left-finite, and left-acyclic. If $G \rightarrow^* H$, then $u(G) \rightarrow_{\mathrm{t}}^{\omega} u(H)$.*

The corollary below shows that there is a natural sense in which term graph rewriting with redex capturing is a sound method for implementing left-linear term rewrite systems (i.e., term rewrite systems having only left-linear rewrite rules).

Corollary 6.2 *Assume that each member of $\mathcal{R}$ is a left-linear term rewrite rule, and let G and H be finite, acyclic graphs. If there is a finite graph rewriting of G to H via $\mathcal{R}$, then there is a finite tree rewriting of $u(G)$ to $u(H)$ via $\mathcal{R}$ composed entirely of terms.*

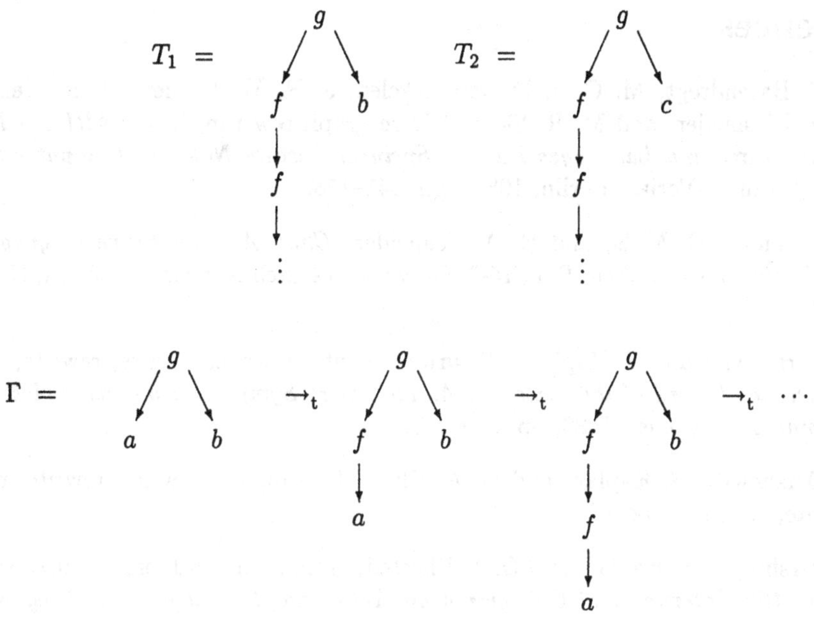

Figure 4: Example 5.4.

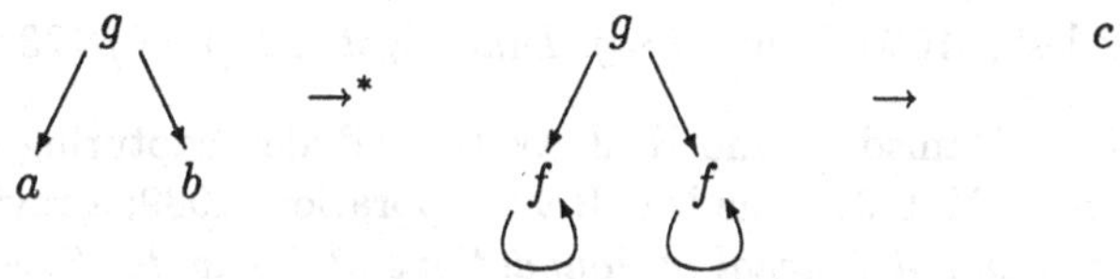

Figure 5: Example 6.3.

The following example shows that the left-linear hypothesis in Corollary 6.2 cannot be discarded:

Example 6.3 [Figure 5] Let $\mathcal{R} = \{\langle a, f(a)\rangle, \langle b, f(b)\rangle, \langle g(x,x), c\rangle\}$. Then $g(a,b) \rightarrow^* g(f^\alpha(f^\alpha), f^\beta(f^\beta))$ (where redex capturing has occurred twice), and hence $g(a,b) \rightarrow^* c$. However, there is no rewriting of $g(a,b)$ to c composed entirely of terms.

We conclude this paper with a corollary about the cyclic Y-rule.

Corollary 6.4 *Let $\mathcal{R}$ be a set of combinator rewrite rules. The cyclic rewrite rule r_Y^c for the fixed-point combinator Y (see Example 4.4) is sound in the sense that, if $G \rightarrow^* H$ using r_Y^c, then $u(G) \rightarrow_t^\omega u(H)$ using r_Y instead of r_Y^c.*

Acknowledgments

The authors wish to express special thanks to Dr. Leonard Monk for many valuable criticisms and suggestions. The authors are also grateful to The MITRE Corporation for supporting this work under the MITRE-Sponsored Research program.

References

[1] H. P. Barendregt, M. C. J. D. van Eekelen, J. R. W. Glauert, J. R. Kennaway, M. J. Plasmeijer, and M. R. Sleep, "Term graph rewriting", in *PARLE - Parallel Architectures and Languages Europe, Springer Lecture Notes in Computer Science 259*, (Springer-Verlag, Berlin, 1987), pp. 141–158.

[2] M. Bickford, C. Mills, and E. A. Schneider, *Clio: An applicative language-based verification system*, Tech. Rep. 15-7, Odyssey Research Associates, Ithaca, NY, June 1989.

[3] N. Dershowitz and S. Kaplan, "Rewrite, rewrite, rewrite, rewrite, rewrite, ...", in *Conference Record of the Sixteenth Annual ACM Symposium on Principles of Programming Languages*, 1989, pp. 250–259.

[4] N. Dershowitz, S. Kaplan, and D. A. Plaisted, "Rewrite, rewrite, rewrite, rewrite, rewrite, ...", to appear.

[5] N. Dershowitz, S. Kaplan, and D. A. Plaisted, "Infinite normal forms", in *Proceedings of the 16th International Colloquium on Automata, Languages, and Programming, Lecture Notes in Computer Science 372*, (Springer-Verlag, Berlin, 1989), pp. 249–262.

[6] W. M. Farmer, J. D. Ramsdell, and R. J. Watro, "A correctness proof for combinator reduction with cycles", *ACM Trans. Prog. Lang. Syst.* **12** (1990) 123–134.

[7] W. M. Farmer, J. D. Ramsdell, and R. J. Watro, "Redex capturing in term graph rewriting", Tech. Rep. M89-36, The MITRE Corporation, 1989; a revised version to appear in the *International Journal of Foundations of Computer Science.*

[8] R. Kennaway, "On 'On graph rewritings'", *Theoret. Comp. Sci.* **52** (1987) 37–58.

[9] J. R. Kennaway, J. W. Klop, M. R. Sleep, F. J. de Vries, "Transfinite reductions in orthogonal term rewriting systems," to appear, Center for Mathematics and Computer Science (CWI) Report CS-R9042, Amsterdam, The Netherlands; a shortened version appears in this volume.

[10] S. L. Peyton Jones, *The Implementation of Functional Programming Languages*, (Prentice Hall, New York, 1987).

[11] J. C. Raoult, "On graph rewritings", *Theoret. Comp. Sci.* **32** (1984) 1–24.

[12] J. Staples, "Computation on graph-like expressions", *Theoret. Comp. Sci.* **10** (1980) 171–185.

[13] D. A. Turner, "A new implementation technique for applicative languages", *Soft. Pract. Exper.* **9** (1979) 31–49.

Rewriting, and Equational Unification: the Higher-Order Cases

Extended Abstract

D.A. Wolfram
Oxford University Computing Laboratory
Programming Research Group
11 Keble Road
Oxford OX1 3QD

Synopsis

We give here a general definition of term rewriting in the simply typed λ-calculus, and use it to define higher-order forms of term rewriting systems, and equational unification and their properties. This provides a basis for generalizing the first- and restricted higher-order results for these concepts. As examples, we generalize Plotkin's criteria for building-in equational theories, and show that pure third-order equational matching is undecidable. This approach simplifies computations in applications involving lexical scoping, and equations. We discuss open problems and summarize future research directions.

1 Introduction

We consider here higher-order rewriting in the simply typed λ-calculus, allowing subterms to be rewritten, and not restricting the order of variables, or the forms of left sides of rewrite rules. Huet and Lang [15] have defined a special form of second-order rewriting for transforming programs. For example, the factorial function

$$fact(x) \Leftarrow \text{ if } x = 0 \text{ then } 1 \text{ else } x * fact(x-1)$$

matches the transformation template

$$f(x) \Leftarrow \text{ if } a(x) \text{ then } b(x) \text{ else } h(d(x), f(e(x)))$$

by the following matching substitution:

$$\{\langle f,\ \lambda x.fact(x)\rangle,\ \langle a,\ \lambda x.x = 0\rangle,\ \langle b,\ \lambda x.1\rangle,\\ \langle h,\ \lambda xy.x * y\rangle,\ \langle d,\ \lambda x.x\rangle,\ \langle e,\ \lambda x.x - 1\rangle\}$$

Rewriting here only occurs at the top level of terms. Another form of second-order rewriting allows subterms to be rewritten [26]. However, the left side of a rewrite rule must be a *pattern* of the form $y(x_1, \ldots, x_n)$ where the x_i are distinct bound variables. This approach relies on the decidability of unification for such second-order patterns [24]. Both second-order definitions [15, 26] are subsumed by our higher-order rewriting systems. This makes the definition of rewriting more intricate.

As applications of this definition, we then consider higher-order generalizations of term-rewriting systems, and equational unification and their properties.

Our definition of term rewriting system is introduced as a basis for generalizing the extensive theory of first-order systems [6, 17] to the full higher-order case. Among many potential applications, using higher-order term rewriting systems in first-order rewrite-rule based languages and theorem provers [7, 9, 27] would simplify their treatment of bound variables. Special encodings, and supplementary code for α-conversions and substitutions become unnecessary.

Another application of higher-order rewriting is our definition of higher-order equational unification which merges the approaches of Huet [12] and Plotkin [29]. It subsumes most forms of unification and matching including first-order equational unification [29] and unification of simply typed λ-terms in the presence of a first-order equational theory [31]. We show that Plotkin's criteria [29] for building-in equational theories in resolution theorem provers follow in the higher-order case from our definitions.

Although no definitions were given, higher-order equational unification was identified by Siekmann [30] in his 1984 survey of unification theory when he considered the tractability of proposed higher-order resolution theorem provers:

> *But may be T-unification for ω-order logics is not more* but less complex *than free ω-unification?*

We shall prove that the special case of third order equational matching is undecidable by a reduction from a form of Hilbert's Tenth Problem [11].

The undecidability of higher order unification [8] has been been used to advocate avoiding higher-order functions in ordinary programming. *Ordinary* programming excepts theorem provers such as Isabelle [28], and TPS [25] which use higher-order unification intensively. They show that most time is spent searching for proofs rather than for unifiers. Moreover, there are decidable unification problems of arbitrarily high order, and the set of all decidable unification problems is polynomial time decidable [32]. Defining higher-order equational unification here is a first step towards determining its decidable subcases and its practicality.

The two applications using the definition of higher-order rewriting — higher-order term rewriting systems, and equational unification — in turn lead to open problems and future directions for research.

Acknowledgements

This work is based on a section of the 1988 version of my Ph.D. Dissertation. I should like to thank Larry Paulson, Roger Hindley, and Gérard Huet for their comments on this work. This research was supported by a Research Fellowship at Christ Church,

Oxford, the Rae and Edith Bennett Travelling Scholarship, an ORS Award, University of Cambridge Computer Laboratory, Trinity College, Cambridge, and the Cambridge Philosophical Society.

2 Preliminary Definitions

We assume that the reader has some familiarity with the simply typed λ-calculus, in particular: the definitions of the order of a type symbol and a term, and the set of free variables $\mathcal{F}(t)$ of a term t. We denote types by α, β, the set of individual types by T_0, write variables as u, v, w, x, y, z, denote the set of all variables by $\mathcal{X}$, write constants as $A, B, C, \ldots, T$, and denote terms by p, q, r, s, t. These symbols may appear with numerical subscripts.

The type of t is written $\tau(t)$. We also assume familiarity with α- and β-, and $\beta\overline{\eta}$-conversions[1], the Strong Normalization and typed Church-Rosser Theorems, the $=_\beta$ relation [1, 4] and the $=_{\beta\overline{\eta}}$ relation. We denote that t is a $\beta\overline{\eta}$-normal form of s by $s \rhd_{\beta\overline{\eta}} t$.

We abbreviate terms in $\beta\overline{\eta}$-normal form as follows:

$$(\cdots(t_1\ t_2)t_3)\cdots)t_m) \text{ by } t_1(t_2, \ldots, t_m),$$

and

$$\lambda x_1.(\lambda x_2.(\cdots(\lambda x_n.t)\cdots) \text{ by } \lambda x_1 \cdots x_n.t$$

where the x_i are distinct variables. Such terms have the form

$$\underbrace{\lambda x_1 \cdots x_n}_{binder} . \overbrace{@\underbrace{(t_1, \ldots, t_m)}_{arguments}}^{matrix}$$

where the binder or arguments may be absent, and @ is either a variable or a constant. *binder*(t) denotes the binder of a term t, and *matrix*(t) denotes its matrix.

The definitions of substitution, the domain $D(\sigma)$ of a substitution σ, its set of introduced variables $I(\sigma)$, and a renaming substitution are not repeated here for brevity [12]. We denote substitutions by $\delta, \gamma, \mu, \pi, \rho, \theta$.

We shall be more specific than usual in identifying the subterms of a term. To do this, we generalize the notion of occurrence used in first-order term rewriting[2] [16].

Definition 2.1 An *occurrence* is an element of the smallest set formed using the rules:

- ϵ is an occurrence.
- 0 and 1 are occurrences.
- If i and j are occurrences, then $i.j$ is an occurrence.
- If i is an occurrence, then $i.\epsilon = \epsilon.i = i$.
- If i, j, and k are occurrences, then $(i.j).k = i.(j.k)$.

[1]This is Huet's $\beta\eta$-normal form [12].
[2]Section 3.1, page 807.

We can now specify the occurrences of a term, and identify its subterms exactly.

Definition 2.2 The *set of occurrences* of a term t, $\mathcal{O}(t)$, and the *subterm of t at occurrence i*, t/i, where $i \in \mathcal{O}(t)$ are defined by:

1. For every term t, $\epsilon \in \mathcal{O}(t)$, and $t/\epsilon = t$.
2. If t is a variable or a constant, then $\mathcal{O}(t) = \{\epsilon\}$.
3. If t has the form $\lambda x.r$, then $\mathcal{O}(t) = \{\epsilon\} \cup \{0.l \mid l \in \mathcal{O}(r)\}$, and $t/0.l = r/l$.
4. If t has the form $(s_0\ s_1)$, then
$$\mathcal{O}(t) = \{\epsilon\} \cup \{0.l \mid l \in \mathcal{O}(s_0)\} \cup \{1.l \mid l \in \mathcal{O}(s_1)\},$$
$t/0.l = s_0/l$, and $t/1.l = s_1/l$.

We define higher-order replacement, which continues the generalization from the first-order case.

Definition 2.3 For $i \in \mathcal{O}(t)$, $t[i \leftarrow s]$ where $\tau(t/i) = \tau(s)$ is

1. $t[\epsilon \leftarrow s] = s$, when $i = \epsilon$.
2. $\lambda x_1.(\lambda x_2.(\cdots(\lambda x_n.((\cdots((t_1\ t_2)t_3)\cdots t_j)\cdots t_m)\cdots)[j.k \leftarrow s] =$
$\lambda x_1.(\lambda x_2.(\cdots(\lambda x_n.((\cdots((t_1\ t_2)t_3)\cdots t_j[k \leftarrow s])\cdots t_m)\cdots),$
when $i = j.k$ and $1 \le j \le m$.

Free variables in the replacement term s can become bound after the replacement.

3 Higher-Order Rewriting

In keeping with the definitions of first-order rewriting [17], we denote an *equation* by $q = r$ for example, where $\tau(=) = (\alpha, \alpha \to o)$ and $\tau(q)$ and $\tau(r)$ are α. An *equational theory* is a set of equations denoted by E.

Higher-order rewriting uses higher-order replacement. We shall rewrite a term p to $p[i \leftarrow r\theta]$ where $i \in \mathcal{O}(p)$, and $q = r$ is an equation. The free variables in the replacement term $r\theta$ for p/i can become bound after the replacement. We use the non-interference property, defined below, to prevent this from happening.

Definition 3.1 Let p be a term and $i \in \mathcal{O}(p)$. A set of variables X is *non-interfering* for p at occurrence i if and only if for every $x \in X$, there is no $j \in \mathcal{O}(p)$ such that p/j has the form $\lambda x.t$ and $i = j.k$ where $k \neq \epsilon$.

Example 3.2 The free variables of the equation
$$\lambda t.h(t, \lambda y_1 y_2.B(y_1, y_2), z) = \lambda v.w(\lambda u.h(x, \lambda x_1 x_2.x_1, u))$$
are $\{h, z, w, x\}$. They are interfering for the term $\lambda xy.f(z, \lambda u.y(B(u, z)))$ at occurrence 0.0.1, which is the subterm $\lambda u.y(B(u, z))$.

This is because the condition of Definition 3.1 is not met with $j = \epsilon$ and $k = 0.0.1$ for the free variable x of the equation.

To prevent this, we can apply the renaming substitution $\{\langle x, x_1\rangle\}$ to the equation to obtain

$$\lambda t.h(t, \lambda y_1 y_2.B(y_1, y_2), z) = \lambda v.w(\lambda u.h(x_1, \lambda x_1 x_2.x_1, u)).$$

We can now define higher-order rewriting using Definition 3.1 of non-interference.

Definition 3.3 Let p and s be closed terms of the same type in $\beta\overline{\eta}$-normal form and E be an equational theory. The *higher-order rewrite relation* $p \longrightarrow_{[i,q=r,\pi]} s$ holds if and only if

- $q = r$ or $r = q$ is an equation in E, and $q = r$ has a $\beta\overline{\eta}$-normal form

$$\lambda x_1 \cdots x_n.t_1 = \lambda x_1 \cdots x_n.t_2$$

- i is an occurrence in $\mathcal{O}(p)$ and $\tau(matrix(p/i))$ is $\tau(t_1)$.

- π is a substitution such that $t_1\pi \rhd_\alpha matrix(p/i)$, and

$$\mathcal{F}(binder(p/i).(t_1\pi) = binder(p/i).(t_2\pi))$$

  is non-interfering for p at i.

- $p[i \leftarrow binder(p/i).(t_2\pi)] \rhd_{\beta\overline{\eta}} s$.

Example 3.4 Here is an example of higher-order rewriting. Let

$$\lambda t.h(t, \lambda y_1 y_2.B(y_1, y_2), z_1) = \lambda t.w(\lambda u.h(x_1, \lambda x_1 x_2.x_1, u))$$

be a $\beta\overline{\eta}$-normal form of an equation in E, and $\lambda xy.f(z, \lambda t.y(B(C, z)))$ be a term.

The substitution π is $\{\langle h, \lambda w_1 w_2 w_3.y(w_2(w_1, z))\rangle, \langle t, C\rangle\}$. The $\beta\overline{\eta}$-normal form of the matrix of the left side of the equation after π has been applied is $y(B(C, z))$.

The matrix of the right side of the equation after π has been applied is the term $w(\lambda u.y(x_1))$, so that $\lambda xy.f(z, \lambda t.w(\lambda u.y(x_1)))$ is the rewritten term.

4 Applications of Higher-Order Rewriting

4.1 Higher-Order Term Rewriting Systems

We define below a higher-order term rewriting system. This generalizes the first-order case [17].

Definition 4.1 A *higher-order term rewriting system* is a set of directed equations $\mathcal{R} = \{q_i \rightarrow r_i \mid i \in I\}$ such that for all $q \rightarrow r \in \mathcal{R}$, $\mathcal{F}(q) \supseteq \mathcal{F}(r)$, and $\tau(q)$ is $\tau(r)$.

We now use Definition 3.3 of the higher-order rewrite relation to define reduction.

Definition 4.2 Given a higher-order term rewriting system $\mathcal{R}$, the *reduction relation* $p \rightarrow_{\mathcal{R}} s$ holds if and only if there is $q \rightarrow r \in \mathcal{R}$ such that $p \rightarrow_{[i,q=r,\pi]} s$ holds.

By continuing the analogy with the first-order case, we could define the confluence, local confluence, and termination properties of $\mathcal{R}$, critical pairs, and normal forms relative to $\mathcal{R}$. The definitions provide a basis for determining whether the Knuth-Bendix Theorem and Completion Algorithm [21] can be similarly generalized.

In Theorem 4.3 below, we relate higher-order equational deduction to higher-order rewriting systems. We shall use the definition of derivation of an equation from hypotheses E in the higher-order logic, $\mathcal{Q}_0$, in the case where E is an equational theory [1]. Derived equality rules of $\mathcal{Q}_0$ are given in Figure 1. Substitution, α-, β-, and η-conversions are theorems of $\mathcal{Q}_0$ which are expressed equationally.

Reflexivity.

$$\frac{}{E \vdash s = s}$$

Symmetry.

$$\frac{E \vdash s = t}{E \vdash t = s}$$

Transitivity.

$$\frac{E \vdash r = s, \quad E \vdash s = t}{E \vdash r = t}$$

Congruence.

$$\frac{E \vdash p = q \quad E \vdash r = s}{E \vdash (p\, r) = (q\, s)}$$

Abstraction. Provided that $x \notin \mathcal{F}(E)$,

$$\frac{E \vdash s = t}{E \vdash \lambda x.s = \lambda x.t}$$

Figure 1: Equality Rules of $\mathcal{Q}_0$

We also recall [16] that $\stackrel{*}{\rightarrow}_{\mathcal{R}}$ is the transitive and reflexive closure of $\rightarrow_{\mathcal{R}}$, and that for all terms s and t, $s \downarrow_{\mathcal{R}} t$ if and only if there is a term p such that $s \stackrel{*}{\rightarrow}_{\mathcal{R}} p$ and $t \stackrel{*}{\rightarrow}_{\mathcal{R}} p$.

The following theorem can be seen as showing that Definition 3.3 is an appropriate one for higher-order terms.

Theorem 4.3 *Let* $\mathcal{R} = \{q_i \rightarrow r_i \mid i \in I\}$ *be a confluent higher-order rewriting system, and* $E = \{q_i = r_i \mid i \in I\}$. *Then* $E \vdash s = t \Leftrightarrow s \downarrow_{\mathcal{R}} t$.

If $\mathcal{R}$ is also a terminating higher-order rewriting system, then this theorem implies that higher-order rewriting can be used to decide the derivability of an equation from a higher-order equational theory. We also note that Henkin's completeness theorem [10] as applied to $\mathcal{Q}_0$ [1] is a suitable generalization to higher-order equational logic of Birkhoff's completeness theorem for first-order equational logic [2].

4.2 Higher-Order Equational Unification

The possibility of higher-order equational unification when the equational theory is a higher-order one does not seem to have been considered in any detail before [31]. We now use Definition 3.3 of higher-order rewrite relation to define higher-order equational unifiers.

Definition 4.4 A substitution θ is a *higher-order equational unifier* or *E-unifier* of two terms p and s of the same type for an equational theory E if and only if

$$p\theta \longrightarrow_{[i_1,\, q_1=r_1,\, \pi_1]} p_1 \longrightarrow_{[i_2,\, q_2=r_2,\, \pi_2]} \cdots$$
$$\longrightarrow_{[i_{n-1},\, q_{n-1}=r_{n-1},\, \pi_{n-1}]} p_{n-1} \longrightarrow_{[i_n,\, q_n=r_n,\, \pi_n]} s\theta$$

Definition 4.5 The relation between $p\theta$ and $s\theta$ is denoted $p\theta \stackrel{*}{\longleftrightarrow}_E s\theta$ and it is the reflexive, symmetric, and transitive closure of the higher-order rewrite relation.

The definition of higher-order equational unifier is now extended to sets of pairs of terms.

Definition 4.6 A *disagreement pair* is a pair of terms $\langle p, s\rangle$ where $\tau(p) = \tau(s)$. A *disagreement set* W is a finite set of disagreement pairs. The set $\mathcal{F}(W)$ is $\bigcup_{\langle p,s\rangle \in W} \mathcal{F}(p) \cup \mathcal{F}(s)$.

To define general properties of E-unifiers, we shall use the following definition [12] of ordering of substitutions.

Definition 4.7 Given a set V of variables, and σ and ρ, we have

- $\rho \leq_V \sigma$ if and only if $\exists \mu : \forall x \in V : x\sigma \rhd_\alpha x\rho\mu$.
- $\sigma =_V \rho$ if and only if $\forall x \in V : x\sigma \rhd_\alpha x\rho$.
- $\sigma = \rho$ if and only if $\sigma =_{\mathcal{X}} \rho$.

It is well-known that for first-order E-unification, the cardinality of the set of unifiers of a disagreement set can be infinite [29].

Definition 4.8 Let E be an equational theory, and W be a disagreement set.

- An E-unifier of W is a substitution which is an E-unifier of every disagreement pair in W. The set of all E-unifiers of W is denoted $\mathcal{U}_E W$.
- $\mathcal{T}_E W$ is the set of all recursively enumerable E-unifiers of W by an E-unification procedure $\mathcal{T}_E$.

- A substitution γ is a *semi-E-unifier* or *E-matcher* of a disagreement set W if and only if for every $\langle s,t\rangle \in W$ we have $s\gamma \overset{*}{\longleftrightarrow}_E t$. The set of all such semi-E-unifiers is denoted by $\mathcal{M}_E W$.

Here are general properties of sets of substitutions for classifying those in Definition 4.8. In the definition, Γ stands uniformly for either $\mathcal{U}_E W$, or $\mathcal{M}_E W$.

Definition 4.9 Let Γ and $\mathcal{S}$ be a sets of substitutions, E be an equational theory, W a disagreement set, and $\mathcal{F}(W) \subset L$. The set Γ may satisfy some of the following properties:

Soundness. $\Gamma \subseteq \mathcal{S}$.

Completeness. $(\forall\sigma \in \mathcal{S}) : (\exists\gamma \in \Gamma) : \gamma \leq_L \sigma$.

Minimality. $\forall\gamma_1, \gamma_2 \in \Gamma : (\gamma_1 \neq \gamma_2) \Rightarrow \gamma_2 \not\leq_L \gamma_1$.

Independence. $\forall\gamma_1, \gamma_2 \in \Gamma : (\gamma_1 \neq \gamma_2) \Rightarrow \neg\exists\theta\theta' : \gamma_1\theta =_L \gamma_2\theta'$.

For example, in Definition 4.9, when $\mathcal{S}$ is $\mathcal{M}_E W$ and $E = \emptyset$, sound and complete sets of minimal matchers can be enumerated [13].

The main case of interest for automated theorem proving is when Γ is $\mathcal{T}_E W$, and $\mathcal{S}$ is $\mathcal{U}_E W$. We consider this case in more detail in the next section. For first-order E-unification, the soundness and completeness conditions are equivalent to Plotkin's [29]. Even in this first-order case the minimality property does not hold in general [14]. Classifying higher-order equational unifiers and matchers using these criteria would supplement the previous results [13, 30].

4.2.1 Logical Soundness and Completeness

In this section, we relate the higher-order equational unifiability of a disagreement set to the validity of equations formed from its disagreement pairs. This result generalizes Plotkin's criterion for building-in equational theories in first-order resolution theorem provers [29] to the higher-order case [33].

The reader is referred to Andrews [1] for a recent definition of a general model. We shall assume that constants in terms and substitutions are from a fixed signature, and each constant in that signature has a denotation in each general model considered.

Definition 4.10 Let $\sim$ be an equivalence relation on closed terms of the same type. If there is a general model $\mathcal{M}$ such that for all terms s and t, $s \sim t$ implies $\mathcal{M} \models s = t$, then $\sim$ is a *logical term relation*, and we call $\mathcal{M}$ a *general $\sim$-model*

Example 4.11 An example of a logical term relation is $\beta\overline{\eta}$ equality on closed terms of the same type.

Definition 4.12 Given a logical term relation $\sim$, if for every general $\sim$-model $\mathcal{M}$, we have $\mathcal{M} \models s = t$ where s and t are terms, we write $\models_\sim s = t$.

An equational theory E abbreviates the formula which is the conjunction of equations in E all of whose free variables are universally quantified at the outermost level. We now relate higher-order equational unification to general models. We define the logical completeness property of $\mathcal{T}_E$ using Definition 4.7, as follows:

Definition 4.13 (*Logical Completeness*). Given a disagreement set W, if $\models_{\sim} E$ and if for every $\langle p, s\rangle \in W$ there is θ such that $\models_{\sim} p\theta = s\theta$, then there is $\sigma \in \mathcal{T}_E W$ such that $\sigma \leq_{\mathcal{F}(W)} \theta$.

We conjecture that given any E where $\models_{\sim} E$ there is a general way to form $\mathcal{T}_E$ with the completeness property. Generalizing narrowing [19] may provide such a procedure.

We shall assume that for some logical term relations $\sim$, E can be constructed such that $\models_{\sim} E$. Using this assumption, we can show that $\mathcal{T}_E$ satisfies the *logical soundness* property:

Theorem 4.14 (*Logical Soundness*). Given a disagreement set W, for every $\sigma \in \mathcal{T}_E W$ and for every $\langle p, s\rangle \in W$, $\models_{\sim} p\theta = s\theta$ where $\sigma \leq_{\mathcal{F}(W)} \theta$.

The next theorem generalizes Plotkin's criteria [29]. Its proof is omitted here for brevity.

Theorem 4.15 *If $\mathcal{T}_E W$ is a sound and complete complete set of E-unifiers for the disagreement set W then $\mathcal{T}_E$ has the logical soundness and logical completeness properties.*

Theorems 4.14 and 4.15 also apply to equational matching problems for simply typed λ-terms of all orders. This is because without loss of generality, we can define an equational matching problem as an equational unification problem in which at most one term in each disagreement pair can contain free variables[3].

4.2.2 Pure Third-Order Equational Matching

First-order equational matching with constant symbols is undecidable [3], but the pure third-order case has not been defined before. Here, terms do not contain constants, and the order of free variables is at most three. We now show that pure third-order equational matching is undecidable.

Hilbert's Tenth Problem [11] is to find an algorithm to determine whether or not a polynomial $P(x_1, \ldots, x_n) = 0$ has a solution in integers. This is not possible because finding whether or not such integers exist is undecidable [23]. The problem for non-negative integers is also undecidable [5]. We reduce this problem to pure third-order equational matching.

Without loss of generality, we only consider equations of the form

$$\exists x_1 \cdots x_n (p(x_1, \ldots, x_n) = q(x_1, \ldots, x_n)) \qquad (1)$$

where $p(x_1, \ldots, x_n)$ and $q(x_1, \ldots, x_n)$ are polynomials with coefficients which are non-negative integers, and whose variables $x_1, \ldots, x_n$ range over non-negative integers.

The equation can be reduced to a pure third-order equational matching problem by using the following encodings which are simply typed versions of their untyped counterparts [20].

Definition 4.16 A non-negative integer constant, a numerical variable, $\times$, and $+$ are λ-definable as follows:

[3]This follows from Lemma 3.16 of Huet [13].

- A coefficient or constant m is represented by the simply typed Church numeral $\overline{m} = \lambda xy.\underbrace{x(x(\cdots x(}_{m}y)\cdots))$ with type restricted to

$$N = ((\iota \rightarrow \iota), \iota \rightarrow \iota).$$

- A variable x is represented by the term $\overline{x}$ which is a variable with type N.

- Multiplication of terms of type N is represented by $\overline{\times} = \lambda xyz.x(y(z))$ with type $(N, N \rightarrow N)$, where $\tau(z) = (\iota \rightarrow \iota)$.

- Addition of terms of type N is represented by $\overline{+} = \lambda xyzt.x(z, y(z,t))$ of type $(N, N \rightarrow N)$, where $\tau(z) = (\iota \rightarrow \iota)$ and $\tau(t) = \iota$.

Using these representations, the polynomials p and q can be encoded uniquely as pure simply typed λ-terms, $\overline{p}$, and $\overline{q}$, respectively. The matching problem has the form $\{\langle \lambda e.e(\overline{p}, \overline{q}),\ \lambda e.e(\lambda xy.y, \lambda xy.y)\rangle\}$ where $\tau(e) = (N, N \rightarrow \iota)$.

The equality theory E is

$$\{\lambda e.e(\lambda xy.x(f), \lambda xy.x(g)) = \lambda e.e(\lambda xy.f, \lambda xy.g)\}$$

where $\tau(f) = N$ and $\tau(g) = N$.

The proof of undecidability uses the observation that when the values of $x_1, \ldots x_n$ have been determined, rewriting with the equation in E successively decrements the representations of the evaluations of p and q until at least one has been reduced to the representation of zero. We have the following result:

Theorem 4.17 *Pure third-order equational matching is undecidable.*

5 Future Directions

Apart from the open problems identified above, higher-order rewriting, its applications, and their properties could be further generalized to other type systems, such as dependent types [22], and their polymorphic versions. Problems requiring these generalizations could occur in currently proposed logical frameworks [18].

Exploring properties, and providing practical procedures for the monomorphic simply typed versions of the applications of higher-order rewriting identified here may give preliminary insights for these more general cases.

References

[1] P.B. Andrews, *An Introduction to Mathematical Logic and Type Theory: To Truth Through Proof*, Academic, Orlando, 1986.

[2] G. Birkhoff, On the Structure of Abstract Algebras, *Proceedings of the Cambridge Philosophical Society* (1935) *31* 433–454.

[3] A. Bockmayr, A Note on a Canonical Theory with Undecidable Unification and Matching Problem, *Journal of Automated Reasoning* **3** (1987) 379–381.

[4] A. Church, A Formulation of the Simple Theory of Types, *Journal of Symbolic Logic* **5** (1940) 56–68. D.C., 1988, 82–90.

[5] M. Davis, *Computability & Unsolvability*, McGraw-Hill, New York, 1958.

[6] N. Dershowitz, Computing with Rewrite Rules, *Information and Control* **65** (1985) 122-157.

[7] S.J. Garland and J.V. Guttag, An Overview of LP: The Larch Prover, *Proceedings of the Third International Conference on Rewriting Techniques and Applications*, Lecture Notes in Computer Science **355**, Springer, Berlin, 1989, 137–151.

[8] J.A. Goguen, Higher Order Functions Considered Unnecessary for Higher Order Programming, in: (D.A. Turner, Ed.), *Proceedings of the University of Texas Year of Programming Institute on Declarative Programming*, Addison-Wesley, 1988.

[9] J.A. Goguen and T. Winkler, Introducing OBJ3, Report SRI-CSL-88-9, SRI International, Menlo Park, 1988.

[10] L. Henkin, Completeness in the Theory of Types, *Journal of Symbolic Logic* **15** (1950) 81–91.

[11] D. Hilbert, Mathematische Probleme, Vortrag gehalten auf dem internationalen Mathematiker-Kongreß zu Paris, 1900. *Nachr. Ges. Wiss. Göttingen*, math.-phys. Kl., (1900) 253–297. English translation by H.W. Newsom in: *Bulletin of the American Mathematical Society* (1901–1902) 437–479.

[12] G.P. Huet, A Unification Algorithm for Typed λ-Calculus, *Theoretical Computer Science* **1** (1975) 27–57.

[13] G.P. Huet, *Résolution d'équations dans des Langages d'Ordre* $1, 2, \ldots, \omega$, Thèse de Doctorat d'Etat, Université Paris VII, Paris, (1976).

[14] G.P. Huet and F. Fages, Complete Sets of Unifiers and Matchers in Equational Theories, *Theoretical Computer Science* **43** 2,3 (1986) 189–200.

[15] G.P. Huet and B. Lang, Proving and Applying Program Transformations Expressed with Second-Order Patterns, *Acta Informatica* **11** (1978) 31–55.

[16] G.P. Huet, Confluent Reductions: Abstract Properties and Applications to Term Rewriting Systems, *Journal of the ACM* **27**, 4 (1980) 797–821.

[17] G.P. Huet and D.C. Oppen, Equations and Rewrite Rules: A Survey, in: *Formal Languages: Perspectives and Open Problems*, R. Book, (Ed.), Academic, 1980, 349–405.

[18] G. Huet and G. Plotkin, Eds., *Proceedings of the First Workshop on Logical Frameworks*, INRIA, 1990.

[19] J.M. Hullot, Canonical Forms and Unification, *Proceedings of the Fifth Conference on Automated Deduction*, Lecture Notes in Computer Science **87**, Springer, Berlin, 1980, 318–334.

[20] S.C. Kleene, λ-definability and Recursiveness, *Duke Math. J.* **2** (1936) 340–353.

[21] D. Knuth and P. Bendix, Simple Word Problems in Universal Algebra, in: *Computational Problems in Abstract Algebra*, (J. Leech, Ed.), Pergamon Press, 1970, 263–297.

[22] P. Martin-Löf, Constructive Mathematics and Computer Programming, in: (C.A.R. Hoare and J.C. Shepherdson, Eds.), *Mathematical Logic and Programming Languages*, Prentice-Hall, 1985, 167–184.

[23] Ju.V. Matiyasevič, Diophantine Representation of Enumerable Predicates, (In Russian), *Izvestija Akademii Nauk SSSR*, Serija Matematika **35** (1971) 3–30. English translation in: *Mathematics of the USSR — Izvestija* **5** (1971) 1–28.

[24] D. Miller, A Logic Programming Language with Lambda-Abstraction, Function Variables, and Simple Unification. *Extensions of Logic Programming*, Lecture Notes in Computer Science, Springer, Berlin. To appear.

[25] D. Miller, E.L. Cohen, and P.B. Andrews, A Look at TPS, *Proceedings of the Sixth Conference on Automated Deduction*, Lecture Notes in Computer Science **138**, Springer, New York, 1982, 50–69.

[26] T. Nipkow, A Critical Pair Lemma for Higher-Order Rewrite Systems and its Application to λ^*, Draft version, in: [18] (1990) 361–376.

[27] M. O'Donnell, *Equational Logic as a Programming Language*, MIT Press, 1985.

[28] L.C. Paulson, The Foundation of a Generic Theorem Prover, *Journal of Automated Reasoning* **5** (1989) 363–397.

[29] G.D. Plotkin, Building-in Equational Theories, in: (B. Meltzer and D. Michie, Eds.), *Machine Intelligence* **7**, Edinburgh, 1972, 73–90.

[30] J.H. Siekmann, Universal Unification, *Proceedings of the Seventh Conference on Automated Deduction*, Lecture Notes in Computer Science **170**, Springer, New York, 1984, 1–42.

[31] W. Snyder, Higher-Order E-Unification, *Proceedings of the Tenth Conference on Automated Deduction*, Lecture Notes in Computer Science, Springer, Berlin, 1990.

[32] R. Statman, On the Existence of Closed Terms in the Typed λ-Calculus II: Transformations of Unification Problems, *Theoretical Computer Science* **15** (1981) 329–338.

[33] D.A. Wolfram, *The Clausal Theory of Types*, Cambridge University Press. *To appear.*

Adding Algebraic Rewriting to the Untyped Lambda Calculus (Extended Abstract)

Daniel J. Dougherty
Wesleyan University
Middletown, CT 06457
ddougherty@eagle.wesleyan.edu

Abstract

We investigate the system obtained by adding an algebraic rewriting system R to the untyped lambda calculus. On certain classes of terms, called here "stable", we prove that the resulting calculus is confluent if R is confluent, and terminating if R is terminating. The termination result has the corresponding theorems for several typed calculi as corollaries. The proof of the confluence result yields a general method for proving confluence of typed β reduction plus rewriting; we sketch the application to the polymorphic calculus F_ω.

1 Introduction

Term rewriting systems and the untyped lambda calculus are each universal models of computation. Algebraic reduction is a natural technique for computing with standard functions such as successor and addition and with operations defined by equations over an abstract data type, while the lambda calculus has proven to be a powerful model of several aspects of modern programming languages (e.g., programmer-defined functions and their parameter passing mechanisms). It would seem profitable to combine the two modes, allowing each to do what it does best. For instance, as pointed out in [Bre88], algebraic rules such as rewriting $x - x$ to 0 could be treated as code optimizations in a functional language. From the point of view of the logic of programming, the equations from which rewriting rules are defined should allow the use of first order properties of the data to be involved in the higher order reasoning about programs.

The following example (from [BM87]) shows that the combination of algebra and untyped lambda calculus is problematic. Suppose we have a language which allows any term $x - x$ to be rewritten to 0, and a term $succ(x) - x$ to be rewritten to 1, and further suppose that terms have fixed points, so that there is a term X with X evaluating to $succ(X)$. Then $X - X$ evaluates to 0 and to 1.

The insight in Val Breazu-Tannen's [Bre88] is that restriction to various type disciplines should allow lambda terms to inherit nice properties from the algebraic system. (See also [BM87], [Bre87]). In [Bre88], it was shown that if a confluent algebraic system is added to the simply typed lambda calculus, the resulting system combining β and algebraic reductions is confluent; the question of preservation of termination was left open. Jean Gallier and Breazu-Tannen have shown that the polymorphic lambda calculus remains confluent when enriched by confluent algebraic rewriting [BG9?a] and remains terminating when combined with terminating algebraic rewriting [BG9?b]. Franco Barbanera [Bar90] has given a proof of many-sorted termination preservation for Girard's F_ω calculus [Gir72] and shown how to lift this result to the Calculus of Constructions [CH88].

In this paper, inspired by [Bre88] but independent of the other work cited above, we explore the pure combinatorics of the interaction between algebraic reduction and β-reduction and attempt to isolate those features of typed systems which allow the smooth interaction between algebraic and higher-order computation. Our strategy is to work in the untyped lambda calculus and to avoid making any use of particular type mechanisms. Our claim is that the relevant virtues of typing are embodied in the notion of ***stability*** below, and further, that the constructions given here provide an elementary explanation for the preservation of confluence and of termination.

We will restrict attention to β-strongly normalizing terms to avoid the difficulties arising from the existence of fixed points, and we will assume that the algebraic signature Σ prescribes arities for its function symbols. We will not want to insist that function-arity is respected in the strictest sense, since we certainly wish to allow function symbols to occur (say, as arguments to higher-order procedures) without being instantiated by their arguments. But when the rewrite system R is thought of as rewriting terms of base type, no function symbol should be presented with *more* arguments than its arity prescribes. Somewhat surprisingly, in the presence of β-strong normalization this very elementary form of type-checking, which we may call "arity-checking", will suffice to ensure inheritance of confluence or termination, without a commitment to a specific type discipline.

Definition 1.1 A set S of terms is *R-stable* if

1. S is closed under taking subterms and under $\xrightarrow{\beta R}$, and
2. each term in S is strongly normalizing under β-reduction and contains no subterm of the form $(f A_1 \cdots A_n)$ where n is greater than the arity of f.

Examples of R-stable sets include the sets of $\Lambda(\Sigma)$ terms which receive a type in the simply typed lambda calculus, polymorphic lambda calculi ([Gir71], [Gir72], [Rey74]), certain systems of dependent types ([Mac86]), and the Calculus of Constructions ([CH88]). When an R and an R-stable set are available from the context we may abuse notation and speak of "stable terms".

Our main results are:

- If R is confluent then $\xrightarrow{\beta R}$ is confluent on R-stable terms.
- If R is terminating then $\xrightarrow{\beta R}$ is terminating on R-stable terms.

The constructions can be outlined as follows. To show preservation of confluence, we follow Breazu-Tannen [Bre88] in projecting βR-reduction to (confluent) R-reduction on β-normal forms, but simplify and generalize his technique by passing to a bottom- up/parallel version of R-reduction which almost commutes, in a technical sense, with β reduction. To analyze termination, we show that any βR-reduction out of a non-β-normal form M can be projected along a properly chosen β reduction in such a way that if the first reduction is infinite then so is the projection. Under the conditions of stability, we will be able to conclude that if M allows an infinite βR-reduction then the β-normal form of M will allow an infinite R-reduction – a contradiction if R is assumed to be a terminating rewrite system.

A consequence of the approach we have adopted is that in order to derive theorems about typed systems, the statements of the theorems above are not sufficient. This is a familiar phenomenon: consider the confluence of the simply typed lambda calculus, which will not follow from the fact that the untyped lambda calculus is confluent, but which submits to exactly the same proof. By making essentially trivial modifications to the proofs of our theorems, the reader may derive termination- and confluence-preservation results for simply typed and polymorphically typed lambda calculi. We outline the modifications needed to address Girard's system F_ω.

Some previous similar work: in [Klo80], Klop considers the addition of new rewriting rules to untyped lambda calculus, with restrictions on the form of the new rules (for example, that variables may not occur twice on the left side of a rule). We treat arbitrary algebraic rules. Toyama, in [Toy87], shows that the direct sum of confluent term rewriting systems is confluent, but the purely algebraic setting is very different from the present framework. The termination of a combination of terminating algebraic rewrite systems is a very delicate issue – Toyama presents several counter-examples in [Toy87]. Termination is known to be preserved in combinations of algebraic systems under various, somewhat restrictive, hypotheses ([Mid89], [Rus87], [TKB89]).

Notation

For definitions and notation not explained below, see [Bar84] for the lambda calculus, and [HO80] for term rewriting. Fix a set $\mathit{Vars} = \{v_i | i \in \omega\}$ of variables, let Σ be set of constant symbols, and let $\Lambda(\Sigma)$ be the set of lambda terms over $\mathit{Vars} \cup \Sigma$. Following a convention of algebraic rewriting, we index subterms of a term by sequences called ***occurrences***; we write A/u for the subterm of A at occurrence u. If a and b are sequences with a an initial segment of b, we say that a ***precedes*** b and that b ***extends*** a; a and b are ***incomparable*** if neither extends the other.

A Σ ***term*** is either a variable or a term $(fA_1 \cdots A_m)$, $f \in \Sigma, A_i \in \Lambda(\Sigma), (0 \leq i \leq m)$. An ***algebraic term*** is either a variable or a term $(fA_1 \cdots A_m)$, $f \in \Sigma$ with arity m, A_i algebraic, $(0 \leq i \leq m)$. We will reserve S and T to stand for algebraic terms.

Substitution into an algebraic term T is particularly simple since there is no variable binding in T. Since we will often have occasion to refer to substitution instances of algebraic terms T, we adopt the following notational convention: If T is an algebraic term whose free variables are among the set $\{v_1, \ldots, v_k\}$, and $\vec{Q} \equiv \{Q_1, \ldots, Q_k\}$ is any multiset of $\Lambda(\Sigma)$ terms, then the result of simultaneously substituting Q_i for v_i in T is denoted $T^{\vec{Q}}$.

An ***algebraic rewrite system*** R is a set of pairs $\langle S, T\rangle$ of algebraic terms such that S is not a variable and $\mathit{Vars}(T) \subseteq \mathit{Vars}(S)$. If P and Q are relations, we often write PQ for $P \cup Q$.

2 Descendants and Projections

The proofs of our preservation theorems proceed by isolating certain subterms of a term and analyzing reductions into steps which take place inside the given terms and other steps which are blind to the internal structure of those terms. In this section we develop some machinery enabling us to track the progress of subterms during a reduction. We use the notions of *descendant* of an occurrence with respect to an algebraic reduction (essentially as in [HL79]).

Definition 2.1 Let $\rho : M \xrightarrow{R} N$. For an occurrence d of M, the set d/ρ of *descendants of d with respect to* ρ is the set of occurrences in N defined as follows.

If d does not extend the redex u of ρ then $d/\rho = \{d\}$. If d is uw, w a non-variable address of the source term S, then $d/\rho = \emptyset$. Otherwise, writing d as uac, where $S/a \equiv v_i$, and writing T for the target term of ρ, d/ρ is $\{ua'c | T/a' = v_i\}$.

If $\mathcal{D}$ is a set of occurrences in M then $\mathcal{D}/\rho$ is $\bigcup\{d/\rho | d \in \mathcal{D}\}$.

If $\rho = \rho_n \circ \cdots \circ \rho_0$ is a several step reduction then $\mathcal{D}/\rho$ is $(\cdots(\mathcal{D}/\rho_0)/\cdots/\rho_n)$.

When the descendant of a certain occurrence of subterm X is under consideration, we will often simply say "descendant of X". For example, suppose $\langle fx, gxx\rangle$ is a rule, and consider the reduction $M \equiv h(f(ky)) \xrightarrow{R} h(g(ky)(ky)) \equiv N$. Then ky has two descendants, the two occurrences of ky in N, and $f(ky)$ has one descendant, *viz.*, $g((ky)(ky))$ in N.

Definition 2.2 The Σ-*boundary* of M, $O_\Sigma(M)$, is the set of occurrences which are minimal (as sequences) among the non-Σ occurrences in M. (So, the corresponding subterms are maximal non-Σ subterms of M.)

Lemma 2.3 *Let* $\rho : M \xrightarrow{R} N$, *and let* d *be an occurrence in* M.

1. *For each* $e \in d/\rho$, $M/d \xrightarrow{R}_{\equiv} N/e$.
2. *If* d *precedes the redex of* ρ *then* $M/d \xrightarrow{R} N/d$.
3. *If* M *is stable and* $\mathcal{D}$ *is* $O_\Sigma(M)$ *then* $\mathcal{D}/\rho$ *is* $O_\Sigma(N)$.

Proof. The first two assertions follow easily from the definition of descendant; stability (arity-checking in particular) is used in the proof of the third statement. □

To isolate the steps of a reduction which are independent of some particular subterms, we consider the term obtained by replacing those subterms by variables. We must do this with some care in order to preserve the rewriting relation.

Definition 2.4 An *R-projection* (or just *projection*, if R is available from the context) is any function π from terms to variables such that if M and N have a common R-reduct then they are assigned the same variable.

Given a set $\mathcal{D}$ of pairwise incomparable occurrences in a term M, and a projection π whose range is disjoint from the variables of M, write M^π for the term obtained from M by replacing M/d by $\pi(M/d)$ for each $d \in \mathcal{D}$, and say that M^π is a *projection of M at* $\mathcal{D}$.

If $\rho : M \xrightarrow{R} N$, π is a projection at $\mathcal{D}$, and $\mathcal{D}/\rho$ is pairwise incomparable, then $\rho\pi$ is the projection at $\mathcal{D}/\rho$ given by $\rho\pi(N/e) \equiv \pi(M/d)$, for each $d \in \mathcal{D}$ and $e \in d/\rho$.

We need Lemma 2.3.1 in order to justify the definition of $\rho\pi$ above. In order to ensure that $\mathcal{D}/\rho$ is pairwise incomparable, it will suffice (by Lemma 2.3.3) to choose $\mathcal{D}$ to be the Σ-boundary of M (it is clear that a Σ-boundary is a pairwise incomparable set of occurrences). To go further and have the projection of a reduction induce a reduction on the projections, we must be careful to project on a sufficiently full set of occurrences, in the following sense:

Definition 2.5 If M/u is of the form $T^{\vec{Q}}$ and $\mathcal{D}$ is a set of occurrences, then $\mathcal{D}$ is (T,u) *full* if no $d \in \mathcal{D}$ is uw with w a non-variable occurrence in T, and for every $d \in \mathcal{D}$ which is uac with T/a a variable, $\mathcal{D}$ contains each $ua'c$ for which T/a' is the same variable.

Lemma 2.6 *Let $\rho : M \xrightarrow{R} N$, have redex u and source term S, let $\mathcal{D}$ be pairwise incomparable and (S,u) full, and suppose that M^π is a projection of M at $\mathcal{D}$. Then*

1. $M^\pi \xrightarrow{R}_{\equiv} N^{\rho\pi}$.
2. *If no $d \in \mathcal{D}$ precedes u, then $M^\pi \xrightarrow{R} N^{\rho\pi}$.*

□

Finally, in order to project an R reduction of several steps on $\mathcal{D}$ we must guarantee that its descendants will be full for the next step. This motivates the next result.

Lemma 2.7 *If $\mathcal{D}$ is $O_\Sigma(M)$ then for any u such that M/u is of the form $T^{\vec{Q}}$, $\mathcal{D}$ is (T,u)-full.*

□

Thus Σ-boundaries are always sets of non-Σ occurrences which are sufficiently full, and in the stable case their descendants inherit this property. These facts will enable us to iterate applications of Lemma 2.6 when we start with a projection of a Σ-boundary.

It will be important to isolate β-redex subterms of a term which are contained in no other β-redexes, and whose descendants are similarly maximal. Leftmost redexes have these properties under β-reduction alone, but algebraic reduction can spoil leftmost-ness. So we need a generalization:

Definition 2.8 An occurrence d of β-redex of M is *outermost* if it is minimal (as a sequence) among the β-redex occurrences in M. (So, the corresponding subterms are maximal β-redexes of M.)

Lemma 2.9 *1. When M is stable, each R-descendant of an outermost β-redex is an outermost β-redex.*

2. When M is a stable β-normal form each R-reduct of M is a β-normal form.

□

3 Confluence

This section shows that when confluent algebraic rewriting is combined with β reduction, confluence is inherited by stable terms.

As pointed out in [Bre88], we cannot expect confluence in the presence of η reduction: if $fx \xrightarrow{R} a$, then $\lambda x.fx$ has the two ηR normal forms $\lambda x.a$ and f.

We first verify that a confluent algebraic system R remains confluent when extended to the expanded set of terms $\Lambda(\Sigma)$. The global strategy in the proof of Theorem 3.1 (projecting βR-reductions to R-reductions on β-normal forms) was used in [Bre88] in the simply typed setting; we avoid the use of types in the argument.

Theorem 3.1 *If R is confluent on algebraic terms, then R is confluent on R-stable $\Lambda(\Sigma)$ terms.*

Proof. We show by induction on stable terms M that if $X \overset{R}{\twoheadleftarrow} M \overset{R}{\twoheadrightarrow} Y$ then there exists N such that $X \overset{R}{\twoheadrightarrow} N \overset{R}{\twoheadleftarrow} Y$. If M is algebraic, confluence holds by hypothesis. If M is itself not a Σ term, then M can be written as one of $xP_1 \cdots P_n$ or $(\lambda x.P_1)P_2 \cdots P_n$, $(n > 0)$, and X and Y must have the same shape, so we can build N using the induction hypothesis on the P_i.

So suppose M is a non-algebraic Σ term, let $\mathcal{D}$ be $O_\Sigma(M)$, and let R^- be the relation R restricted to $\{A | \exists d \in \mathcal{D}, M/d \overset{R}{\twoheadrightarrow} A\}$. Since M is a Σ term, each M/d for $d \in \mathcal{D}$ is smaller than M. By the induction hypothesis, R confluence holds out of M/d when $d \in \mathcal{D}$, and it follows that R confluence holds out of every term in the domain of R^-. So R^- is a confluent relation.

Let π be defined over $\mathcal{D}$ so that terms M/d and M/e are replaced by the same variable if and only if $M/d \overset{R^-}{\twoheadleftarrow\!\!\twoheadrightarrow} M/e$. This is an R-projection. Since M is not algebraic, M^π is smaller than M.

By iterating Lemma 2.6 we can project the two reductions on $\mathcal{D}$ and its descendants, obtaining $M^\pi \overset{R}{\twoheadrightarrow} X'$ and $M^\pi \overset{R}{\twoheadrightarrow} Y'$. By the induction hypothesis applied to M^π there exists N' with $X' \overset{R}{\twoheadrightarrow} N' \overset{R}{\twoheadleftarrow} Y'$.

The terms M^π, X', and Y' are obtained from M, X, and Y respectively by replacing subterms by new variables. We build our desired N by finding appropriate terms to substitute for these variables in N'.

Consider one of the new variables z and let $A_1, \ldots A_l$ be the subterms of X replaced by z to give X', and $B_1, \ldots B_m$ the subterms of Y replaced by z to give Y'. Each A_i is a reduct of some M/d with $d \in \mathcal{D}$ and $\pi(M/d) \equiv z$, and the same holds for each B_j. By the confluence of R^-, we can produce a term $C_{(z)}$ which is a common R-reduct of all the A_i and B_j.

When this has been done for each z, take N to be N' with each z replaced by $C_{(z)}$.

Now we have $X \overset{R}{\twoheadrightarrow} X'[\vec{z} := \vec{C}_{(z)}]$ by rewriting the various A_i to $C_{(z)}$ for each z and A_i as above. Similarly, $Y \overset{R}{\twoheadrightarrow} Y'[\vec{z} := \vec{C}_{(z)}]$. Finally, $X'[\vec{z} := \vec{C}_{(z)}] \overset{R}{\twoheadrightarrow} N'[\vec{z} := \vec{C}_{(z)}]$ and $Y'[\vec{z} := \vec{C}_{(z)}] \overset{R}{\twoheadrightarrow} N'[\vec{z} := \vec{C}_{(z)}]$ by substitutivity of R. Thus $N'[\vec{z} := \vec{C}_{(z)}]$ is the desired N.

□

To lift this result to full βR reduction, we attempt to project reductions to reductions on β normal forms (the latter reductions will be purely algebraic if the original term is stable). This is the key step in the proof of confluence-preservation in [Bre88]; we give a treatment here which separates the relationship between β and R from the existence of β normal forms.

The key step is to introduce a relation with the same transitive closure as R-reduction, but which interacts more smoothly with β-reduction. This technique is inspired by a proof of the confluence of $\xrightarrow{\beta}$ due to Tait and Martin-Löf.

Definition 3.2 The relation $\xrightarrow{R}\!\!\!\!\rightarrow_1$ is defined inductively as follows:

1. $M \xrightarrow{R}\!\!\!\!\rightarrow_1 M$.
2. If $M \xrightarrow{R}\!\!\!\!\rightarrow_1 M'$ and $N \xrightarrow{R}\!\!\!\!\rightarrow_1 N'$ then $MN \xrightarrow{R}\!\!\!\!\rightarrow_1 M'N'$.
3. If $M \xrightarrow{R}\!\!\!\!\rightarrow_1 M'$ then $\lambda x.M \xrightarrow{R}\!\!\!\!\rightarrow_1 \lambda x.M'$.
4. If $\langle S,T\rangle \in R$, and for $1 \le i \le n, P_i \xrightarrow{R}\!\!\!\!\rightarrow_1 Q_i$, then $S^{\vec{P}} \xrightarrow{R}\!\!\!\!\rightarrow_1 T^{\vec{Q}}$.

Lemma 3.3 *1.* $\xrightarrow{R} \subseteq \xrightarrow{R}\!\!\!\!\rightarrow_1 \subseteq \xrightarrow{R}\!\!\!\!\rightarrow$

2. $A \xrightarrow{R}\!\!\!\!\rightarrow_1 B$ *implies* $M[x := A] \xrightarrow{R}\!\!\!\!\rightarrow_1 M[x := B]$.

3. $A \xrightarrow{R}\!\!\!\!\rightarrow_1 B$ *and* $M \xrightarrow{R}\!\!\!\!\rightarrow_1 N$ *imply* $M[x := A] \xrightarrow{R}\!\!\!\!\rightarrow_1 N[x := B]$.

□

The important feature of $\xrightarrow{R}\!\!\!\!\rightarrow_1$ is that we can project and develop a single step $\xrightarrow{R}\!\!\!\!\rightarrow_1$ reduction to a *single step* $\xrightarrow{R}\!\!\!\!\rightarrow_1$ reduction, as follows.

Proposition 3.4 *Let* M *be any* $\Lambda(\Sigma)$ *term. If* $\rho : M \xrightarrow{\beta} X$ *and* $M \xrightarrow{R}\!\!\!\!\rightarrow_1 N$*, then there are* X' *and* Z *such that* $X \xrightarrow{\beta}\!\!\!\!\rightarrow X' \xrightarrow{R}\!\!\!\!\rightarrow_1 Z$ *and* $N \xrightarrow{\beta}\!\!\!\!\rightarrow Z$.

Proof. By induction on the derivation of $M \xrightarrow{R}\!\!\!\!\rightarrow_1 N$. The only interesting case is the last: suppose $M \equiv S^{\vec{P}} \xrightarrow{R}\!\!\!\!\rightarrow_1 T^{\vec{Q}} \equiv N$, with each $P_i \xrightarrow{R}\!\!\!\!\rightarrow_1 Q_i$: If $(\lambda x A)B$ is the redex term in ρ then there is an i such that $(\lambda x A)B \sqsubseteq P_i$. For this i, we have a β-reduction out of P_i and $P_i \xrightarrow{R}\!\!\!\!\rightarrow_1 Q_i$, so by induction there are P_i' and Q_i', with $P_i \xrightarrow{\beta}\!\!\!\!\rightarrow P_i'$, $Q_i \xrightarrow{\beta}\!\!\!\!\rightarrow Q_i'$, and $P_i' \xrightarrow{R}\!\!\!\!\rightarrow_1 Q_i'$. Let $\vec{P'}$ denote the sequence of terms obtained from $\vec{P}$ by replacing P_i by P_i', and take X' to be $S^{\vec{P'}}$. Let $\vec{Q'}$ denote the sequence of terms obtained from $\vec{Q}$ by replacing Q_i by Q_i', and take Z to be $T^{\vec{Q'}}$. Then $M \equiv S^{\vec{P}} \xrightarrow{\beta} X \xrightarrow{\beta}\!\!\!\!\rightarrow S^{\vec{P'}} \equiv X'$ by suitably reducing *all* occurrences of P_i to P_i', and $X' \xrightarrow{R}\!\!\!\!\rightarrow_1 Z$ since each element of P' reduces via $\xrightarrow{R}\!\!\!\!\rightarrow_1$ to the corresponding element of Q'. □

Preservation of confluence now follows.

Theorem 3.5 *If* R *is confluent on algebraic terms, then* βR *is confluent over* R*-stable* $\Lambda(\Sigma)$ *terms.*

Proof. Write $\beta nf(A)$ for the β-normal form of a term A. We first show, by induction along $\xrightarrow{\beta}$, that when M is stable and $M \xrightarrow{R}\!\!\!\rightarrow_1 N$ then $\beta nf(M) \xrightarrow{R}\!\!\!\rightarrow_1 \beta nf(N)$. If M is a β normal form then by Lemma 2.9 .2 so is N. Otherwise, let $M \xrightarrow{\beta} X$ be any reduction, define X' and Z as in Proposition 3.4, and apply the induction hypothesis to the instance $X' \xrightarrow{R}\!\!\!\rightarrow_1 Z$.

It follows that M stable and $M \xrightarrow{R} N$ imply that $\beta nf(M) \xrightarrow{R}\!\!\!\rightarrow \beta nf(N)$.

Now, to show confluence, suppose M is R stable, with $A \twoheadleftarrow^{\beta R} M \xrightarrow{\beta R}\!\!\!\rightarrow B$. Then $\beta nf(A) \twoheadleftarrow^{R} \beta nf(M) \xrightarrow{R}\!\!\!\rightarrow \beta nf(B)$. Confluence of R on $\Lambda(\Sigma)$ yields P such that $\beta nf(A) \xrightarrow{R}\!\!\!\rightarrow P \twoheadleftarrow^{R} \beta nf(B)$, so that $A \xrightarrow{\beta R}\!\!\!\rightarrow P \twoheadleftarrow^{\beta R} B$ as desired. □

Application to typed systems

The proofs above provide a method for proving inheritence of confluence in β strongly normalizing typed calculi. We will outline such a proof for Girard's [Gir72] system F_ω.

The types of F_ω form a simply typed lambda calculus, where the types of the types are called *kinds.* The *raw terms* of F_ω are objects of the form x, f, MN, $M\sigma$, $\lambda x : \sigma.M$, or $\lambda t : K.M$, where x is a term variable, f is from Σ, M and N are terms, t is a type variable, and K is a kind. We allow β reductions on types as well as terms, so that $(\lambda t : K.M)\sigma \xrightarrow{\beta} M[t := \sigma]$.

The *well-typed* terms of F_ω are given by triples $\Delta \vdash M : \sigma$ provable in the type inference system of F_ω, where Δ is a type assignment to variables. We will not present the type inference system here; part of our message is that we do not need to know the type system in detail in order to derive preservation of confluence. All we need to know is that for each Δ and each many-sorted algebraic rewrite system R, the set of terms which are well-typed under Δ is stable.

The results in Section 3 which mention only R-reduction are true in the F_ω setting, with precisely the same proofs – we need only be careful to respect types when replacing subterms by variables. Now suppose that R is confluent as a rewrite system. R-confluence on well-typed F_ω terms has the same proof, verbatim, as Theorem 3.1. The confluence of βR on F_ω is proved just as in Theorem 3.5, once we have proved a version of Proposition 3.4 using an expanded definition of $\xrightarrow{R}\!\!\!\rightarrow_1$ in which:

- if $M \xrightarrow{R}\!\!\!\rightarrow_1 M'$ then $M\sigma \xrightarrow{R}\!\!\!\rightarrow_1 M'\sigma$
- if $M \xrightarrow{R}\!\!\!\rightarrow_1 M'$ then $\lambda t : K.M \xrightarrow{R}\!\!\!\rightarrow_1 \lambda t : K.M'$.

4 Termination

In this section it will be shown that if a terminating algebraic rewriting system is added to the $\Lambda(\Sigma)$ calculus of β-reduction, the resulting system is terminating on stable terms. We assume that all terms under consideration are stable.

The first step is to record some well-known results on β-reduction which parallel some of the results of Section 2. The notions of *residual* of a β-redex and of a *development* of a specified set of redexes are standard, and it turns out that we can confine our attention to developing sets of incomparable β-redexes. In the interest of maintaining a uniform terminology we will use "descendant" to refer to the image of an occurrence under either type of reduction. Hence, if $\rho : M \xrightarrow{\beta}\!\!\!\!\rightarrow \cdots$, and $\mathcal{D}$ is a set of β-redex occurrences in M, then the set $\mathcal{D}/\rho$, of descendants of of $\mathcal{D}$ with respect to ρ is the set of occurrences of residuals of the terms at $\mathcal{D}$, and when $\mathcal{D}$ is a set of pairwise incomparable β-redexes in M then $\varphi(\mathcal{D}\ , M)$ is the term obtained from M by contracting those redexes. We say that $\varphi(\mathcal{D}\ , M)$ is a *development* of M.

We have the following facts, parallel to the earlier results about R-descendants. Let $\rho : M \xrightarrow{\beta} N$, and let d be an outermost β-redex occurrence in M. If d is the redex of ρ then $d/\rho = \emptyset$, otherwise $d/\rho = \{d\}$. We always have $M/d \xrightarrow{\beta}_{\equiv} N/d$, and if d precedes the redex of ρ, then $M/d \xrightarrow{\beta} N/d$. Finally, each βR descendant of a outermost β-redex is a outermost β-redex.

The construction in the proof of termination for βR involves choosing an outermost β-redex from the initial term of a reduction and developing it and all of its descendants. The next two results show that under the right conditions, such a development preserves $\xrightarrow{\beta}\!\!\!\!\rightarrow$ and $\xrightarrow{R}\!\!\!\!\rightarrow$.

In a β-redex term $(\lambda x P)Q$, call Q the *argument term*. Note that if D is a β-redex subterm of Q in $(\lambda x P)Q$, and if furthermore $x \notin FV(P)$, then a contraction of D is rendered moot by a subsequent contraction of $(\lambda x P)Q$. This possibility plays a role in the next lemma.

Lemma 4.1 *Let $\rho : M \xrightarrow{\beta R} N$, let $\mathcal{D}$ be a set of pairwise incomparable β redexes in M, and let $\mathcal{E}$ be $\mathcal{D}/\rho$.*

1. $\varphi(\mathcal{D}, M) \xrightarrow{\beta R}\!\!\!\!\rightarrow \varphi(\mathcal{E}, N)$.
2. *If the redex of ρ is neither an element of $\mathcal{D}$ nor an occurrence in an argument term of a redex from $\mathcal{D}$ then at least one reduction is done in $\varphi(\mathcal{D}, M) \xrightarrow{\beta R}\!\!\!\!\rightarrow \varphi(\mathcal{E}, N)$.*

Proof. For β-reduction, the results are Lemmas 11.1.7 and 11.3.3 of [Bar84], so let ρ be an R-reduction, with redex u and source term S.

For the first assertion, we isolate two cases, defined by the position of u with respect to $\mathcal{D}$ (of course $u \notin \mathcal{D}$). If no d in $\mathcal{D}$ precedes u, expand $\mathcal{D}$ to the smallest (S, u) full set $\mathcal{D}^+$ containing $\mathcal{D}$. Then $\mathcal{D}^+$ is still a set of outermost β redexes and $\mathcal{D}^+/\rho$ is $\mathcal{E}$. Now if π is any projection on $\mathcal{D}^+$, Lemma 2.6, implies that $M^\pi \xrightarrow{R} N^{\rho\pi}$. Therefore we can perform β reductions in M before the R reduction yielding $\varphi(\mathcal{D}, M) \xrightarrow{\beta}\!\!\!\!\rightarrow \varphi(\mathcal{D}^+, M) \xrightarrow{R} \varphi(\mathcal{E}, N)$.

If the redex u extends some $d_0 \in \mathcal{D}$, then no element of $\mathcal{D}$ extends u, since they are all incomparable with d_0. Therefore $\mathcal{D}$ is trivially (S, u) full and each element of $\mathcal{D}$ is its own descendant. Write $M/d_0 \equiv (\lambda x A)B$, and $M/u \equiv S^{\vec{Q}}$. When $S^{\vec{Q}}$ is a subterm of A, substitutivity of R implies that $\varphi(\mathcal{D}, M) \xrightarrow{R} \varphi(\mathcal{E}, N)$. When $S^{\vec{Q}}$ is a subterm of B, we can mimic ρ at the address of each free occurrence of x in A to obtain $\varphi(\mathcal{D}, M) \xrightarrow{R}\!\!\!\!\rightarrow \varphi(\mathcal{E}, N)$.

The second assertion can be seen by examining the cases in (1) - the only case where collapsing might occur is in the last case, when x is not free in A. □

We are now in a position to see that βR-reduction is terminating on stable terms. It is convenient to treat pure R-reduction first.

Theorem 4.2 *If R is terminating on algebraic terms, then R is terminating on R-stable $\Lambda(\Sigma)$ terms.*

Proof. For the sake of contradiction, let M be a stable term of minimal size among those which are R-infinite.

By hypothesis, M cannot be algebraic. Suppose M were not a Σ term. Then M would be one of $xP_1\cdots P_n$, or $(\lambda x.P_1)P_2\cdots P_n$, $(n > 0)$, each R-reduct would be of the same shape, so that some P_i would be R-infinite, contradicting the minimality of M.

So let $\mathcal{D}$ be $O_\Sigma(M)$ and let π be a projection on $\mathcal{D}$ which replaces all subterms by the same variable. Since M is a Σ term, each subterm represented in $\mathcal{D}$ is smaller than M, and since M is not algebraic, M^π is smaller than M.

Now let $\rho : M \equiv M_0 \xrightarrow{R} M_1 \xrightarrow{R} \cdots$ be an infinite R-reduction, set $\mathcal{D}_0 \equiv \mathcal{D}$, $\mathcal{D}_{n+1} \equiv \mathcal{D}_n/\rho_n$, set $\pi_0 \equiv \pi$, $\pi_{n+1} \equiv \rho_n\pi_n$ and construct the sequence of terms $M_n^{\pi_n}$. By Lemma 2.6.1, $M_n^{\pi_n} \xrightarrow{R}_{\equiv} M_{n+1}^{\pi_{n+1}}$ for each n.

Since M^π is smaller than M, the sequence above is finite as a reduction sequence, so that for some k, $M_n^{\pi_n} \xrightarrow{R} M_{n+1}^{\pi_{n+1}}$ fails for all $n \geq k$.

For $n \geq k$, Lemma 2.6.2 applied to the reduction $\rho_n : M_n \xrightarrow{R} M_{n+1}$ yields a $d_n \in \mathcal{D}_n$ preceding the redex of ρ_n. It follows that for $n \geq k$, $\mathcal{D}_n/\rho_n = \mathcal{D}_k$. Furthermore, there must be a particular $d_k \in \mathcal{D}_k$ such that for infinitely many n, d_k precedes the redex of ρ_n. Choose $d_0 \in \mathcal{D}_0$ such that d_k is a descendant of d_0; M/d_0 must be R-infinite, contradicting the minimality of M. □

Theorem 4.3 *If R is terminating on algebraic terms, then βR is terminating on R-stable $\Lambda(\Sigma)$ terms.*

Proof. The proof is by induction on the maximum number of steps which can occur in a β-reduction of a stable term M. For the sake of contradiction, let $\rho : M \equiv M_0 \xrightarrow{\beta R} M_1 \xrightarrow{\beta R} \cdots$ be an infinite reduction.

When M is a β-normal form, Lemma 2.9 .2 implies that each ρ_n is an R-reduction, so ρ is finite by Theorem 4.2.

So let d_0 be any outermost β-redex in M_0, $M_0/d_0 \equiv (\lambda x P_0)Q_0$. Since stability is inherited by subterms, the induction hypothesis applies to Q_0, so Q_0 is βR terminating.

Set $\mathcal{D}_0 = \{d_0\}$, $\mathcal{D}_{n+1} = \mathcal{D}_n/\rho_n$. Each $\mathcal{D}_n$ is a set of outermost β redexes, hence is pairwise incomparable. Lemma 4.1.1 implies that $\varphi(\mathcal{D}_n, M_n) \overset{\beta R}{\twoheadrightarrow} \varphi(\mathcal{D}_{n+1}, M_{n+1})$ for each n, but by induction, $\varphi(\{d_0\}, M)$ is βR-terminating, so this is finite as a $\xrightarrow{\beta R}$ reduction.

By Lemma 4.1.2, from some point on each ρ_n-redex term is either equal to some β-redex term from $\mathcal{D}_n$, or is a subterm of the argument part of such a term. A reduction ρ_n of the first type results in $\mathcal{D}_{n+1}$ being smaller than $\mathcal{D}_n$, while one of the second type yields $\mathcal{D}_{n+1}$ the same size as $\mathcal{D}_n$, so eventually every reduction is of the second type. That is, there is a k such that for $n \geq k$ each ρ_n has its redex term inside the Q of some term $(\lambda x P)Q$ occurring in $\mathcal{D}_n$. Just as in the previous theorem, for $n \geq k$, $\mathcal{D}_n/\rho_n = \mathcal{D}_k$, and there is a particular $d_k \in \mathcal{D}_k$ such that for infinitely many n, d_k precedes the redex of ρ_n.

Now, M_k/d_k is of the form $(\lambda x P_k)Q_k$ and $(\lambda x P_0)Q_0 \overset{\beta R}{\twoheadrightarrow} (\lambda x P_k)Q_k$. For all $n < k$, $\{d_n\}/\rho_n \neq \emptyset$, since $d_{n+1} \in \{d_n\}/\rho_n$. Thus for all $n < k$, d_n is not the redex of ρ_n, so in fact $Q_0 \xrightarrow{\beta R}\!\!\!\!\rightarrow Q_k$. The previous paragraph showed that Q_k is βR-infinite, so we have a contradiction of the assumption that Q_0 is βR terminating. □

Application to typed systems

Often, a typed lambda calculus admits a notion of "type erasure". If the type system is such that well-typed terms are β-strongly normalizing and if there cannot be an infinite sequence of reductions between terms with the same type erasure, then the theorems of this section will yield preservation of termination when terminating algebraic reduction is added.

For example, if a terminating algebraic rewrite system defined over a one-sorted signature Σ is added to Girard's polymorphic F_ω calculus, the augmented system is terminating. The fact that there are no infinite sequences of type-reductions in F_ω is crucial for this argument.

Acknowledgements

I am indebted to Val Breazu-Tannen and Jean Gallier for illuminating discussions about the topics of this work, and to Patricia Johann for careful readings and suggestions for improvement. The referees helped to clarify the notion of stability and provided a substantial number of corrections.

References

[Bar84] H. P. Barendregt. *The Lambda Calculus: Its Syntax and Semantics.* North-Holland, Amsterdam. 1981, revised 1984.

[Bar90] F. Barbanera. Adding Algebraic Rewriting to the Calculus of Constructions: Strong Normalization Preserved. In *Proceedings of the Second International Workshop on Conditional and Typed Rewriting Systems*, Concordia University, 1990.

[Bre87] V. Breazu-Tannen. *Conservative extensions of type theories.* dissertation, Massachusetts Institute of Technology 1987.

[Bre88] V. Breazu-Tannen. Combining algebra and higher-order types. In *Proceedings of the Third Annual Symposium on Logic in Computer Science*, pp. 82- 90. 1988.

[BG9?a] V. Breazu-Tannen and J. Gallier. Polymorphic rewriting conserves algebraic strong normalization. *Theoretical Computer Science*, to appear.

[BG9?b] V. Breazu-Tannen and J. Gallier. Polymorphic rewriting conserves algebraic confluence. *Information and Computation*, to appear.

[BM87] V. Breazu-Tannen and A. R. Meyer. Computable values can be classical. In *Proceedings of the Fourteenth Symposium on Principles of Programming Languages* pp. 238-245, ACM, 1987.

[CH88] T. Coquand and G. Huet. The Calculus of Constructions. *Information and Control*, v.76, no.2/3, pp. 95-120, 1988.

[Der87] N. Dershowitz. Termination of rewriting. *J. Symbolic Computation* **3**, pp. 69-116, 1987.

[Gir71] J-Y. Girard. Une extension de l'interprétation de Gödel à l'analyse, et son application à l'elimination des coupures dans l'analyse et la théorie des types. In *Proc. Second Scandinavian Logic Symposium*, ed. J.E. Fenstad. North-Holland, Amsterdam, 1971.

[Gir72] J-Y. Girard. *Interprétation functionelle et élimination des coupures de l'arithmétique d'ordre supérieur*. These D'Etat, Universite Paris VII, 1972.

[HL79] G. Huet, J.J. Lévy. Call by need computations in non-ambiguous linear term rewriting systems. Rapport Laboria 359, INRIA, 1979

[HO80] G. Huet, D. Oppen. Equations and rewrite rules: a survey. In Formal Languages: Perspectives and Open Problems, ed. R. Book. Academic Press, New York 1980.

[Klo80] J. W. Klop. *Combinatory Reduction Systems*. Mathematical Center Tracts 127, Amsterdam, 1980.

[Mac86] D. B. MacQueen. Using dependent types to express modular structure. In *Conference Record of the Thirteenth Annual ACM Symposium on Principles of Programming Languages*, pp. 277-286, 1986.

[Mid89] A. Middeldorp. Modular aspects of properties of term rewriting systems related to normal forms. In *Proc. Third International Conference on Rewriting Techniques and Applications*, Springer-Verlag LNCS 355, pp. 263-277, 1989.

[Pot81] G. Pottinger. The Church-Rosser theorem for the typed λ calculus with surjective pairing. *Notre Dame Journal of Formal Logic*, v. 22, no. 3, pp. 264-268, 1981.

[Rey74] J. C. Reynolds. Towards a theory of type structure. In *Proc. Colloque sur la Programmation*, Springer-Verlag LNCS 19, pp. 408-425, 1974.

[Rus87] M. Rusinowitch. On termination of the direct sum of term rewriting systems. *Information Processing Letters* **26** pp.65-70, 1987.

[Toy87] Y. Toyama. On the Church-Rosser property for the direct sum of term rewriting systems. *Journal of the ACM*, v.34, no.1, pp.128-143, 1987.

[TKB89] Y. Toyama, J. W. Klop and H. Barendregt. Termination for the direct sum of left-linear term rewriting systems. In *Proc. Third International Conference on Rewriting Techniques and Applications*, Springer-Verlag LNCS 355, pp. 477-491, 1989.

Incremental Termination Proofs and the Length of Derivations

Frank Drewes
Universität Bremen

Clemens Lautemann[1]
Universität Mainz

Abstract. Incremental termination proofs, a concept similar to termination proofs by quasi-commuting orderings, are investigated. In particular, we show how an incremental termination proof for a term rewriting system $\mathcal{T}$ can be used to derive upper bounds on the length of derivations in $\mathcal{T}$. A number of examples show that our results can be applied to yield (sharp) low-degree polynomial complexity bounds.

1 Introduction

Term rewriting systems (TRSs) are a general computational model widely used as a theoretical and practical basis for algebraic specification and functional and logic programming. There are a number of surveys of this field (e.g. [HO80, Kl87]), for a recent survey see [DJ89].

As any other computation, a derivation in a term rewriting system may or may not terminate, and a good deal of research has been devoted to the development of methods for proving that all derivations in a term rewriting system must terminate, i.e., that the TRS is *terminating*. For a comprehensive survey of this area, see [De87].

Obviously, in practical applications, knowing that an algorithm will eventually terminate is not always enough. It is also important to have an estimate for the time it will take to do so, i.e., for its (time) complexity. The time that a derivation takes depends on several factors, the most basic one being the number of derivation steps. It is this complexity measure that we will be concerned with in this paper, more precisely, we will consider $dh_{\mathcal{T}}(t)$, the *derivation height* of term t in the TRS $\mathcal{T}$, i.e., the maximal length of a derivation in $\mathcal{T}$ which starts with t. Little is known about derivation height, but some work has been done to compare it with Turing complexity (cf. [EM80, BO84]) and to derive upper complexity bounds (cf. [HL89]). In this paper, we continue the investigations in [HL89] and present new upper complexity bounds based on termination proofs.

Usually, a termination proof for a TRS $\mathcal{T}$ comprises two major steps:

- The definition of a Noetherian order relation $>$ on terms.
- A proof that $\mathcal{T}$ is reducing for $>$, i.e., that $t \xrightarrow{\mathcal{T}} t'$ implies $t > t'$.

Many different kinds of Noetherian order relations have been developed in this context, with two types of relations being particularly useful: *simplification orderings* (e.g. Knuth–Bendix ordering, or recursive path ordering) and orderings induced by *valuations*[2].

If termination of a TRS $\mathcal{T}$ can be proved in the above fashion, with $>$ being induced by a valuation f, then, for every term t, the value $f(t)$ is an upper bound on $dh_{\mathcal{T}}(t)$. Thus

[1]Correspondence to: Clemens Lautemann, Joh.-Gutenberg Universität, Department of Computer Science, P.O. Box 3980, 6500 Mainz, Germany.

[2]These are just mappings from terms to natural numbers.

termination proofs based on valuations are particularly useful when complexity bounds are looked for.

One major difficulty encountered when trying to prove termination of a TRS $\mathcal{T} = \{R_1, \ldots, R_k\}$ by the method sketched above is that *every* one of the rules has to be reducing for the order relation $>$. Since the rules may be quite different, this might require quite different properties of $>$ for each of them, and thus it can turn out a formidable task to find such a $>$ which accomodates all rules simultaneously. In order to facilitate this task, in [BD86], Bachmair and Dershowitz developed the notion of *quasi-commutativity* of one relation over another. They showed that a termination proof for $\mathcal{T}$ can be obtained from two separate proofs for systems $\mathcal{T}_1, \mathcal{T}_2$ (where $T = \mathcal{T}_1 \cup \mathcal{T}_2$), if one of the order relations $>_1, >_2$, used in these proofs quasi-commutes over the other. The final proof for $\mathcal{T}$ then can, of course, be reformulated in terms of one order relation $>$ obtained from $>_1$ and $>_2$. However, as $>_1$ need not work for $\mathcal{T}_2$, and $>_2$ not for $\mathcal{T}_1$, $>$ may be considerably more complicated than any of $>_1, >_2$.
If the order relations are induced by valuations such a proof might result in smaller valuations for $\mathcal{T}_1$ and $\mathcal{T}_2$, whereas one valuation for all rules of $\mathcal{T}$ may have to take much larger values. It is therefore conceivable that this approach will lead to considerably better complexity bounds, and we will show in this paper that, indeed, low degree polynomial bounds can be obtained in this way, for a number of TRSs. As for quasi-commutativity, the idea behind our method is to define separate termination orderings for the rules of a TRS $\mathcal{T} = \{R_1, \ldots, R_k\}$. More specifically, we take valuations $f_1, \ldots, f_k$ such that, for each i, R_i is reducing for f_i. The only additional requirement then is that, for $j < i$, R_j be consistent with f_i, i.e., that $t \xrightarrow{R_j} t' \Longrightarrow f_i(t) \geq f_i(t')$, whereas *no* restrictions are placed on f_i with respect to any of the R_j for $j > i$. We call such termination proofs *incremental.* They seem particularly well-suited for situations in which the rules of a TRS are not given in one chunk, but generated one by one as is the case, e.g., in applications of the Knuth–Bendix completion procedure. However, we will study them here mainly as a means of obtaining good complexity bounds.

Section 2 contains the formal definition of incremental termination proofs. We also show by means of an example that they can be used in situations where neither RPO nor KBO apply. Section 3 is the core of our paper. We derive upper complexity bounds under the assumption that for $j < i$ there is a function $g_{i,j}$ such that $t \xrightarrow{R_i} t' \Longrightarrow f_j(t') \leq g_{i,j}(f_j(t))$. In order to analyse this situation, we consider certain replacement systems over $\mathbb{N}^k$. The results obtained there are then used to develop a general recurrence for the maximal length of derivations, and we show that the derivation height can be bounded by a linear function of the valuations, if they satisfy $g_{i,j}(n) \leq n + c$, where c is a constant. At the end of this section the use of the main result is illustrated by some examples. In Section 4 two extensions of our method are briefly discussed. The first deals with rewriting modulo AC, and the second is a method which can be used to prove satisfactory upper bounds for TRSs with non-right linear rules. These rules are difficult to cope with because of their ability to double almost the whole term in one step. An unsophisticated method would assume this to be the case in every step of a derivation and thus yield more or less trivial results in cases where — from an intuitive point of view — better ones should be obtainable.

This is the first systematic approach which yields small complexity bounds for term rewriting systems; except for a few odd examples dealt with by ad-hoc methods, the best complexity bounds derived so far were exponential (cf. [HL89]).

We conclude this introduction by compiling the basic terminology that we will use in this paper.

Preliminaries

We assume that the reader is familiar with the basic notions used in the area of TRSs. So we will not define what a replacement relation, a term rewriting system or a derivation is. Normally we will identify a TRS with the set of its rules, without mentioning the underlying signature Σ. If the rules do not contain any constant symbol we assume that there is one such symbol $0 \in \Sigma$. The size of a term t (i.e. the number of symbols in it) is denoted by $|t|$. For a set S, $|S|$ denotes the cardinality of S. With $\#_f(t)$, where f is an arbitrary function symbol and t a term, we denote the number of occurrences of f in t. $t' \sqsubseteq t$ means that t' is an element of the multiset of all subterms of t.
We say that a relation Θ is *Noetherian* if Θ permits no infinite descending chain, i.e. every sequence $a_1\Theta a_2\Theta\ldots$ is finite.[3] The transitive closure of any relation Θ is denoted by Θ^+, the transitive and reflexive closure by Θ^*. The replacement relation associated with a TRS $\mathcal{T}$ is denoted by $\xrightarrow{\mathcal{T}}$.[4] If it is clear what set of rules we have in mind, we simply write $\longrightarrow$. In addition to $\longrightarrow^*$ and $\longrightarrow^+$ we will use the notation $t\longrightarrow^n t'$ to indicate that there is a derivation of length n starting with t and ending with t'.
Assume that $\succsim$ is a binary relation on ground terms.[5] Then a set of rewrite rules $\mathcal{T}$ is *reducing* with respect to $\succsim$ if $\succsim$ contains $\xrightarrow{\mathcal{T}}$, i.e. if $t\xrightarrow{\mathcal{T}}t'$ implies $t\succsim t'$.
A transitive relation $\succsim$ is said to be a *quasi-ordering*. The order relation $\succ$ is then defined by $x \succ y \;:\Longleftrightarrow\; x\succsim y \wedge \neg y\succsim x$.
A TRS $\mathcal{T}$ is called *terminating* if $\xrightarrow{\mathcal{T}}$ is Noetherian.

1.1 Fact

A TRS $\mathcal{T}$ is terminating iff there exists a Noetherian order relation $\succ$ on terms such that $\mathcal{T}$ is reducing with respect to $\succ$. □

A possibility to get Noetherian relations on terms which avoids the need for a proof of well-foundedness is to define $t\gtrsim t' \iff h(t) \geq h(t')$, where $>$ is a Noetherian relation on a set M and $h : T_\Sigma \to M$. If $M = \mathbb{N}$, we call h a *valuation*.
One particular type of valuation for a term rewriting system are *polynomial interpretations* (cf. [Ln75, Ln79]). Here an n-ary (positive) integer polynomial p_f is associated with each n-ary function symbol f, thus each ground term $f(t_1,\ldots,t_n)$ gets the value $p(f(t_1,\ldots,t_n)) = p_f(p(t_1),\ldots,p(t_n))$. If the polynomials used are strongly monotonic in each argument it suffices to prove that the rules are reducing for instances of the left-hand sides.

The complexity measure for TRS used in this paper is the *derivation height*, defined as follows.

---

[3] A Noetherian order relation is also said to be *well-founded*.

[4] As usual, we will omit the braces around sets consisting of one element, i.e. if R is a term rewriting rule we write $\xrightarrow{R}$ instead of $\xrightarrow{\{R\}}$.

[5] We only consider rewriting on ground terms, i.e. terms without variables, throughout this paper.

1.2 Definition (*Derivation height*, cf. [HL89])

Let $\mathcal{T}$ be a TRS and let t be a term.
We define $dh_{\mathcal{T}}(t)$, the *length of the longest derivation starting with* t by

$$dh_{\mathcal{T}}(t) := \max\{n \in \mathbb{N} \;/\; \exists t' : t \longrightarrow^n t'\}.$$

For all $n \in \mathbb{N}$ we define $Dh_{\mathcal{T}}(n)$ by

$$Dh_{\mathcal{T}}(n) := \max\{dh_{\mathcal{T}}(t) \;/\; |t| = n\}.$$ □

2 Incremental Termination Proofs

In this section we introduce a method for proving termination of TRSs. Using this method, we will not only be able to show termination of certain systems, but also obtain a number of low–degree polynomial bounds on the derivation height of these systems. The notion we propose here is closely related to the quasi–commuting notion of Bachmair and Dershowitz, but it is defined in a more operational way. This makes it easier to prove statements concerned with the complexity of TRSs. We call it *incremental*, because it provides the opportunity to extend an already developed incremental termination proof for k_1 rules of a TRS to one for $k_1 + k_2$ rules, which in turn can be extended to one for $k_1 + k_2 + k_3$ rules, and so on. The "atomic pieces" of such an ITP are ordinary valuations which, of course, must satisfy certain constraints.

2.1 Definition (*Incremental termination proof*)

Let $\mathcal{T}_1, \ldots, \mathcal{T}_k$ be TRSs and let $\mathcal{F} = (f_1, \ldots, f_k)$ be a tuple of valuations.
We call $\mathcal{F}$ an *incremental termination proof (ITP)* for $(\mathcal{T}_1, \ldots, \mathcal{T}_k)$ if for all $i, 1 \leq i \leq k$, and terms t, t',

1. $\mathcal{T}_i$ is reducing with respect to f_i, i.e. if $t \xrightarrow{\mathcal{T}_i} t'$ then $f_i(t) > f_i(t')$, and
2. for all $j, 1 \leq j < i$, $\xrightarrow{\mathcal{T}_j}$ does not increase f_i, i.e., if $t \xrightarrow{\mathcal{T}_j} t'$ then $f_i(t) \geq f_i(t')$. □

So, when choosing f_i we do not have to pay attention to any rule from $\mathcal{T}_{i+1}, \ldots, \mathcal{T}_k$ because the two conditions given in Definition 2.1 involve only $\mathcal{T}_1, \ldots, \mathcal{T}_i$. Moreover, only the rules of $\mathcal{T}_i$ have to be strictly reducing with respect to f_i. In fact, requiring $\mathcal{T}_j$ to be strictly reducing with respect to f_i for $j < i$ would of course result in f_i being a termination proof for the whole system $\mathcal{T}_1 \cup \ldots \cup \mathcal{T}_i$. In contrast to this, our definition allows $t \xrightarrow{\mathcal{T}_j} t'$ for terms t and t' which are *equivalent* with respect to f_i (i.e., $f_i(t) = f_i(t')$).

2.2 Proposition

If $(\mathcal{T}_1, \ldots, \mathcal{T}_k)$ has an ITP then $\mathcal{T} = \mathcal{T}_1 \cup \ldots \cup \mathcal{T}_k$ is terminating.

Proof
Let $(f_1, \ldots, f_k)$ be an ITP for $(\mathcal{T}_1, \ldots, \mathcal{T}_k)$.
Then, clearly, $t \xrightarrow{\mathcal{T}} t'$ implies $(f_k(t), \ldots, f_1(t)) >^k_{lex} (f_k(t'), \ldots, f_1(t'))$, where $>^k_{lex}$ is the lexicographic ordering on $\mathbb{N}^k$, so we are ready since $>^k_{lex}$ is Noetherian. □

ITPs are closely connected to the notion of quasi-commutativity of [BD86]. For a detailled study of these connections the reader is refered to [DL90].

Our first example is derived from a TRS found in [Kl87].

2.3 Example

$$\text{Let } \mathcal{T} = \{ \begin{array}{lcll} h(x,y) & \longrightarrow & g(x,y), & (R_1) \\ g(f(g(x,y)),z) & \longrightarrow & f(h(g(x,y),z), & (R_2) \\ g(f(g(x,y)),z) & \longrightarrow & f(f(g(g(x,y),z))) \ \} & (R_3). \end{array}$$

As valuations we choose

$f_1(t) = \#_h(t)$

$f_2(t) =$ the sum of all $\#_f(t')$ over all $t' \sqsubseteq t$ having g or h as their topmost symbol

$f_3(t) = |\{t' \sqsubseteq t \ / \ t' = G(f(H(...)),...),$ where $G, H \in \{g,h\}\}|$

These valuations form an ITP for (R_1, R_2, R_3):[6]

1. Obviously, R_i reduces f_i, for $i = 1, \dots, 3$.
2. An application of R_1 touches neither f_2 nor f_3 since g's and h's are treated equally by f_2.
3. R_2 does not increase f_3, because the f on the left hand side of R_2 is counted by f_3 (it stands between two g's) and so the worst case is that the right hand side gets the same value.

This example shows that ITPs suffice to get easy termination proofs for systems that other methods cannot handle, because $\mathcal{T}$, for example, cannot be proved terminating using KBO[7], because R_3 cannot be ordered unless f gets the weight 0, in which case f has to be the greatest symbol, and the right hand side must be greater than the left hand side. RPO cannot be used either. Because of R_1, h must be greater than g and so R_2 cannot be reducing since for that we eventually would have to prove $g(x,y) >_{RPO} h(g(x,y))$ which is wrong because RPO satisfies the subterm property. (Observe that even "RPO with status" does not help in any way because in comparing the two sides of R_2 RPO with status does not behave different from the usual RPO.) □

3 Upper bounds on derivation height

We now turn to the main question of this paper: If the termination of some TRS $\mathcal{T}$ can be shown with an ITP, what can be said about upper bounds on $Dh_{\mathcal{T}}$ for this particular system $\mathcal{T}$? As one can easily see, there is a trivial ITP for each terminating TRS. Simply setting $f_1 = \dots = f_k = dh_{\mathcal{T}}$ yields a correct ITP for $\mathcal{T}$, because "$\mathcal{T}$ terminates" means the same as "$\longrightarrow^+$ is well founded". Therefore we cannot hope to get non-trivial results in general and so we will look for a more special (but as general as possible) situation to deal with.

The main idea leading to upper bounds on $Dh_{\mathcal{T}}$ arises from the following observation. Derivations starting with a term t cannot have more than $f_k(t)$ occurrences of rules from

[6] The termination of the original system from [Kl87] can be proved using f_1 and f_2.

[7] See [KB70], [De82], and [St88] for the defintions of KBO, RPO, and RPO with status, respectively.

$\mathcal{T}_k$, because each application of some $R \in \mathcal{T}_k$ reduces the kth value and no other rule is able to increase it. It would be nice to have a similar upper bound on the number of occurrences of the other rules as well, but at this point our problems arise, because we do not have any information about how big the increase of $f_1, \ldots f_{k-1}$, caused by an application of R, is. Therefore, in general, it is not possible to give upper bounds for these values (in terms of $f_1(t), \ldots, f_{k-1}(t)$) after any one application of R has taken place. So what we need are functions $g_{i,j} : \mathbb{N} \to \mathbb{N}$ for all $i, j : 1 \leq j < i \leq k$ which serve as upper bounds on f_j after one application of $R \in \mathcal{T}_i$. In order to prove a general result for cases in which such $g_{i,j}$ are given we will first abstract from our concrete topic and deal with properties of replacement systems over $\mathbb{N}^k$.

3.1 Definition (k–bound)

Let $k \geq 1$ be a natural number and, for each pair $(i,j) \in \mathbb{N}^2$ with $1 \leq j < i \leq k$, let $g_{i,j} : \mathbb{N} \to \mathbb{N}$ be a strongly monotonic growing function.

We call $\mathcal{G} = (g_{i,j})_{1 \leq j < i \leq k}$ a *k-bound* and associate the following replacement relations with it:

1. For each $i \in \mathbb{N}, 1 \leq i \leq k$, and $(x_1, \ldots, x_k), (y_1, \ldots, y_k) \in \mathbb{N}^k$:
   $(x_1, \ldots, x_k) \searrow^{\mathcal{G},i} (y_1, \ldots, y_k) : \Longleftrightarrow$ • $y_i < x_i$ and
   • $\forall j, 1 \leq j < i : y_j \leq g_{i,j}(x_j)$ and
   • $\forall j, i < j \leq k : y_j \leq x_j$
2. $\searrow^{\mathcal{G}} := \bigcup_{1 \leq l \leq k} \searrow^{\mathcal{G},l}$ □

3.2 Example

Let $k = 3$ and let $\mathcal{G}$ consist of
$$g_{2,1}(x) = x + 3, \quad g_{3,1}(x) = x + 1, \text{ and} \quad g_{3,2}(x) = x * 2.$$

Then $(n_1, n_2, n_3) \searrow^{\mathcal{G},1} (n'_1, n'_2, n'_3)$ if $n_1 > n'_1$, $n_2 \geq n'_2$, and $n_3 \geq n'_3$,
$(n_1, n_2, n_3) \searrow^{\mathcal{G},2} (n'_1, n'_2, n'_3)$ if $n_1 + 3 \geq n'_1, n_2 > n'_2$, and $n_3 \geq n'_3$, and
$(n_1, n_2, n_3) \searrow^{\mathcal{G},3} (n'_1, n'_2, n'_3)$ if $n_1 + 1 \geq n'_1, 2{\cdot}n_2 \geq n'_2$, and $n_3 > n'_3$. □

It is clear that $\searrow^{\mathcal{G}}$ is Noetherian, because without the second constraint on the $\searrow^{\mathcal{G},i}$ (i.e., $\forall j, 1 \leq j < i : y_j \leq g_{i,j}(x_j)$)) it compares to the lexicographic ordering $>^k_{lex}$ on $\mathbb{N}^k$ where the tuples are read backwards, thus $(x_1, \ldots, x_k) \searrow^{\mathcal{G}} (y_1, \ldots, y_k)$ implies $(x_k, \ldots, x_1) >^k_{lex} (y_k, \ldots, y_1)$. We now want to bound the length of such $\searrow^{\mathcal{G}}$–chains. Since we actually know how many $\searrow^{\mathcal{G},k}$–steps can at most occur, namely x_k, and furthermore the maximum number of possible $\searrow^{\mathcal{G},j}$–steps cannot be increased through steps of an index smaller than j, i.e., does at most depend on x_j and the maximum possible number of occurrences of $\searrow^{\mathcal{G},j+1}, \ldots, \searrow^{\mathcal{G},k}$, it makes sense to think of a recurrence for this.
For a derivation $(x_1, \ldots, x_k) \searrow^{\mathcal{G},i_1} \ldots \searrow^{\mathcal{G},i_n} (y_1, \ldots, y_k)$, for simplicity containing steps from $\searrow^{\mathcal{G},i+1}, \ldots, \searrow^{\mathcal{G},k}$ only, the increase of the ith component can of course be estimated by $g_{i_n,i}(g_{i_{n-1},i} \cdots g_{i_1,i}(x_i) \ldots)$. So, at first, think of x_j as the maximum possible

number of occurrences of $\searrow^{\mathcal{G},j}$ –steps for all $j, i < j \leq k$. Then, throughout the derivation the ith component is increased at most x_j times through $\searrow^{\mathcal{G},j}$ for each $j, i < j \leq k$. Hence, $i_l = j$ for at most x_j of the i_l. The order in which the different steps occur may of course influence the result, so we have to take the maximum over all these possibilities. Since every $\searrow^{\mathcal{G},i}$ –step decreases the ith value, this gives us the upper bound on the number of occurrences of $\searrow^{\mathcal{G},i}$ (under the assumption that each $\searrow^{\mathcal{G},j}$ occurs at most x_j times). Furthermore, if we now replace each of these x_j, $i < j \leq k$, by a recursion, we obtain a recursive definition of functions $\Pi_i^{\mathcal{G}}$ estimating the maximum number of $\searrow^{\mathcal{G},i}$ –steps in a derivation starting with some $X \in \mathbb{N}^k$. (This, of course, remains to be proven.) The formal definition given below follows the line of this intuition.

3.3 Definition

Let k be a natural number and let $\mathcal{G} = (g_{i,j})_{1 \leq j < i \leq k}$ be a k–bound.

For each $i \in \mathbb{N}, 1 \leq i \leq k$ and every k–tuple $X = (x_1, \ldots, x_k) \in \mathbb{N}^k$ we define:

1. $\pi^{\mathcal{G}}_{x_{i+1},\ldots,x_k}(x_i) := \max\{g_{i_1,i}(g_{i_2,i}(\cdots g_{i_q,i}(x_i)\ldots)) \;/\; q = \sum_{l=i+1}^{k} x_l$ and each $j > i$ occurs exactly x_j times among the $i_l\}$
2. $\Pi_i^{\mathcal{G}}(X) := \pi^{\mathcal{G}}_{\Pi_{i+1}(X),\ldots,\Pi_k(X)}(x_i)$. □

For the rest of this section we will simply write $\searrow$, $\searrow^{i}$, $\pi_{x_{i+1},\ldots,x_k}$ and Π_i as long as it is clear which k–bound we have in mind.

Because of the fact that the $g_{i,j}$ are strongly monotonic, we have the following fact.

3.4 Fact

Let $\mathcal{G}$ be a k–bound and let $i \in \mathbb{N}, 1 \leq i \leq k$.

Then $\pi_{x_{i+1},\ldots,x_k}(x_i)$ and $\Pi_i((x_1, \ldots, x_k))$ are monotonic with respect to each of $x_{i+1}, \ldots, x_k$ and are strongly monotonic with respect to x_i. □

3.5 Example

Consider the situation of Example 3.2. For $N = (n_1, n_2, n_3)$ we have

$$\begin{aligned}
\pi(n_3) &= n_3 \\
\pi_{n_3}(n_2) &= \underbrace{g_{3,2}(\cdots g_{3,2}}_{n_3}(n_2)\cdots) = 2^{n_3}\cdot n_2, \text{ and} \\
\pi_{n_2,n_3}(n_1) &= \max\left\{ g_{i_1,1}(\cdots g_{i_{n_2+n_3},1}(n_1)\cdots) \;\middle/\; \begin{array}{l} i_j = 2 \text{ for } n_2 \text{ of the } i_j \text{ and} \\ i_j = 3 \text{ for } n_3 \text{ of the } i_j \end{array} \right\} \\
&= n_1 + 3\cdot n_2 + n_3.
\end{aligned}$$

Therefore, for Π we get

$\Pi_3(N) = \pi(n_3) = n_3$,
$\Pi_2(N) = \pi_{\Pi_3(N)}(n_2) = \pi_{n_3}(n_2) = 2^{n_3}\cdot n_2$, and
$\Pi_1(N) = \pi_{\Pi_2(N),\Pi_3(N)}(n_1) = \pi_{2^{n_3}\cdot n_2, n_3}(n_1) = n_1 + 3\cdot 2^{n_3}\cdot n_2 + n_3$. □

The following lemma says that $\Pi_i(X)$ provides an upper bound on the number of occurrences of $\searrow^{i}$ –steps in derivations starting with some k–tuple X. It is the technical core of our paper.

3.6 Lemma

Let $\mathcal{G}$ be a k–bound, let $N \in \mathbb{N}^k$, and let $i \in \mathbb{N}, 1 \leq i \leq k$.
In each $\searrow$ –derivation beginning with N there are at most $\Pi_i(N)$ occurrences of $\searrow^i$.

Proof
We proceed by induction on i, with i going down from k to 1. Let $N = (n_1^0, \ldots, n_k^0)$.

For $i = k$ the claim holds by definition.

If $i < k$, let $(n_1^0, \ldots, n_k^0) \searrow^{i_1} (n_1^1, \ldots, n_k^1) \searrow^{i_2} \cdots \searrow^{i_m} (n_1^m, \ldots, n_k^m)$ be a $\searrow$ –derivation.

Since, for $j_1 < i \leq j_2$, by definition $(m_1^0, \ldots, m_k^0) \searrow^{j_1} (m_1^1, \ldots, m_k^1) \searrow^{j_2} (m_1^2, \ldots, m_k^2)$ implies $(0, \ldots, 0, m_i^0, \ldots, m_k^0) \searrow^{j_2} (0, \ldots, 0, m_i^2, \ldots, m_k^2)$, we can assume w.l.o.g. that $n_1^j = \ldots = n_{i-1}^j = 0$ for $1 \leq j \leq m$, and that $i_1, \ldots, i_m \geq i$.

Furthermore, we can assume that $\searrow^i$ –steps occur only at the end of the derivation, since, for $j \geq i$, $(0, \ldots, 0, m_i^0, \ldots, m_k^0) \searrow^i (0, \ldots, 0, m_i^1, \ldots, m_k^1) \searrow^j (0, \ldots, 0, m_i^2, \ldots, m_k^2)$ implies $(0, \ldots, 0, m_i^0, \ldots, m_k^0) \searrow^j (0, \ldots, 0, m_i^2 + 1, m_{i+1}^2, \ldots, m_k^2) \searrow^i (0, \ldots, 0, m_i^2, \ldots, m_k^2)$ by definition of the $\searrow^i$ and by strong monotonicity of the $g_{i,j}$.

So let $\searrow^{j_{n+1}}$ be the first $\searrow^i$ –step of this derivation. By induction hypothesis, every $\searrow^j$, $i < j \leq k$, occurs at most $\Pi_j(N)$ times in the derivation. Thus we have (see Definition 3.1)

$$\begin{aligned} n_i^{n+1} &\leq g_{j_1,i}(g_{j_2,i}(\cdots g_{j_n,i}(n_i^0)\ldots)) \\ &\leq \pi_{\Pi_{i+1}(n^0),\ldots,\Pi_k(n^0)}(n_i^0) \qquad \text{by strong monotonicity} \\ &= \Pi_i(N). \end{aligned}$$ □

In order to bound the length of $\searrow$ –derivations, we now need estimates on Π_i. Of course, such estimates can only be given under certain restrictions on the $g_{i,j}$; of particular interest for us is the case where $g_{i,j}(x) = x + c_{i,j}$, for some constants $c_{i,j}$. By induction on i (from k down to 1) we can show:[8]

3.7 Theorem

Let $\mathcal{G}$ be a k–bound, and assume that there are constants $c_{i,j}$, $1 \leq j < i \leq k$, such that $g_{i,j}(x) = x + c_{i,j}$ for all $x \in \mathbb{N}$.

If $c_i = max\left\{ \prod_{p=1}^{n-1} c_{m_p,m_{p+1}} \Big/ i = m_1 > \ldots > m_n \geq 1 \right\}$ then, for all $N = (n_1, \ldots, n_k) \in \mathbb{N}^k$,

$$\sum_{i=1}^{k} \Pi_i(N) \leq 2^k \sum_{l=1}^{k} c_l n_l.$$ □

We now go back to our original problem of deriving upper bounds on the derivation height of term rewriting systems.

[8] The case $g_{i,j}(x) = c_{i,j} \cdot x$ can be handled similarly. For this and the proof of Theorem 3.7 consider [DL90].

3.8 Definition (*G–boundedness of derivations*)

Let $\mathcal{T} = \bigcup_{1 \le l \le k} \mathcal{T}_l$ be a TRS, $(f_1, \ldots, f_k)$ an ITP for $(\mathcal{T}_1, \ldots, \mathcal{T}_k)$ and let $\mathcal{G} = (g_{i,j})_{1 \le j < i \le k}$ be a k–bound.

We say that a derivation $t_1 \longrightarrow t_2 \longrightarrow \ldots$ in $\mathcal{T}$ is *$\mathcal{G}$–bounded* if, for every step $t_i \xrightarrow{\mathcal{T}_j} t_{i+1}$ of this derivation, we have $(f_1(t_i), \ldots, f_k(t_i)) \searrow^j (f_1(t_{i+1}), \ldots, f_k(t_{i+1}))$. □

Thus, a derivation is $\mathcal{G}$–bounded, if for all steps $t_i \xrightarrow{\mathcal{T}_j} t_{i+1}$ of this derivation, and for all $l, 1 \le l < j$, the lth value of t_{i+1} is bounded from above by $g_{j,l}$, that is $f_l(t_{i+1}) \le g_{j,l}(f_l(t_i))$. The following is a direct consequence of Lemma 3.6 and the definition of $\mathcal{G}$–boundedness.

3.9 Theorem

Let $\mathcal{T} = \bigcup_{1 \le l \le k} \mathcal{T}_l$ be a TRS, $(f_1, \ldots, f_k)$ an ITP for $(\mathcal{T}_1, \ldots, \mathcal{T}_k)$ and let $\mathcal{G} = (g_{i,j})_{1 \le j < i \le k}$ be a k–bound.

For every $\mathcal{G}$–bounded derivation $D = t_1 \xrightarrow{\mathcal{T}_{j_1}} t_2 \xrightarrow{\mathcal{T}_{j_2}} \ldots \xrightarrow{\mathcal{T}_{j_m}} t_{m+1}$ the following hold.

1. D contains at most $\Pi_i((f_1(t_1), \ldots, f_k(t_1)))$ applications of rules from $\mathcal{T}_i$, for $i = 1, \ldots, k$.
2. The length of D is bounded from above by $\sum_{i=1}^{k} \Pi_i((f_1(t_1), \ldots, f_k(t_1)))$. □

Furthermore, from Theorem 3.7 we immediately get the following.

3.10 Corollary

Let $\mathcal{T} = \bigcup_{1 \le l \le k} \mathcal{T}_l$ be a TRS, let $(f_1, \ldots, f_k)$ be an indirect ITP for $(\mathcal{T}_1, \ldots, \mathcal{T}_k)$, and assume that, for some k–bound $\mathcal{G}$, there are constants $c_{i,j}$, $1 \le j < i \le k$, such that $g_{i,j}(x) = x + c_{i,j}$ for all $x \in \mathbb{N}$.

If $c_i = max \left\{ \prod_{p=1}^{n-1} c_{m_p, m_{p+1}} \Big/ i = m_1 > \ldots > m_n \ge 1 \right\}$, then every $\mathcal{G}$–bounded derivation is of length at most $2^k \sum_{l=1}^{k} c_l f_l(t)$, where t is the starting term of the derivation □

We conclude this section with some examples showing that the results obtained so far can indeed be used to prove polynomial bounds on the derivation height of some well-known TRSs.

3.11 Example (cf. [Ch86], p. 46)

Consider $\mathcal{T} = \{$

$$\begin{array}{llll} x+0 & \longrightarrow & x, & (R_1)\\ x+(-x) & \longrightarrow & 0, & (R_2)\\ x+(-y)+y & \longrightarrow & x, & (R_3)\\ -0 & \longrightarrow & 0, & (R_4)\\ -(-x) & \longrightarrow & x, & (R_5)\\ -(x+y) & \longrightarrow & (-x)+(-y)\ \} & (R_6). \end{array}$$

Since the first five rules are length–decreasing, we can choose $f_1(t) = |t|$ as a valuation for $\{R_1, \dots, R_5\}$. For R_6 we use $f_2(t) = \sum_{-t_1 \sqsubseteq t} \#_+(t_1)$.

Obviously, f_2 is not increased by $R_1, \dots, R_5$ and if $t \xrightarrow{R_6} t'$ then

- $f_2(t') = f_2(t) - 1$, so R_6 is reducing for f_2 as required, and
- $f_1(t') = f_1(t) + 1$.

Hence every derivation in $\mathcal{T}$ is $(g_{2,1})$–bounded, if $g_{2,1}(x) = x+1$ and, using Corollary 3.10, we obtain the upper bound $dh_{\mathcal{T}}(t) \in O(f_1(t) + f_2(t)) = O(|t|^2)$, so

$$Dh_{\mathcal{T}}(n) \in O(n^2).$$

This bound is essentially sharp, as each term of the form $\underbrace{-(\dots-(}_{n}\underbrace{+(\dots+(}_{n}0,0)\dots,0))\dots)$ makes n^2 applications of rule R_6 possible. □

3.12 Example

Let us have a second look at the system $\mathcal{T} =$

$$\begin{array}{llll} \{\ h(x,y) & \longrightarrow & g(x,y), & (R_1)\\ g(f(g(x,y)),z) & \longrightarrow & f(h(g(x,y),z), & (R_2)\\ g(f(g(x,y)),z) & \longrightarrow & f(f(g(g(x,y),z)))\ \} & (R_3) \end{array}$$

from Example 2.3, where the valuations were

$f_1(t) =\ \#_h(t)$
$f_2(t) =$ the sum of all $\#_f(t')$ over all $t' \sqsubseteq t$ having g or h as their topmost symbol
$f_3(t) =\ |\{t' \sqsubseteq t / t' = G(f(H(\dots)),\dots)$, where $G, H \in \{g,h\}\}|$

For each term t let $\mathcal{G}^t = \{g^t_{2,1}, g^t_{3,1}, g^t_{3,2}\}$ with

$$g^t_{2,1}(x) = x+1, \quad g^t_{3,1}(x) = x, \quad \text{and } g^t_{3,2}(x) = x + |t|.$$

Then every derivation starting with a term t is $\mathcal{G}^t$–bounded, and we can use Corollary 3.10, where $c_1 = c_2 = 1$ and $c_3 = |t|$ to obtain

$$dh_{\mathcal{T}}(t)\ \in\ O\left(1\cdot|t| + 1\cdot|t|^2 + |t|\cdot|t|\right)\ =\ O\left(|t|^2\right)$$

since $f_1(t) \leq |t|$, $f_2(t) \leq |t|^2$ *and* $f_3(t) \leq |t|$.

Again, this bound is essentially sharp since a term of the form $g(f(g(f(\dots),0)),0)$ can, using R_1 and R_2, be rewritten to $f(f(\dots g(g(\dots),0)\dots))$ in $\Omega(|t|^2)$ steps. □

3.13 Example (cf. [CL87])

Consider the system $\mathcal{T} = \{ \begin{array}{lcll} (x*y)*z & \longrightarrow & x*(y*z), & (R_1) \\ f(x)*f(y) & \longrightarrow & f(x*y), & (R_2) \\ f(x)*(f(y)*z) & \longrightarrow & f(x*y)*z \;\} & (R_3). \end{array}$

For the first rule we use $f_1(t) = \begin{cases} \#_*(t_1) + f_1(t_1) + f_1(t_2) & if\ t = t_1 * t_2 \\ f_1(t_1) & if\ t = f(t_1) \\ 0 & \text{otherwise.} \end{cases}$

R_1 is reducing for f_1, and since R_2 and R_3 are length–decreasing (and R_1 is not length–increasing) we are allowed to take $f_2(t) = |t|$ for $\{R_2, R_3\}$.

Obviously, R_2 does not change f_1, and R_3 increases it by at most $\#_*(t)$. This together with the observation that no rule increases the number of $*$–symbols a term contains, implies that each derivation starting with a term t is $(g^t_{2,1})$–bounded, if $g^t_{2,1}(n) = n + |t|$.

Thus, with Corollary 3.10 we get $dh_{\mathcal{T}}(t) \in O(|t|{\cdot}f_2(t) + f_1(t)) = O(|t|^2)$, and so

$$Dh_{\mathcal{T}}(n) \in O(n^2).$$

Again, this bound is sharp, since R_1 alone is able to generate chains of length $\Omega(n^2)$ when applied to terms of the form $(\ldots(0+0)+\ldots+0)$. □

3.14 Example

An example which is already given in [HL89] (and from which the general notion of ITPs was derived) deals with the famous rewrite system for group theory, first considered in [KB70, p.279]. This system $\mathcal{G}$ consists of the 10 rules

$$\begin{array}{llcllllcl} (G1) & x+0 & \longrightarrow & x & \quad & (G6) & -x+(x+y) & \longrightarrow & y \\ (G2) & 0+x & \longrightarrow & x & & (G7) & -0 & \longrightarrow & 0 \\ (G3) & x+-x & \longrightarrow & 0 & & (G8) & --x & \longrightarrow & x \\ (G4) & -x+x & \longrightarrow & 0 & & (G9) & (x+y)+z & \longrightarrow & x+(y+z) \\ (G5) & x+(-x+y) & \longrightarrow & y & & (G10) & -(x+y) & \longrightarrow & -y+-x \end{array}$$

As a valuation f_1 for $\{G1, \ldots, G8\}$ we use $|t|$ and for $G9$ we choose the valuation already used in Example 3.13, i.e., $f_2(t)$ is the sum of all $\#_+(t')$ over all subterms t' that occur as a left son of a $+$–symbol in t. Similarly we define $f_3(t)$ to be the sum of all $\#_+(t')$ over all subterms $-t'$ of t.

Obviously, $f_1(t)$ is bounded from above by $|t|$ and $G1, \ldots, G8$ do not increase f_2 and f_3. Moreover, $G9$ does not influence f_1 at all and $G10$ increases it by one, so $g_{2,1}(x) = x$ and $g_{3,1}(x) = x + 1$.

As stated in [HL89], the following holds for the relation of f_2 and f_3.

- $f_3(t)$ is not increased by $G9$,
- if $t_1 \longrightarrow^* t_2 \xrightarrow{G10} t_3$ then $f_2(t_3) \le f_2(t_2) + |t_1|$ and
- both $f_2(t)$ and $f_3(t)$ are bounded from above by n^2.

So we are able to apply Corollary 3.10 to a derivation with starting term t by setting $c_1^t = c_2^t = 1$ and $c_3^t = \max\{1, 0, |t|\} = |t|$, which leads to the result

$$dh_{\mathcal{G}}(t) \in O(|t| + |t|^2 + |t|^2 * |t|) = O(|t|^3).$$

This bound is the same as the one given in [HL89] (where it is shown to be sharp). □

4 Extensions

First note that the derivation height of a TRS $\mathcal{T}$ might become considerably larger when rewriting modulo AC is considered. If however $Dh_{\mathcal{T}}$ can be bounded by means of an ITP $(f_1, \dots, f_k)$ whose valuations are AC–invariant (i.e. $t \equiv_{AC} t' \Longrightarrow f_i(t) = f_i(t')$) then this bound is also valid for $\mathcal{T}/\mathrm{AC}$. This is the case, e.g., in Example 3.11. Under AC, the TRS studied there is canonical for Abelian groups.

Our second remark concerns non–right linear rules. So far, all our examples were right–linear, and this is no coincidence, as doubling of subterms by non–right linear rules will, in general, also double the value of valuations, making a direct application of Corollary 3.10 impossible. Often, however, we can find a way around this problem by means of the following result from [DL90].

4.1 Theorem

Let $(f_1, \dots, f_k)$ be an indirect ITP for $(\mathcal{T}_1, \dots, \mathcal{T}_k)$ and V a valuation such that for all terms t, t'

- $t \xrightarrow{\mathcal{T}_i} t' \Longrightarrow V(t') \leq V(t)$,
- $V(t) \geq |t|$

and $f_i(n) > 0$ for all $n > 0, 1 \leq i \leq k$. [9]

Then every derivation starting with a term t is $\mathcal{G}^t$–bounded, where $\mathcal{G}^t = (g_{i,j}^t)_{1 \leq j < i \leq k}$ and

$$g_{i,j}^t(x) := x + f_j(V(t)).$$ □

As an application of Theorem 4.1 we can bound the complexity of systems containing both associativity and distributivity laws.

4.2 Example

$$\text{Let } \mathcal{M} = \left\{ \begin{array}{rcll} (x*y)*z & \longrightarrow & x*(y*z), & (M_1) \\ (x+y)*z & \longrightarrow & (x*z)+(y*z), & (M_2) \\ x*(y+z) & \longrightarrow & (x*y)+(x*z), & (M_3) \\ (x+y)+z & \longrightarrow & x+(y+z) & \} \ (M_4). \end{array} \right.$$

For M_1 we can take the valuation f_1 for associativity rules from Example 3.13, and the last three rules can be handled with a linear interpretation f_2, defined by (see [De87])

$$f_2(t) = \begin{cases} f_2(t_1) \cdot f_2(t_2) & \text{if } t = t_1 * t_2 \\ 2 \cdot f_2(t_1) + f_2(t_2) + 1 & \text{if } t = t_1 + t_2 \\ 2 & \text{if } t \text{ is a constant.} \end{cases}$$

[9] For $f : T_\Sigma \to \mathbb{N}$ we define $f : \mathbb{N} \to \mathbb{N}$ by $f(n) = \max\{f(t) \mid |t| \leq n\}$.

We let V be the standard interpretation, i.e. we interpret "$*$" and "$+$" as multiplication and addition, respectively. Constants are interpreted by 2. Then each application of a rule from $\mathcal{M}$ preserves $V(t)$, and $|t| < V(t)$. So the applicability conditions for Theorem 4.1 are fulfilled and we can apply Corollary 3.10 with $g_{i,j}(x) = x + f_j(V(t))$ to obtain

$$dh_{\mathcal{M}}(t) \leq 4\cdot\sum_{l=1}^{2}\left(f_l(t)\cdot\prod_{m=1}^{l-1} f_m(V(t))\right) \leq 4\cdot(|t|^2 + 2^{|t|}.2^{2\cdot|t|}) \in O(2^{3\cdot|t|}),$$

since $f_1(t) \leq |t|^2$ and $f_2(t) \leq 2^{|t|}$. □

Again, this upper bound is essentially sharp, because the distributivity rules alone make derivations of length $2^{\Omega(n)}$ possible.

References

[BD86] L. Bachmair, N. Dershowitz: *Commutation, transformation, and termination.* Proc. 8th Conf. on Automated Deduction, LNCS 230, pp. 5–20.

[BO84] G. Bauer, F. Otto: *Finite complete rewriting systems and the complexity of the word problem.* Acta Informatica 21, pp. 521–540.

[Ch86] Ph. le Chenadec: *Canonical Forms in Finitely Presented Algebras.* J. Wiley & Sons

[CL87] A. Ben Cherifa, P. Lescanne: *Termination of Rewriting Systems by Polynomial Interpretations and its Implementation.* Sci. of Comp. Prog. 9, pp. 137–159.

[De82] N. Dershowitz: *Orderings for Term-rewriting systems.* TCS 17, pp. 279–301.

[De87] N. Dershowitz: *Termination of rewriting.* J. Symbolic Computation 3, pp. 69–116.

[DJ89] N. Dershowitz, Jean-Pierre Jouannaud: *Rewrite systems.* Rapport de Recherche n° 478, Université de Paris-Sud.

[DL90] F. Drewes, C. Lautemann: *Incremental Termination Proofs and the Length of Derivations.* Universität Bremen, report 7/90.

[EM80] H. Ehrig, B. Mahr: *Complexity of Implementations on the Level of Algebraic Specifications.* Proc. ACM Symp. Theory of Computing.

[HL89] D. Hofbauer, C. Lautemann: *Termination proofs and the length of derivations.* Proc. 3rd Intern. Conf. on Rewriting Techn. and Appl., LNCS 355, pp. 167–177.

[HO80] G. Huet, D. Oppen: *Equations and rewrite rules: a survey.* In Ronald V. Book, Ed.: *Formal languages, perspectives and open problems*, Academic Press.

[KB70] D. E. Knuth, P. B. Bendix: *Simple Word Problems in Universal Algebras.* In: J. Leech, Ed.: *Computational Problems in Abstract Algebra*, Oxford, Pergamon Press.

[Kl87] J.-W. Klop: *Term Rewriting Systems: A Tutorial*, EATCS Bulletin 32, pp. 143.

[Ln75] D. Lankford: *Canonical algebraic simplification in computational logic.* Report ATP-25, University of Texas.

[Ln79] Dallas Lankford: *On proving term rewriting systems are Noetherian.* Report MTP-3, Louisiana Tech University.

[St88] J. Steinbach: *Extension and comparison of simplification orderings.* Proc. 3rd Intern. Conf. on Rewriting Techniques and Applications, LNCS 355, pp. 434–448.

Time Bounded Rewrite Systems and Termination Proofs by Generalized Embedding

Dieter Hofbauer
Technische Universität Berlin *

Abstract

It is shown that term rewriting systems with primitive recursively bounded derivation heigths can be simulated by rewriting systems that have termination proofs using generalized embedding, a very restricted class of simplification orderings. As a corollary we obtain a characterization of the class of relations computable by rewrite systems having primitive recursively bounded derivation heights using recent results on termination proofs by multiset path orderings.

1. Introduction

The *derivation height* of a term t w.r.t. a (finite) rewrite system R (Hofbauer, Lautemann [HL89]) is defined as the length of a longest derivation sequence using R starting with t:

$$dh_R(t) := \max\{\, m \in \mathbb{N} \mid \text{there is a term } t' \text{ such that } t\, (\longrightarrow_R)^m\, t' \,\}$$

Taking the maximum over all terms of bounded size we get a unary function on nonzero natural numbers:

$$Dh_R(k) := \max\{\, dh_R(t) \mid size(t) \le k \,\}$$

Thus the derivation height can be seen as a time complexity measure for term rewriting systems as a model of computation. If the derivation height of a rewrite system is primitive recursively bounded as for the systems we are concerned with in this paper, all natural computation models will be able to compute normal forms of terms within primitive recursive time bounds if terms are coded appropriately.

Different termination proof methods for term rewriting systems put different types of bounds on the length of derivation sequences. In [HL89] termination proofs via *polynomial interpretation* and via the *Knuth-Bendix ordering* are investigated, whereas Hofbauer [Hof90] and Cichon [Cich90] treat *path orderings* independently. There it is shown that termination proofs using *multiset path orderings* yield a primitive recursive bound on the derivation length. Multiset path orderings have been introduced by Plaisted [Pla78] and Dershowitz [Der82] (alternative definitions can be found in Klop [Klop87], Dershowitz, Jouannaud [DJ90] and Geser [Ges90]). For related results see Dershowitz, Okada [DO89], where *ordinal types* of reduction orderings are investigated, and [Cich90], who shows a close connection between ordinal types and derivation heights. Drewes, Lautemann [DL90] treat termination methods that yield polynomial bounds on the derivation height.

---

* Technische Universität Berlin, Fachbereich 20 (Informatik), FR 6-2, Franklinstr. 28/29, D - 1000 Berlin 10, Germany

In this paper we start with systems having primitive recursively bounded derivation heights and *simulate* them by systems that have termination proofs of a very restricted form; the termination ordering used is *generalized embedding*, a concept generalizing the homeomorphic embedding relation but weaker than multiset path orderings.

As a corollary we obtain a characterization of the class of relations computable by rewrite systems having primitive recursively bounded derivation heights: There is a rewrite system that computes a relation P with primitive recursively bounded derivation height iff there is a rewrite system that computes P with termination proof by generalized embedding iff there is a rewrite system that computes P with termination proof by multiset path ordering.

Although our results hold for many sorted signatures too, we state and prove them only for the one sorted case to simplify the matter.

2. Preliminaries

We assume the reader to be familiar with basic notations of *term rewriting systems* (see Dershowitz, Jouannaud [DJ90] or Klop [Klop87] for surveys). We just recall some of them. A *term rewrite system* R consists of a set of term pairs $l \rightarrow r$, the *rules* of R, where $l, r \in T(\Sigma,X)$, i.e. they are terms over a (one sorted) signature Σ, possibly containing variables from a set X; since we are concerned with *terminating* systems, i.e. systems not allowing infinite rewrite sequences, all variables occurring in r also occur in l and l itself is not a variable. $T(\Sigma)$ denotes the set of *ground* terms over signature Σ.
Let t be a term; t / u denotes the *subterm* of t at occurrence u where the set Occ(t) of *occurrences* consists of lists of natural numbers, denoting positions in t; the outermost symbol of a term occurs at *empty list* λ. Occurrences are partially ordered by $v < u$ iff $\exists w{\neq}\lambda: vw = u$.
Let $|t| := |Occ(t)|$ be the *size* of t and $var(t) := \{ x \in X \mid \exists u \in Occ(t): t/u = x \}$ the set of variables occurring in t. We use $top(t) = f$ iff $t = f(t_1,\ldots, t_n)$ $(n \geq 0)$ and $t // u := top(t / u)$ if occurrence u denotes a function symbol in t. Applying a rule $l \rightarrow r$ in term t means choosing a substitution σ and an occurrence u in t such that $t / u = l\sigma$ and replacing the subterm at this position by $r\sigma$; that *rewrite step* thus results in $t[u \leftarrow r\sigma]$. The one step rewrite relation is $\longrightarrow_R$, $\xrightarrow{+}_R$ is its transitive and $\xrightarrow{*}_R$ its reflexive and transitive closure. $NF_R(t)$ denotes the *set of all normal forms* of term t w.r.t. R; if t has a *unique* normal form w.r.t. R we denote it by $t\downarrow_R$.

3. Generalized Embedding

In this section we define the generalized embedding relation $\succ_{GE}$, a precedence based simplification ordering. It generalizes the homeomorphic embedding $\succ_{HE}$ (see [Klop87], [DJ90]) but is still weaker than the multiset path ordering $\succ_{MPO}$ (see [DJ90]). As [Klop87], [DJ90] and [Ges90], we use rewrite relations for this purpose.

Definition

For a given precedence $\succ \subseteq \Sigma\times\Sigma$, we define a rewrite relation $\longrightarrow_{GE}$ through the following two rule schemes $(n, m \geq 0, t_1,\ldots, t_n, s_1,\ldots, s_m \in T(\Sigma,X))$:

(i) $f(t_1,...,t_n) \longrightarrow_{GE} t_i$ for all i $(1 \le i \le n)$

(ii) $f(t_1,...,t_n) \longrightarrow_{GE} g(s_1,...,s_m)$ if $f \succ g$ and $f(t_1,...,t_n) \xrightarrow{+}_{GE} s_i$ for all i $(1 \le i \le m)$

$\succ_{GE} := \xrightarrow{+}_{GE}$ is called *generalized embedding*. □

Lemma

(1) For a fixed precedence $\succ$ we have $\succ_{HE} \subsetneq \succ_{GE} \subsetneq \succ_{MPO}$.

(2) There is a signature Σ, a precedence $\succ$ on Σ and (ground) terms $t_1,..., t_4 \in T(\Sigma)$ such that (i) $t_1 \succ_{GE} t_2$ and $t_1 \not\succ_{HE} t_2$ and (ii) $t_3 \succ_{MPO} t_4$ and $t_3 \not\succ_{GE} t_4$.

Proof

(1) is trivial. For (2) choose $\Sigma = \{a, b, c, f, g\}$ where $f \succ a \succ b \succ c \succ g$, $t_1 = a$, $t_2 = g(b)$ and $t_3 = f(a, c)$, $t_4 = f(b, b)$. □

Thus $\succ_{HE}$, $\succ_{GE}$ and $\succ_{MPO}$ in general differ even on ground terms and if the underlying precedence is total; of course $\succ_{GE} = \succ_{MPO}$ if Σ contains only monadic symbols and constants.

4. Simulating Primitive Recursively Time Bounded Systems

In this section we state the main theorem of the paper, explain the simulation used and define the simulating system; section 5 will be devoted to the proof.

Theorem

Let R be a terminating rewrite system over signature Σ such that Dh_R is primitive recursively bounded. Then there exists a rewrite system S over an extension Σ_0 of Σ (i.e. $\Sigma \subset \Sigma_0$) and a unary symbol normalize $\in \Sigma_0 \setminus \Sigma$ such that

(1) $NF_S(\text{normalize}(t)) = NF_R(t)$ for all $t \in T(\Sigma)$

(2) $l \succ_{GE} r$ for all rules $l \to r \in S$

(3) If R is finite then S is finite too. □

If we allow S to be infinite, the claim becomes trivial: just choose

$S := \{ \text{normalize}(t) \to t' \mid t, t' \in T(\Sigma) \text{ and } t' \in NF_R(t) \}$.

Termination via generalized embedding is easily obtained by choosing normalize $\succ$ f for all symbols $f \in \Sigma$. Thus for the rest of the paper let $R = \{ l_1 \to r_1,..., l_k \to r_k \}$ be a finite rewrite system over signature Σ and B: $\mathbb{N} \to \mathbb{N}$ be a primitive recursive function that puts an upper bound on Dh_R, i.e. $B(n) \ge Dh_R(n)$ for all $n > 0$.

Σ_0 will consist of the following symbols (binary symbols will sometimes be used as infix symbols):

(i) $\Sigma \cup \Sigma' \cup \Sigma''$ where $\Sigma' := \{ f' \mid f \in \Sigma \}$ and $\Sigma'' := \{ f'' \mid f \in \Sigma \}$ are disjoint copies of Σ,

(ii) unary symbols normalize, size, s, code, decode, reducible, reduce, rule$_i$ $(1 \le i \le k)$, test, $\Diamond$; binary symbols $\otimes$, $\oplus$, +, equal, $\wedge$, $\vee$; constant symbols 0, $\triangleright$, $\perp$ and a symbol test' of arity k+1 ,

(iii) symbols for computing the primitive recursive time bound (see below).

As a tool to switch from signature Σ to signature Σ' and vice versa, we define (canonical) rewrite systems C ("*code*") and D ("*decode*") by

$C := \{ f(x_1,...,x_n) \to f'(x_1,...,x_n) \mid f \in \Sigma \}$, $D := \{ f'(x_1,...,x_n) \to f(x_1,...,x_n) \mid f \in \Sigma \}$.

Note that C and D will not be part of S ; some useful properties of C and D are stated in lemma 5.1.

S will consist of the rules of type (1) - (14) defined below. Simulation of R using S can be devided into three phases, starting with the term normalize(t) for $t \in T(\Sigma)$:

(I) computing the time bound $B(|t|)$,

(II) nondeterministically simulating R-steps on a copy $t{\downarrow_C}$ of t and testing the resulting terms for irreducibility ,

(III) decoding the resulting term if normal form is reached.

Phase (I) starts using rule (1), then computes the size of t (i.e. $s^{|t|}(0)$ where s denotes the successor on natural numbers) using rules of type (3), then computes the time bound (i.e. $s^{B(|t|)}(0)$) using rules of type (4) and finally uses the rules of type (2) to get $s^{B(|t|)}(\text{code}(t) \oplus \text{test}(t))$.

$\text{normalize}(x) \rightarrow b(\text{size}(x)) \otimes x$		(1)
$s(x) \otimes y \rightarrow s(x \otimes y)$		(2a)
$0 \otimes y \rightarrow \text{code}(y) \oplus \text{test}(y)$		(2b)
$\text{size}(f(x_1,\dots,x_n)) \rightarrow s(\text{size}(x_1) + (\dots + (\text{size}(x_{n-1}) + \text{size}(x_n))\dots))$	(for $f \in \Sigma$, $n>0$)	(3a)
$\text{size}(f) \rightarrow s(0)$	(for constant $f \in \Sigma$)	(3b)
$s(x) + y \rightarrow s(x + y)$		(3c)
$0 + y \rightarrow y$		(3d)

The set of rules of type (4) is defined by transforming the defining equations of the primitive recursive definition of B into rewrite rules such that $b(s^n(0)) \xrightarrow{*}_S s^k(0)$ iff $B(n) = k$ (see [Börg89] or [Rose84]). s and 0 will be used to denote natural numbers, u^n_i are projection symbols. Thus there are rules according to the following schemes:

$u^n_i(x_1,\dots,x_n) \rightarrow x_i$	*(projection)*	(4a)
$d(x_1,\dots,x_n) \rightarrow h(g_1(x_1,\dots,x_n),\dots,g_m(x_1,\dots,x_n))$	*(composition)*	(4b)
$d(0,x_1,\dots,x_n) \rightarrow g(x_1,\dots,x_n)$	*(primitive*	(4c)
$d(s(x_0),x_1,\dots,x_n) \rightarrow h(d(x_0,x_1,\dots,x_n), x_0,x_1,\dots,x_n)$	*recursion)*	(4d)

Phase (II) works on terms of the form $s^j(t' \oplus t'')$. Rules of type (7a) allow to replace a subterm $(l_i{\downarrow_C})\sigma$ corresponding to an R-redex by reducible($\text{rule}_i(x_1\sigma,\dots,x_n\sigma)$) where $\{x_1,\dots,x_n\} = \text{var}(l_i)$. Type (7b) rules propagate information of the existence of a reducible subterm to the root of the term by replacing symbols f' by f'' and marking the path leading to the reducible subterm by $\triangleright$. If the symbol "reducible" has reached the root of t' , rule (8) allows to initiate the actual reduction step by replacing it by "reduce" and (using rules of type (9)) unambiguously do the corresponding reduction step at the right occurrence. Each time rule (8) is used, one of the s-symbols vanishes; if the precedence for proving termination is chosen appropriately, this guarantees termination of S .

Define $\text{RULE}_i := \text{rule}_i(x_1,\dots,x_n)$ where $\{x_1,\dots,x_n\} = \text{var}(l_i)$, $1 \le i \le k$. RULE_i thus is uniquely determined up to permutation of $\{x_1,\dots,x_n\}$; choose the same fixed order for rules of type (7a) and (9b).

$$\text{code}(\ f(x_1,...,x_n)\ ) \ \rightarrow\ f'(\ \text{code}(x_1),...,\text{code}(x_n)\ ) \qquad \text{(for } f\in\Sigma,\ n\geq 0) \qquad (5)$$

$$\text{decode}(\ f'(x_1,...,x_n)\ ) \ \rightarrow\ f(\ \text{decode}(x_1),...,\text{decode}(x_n)\ ) \qquad \text{(for } f\in\Sigma,\ n\geq 0) \qquad (6)$$

$$l_i \downarrow_C \ \rightarrow\ \text{reducible}(\ \text{RULE}_i\ ) \qquad \text{(for } 1\leq i\leq k) \qquad (7a)$$

$$f'(x_1,...,\text{reducible}(x_i),...,x_n) \ \rightarrow\ \text{reducible}(\ f''(x_1,...,\triangleright(x_i),...,x_n)\ ) \qquad \text{(for } f\in\Sigma,\ n>0) \qquad (7b)$$

$$s(\ \text{reducible}(x) \oplus y\ ) \ \rightarrow\ \text{reduce}(x) \oplus \text{test}(\ \text{decode}(\ \text{reduce}(x)\ )\ ) \qquad (8)$$

$$\text{reduce}(\ f''(x_1,...,\triangleright(x_i),...,x_n)\ ) \ \rightarrow\ f'(x_1,...,\text{reduce}(x_i),...,x_n) \qquad \text{(for } f\in\Sigma,\ n>0) \qquad (9a)$$

$$\text{reduce}(\ \text{RULE}_i\ ) \ \rightarrow\ r_i \downarrow_C \qquad \text{(for } 1\leq i\leq k) \qquad (9b)$$

Example

A typical term t' would look like that:

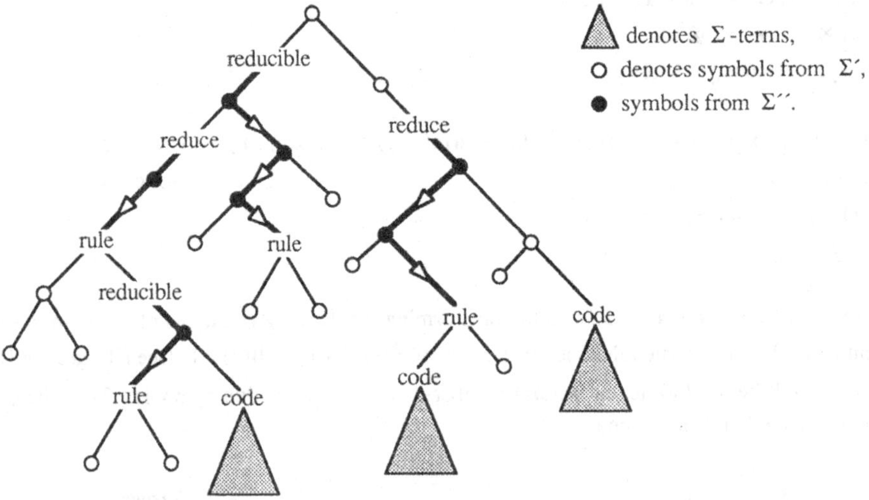

□

Testing for irreducibility is done on the second argument of $\oplus$, using rules of type (10), (11) and (12). For $t\in T(\Sigma)$, the i-th rule of R is not applicable to t at occurrence λ iff $\text{test}'(t,y_1,...,\Diamond(t),...,y_k) \xrightarrow{*}_S \text{test}'(t,y_1,...,\bot,...,y_k)$ due to the rules of type (10b) and (10c); (10b) takes care of function symbols and (10c) of nonlinear variables in l_i. Finally t is in normal form w.r.t. R iff $\text{test}(t) \xrightarrow{*}_S \bot$. For defining rules of type (10b) and (10c), the following notations are used:

Definition

(1) For $t, t'\in T(\Sigma,X)$ and $u\in Occ(t)$ such that $t/u \notin X$ let $PATH(t,u,t') \in T(\Sigma,X)$ such that

$PATH(t, \lambda, t') := t'$

$PATH(f(t_1,...,t_n), i.u, t') := f(x_1,...,x_{i-1},PATH(t_i, u, t'),x_{i+1},...,x_n)$

where $x_1,..., x_{i-1}, x_{i+1},..., x_n$ are distinct variables not occurring in $PATH(t_i, u, t')$.

(2) For $t\in T(\Sigma,X)$ let $LIN(t)\in T(\Sigma,X)$ such that $LIN(t)$ is linear, $Occ(LIN(t)) = Occ(t)$ and for all $u\in Occ(t)$: $t/u \in X$ iff $LIN(t)/u \in X$ and if $t/u \notin X$ then $LIN(t)//u = t//u$.

(Thus $PATH(t, u, t')$ and $LIN(t)$ are uniquely determined up to variable renaming.)

(3) For nonlinear $t \in T(\Sigma,X)$ and a fixed term LIN(t) define $EQ(t) \in T(\{equal, \wedge\},X)$ as follows. Let $U_x := \{ u \in Occ(t) \mid t/u = x \}$ for $x \in X$ and let $\{z_1,\dots,z_p\} = \{ x \in var(t) \mid |U_x| > 1 \}$ be the set of nonlinear variables in t (thus $p > 1$). For $\{u_1,\dots,u_{n_i}\} = U_{z_i}$ define

$EQ_{z_i}(t) := (LIN(t)/u_1 \text{ equal } LIN(t)/u_2) \wedge (\dots \wedge (LIN(t)/u_{n_i-1} \text{ equal } LIN(t)/u_{n_i})\dots)$ and finally

$EQ(t) := EQ_{z_1}(t) \wedge (\dots \wedge EQ_{z_p}(t)\dots)$. □

Example

Let $R = \{ f(a, x) \to \dots, f(f(x, a), f(x, f(y, x))) \to \dots \}$ be a two rule system over signature $\Sigma = \{ a, f \}$; thus test' is of arity 3. Rules in S of type (10b) are

$test'(a, \Diamond(y_1), y_2) \to test'(a, \perp, y_2)$ and

$test'(f(f(x_1, x_2), x_3), \Diamond(y_1), y_2) \to test'(f(f(x_1, x_2), x_3), \perp, y_2)$ from the first rule,

$test'(a, y_1, \Diamond(y_2)) \to \dots,$ $test'(f(a, x_1), y_1, \Diamond(y_2)) \to \dots,$

$test'(f(f(x_1, f(x_2, x_3)), x_4), y_1, \Diamond(y_2)) \to \dots,$ $test'(f(x_1, a), y_1, \Diamond(y_2)) \to \dots$ and

$test'(f(x_1, f(x_2, a)), y_1, \Diamond(y_2)) \to \dots$ from the second rule (right hand sides omitted).

There is only one rule of type (10c) in S since the first rule is left linear; the second rule yields

$test'(x, y_1, \Diamond(f(f(x_1, a), f(x_2, f(x_3, x_4))))) \to test'(x, y_1, \Diamond((x_1 \text{ equal } x_2) \wedge (x_2 \text{ equal } x_4)))$. □

$test(x) \to test'(x, \Diamond(x),\dots,\Diamond(x))$ (10a)

$test'(PATH(l_i, u, g(x_1,\dots,x_n)), y_1,\dots,\Diamond(y_i),\dots,y_k) \to test'(PATH(l_i, u, g(x_1,\dots,x_n)), y_1,\dots,\perp,\dots,y_k)$
(for $1 \le i \le k$, $u \in Occ(l_i)$ such that $l_i/u \notin X$, $g \in \Sigma$, $n \ge 0$ such that $g \ne l_i//u$, $y_1,\dots,y_k \notin var(PATH(l_i, u, g(x_1,\dots,x_n)))$) (10b)

$test'(x, y_1,\dots,\Diamond(LIN(l_i)),\dots,y_k) \to test'(x, y_1,\dots,\Diamond(EQ(l_i)),\dots,y_k)$
(for $1 \le i \le k$ if l_i is nonlinear) (10c)

$test'(f(x_1,\dots,x_n), \perp,\dots,\perp) \to test(x_1) \vee (\dots \vee (test(x_{n-1}) \vee test(x_n))\dots)$
(for $f \in \Sigma$, $n>0$) (10d)

$test'(f, \perp,\dots,\perp) \to \perp$ (for constant $f \in \Sigma$) (10e)

$\perp \vee \perp \to \perp$ (11a) $\Diamond(\perp) \to \perp$ (11b)

$x \wedge \perp \to \perp$ (11c) $\perp \wedge x \to \perp$ (11d)

$f(x_1,\dots,x_n) \text{ equal } g(y_1,\dots,y_m) \to \perp$ (for $f,g \in \Sigma$, $n,m \ge 0$, $f \ne g$) (12a)

$f(x_1,\dots,x_n) \text{ equal } f(y_1,\dots,y_n) \to (x_1 \text{ equal } y_1) \wedge (\dots \wedge (x_n \text{ equal } y_n)\dots)$ (for $f \in \Sigma$, $n>0$) (12b)

Finally, rules (13) and (14) (together with (6)) allow to produce a term of signature Σ again in phase (III):

$x \oplus \perp \to decode(x)$ (13)

$s(f(x_1,\dots,x_n)) \to f(x_1,\dots,x_n)$ (for $f \in \Sigma$, $n \ge 0$) (14)

5. Proof of the Main Theorem

The proof of our main theorem is separated into three parts:

(1) *completeness* (i.e. $NF_R(t) \subseteq NF_S(\text{normalize}(t))$ for all $t \in T(\Sigma)$)

(2) *correctness* (i.e. $NF_S(\text{normalize}(t)) \subseteq NF_R(t)$ for all $t \in T(\Sigma)$)

(3) *termination* (i.e. $l >_{GE} r$ for all rules $l \rightarrow r \in S$)

Some proofs of lemmas are omitted due to space limitations. For details we refer to [Hof91]. First some trivial properties of C and D resp. as defined in section 4 are stated in lemma 5.1:

Lemma 5.1

Let $t, t' \in T(\Sigma,X)$, $u \in Occ(t)$, $\sigma: X \rightarrow T(\Sigma,X)$ be a substitution. Then $t\downarrow_C \in T(\Sigma',X)$, $t\downarrow_C\downarrow_D = t$, $t[u \leftarrow t']\downarrow_C = t\downarrow_C[u \leftarrow t'\downarrow_C]$, $(t/u)\downarrow_C = t\downarrow_C/u$, $(t\sigma)\downarrow_C = (t\downarrow_C)(\sigma\downarrow_C)$ (for $\downarrow_D$ analogously).
Let $t \in T(\Sigma)$. Then $\text{code}(t) \xrightarrow{*}_S t\downarrow_C$. Let $t \in T(\Sigma')$. Then $\text{decode}(t) \xrightarrow{*}_S t\downarrow_D$.

5.1. Completeness

Lemma 5.2

Let $t \in T(\Sigma)$. Then $b(\text{size}(t)) \xrightarrow{*}_S s^{B(|t|)}(0)$.

Lemma 5.3

Let $t \in T(\Sigma)$, $u \in Occ(t)$, $x, y \in X$. Then
$s(t\downarrow_C[u \leftarrow \text{reducible}(x)] \oplus y) \xrightarrow{*}_S t\downarrow_C[u \leftarrow \text{reduce}(x)] \oplus \text{test}(t[u \leftarrow \text{decode}(\text{reduce}(x))])$.

Proof

Structural induction on u:

<u>u = λ:</u> $s(t\downarrow_C[\lambda \leftarrow \text{reducible}(x)] \oplus y) = s(\text{reducible}(x) \oplus y) \longrightarrow_{(8)}$
$\text{reduce}(x) \oplus \text{test}(\text{decode}(\text{reduce}(x))) = t\downarrow_C[\lambda \leftarrow \text{reduce}(x)] \oplus \text{test}(t[\lambda \leftarrow \text{decode}(\text{reduce}(x))])$

<u>u = vi:</u> Let $t\downarrow_C//v = f'$. Thus

$s(t\downarrow_C[vi \leftarrow \text{reducible}(x)] \oplus y) =$

$s(t\downarrow_C[v \leftarrow f'(t\downarrow_C/v1,\ldots,\text{reducible}(x),\ldots,t\downarrow_C/vn)] \oplus y) \longrightarrow_{(7b)}$

$s(t\downarrow_C[v \leftarrow \text{reducible}(f''(t\downarrow_C/v1,\ldots,\triangleright(x),\ldots,t\downarrow_C/vn))] \oplus y) \xrightarrow{*}_S$ by induction hypothesis

$t\downarrow_C[v \leftarrow \text{reduce}(f''(t\downarrow_C/v1,\ldots,\triangleright(x),\ldots,t\downarrow_C/vn))] \oplus$
$\quad \text{test}(t[v \leftarrow \text{decode}(\text{reduce}(f''(t\downarrow_C/v1,\ldots,\triangleright(x),\ldots,t\downarrow_C/vn)))]) \xrightarrow{*}_S$

$t\downarrow_C[v \leftarrow f'(t\downarrow_C/v1,\ldots,\text{reduce}(x),\ldots,t\downarrow_C/vn)] \oplus$
$\quad \text{test}(t[v \leftarrow f(t/v1,\ldots,\text{decode}(\text{reduce}(x)),\ldots,t/vn)]) =$

$t\downarrow_C[vi \leftarrow \text{reduce}(x)] \oplus \text{test}(t[vi \leftarrow \text{decode}(\text{reduce}(x))])$

since $\text{reduce}(f''(t\downarrow_C/v1,\ldots,\triangleright(x),\ldots,t\downarrow_C/vn)) \longrightarrow_{(9a)} f'(t\downarrow_C/v1,\ldots,\text{reduce}(x),\ldots,t\downarrow_C/vn)$

and $\text{decode}(f'(t\downarrow_C/v1,\ldots,\text{reduce}(x),\ldots,t\downarrow_C/vn)) \longrightarrow_{(6)}$
$f(\text{decode}(t\downarrow_C/v1),\ldots,\text{decode}(\text{reduce}(x)),\ldots,\text{decode}(t\downarrow_C/vn)) \xrightarrow{*}_S$
$f((t\downarrow_C/v1)\downarrow_D,\ldots,\text{decode}(\text{reduce}(x)),\ldots,(t\downarrow_C/vn)\downarrow_D) =$
$f(t/v1,\ldots,\text{decode}(\text{reduce}(x)),\ldots,t/vn)$ using lemma 5.1. □

Lemma 5.4

Let $t, \bar{t} \in T(\Sigma)$, $x \in X$.

(1) If $t \longrightarrow_R \bar{t}$ then $s(t\downarrow_C \oplus x) \xrightarrow{*}_S \bar{t}\downarrow_C \oplus \mathrm{test}(\bar{t})$.

(2) If $t\,(\longrightarrow_R)^n\,\bar{t}$ then $s^n(t\downarrow_C \oplus x) \xrightarrow{*}_S \bar{t}\downarrow_C \oplus \mathrm{test}(\bar{t})$.

Proof

(1) Let $u \in Occ(t)$, $t/u = l_i\sigma$ and $\bar{t} = t[u \leftarrow r_i\sigma]$ where $l_i \rightarrow r_i$ is a rule from R. Then

$s(t\downarrow_C \oplus x) =$ by lemma 5.1

$s(t\downarrow_C[u \leftarrow (l_i\sigma)\downarrow_C] \oplus x) =$ by lemma 5.1

$s(t\downarrow_C[u \leftarrow (l_i\downarrow_C)(\sigma\downarrow_C)] \oplus x) \longrightarrow_{(7a)}$

$s(t\downarrow_C[u \leftarrow \mathrm{reducible}(\mathrm{RULE}_i(\sigma\downarrow_C))] \oplus x) \xrightarrow{*}_S$ by lemma 5.3

$t\downarrow_C[u \leftarrow \mathrm{reduce}(\mathrm{RULE}_i(\sigma\downarrow_C))] \oplus \mathrm{test}(t[u \leftarrow \mathrm{decode}(\mathrm{reduce}(\mathrm{RULE}_i(\sigma\downarrow_C)))]) \xrightarrow{*}_S$

$t\downarrow_C[u \leftarrow (r_i\sigma)\downarrow_C] \oplus \mathrm{test}(t[u \leftarrow r_i\sigma]) =$ by lemma 5.1

$t[u \leftarrow r_i\sigma]\downarrow_C \oplus \mathrm{test}(t[u \leftarrow r_i\sigma]) = \bar{t}\downarrow_C \oplus \mathrm{test}(\bar{t})$

since $\mathrm{reduce}(\mathrm{RULE}_i(\sigma\downarrow_C)) \longrightarrow_{(9b)} (r_i\downarrow_C)(\sigma\downarrow_C) = (r_i\sigma)\downarrow_C$

and $\mathrm{decode}((r_i\sigma)\downarrow_C) \xrightarrow{*}_S (r_i\sigma)\downarrow_C\downarrow_D = r_i\sigma$ by lemma 5.1

(2) Easy induction using (1). □

Lemma 5.5

Let $t, t' \in T(\Sigma)$. If $t \neq t'$ then $t \text{ equal } t' \xrightarrow{*}_S \bot$.

Proof

Use rules of type (11c,d) and (12a,b). □

Lemma 5.6

Let $t \in T(\Sigma)$, $l_i \rightarrow r_i \in R$ $(1 \leq i \leq k)$. If there is no substitution σ such that $t = l_i\sigma$ then $\mathrm{test}'(t, y_1,\ldots,y_{i-1},\Diamond(t),y_{i+1},\ldots,y_k) \xrightarrow{*}_S \mathrm{test}'(t, y_1,\ldots,y_{i-1},\bot,y_{i+1},\ldots,y_k)$.

Proof

If l_i does not match t then (case 1) there is a nonvariable occurrence u in l_i that is also an occurrence in t such that $t//u \neq l_i//u$ or (case 2) there is a nonlinear variable in l_i such that the corresponding subterms in t differ.

<u>case 1</u> Choose an outermost occurrence $u \in Occ(l_i) \cap Occ(t)$ such that $l_i/u \;\; X$ and $t//u \neq l_i//u$ (i.e. $t//v = l_i//v$ for all $v < u$).
Then t is an instance of $\mathrm{PATH}(l_i, u, t/u)$, thus an instance of $\mathrm{PATH}(l_i, u, (t//u)(x_1,\ldots,x_n))$ if $t//u$ is of arity n. Thus $\mathrm{test}'(t, y_1,\ldots,y_{i-1},\Diamond(t),y_{i+1},\ldots,y_k) \longrightarrow_{(10b)} \mathrm{test}'(t, y_1,\ldots,y_{i-1},\bot,y_{i+1},\ldots,y_k)$.

<u>case 2</u> If there is no match from l_i to t but case 1 does not apply then $Occ(l_i) \subseteq Occ(t)$ and for all nonvariable occurrences u in l_i we have $l_i//u = t//u$. Hence t is an instance of $\mathrm{LIN}(l_i)$, $t = \mathrm{LIN}(l_i)\sigma$ say, and $\mathrm{test}'(t, y_1,\ldots,y_{i-1},\Diamond(t),y_{i+1},\ldots,y_k) \longrightarrow_{(10c)} \mathrm{test}'(t, y_1,\ldots,y_{i-1},\Diamond(\mathrm{EQ}(l_i)\sigma),y_{i+1},\ldots,y_k)$.
There are different occurrences u_1, u_2 in l_i such that $l_i/u_1 = l_i/u_2 \in X$ and $t/u_1 \neq t/u_2$. W.l.o.g. choose u_1, u_2 such that $(\mathrm{LIN}(l_i)/u_1 \text{ equal } \mathrm{LIN}(l_i)/u_2)$ is a subterm of $\mathrm{EQ}(l_i)$. Since $(\mathrm{LIN}(l_i)/u_1)\sigma = t/u_1$ and $(\mathrm{LIN}(l_i)/u_2)\sigma = t/u_2$, $(t/u_1 \text{ equal } t/u_2)$ is a subterm of $\mathrm{EQ}(l_i)\sigma$. From lemma 5.5 we get $(t/u_1 \text{ equal } t/u_2) \xrightarrow{*}_S \bot$, thus $\mathrm{EQ}(l_i)\sigma \xrightarrow{*}_S \bot$ using rules of type (11c,d). Hence

$\mathrm{test}'(t, y_1,\ldots,y_{i-1},\Diamond(\mathrm{EQ}(l_i)\sigma),y_{i+1},\ldots,y_k) \xrightarrow{*}_S \mathrm{test}'(t, y_1,\ldots,y_{i-1},\Diamond(\bot),y_{i+1},\ldots,y_k) \longrightarrow_{(11b)}$
$\mathrm{test}'(t, y_1,\ldots,y_{i-1},\bot,y_{i+1},\ldots,y_k)$. □

Lemma 5.7

Let $t \in T(\Sigma)$ be in normal form w.r.t. R. Then $\text{test}(t) \xrightarrow{*}_S \perp$.

Proof

Structural induction on t: $\text{test}(t) \longrightarrow_{(10a)} \text{test}'(t, \Diamond(t),\dots,\Diamond(t)) \xrightarrow{*}_S \text{test}'(t, \perp,\dots,\perp)$ by lemma 5.6. If t is a constant then $\text{test}'(t, \perp,\dots,\perp) \longrightarrow_{(10e)} \perp$. If $t = f(t_1,\dots,t_n)$ then $\text{test}'(t, \perp,\dots,\perp) \longrightarrow_{(10d)}$ $\text{test}(t_1) \vee (\dots \vee (\text{test}(t_{n-1}) \vee \text{test}(t_n))\dots) \xrightarrow{*}_S \perp \vee (\dots \vee (\perp \vee \perp)\dots) \xrightarrow{*}_{(11a)} \perp$ by induction hypothesis since subterms of normal form terms are in normal form. □

Main Lemma 5.8 (Completeness)

Let $t, \bar{t} \in T(\Sigma)$, $\bar{t} \in NF_R(t)$. Then $\text{normalize}(t) \xrightarrow{*}_S \bar{t}$ and $\bar{t}$ is in normal form w.r.t. S.

Proof

Since left hand sides of rules of S are not in $T(\Sigma,X)$, every term in $T(\Sigma)$ is in normal form w.r.t. S. $\text{normalize}(t) \xrightarrow{*}_S \bar{t}$ is shown by combining lemma 5.2, lemma 5.4 and lemma 5.7:

normalize(t)	$\longrightarrow_{(1)}$	$b(\text{size}(t)) \otimes t$	
	$\xrightarrow{*}_S$	$s^k(0) \otimes t$	where $k = B(\lvert t\rvert) \geq Dh_R(\lvert t\rvert) \geq dh_R(t)$ by lemma 5.2
	$\xrightarrow{*}_{(2a)}$	$s^k(0 \otimes t)$	
	$\longrightarrow_{(2b)}$	$s^k(\text{code}(t) \oplus \text{test}(t))$	
	$\xrightarrow{*}_{(5)}$	$s^k(t\downarrow_C \oplus \text{test}(t))$	by lemma 5.1
	$\xrightarrow{*}_S$	$s^j(\bar{t}\downarrow_C \oplus \text{test}(\bar{t}))$	where $j \geq k - dh_R(t) \geq 0$ by lemma 5.4
	$\xrightarrow{*}_S$	$s^j(\bar{t}\downarrow_C \oplus \perp)$	by lemma 5.7
	$\longrightarrow_{(13)}$	$s^j(\text{decode}(\bar{t}\downarrow_C))$	
	$\xrightarrow{*}_{(6)}$	$s^j(\bar{t}\downarrow_C\downarrow_D)$	
	$=$	$s^j(\bar{t})$	by lemma 5.1
	$\xrightarrow{*}_{(14)}$	$\bar{t}$	□

5.2. Correctness

Lemma 5.9

Let $t \in T(\Sigma)$ and $\text{normalize}(t) \longrightarrow_S t_1 \longrightarrow_S \dots \longrightarrow_S t_n$ be a terminating derivation sequence. Then there exists i $(1 \leq i \leq n)$ such that $t_i = s^k(\text{code}(t) \oplus \text{test}(t))$ and $k \geq dh_R(t)$.

Proof

Use confluence of the set of rules of type (1), (2), (3) and (4) and lemma 5.2. □

Define N to be the rewrite system consisting of all rules according to the following schemes:

$\text{reduce}(f''(x_1,\dots,\triangleright(x_i),\dots,x_n)) \rightarrow f'(x_1,\dots,\text{reduce}(x_i),\dots,x_n)$		(for $f \in \Sigma$, $n>0$)
$\text{reduce}(\text{RULE}_i) \rightarrow r_i\downarrow_C$		(for $1 \leq i \leq k$)
$\text{reducible}(f''(x_1,\dots,\triangleright(x_i),\dots,x_n)) \rightarrow f'(x_1,\dots,\text{reducible}(x_i),\dots,x_n)$		(for $f \in \Sigma$, $n>0$)
$\text{reducible}(\text{RULE}_i) \rightarrow l_i\downarrow_C$		(for $1 \leq i \leq k$)
$\text{code}(f(x_1,\dots,x_n)) \rightarrow f'(\text{code}(x_1),\dots,\text{code}(x_n))$		(for $f \in \Sigma$, $n \geq 0$)

Lemma 5.10

Let $t \in T(\Sigma)$. If $s^k(\text{code}(t) \oplus ...) \xrightarrow{*}_S s^j(t_0 \oplus ...)$ then

(1) t_0 has a unique normal form w.r.t. N and $t_0 \downarrow_N \in T(\Sigma')$

(2) if $s(t_0 \oplus x)$ is in normal form w.r.t. S then $t_0 \in T(\Sigma')$ and $t_0 \downarrow_D$ is in normal form w.r.t. R .

Proof

Use the fact that for $t_0 \xrightarrow{*}_N t'$ the following holds (cf. the figure in section 4):

- t' is a term over signature $\Sigma \cup \Sigma' \cup \Sigma'' \cup \{\text{reducible, reduce, rule}_i, \triangleright, \text{code}\}$.
- no subterm of t' has the form $f''(...,\triangleright(...),...,\triangleright(...),...)$.
- for every subterm $f''(t_1,...,t_n)$ of t' there exists i such that $\text{top}(t_i) = \triangleright$.
- for every subterm reduce(t") or reducible(t") of t' we have $\text{top}(t'') \in \Sigma''$ or t" is an instance of a term RULE_i .
- for every subterm code(t") of t' we have $t'' \in T(\Sigma)$.
- $t'/\!/\lambda \neq \triangleright$ and if $t'/\!/ui = \triangleright$ then $t'/\!/u \in \Sigma''$.
- $t'/\!/\lambda \notin \Sigma''$ and if $t'/\!/ui \in \Sigma''$ then $t'/\!/u \in \{\text{reduce, reducible}, \triangleright\}$.

(Proof by induction on the length of the $S \cup N$ derivation.) □

Lemma 5.11

Let $t \in T(\Sigma)$. If $s^k(\text{code}(t) \oplus ...) \xrightarrow{*}_S s^k(t' \oplus ...)$ then $t = t'\downarrow_N\downarrow_D$ and if

$s^k(\text{code}(t) \oplus ...) \xrightarrow{*}_S s^{j+1}(\text{reducible}(t_0) \oplus ...) \longrightarrow_S s^j(\text{reduce}(t_0) \oplus ...) \xrightarrow{*}_S s^j(t' \oplus ...)$

(thus $k \geq j+1$) then $t\ (\longrightarrow_R)^{k-j-1}\ \text{reducible}(t_0)\downarrow_N\downarrow_D \longrightarrow_R \text{reduce}(t_0)\downarrow_N\downarrow_D = t'\downarrow_N\downarrow_D$.

Lemma 5.12

Let $t \in T(\Sigma)$. If $s^k(\text{code}(t) \oplus \text{test}(t)) \xrightarrow{*}_S s^j(t_0 \oplus \bot)$ then $t_0 \downarrow_N\downarrow_D$ is in normal form w.r.t. R and $t_0 \downarrow_N\downarrow_D$ is the unique normal form of $s^j(\text{decode}(t_0))$ w.r.t. S .

Main Lemma 5.13 (Correctness)

Let $t, \bar{t} \in T(\Sigma)$, $\bar{t} \in NF_S(\text{normalize}(t))$. Then $t \xrightarrow{*}_R \bar{t}$ and $\bar{t}$ is in normal form w.r.t. R .

Proof

By lemma 5.9 $\bar{t} \in NF_S(\ s^k(\text{code}(t) \oplus \text{test}(t))\ )$ where $k \geq dh_R(t)$. If $s^k(\text{code}(t) \oplus \text{test}(t)) \xrightarrow{*}_S \bar{t}$ without using rule (13), $\bar{t}$ would be of the form $s^j(t' \oplus t'')$ with $t'' \neq \bot$. By lemma 5.11 $t\ (\longrightarrow_R)^{k-j}\ t'\downarrow_N\downarrow_D$. If $j = 0$ we get $k = dh_R(t)$, thus $t'\downarrow_N\downarrow_D$ is in normal form w.r.t. R . If $j > 0$ then by lemma 5.10 $t'\downarrow_N\downarrow_D = t'\downarrow_D$ is in normal form w.r.t. R too.

<u>case 1</u> $k = j$: Thus $\text{code}(t) \xrightarrow{*}_S t'$ and $\text{test}(t) \xrightarrow{*}_S t''$. By lemma 5.11 $t = t'\downarrow_N\downarrow_D$, thus by lemma 5.7 $\text{test}(t) \xrightarrow{*}_S \bot$, a contradiction to the fact that test(t) has a unique normal form w.r.t. S for $t \in T(\Sigma)$ (easily proven using confluence of the set of rules of type (10), (11) and (12)).

<u>case 2</u> $k > j$: Thus $s^k(\text{code}(t) \oplus \text{test}(t)) \xrightarrow{*}_S s^{j+1}(\text{reducible}(t_0) \oplus ...)$, $\text{reduce}(t_0) \xrightarrow{*}_S t'$ and $\text{test}(\text{decode}(\text{reduce}(t_0))) \xrightarrow{*}_S t''$ for a term t_0. We get a contradiction similar to case 1 using $\text{reduce}(t_0)\downarrow_N = t'\downarrow_N$, the fact that test(decode(reduce(t_0))) has a unique normal form w.r.t. S and $\text{decode}(\text{reduce}(t_0)) \xrightarrow{*}_S \text{decode}(\text{reduce}(t_0)\downarrow_N) \xrightarrow{*}_S \text{reduce}(t_0)\downarrow_N\downarrow_D = t'\downarrow_N\downarrow_D$.

Hence rule (13) has been used and there is a term t_1 such that

$s^k(\text{code}(t) \oplus \text{test}(t)) \xrightarrow{*}_S s^j(t_1 \oplus \bot) \longrightarrow_{(13)} s^j(\text{decode}(t_1)) \xrightarrow{*}_S \bar{t}$.

Lemma 5.11 and lemma 5.12 yield $t_1\downarrow_N\downarrow_D \in NF_R(t)$; since $\bar{t} \in NF_S(\ s^j(\text{decode}(t_1))\ )$, the uniqueness property from lemma 5.12 gives us $t_1\downarrow_N\downarrow_D = \bar{t}$. □

5.3. Termination

Termination of the system S can be proven using generalized embedding by showing $l \succ_{GE} r$ for all rules $l \rightarrow r$ in S. Choose a (partial) precedence $\succ$ on Σ_0 according to the diagram below (from left to right in decreasing order w.r.t. $\succ$). f (f´, f´´) denotes symbols in Σ (Σ', Σ''). b denotes symbols used for computing the primitive recursive time bound, i.e. symbols occurring in rules of type (4) except for 0 and s. These symbols are ordered by $\succ$ according to their use in the rules: $d \succ d'$ if $d \neq d'$ and there is a rule $l \rightarrow r$ of type (4) such that $top(l) = d$ and d' occurs in r; additionally the projection symbols u^n_i are less than all other such symbols w.r.t. $\succ$. This implies $d \succ h, g_1,..., g_m$ in rules of type (4b) and $d \succ g$ and $d \succ h$ in (4c) and (4d) respectively.

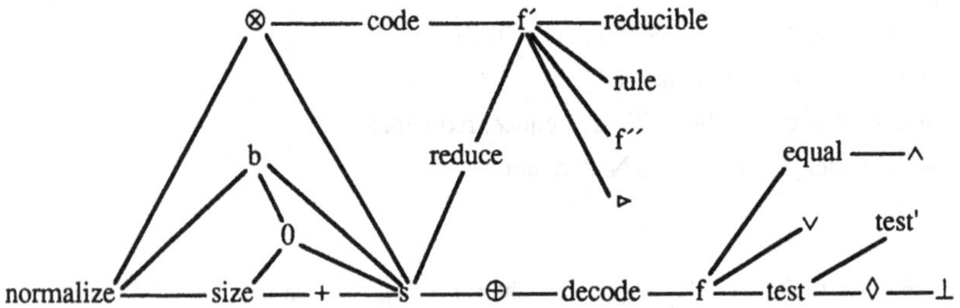

6. Relations Computed by Rewrite Systems

Finally we just sketch a corollary to the main theorem concerning relations that are computable by rewrite systems.

E.g. let $\Sigma = \{s, 0\}$ and $ADD : T(\Sigma)^2 \rightarrow T(\Sigma)$ be the function defined by $ADD(\, s^m(0), s^n(0)\,) := s^{m+n}(0)$. Then $R = \{\, 0+y \rightarrow y,\ s(x)+y \rightarrow s(x+y)\,\}$ *computes* ADD via the new symbol + in a natural sense.

Definition

Let $P \subseteq T(\Sigma)^{n+1}$, n>0 be a relation on ground terms over signature Σ, R be a finite rewrite system over an extension Γ of Σ (i.e. $\Sigma \subset \Gamma$) and p be a n-ary symbol in $\Gamma \setminus \Sigma$ such that

(1) R is terminating and

(2) for all $t_1,...,t_n \in T(\Sigma)$ we have $(\, t_1,...,t_n, \bar{t}\,) \in P$ iff $\bar{t} \in NF_R(\, p(t_1,...,t_n)\,)$.

Then R *computes* P *via* p. □

Corollary

For $P \subseteq T(\Sigma)^{n+1}$, n>0 are equivalent:

(1) There is a rewrite system that computes P with primitive recursively bounded derivation height.

(2) There is a rewrite system that computes P with termination proof by generalized embedding.

(3) There is a rewrite system that computes P with termination proof by multiset path ordering.

Proof

(2) implies (3) by the lemma from section 3 and (3) implies (1) by results of [Cich90] and [Hof90].

To show that (1) implies (2) let R be a rewrite system over Γ that computes P via p. Define S over Σ_0 to be the system constructed from R as in section 4 (note that Γ now plays the role of Σ in the construction) together with the rule $q(x_1,...,x_n) \rightarrow$ normalize($p(x_1,...,x_n)$) where q is added to Σ_0. Thus for all $t_1,...,t_n \in T(\Sigma)$ we have

$$NF_S(\ q(t_1,...,t_n)\ ) \ = \ NF_S(\ \text{normalize}(p(t_1,...,t_n))\ ) \ = \ NF_R(\ p(t_1,...,t_n)\ )$$

by the main theorem, hence S computes P via q. □

Using results from Rusinowitch [Rus87] and Steinbach [Stei89], in the corollary multiset path ordering can be replaced by PSO (path of subterms ordering [Pla78]), RDO (recursive decomposition ordering [JLR82]), KNS (the path ordering of Kapur, Narendran, Sivakumar [KNS85]), IRD (improved recursive decomposition ordering [Rus87]) or PSD (path of subterms ordering on decompositions [Stei89]) if lexicographic status is not incorporated.

Acknowledgement

I want to thank Frank Drewes, Maria Huber and an anonymous referee for helpful comments.

References

[Börg89] Börger, E. *Computability, Complexity, Logic*, North Holland (1989).

[Cich90] Cichon, E.A. *Bounds on Derivation Lengths from Termination Proofs*, Report CSD-TR-622, University of London, Egham (1990). To appear in Proc. Conf. on Proof Theory, Leeds (1990).

[Der82] Dershowitz, N. *Orderings for term rewriting systems*, TCS 17 (3) (1982), pp. 279-301.

[DJ90] Dershowitz, N., Jouannaud, J.-P. *Rewrite systems*, in: Handbook of Theoretical Computer Science, Vol. 2, Ed. J. van Leeuwen, North Holland (1990).

[DO89] Dershowitz, N., Okada, M. *Proof-theoretic techniques for term rewriting theory*, Proc. 3rd LICS (1989), pp. 104-111.

[DL90] Drewes, F., Lautemann, C. *Incremental Termination Proofs and the Length of Derivations*, Report 7/90, Universität Bremen (1990).

[Ges90] Geser, A. *Termination relative*, Ph.D. thesis, Univ. Passau (1990).

[Hof90] Hofbauer, D. *Termination proofs by multiset path orderings imply primitive recursive derivation lengths*, Proc. 2nd ALP, LNCS 463 (1990), pp. 347-358.

[Hof91] Hofbauer, D. *Time bounded rewrite systems and termination proofs by generalized embedding*, Report, Technische Universität Berlin (1991).

[HL89] Hofbauer, D., Lautemann, C. *Termination proofs and the length of derivations*, Proc. 3rd RTA, LNCS 355 (1989), pp. 167-177.

[JLR82] Jouannaud, J.-P., Lescanne, P., Reinig, F. *Recursive decomposition ordering*, IFIP Working Conf., Ed. Bjørner, North Holland (1982), pp. 331-348.

[Klop87] Klop, J.W. *Term rewriting systems: A tutorial*, Bulletin of the EATCS 32 (1987), pp. 143-183.

[KNS85] Kapur, D., Narendran, P., Sivakumar, G. *A path ordering for proving termination of term rewriting systems*, Proc. 10th CAAP, LNCS 185 (1985), pp. 173-187.

[Pla78] Plaisted, D. *A recursively defined ordering for proving termination of term rewriting systems*, Report UIUCDCS-R-78-943, University of Illinois, Urbana (1978).

[Rose84] Rose, H.E. *Subrecursion: Functions and hierarchies*, Oxford University Press (1984).

[Rus87] Rusinowitch, M. *Path of subterms ordering and recursive decomposition ordering revisited*, J. Symbolic Comp. 3 (1&2) (1987), pp. 117-131.

[Stei89] Steinbach, J. *Extensions and comparison of simplification orderings*, Proc. 3rd RTA, LNCS 355, pp. 434-448.

Detecting Redundant Narrowing Derivations by the LSE-SL Reducibility Test

Stefan Krischer, Alexander Bockmayr
Sonderforschungsbereich 314
Institut für Logik, Komplexität und Deduktionssysteme
Universität Karlsruhe, P.O.Box 6980
D-7500 Karlsruhe 1, F.R.Germany

Abstract
Rewriting and narrowing provide a nice theoretical framework for the integration of logic and functional programming. For practical applications however, narrowing is still much too inefficient. In this paper we show how reducibility tests can be used to detect redundant narrowing derivations. We introduce a new narrowing strategy, LSE-SL left-to-right basic normal narrowing, prove its completeness for arbitrary canonical term rewriting systems, and demonstrate how it increases the efficiency of the narrowing process.

1. Introduction
Rewriting and narrowing provide a nice theoretical framework for the integration of logic and (first order) functional programming. In the rewriting process the rewriting rules are employed for the simplification or evaluation of terms ("functional programming"), whereas in the narrowing process, which is closely related to resolution, the same rules are used for the solution of goals or equations ("logic programming") [Dershowitz/Plaisted 85, Bockmayr 86a].
From a theoretical point of view, narrowing provides a complete unification procedure for any equational theory that can be defined by a canonical term rewriting system. For practical applications however, narrowing in its original form is much too inefficient [Bockmayr 86b]. Therefore many optimizations have been proposed during the last years [Hullot 80, Réty et al. 85, Fribourg 85, Herold 86, Réty 87, Nutt/Réty/Smolka 89, Bosco/Giovannetti/Moiso 88, Echahed 88, Réty 88, Bockmayr 88, You 88, Padawitz 88, Hölldobler 89, Darlington/Guo 89, You 90].
In this paper we consider narrowing for arbitrary canonical term rewriting systems. We don't impose any restriction on the canonical term rewriting system such as constructor discipline [Fribourg 85] or left-linearity and non-overlapping left-hand sides [You 88, 90, Darlington/Guo 89]. For arbitrary canonical systems the most efficient complete narrowing strategy known so far is the SL left-to-right basic normal narrowing [Réty 88]. However, a thorough analysis of Réty's approach shows that it can be considerably improved if the term rewriting system has non-regular rules and overlapping left-hand sides. In this case various redundancies in the narrowing process can be avoided. In our paper we introduce three reducibility tests that allow to detect redundant narrowing derivations. They yield a new narrowing strategy, the LSE-SL left-to-right basic normal narrowing. We prove the completeness of this strategy, show how it can be implemented, and illustrate it by various examples.

The organization of the paper is as follows. After some preliminaries in section 2, we present in section 3 the most important narrowing strategies for arbitrary canonical term rewriting systems known today. For the first time we give quantitative results to compare the efficiency of the various approaches. In section 4 we show how SL left-to-right basic normal narrowing which is the most efficient general narrowing strategy known so far may be considerably improved by the LSE-SL test which is a substantial extension of Réty's SL test. In section 5 we describe how the LSE-SL test can be efficiently implemented. Finally, in section 6, we demonstrate the utility of our method by a number of examples and again give some quantitative results.

2. Preliminaries

We recall briefly some basic notions that are needed in the sequel. More details can be found in the survey of [Huet/Oppen 80].

$\Sigma = (S,F)$ denotes a *signature* (S a set of sort symbols and F a set of function symbols together with an arity function).

A *Σ-algebra* A consists of a family of sets $(A_s)_{s \in S}$ and a family of functions $(f^A)_{f \in F}$ such that if $f: s_1 \times ... \times s_n \to s$ then $f^A: A_{s1} \times ... \times A_{sn} \to A_s$.

X represents a family $(X_s)_{s \in S}$ of countably infinite sets X_s of *variables* of sort s.

T(F,X) is the algebra of *terms* (with variables) over Σ.

Occ(t) denotes the set of *occurrences* and FuOcc(t) the set of *non-variable occurrences* in the term t.

t/ω is the *subterm* of t at position $\omega \in Occ(t)$, $t[\omega \leftarrow s]$ the term obtained from t by *replacement* of the subterm t/ω by s. Var(t) is the set of variables occurring in t.

A *substitution* is a mapping $\sigma: X \to T(F,X)$ which is different from the identity only for a finite subset $D(\sigma)$ of X. We don't distinguish σ from its canonical extension to T(F,X).

$I(\sigma) =_{def} \bigcup_{x \in D(\sigma)} Var(\sigma(x))$ is the set of *variables introduced* by σ.

A *precongruence* is a binary relation $\to$ on T(F,X) or T(F) such that $s_1 \to t_1, \ldots, s_n \to t_n$ implies $f(s_1,\ldots,s_n) \to f(t_1,\ldots,t_n)$ for all terms s_i,t_i and all $f \in F$. By $\to^+$, $\to^*$ and $\leftrightarrow^*$ we denote the transitive, the reflexive-transitive and the reflexive-transitive- symmetric closure of $\to$ respectively. A *congruence* is a precongruence which is also an equivalence relation.

An *equation* G is an expression of the form $t \approx u$ where t and u are terms of T(F,X) belonging to the same sort. Let E be a set of equations. The *equational theory* $\equiv_E$ associated with E is the smallest congruence $\equiv$ on T(F,X) such that $\sigma(l) \equiv \sigma(r)$ for all equations $l \approx r$ in E and all substitutions σ.

The *E-subsumption* preorder $\leq_E$ on T(F,X) is defined by $s \leq_E t$ iff there is a substitution $\sigma: X \to T(F,X)$ with $t \equiv_E \sigma(s)$. For two substitutions $\sigma,\tau: X \to T(F,X)$ and a set of variables V we say $\sigma \leq_E \tau$ [V] iff there is a substitution λ with $\tau(x) \equiv_E \lambda(\sigma(x))$ for all $x \in V$.

A *E-unifier* of two terms s and t is a substitution $\sigma: X \to T(F,X)$ with $\sigma(s) \equiv_E \sigma(t)$.

Let W be a finite set of variables containing $Var(s \approx t)$.

A set $cU_E(s,t,W)$ of substitutions is called a *complete set of E–unifiers of s and t away from W iff*

- every $\sigma \in cU_E(s,t,W)$ is a E-unifier of s and t
- for any E-unifier τ of s and t there is $\sigma \in cU_E(s,t,W)$ such that $\sigma \leq_E \tau$ [V].
- for all $\sigma \in cU_E(s,t,W)$: $D(\sigma) \subseteq V$ and $I(\sigma) \cap W = \emptyset$

$cU_E(s,t,W)$ is called *minimal* iff it satisfies further the condition

- for all $\sigma, \sigma' \in cU_E(s,t,W)$: $\sigma \leq_E \sigma'$ [V] implies $\sigma = \sigma'$.

A *rewriting rule* is an expression of the form $l \to r$ with terms $l, r \in T(F, X)$ of the same sort such that

$Var(r) \subseteq Var(l)$. The rule is *regular* iff $Var(l) = Var(r)$.

A *term rewriting system* R is a set of rewriting rules.

The *rewriting relation* $\rightarrow_R$ associated with R is defined as follows:

For terms $s, t \in T(F, X)$ we have $s \rightarrow_{[\omega, l\rightarrow r]}\rightarrow_R t$ iff there is an occurrence $\omega \in Occ(s)$, a rule $l\rightarrow r$ in R, and a substitution $\sigma: X \rightarrow T(F, X)$ such that $\sigma(l) = s/\omega$ and $t = s[\omega \leftarrow \sigma(r)]$.

The *narrowing relation* $\rightsquigarrow_R$ associated with R is defined as follows:

For terms $s, t \in T(F, X)$ we have $s \rightsquigarrow_{[\omega, l\rightarrow r, \sigma]}\rightarrow_R t$ iff there is a non-variable occurrence $\omega \in Occ(s)$, a rule $l\rightarrow r$ in R, and a substitution $\sigma: X \rightarrow T(F, X)$ such that $\sigma(l) = \sigma(s/\omega)$ and $t = \sigma(s)[\omega \leftarrow \sigma(r)]$.

Here σ is called the *narrowing substitution* .

The term rewriting system R is *canonical* iff the rewriting relation $\rightarrow_R$ is confluent and noetherian.

Given a canonical term rewriting system R and terms $s_o, t_o \in T(F,X)$ the set of all substitutions σ such that

- there exists a narrowing derivation

  $$s_o \approx t_o \rightsquigarrow_{\delta_o}\rightarrow \ldots \rightsquigarrow_{\delta_{n-1}}\rightarrow s_n \approx t_n, n \geq 0,$$

  such that s_n and t_n are unifiable by a most general unifier μ
- $\sigma = \mu \circ \delta_{n-1} \circ \ldots \circ \delta_o$

is a complete set of R-unifiers of s_o and t_o.

3. Optimizing Narrowing Strategies

One of the most important optimizations of naive narrowing is *normal narrowing*: after every narrowing step the goal is normalized with respect to the given canonical term rewriting system. This allows us to take advantage of the special properties of rewriting steps compared to narrowing steps. Any rewriting step is also a special narrowing step. Naive narrowing does not distinguish rewriting and narrowing steps. Every rewriting step leads to a new path in the search space ("don't know indeterminism"). However, we know that in a canonical term rewriting system the rewriting steps may be executed in an arbitrary ordering ("don't care indeterminism").

A second fundamental approach for optimization is to restrict the set of occurrences at which a narrowing step is performed. Naive narrowing considers any non-variable occurrence in the goal. The idea of *basic narrowing* [Hullot 80] is to discard those occurrences which have been introduced by the narrowing substitution in a previous narrowing step.

Definition

The sets B_i, $i = 1,\ldots,n$, of *basic occurrences* in a narrowing derivation

$$G_1 \rightsquigarrow_{[\omega_1, l_1\rightarrow r_1, \sigma_1]}\rightarrow G_2 \rightsquigarrow_{[\omega_2, l_2\rightarrow r_2, \sigma_2]}\rightarrow \ldots\ldots \rightsquigarrow_{[\omega_{n-1}, l_{n-1}\rightarrow r_{n-1}, \sigma_{n-1}]}\rightarrow G_n$$

are inductively defined as follows

- $B_1 =_{def} \{ \omega \in Occ(G_1) \mid G_1/\omega \text{ is not a variable} \}$
- $B_{i+1} =_{def} (B_i \setminus \{ \upsilon \in B_i \mid \upsilon \geq \omega_i \}) \cup \{ \omega_i.\upsilon \mid \upsilon \in Occ(r_i), r_i/\upsilon \text{ is not a variable} \}, i > 1$

For a *basic narrowing derivation* we require that $\omega_i \in B_i$, for all $i = 1,\ldots,n-1$.

It is not difficult to show that basic narrowing is complete for arbitrary canonical term rewriting systems [Hullot 80]. However, it is possible to restrict the set of narrowing occurrences further without loosing completeness. After a narrowing step $G \rightsquigarrow_{[\omega, l\rightarrow r, \sigma]}\rightarrow G'$, we may also discard those narrowing occurrences which are left of ω [Herold 86].

Definition

An occurrence ω is *left* of an occurrence ω', $\omega \triangleleft \omega'$, iff there exist occurrences o, υ, υ' and natural numbers i, i' such that $i < i'$, $\omega = o.i.\upsilon$ and $\omega' = o.i'.\upsilon'$

Definition

The sets LRB_i, i = 1,...,n, of *left-to-right basic occurrences* in a narrowing derivation

$$G_1 \rightsquigarrow[\omega_1, l_1 \to r_1, \sigma_1] \to G_2 \rightsquigarrow[\omega_2, l_2 \to r_2, \sigma_2] \to \ldots\ldots \rightsquigarrow[\omega_{n-1}, l_{n-1} \to r_{n-1}, \sigma_{n-1}] \to G_n$$

are inductively defined as follows

- $LRB_1 =_{def} \{ \omega \in Occ(G_1) \mid G_1/\omega \text{ is not a variable} \}$
- $LRB_{i+1} =_{def} (LRB_i \setminus \{\upsilon \in LRB_i \mid \upsilon \geq \omega_i \text{ or } \upsilon \triangleleft \omega_i\})$
  $\cup \{\omega_i\upsilon \mid \upsilon \in Occ(r_i), r_i/\upsilon \text{ is not a variable} \}, i > 1$

For a *left-to-right basic narrowing derivation* we require that $\omega_i \in LRB_i$, for all i = 1,...,n-1.

We could also define a right-to-left basic narowing derivation. If we allow arbitrary selection strategies we obtain the *basic selection narrowing* of [Bosco/Giovannetti/Moiso 88], which includes the left-to-right and the right-to-left basic narrowing as special cases.

If we try to combine in a naive way the two fundamental approaches for optimization discussed so far (normalization and restriction of the narrowing occurrences) we loose completeness. For rewriting derivations the computation of the sets of basic occurrences is more complicated than for narrowing derivations. We need the notion of weakly basic rewriting derivation [Réty 87].

Definition

Let $t \dashv[\omega, l \to r] \to t'$ be a rewriting step.

We say that the occurrence υ in t is an *antecedent* of the occurrence υ' in t' iff

- $\upsilon = \upsilon'$ and υ is not comparable to ω or
- there exists an occurrence ρ' of a variable x in r such that $\upsilon' = \omega.\rho'.o$ and
  $\upsilon = \omega.\rho.o$ where ρ is an occurrence of the (same) variable x in l

Definition

Given a rewriting derivation

$$G_1 \dashv[\omega_1, l_1 \to r_1] \to G_2 \dashv[\omega_2, l_2 \to r_2] \to \ldots\ldots \dashv[\omega_{n-1}, l_{n-1} \to r_{n-1}] \to G_n$$

and a set $WB \subseteq Occ(G_1)$ of occurrences in G_1 the sets WB_i, i = 1,...,n, of *weakly basic occurrences* are inductively defined as follows

- $WB_1 =_{def} WB$
- $WB_{i+1} =_{def} (WB_i \setminus \{\upsilon \in WB_i \mid \upsilon \geq \omega_i \}) \cup \{\omega_i\upsilon \mid \upsilon \in Occ(r_i), r_i/\upsilon \text{ is not a variable} \}$
  $\cup \{ \upsilon \in Occ(G_{i+1}) \mid \upsilon = \omega_i.o, o \notin FuOcc(r_i)$ and all antecedents of υ in G_i are in WB_i }. The derivation in *weakly based on WB* iff $\omega_i \in WB_i$, for all i = 1,...,n-1.

The main difference compared to the computation of the set B_{i+1} of basic occurrences is that occurrences under ω_i may belong to WB_{i+1}.

Definition

A *left-to-right basic normal narrowing step*

$$(G, U) \rightsquigarrow\!\sigma\!\rightarrow_n (G', U')$$

with goals G, G' and sets of occurrences U, U'

is given by a left-to-right basic narrowing step $(G, U) \rightsquigarrow\!\sigma\!\rightarrow (G_1, U_1)$

followed by a normalization $(G_1, U_1) \rightarrow (G_2, U_2) \rightarrow \ldots \rightarrow (G_n, U_n) = (G', U')$

that is weakly based on U_1.

We now illustrate the various narrowing strategies by an example. For the first time in the literature we give quantitative results. The computations have been done in the Karlsruhe Narrowing Laboratory KANAL [Krischer 90].

Example

Let $R = \{0 + x \rightarrow x,\ s(x) + y \rightarrow s(x+y),$
$0 * x \rightarrow 0,\ s(x) * y \rightarrow y + x * y\ \}$

be a canonical term rewriting system for the addition and multiplication of natural numbers.

We would like to answer the query

$$?\text{- } x*x + y*y \approx s(0)$$

which has two solutions

$$\sigma_1 = [\ x/0,\ y/s(0)\ ] \text{ und } \sigma_2 = [\ x/s(0),\ y/0\ ].$$

First we consider the narrowing strategies without normalization.

The solution σ_1 is found in depth 6, the solution σ_2 in depth 7 of the narrowing tree.

The number of nodes in the narrowing tree is as follows

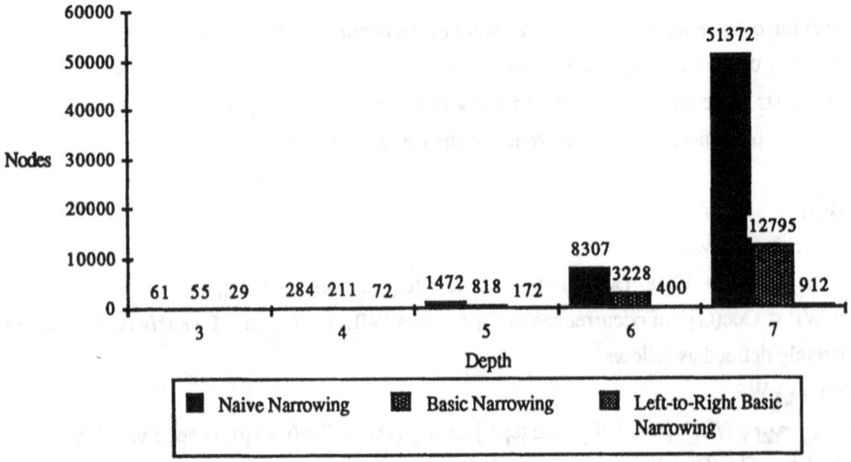

If we do narrowing with normalization both solutions are found in depth 3 and much less narrowing steps are needed. The naive narrowing tree contains 51372 nodes at depth 7 whereas in the normal tree at depth 3 there are only 72. Although normal narrowing steps are more costly than naive narrowing steps, this is an enormous gain of efficiency.

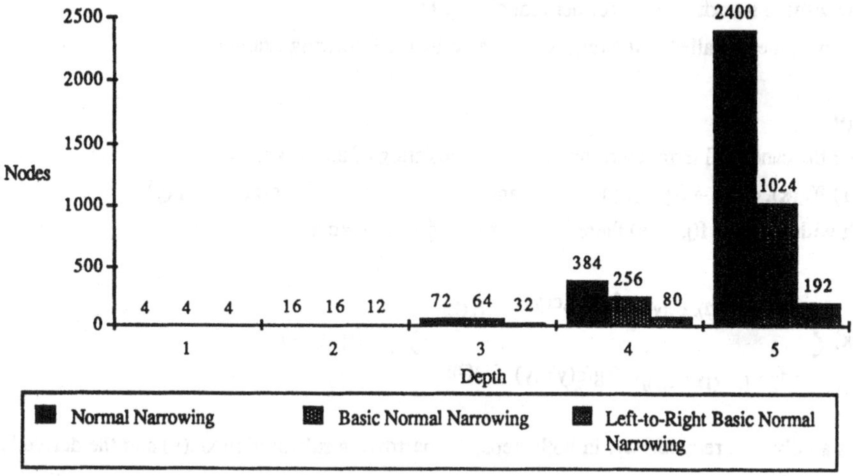

4. LSE-SL Basic Normal Narrowing

The two main approaches for optimizing the narrowing procedure considered so far have been

- normalization
- restricting the set of narrowing occurrences

Now we exploit a third fundamental idea

- reducibility tests after a narrowing step

A first step in this direction was Réty's test for sufficient largeness [Réty 87].

Definition

A set U of occurrences of a term t is said to be *sufficiently large* on t, iff t/ω is in normal form for all $\omega \in Occ(t) \setminus U$.

Theorem (Completeness of left-to-right SL-basic normal narrowing, Réty 88)
Let R be a canonical term rewriting system and s_o, t_o be two terms. Let $U_o =_{def} FuOcc(s_o \approx t_o)$.
The set of substitutions σ such that

- there exists a left-to-right basic normal narrowing derivation
  $(s_o \approx t_o, U_o) \leadsto \delta_o \rightarrow_n \ldots \leadsto \delta_{n-1} \rightarrow_n (s_n \approx t_n, U_n), n \geq 0,$
  with U_i sufficiently large on $s_i \approx t_i$, $i = 1,\ldots,n$
  such that s_n and t_n are unifiable by a most general unifier μ
- $\sigma = \mu \circ \delta_{n-1} \circ \ldots \circ \delta_o$ is normalized

is a complete set of R-unifiers of s_o and t_o.

If U_i is not sufficiently large on $s_i \approx t_i$ this means that a subterm of $s_i \approx t_i$ at a non-basic occurrence is reducible. This indicates that the previous narrowing step was superfluous. Either the narrowing substitution is reducible (and thus would yield a reducible solution) or it is more special than another narrowing substitution.

We now want to introduce three further reducibility tests.

The first one, the so-called *L-sub test*, is motivated by the following example.

Example

Consider the canonical term rewriting system R consisting of the two rules

(1) $f(g(x), c(y)) \rightarrow f(g(x), y)$ and (2) $g(c(y)) \rightarrow g(y)$.

Starting with the term f(g(x),x) there are two narrowing derivations

$$f(g(x),x) \rightsquigarrow_{[1,\,(2),\,x/c(y)]} f(g(y),c(y)) \rightsquigarrow_{[\varepsilon,\,(1)]} f(g(y),y)$$

$$f(g(x),x) \rightsquigarrow_{[\varepsilon,\,(1),\,x/c(y)]} f(g(c(y)),y) \rightsquigarrow_{[1,\,(2)]} f(g(y),y)$$

There is an obvious redundancy. In both steps, the narrowing substitution x/c(y) and the derived term f(g(y),y) are the same.

Our second test detects redundancies by testing the narrowing occurrences themselves for reducibility. We will call it the *L-epsilon test*. While all basic narrowing strategies restrict the indeterminism in the selection of the narrowing occurrence, the L-epsilon test allows to restrict the indeterminism in the selection of the rewriting rule. We assume that the term rewriting system is numbered, i.e. we associate with every rule a natural number *R*. The ordering of the rules is arbitrary. In fact, it is only necessary to order those rules which have a critical pair at top.

Definition

In a left-to-right basic narrowing derivation

$$(G_0,U_0) \rightsquigarrow_{R\,[\omega_0, R_0, \delta_0]} (G_1,U_1) \rightsquigarrow_{R\,[\omega_1, R_1, \delta_1]} \cdots \rightsquigarrow_{R\,[\omega_{n-2}, R_{n-2}, \delta_{n-2}]} (G_{n-1},U_{n-1}) \rightsquigarrow_{R\,[\omega_{n-1}, R_{n-1}, \delta_{n-1}]} (G_n,U_n)$$

with $\sigma = \delta_{n-1} \circ \delta_{n-2} \circ \ldots \circ \delta_1 \circ \delta_0$

the step $(G_{n-1},U_{n-1}) \rightsquigarrow_R (G_n,U_n)$ is *LSE-SL*, iff

- U_n is sufficiently large on G_n *(SL test)*
- $\forall\, i \in \{0,\ldots,n-1\}$: all proper subterms of $\sigma(G_i/\omega_i)$ are in normal form *(L-sub test)*
- $\forall\, i \in \{0,\ldots,n-1\}$: $\sigma(G_i/\omega_i)$ is not reducible at the top with a rule whose number is strictly less than R_i. *(L-epsilon test)*

Remarks

1. Note that $\sigma(G_i/\omega_i) = \sigma(l)$ where R_i is the rule number of $l \rightarrow r$.
   We apply the narrowing substitution σ of the whole derivation to the left-hand sides l of *all* rules involved in the previous narrowing steps and then test them for reducibility.
2. From the left-hand sides l, the "L" in LSE-SL stems. The "S" stands for sub (from L-sub test) and the "E" for epsilon (from L-epsilon test). "SL" means sufficient largeness.

The LSE-SL property is operationally a *test* that has to be done after every narrowing step. According to the definition above, there are terms that will be tested multiple times. This can be avoided as will be shown in section 5.

Example (continued)
The upper derivation is allowed but the lower not because of the L-sub test: the instantiation of the starting term $\sigma(\ f(g(x),x)\ ) = f(g(c(y)),c(y))$ is reducible at occurrence 1 with rule (2).
Without the L-sub test the narrowing tree is exponentially increasing (breadths 2, 4, 8, 16, ...). With the L-sub test, the breadth of the tree is constant (1, 1, ...).

The following lemma is crucial to prove the completeness of LSE-SL left-to-right basic narrowing.

Lemma
Let

$$(G_0,U_0) \underset{[\omega_0,R_0,\delta_0]}{\rightsquigarrow_R} (G_1,U_1) \underset{[\omega_1,R_1,\delta_1]}{\rightsquigarrow_R} \cdots \underset{[\omega_{n-2},R_{n-2},\delta_{n-2}]}{\rightsquigarrow_R} (G_{n-1},U_{n-1}) \underset{[\omega_{n-1},R_{n-1},\delta_{n-1}]}{\rightsquigarrow_R} (G_n,U_n)$$

be a left-to-right basic narrowing derivation.
Let $\sigma =_{def} \delta_{n-1} \circ \delta_{n-2} \circ \ldots \circ \delta_1 \circ \delta_0$.
If all narrowing steps except the last one are LSE-SL (and the last one is not LSE-SL) then

- $\sigma \mid Var(G_0)$ is reducible or
- there exists $i \in \{0,\ldots,n-1\}$ and a LSE-SL basic narrowing step
  $(G_i, U_i) \rightsquigarrow_{[\upsilon, R, \delta]} (G, U)$
  such that $\delta \circ \delta_{i-1} \circ \ldots \circ \delta_1 \circ \delta_0 \leq_{\emptyset} \sigma\ [Var(G_0)]$.
  Moreover we have $\omega_i < \upsilon$ or if $\omega_i = \upsilon$ then $R < R_i$.

Sketch of Proof: If the derivation in the lemma is not LSE-SL because of the L-epsilon test or the L-sub test, then by definition there exists $i \in \{0,\ldots,n-1\}$ such that $\sigma(G_i/\omega_i)$ is reducible with a rule R either at the top or at a deeper occurrence ω. These are the index and the rule mentioned in the lemma. The occurrence υ is ω_i in the case of the L-epsilon test and $\omega_i.\omega$ in the case of the L-sub test.
If the derivation is not LSE-SL because of the SL test, then one has to examine why the occurrence in G_n, which is reducible by a rule R, is not basic: either it is an instance of a variable in G_0 or it is an instance of a variable in a rule R_i, $i \in \{0,\ldots,n-1\}$. In the first case, $\sigma|Var(G_0)$ is reducible and in the second case the rule R can be applied in a narrowing step $(G_i, U_i) \rightsquigarrow_{[\upsilon, R, \delta]} (G, U)$ with $\omega_i < \upsilon$.

Theorem (Completeness of LSE-SL left-to-right basic narrowing)
Let R be a canonical term rewriting system and s_0, t_0 be two terms. Let $U_0 =_{def} FuOcc(s_0 \approx t_0)$.
The set of all substitutions σ such that

- there exists a LSE-SL left-to-right basic narrowing derivation
  $(s_0 \approx t_0, U_0) \rightsquigarrow_{\delta_0} \ldots \rightsquigarrow_{\delta_{n-1}} (s_n \approx t_n, U_n)$, $n \geq 0$,
  such that s_n and t_n are unifiable by a most general unifier μ
- $\sigma = \mu \cdot \delta_{n-1} \cdot \ldots \cdot \delta_0$

is a complete set of R-unifiers of s_0 and t_0.

If we want to do left-to-right basic *normal* narrowing we need our third test, the *L-reduction test.* The aim of this test is to examine those non-basic occurrences that may be lost during normalization (see example 3 in section 6).

Definition

In a left-to-right basic normal narrowing derivation ($\xrightarrow{*}_{\downarrow R}$ denotes the normalization)

$$(G_0',U_0') \xrightarrow{*}_{\downarrow R} (G_0,U_0) \underset{[\omega_0,R_0,\delta_0]}{\rightsquigarrow_R} (G_1',U_1') \xrightarrow{*}_{\downarrow R} (G_1,U_1) \underset{[\omega_1,R_1,\delta_1]}{\rightsquigarrow_R} \cdots$$

$$\cdots \underset{[\omega_{n-2},R_{n-2},\delta_{n-2}]}{\rightsquigarrow_R} (G_{n-1}',U_{n-1}') \xrightarrow{*}_{\downarrow R} (G_{n-1},U_{n-1}) \underset{[\omega_{n-1},R_{n-1},\delta_{n-1}]}{\rightsquigarrow_R} (G_n',U_n') \xrightarrow{*}_{\downarrow R} (G_n,U_n)$$

with $\sigma = \delta_{n-1} \circ \delta_{n-2} \circ \ldots \circ \delta_1 \circ \delta_0$

the step $(G_{n-1},U_{n-1}) \rightsquigarrow_R (G_n',U_n') \xrightarrow{*}_{\downarrow R} (G_n,U_n)$ is *LSE-SL*, iff

- U_n' is sufficiently large on G_n' *(SL test)*
- $\forall i \in \{0,\ldots,n-1\}$: all proper subterms of $\sigma(G_i/\omega_i)$ are in normal form *(L-sub test)*
- $\forall i \in \{0,\ldots,n-1\}$: $\sigma(G_i/\omega_i)$ is not reducible at the top with a rule whose number is strictly less than R_i *(L-epsilon test)*
- $\forall i \in \{0,\ldots,n-1\}$: U_i' is sufficiently large on $\sigma(G_i')$ *(L-reduction test)*

The L-reduction test seems to be very costly, but the last condition "U_i' is sufficiently large on $\sigma(G_i')$" is too strong. It's enough to check those non-basic subterms (= whose occurrence lies not in U_i') which are situated under an occurrence of a reduction that follows (see next section).

The completeness theorem and the preceding lemma remain true if we replace the narrowing relation $\rightsquigarrow$ with the normal narrowing relation $\rightsquigarrow_n$. In the proof of the lemma the argumentation for the L-reduction test is similar to the one for the SL test.

5.Realization

For not testing the reducibility of a subterm more than once, one has to take into consideration that a subterm of an equation situated under a narrowing (or reduction) occurrence will be transported into the derived equation if a rule variable (that occurs in the rhs of the rule) is instantiated to it. It then will be tested by the SL test and does not need to be tested by the L-sub test. A lhs rule variable that does not occur in the rhs is called a *disappearing variable.* So we have to apply the *L-sub test* only to the non-variable occurrences of the lhs and the instances of the disappearing variables of the rule.

For the *L-reduction test* we have to check the reducibility of those non-basic subterms of the equation that are matched by a function symbol of the lhs or by a disappearing variable of the rule. This way every non-basic occurrence will be tested exactly once.

Further points for optimization of the testing procedure are:

- the detection of variables or subterms that occur several times in the first equation or in the rhs of a rule. For the computation of the basic occurrence sets in a narrowing or reduction step and for the LSE-SL tests, these occurrences have to be considered only once.

- only those subterms have to be tested for reducibility that have been extended by the last increment δ_{n-1} of σ, i.e. that contain a variable which is instantiated by δ_{n-1}
- ground subterms in normal form never need to be tested for reducibility
- the success of the L-epsilon and the L-sub test can be detected in advance in the case of an overlapping application of two or more rules in a narrowing derivation. More precisely: Let R: $l \rightarrow r$ and R': $l' \rightarrow r'$ be two rules of the canonical term rewriting system. If σ is the mgu between r/υ and l', and if $\sigma(l)$ is reducible at the top with a rule whose number is less than R (L-epsilon test) or a subterm of $\sigma(l)$ is reducible (L-sub test), then store at r/υ (in the derived equation) that the rule R' need not to be applied for narrowing (see Example 1 in the next section, anticipated evaluation).

LSE-SL left-to-right basic normal narrowing has been implemented in the Karlsruhe Narrowing Laboratory KANAL [Krischer 90]. Some small instructive and one large practical example which has been solved using KANAL will be presented in the next section.

6. Examples

Example 1

Let the following two rules belong to a canonical term rewriting system R

(1) s(x)+y→ s(x+y) and (2) x+p(y) → p(x+y).

Starting with the term x+y there are two narrowing derivations:

$$x+y \begin{cases} \rightsquigarrow_{[\varepsilon,\,(1),\,x/s(x')]} s(x'+y) \rightsquigarrow_{[1,\,(2),\,y/p(y')]} s(p(x'+y')) \\ \rightsquigarrow_{[\varepsilon,\,(2),\,y/p(y')]} p(x+y') \rightsquigarrow_{[1,\,(1),\,x/s(x')]} p(s(x'+y')) \end{cases}$$

The upper derivation is LSE-SL but the lower not because σ(x+y) = s(x')+p(y') is reducible with rule (1) and 1 < 2 (L-epsilon test).

Anticipated evaluation of the test: We superpose the rhs of rule (2) p(x+y) at occurrence 1 with the lhs of rule (1) s(x')+y' and see that the instantiated lhs of rule (2) s(x')+p(y) is reducible by rule (1) where 1 < 2. So we have to store in the rhs of rule (2) p(x+y) that rule (1) need not to be applied for narrowing at occurrence 1 after rule (2) has been applied.

Example 2

Let the following two rules belong to a canonical term rewriting system R

(1) s(x)+y→ s(x+y) and (2) x+s(y) → s(x+y).

There are two narrowing derivations starting with the term z+z

$$z+z \begin{cases} \rightsquigarrow_{[\varepsilon,\,(1),\,z/s(x)]} s(x+s(x)) \\ \rightsquigarrow_{[\varepsilon,\,(2),\,z/s(y)]} s(s(y)+y) \end{cases}$$

The upper derivation is LSE-SL but the lower not, because σ(z+z) = s(x)+s(y) is reducible with rule (1) and 1 < 2 (L-epsilon test).

Example 3

We take the canonical term rewriting system

(1) $h(g(x)) \to h(x)$

(2) $f(g(g(x)), y) \to f(x, y)$

(3) $g(g(g(x))) \to g(x)$

There are two narrowing derivations starting with the term f(g(x),h(x))

$$f(g(x),h(x)) \begin{cases} \rightarrowtail_{[2,\ (1),\ x/g(y)]} f(\underline{g(g(y))},h(y)) \to f(y,h(y)) \rightarrowtail_{[2,\ (1),\ y/g(y')]} f(g(y'),h(y')) \\ \rightarrowtail_{[1,\ (3),\ x/g(g(y'))]} f(g(y'),h(g(g(y')))) \to^{*} f(g(y'),h(y')) \end{cases}$$

The L-reduction test detects that the upper derivation is superfluous. The underlined non-basic subterm g(g(y)) disappears during the reduction. It becomes reducible by the following narrowing substitution y/g(y').

Finally we give a large practical example to show how the efficiency of narrowing is improved by our method.

Example 4

Consider the canonical term rewriting system for the integers from [Réty et al. 85] which contains 24 rules and take the goal ? - x*x + y*y ≈ s(0).

The number of nodes in the narrowing tree is as follows:

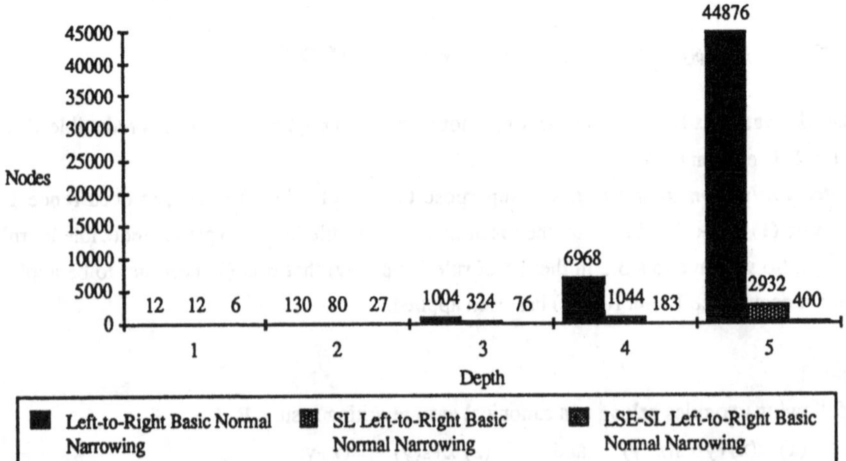

To compute the narrowing tree of depth 5, we needed 1424 sec. for ordinary left-to-right basic normal narrowing, 278 sec. for SL left-to-right basic normal narrowing, and 73 sec. for LSE-SL left-to-right basic normal narrowing. The computations were done in the Karlsruhe narrowing laboratory KANAL, which is implemented in the Prolog dialect KA-Prolog, on a SUN 4/60.

References

[Bockmayr 86a] A. Bockmayr: Conditional Rewriting and Narrowing as a Theoretical Framework for Logic-Functional Programming: A Survey. Interner Bericht 10/86, Fakultät für Informatik, Universität Karlsruhe, 1986

[Bockmayr 86b] A. Bockmayr: Narrowing with Inductively Defined Functions. Interner Bericht 25/86, Fakultät für Informatik, Universität Karlsruhe, 1986

[Bockmayr 88] A. Bockmayr: Narrowing with Built-in Theories. First Int.Workshop Logic and Algebraic Programming, Gaußig, DDR, 1988, Springer LNCS 343

[Bosco/Giovannetti/Moiso 88] P. G. Bosco, E. Giovannetti, C. Moiso: Narrowing vs. SLD-Resolution. Theoretical Computer Science 59 (1988), 3 - 23

[Darlington/Guo 89] J. Darlington, Y. Guo: Narrowing and Unification in Functional Programming - an Evaluation Mechanism for Absolute Set Abstraction. Rewriting Techniques and Applications, Chapel Hill, 1989, Springer LNCS 355

[Dershowitz/Plaisted 85] N. Dershowitz, D. A. Plaisted: Logic Programming cum Applicative Programming. Symp. on Logic Programming, Boston 1985, IEEE

[Dershowitz/Plaisted 88] N. Dershowitz, D. A. Plaisted: Equational Programming. Mach. Intell. 11 (Ed. J. Richards), Oxford, 1988

[Echahed 88] R. Echahed: On Completeness of Narrowing Strategies. CAAP 88, Nancy, Springer LNCS 299

[Fribourg 85] L. Fribourg: Handling Function Definitions Through Innermost Superposition and Rewriting. Rewriting Techniques and Applications, Dijon, 1985, Springer LNCS 202

[Herold 86] A. Herold: Narrowing Techniques Applied to Idempotent Unification. SEKI Report SR-86-16, Universität Kaiserslautern, 1986

[Hölldobler 89] S. Hölldobler: Foundations of Equational Logic Programming. Springer, LNCS 353, 1989

[Huet/Oppen 80] G. Huet, D. C. Oppen: Equations and Rewrite Rules, A Survey. Formal Language Theory (Ed. R. V. Book), Academic Press 1980

[Hullot 80] J. M. Hullot: Canonical Forms and Unification. 5th Conference on Automated Deduction, Les Arcs 1980, Springer LNCS 87

[Krischer 90] S. Krischer: Vergleich und Bewertung von Narrowing-Strategien. Diplomarbeit, Fakultät für Informatik, Universität Karlsruhe, 1990

[Nutt/Réty/Smolka 89] W. Nutt, P. Réty, G. Smolka: Basic Narrowing Revisited. J. Symb. Comput. 7 (1989), 295-317

[Padawitz 88] P. Padawitz: Computing in Horn Clause Theories. EATCS Monograph Vol. 16, Springer, 1988

[Réty et al. 85] P. Réty, C. Kirchner, H. Kirchner, P. Lescanne: NARROWER: a New Algorithm for Unification and its Application to Logic Programming. Rewriting Techniques and Applications, Dijon, 1985, Springer LNCS 202

[Réty 87] P. Réty: Improving Basic Narowing Techniques. Rewriting Techniques and Applications, Bordeaux, 1987, Springer LNCS 256

[Réty 88] P. Réty: Méthodes d'Unification par Surréduction. Thèse. Univ. Nancy, 1988

[You 88] J. You: Solving Equations in an Equational Language. First Int.Workshop Logic and Algebraic Programming, Gaußig, DDR, 1988, Springer LNCS 343

[You 90] J. You: Unification Modulo an Equality Theory for Equational Logic Programming. to appear in: J. Comp. Syst. Sc. 40 (1990)

Unification, Weak Unification, Upper Bound, Lower Bound, and Generalization Problems

Franz Baader
German Research Center for Artificial Intelligence
DFKI, Projektgruppe WINO, Postfach 2080
W-6750 Kaiserslautern, Germany
email: baader@dfki.uni-kl.de

Abstract

We introduce E-unification, weak E-unification, E-upper bound, E-lower bound, and E-generalization problems, and the corresponding notions of unification, weak unification, upper bound, lower bound, and generalization type of an equational theory. When defining instantiation preorders on solutions of these problems, one can compared substitutions w.r.t. their behaviour on all variables or on finite sets of variables. We shall study the effect which these different instantiation preorders have on the existence of most general or most specific solutions of E-unification, weak E-unification, and E-generalization problems. In addition, we shall elucidate the subtle difference between most general unifiers and coequalizers. and we shall consider generalization in the class of commutative theories.

1 Introduction

The research which will be presented in this paper was motivated by the following observations:
(1) Generalization of terms is often called the dual problem to unification. However, this is not true if the variables of the terms in question are not disjoint. In that case, generalization is dual to weak unification. The name "weak unification" was introduced by Eder (1985); however, this kind of unification has already been considered before (see e.g. Huet (1980)).
Weak unification means that, for given terms s, t, we want to find *two* substitutions σ, τ such that $s\sigma = t\tau$, whereas unification requires *one* substitution δ such that $s\delta = t\delta$. The weakly unified terms $s\sigma = t\tau$ are just the upper bounds of s, t in the instantiation lattice of first order terms (see Huet (1980)).
On the other hand, generalization of terms means that, for given terms s, t, we want to find a term g and *two* substitutions σ, τ with $s = g\sigma$ and $t = g\tau$ (see e.g. Plotkin (1970), Huet (1980)). The generalized terms g are the lower bounds of s, t in the instantiation lattice of first order terms. In this sense weak unification and generalization are duals of each other.
The problem of weak unification naturally arises in some applications, e.g., when computing critical pairs of term rewriting rules. Of course, by renaming of variables, weak unification can be reduced to unification. But these are nevertheless two different notions, and we shall see that they have different behaviour w.r.t. existence of most general solutions – at least if one takes the wrong instantiation ordering (see point (3) below).
(2) In many applications one is not only interested in solvability of weak unification problems (generalization problems), but also in most general (most specific) solutions. In this case it makes a difference whether one only considers the weakly unified term (generalized term) as solution, or whether the substitutions which yield this term are also taken as part of the solution. Hence we have to

distinguish between lower bound problems and generalization problems, and in the dual case, between weak unification problems and upper bound problems.
(3) In Robinson's seminal paper (Robinson (1965)), and in many subsequent papers on unification, the instantiation preorder $\leq$ on substitutions is defined by $\sigma \leq \tau$ iff there exists a substitution λ such that $\sigma\lambda = \tau$.[1] If unification is generalized to equational unification, most authors (e.g., Plotkin (1972), Siekmann (1989)) use the restricted instantiation ordering, i.e., they just require $x\sigma\lambda = x\tau$ for all variables x occurring in some term of the unification problem. In order to account for this fact, we shall give an example of an equational theory – namely, the theory of commutative idempotent monoids – where unification problems always have a most general unifier w.r.t. restricted instantiation, but not w.r.t. unrestricted instantiation. Eder's unpleasant results for weak unification (see Eder (1985)) are also due to the fact that he uses unrestricted instantiation.
(4) Generalization of terms can also be done modulo an equational theory. But then a pair of terms may have more than one most specific generalizer (see e.g., Pottier (1989)). In fact, all the phenomena known from equational unification are conceivable in this case. One can thus also define generalization types of equational theories, analogously to the definition of unification types. In the present paper, we shall give – within the framework of preordered sets – a uniform definition of unification, weak unification, upper bound, lower bound, and generalization types of equational theories.
(5) Unification and equational unification can be formulated in a categorical framework where the term algebras are the objects and the substitutions are the morphisms of the category. However, most general unifiers are in general not coequalizers (as sometimes claimed in this context), but only weak coequalizers (see McLane (1971) for the definition of weak limits and colimits). The reason is that they do not satisfy a certain uniqueness condition which is required in the definition of coequalizers (see e.g., Goguen (1988)). Please note that this uniqueness condition is completely meaningless for the applications of unification in term rewriting or resolution theorem proving.
For the empty theory, most general unifiers w.r.t. restricted instantiation can always be chosen as coequalizers. For this reason, the definition of most general unifier is sometimes modified in order to match the definition of coequalizer (see e.g., Rydeheard-Burstall (1985)).[2] However, this modified definition should not be used together with unrestricted instantiation. In fact, one can show that only trivial unification problems have a coequalizer w.r.t. unrestricted instantiation.
For equational unification, the modified definition is not appropriate, even if one takes the restricted instantiation preorder. We shall show that there exists an equational theory E and an E-unification problem which has a most general E-unifier (w.r.t. restricted instantiation), but which does not have a coequalizer (w.r.t. restricted instantiation).
(6) For the class of commutative theories (see Baader (1989)), unification can be reduced to solving linear equations in the corresponding semiring. This correspondence also yields algebraic characterizations of the unification types of commutative theories (see Nutt (1990), Baader (1990a)). We shall show that this fact can also be used to characterize the existence of coequalizers in an algebraic way. In addition, we shall show that, for the class of commutative theories, there always exist most specific generalizers. That means that all commutative theories are of generalization type "unitary".

2 Definitions and Notations

Let Ω be a signature, i.e., a set of function symbols with fixed arity, and let V be a countable set of variables. For any subset X of V we denote the set of all Ω-terms with variables in X by F(X). This set is the carrier of the free Ω-algebra with generators X, which will also be denoted by F(X). Any mapping of X into an Ω-algebra $\mathcal{A}$ can be uniquely extended to a homomorphism of F(X) into $\mathcal{A}$. We write homomorphisms in suffix notation, i.e., $s\theta$ instead of $\theta(s)$. Consequently, composition is written from

[1]This preorder will be called unrestricted instantiation preorder.
[2]In the present paper, we shall not introduce the catagorical reformulation of unification. Coequalizers will be defined as most general unifiers satisfying an additional uniqueness condition.

left to right, i.e., $\sigma\theta$ means first σ and then θ. An endomorphism θ of F(V) is called substitution iff it has finite domain, where the domain of θ is defined as $D(\theta) := \{x;\ x\theta \neq x\}$. A substitution υ is called *variable renaming* if its restriction to V is a permutation of V. Since υ is a permutation, there also exists an *inverse variable renaming* υ^{-1} such that $\upsilon\upsilon^{-1}$ and $\upsilon^{-1}\upsilon$ are the identity.

Let s be a term, θ be a substitution, and X be a subset of V. The set of all variables occurring in s is denoted by V(s). The set $\{y;$ There is $x \in X$ with $y \in V(x\theta)\}$ is denoted by $V(X\theta)$. For an n-tuple $\underline{s} = (s_1,\ldots,s_n)$, $V(\underline{s})$ denotes the union of the sets $V(s_i)$ for $i = 1, \ldots, n$.

For a set of identities (equational theory) E, let $=_E$ denote the equality of terms, induced by E. The equational theory E defines a variety V(E), i.e., the class of all algebras (over the given signature Ω) which satisfy each identity of E. For any subset X of V, the quotient algebra $F(X)/{=_E}$ is the E-free Ω-algebra with generators X, which is an element of V(E), and which will be denoted by $F_E(X)$. The relation $=_E$ can be extended to substitutions in the obvious way, namely $\sigma =_E \tau$ iff $x\sigma =_E x\tau$ for all variables $x \in V$.

The solutions of unification, weak unification, upper bound, lower bound, and generalization problems (see Definition 2.3 below) will be tuples of terms or substitutions. Since usually one is not interested in all solutions, but only in most general (most specific) ones, one needs an appropriate notion of instantiation order with respect to which "most general" or "most specific" is meant. In order to have a succinct notation, we shall now define these orderings on n-tuples of terms and substitutions.

Definition 2.1 For an n-tuple of terms $\underline{s} = (s_1,\ldots,s_n)$, an n-tuple of substitutions $\underline{\sigma} = (\sigma_1,\ldots,\sigma_n)$, and a substitution λ, let $\underline{s}\lambda := (s_1\lambda,\ldots,s_n\lambda)$, $\underline{\sigma}\lambda := (\sigma_1\lambda,\ldots,\sigma_n\lambda)$, and $\lambda\underline{\sigma} := (\lambda\sigma_1,\ldots,\lambda\sigma_n)$.
(1) Let $\underline{s} = (s_1,\ldots,s_n)$ and $\underline{t} = (t_1,\ldots,t_n)$ be n-tuples of terms. Then we define $\underline{s} \leq_E \underline{t}$ iff there exists a substitution λ such that $\underline{s}\lambda =_E \underline{t}$.
(2) Let $\underline{\sigma} = (\sigma_1,\ldots,\sigma_n)$ and $\underline{\tau} = (\tau_1,\ldots,\tau_n)$ be n-tuples of substitutions, and let $\underline{X} = (X_1,\ldots,X_n)$ be an n-tuple of finite sets of variables. We define the *restricted instantiation preorder* as $\underline{\sigma} \leq_E \underline{\tau}$ $\langle\underline{X}\rangle$ iff there exists a substitution λ such that, for all i and all x in X_i, we have $x\sigma_i\lambda =_E x\tau_i$.
(3) Let $\underline{\sigma} = (\sigma_1,\ldots,\sigma_n)$ and $\underline{\tau} = (\tau_1,\ldots,\tau_n)$ be n-tuples of substitutions. We define the *unrestricted instantiation preorder* as $\underline{\sigma} \leq_E \underline{\tau}$ iff there exists a substitution λ such that $\underline{\sigma}\lambda =_E \underline{\tau}$.

If E is the empty theory, we shall sometimes omit the index. For $n = 1$ we have the usual instantiation preorders (see e.g. Huet (1980), Eder (1985), Siekmann (1989)); but for $n > 1$, the n-tuples are *not* ordered componentwise w.r.t. the usual instantiation preorder because we require the same substitution λ for all components. Recall that $\underline{\sigma}$ is less or equal $\underline{\tau}$ w.r.t. the unrestricted *componentwise* E-instantiation preorder iff there exist substitutions $\lambda_1, \ldots, \lambda_n$ such that $\sigma_1\lambda_1 = \tau_1, \ldots, \sigma_n\lambda_n = \tau_n$.

In order to define unification, weak unification, etc. types of equational theories in a uniform, order theoretic framework, we need some definitions and facts concerning preorders. Let $\leq$ be a preorder, i.e., a reflexive, transitive relation, on a set Q. This preorder defines an equivalence relation $\equiv$ in the usual way: $a \equiv b$ iff $a \leq b$ and $b \leq a$. Now $\leq$ induces a partial order on the equivalence classes $[a] = \{b;\ a \equiv b\}$ of $\equiv$ by $[a] \leq [b]$ iff $a \leq b$.

A non-empty subset A of Q is a *lower set (upper set)* iff $a \in A$ and $b \leq a$ implies $b \in A$ ($a \in A$ and $a \leq b$ implies $b \in A$). The lower set (upper set) A is generated by $B \subseteq A$ iff $A = \{a \in Q;$ There is $b \in B$ such that $a \leq b\}$ ($A = \{a \in Q;$ There is $b \in B$ such that $b \leq a\}$). Let A be a lower set (upper set) which is generated by B. Then B is called a *base* of A iff two different elements of B are not comparable w.r.t. $\leq$.

LEMMA 2.2 Let $\leq$ be a preorder on the set Q and let [Q] be the set of all $\equiv$-classes. Let A be a lower set (upper set) in Q, and let M be the set of all maximal (minimal) elements of $[A] = \{[a];\ a \in A\}$.
(1) A has a base (w.r.t. $\leq$ on Q) iff M generates [A] (w.r.t. $\leq$ on [Q]).
(2) If B is a base of A, then $M = \{[b];\ b \in B\}$.
(3) If M generates [A], then any set of representatives for M is a base of A.
Proof. See Baader (1989b), Lemma 2.2 and Proposition 2.3. ❑

Evidently, the lower set (upper set) A may have four possible types with respect to existence and cardinality of a base:

(1) M generates [A] and is a singleton. (type unitary)
(2) M generates [A] and is finite, but not a singleton. (type finitary)
(3) M generates [A] and is infinite. (type infinitary)
(4) A does not have a base, i.e., M does not generate [A]. (type zero)

These types are ordered as follows: unitary < finitary < infinitary < zero.

We are now ready to define the problems mentioned in the title of the paper. Choosing the appropriate instantiation preorders on their solutions is straightforward for unification, weak unification, upper bound, and lower bound problems; but it is a bit tricky for generalization problems. The problems are all of the form $\Gamma = \langle \underline{s}, \underline{t} \rangle_E$ for n-tuples of terms $\underline{s} = (s_1,\ldots,s_n)$ and $\underline{t} = (t_1,\ldots,t_n)$.

Definition 2.3 (problems, solutions, and preorders on the solutions)

(1) The solutions of an *E-unification problem* $\Gamma = \langle \underline{s}, \underline{t} \rangle_E$ are E-unifiers of Γ, i.e., substitutions σ such that $\underline{s}\sigma =_E \underline{t}\sigma$. E-unifiers of Γ can be preordered by the unrestricted instantiation preorder (see Definition 2.1, (3) for n = 1), or by the restricted instantiation preorder $\leq_E \langle X \rangle$ where $X := V(\underline{s}) \cup V(\underline{t})$ is the set of variables occurring in Γ. The *set of all E-unifiers of* Γ will be denoted by $U_E(\Gamma)$.

(2) The solutions of a *weak E-unification problem* $\Gamma = \langle \underline{s}, \underline{t} \rangle_E$ are weak E-unifiers of Γ, i.e., pairs $\underline{\sigma} = (\sigma_1,\sigma_2)$ of substitutions such that $\underline{s}\sigma_1 =_E \underline{t}\sigma_2$. Weak E-unifiers of Γ can be preordered by the unrestricted instantiation preorder (see Definition 2.1, (3) for n = 2), or by the restricted instantiation preorder $\leq_E \langle X_1,X_2 \rangle$ where $X_1 := V(\underline{s})$ is the set of variables occurring in $\underline{s}$, and $X_2 := V(\underline{t})$ is the set of variables occurring $\underline{t}$. The *set of all weak E-unifiers of* Γ will be denoted by $W_E(\Gamma)$.

(3) The solutions of an *E-upper bound problem* $\Gamma = \langle \underline{s}, \underline{t} \rangle_E$ are E-upper bounds of Γ, i.e., n-tuples $\underline{u}$ of terms such that $\underline{s} \leq_E \underline{u}$ and $\underline{t} \leq_E \underline{u}$. E-upper bounds of Γ are preordered by the E-instantiation preorder on tuples of terms (see Definition 2.1, (1)). The *set of all E-upper bounds of* Γ will be denoted by $UB_E(\Gamma)$.

(4) The solutions of an *E-lower bound problem* $\Gamma = \langle \underline{s}, \underline{t} \rangle_E$ are E-lower bounds of Γ, i.e., n-tuples $\underline{g}$ of terms such that $\underline{g} \leq_E \underline{s}$ and $\underline{g} \leq_E \underline{t}$. E-lower bounds of Γ are also preordered by the E-instantiation preorder on tuples of terms. The *set of all E-lower bounds of* Γ will be denoted by $LB_E(\Gamma)$.

(5) The solutions of an *E-generalization problem* $\Gamma = \langle \underline{s}, \underline{t} \rangle_E$ are E-generalizers of Γ, i.e., pairs $\underline{\sigma} = (\sigma_1,\sigma_2)$ of substitutions together with an n-tuple g of terms such that $g\sigma_1 =_E \underline{s}$ and $g\sigma_2 =_E \underline{t}$. E-generalizers of Γ can be preordered by the unrestricted instantiation preorder, or by the restricted instantiation preorder (see Definition 2.4 below). $G_E(\Gamma)$ denotes the *set of all E-generalizers of* Γ.

Definition 2.4 (instantiation preorders for solutions of generalization problems)

The elements of $G_E(\Gamma)$ are of the form $(\underline{\sigma},g)$ where $\underline{\sigma}$ is a pair of substitutions and g is an n-tuple of terms. Let $Q := \{(\underline{\sigma},g);\ \underline{\sigma}$ is a pair of substitutions and g is an n-tuple of terms$\}$.

(1) The unrestricted instantiation preorder on Q is defined by $(\underline{\sigma},\underline{g}) \leq_E (\underline{\sigma}',\underline{g}')$ iff there exists a substitution λ such that $\underline{\sigma} =_E \lambda\underline{\sigma}'$ and $g\lambda =_E g'$.

(2) The restricted preorder on Q is defined by $(\underline{\sigma},\underline{g}) \leq_E (\underline{\sigma}',\underline{g}')$ iff there exists a substitution λ such that $x\underline{\sigma} =_E x\lambda\underline{\sigma}'$ for all variables x occurring in g and $g\lambda =_E g'$.

The weakly unified term $\underline{s}\sigma_1 =_E \underline{t}\sigma_2$ need not be included into the solution of a weak unification problem $\langle \underline{s}, \underline{t} \rangle_E$ because it can be obtained by applying the pair of weak unifiers to the problem. Since this is not the case for generalization, we have to incorporate the generalized term into the solution of the generalization problem.

It is easy to see that the sets $U_E(\Gamma)$, $W_E(\Gamma)$ and $UB_E(\Gamma)$ are either empty (if Γ is not solvable) or upper sets w.r.t. the corresponding preorders; whereas the sets $LB_E(\Gamma)$ and $G_E(\Gamma)$ are always lower sets w.r.t. the corresponding preorders. Please note that $W_E(\Gamma)$ would not be an upper set w.r.t. the componentwise instantiation preorder, which is used in Eder (1985) to compare weak unifiers.

Definition 2.5 (Types of problems and equational theories)

(1) Let Γ be a solvable E-unification (weak E-unification, E-upper bound, E-lower bound, E-generalization) problem, and let A be the set of solutions of Γ. Then A is an upper set or lower set w.r.t.

the appropriate restricted (unrestricted) E-instantiation preorder. The restricted (unrestricted) *type of* Γ is defined to be the type of A.

(2) Let E be an equational theory. Then the restricted (unrestricted) *unification (weak unification, upper bound, lower bound, generalization) type of E* is defined as

max{T; T is the restricted (unrestricted) type of a solvable E-unification (weak E-unification, E-upper bound, E-lower bound, E-generalization) problem}.

In the present paper, we shall mostly be interested in the question whether a given problem or theory is unitary or not. Let Γ be a unitary E-unification (weak E-unification, E-upper bound, E-lower bound, E-generalization) problem. Then all solutions of Γ can be generated from a single solution. This solution is unique up to equivalence, and is called *most general E-unifier (most general weak E-unifier, E-supremum, E-infimum, most specific E-generalizer)* of Γ.

When working with a categorical reformulation of E-unification (see Rydeheard-Burstall (1985), Baader (1989,1989a,1990)), it turns out that the concept "most general unifier" is very similar to the categorical notion "coequalizer". However, there is a subtle difference between these two concepts. We shall now give a definition of coequalizers within our framework, and without explicit reference to categories.

To that purpose, let us consider the definition of most general unifier more closely. Obviously, the substitution μ is a most general E-unifier of Γ w.r.t. unrestricted instantiation iff μ is an E-unifier of Γ, and for any other E-unifier δ of Γ there exists a substitution λ such that $\delta =_E \mu\lambda$. If μ satisfies the additional property that the substitution λ is unique (modulo $=_E$), then μ is called *coequalizer of* Γ *w.r.t. unrestricted instantiation.* Requiring λ to be unique (as substitution on all variables) is obviously not appropriate for restricted instantiation because λ may always be arbitrary on $V \setminus V(V_0\mu)$ where V_0 is the set of all variables occurring in Γ. In this case, the restriction of λ to $V(V_0\mu)$ has to be unique modulo $=_E$ in order to let μ be a *coequalizer w.r.t. restricted instantiation.*

In the remainder of this section we shall exhibit some easy facts concerning the connection between the notions introduced above.

It is an immediate consequence of the definitions that an E-unifier which is most general w.r.t. unrestricted instantiation is also most general w.r.t. restricted instantiation. The same is true for weak unification and generalization. But the other direction need not hold (see Section 3.3, 3.4, and 4). Therefore we have

Proposition 2.6 If E is of unification (weak unification, generalization) type unitary w.r.t. unrestricted instantiation, then E is also of type unitary w.r.t. restricted instantiation.

Next we shall show that, with respect to the restricted instantiation preorder, the existence of most general unifiers implies the existence of most general weak unifiers. However, with respect to the unrestricted instantiation preorder, this need not be the case (see Section 3.3 below).

Proposition 2.7 If the equational theory E is of unification type unitary w.r.t. restricted instantiation, then it is also of weak unification type unitary w.r.t. restricted instantiation.

Proof. Let $\Gamma = <\underline{s}, \underline{t}>_E$ be a solvable weak E-unification problem, and let υ_1, υ_2 be variable renamings such that the variables occurring in $\underline{s}\upsilon_1$ are distinct from those occurring in $\underline{t}\upsilon_2$. We consider the unification problem $\Delta := <\underline{s}\upsilon_1, \underline{t}\upsilon_2>_E$. Since Γ is solvable, Δ is also solvable. In fact, a pair of weak unifiers (σ,τ) of Γ can be used to define a unifier δ of Δ as follows: $x\delta := (x\upsilon_1^{-1})\sigma$ for all x in $V(\underline{s}\upsilon_1)$, $x\delta := (x\upsilon_2^{-1})\tau$ for all x in $V(\underline{t}\upsilon_2)$, and $x\delta := x$ otherwise.

Let μ be a most general E-unifier of Δ w.r.t. restricted instantiation (which exists since Δ is solvable, and E is unitary w.r.t. restricted instantiation). Obviously, $(\upsilon_1\mu, \upsilon_2\mu)$ is a pair of weak E-unifiers of Γ. It remains to be shown that this pair is most general.

Let (σ,τ) be a pair of weak E-unifiers of Γ. We have to show that $(\upsilon_1\mu, \upsilon_2\mu) \leq_E (\sigma,\tau) <V(\underline{s}), V(\underline{t})>$.

Let δ be the E-unifier of Δ defined by σ, τ as described above. Please note that the substitution δ is

defined in a way such that, for all y in $V(\underline{s})$ we have $y\sigma = y\upsilon_1\delta$, and for all y in $V(\underline{t})$ we have $y\tau = y\upsilon_2\delta$.
Since μ is a most general E-unifier of Δ, we have $\mu \leq_E \delta <V(\underline{s}\upsilon_1) \cup V(\underline{t}\upsilon_2)>$, i.e., there exists a substitution λ such that, for all x in $V(\underline{s}\upsilon_1) \cup V(\underline{t}\upsilon_2)$, we have $x\mu\lambda =_E x\delta$.
Let y be a variable in $V(\underline{s})$. Then $y\upsilon_1$ is a variable in $V(\underline{s}\upsilon_1)$, and thus we have $y\sigma = y\upsilon_1\delta =_E y\upsilon_1\mu\lambda$. Similarly, we can show for any y in $V(\underline{t})$ that $y\tau = y\upsilon_2\delta =_E y\upsilon_2\mu\lambda$. Thus we have $(\upsilon_1\mu, \upsilon_2\mu) \leq_E (\sigma,\tau) <V(\underline{s}), V(\underline{t})>$. ❑

The construction we have used in this proof is an instance of the general categorical construction of pushouts from coequalizers and coproducts (see Baader (1990), Proposition 3.8). It cannot be used for unrestricted instantiation because the category which has to be used there need not have all binary coproducts.

The connection between E-suprema (E-infima) and most general weak E-unifiers (most specific E-generalizers) is as follows. It is easy to see that, for a most general weak E-unifier (σ,τ) (most specific E-generalizer $((\sigma_1,\sigma_2),\underline{g})$) of the problem $\Gamma = <\underline{s}, \underline{t}>_E$, the tuple $\underline{s}\sigma =_E \underline{t}\tau$ (the tuple $\underline{g}$) is an E-supremum (E-infimum) of Γ. This yields

Proposition 2.8 If E is of weak unification type (weak generalization type) unitary w.r.t. restricted or unrestricted instantiation, then E is also of upper bound (lower bound) type unitary.

However, we shall see in Section 3 that the other direction need not be true. The connection between E-suprema (E-infima) and most general weak E-unifiers (most specific E-generalizers) can also be neatly formulated and explained in the categorical framework (see Baader (1990), Proposition 3.5 and the remark following that proposition).

3 Results for the Empty Theory

The problems of Definition 2.3 are now considered for the empty theory $E = \emptyset$. We shall usually omit the prefix or index "Ø", and write e.g., "unifier" and "=" instead of "Ø-unifier" and "$=_\emptyset$". It will turn out that the unrestricted instantiation preorder should only be used for unification. For the other problems the restricted preorder yields much better results.

3.1 Upper Bound and Lower Bound Problems

It is well-known that any pair of terms has an infimum, and that any weakly unifiable pair of terms has a supremum (see e.g., Huet (1980)). Let $\underline{s} = (s_1,\ldots,s_n)$ and $\underline{t} = (t_1,\ldots,t_n)$ be n-tuples of terms, and let f be a binary function symbol.
We define $s := f(s_1,f(s_2,\ldots f(s_{n-1},s_n)\ldots))$ and $t := f(t_1,f(t_2,\ldots f(t_{n-1},t_n)\ldots))$. Then $\underline{u} = (u_1,\ldots,u_n)$ is an infimum (supremum) of the lower bound (upper bound) problem $\Gamma = <\underline{s}, \underline{t}>$ iff $u := f(u_1,f(u_2,\ldots f(u_{n-1},u_n)\ldots))$ is an infimum (supremum) of s, t. As a consequence, we get

Proposition 3.1 Any Ø-lower bound problem has an infimum, and any solvable Ø-upper bound problem has a supremum.

3.2 Unification Problems

It is well-known (see e.g., Robinson (1965), Eder (1985)) that any solvable unification problem $\Gamma = <\underline{s}, \underline{t}>$ has a most general unifier δ w.r.t. unrestricted instantiation which satisfies the following properties:
(P1) $D(\delta) \cup V(D(\delta)\delta) \subseteq V_0$ where $V_0 := V(\underline{s}) \cup V(\underline{t})$ is the set of all variables occurring in $\underline{s}$ or $\underline{t}$.
(P2) δ is idempotent, i.e., $D(\delta) \cap V(D(\delta)\delta) = \emptyset$.

Proposition 3.2 Any solvable Ø-unification problem has a most general unifier w.r.t. unrestricted instantiation, and thus also w.r.t. restricted instantiation.

Most general unifiers w.r.t. unrestricted instantiation are unique up to ≡-equivalence (where ≡

denotes the equivalence induced by the unrestricted instantiation preorder). The equivalence relation $\equiv$ can be described as follows: $\sigma \equiv \tau$ iff there exists a variable renaming π which satisfies $\sigma = \tau\pi$ (see Eder (1985)).

We shall now show that, w.r.t. unrestricted instantiation, only trivial problems have coequalizers. On the other hand, all solvable unification problems have a coequalizer w.r.t. restricted instantiation.

Proposition 3.3 (1) The $\emptyset$-unification problem $\Gamma = \langle \underline{s}, \underline{t} \rangle$ has a coequalizer w.r.t. unrestricted instantiation iff $\underline{s} = \underline{t}$.
(2) With respect to restricted instantiation, all solvable $\emptyset$-unification problems have a coequalizer.
Proof. (1) If $\underline{s} = \underline{t}$, then the identity (i.e., the substitution with empty domain) is a coequalizer of Γ. On the other hand, assume that $\underline{s} \neq \underline{t}$. If Γ is not solvable, then it trivially does not have a coequalizer. Now let Γ be solvable, and let δ be a most general unifier w.r.t. unrestricted instantiation satisfying the properties (P1) and (P2). Obviously, (P2) implies that the variables of $D(\delta)$ do not occur in $V(V\delta)$; and $\underline{s} \neq \underline{t}$ yields that $D(\delta) \neq \emptyset$.
Assume that γ is a coequalizer of Γ w.r.t. unrestricted instantiation. In particular, γ is also a most general unifier of Γ, and thus equivalent to δ. That means that there exists a variable renaming π which satisfies $\gamma = \delta\pi$. Since $V(V\delta)$ is a strict subset of V, this property also holds for $V(V\delta\pi)$, i.e., there exists a variable x in V such that x is not contained in $V(V\delta\pi) = V(V\gamma)$. But then the substitutions λ can never be unique since their values on x are arbitrary.
(2) Let μ be a most general unifier of Γ w.r.t. restricted instantiation, and let V_0 denote the set of all variables occurring in Γ. In order to prove that μ is a coequalizer of Γ[3], it suffices to show that $x\mu\lambda_1 = x\mu\lambda_2$ for all x in V_0 implies that $y\lambda_1 = y\lambda_2$ for all y in $V(V_0\mu)$. This is a consequence of the following trivial fact: if t is a term containing the variable y, and if the substitutions σ, τ satisfy $y\sigma \neq y\tau$, then $t\sigma \neq t\tau$. ❑

3.3 Weak Unification Problems

By Proposition 3.2 and 2.7, we know that any solvable weak $\emptyset$-unification problem has a most general weak $\emptyset$-unifier w.r.t. restricted instantiation. But if we compare weak unifiers w.r.t. the unrestricted instantiation preorder, the situations becomes completely different.

Proposition 3.4 (1) Any solvable weak $\emptyset$-unification problem has a most general weak $\emptyset$-unifier w.r.t. restricted instantiation.
(2) Weak $\emptyset$-unification problem never have a most general weak $\emptyset$-unifiers w.r.t. the unrestricted instantiation preorder
Proof. To prove (2), assume that $\Gamma = \langle \underline{s}, \underline{t} \rangle$ with $\underline{s} = (s_1,...,s_n)$ and $\underline{t} = (t_1,...,t_n)$ is a weak unification problem. Let $X := V(\underline{s}) \cup V(\underline{t})$ be the set of all variables occurring in $\underline{s}$ or $\underline{t}$. Now assume that (γ,δ) is a most general weak unifier of Γ. For $z \notin X \cup D(\gamma) \cup D(\delta)$ we have $z\gamma = z = z\delta$, and hence $z\gamma\lambda = z\delta\lambda$ for all substitutions λ. Now any pair (γ,δ') with $x\gamma = x\gamma$ and $x\delta' = x\delta$ for $x \neq z$ and $z\gamma \neq z\delta'$ is a pair of weak unifiers of Γ which is not an instance of γ, δ. ❑

This proof depends on the fact that our preorder requires a common right factor λ to obtain γ and δ'. Nevertheless, we prefer this preorder to the componentwise instantiation preorder because:
(1) Instances of weak unifiers should also be weak unifiers; otherwise $W_E(\Gamma)$ would not be an upper set.
(2) Even with the componentwise instantiation preorder, the terms $s = x$ and $t = f(x,y)$ have weak $\emptyset$-unifiers, but they do not have a most general weak $\emptyset$-unifier w.r.t. unrestricted instantiation (see Eder (1985), Example 5.5).

[3]The fact that – w.r.t. restricted instantiation – all most general $\emptyset$-unifiers are coequalizers depends on our definition of coequalizer w.r.t. restricted instantiation, which corresponds to Rydeheard-Burstall's reformulation of $\emptyset$-unification. With respect to the categorical reformulation of E-unification given in Baader (1989), not all most general $\emptyset$-unifiers are coequalizers, but one can always find a coequalizer among the most general $\emptyset$-unifiers (see Baader (1990)).

3.4 Generalization Problems

As for weak unification, we get different behaviour w.r.t. existence of most specific solutions, depending on whether we take the restricted or the unrestricted instantiation preorder.

Proposition 3.5 (1) Any Ø-generalization problem has a most specific Ø-generalizer w.r.t. restricted instantiation.
(2) There exist solvable Ø-generalization problems which do not have a most specific Ø-generalizer w.r.t. unrestricted instantiation.

Proof. (1) Let $\Gamma = \langle \underline{s}, \underline{t} \rangle$ be a Ø-generalization problem, and let $\underline{g}$ be an infimum of $\underline{s}, \underline{t}$ (which exists by Proposition 3.1). Since $\underline{g}$ is a lower bound of $\underline{s}, \underline{t}$, there exist substitutions σ_1, σ_2 such that $\underline{g}\sigma_1 = \underline{s}$ and $\underline{g}\sigma_2 = \underline{t}$. We shall show that $((\sigma_1,\sigma_2),\underline{g})$ is a most specific generalizer of Γ w.r.t. restricted instantiation.
Assume that $((\tau_1,\tau_2),\underline{h})$ is an arbitrary generalizer of Γ. We have to show that $((\sigma_1,\sigma_2),\underline{g})$ is an instance of $((\tau_1,\tau_2),\underline{h})$ w.r.t. the restricted instantiation preorder as defined in Definition 2.4.
We have $\underline{g}\sigma_1 = \underline{s}$, $\underline{g}\sigma_2 = \underline{t}$, $\underline{h}\tau_1 = \underline{s}$, $\underline{h}\tau_2 = \underline{t}$ because $((\sigma_1,\sigma_2),\underline{g})$ and $((\tau_1,\tau_2),\underline{h})$ are generalizers of Γ. Since $\underline{g}$ is an infimum of Γ, there exists a substitution λ such that $\underline{h}\lambda = \underline{g}$. It remains to be shown that, for all variables x occurring in $\underline{h}$, $x\lambda\sigma_i = x\tau_i$ holds for $i = 1, 2$.
For $i = 1$, we have $\underline{h}\tau_1 = \underline{s} = \underline{g}\sigma_1 = \underline{h}\lambda\sigma_1$. Since x occurs in $\underline{h}$, there exists j with $x \in V(h_j)$. But then $h_j\tau_1 = s_j = h_j\lambda\sigma_1$ implies $x\tau_1 = x\lambda\sigma_1$ (see proof of part (2) of Proposition 3.3). Similarly, one can show $x\tau_2 = x\lambda\sigma_2$.
(2) Lemma 3.6 below will give an example of such a problem. Please recall that any Ø-generalization problem is solvable. ❑

Lemma 3.6 Let the signature consist of the two unary function symbols f and g. We define $s := f(x)$ and $t := g(x)$. Then the Ø-generalization problem $\Gamma := \langle s, t \rangle$ does not have a most specific Ø-generalizer w.r.t. unrestricted instantiation.

Proof. Assume that the term h together with the substitutions π_1, π_2 is a most specific generalizer of Γ w.r.t. unrestricted instantiation. Obviously, $h\pi_1 = f(x)$ and $h\pi_2 = g(x)$ implies that h is a variable z. Let $Z = \{ z, z_1, ..., z_m \}$ be the set $D(\pi_1) \cup D(\pi_2)$. We have $z\pi_1 = f(x)$, $z\pi_2 = g(x)$, and $z_i\pi_1 = s_i$, $z_i\pi_2 = t_i$ for terms s_i and t_i.
Let $r > 1$ be a positive integer such that $f^r(x) \neq s_i$ for all i, $1 \leq i \leq m$. We define the term k, and substitutions δ_1, δ_2 as follows:

$$k := y,\ D(\delta_1) := D(\delta_2) := \{y, y_1\}, \text{ and } y\delta_1 := f(x),\ y_1\delta_1 := f^r(x),$$
$$y\delta_2 := g(x),\ y_1\delta_2 := g^r(x).$$

Obviously, $((\delta_1,\delta_2),k)$ is a Ø-generalizer of Γ, and thus there exists a substitution λ with $k\lambda = h$, $\lambda\pi_1 = \delta_1$, and $\lambda\pi_2 = \delta_2$.
Now $y_1\lambda\pi_1 = y_1\delta_1 = f^r(x)$ and $y_1\lambda\pi_2 = y_1\delta_2 = g^r(x)$ imply that $y_1\lambda$ is a variable in $D(\pi_1) \cap D(\pi_2)$. This variable is different from z because $z\pi_1 = f(x)$, $(y_1\lambda)\pi_1 = f^r(x)$, and $r > 1$. Hence $y_1\lambda = z_i$ for some i with $1 \leq i \leq m$. But that means that $s_i = z_i\pi_1 = y_1\lambda\pi_1 = y_1\delta_1 = f^r(x)$, which is a contradiction to our choice of r. ❑

4 The Unification Type of a Theory Depends on the Instantiation Preorder

Until now we have seen that the weak unification (generalization) type of an equational theory may depend on the chosen instantiation preorder: the empty theory has weak unification (generalization) type "unitary" if we use the restricted Ø-instantiation preorder; with respect to the unrestricted Ø-instantiation preorder, the empty theory does not have weak unification (generalization) type "unitary".

In this section, we shall give an example of an equational theory E which has unification type "unitary" w.r.t. the restricted E-instantiation preorder, but not w.r.t. the unrestricted E-instantiation preorder.

Let CIM be the theory of commutative idempotent monoids, i.e., the signature consists of a binary function symbol "+" and a constant symbol "0", and the equational theory is is defined by CIM $:= \{ x+0=x, x+y=y+x, x+(y+z)=(x+y)+z, x+x=x \}$.

Terms s, t are equal w.r.t. CIM iff $V(s) = V(t)$ and $s =_{CIM} 0$ iff $V(s) = \emptyset$. Baader-Büttner (1988) show that CIM is unitary w.r.t. restricted instantiation, and they describe a unification algorithm for this theory.

Let s, t be terms, $\Gamma = \langle s, t\rangle_{CIM}$ be a CIM-unification problem, and $V_0 := V(s) \cup V(t)$ be the set of all variables occurring in s or t. Assume that the substitution σ is a most general CIM-unifier of Γ where "most general" is meant w.r.t. the unrestricted instantiation preorder. We define $W_0 := V(V_0\sigma)$.

From Baader-Büttner (1988) one can easily derive that there exist CIM-unification problems such that σ must satisfy $|W_0| > |V_0|$. Assume that Γ is such a unification problem. Without loss of generality we may also assume that $V_0 \subseteq W_0$. Otherwise, let W be a subset of W_0 of cardinality V_0, and let π be a variable renaming with $W\pi = V_0$. Then $\sigma\pi$ is a most general CIM-unifier of Γ which satisfies $V_0 \subseteq V(V_0\sigma\pi)$.

Let x_0 be an element of W_0 which is not contained in V_0. We define a substitution τ_0 as follows: $x\tau_0 := 0$ for $x \in V_0$, and $x\tau_0 := x$ for $x \notin V_0$. Obviously, $s\tau_0 =_{CIM} 0 =_{CIM} t\tau_0$, which shows that τ_0 is a CIM-unifier of Γ. Thus there exists a substitution λ_0 with $\tau_0 =_{CIM} \sigma\lambda_0$. We consider the term $t_0 := x_0\sigma$.

Lemma 4.1 The set $V(t_0)$ contains a variable x_1 which is not contained in W_0, and thus $x_0 \in D(\sigma)$.
Proof. (1) Assume that $V(t_0) \subseteq W_0$, and let y be an element of $V(t_0)$. Then y is an element of $W_0 = V(V_0\sigma)$, and thus there exists $x \in V_0$ with $y \in V(x\sigma)$. But now $x\sigma\lambda =_{CIM} x\tau = 0$ implies $y\lambda =_{CIM} 0$. Since this holds for any element y of $V(t_0)$, we get $x_0\sigma\lambda = t_0\lambda =_{CIM} 0$. This is a contradiction to $x_0\sigma\lambda =_{CIM} x_0\tau_0 = x_0$.
(2) Now $x_0 \in D(\sigma)$ is an immediate consequence of the fact that $x_0 \in W_0$. ❑

Let $n \geq 1$, and assume that we already have n+1 distinct variables $x_0, x_1, \ldots, x_n$ which satisfy the following conditions:

(C1) $x_n \notin V_0$.
(C2) $x_0, x_1, \ldots, x_{n-1} \in D(\sigma)$.
(C3) For all i, $1 \leq i \leq n$, $x_i \in V(x_{i-1}\sigma)$.

By Lemma 4.1, the sequence x_0, x_1 satisfies these conditions for n = 1. We define $V_n := V_0 \cup \{ x_0, x_1, \ldots, x_{n-1}\}$ and $W_n := V(V_n\sigma)$. Please note that (C3) together with $V_0 \subseteq W_0 \subseteq W_n$ implies that $V_n \subseteq W_n$. In addition, we have $x_n \in W_n \setminus V_n$.

The substitution τ_n is defined as $x\tau_n := 0$ for $x \in V_n$, and $x\tau_n := x$ for $x \notin V_n$. Because $V_0 \subseteq V_n$, we have $s\tau_n =_{CIM} 0 =_{CIM} t\tau_n$, which shows that τ_n is a CIM-unifier of Γ. Thus there exists a substitution λ_n with $\tau_n =_{CIM} \sigma\lambda_n$. We consider the term $t_n := x_n\sigma$.

Lemma 4.2 The set $V(t_n)$ contains a variable x_{n+1} which is not contained in W_n, and thus $x_n \in D(\sigma)$.
Proof. (1) The existence of a variable $x_{n+1} \in V(t_n) \setminus W_n$ is proved as in Lemma 4.1.
(2) Then $x_n \in D(\sigma)$ is an immediate consequence of $x_n \in W_n$. ❑

We want to show that the sequence $x_0, x_1, \ldots, x_n, x_{n+1}$ also satisfies the conditions (C1), (C2), (C3). Condition (C1), i.e., $x_{n+1} \notin V_0$, follows from $x_{n+1} \notin W_n$ and $V_0 \subseteq W_n$. The conditions (C2) and (C3) are also immediate consequences of Lemma 4.2.

By induction, we thus get an infinite chain $x_0, x_1, x_2, \ldots$ of different variables such that $x_i \in D(\sigma)$ for all $i \geq 0$; so the domain $D(\sigma)$ of σ cannot be finite. This is a contradiction since most general unifiers must be substitutions, and substitutions are defined to have finite domain. Thus we have shown

Proposition 4.3 The theory CIM of commutative idempotent monoids is unitary w.r.t. restricted instantiation, but it is not unitary w.r.t. unrestricted instantiation.

5 Unification and Generalization in Commutative Theories

In this section we shall *only use the restricted instantiation preorder*. Unification and generalization types, as well as most general unifiers, coequalizers, and most specific generalizers are always meant w.r.t. restricted instantiation.

Motivated by the categorical reformulation of unification, the class of commutative theories was defined in Baader (1989) within a categorical framework. Baader (1989) also contains an algebraic characterization of commutative theories. In order to avoid introducing notions from category theory and universal algebra, we shall now give a very short definition of commutative theories. The reader is referred to Baader (1989,1989a,1990a) for more information about commutative theories.

Definition 5.1 An equational theory E is called *commutative* iff it satisfies the following properties:
(1) The E-free Ω-algebra without generators, $F_E(\emptyset)$, is of cardinality 1.
(2) The E-free Ω-algebra in two generators is the direct product of two E-free Ω-algebras in one generator, i.e., $F_E(\{x,y\})$ is isomorphic to $F_E(\{x\}) \times F_E(\{y\})$.

The class of commutative theories is interesting for at least two reasons. First, it contains equational theories of practical importance, such as the theory CM of commutative monoids, the theory CIM of commutative idempotent monoids, and the theory AG of abelian groups. Second, well-known algebraic methods for solving systems of linear equations can be used to get unification algorithms for certain commutative theories (see e.g., Nutt (1990), Baader (1990a)).

The reason for the second fact is that one can associate a semiring S(E) to any commutative theory E.[4] For the examples from above we get the following semirings: S(CM) is the semiring **N** of all nonnegative integers, S(CIM) is the two-element boolean semiring, and S(AG) is the ring **Z** of all integers.

Unification in the commutative theory E can be reduced to solving systems of linear equations in S(E) (see Nutt (1990), Baader (1990a)). In fact, an E-unification problem $\Gamma = \langle \underline{s}, \underline{t} \rangle_E$ corresponds to a pair M_σ, M_τ of matrices over S(E), and an E-unifier δ of Γ corresponds to a matrix M_δ with $M_\sigma M_\delta = M_\tau M_\delta$.

Unification problems in commutative theories are either of type unitary or zero, and these types can be characterized in an algebraic way (see Nutt (1990), Baader (1990a)): Let $\Gamma = \langle \underline{s}, \underline{t} \rangle_E$ be an E-unification problem for the commutative theory E, and let M_σ, M_τ be the corresponding matrices over S(E). Then Γ is unitary iff the right S(E)-module

$$U(M_\sigma,M_\tau) := \{a \in S(E)^n;\ M_\sigma a = M_\tau a\}$$

is finitely generated, i.e., there exist finitely many generators $a_1, \ldots, a_r \in S(E)^n$ such that $U(M_\sigma,M_\tau) = \{a_1s_1 + \ldots + a_rs_r;\ s_1, \ldots, s_r \in S(E)\}$. The substitution δ is a most general E-unifier of Γ iff the columns of M_δ generate $U(M_\sigma,M_\tau)$.

The theories CM, CIM, and AG are unitary (see Baader (1989), Nutt (1990)). The theory CIMH of commutative idempotent monoids with a homomorphism is an example of a commutative theory of type zero (see Baader (1989)).

5.1 The Existence of Coequalizers

Before we can characterize the existence of coequalizers in an algebraic way, we need one more definition from algebra.

Definition 5.2 Let S be a semiring, and U be a right S-module.
The multiset $B = \{b_1, \ldots, b_r\}$ is a *base* of U if and only if
(1) $U = \{b_1s_1 + \ldots + b_rs_r;\ s_1, \ldots, s_r \in S\}$ and
(2) $b_1s_1 + \ldots + b_rs_r = b_1s_1' + \ldots + b_rs_r'$ implies $s_1 = s_1', \ldots, s_r = s_r'$.

[4]This was first observed by W. Nutt (see Nutt (1990)) for his class of "monoidal theories", which is – modulo a translation of the signature – the same as the class of commutative theories.

The characterization of most general E-unifiers mentioned above can now easily be modified to a characterization of coequalizers.

Proposition 5.3 Let E be a commutative theory, and let M_σ, M_τ be the matrices over S(E) corresponding to the E-unification problem $\Gamma = \langle \underline{s}, \underline{t} \rangle_E$. Then δ is a coequalizer of Γ iff the columns of M_δ are a base of $U(M_\sigma, M_\tau)$.

In general, finitely generated right S(E)-modules need not have a base. This is demonstrated by the following example, which together with Proposition 5.3 proves Proposition 5.5 below.

Example 5.4 We consider the theory CM of commutative monoids. Since S(CM) is isomorphic to the semiring N of all nonnegative integers, we have to deal with matrices over N. Let $\Gamma = \langle s, t \rangle_{CM}$ be a CM-unification problem such that the corresponding matrices are $M_\sigma = (2\ 3\ 0)$ and $M_\tau = (0\ 0\ 5)$. The elements of $U := U(M_\sigma, M_\tau)$ can be ordered by the componentwise $\leq$-ordering on natural numbers. The module U is generated by the minimal elements of $U \setminus \{\underline{0}\}$, and any set that generates U must contain these minimal elements.
It is easy to see that $(5\ 0\ 2)^T$, $(0\ 5\ 3)^T$ and $(1\ 1\ 1)^T$ are minimal elements of $U \setminus \{\underline{0}\}$. Since $(5\ 0\ 2)^T \cdot 1 + (0\ 5\ 3)^T \cdot 1 = (1\ 1\ 1)^T \cdot 5$, the module U does not have a base.

Proposition 5.5 There exist CM-unification problems which have a most general CM-unifier, but which do not have a coequalizer.

However, if S is a principal ideal domain, then any finitely generated S-module has a base (see e.g. Oeljeklaus-Remmert (1974)). As a consequence we get

Proposition 5.6 Any solvable AG-unification problem $\Gamma = \langle \underline{s}, \underline{t} \rangle_{AG}$ has a coequalizer.

5.2 Generalization

Let E be a commutative theory, and let S(E) be the corresponding semiring. As for unification, an E-generalization problem $\Gamma = \langle \underline{s}, \underline{t} \rangle_E$ corresponds to a pair M_σ, M_τ of matrices over S(E). An E-generalizer $((\gamma_1, \gamma_2), \underline{g})$ of Γ corresponds to the triple of matrices $((M_{\gamma_1}, M_{\gamma_2}), M_\gamma)$ over S(E) with $M_\gamma M_{\gamma_1} = M_\sigma$ and $M_\gamma M_{\gamma_2} = M_\tau$.

The restricted instantiation preorder on E-generalizers (see Definition 2.4.2) can be expressed with the corresponding matrices as follows:
Let $((\gamma_1, \gamma_2), \underline{g})$ and $((\eta_1, \eta_2), \underline{h})$ be E-generalizers of Γ, and let $((M_{\gamma_1}, M_{\gamma_2}), M_\gamma)$ and $((M_{\eta_1}, M_{\eta_2}), M_\eta)$ be the corresponding triples of matrices. Then

$((\eta_1, \eta_2), \underline{h}) \leq_E ((\gamma_1, \gamma_2), \underline{g})$ iff there exists a matrix M_λ over S(E) such that $M_\eta M_\lambda = M_\gamma$, $M_{\eta_1} = M_\lambda M_{\gamma_1}$, and $M_{\eta_2} = M_\lambda M_{\gamma_2}$.

Proposition 5.7 Let E be a commutative theory, and let $\Gamma = \langle \underline{s}, \underline{t} \rangle_E$ be an E-generalization problem. Then Γ has a most specific E-generalizer.
Proof. Let M_σ, M_τ be the pair of matrices corresponding to Γ, where M_σ has dimension k×n and M_τ has dimension k×m. Let $a_1, \ldots, a_n$ be the columns of M_σ, and $b_1, \ldots, b_m$ be the columns of M_τ.
We define $G := (M_\sigma\ M_\tau)$ to be the k×(n+m)-matrix with columns $a_1, \ldots, a_n, b_1, \ldots, b_m$. The matrices G_1, G_2 are defined as

$$G_1 := \begin{pmatrix} E_{n\times n} \\ Z_{m\times m} \end{pmatrix} \text{ and } G_2 := \begin{pmatrix} Z_{n\times n} \\ E_{m\times m} \end{pmatrix}.$$

where the matrix $Z_{m\times m}$ ($Z_{n\times n}$) is the m×m (n×n) zero matrix, and the matrix $E_{n\times n}$ ($E_{m\times m}$) is the n×n (m×m) identity matrix.
Let $((\gamma_1, \gamma_2), \underline{g})$ be the triple (consisting of substitutions γ_1, γ_2 and a tuple of terms $\underline{g}$) which corresponds to the triple of matrices $((G_1, G_2), G)$.
It is not hard to show that $((\gamma_1, \gamma_2), \underline{g})$ is a most specific E-generalizer of Γ. ❑

Theorem 5.8 Any commutative theory E is of lower bound and generalization type unitary.

6 Conclusion

We have presented unification, weak unification, upper bound, lower bound, and generalization problems within a uniform framework. The results for the empty theory demonstrate that using the unrestricted instantiation preorder is only admissible for unification. When working modulo an equational theory, the unrestricted instantiation preorder should not even be used for unification (see Section 4).

For the empty theory a most specific generalizer of two terms yields a shorter description of these terms (see Ohlbach (1990)). In Section 5 we have seen that any commutative theory E is of generalization type "unitary". In this case however, a most specific E-generalizer of two terms does not give a shorter description of the terms (see the proof of Proposition 5.7).

The difference between most general unifiers and coequalizers has been demonstrated for the empty theory w.r.t. unrestriced instantiation, and for commutative theories w.r.t. restricted instantiation.

In Baader (1990), the problems are also considered in a categorical framework. The categories are used to find the correct definitions (e.g., of the instantiation preorders on generalizers) and to clarify the connection between different notions (such as unification and weak unification, or lower bound and generalization problems). Unification, weak unification, etc problems are not only considered for terms, but also for substitutions.

7 References

Baader, F. (1989). Unification in Commutative Theories. *J. Symbolic Computation* **8**.

Baader, F. (1989a). Unification Properties of Commutative Theories: A Categorical Treatment. Proceedings of the Summer Conference on Category Theory and Computer Science, *LNCS* **389**.

Baader, F. (1989b). Characterizations of Unification Type Zero. Proceedings of the 3rd International Conference on Rewriting Techniques and Applications, RTA 89, *LNCS* **355**.

Baader, F. (1990). Unification, Weak Unification, Upper Bound, Lower Bound, and Generalization Problems. SEKI Report **SR-90-02**, Universität Kaiserslautern.

Baader, F. (1990a). Unification in Commutative Theories, Hilbert's Basis Theorem and Gröbner Bases. SEKI Report **SR-90-01**, Universität Kaiserslautern.

Baader, F., Büttner, W. (1988). Unification in Commutative Idempotent Monoids. *Theor. Comp. Sci.* **56**.

Eder, E. (1985). Properties of Substitutions and Unifications. *J. Symbolic Computation* **1**.

Goguen, J.A. (1988). What is Unification? Technical Report, Computer Science Laboratory SRI International, **SRI-CSL-88-2**.

Huet, G. (1980). Confluent Reductions: Abstract Properties and Applications to Term Rewriting Systems. *J. ACM* **27**.

Mac Lane, S. (1971). *Categories for the Working Mathematician.* Heidelberg: Springer-Verlag.

Nutt, W. (1990). Unification in Monoidal Theories. Proceedings of the 10th International Conference on Automated Deduction, CADE 90, *LNCS* **449**.

Oeljeklaus, E., Remmert, R. (1974). *Lineare Algebra* I. Berlin: Springer-Verlag.

Ohlbach, H.J. (1990). Abstraction Tree Indexing for Terms. Proceedings of the 9th European Conference on Artificial Intelligence, ECAI 90.

Plotkin, G. (1970). A Note on Inductive Generalization. *Machine Intelligence* **5**.

Plotkin, G. (1972). Building in Equational Theories. *Machine Intelligence* **7**.

Pottier, L. (1989). Algorithmes de completion et generalisation en logique du premier ordre. These de Doctorat Sciences, Universite de Nice-Sophia Antipolis LISAN-CNRS.

Robinson, J.A. (1965). A Machine-Oriented Logic Based on the Resolution Principle. *J. ACM* **12**.

Rydeheard, D.E., Burstall, R.M. (1985). A Categorical Unification Algorithm. Proceedings of the Workshop on Category Theory and Computer Programming, *LNCS* **240**.

Siekmann, J. (1989). Unification Theory. *J. Symbolic Computation* **7**, Special Issue on Unification.

AC Unification Through Order-Sorted AC1 Unification*

Eric Domenjoud
CRIN & INRIA Lorraine
BP 239 F-54506 Vandœuvre-Lès-Nancy Cedex
FRANCE

Abstract

We design in this paper a new algorithm to perform unification modulo Associativity and Commutativity. This problem is known to be NP-Complete, and none of the solutions proposed until now is very satisfying because of the huge amount of minimal unifiers of some equations. Unlike many authors, we did not try to speed up computations by optimizing some parts of the algorithm, but we tried to design an extension of the algebra in which unification would be less complex. This goal is achieved by adding axioms expressing the existence of an identity for every AC-operator and working in the AC1 theory. In order to get a conservative extension of the quotient algebra, we work in an order-sorted framework, and thus have to deal with order-sorted unification in a collapsing theory.

1 Introduction

Unification in the Associative Commutative theory is a major topic in automated deduction since many mathematical operators have these properties. The first AC-unification algorithm was discovered concurrently M. Stickel [22] and M. Livesey & J. Siekmann [19] in 1975, and proved terminating by F. Fages [9] in 1984. After them, many authors studied this problem, including A. Herold & J. Siekmann [10], C. Kirchner [14], J. Christian & P. Lincoln [18] and many other.

Unifying modulo AC usually requires the solving of linear diophantine equations followed by a combination step of their solutions. Solving linear diophantine equations has been successfully addressed by many authors including G. Huet [11], J.-L. Lambert [17], M. Clausen & A. Fortenbacher [4], E. Contejean & H. Devie [5], E. Domenjoud [8]. Conversely, the only significant improvements of the combination step were proposed by J.-M. Hullot in 1979 [12] and J. Christian & P. Lincoln in 1987 [18]. J.-M. Hullot gave criteria to eliminate early some combinations which do not lead to a solution, while J. Christian & P. Lincoln proposed a particular very fast algorithm for the case of linear

*This work was partly supported by the GRECO de programmation (CNRS)

equations modulo AC. This last solution may however not be used as a general AC-unification algorithm. In the general case, even simple equations may be solved by no current algorithm because of the huge amount of their minimal AC-unifiers. For instance, if $+$ is associative and commutative, then $x+x+x+x = y_1+y_2+y_3+y_4$ has 34 359 607 481 minimal unifiers as shown in [7]. A. Herold & J. Siekmann [10] proposed to introduce an identity for $+$, and to perform unification modulo Associativity, Commutativity and Identity. The theory defined by these axioms, called AC1, is then unitary unifying when only variables and one AC1-operator are involved in terms to unify. This works well as long as no other operators are involved and no other axioms are taken into account but in general, we do not get a conservative extension of the quotient algebra. For example, consider the theory TH defined by the following axioms

$$
\begin{aligned}
&+ \text{ AC} \\
x + x &= \Lambda \\
x + \Lambda &= \Lambda
\end{aligned}
$$

where Λ is a constant. If we add the axiom $x + 0 = x$, where 0 is a new constant, the new theory $(TH1)$ is inconsistent. Indeed, $u = u + 0 = u + (0 + 0) = u + \Lambda = \Lambda$ proves $u = \Lambda$ for any term u. The trouble is due to the fact that the new axiom $x + 0 = x$ may be overlapped with the previous ones, if some variables are instanciated with 0. Such instanciations should thus be forbidden. The solution proposed by C. Kirchner & H. Kirchner [16] and W. L. Buntine & H.-J. Bürckert [3] is to handle constraints expressing that some variables should not be instanciated with the identity. Unfortunately, this does not really solve the problem. Consider again the theory TH: the equation $x =_{TH} y$ would be transformed into $x =_{TH1} y \wedge x \neq_{TH1} 0 \wedge y \neq_{TH1} 0$ which has no solution because disequations have no solution in an inconsistent theory.

A careful study of the combination step in Stickel's algorithm leads to another solution which avoids inconsistency. When solving $x + y =_{AC} z + t$ for example, we first solve the diophantine equation $X_1 + X_2 = X_3 + X_4$. The minimal solutions of this equation are

$$S^1 = (1,0,1,0),\ S^2 = (1,0,0,1),\ S^3 = (0,1,1,0),\ S^4 = (0,1,0,1)$$

and we have to find all subsets of $\{S^1, S^2, S^3, S^4\}$ containing for each j some S^i such that $S^i_j \neq 0$. The minimal solutions of $x + y =_{AC} z + t$ are then linear combinations, with new variables as coefficients, of S^i's in these subsets. In our case, there are 7 such subsets, leading to 7 minimal solutions of $x + y =_{AC} z + t$. But considering each component, we see that a subset fulfills the requirement if and only if it contains at least S^1 and S^4 or S^2 and S^3, other S^i's being *"free"* in each case. For each of these two possibilities, we take *"normal"* new variables u_i as coefficients for constrained S^i's, and *"special"* new variables u_i^+ which may be instanciated with the identity as coefficients for *"free"* ones . We get then two solutions

$$
\begin{cases}
x = u_1 + u_2^+ \\
y = u_3^+ + u_4 \\
z = u_1 + u_3^+ \\
t = u_2^+ + u_4
\end{cases}
\quad \text{and} \quad
\begin{cases}
x = u_1^+ + u_2 \\
y = u_3 + u_4^+ \\
z = u_1^+ + u_3 \\
t = u_2 + u_4^+
\end{cases}
$$

instead of seven. We see on this example that we have variables of two sorts: *"special"* variables may be instanciated with the identity while *"normal"* ones may not. The second sort is embedded in the first one and this leads us to work in an order-sorted framework with the semantics of G. Smolka et al. [21]. Applying the same method, $x+x+x+x = y_1+y_2+y_3+y_4$ has 437 minimal unifiers and this may now be used for practical purposes.

All the proofs that are missing in this paper may be found in [6].

2 The order-sorted extended algebra

According to the idea presented in the previous section, we introduce for each AC-operator a new sort and an identity. Formally, we consider:

- a finite set of operator symbols $\mathcal{F} = F \uplus F_{AC} \uplus \mathcal{I}$ where $F_{AC} = \{+_1, \ldots, +_N\}$ is the set of AC-operators and $\mathcal{I} = \{I_{+_1}, \ldots, I_{+_N}\}$ is the set of identities. For any operator symbol f, $|f|$ is the arity of f.
- a set of sort symbols $\mathcal{S} = \{s, s_{+_1}, \ldots, s_{+_N}\}$, with the subsorts declarations: $s \leq s_{+_i}$
- the operator declarations:

$$\begin{array}{llcccl} \text{for each } f \in F, & f: & & s^{|f|} & & s \\ \text{for each } + \in F_{AC}, & +: & s_+ & \times & s_+ & \longrightarrow s_+ \\ & & s_+ & \times & s & \longrightarrow s \\ & & s & \times & s_+ & \longrightarrow s \\ & I_+: & & & & s_+ \end{array}$$

- The order-sorted signature $\Sigma = (\mathcal{F}, \mathcal{S}, \leq)$, and the one-sorted signature $\Sigma_0 = (F \cup F_{AC}, \{s\})$
- a set of sorted variables $\mathcal{X} = X \uplus X_{+_1} \uplus \cdots \uplus X_{+_N}$ such that X and each X_{+_i} are countable. X is the set of variables of sort s and X_{+_i} is the set of variables of sort s_{+_i}.

$\mathcal{T} = T(\Sigma, \mathcal{X})$ is the order-sorted free Σ-algebra over $\mathcal{X}$, and $T = T(\Sigma_0, X)$ is the one-sorted free Σ_0-algebra over X.

In all the paper, x^+ denotes a variable of sort s_+, sort(t) denotes the least sort of a term t, which exists because Σ is a regular signature, and if S is a set or a multiset of terms, sort(S) denotes the greatest common subsort of sorts of terms in S, which also exist and is unique.

For any equational theory $E_0 \uplus AC$ on T, where

$$AC = \{(x+y)+z = x+(y+z), x+y = y+x \mid + \in F_{AC}\}$$

we define on $\mathcal{T}$ the equational theory $\mathcal{E} \cup \mathcal{AC}$, where $\mathcal{E} = E_0 \cup \mathbf{1}$ with

$$\begin{array}{rcl} \mathcal{AC} & = & \{(x^+ + y^+) + z^+ = x^+ + (y^+ + z^+), x^+ + y^+ = y^+ + x^+ \mid + \in F_{AC}\} \\ \mathbf{1} & = & \{x^+ + I_+ = x^+ \mid + \in F_{AC}\} \end{array}$$

$=_{(\Sigma_0,E_0\cup AC)}$ and $=_{(\Sigma,\mathcal{E}\cup AC)}$ denote the corresponding congruences, and $AC1$ denotes the set of axioms $AC \cup 1$. We take an arbitrary theory for E_0 in order to get results which remain true if the AC theory is combined with other ones. This is especially important if we want to use our unification algorithm in an equational completion procedure since we must take into account the theory defined by the rewrite rules. The relationship between $=_{(\Sigma_0,E_0\cup AC)}$ and $=_{(\Sigma,\mathcal{E}\cup AC)}$ is clarified by the following theorem:

Theorem 1 $\mathcal{T}/_{=_{(\Sigma,\mathcal{E}\cup AC)}}$ *is a conservative extension of* $T/_{=_{(\Sigma_0,E_0\cup AC)}}$*, i.e.:*

$$\forall t,t' \in T, t =_{(\Sigma,\mathcal{E}\cup AC)} t' \Longleftrightarrow t =_{(\Sigma_0,E_0\cup AC)} t'$$

This result allows us to perform in $\mathcal{T}/_{=_{(\Sigma,\mathcal{E}\cup AC)}}$ all the computations we would perform in $T/_{=_{(\Sigma_0,E_0\cup AC)}}$, especially unification the complexity of which drastically decreases.

3 The unification algorithm

3.1 Main definitions

Since we want to perform unification in $\mathcal{T}/_{=_{(\Sigma,\mathcal{E}\cup AC)}}$, we suppose that E defines on $\mathcal{T}$ a simple theory, disjoint from $AC1$, for which a finite and complete unification algorithm is known. We recall that a theory is simple if a term is never equal to one of its proper subterms. This is required because $AC1$ is a collapsing theory, hence is not simple, and combining non simple theories is a difficult problem (see M. Schmidt-Schauß[20] or A. Boudet [2] for more details). Let us first fix our main notations and terminology.

In all the rest of this paper we consider terms of $\mathcal{T}/_{=_{(\Sigma,AC)}}$ rather than terms of $\mathcal{T}$. Terms of $\mathcal{T}/_{=_{(\Sigma,AC)}}$ are represented as flattened terms. $\mathcal{E}$ and E_0 denote then the equational theories induced on $\mathcal{T}/_{=_{(\Sigma,AC)}}$ by $\mathcal{E} \cup AC1$ and $E_0 \cup AC$ respectively, and $\Im$ denotes the rewriting system $\{x^+ + I_+ \rightarrow x^+ \mid + \in F_{AC}\}$ on $\mathcal{T}/_{=_{(\Sigma,AC)}}$.

Definition 1

- *if σ is an idempotent substitution in $\Im$-normal form, σ is a* **$\mathcal{I}$-substitution** *iff $Codom(\sigma) \subset \mathcal{I}$ and σ is* **out of $\mathcal{I}$** *iff $Codom(\sigma) \cap \mathcal{I} = \emptyset$. If $Dom(\sigma) = \{x_1, \ldots, x_n\}$, σ is represented as $\{x_1 \mapsto \sigma(x_1), \ldots, x_n \mapsto \sigma(x_n)\}$ SUBST is the set of all idempotent substitutions in $\Im$-normal form, and $SUBST_{\mathcal{I}}$ is the set of $\mathcal{I}$-substitutions.*
- *A* **multiequation** *is a multiset $\{t_1, \ldots, t_n\}$ of terms, also written $t_1 \doteq \cdots \doteq t_n$. For any multiequation e, $|e|$ is the cardinality of e, $V(e)$ is the multiset of variable terms of e and $T(e)$ is $e - V(e)$. If e and e' are multiequations, we write $e \doteq e'$ for $e \cup e'$. If $|e| = 2$, e is an* **equation**. *$x \doteq t$ is* **solved** *if x is a variable and $sort(x) \geq sort(t)$.*
- *A* **system of multiequations** *(or system for short) is a multiset $\{e_1, \ldots, e_n\}$ of multiequations, also written $e_1 \wedge \cdots \wedge e_n$. $|S|$ is the cardinality of S, and if S_1 and S_2 are systems, we write $S_1 \wedge S_2$ for $S_1 \cup S_2$.*

- *A* **separated system** *is a pair* (σ, S)*, also written* $\sigma \,\&\, S$*, where* S *is a system of multiequations and* σ *is an idempotent substitution, the domain of which is disjoint from the set of variables occurring in* S*. We also say system when no ambiguity arises.* $\sigma \,\&\, S$ *is* **almost solved** *if* S *contains only solved equations and does not contain two equations* $x \doteq t$ *and* $x \doteq t'$ *with* $x \in \mathcal{X}$*, and it is* **solved** *if* S *is empty.*
- *A* **disjunction** *is a multiset* $\{S_1, \cdots, S_n\}$ *of separated systems, also written* $S_1 \vee \cdots \vee S_n$*.*

3.2 Standard rules

Our first transformation rules are very similar to the ones for standard unification:

VAR-DEL	$\dfrac{\sigma \,\&\, (S \wedge x \doteq x \doteq e)}{\sigma \,\&\, (S \wedge x \doteq e)}$	if x is a variable.
SUP-EQ	$\dfrac{\sigma \,\&\, (S \wedge e)}{\sigma \,\&\, S}$	if $\lvert e \rvert = 1$.
SORT-UNIF	$\dfrac{\sigma \,\&\, (S \wedge e)}{\{x \mapsto z\}(\sigma \,\&\, (S \wedge e))}$	if $x \in V(e)$, $V(e)$ contains no variable of sort sort$(V(e))$, z is a new variable of sort s.
VAR-REP	$\dfrac{\sigma \,\&\, (S \wedge x \doteq y \doteq e)}{\{x \mapsto y\}(\sigma \,\&\, (S \wedge y \doteq e))}$	if x and y are distinct variables, sort$(x) \geq$ sort(y).
MERGE	$\dfrac{\sigma \,\&\, (S \wedge x \doteq e \wedge x \doteq e')}{\sigma \,\&\, (S \wedge x \doteq e \doteq e')}$	if x is a variable.
TERM-REP	$\dfrac{\sigma \,\&\, (S \wedge x \doteq t)}{(\{x \mapsto t\}(\sigma \,\&\, S))\!\downarrow_{\Im}}$	if x is a variable, x does not occur in t, sort$(x) \geq$ sort(t).
$\mathcal{I}$-REP	$\dfrac{\sigma \,\&\, (S \wedge x^+ \doteq I_+)}{(\{x^+ \mapsto I_+\}(\sigma \,\&\, S))\!\downarrow_{\Im}}$	

The only differences with standard unification are on one hand, the rule SORT-UNIF and the precondition of VAR-REP because of the order-sorted framework, and on the other hand, the fact that terms are always maintained in $\Im$-normal form. The rule $\mathcal{I}$-REP is simply an instance of TERM-REP. We also introduce the decomposition rule for which we need some additional definitions.

Definition 2 *A symbol* $f \in \mathcal{F}$ *is decomposable iff* $f(t_1, \ldots, t_n) =_{(\Sigma,\mathcal{E})} f(t'_1, \ldots, t'_n) \iff t_1 =_{(\Sigma,\mathcal{E})} t'_1 \wedge \cdots \wedge t_n =_{(\Sigma,\mathcal{E})} t'_n$*. A multiequation* e *is* **decomposable** *iff* $\lvert e \rvert \geq 2$ *and the same decomposable symbol appears at the top of all terms in* e*. The* **common part (CP)** *and the* **frontier (FR)** *of a multiequation* $e = (t_1 \doteq \cdots \doteq t_n)$ *are recursively defined by:*

1. *if* $\forall\, i = 1, \ldots, n$, $t_i = f(t_{i,1}, \ldots, t_{i,m})$ *with* f *decomposable, then:*

$$CP(e) = f(CP(t_{1,1} \doteq \cdots \doteq t_{n,1}), \ldots, CP(t_{1,m} \doteq \cdots \doteq t_{n,m}))$$
$$FR(e) = FR(t_{1,1} \doteq \cdots \doteq t_{n,1}) \wedge \ldots \wedge FR(t_{1,m} \doteq \cdots \doteq t_{n,m})$$

2. *else, if t_i is a variable for some i, then $CP(e) = t_i$ and $FR(e) = e$*

3. *else, $CP(e) = x$ and $FR(e) = (x \doteq e)$ where x is a new variable of sort s.*

The transformation rule for decomposition is then:

$$\textbf{DEC} \quad \frac{\sigma \,\&\, (S \wedge e)}{\sigma \,\&\, (S \wedge V(e) \doteq \mathrm{CP}(T(e)) \wedge \mathrm{FR}(T(e)))} \quad \text{if } T(e) \text{ is decomposable}$$

3.3 Homogenization

We present now a transformation of a system that allows us to reuse as much as possible the simplification rules of each subtheory, resulting in a more modular procedure. The idea is to use the collapsing axioms only by need to make *"compatible"* symbols appear at the top of all terms in a multiequation. Roughly speaking, two symbols are compatible if they are constrained by the same subtheory. From now on, $+$ and $*$ are AC-operators

Definition 3 *$\forall f, g \in \mathcal{F} \cup \mathcal{X}$, f and g are compatible, written $f \sim g$, iff*

$$\exists t, t', \sigma \text{ out of } \mathcal{I} \text{ s.t. } top(t) = f,\ top(t') = g,\ \sigma(t) =_{(\Sigma, E_0)} \sigma(t')$$

A multiequation e is **homogeneous** *if $\forall t_1, t_2 \in e$, $top(t_1{\downarrow_{\mathfrak{I}}}) \sim top(t_2{\downarrow_{\mathfrak{I}}})$. A substitution σ* **homogenizes** *a multiequation e if $\sigma(e)$ is homogeneous, and it homogenizes a system S if it homogenizes each multiequation in S. A* **minimal complete set of homogenizing substitutions** *of a system S (MCSHS for short), is a set of substitutions such that*

1. *$\forall\ \sigma \in MCSHS(S)$, σ homogenizes S*

2. *$\forall\ \sigma \in SUBST$, σ homogenizes $S \Rightarrow \exists \sigma' \in MCSHS(S)$, $\sigma' \leq_{\mathcal{E}} \sigma$*

3. *$\forall\ \sigma, \sigma' \in MCSHS(S)$, neither $\sigma \leq_{\mathcal{E}} \sigma'$, nor $\sigma' \leq_{\mathcal{E}} \sigma$ holds.*

In order to compute MCSHS(S), we associate with each term t a finite set of pairs (f_i, c_i), noted c-top(t), where $f_i \in \mathcal{F} \cup \mathcal{X}$ and c_i is a logical formula. c-top(t) is recursively defined as follows:

- If $t \in X_+$, c-top$(t) = \{(t, t \neq I_+), (I_+, t = I_+)\}$
- if top$(t) \in X \cup F \cup \mathcal{I}$, c-top$(t) = \{(\mathrm{top}(t), \mathrm{True})\}$
- if top$(t) = + \in F_{AC}$, $t = t_1 + t_2$ with c-top$(t_1) = \{(f_1, c_1), \ldots, (f_n, c_n)\}$ and c-top$(t_2) = \{(g_1, c'_1), \ldots, (g_m, c'_m)\}$,

$$\begin{aligned} \text{c-top}(t) \;=\; & \{(+, \neg\text{top-c}(t_1, I_+) \wedge \neg\text{top-c}(t_2, I_+)), \\ & (f_1, c_1 \wedge \text{top-c}(t_2, I_+)), \ldots, (f_n, c_n \wedge \text{top-c}(t_2, I_+)) \\ & (g_1, c'_1 \wedge \text{top-c}(t_1, I_+)), \ldots, (g_m, c'_m \wedge \text{top-c}(t_1, I_+))\} \end{aligned}$$

where $\text{top-c}(t, f) = \displaystyle\bigvee_{(f,c) \in \text{c-top}(t)} c$

Example 1 $F = \{f\}$, $u, z \in X$, $t_1 = x^+ + (y^* * f(u + x^+))$, $t_2 = y^* * (z + x^+)$

$$\begin{aligned} c\text{-}top(t_1) &= \{(+, \neg c_{x^+}), (*, c_{x^+} \wedge \neg c_{y^*}), (f, c_{x^+} \wedge c_{y^*})\} \\ c\text{-}top(t_2) &= \{(*, \neg c_{y^*}), (+, c_{y^*} \wedge \neg c_{x^+}), (z, c_{y^*} \wedge c_{x^+})\} \\ top\text{-}c(t_1, f) &= c_{x^+} \wedge c_{y^*} \end{aligned}$$

where c_{x^+} *stands for* $x^+ = I_+$.

For each $(f, c) \in$ c-top(t), c is interpreted as the set of $\mathcal{I}$-substitutions, noted $\text{subst}_{\mathcal{I}}(c)$, defined by:

- $\text{subst}_{\mathcal{I}}(c_x) = \{\sigma \mid \sigma \in \text{SUBST}_{\mathcal{I}},\ x \in \text{Dom}(\sigma)\}$
- $\text{subst}_{\mathcal{I}}(\text{False}) = \emptyset$
- $\text{subst}_{\mathcal{I}}(c \wedge c') = \text{subst}_{\mathcal{I}}(c) \cap \text{subst}_{\mathcal{I}}(c')$
- $\text{subst}_{\mathcal{I}}(\neg c) = \text{SUBST}_{\mathcal{I}} - \text{subst}_{\mathcal{I}}(c)$

By induction on the structure of t, we prove then easily:

Theorem 2 $\forall f \in \mathcal{F} \cup \mathcal{X}:$ *$subst_{\mathcal{I}}(top\text{-}c(t, f)) = \{\sigma \mid \sigma \in SUBST_{\mathcal{I}},\ top(\sigma(t)\downarrow_{\Im}) = f\}$*

This allows us to represent the set of all $\mathcal{I}$-substitutions homogenizing a system of multiequations S by

$$\text{hom}(S) = \bigwedge_{e \in S} \bigwedge_{t_1, t_2 \in e} \bigvee_{\substack{(f,c) \in \text{c-top}(t_1) \\ (f',c') \in \text{c-top}(t_2) \\ f \sim f'}} (c \wedge c')$$

In order to get a basis of the set represented by hom(S), we take its **positive part**, obtained by removing from its disjunctive form all conjunctions containing a literal and its negation, and replacing with True all negative literals in the resulting formula. We note PP(hom(S)) the normal form of this positive part, which is well defined because all literals are positive. We get then MCSHS(S) in the following way:

1. if PP(hom(S)) = False then MCSHS$(S) = \emptyset$
2. else if PP(hom(S)) = True then MCSHS$(S) = \{\text{Id}\}$
3. else $\text{PP}(\text{hom}(S)) = \bigvee_{i=1,\dots,n} (x_{i,1} = I_{+_{i,1}}) \wedge \cdots \wedge (x_{i,n_i} = I_{+_{i,n_i}})$ and

$$\text{MCSHS}(S) = \{\{x_{i,1} \mapsto I_{+_{i,1}}, \dots x_{i,n_i} \mapsto I_{+_{i,n_i}}\} \mid i = 1, \dots, n\}$$

Example 2 (Example 1 continued) *$hom(x^+ + (y^* * f(u + x^+)) \doteq y^* * (z + x^+)) =$*

$$\begin{aligned} &(\neg c_{x^+} \wedge c_{y^*} \wedge \neg c_{x^+}) \vee (\neg c_{x^+} \wedge c_{y^*} \wedge c_{x^+}) \vee (c_{x^+} \wedge \neg c_{y^*} \wedge \neg c_{y^*}) \\ \vee\ &(c_{x^+} \wedge \neg c_{y^*} \wedge c_{y^*} \wedge c_{x^+}) \vee (c_{x^+} \wedge c_{y^*} \wedge c_{y^*} \wedge c_{x^+}) \end{aligned}$$

which reduces to $(\neg c_{x^+} \wedge c_{y^*}) \vee (c_{x^+} \wedge \neg c_{y^*}) \vee (c_{x^+} \wedge c_{y^*})$. *Its positive part is* $(True \wedge c_{y^*}) \vee (c_{x^+} \wedge True) \vee (c_{x^+} \wedge c_{y^*})$ *the normal form of which is* $c_{y^*} \vee c_{x^+}$. *We get then*

$$MCSHS(x^+ + (y^* * f(u + x^+)) \doteq y^* * (z + x^+)) = \{\{y^* \mapsto I_*\}, \{x^+ \mapsto I_+\}\}$$

The transformation rule for homogenizing is:

$$\text{HOM}\quad \frac{\sigma \,\&\, S}{\bigvee_{\tau \in \mathrm{MCSHS}(S)} (\tau(\sigma \,\&\, S))\downarrow_{\Im}} \quad \text{if } \mathrm{MCSHS}(S) \neq \{Id\}$$

3.4 Solving cycles

A cycle is a system $C = (x_1 \doteq t_1[x_2] \wedge \cdots \wedge x_{n-1} \doteq t_{n-1}[x_n] \wedge x_n \doteq t_n[x_1])$ where x_i's are variables. C is trivial if all t_i's are variables. Unlike in a simple theory, a non trivial cycle may have a solution modulo $(\Sigma, \mathcal{E})$ because it is a collapsing theory. Solving cycles modulo $(\Sigma, \mathcal{E})$ is actually very easy, and whenever a solution exists, there is a unique most general one.

Let Sym(t) be the multiset of all symbols (variables and operators) occurring in a term t and Sym(C) be $\bigcup_{i=1}^{n} \mathrm{Sym}(t_i) - \{x_1, \ldots, x_n\}$ where $\{x_1, \ldots, x_n\}$ is a multiset.

Theorem 3 *C has a solution only if Sym(C) contains only AC-operators and variables in $\mathcal{X} - X$. In this case, if a solution exists, it is of the form $\theta \cdot \tau_C$ where τ_C is the $\mathcal{I}$-substitution the domain of which is the set of variables occurring in Sym(C).*

The transformation rule for solving cycles is then:

$$\text{CYCLE}\quad \frac{\sigma \,\&\, (S \wedge C)}{\bigvee_{\tau \in \mathrm{SOL}(C)} (\tau(\sigma \,\&\, (S \wedge C)))\downarrow_{\Im}} \quad \text{if } C \text{ is a non trivial cycle}$$

where SOL(C) is $\{\tau_C\}$ if the condition of the previous theorem holds, $\emptyset$ otherwise. In this last case, we introduce an empty disjunction rather than failure because our procedure never explicitly fails.

Example 3 *Let us consider the cycles*

$$C_1 = \begin{cases} x_1 \doteq u^+ + (v^* * x_2) \\ x_2 \doteq w^* * (u^+ + z^+ + x_3) \\ x_3 \doteq y^* * x_1 \end{cases} \qquad C_2 = \begin{cases} x_1 \doteq x_4 + x_2 \\ x_2 \doteq w^* * x_1 \end{cases}$$

where x_i's are variables of sort s.

Sym(C_1) $= \{u^+, +, v^*, *, w^*, *, u^+, +, z^+, +, y^*, *\}$ and $\tau_{C_1} = \{u^+ \mapsto I_+, v^* \mapsto I_*, w^* \mapsto I_*, z^+ \mapsto I_+, y^* \mapsto I_*\}$. The solution of C_1 is then $\theta \cdot \tau_{C_1}$ where θ is the solution of $x_1 \doteq x_2 \wedge x_2 \doteq x_3 \wedge x_3 \doteq x_1$.

Sym(C_2) $= \{x_4, +, w^*, *\}$ and C_2 has no solution because sort(x_4) $= s$.

3.5 Mutation

The mutation, introduced by C. Kirchner in [13] is intended to solve equations which may not be solved by decomposition. In [15], he investigates how mutation may be reused in an order-sorted framework. In this section, we shall first describe briefly how the mutation for the E_0 theory operate, and then detail the mutation for the $\mathcal{A}C1$ theory.

Let us call F_{E_0} the set of all operators occurring in some axiom of E_0. We suppose that a process is known, which returns a complete set of solutions $\mathrm{SOL}_{E_0}(S)$ for any system S pure in E_0, that is a system where all operators belong to F_{E_0}. We consider then for any system S the subsystem S_{E_0} of all homogeneous multiequations $e \in S$ such that $\forall t \in T(e), \mathrm{top}(t) \in F_{E_0}$. S_{E_0} may be written as $\mu(\overline{S}_{E_0})$ where $\overline{S}_{E_0}$ is a pure in E_0 and $\forall t \in \mathrm{Codom}(\mu), \mathrm{top}(t) \notin F_{E_0}$. This is known as variable abstraction. The mutation rule for the theory E_0 is then:

$$\textbf{MUTATE}_{E_0} \quad \frac{\sigma \,\&\, S}{\bigvee_{\tau \in \mathrm{SOL}_{E_0}(\overline{S}_{E_0})} \sigma \,\&\, ((\bigwedge_{x \in \mathrm{Dom}(\mu\cdot\tau)} x \doteq \mu \cdot \tau(x)) \wedge (S - S_{E_0}))} \quad \text{if } S_{E_0} \text{ is unsolved}$$

We describe now the mutation for the $\mathcal{AC}1$ theory. Let us first specify the part of a system mutation applies to:

Definition 4 *The* **maximal mutable subsystem (for +)** *of a system S (MMSS$(S,+)$ for short) is defined as follows:*

1. *$MSS_0(S,+)$ is the multiset of all homogeneous unsolved multiequations $e \in S$ such that $\forall t \in T(e)$, $top(t) = +$.*
2. *for all $i \geq 0$, let*
   - *$SV_i(S,+)$ be the set of variables appearing as a variable term, or as an immediate subterm of a term in some multiequation in $MSS_i(S,+)$ where the immediate subterms of $t_1 + \cdots + t_n$ are t_i's if their top symbol is not +*
   - *$MSS_{i+1}(S,+) = MSS_i \cup \{e \in S - MSS_i(S,+) \mid e$ is homogeneous, $V(e) \cap SV_i(S,+) \neq \emptyset$ and $\forall t \in T(e)$, $top(t) = +\}$*
3. *$\exists i_0 \geq 0$, $MSS_{i_0+1}(S,+) = MSS_{i_0}(S,+)$ and then, $MMSS(S,+) = MSS_{i_0}(S,+)$*

Like in Stickel's algorithm [22], we associate with MMSS$(S,+)$ a linear homogeneous diophantine system defined as follows: we consider the vector $\mathrm{VIM}(S,+) = [t_1, \ldots, t_n]$ of all distinct variable terms and immediate subterms of non variable terms in all multiequations of MMSS$(S,+)$. There exists a system of equations equivalent to MMSS$(S,+)$, and such a system may be written as $\mathrm{VIM}(S,+)D_1 \doteq \mathrm{VIM}(S,+)D_2$ where D_1 and D_2 are matrices of naturals. The matrix D of the diophantine system is then $D_1 - D_2$.

Let us write $\mathrm{hom}_+(t)$ for $\mathrm{hom}(t \doteq v_1 + v_2)$ where v_1 and v_2 are new variables of sort s. B is a matrix the rows of which are all minimal solutions of the diophantine system $\vec{Y}D = \vec{0}$ over $\mathbb{N}^n$, except the ones satisfying either $X_j \geq 2$ for some j such that $\mathrm{hom}_+(t_j) = \mathrm{False}$, or $B_{i,j_1} + B_{i,j_2} \geq 2$ for some $j_1 \neq j_2$ such that $\mathrm{hom}_+(t_{j_1}) = \mathrm{hom}_+(t_{j_2}) = \mathrm{hom}(t_{j_1} \doteq t_{j_2}) = \mathrm{False}$. The mutation of MMSS$(S,+)$ is then a disjunction of systems of the form $\mathrm{VIM}(S,+) \doteq \vec{u}B$ where $\vec{u}$ is a vector of new variables of sort s or s_+. In order to determine the sorts of u_i's, let us write $u = [u_1, \ldots, u_m]$ where m is the number of rows of B. Since we generate equations of the form $t_j \doteq u_1 B_{1,j} + \cdots + u_m B_{m,j}$, and every solution σ of S must satisfy $\mathrm{sort}(\sigma(t_j)) = \mathrm{sort}(\sigma(u_1 B_{1,j} + \cdots + u_m B_{m,j}))$, we may express constraints on sorts of t_j's by constraints on sorts of u_i's.

If $\mathrm{sort}(t_j) \neq s_+$ then $\mathrm{sort}(\sigma(t_j)) = s$ and it is sufficient to satisfy $\mathrm{sort}(u_1B_{1,j} + \cdots + u_mB_{m,j}) = s$. This holds only if $\mathrm{sort}(u_iB_{i,j}) = s$ for at least one i, that is to say, $B_{i,j} \neq 0$ and $\mathrm{sort}(u_i) = s$.

Now suppose that t_j is not a variable. If $\sum_{j=1}^{m} B_{i,j} > 1$, the generated equation is inhomogeneous and must be later homogenized. If $\mathrm{hom}_+(t_j) = \mathrm{False}$ then $u_1B_{1,j} + \cdots + u_mB_{m,j}$ must be reduced to a variable. This will be possible only if $\sum_{\mathrm{sort}(u_i)=s} B_{i,j} \leq 1$.

For each t_j, we have then to consider the constraints:

1. $\bigvee_{B_{i,j}\neq 0} \mathrm{sort}(u_i) = s$ if $\mathrm{sort}(t_j) \neq s_+$

2. $\bigwedge_{\substack{B_{i_1,j}=B_{i_2,j}=1\\ i_1\neq i_2}} (\mathrm{sort}(u_{i_1}) \neq s \vee \mathrm{sort}(u_{i_2}) \neq s)$ if $\mathrm{hom}_+(t_j) = \mathrm{False}$

We reduce the conjunction of the above constraints in normal form (disjunction of conjunctions of literals) $C_1 \vee \cdots \vee C_p$ and for each C_k, we form the system $\mathrm{Mut}_i(S,+) = (\mathrm{VIM}(S,+) \doteq \vec{u}B)$ where

$$u_i \in \begin{cases} X & \text{if } C_k = (C'_k \wedge \mathrm{sort}(u_i) = s) \\ X_+ & \text{otherwise} \end{cases}$$

The mutation of the system S is then defined as $\mathrm{Mut}(S,+) = \{(S - \mathrm{MMSS}(S,+)) \wedge \mathrm{Mut}_i(S,+) \mid i = 1,\ldots,p\}$ and the transformation rule for the AC-mutation is:

$$\textbf{MUTATE}_+ \quad \frac{\sigma \,\&\, S}{\bigvee_{S_i\in\mathrm{Mut}(S,+)} \sigma \,\&\, S_i} \quad \text{if } \mathrm{MMSS}(S,+) \neq \emptyset$$

Example 4 *We want to mutate the equation* $x + x + y^+ \doteq a + (u * v^*)$ *where* a *is a constant.* $VIM = [x, y^+, a, u*v^*]$ *and* ${}^tD = [\,2\;1\;-1\;-1\,]$. *We do not consider the solution* $[1\;0\;2\;0]$ *of the diophantine equation because* $hom_+(a) = False$. *The matrix of remaining minimal solutions is*

$$B = \begin{bmatrix} 1 & 0 & 1 & 1 \\ 1 & 0 & 0 & 2 \\ 0 & 1 & 1 & 0 \\ 0 & 1 & 0 & 1 \end{bmatrix}$$

The constraints are

- $sort(u_1) = s \vee sort(u_2) = s$ *because* $sort(x) \neq s_+$
- $sort(u_1) = s \vee sort(u_3) = s$ *because* $sort(a) \neq s_+$
- $sort(u_1) = s \vee sort(u_2) = s \vee sort(u_4) = s$ *because* $sort(u * v^*) \neq s_+$
- $sort(u_1) \neq s \vee sort(u_3) \neq s$ *because* $hom_+(a) = False$

and their conjunction reduces to $(sort(u_1) = s \wedge sort(u_3) \neq s) \vee (sort(u_1) \neq s \wedge sort(u_2) = s \wedge sort(u_3) = s)$. *The mutation of the equation contains then two systems:*

$$\begin{cases} x & \doteq u_1 + u_2^+ \\ y^+ & \doteq u_3^+ + u_4^+ \\ a & \doteq u_1 + u_3^+ \\ u * v^* & \doteq u_1 + u_2^+ + u_2^+ + u_4^+ \end{cases} \quad \text{and} \quad \begin{cases} x & \doteq u_1^+ + u_2 \\ y^+ & \doteq u_3 + u_4^+ \\ a & \doteq u_1^+ + u_3 \\ u * v^* & \doteq u_1^+ + u_2 + u_2 + u_4^+ \end{cases}$$

4 Control and Termination

Let us now give the control on our transformation rules and sketch the proof of termination for this control. The termination proof may not hold for the case of a non empty theory E_0. In this case, the whole procedure remains nevertheless sound and complete. One may see M. Schmidt-Schauß[20] or A. Boudet [2] for more details about the combination of equational theories and the termination.

The unification procedure is described by the following control:

(RED* MUTATE)* TERM-REP*

where

- RED=MERGE | VAR-REP | VAR-DEL | SUP-EQ | $\mathcal{I}$-REP | SORT-UNIF | HOM | DEC | CYCLE
- MUTATE=MUTATE$_{+_1}$ | $\cdots$ | MUTATE$_{+_\mathcal{N}}$ | MUTATE$_{E_0}$
- $R \mid R'$ means apply R or R', and R^* means apply R as long as possible.

It is easy to see that the procedure returns solved separated systems if it terminates. We suppose now that E_0 is the empty theory. In order to prove the termination, we first consider for each separated system σ & S, the multiset H of heights of all terms in all multiequations in S, which are not of the form (F) $x_1 + \cdots + x_n$ where x_i's are variables and $n \geq 0$ (by convention, such a term is x_1 if $n = 1$ and I_+ if $n = 0$). H never increases, and strictly decreases if either DEC is applied to a multiequation which is not $V(e) \doteq I_+ \doteq \cdots \doteq I_+$ or $MUTATE_+$ is applied to a subsystem where appears a term which is not of the form (F). Since H cannot decrease indefinitely, terms which are not of the form (F) ultimately disappear or become *"passive"* in the sense that they may still be modified but do not act upon the rest of the system any more. We may thus assume that only terms of the form (F) occur in the system.

We prove the termination of RED* using the complexity $<$ SV, $|S|$, CM $>$ where SV is the set of all variables occurring in S and CM is the multiset $\{|e|,\ e \in S\}$. The second and third components are compared with the usual ordering $\leq$ on naturals and its multiset extension $\lll$, and the first components are compared with the ordering defined by:

1. $\forall x \in \mathcal{X}$, $\mathrm{SV} \cup \{x\} > \mathrm{SV}$ if $x \notin \mathrm{SV}$
2. $\forall x \in X$, $\forall y^+ \in X_+$, $\mathrm{SV} \cup \{y^+\} > \mathrm{SV} \cup \{x\}$ if $y^+ \notin \mathrm{SV}$

Since these orderings are well founded, so is their lexicographic combination. This complexity behaves as shown in the table below where $\downarrow$ (resp. $\uparrow$) means that the component strictly decreases (resp. increases), $\searrow$ (resp. $\nearrow$) means that the component may decrease (resp. increase) or remain the same, and = means that the component does not change.

	SV	$\|S\|$	CM
MERGE	$=$	$\downarrow$	$\uparrow$
SUP-EQ	$\searrow$	$\downarrow$	$\downarrow$
$\mathcal{I}$-REP	$\downarrow$	$\downarrow$	$\downarrow$
VAR-REP	$\downarrow$	$=$	$\downarrow$
VAR-DEL	$\searrow$	$\searrow$	$\downarrow$
SORT-UNIF	$\downarrow$	$=$	$=$
HOM	$\downarrow$	$=$	$=$
DEC	$=$	$=$	$\downarrow$
CYCLE	$\downarrow$	$=$	$=$

For the last step of the proof, we notice that when MUTATE is applied, S may be written as $S_{+_1} \wedge \ldots \wedge S_{+_n}$ where the top symbol of all non variable terms in S_{+_i} is $+_i$. We consider then the complexity $<$ SHV, USS $>$ where SHV is the number of shared variables, that is to say variables that occur in S_{+_i} and S_{+_j} for $i \neq j$, and USS is the number of unsolved subsystems, that is to say subsystems containing an unsolved multiequation. This complexity strictly decreases between two applications of MUTATE.

5 Conclusion and perspectives

Let us first give some comparisons between the number of unifiers returned by our procedure and the number of minimal AC-unifiers. Consider equations of the form

$$\underbrace{x + \cdots + x}_{p \text{ times}} \doteq y_1 + \cdots y_q$$

The table below displays the number of minimal unifiers in both cases for some values of p and q. The first values correspond to the $\mathcal{AC}1$ case.

$q \backslash p$	2	3	4
2	2 / 5	3 / 13	4 / 29
3	7 / 45	20 / 981	40 / 32 677
4	26 / 809	139 / 1 044 569	437 / 34 359 607 481

This shows clearly that $\mathcal{AC}1$-unification saves time and space and could be a good alternative to of AC-unification. In addition, using the $\mathcal{AC}1$ theory could be interesting in an implementation of completion modulo AC since we often have to add AC-extensions of rules. The AC-extension of a rule $t_1 + t_2 \rightarrow t_3$ is the rule $t_1 + t_2 + x \rightarrow t_3 + x$ where x is a new variable. Modulo $\mathcal{AC}1$, $t_1 + t_2 + x^+ \rightarrow t_3 + x^+$ *"modelizes"* both the rule and its extension. Rewriting has then to be performed modulo $\mathcal{AC}1$ and we may use results of T. B. Baird et al. [1] for completion and rewriting modulo a collapsing theory.

Some questions arise after this work:

- Do similar methods apply to other theories? Do they improve the complexity of unification? It is actually possible to design such an order-sorted extension for a theory, as soon as some operators are Associative. One can introduce an identity for such operators if terms are sorted. The complexity of unification should then be carefully studied.
- Specific tools are necessary for unification modulo $\mathcal{AC}1$ because it is a collapsing theory. In our case, homogenization works well because a minimal set of closed homogenizing substitutions may be associated with each system of multiequations. It would be interesting to reuse this idea for other collapsing theories, including AC1 in the unsorted case or idempotency.

Acknowledgements: I am very grateful to Claude Kirchner for numerous discussions about this paper, and to Pierre Lescanne who read a previous version of it and suggested many improvements.

References

[1] T.B. Baird, G.E. Peterson, and R.W. Wilkerson. Complete sets of reductions modulo associativity, commutativity and identity. In N. Dershowitz, editor, *Proceedings of RTA'89, Chapel Hill, (North Carolina, USA)*, volume 355 of *LNCS*, pages 29–44. Springer-Verlag, 1989.

[2] A. Boudet. Unification dans les mélanges de théories équationelles. application aux axiomes d'associativité, commutativité, identité et idempotence, aux anneaux booléens, et aux groupes abéliens. Thèse de l'Université d'Orsay, 1990.

[3] W. Buntine and H.-J. Bürckert. On solving equations and disequations. Technical Report SR-89-03, Universität Kaiserslautern, 1989.

[4] M. Clausen and A. Fortenbacher. Efficient solution of linear diophantine equations. *JSC*, 8:201–216, 1989. Special issue on unification.

[5] E. Contejean and H. Devie. Solving systems of linear diophantine equations. In H.-J. Bürckert and W. Nutt, editors, *Proceedings of UNIF'89, Lambrecht (Germany)*, 1989.

[6] E. Domenjoud. AC-unification through order-sorted AC1-unification. Research Report 89-R-67, CRIN, Nancy (France), 1989.

[7] E. Domenjoud. Number of minimal unifiers of the equation $\alpha x_1 + \cdots + \alpha x_p \doteq_{AC} \beta y_1 + \cdots + \beta y_q$. Research Report 89-R-2, CRIN, Nancy (France), 1989. To appear in *JAR* (1990).

[8] E. Domenjoud. Solving systems of linear diophantine equations: an algebraic approach. In *Proceedings of UNIF'90*, Leeds (U.K.), 1990.

[9] F. Fages. Associative-commutative unification. In R. Shostak, editor, *Proceedings of CADE'84, Napa Valley (California, USA)*, volume 170 of *LNCS*, pages 194–208. Springer-Verlag, 1984.

[10] A. Herold and J. Siekmann. Unification in abelian semigroups. *JAR*, 3:247–283, 1987.

[11] G. Huet. An algorithm to generate the basis of solutions to homogenous linear diophantine equations. *Information Processing Letters*, 7:144–147, 1978.

[12] J.-M. Hullot. Associative-commutative pattern matching. In *Proceedings 9th International Joint Conference on Artificial Intelligence*, 1979.

[13] C. Kirchner. Méthodes et outils de conception systématique d'algorithmes d'unification dans les théories équationnelles. Thèse d'état, Université de Nancy I, 1985.

[14] C. Kirchner. From unification in combination of equational to a new AC-unification algorithm. Technical Report 87-R-132, Centre de Recherche en Informatique de Nancy, 1987.

[15] C. Kirchner. Order-sorted equational unification. Presented at the fifth International Conference on Logic Programming (Seattle, USA), 1988. Also as rapport de recherche INRIA 954.

[16] C. Kirchner and H. Kirchner. Constrained equational reasoning. In *Proceedings of the ACM-SIGSAM 1989 International Symposium on Symbolic and Algebraic Computation*, pages 382–389, Portland (Oregon), 1989. ACM Press. Report CRIN 89-R-220.

[17] J.-L. Lambert. Une borne pour les générateurs des solutions entières positives d'une équation diophantienne linéaire. *Compte-rendu de L'Académie des Sciences de Paris*, 305:39–40, 1987.

[18] P. Lincoln and J. Christian. Adventures in associative-commutative unification. *JSC*, 8:217–240, 1989. Special issue on unification.

[19] M. Livesey and J. Siekmann. Unification of bags and sets. Technical report, Institut für Informatik I, Universität Karlsruhe, 1976.

[20] M. Schmidt-Schauß. Combination of unification algorithms. *JSC*, 8:51–100, 1989. Special issue on unification.

[21] G. Smolka, W. Nutt, J.A. Goguen, and J. Meseguer. Order-sorted equational computation. In H. Ait-Kaci and M. Nivat, editors, *Resolution of Equations in Algebraic Structures, Volume 2: Rewriting Techniques*, pages 297–367. Academic Press, 1989.

[22] M.E. Stickel. A complete unification algorithm for associative-commutative functions. *Proceedings 4th International Joint Conference on Artificial Intelligence, Tbilissi*, 1975.

Narrowing directed by a graph of terms

Jacques Chabin and Pierre Réty

LIFO & LRI, Dépt. math-info, Université d'Orléans
BP 6759, 45067 Orléans cedex 2, France
E-mail : {chabin, rety}@univ-orleans.fr

Abstract

Narrowing provides a complete procedure to solve equations modulo confluent and terminating rewriting systems. But it seldom terminates. This paper presents a method to improve the termination. The idea consists in using a finite graph of terms built from the rewriting system and the equation to be solved, which helps one to know the narrowing derivations possibly leading to solutions. Thus, the other derivations are not computed. This method is proved complete. An example is given and some improvements are proposed.

1 Introduction

Solving equations (unification) modulo term rewriting systems is an important problem, which is necessary to combine functional and logic programming, for example as in EQLOG langage [Goguen-Meseguer-86]. In this framework, completeness and termination of unification methods are desirable. Indeed, during a clause superposition attempt, it is interesting to detect the unsatisfiability of the equation to be solved, which implies that the superposition is not possible. Conditional completion procedures also need methods that solve the equations appearing in the conditions. If such an equation is unsatisfiable, the conditional rewrite rule containing it can be deleted, because this rule can't be applied. And to detect that, a complete and terminating procedure is necessary.

This paper presents a method that improves the termination of basic narrowing. Basic narrowing gives a complete set of solutions modulo confluent and terminating rewriting systems, by computing all the narrowing derivations issued from the equation to be solved [Hullot-80]. Actually, only the derivations that lead to a syntactically unifiable equation give solutions. Thus, the others are useless. Our idea consists in using this fact to prune useless paths from the search tree.

Let's look at a simple example. Consider the following confluent and terminating rewriting system :

$$\begin{aligned} r_1: \quad h(f(g(y))) &\rightarrow h(f(y)) \\ r_2: \quad h(0) &\rightarrow 0 \end{aligned}$$

We want to solve the equation $h(x) \doteq_R 0$. By using basic narrowing, there are two derivations:

$$h(x) \doteq_R 0 \;\rightsquigarrow_{[r_1, x\mapsto f(g(x))]} h(f(x)) \doteq_R 0 \rightsquigarrow_{[r_1, x\mapsto g(x)]} h(f(x)) \doteq_R 0 \rightsquigarrow \dots$$
$$\rightsquigarrow_{[r_2, x\mapsto 0]} 0 \doteq_R 0$$

The first derivation does not terminate and gives no solution. The second gives the solution $(x \mapsto 0)$.

To improve the termination, we build a graph (called graph of terms) whose nodes are left-hand-sides and right-hand-sides of the rules, as well as the sides of the equation to be solved; and whose edges are the rewrite arrows $\rightarrow$ and arrows $\xrightarrow{gr}$ (gr means graph) between the syntactically unifiable terms (their variables are implicitly renamed to be disjoint):

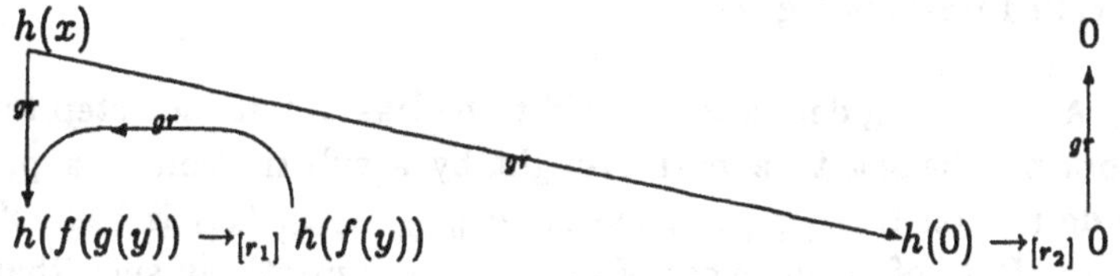

Now we apply the only rewrite rules that occur along the paths going from $h(x)$ to 0. In this graph, there is only one (by using r_2). Then the method provides the solution $(x \mapsto 0)$ and terminates. In general, the graph of terms is more complicated because some arrows may appear between subterms, and the two sides of the equations can be narrowed.

Our method is an extension of [Sivakumar-Dershowitz-87] and [Dershowitz-Sivakumar-87], where a graph of top symbols is used. Their procedure applied on the above example does not terminate, because if we only consider the top symbols, the graph becomes:

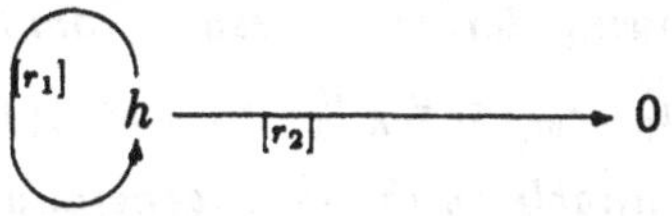

and there are infinitely many paths going from h to 0. In order to get a finite graph, we will suppose the rewriting system is finite.

After introducing preliminary notions in section 2, we present the graph building algorithm as inference rules in section 3. A narrowing procedure is given in section 4 by inference rules and using equation systems. Its completeness is proved. Two variants of an example (about lists) are shown : one which terminates (in section 5) whereas it wouln't, if some known narrowing optimizations are used; the other which doesn't terminate (in section 6), and we show how the method can be improved.

2 Preliminary notions

We assume the reader is familiar with the standard definitions of one-sorted terms, substitutions, equations, rewriting systems, complete sets of unifiers (see [Dershowitz-Jouannaud-90]).

We denote constants by a, b; function symbols by $f, g, h, \ldots$; variables by $x, y, z, \ldots$; terms by $s, t, \ldots$; occurrences (also called positions) by $u, v, w, \ldots$, and ϵ denotes the top occurrence; substitutions by $\sigma, \theta, \ldots$. The set of variables of a term t is denoted by $V(t)$. We denote by $t|_u$ the subterm of t at occurrence u, and by $t[u \leftarrow s]$ the term obtained by replacing $t|_u$ by s in t. $D(t)$ is the set of the occurrences of t, and $O(t)$ is the set of non-variable occurrences. In the whole paper, we consider a rewriting system R which contains finitely many rewrite rules. The rewriting relation is denoted by $\rightarrow$.

The narrowing relations defined below don't normalize resulting terms, in contrast with [Fay-78] and [Réty-80]. The definitions and the theorem come from [Hullot-80].

Definition 1 We say t is narrowable into t' and we write $t \rightsquigarrow_{[u,l\rightarrow r,\sigma]} t'$ if $t|_u$ and l are unifiable by the most general unifier σ, with $u \in O(t)$, and $t' = \sigma(t[u \leftarrow r])$. The relation $\rightsquigarrow$ is called **narrowing.** $\diamond$

Definition 2 A narrowing derivation is said to be **basic** if at each step the narrowing occurrence does not belong to a part brought by a substitution in a previous step. Formally the derivation $t_1 \rightsquigarrow_{[u_1,l_1\rightarrow r_1,\sigma_1]}\rightsquigarrow \cdots \rightsquigarrow_{[u_n,l_n\rightarrow r_n,\sigma_n]} t_{n+1}$ is basic if there exist some sets $U_2, \ldots, U_{n+1}$ of occurrences of $t_2, \ldots, t_{n+1}$ respectively such that for all $i \in \{2, \ldots, n\}$, $u_i \in U_i$ and

$$U_{i+1} = (U_i - \{v \in U_i / \exists w, v = u_i.w\}) \cup \{u_i.v / v \in O(r_i)\}$$

$\diamond$

The equations to be solved modulo the rewriting system R are written in the form $t \doteq_R t'$, and are considered as new terms in the signature.

Theorem 1 *If the rewriting system R is confluent and terminating, the set of substitutions θ such that*

- *there exists a basic narrowing derivation issued from $t_0 \doteq_R t'_0$*

$$t_0 \doteq_R t'_0 \rightsquigarrow_{[\sigma_1]} t_1 \doteq_R t'_1 \rightsquigarrow \ldots \rightsquigarrow_{[\sigma_n]} t_n \doteq_R t'_n$$

such that t_n and t'_n are unifiable by the most general unifier β

- *$\theta = \beta.\sigma_n \ldots \sigma_1$ and θ is normalized on $V(t_0) \cup V(t'_0)$,*

is a complete set of R-unifiers of t_0 and t'_0.

3 Building the graph of terms

3.1 Examples

In this section some graphs are shown to explain the method. They are not completely built, i.e. some arrows may be missing.

Consider a basic narrowing derivation issued from a term t_0. In each narrowing step $t_i \rightsquigarrow_{[u_i,l_i\rightarrow r_i,\sigma_i]} t_{i+1}$, the occurrence u_i does not belong to a part brought by a substitution.

Then it comes from t_0 or from a right-hand-side of a rewrite rule. Moreover $t_i|_{u_i}$ and the left-hand-side l_i are unifiable. Therefore, we can summarize this basic narrowing derivation by considering only t_0 and the rule sides. We see them as the nodes of a graph, whose edges are rewrite arrows and syntactical unification possibilities. The part brought by narrowing substitutions is omitted. This graph is finite since the set of rewrite rules is. For example, consider the rewrite rule :

$$r: \ h(x, g(y)) \to i(x, y)$$

Then the step :

$$h(f(x_1), y_1) \rightsquigarrow_{[\epsilon, r, (x \mapsto f(x_1), y_1 \mapsto g(y))]} i(f(x_1), y)$$

is summarized by the graph :

$$h(f(x_1), y_1) \xrightarrow{gr} h(x, g(y)) \to_{[r]} i(x, y)$$

The arrow $\xrightarrow{gr}$ shows that two terms are unifiable (gr means graph). Since the narrowing substitution is not applied on the terms in the graph, the last term of the narrowing step $i(f(x_1), y)$ is an instance of the term $i(x, y)$ of the graph.

Conversely, if the graph contains $\xrightarrow{gr}$ whenever two terms are unifiable, then it associates a path in the graph with every basic narrowing derivation issued from t_0. Therefore, by considering all the paths that lead to a given term s of the graph, we know all the basic derivations issued from t_0, that may lead to an instance of s.

However, some narrowing steps may be applied on subterms. For example, consider the rewriting system :

$$\begin{array}{rcrcl} r_1 & : & h_1(0, x) & \to & f(g(x)) \\ r_2 & : & g(0) & \to & h(0) \\ r_3 & : & i(f(h(0))) & \to & 0 \end{array}$$

and the basic narrowing derivation :

$$i(h_1(x_1, x_2)) \rightsquigarrow_{[1, r_1, (x_1 \mapsto 0, x_2 \mapsto x)]} i(f(g(x))) \rightsquigarrow_{[2, r_2, x \mapsto 0]} i(f(h(0))) \rightsquigarrow_{[\epsilon, r_3, Id]} 0$$

The two first steps are applied on subterms. In order to get in the graph a path at the top from $i(h_1(x_1, x_2))$ to 0, first we add a $\xrightarrow{gr}$ corresponding to the second step, then a $\xrightarrow{gr}$ corresponding to the first step. Obviously, the arrow $\xrightarrow{gr}$ from $i(h_1(x_1, x_2))$ to $i(f(h(0)))$ does not mean these two terms are unifiable, it only means $i(h_1(x_1, x_2))$ can be narrowed only on subterms into an instance of $i(f(h(0)))$. Actually, the arrows $\xrightarrow{gr}$ have two meanings.

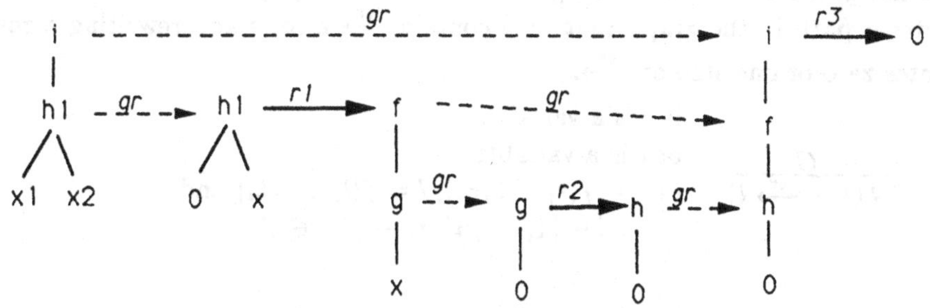

Observe that one of the $\xrightarrow{gr}$ arrows goes from a term to a subterm.

Now, let's try to solve an equation $t \doteq_R t'$. A solution is found whenever $t \doteq_R t'$ is narrowed into a syntactically unifiable equation. Then t and t' can be narrowed into unifiable terms t_2 and t'_2 :

$$t \rightsquigarrow^* \rightsquigarrow_{[\epsilon]} t_1 \rightsquigarrow^*_{[\neq\epsilon]} t_2$$
$$t' \rightsquigarrow^* \rightsquigarrow_{[\epsilon]} t'_1 \rightsquigarrow^*_{[\neq\epsilon]} t'_2$$

t_1, t'_1 are instances of right-hand-sides (or of t, t'), and we add the arrow $\xleftrightarrow{gr}$ between them in the graph to show that they can be narrowed (possibly by 0 step) only on subterms into two unifiable terms. For example consider the following graph, where the terms to be solved are encercled. Remark that adding a $\xleftrightarrow{gr}$ arrow may create another.

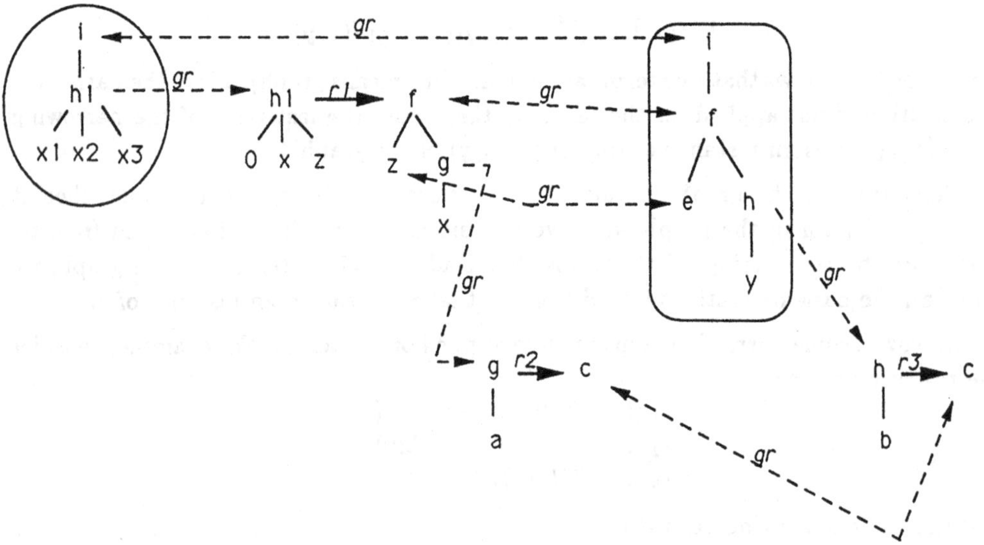

3.2 The algorithm

We present the algorithm in a simple form, via inference rules. It could be optimized to avoid useless arrows.

Notations : G denotes the current graph, l is a (sub)-left-hand-side, and r, r' are (sub)-right-hand-sides. Actually r, r', l are both pointers in the graph and terms. We assume r and r' do not point inside the same right-hand-side of the graph (except in section 6). $\longrightarrow^*$ denotes a path in the graph that may contain $\xrightarrow{gr}$ arrows and rewriting arrows. $\xleftrightarrow{gr}^?$ denotes zero or one step of $\xleftrightarrow{gr}$.

$$(\text{Add}\xrightarrow{gr}) \quad \frac{G}{G \cup \{r \xrightarrow{gr} l\}} \quad \begin{array}{l} \text{if } r \text{ is a variable} \\ \text{or } l \text{ is a variable} \\ \text{or } r = f(r_1, \ldots, r_n),\ l = f(l_1, \ldots, l_n) \text{ and} \\ \quad \forall i \in \{1, \ldots, n\}, r_i \longrightarrow^* l_i \in G \end{array}$$

$$(\text{Add}\overset{gr}{\longleftrightarrow})\quad \frac{G}{G\cup\{r\overset{gr}{\longleftrightarrow}r'\}}\quad \begin{array}{l}\text{if } r \text{ is a variable}\\ \text{or } r' \text{ is a variable}\\ \text{or } r=f(r_1,\ldots,r_n),\ r'=f(r'_1,\ldots,r'_n) \text{ and}\\ \quad \forall i\in\{1,\ldots,n\},\exists t_i,t'_i/\ r_i\longrightarrow^* t_i \overset{gr}{\longleftrightarrow}{}^{?}\ t'_i\ {}^*\!\longleftarrow r'_i\in G\end{array}$$

Let t and t' be the terms to be unified modulo the rewriting system R. The algorithm consists in two steps :

- initialization : $G \leftarrow R \cup \{t, t'$ considered as right-hand-sides$\}$
- adding arrows : apply the inferences rules as long as they add new arrows.

The termination is ensured since the graph can't contain infinitely many arrows (recall R is finite).

The graph depends on the terms to be unified. However, if one wants to unify several pairs of terms modulo the same rewriting system, one can avoid building the whole graph several times :

- Initialize the graph by R and add all the arrows.
- For each pair of terms t, t' to be unified, add t, t' into the graph and add all the arrows between t (respectively t') and others (sub)terms. Once having finished dealing with t and t', remove them out of the graph and delete the arrows that arrive at or start from them.

4 The narrowing procedure

We describe here how to use the graph of terms. For that, we present an equational formulation of narrowing, which is not identical to that of [Martelli-etal-87], because we introduce a further equation symbol $\stackrel{\rightarrow}{=}_R$ in order to avoid some redundancies. We use equational systems, called unificands, as in [Kirchner-84], except that we don't use any multiequations.

We define 5 kinds of equations, and the associated sets of solutions SOL. For terms t, t', consider :

- $t \doteq_R t'$ where $SOL(t \doteq_R t') = \{\theta \text{ normalized } | \theta(t) =_R \theta(t')\}$
- $t \stackrel{\rightarrow}{=}_R t'$ where $SOL(t \stackrel{\rightarrow}{=}_R t') = \{\theta$ normalized $|\theta(t) \rightarrow^*_R \theta(t')$ by a basic rewriting derivation$\}$
- $t \doteq t'$ where $SOL(t \doteq t') = \{\theta \text{ normalized } | \theta(t) = \theta(t')\}$
- T where $SOL(T) = \{\theta \text{ normalized}\}$
- F where $SOL(F) = \emptyset$

Let $\wedge, \vee$ be two new associative and commutative symbols of variable arity. The unificands are defined by :

- any equation $t \doteq t'$ or $t \stackrel{\rightarrow}{=}_R t'$ or $t \doteq_R t'$ or F or T is a unificand.
- if $S_1, \ldots, S_n$ are unificands, then $\wedge S_1 \ldots S_n$ and $\vee S_1 \ldots S_n$ are unificands.

A unificand in the form $\wedge S_1 \ldots S_n$ is called conjunctive factor. SOL is extended in the natural way :

$$SOL(\wedge S_1 \ldots S_n) = \cap_{i=1}^{n} SOL(S_i)$$
$$SOL(\vee S_1 \ldots S_n) = \cup_{i=1}^{n} SOL(S_i)$$

The unificands are viewed as flattened terms in a new signature. It is assumed that considered unificands are completely flattened. $\wedge, \vee$ are prefix symbols, but they may be used as infix symbols.

The procedure is given by some inference rules. They are similar to those of [Chabin-90]. Each inference rule preserves the solutions, thus the choice of the equation to be reduced is "don't care". The following property is preserved by the inference rules, up to a variable renaming :

- In an equation $t \doteq_R t'$, the terms t and t' are (sub)-right-hand-sides of the graph.
- in an equation $t \overset{\rightarrow}{=}_R t'$, the term t is a (sub)-right-hand-side of the graph, and the term t' is a (sub)-left-hand-side of the graph.

In the inference rules, we distinguish variables and non variable terms. x and y denote variables, while s, t denote non-variable terms. $top(s)$ is the top-symbol of s and $s_1, \ldots, s_n$ are its subterms. $t_1, \ldots, t_n$ are subterms of t and $l \rightarrow r$ is a rewrite rule. Recall that $\overset{gr}{\longleftrightarrow}^{?}$ means zero or one step of $\overset{gr}{\longleftrightarrow}$. The big symbol $\bigvee$ means there are several conjunctive factors in the line if there exist several paths in the graph satisfying the condition. The symbol $\longrightarrow^*$ denotes a path in the graph which may contain $\overset{gr}{\longrightarrow}$ and rewrite arrows $\rightarrow$. If an inference rule creates an empty conjunctive factor, it has to be replaced by T, and if a rule creates an empty disjunctive factor (for none of the conditions is satisfied), it has to be replaced by F. When a rewriting rule is used, its variables must be renamed (if necessary) to avoid conflicts of variables. Note that the graph is only used for the equations whose terms are both non-variable.

The narrowing rules

(Narrow-$\doteq_R$)

$$
\begin{array}{llll}
s \doteq_R t \longrightarrow & & s_1 \doteq_R t_1 \wedge \ldots \wedge s_n \doteq_R t_n & \text{if } s \overset{gr}{\longleftrightarrow} t \\
& \bigvee & s_1 \overset{\rightarrow}{=}_R l_1 \wedge \ldots \wedge s_n \overset{\rightarrow}{=}_R l_n \wedge r \doteq_R t & \text{if } s \overset{gr}{\longrightarrow} l \rightarrow r \longrightarrow^* \overset{gr}{\longleftrightarrow}^{?} {}^* \longleftarrow t \\
& \bigvee & t_1 \overset{\rightarrow}{=}_R l_1 \wedge \ldots \wedge t_n \overset{\rightarrow}{=}_R l_n \wedge s \doteq_R r & \text{if } s \overset{gr}{\longleftrightarrow}^{?} {}^* \longleftarrow r \leftarrow l \overset{gr}{\longleftarrow} t
\end{array}
$$

$$
\left.\begin{array}{l} s \doteq_R y \\ y \doteq_R s \end{array}\right. \longrightarrow \quad y \doteq s
$$

$$
\bigvee \quad s|_u \doteq l \wedge s[u \leftarrow r] \doteq_R y \qquad \text{if } u \in O(s) \text{ and } s|_u \text{ and } l \text{ have the same top symbol}
$$

$$
x \doteq_R y \longrightarrow \quad x \doteq y
$$

(Narrow- $\stackrel{\rightarrow}{=}_R$)

$$s \stackrel{\rightarrow}{=}_R t \longrightarrow \quad s_1 \stackrel{\rightarrow}{=}_R t_1 \wedge \ldots \wedge s_n \stackrel{\rightarrow}{=}_R t_n \qquad \text{if } s \xrightarrow{gr} t$$

$$\vee \quad s_1 \stackrel{\rightarrow}{=}_R l_1 \wedge \ldots \wedge s_n \stackrel{\rightarrow}{=}_R l_n \wedge r \stackrel{\rightarrow}{=}_R t \quad \text{if } s \xrightarrow{gr} l \rightarrow r \longrightarrow^* t$$

$$y \stackrel{\rightarrow}{=}_R t \longrightarrow \quad y \doteq t$$

$$s \stackrel{\rightarrow}{=}_R x \longrightarrow \quad x \doteq s$$

$$\vee \quad s|_u \doteq l \wedge s[u \leftarrow r] \stackrel{\rightarrow}{=}_R x \qquad \text{if } u \in O(s) \text{ and } s|_u \text{ and } l \text{ have the same top symbol}$$

$$y \stackrel{\rightarrow}{=}_R x \longrightarrow \quad y \doteq x$$

Now, we need some rules to simplify the unificands.

Definition 3 A conjunctive factor C is said to be in **solved form** if $C = F$ or $C = T$ or $C = \wedge\, x_1 \doteq t_1 \, \ldots \, x_n \doteq t_n$ where $\forall i,j \ \ (i \neq j \Longrightarrow x_i \neq x_j)$ and there is no cycle of variable. ◇

The simplification rules

Transforming the conjunctive factors into solved form (well known problem of syntactical unification) :

(Solve-$\doteq$) $\dfrac{\wedge t_1 \doteq t'_1 \ldots t_n \doteq t'_n}{S'}$ where $\left\{ \begin{array}{l} S' \text{ is a conjunctive factor in solved form} \\ \text{and } SOL(S') = SOL(\wedge t_1 \doteq t'_1 \ldots t_n \doteq t'_n) \end{array} \right.$

Deleting the equations that lead to non normalized solutions (they are redundant):

(Del-$\doteq$) $\dfrac{x \doteq t}{F}$ if t is not in normal form

Deleting trivial equations :

(Del-$\doteq_R$) $\dfrac{a \doteq_R a}{T}$ if a is a constant (Del- $\stackrel{\rightarrow}{=}_R$) $\dfrac{a \stackrel{\rightarrow}{=}_R a}{T}$ if a is a constant

Deleting the equations T and F :

(Del-$\vee$-T) $\dfrac{\vee S_1 \ldots S_{i-1} T S_{i+1} \ldots S_n}{T}$ (Del-$\vee$-F) $\dfrac{\vee S_1 \ldots S_{i-1} F S_{i+1} \ldots S_n}{\vee S_1 \ldots S_n}$

(Del-$\wedge$-T) $\dfrac{\wedge S_1 \ldots S_{i-1} T S_{i+1} \ldots S_n}{\wedge S_1 \ldots S_n}$ (Del-$\wedge$-F) $\dfrac{\wedge S_1 \ldots S_{i-1} F S_{i+1} \ldots S_n}{F}$

The distributivity rule

Transforming the unificands into disjunctive form :

(Distributivity) $\dfrac{\wedge S_1 \ldots S_n (\vee Q_1 \ldots Q_p)}{\vee(\wedge S_1 \ldots S_n Q_1) \ldots (\wedge S_1 \ldots S_n Q_p)}$

Remark : The narrowing relation defined by these inference rules is basic since the part brought by substitutions is inside the $\doteq$ equations (after applying (Solve-$\doteq$)), which are not narrowed. It can even be left-to-right basic [Herold-86] (or right-to-left) because of the "don't care" choice of the equation to be narrowed.

We use the following strategy to apply the rules.

Definition 4 A derivation $S_0 \longrightarrow \ldots \longrightarrow S_n \longrightarrow \ldots$ is said **fair** if it is created by the following procedure :
while the current unificand contains $\doteq_R$ or $\stackrel{\rightarrow}{=}_R$ equations do

- apply the rule (Narrow-$\doteq_R$) on all the $\doteq_R$ equations, and (Narrow-$\stackrel{\rightarrow}{=}_R$) on all the $\stackrel{\rightarrow}{=}_R$ equations,
- apply the simplifications rules.
- if you wish, transform into disjunctive form by applying (Distributivity), and then apply the simplification rules.

◇

Transforming the unificand into disjunctive form is necessary to compute the possibly solutions. The loop must not be run infinitely many times without transforming the unificand into disjunctive form.

The above procedure is complete, i.e. any solution is found in a finite time.

Theorem 2 *Let S_0 be a unificand, $\theta_0 \in SOL(S_0)$, and $S_0 \longrightarrow * \ldots \longrightarrow *S_n \longrightarrow \ldots$ a fair derivation issued from S_0, where $S_1, \ldots, S_n, \ldots$ are the unificands in disjunctive form obtained at the end of some loop steps in the procedure.*

Then there exists a unificand S_i of this derivation and a substitution θ_i such that

- *θ_i is solution of a conjunctive factor of S_i that contains no $\doteq_R$ and $\stackrel{\rightarrow}{=}_R$ equation*
- *$\theta_i = \theta_0\ [V(S_0)]$*

Proof : See [Chabin-Réty-91]. □

Intuitively, we see that if a term can be narrowed infinitely many times, then there is a cycle in the graph of terms.

Definition 5 The graph of terms is said to contain a **cycle by subterm** if there are some right-hand-sides $t_1, \ldots, t_n$ and some occurrences $u_1 \in O(t_1), \ldots, u_n \in O(t_n)$, such that there exist paths in the graph of the form $t_1|_{u_1} \longrightarrow^* t_2$, $t_2|_{u_2} \longrightarrow^* t_3$, ..., $t_n|u_n \longrightarrow^* t_1$. ◇

Conjecture 1 *(termination criterion)*
If the graph of terms doesn't have any cycles by subterm, then the above procedure terminates.

Unfortunately, there is a cycle by subterm if a function symbol is recursively defined.

5 Example

Consider a set $\{a, b\}$ (which could be extended) of constants, strictly ordered by $<$, and the lists with the symbols $\{empty, cons\}$, the booleans with $\{true, false, and\}$, and the function *is_sorted* that says whether a list is strictly sorted. The graph is below. Some arrows are missing, only the usefull arrows are indicated. The terms to be unified are encercled.

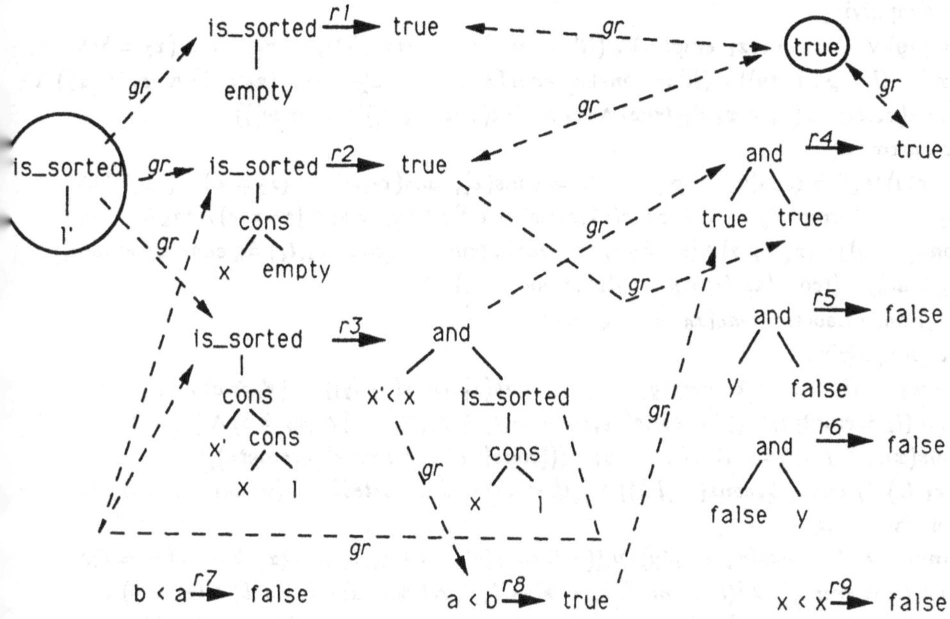

We solve $is\_sorted(l') \doteq_R true$. We obtain :

1. After narrowing

$(((l' \overset{\rightarrow}{=}_R empty) \wedge (true \doteq_R true)) \vee ((l' \overset{\rightarrow}{=}_R cons(x_1, empty)) \wedge (true \doteq_R true))$
$\vee((l' \overset{\rightarrow}{=}_R cons(x_2', cons(x_2, l_2))) \wedge ((x_2' < x_2)\ and\ is\_sorted(cons(x_2, l_2))) \doteq_R true)))$

2. After simplifying

$((l' \overset{\rightarrow}{=}_R empty) \vee (l' \overset{\rightarrow}{=}_R cons(x_1, empty)) \vee ((l' \overset{\rightarrow}{=}_R cons(x_2', cons(x_2, l_2))) \wedge$
$((x_2' < x_2)\ and\ (is\_sorted(cons(x_2, l_2))) \doteq_R true)))$

3. After narrowing

$((l' \doteq empty) \vee (l' \doteq cons(x_1, empty)) \vee ((l' \doteq cons(x_2', cons(x_2, l_2))) \wedge ((x_2' < x_2) \overset{\rightarrow}{=}_R true) \wedge$
$(is\_sorted(cons(x_2, l_2)) \overset{\rightarrow}{=}_R true) \wedge (true \doteq_R true)))$

4. After simplifying

$((l' \doteq empty) \vee (l' \doteq cons(x_1, empty)) \vee ((l' \doteq cons(x_2', cons(x_2, l_2))) \wedge ((x_2' < x_2) \overset{\rightarrow}{=}_R true) \wedge$
$(is\_sorted(cons(x_2, l_2)) \overset{\rightarrow}{=}_R true)))$

5. After narrowing

$((l' \doteq empty) \vee (l' \doteq cons(x_1, empty)) \vee ((l' \doteq cons(x_2', cons(x_2, l_2))) \wedge (x_2' \overset{\rightarrow}{=}_R a) \wedge (x_2 \overset{\rightarrow}{=}_R b) \wedge$
$(true \overset{\rightarrow}{=}_R true) \wedge (((cons(x_2, l_2) \overset{\rightarrow}{=}_R cons(x_3, empty)) \wedge (true \overset{\rightarrow}{=}_R true)) \vee$
$((cons(x_2, l_2) \overset{\rightarrow}{=}_R cons(x_4', cons(x_4, l_4)) \wedge ((x_4' < x_4) \wedge is\_sorted(cons(x_4, l_4))) \overset{\rightarrow}{=}_R true))))$

6. After simplifying and transforming into normal form
$((l' \doteq empty) \vee (l' \doteq cons(x_1, empty)) \vee ((l' \doteq cons(x_2', cons(x_2, l_2))) \wedge (x_2' \stackrel{?}{=}_R a) \wedge (x_2 \stackrel{?}{=}_R b) \wedge$
$(cons(x_2, l_2) \stackrel{?}{=}_R cons(x_3, empty))) \vee ((l' \doteq cons(x_2', cons(x_2, l_2))) \wedge (x_2' \stackrel{?}{=}_R a) \wedge (x_2 \stackrel{?}{=}_R b) \wedge$
$((cons(x_2, l_2) \stackrel{?}{=}_R cons(x_4', cons(x_4, l_4))) \wedge ((x_4' < x_4)\ andis\_sorted(cons(x_4, l_4))) \stackrel{?}{=}_R true))))$
7. After narrowing
$((l' \doteq empty) \vee (l' \doteq cons(x_1, empty)) \vee ((l' \doteq cons(x_2', cons(x_2, l_2))) \wedge (x_2' \doteq a) \wedge (x_2 \doteq b) \wedge$
$(x_2 \stackrel{?}{=}_R x_3) \wedge (l_2 \stackrel{?}{=}_R empty)) \vee ((l' \doteq cons(x_2', cons(x_2, l_2))) \wedge (x_2' \doteq a) \wedge (x_2 \doteq b) \wedge (x_2 \stackrel{?}{=}_R x_4') \wedge$
$(l_2 \stackrel{?}{=}_R cons(x_4, l_4)) \wedge (x_4' < x_4 \stackrel{?}{=}_R true) \wedge (is\_sorted(cons(x_4, l_4))) \stackrel{?}{=}_R true) \wedge (true \stackrel{?}{=}_R true)))$
8. After simplifying
$((l' \doteq empty) \vee (l' \doteq cons(x_1, empty)) \vee ((l' \doteq cons(x_2', cons(x_2, l_2))) \wedge (x_2' \doteq a) \wedge (x_2 \doteq b) \wedge$
$(x_2 \stackrel{?}{=}_R x_3) \wedge (l_2 \stackrel{?}{=}_R empty)) \vee ((l' \doteq cons(x_2', cons(x_2, l_2))) \wedge (x_2' \doteq a) \wedge (x_2 \doteq b) \wedge (x_2 \stackrel{?}{=}_R x_4') \wedge$
$(l_2 \stackrel{?}{=}_R cons(x_4, l_4)) \wedge (x_4' < x_4 \stackrel{?}{=}_R true) \wedge (is\_sorted(cons(x_4, l_4))) \stackrel{?}{=}_R true)))$
9. After narrowing
$((l' \doteq empty) \vee (l' \doteq cons(x_1, empty)) \vee ((l' \doteq cons(x_2', cons(x_2, l_2))) \wedge (x_2' \doteq a) \wedge (x_2 \doteq b) \wedge$
$(x_2 \doteq x_3) \wedge (l_2 \doteq empty)) \vee ((l' \doteq cons(x_2', cons(x_2, l_2))) \wedge (x_2' \doteq a) \wedge (x_2 \doteq b) \wedge (x_2 \doteq x_4') \wedge$
$(l_2 \doteq cons(x_4, l_4)) \wedge (x_4' \stackrel{?}{=}_R a) \wedge (x_4 \stackrel{?}{=}_R b) \wedge (true \stackrel{?}{=}_R true) \wedge (((cons(x_4, l_4) \stackrel{?}{=}_R cons(x_5, empty)) \wedge$
$(true \stackrel{?}{=}_R true)) \vee ((cons(x_4, l_4) \stackrel{?}{=}_R cons(x_6', cons(x_6, l_6))) \wedge$
$((x_6' < x_6)\ and\ is\_sorted(cons(x_6, l_6)) \stackrel{?}{=}_R true)))$
10. After simplifying
$((l' \doteq empty) \vee (l' \doteq cons(x_1, empty)) \vee ((l' \doteq cons(x_2', cons(x_2, l_2))) \wedge (x_2' \doteq a) \wedge (x_2 \doteq b) \wedge$
$(x_3 \doteq b) \wedge (l_2 \doteq empty)) \vee ((l' \doteq cons(x_2', cons(x_2, l_2))) \wedge (x_2' \doteq a) \wedge (x_2 \doteq b) \wedge (x_4' \doteq b) \wedge$
$(l_2 \doteq cons(x_4, l_4)) \wedge (x_4' \stackrel{?}{=}_R a) \wedge (x_4 \stackrel{?}{=}_R b) \wedge (((cons(x_4, l_4) \stackrel{?}{=}_R cons(x_5, empty))) \vee$
$((cons(x_4, l_4) \stackrel{?}{=}_R cons(x_6', cons(x_6, l_6))) \wedge ((x_6' < x_6)\ and\ is\_sorted(cons(x_6, l_6)) \stackrel{?}{=}_R true)))$
11. After narrowing
$((l' \doteq empty) \vee (l' \doteq cons(x_1, empty)) \vee ((l' \doteq cons(x_2', cons(x_2, l_2))) \wedge (x_2' \doteq a) \wedge (x_2 \doteq b) \wedge$
$(x_3 \doteq b) \wedge (l_2 \doteq empty)) \vee ((l' \doteq cons(x_2', cons(x_2, l_2))) \wedge (x_2' \doteq a) \wedge (x_2 \doteq b) \wedge (x_4' \doteq b) \wedge$
$(l_2 \doteq cons(x_4, l_4)) \wedge (x_4' \doteq a) \wedge (x_4 \doteq b) \wedge (((x_4 \stackrel{?}{=}_R x_5) \wedge (l_4 \stackrel{?}{=}_R empty)) \vee ((x_4 \stackrel{?}{=}_R x_6') \wedge$
$(l_4 \stackrel{?}{=}_R cons(x_6, l_6)) \wedge (x_6' < x_6 \stackrel{?}{=}_R true) \wedge (is\_sorted(cons(x_6, l_6)) \stackrel{?}{=}_R true) \wedge (true \stackrel{?}{=}_R true)))$
12. After simplifying
$((l' \doteq empty) \vee (l' \doteq cons(x_1, empty)) \vee ((l' \doteq cons(x_2', cons(x_2, l_2))) \wedge (x_2' \doteq a) \wedge (x_2 \doteq b) \wedge$
$(x_3 \doteq b) \wedge (l_2 \doteq empty)))$

The procedure terminates, although there is a cycle by subterm in the graph, and we obtain 3 solutions, from which 4 closed solutions can be deduced.

Consider the infinite derivation :

$$is\_sorted(l') \leadsto_{[r_3]} (x' < x)\ and\ is\_sorted(cons(x, l))$$
$$\leadsto (x' < x)\ and\ (x_1' < x_1\ and\ is\_sorted(cons(x_1, l_1))) \leadsto \ldots$$

It is left-to-right basic, and every term is in normal form. Therefore none among the ordinary, basic, left-to-right basic, normalizing, basic normalizing narrowings terminates.

6 Improvements and conclusion

The previous example works well because most rewriting rules have closed terms as right-hand-side. Unfortunately, if we replace the rule r_4 by $(true\ and\ y) \rightarrow y$, the unificand obtained at step 3 contains the conjunctive factor :

$$((l' \doteq cons(x_2', cons(x_2, l_2))) \wedge ((x_2' < x_2) \stackrel{\rightarrow}{=}_R true) \wedge (is\_sorted(cons(x_2, l_2)) \stackrel{\rightarrow}{=}_R y) \wedge (y \doteq_R true)))$$

Then $y \doteq_R true$ is transformed into $y \doteq true$, and $is\_sorted(cons(x_2, l_2)) \stackrel{\rightarrow}{=}_R y$ can be narrowed infinitely many times. Therefore our procedure does not terminate.

Actually, we need to take into account the fact that y is equal to $true$, i.e. by solving $is\_sorted(cons(x_2, l_2)) \stackrel{\rightarrow}{=}_R true$, which terminates. For that, we propose to introduce merging rules, like

$$s \stackrel{\rightarrow}{=}_R x \wedge x \doteq t \longrightarrow s \stackrel{\rightarrow}{=}_R t \wedge x \doteq t$$
$$s \doteq_R x \wedge x \doteq t \longrightarrow s \doteq_R t \wedge x \doteq t$$
$$s \stackrel{\rightarrow}{=}_R x \wedge x \doteq_R t \longrightarrow s \doteq_R t \wedge x \doteq_R t$$

and so on. Thus, we don't have to narrow equations in the form $s \stackrel{\rightarrow}{=}_R x$ or $x \doteq_R t$ if $x \notin s$. So, we hope the method will often terminate.

Building the graph of terms obviously spends a long time. By using a graph of top symbols, as in [Dershowitz-Sivakumar-87], it would spend less time. But it would terminate less often.

7 References

[Chabin-90] Jacques Chabin. Surréduction dirigée par un graphe d'opérateurs. Formulation équitable par des systèmes d'équations et implantation. Rapport de DEA, université d'Orléans, Septembre 1990. (In french).

[Chabin-Réty-91] J. Chabin and P. Réty. Narrowing directed by a graph of terms (including proofs). Internal report 91-1, LIFO, Université d'Orléans. 1991.

[Dershowitz-Sivakumar-87] N. Dershowitz and G. Sivakumar. Solving goals in equational langages. Proceedings of the first workshop CTRS. Springer-Verlag, vol 308. 1987.

[Dershowitz-Jouannaud-90] N. Dershowitz and J.P. Jouannaud. Rewrite system. Handbook of Theoretical Computer Science, Vol B, North-Holland, 1990.

[Fay-78] M. Fay. First-Order Unification in an Equational Theory. Master Thesis 78-5-002. University of Santa Cruz. 1978.

[Goguen-meseguer-86] J.A. Goguen and J. Meseguer. EQLOG : Equality, Types and generic Modules for logic Programming. In logic programming: functions, relations and equations. D. Degroot and G. Lindstrom, eds. Prentice-Hall, Englewood Cliffs, NJ, pp 295-363, 1986.

[Herold-86] A. Herold. Narrowing Techniques applied to idempotent Unification. Internal SEKI Report SR-86-16. August 1986.

[Hullot-80] J.M. Hullot. Canonical forms and unification. Proceedings of the fifth Conference on Automated Deduction, Springer-Verlag, vol 87, July 1980.

[Kirchner-84] C. Kirchner. A new equational unification method : a generalization of Martelli-Montanari's algorithm. Proceedings of the 7th CADE. Springer-Verlag, vol 170. 1984.

[Martelli-etal-87] A. Martelli, C. Moiso, and G-F. Rossi. Lazy unification algorithms for canonical rewrite systems. Proceedings of CREAS, Austin, Texas. May 1987.

[Réty-87] P. Réty. Improving basic Narrowing Techniques. Proceedings of the second conference on Rewriting Techniques and Applications, Springer-Verlag, vol 256. May 1987.

[Sivakumar-Dershowitz-87] G. Sivakumar and N. Dershowitz. Goal directed equation solving. Technical report. University of Illinois, USA. 1987.

Adding Homomorphisms to Commutative/Monoidal Theories
or
How Algebra Can Help in Equational Unification

Franz Baader Werner Nutt
German Research Center for Artificial Intelligence (DFKI)
Postfach 2080, D-6750 Kaiserslautern, Germany
e-mail: {baader, nutt}@dfki.uni-kl.de

Abstract

In this paper we consider the class of theories for which solving unification problems is equivalent to solving systems of linear equations over a semiring. This class has been introduced by the authors independently of each other as commutative theories (Baader) and monoidal theories (Nutt). The class encompasses important examples like the theories of abelian monoids, idempotent abelian monoids, and abelian groups.

We identify a large subclass of commutative/monoidal theories that are of unification type zero by studying equations over the corresponding semiring. As a second result, we show with methods from linear algebra that unitary and finitary commutative/monoidal theories do not change their unification type when they are augmented by a finite monoid of homomorphisms, and how algorithms for the extended theory can be obtained from algorithms for the basic theory. The two results illustrate how using algebraic machinery can lead to general results and elegant proofs in unification theory.

1 Introduction

In unification theory, a common approach to devising algorithms is to consider classes of theories which are defined by syntactic properties of the set of axioms (see e.g., [KK90]). Such algorithms mostly depend on transformations of terms, but do not take into account the properties of the algebras defined by the theory. On the other hand, special purpose algorithms designed for theories of practical importance—such as the theories of boolean rings, abelian monoids (AM), idempotent abelian monoids (AIM), and abelian groups (AG)—usually depend on algebraic properties of these theories.

The theories AM, AIM, and AG belong to the class of commutative theories, that has been defined by categorical properties of the corresponding free algebras [Ba89a, Ba89b, Ba90]. It turns out (see Section 3 below) that the class of commutative theories is—modulo a translation of the signature—the same as the class of monoidal theories [Nu88, Nu90].

Unification in these theories can always be reduced to solving linear equations in semirings [Nu88]. On the one hand, this fact can be used to derive general results on unification in commutative/monoidal theories. For example, it can be shown that constant free unification problems are either unitary or of type zero, and the unification type of a theory can be characterized by algebraic properties of the corresponding semiring. These characterizations were used in [Nu88, Ba89b, Nu90] to determine the unification types of several commutative/monoidal theories. On the other hand, unification algorithms for certain commutative/monoidal theories—for example, the theory of abelian groups with n commuting homomorphisms—can be derived with the help of well-known algebraic methods for the corresponding semiring—for instance, Buchberger's algorithm for the ring $\mathbf{Z}[X_1, \ldots, X_n]$ of integer polynomials in n indeterminates [Ba90].

Let us now reconsider two of the examples in [Ba89b, Ba90]. Using algebraic properties of the semiring of polynomials with nonnegative integer coefficients, $\mathbf{N}[X]$, it was shown in [Ba90] that the corresponding theory, i.e., the theory of abelian monoids with a homomorphism, is of unification type zero. In contrast, the theory of abelian monoids with an involution[1] is unitary (finitary w.r.t. unification with constants). In both cases, the corresponding semiring has a specific structure: it is a monoid semiring $\mathcal{S}\langle H\rangle$, i.e., a semiring $\mathcal{S}$ with an adjoint monoid H. In the first example, the monoid H is the free monoid in one generator, which is an infinite monoid, while in the second example, we have the cyclic group of order two, which is finite. In both examples, the semiring $\mathcal{S}$ is the semiring $\mathbf{N}$ of all nonnegative integers. This semiring corresponds to the theory AM of commutative monoids, which is a finitary commutative/monoidal theory.

In the present paper we shall consider commutative/monoidal theories where the corresponding semiring is a monoid semiring $\mathcal{S}\langle H\rangle$ more closely. The result for the theory of abelian monoids with a homomorphism can now be generalized to a whole class of theories as follows. If $\mathcal{S}$ is a *strict semiring*—i.e., a semiring which is not a ring—and H is a *free monoid* then the corresponding commutative/monoidal theory is of unification type zero. On the other hand, assume that $\mathcal{S}$ is a semiring such that unification in the corresponding commutative/monoidal theory is unitary (finitary w.r.t. unification with constants), and let H be a *finite monoid.* In that case, the theory corresponding to the semiring $\mathcal{S}\langle H\rangle$ is also of unification type unitary (finitary w.r.t. unification with constants). This generalizes the result for the theory of abelian monoids with an involution. Moreover, a finite unification algorithm for the theory corresponding to $\mathcal{S}$ can be used to derive a finite unification algorithm for the theory corresponding to $\mathcal{S}\langle H\rangle$.

The paper is organized as follows. After recalling some basic definitions concerning equational theories, unification, and semirings in Section 2, we shall introduce in Section 3 commutative and monoidal theories and prove their equivalence. In Section 4 we shall recall the algebraic characterizations of the unification types for these theories. The last two sections contain the exact formulations and the proofs of the two general results mentioned above. An extended version of this paper has appeared as technical report [BN90].

2 Basic Definitions

We assume that the reader is familiar with the basic notions of unification theory (see [Si89]). The composition of mappings is written from left to right, that is, $\phi\psi$ means first ϕ and then ψ. Consequently, we use suffix notation for mappings (but not for function symbols in terms).

2.1 Equational Theories and Unification

We assume that two disjoint infinite sets of symbols are given, a set of function symbols and a set of variables. A *signature* Σ is a finite set of function symbols each of which is associated with its arity. Every signature Σ determines a class of Σ-*algebras* and Σ-*homomorphisms*. We define Σ-terms and Σ-substitutions as usual. By $[x_1/t_1, \ldots, x_n/t_n]$ we denote the substitution which replaces the variables x_i by the terms t_i.

An *equational theory* $\mathcal{E} = (\Sigma, E)$ is a pair consisting of a signature Σ and a set of identities E. The equality of Σ-terms induced by $\mathcal{E}$ will be denoted by $=_{\mathcal{E}}$. If X is a set of variables then we denote the free $\mathcal{E}$-algebra with generators X by $\mathcal{F}_{\mathcal{E}}(X)$.

[1] An involution is a homomorphism h satisfying $h^2(x) = x$.

Let $\mathcal{E} = (\Sigma, E)$ be an equational theory. An *$\mathcal{E}$-unification problem* is a finite sequence of equations $\Gamma = \langle s_i \doteq t_i \mid 1 \le i \le n \rangle$, where s_i and t_i are Σ-terms. A Σ-substitution θ is called an *$\mathcal{E}$-unifier* of Γ if $s_i\theta =_{\mathcal{E}} t_i\theta$ for each i. The set of all $\mathcal{E}$-unifiers of Γ is denoted by $U_{\mathcal{E}}(\Gamma)$. In general one does not need the set of all $\mathcal{E}$-unifiers. A complete set of $\mathcal{E}$-unifiers, i.e., a set of $\mathcal{E}$-unifiers from which all unifiers may be generated by $\mathcal{E}$-instantiation, is usually sufficient. More precisely, for every set of variables V we extend $=_{\mathcal{E}}$ to a relation $=_{\mathcal{E},V}$ between substitutions, and introduce the *$\mathcal{E}$-instantiation quasi-ordering* $\le_{\mathcal{E},V}$ as follows: we write $\sigma =_{\mathcal{E},V} \theta$ if $x\sigma =_{\mathcal{E}} x\theta$ for all $x \in V$, and we write $\sigma \le_{\mathcal{E},V} \theta$ if there exists a substitution λ such that $\theta =_{\mathcal{E},V} \sigma \circ \lambda$. A set $C \subseteq U_{\mathcal{E}}(\Gamma)$ is a *complete set of $\mathcal{E}$-unifiers* of Γ if for every unifier θ of Γ there exists $\sigma \in C$ such that $\sigma \le_{\mathcal{E},V} \theta$, where V is the set of variables occurring in Γ. For reasons of efficiency, this set should be as small as possible. Thus one is interested in *minimal* complete sets of $\mathcal{E}$-unifiers. In minimal complete sets two different elements are not comparable w.r.t. $\mathcal{E}$-instantiation.

The *unification type* of a theory $\mathcal{E}$ is defined with reference to the existence and cardinality of minimal complete sets. The theory $\mathcal{E}$ is *unitary* (*finitary*, *infinitary*, respectively) if minimal complete sets of $\mathcal{E}$-unifiers always exist, and their cardinality is at most one (always finite, at least once infinite, respectively). The theory $\mathcal{E}$ is of *unification type zero* if there exists an $\mathcal{E}$-unification problem without a minimal complete set of $\mathcal{E}$-unifiers.

If the terms in the unification problems may contain free constants, we talk about *unification with constants*, otherwise we talk about unification without constants. In this version of the paper we consider only unification without constants. The constant case is treated in the report version [BN90].

2.2 Semirings

A *semiring* $\mathcal{S}$ is a tuple $(\mathcal{S}, +, 0, \cdot, 1)$ such that $(\mathcal{S}, +, 0)$ is an abelian monoid, $(\mathcal{S}, \cdot, 1)$ is a monoid, and all q, r, $s \in \mathcal{S}$ satisfy the equalities

1. $(q + r) \cdot s = q \cdot s + r \cdot s$
2. $q \cdot (r + s) = q \cdot r + q \cdot s$
3. $0 \cdot s = s \cdot 0 = 0.$

The elements 0 and 1 are called *zero* and *unit*. Semirings are different from rings in that they need not be groups w.r.t. addition. Obviously, any ring is a semiring. A prominent example for a semiring which is not a ring is the semiring $\mathbf{N}$ of nonnegative integers.

Similar to the construction of polynomial rings over a given ring, one can use a semiring $\mathcal{S}$ and a monoid H to construct a new semiring, namely the *monoid semiring* $\mathcal{S}\langle H \rangle$. As for polynomials, the elements of the monoid semiring may be represented as sums of the form $\sum_{h \in H} s_h \cdot h$ where only finitely many of the coefficients $s_h \in \mathcal{S}$ are nonzero. The zero element of $\mathcal{S}\langle H \rangle$ is the sum where all the coefficients are zero, and the unit element is the sum where only the unit of H has a coefficient different from zero and this coefficient is the unit element of $\mathcal{S}$. Addition and multiplication in $\mathcal{S}\langle H \rangle$ are defined as follows:

$$\sum_{h \in H} s_h \cdot h + \sum_{h \in H} t_h \cdot h = \sum_{h \in H} (s_h + t_h) \cdot h$$
$$\sum_{f \in H} s_f \cdot f \cdot \sum_{g \in H} t_g \cdot g = \sum_{h \in H} \Big(\sum_{h = fg} s_f \cdot t_g \Big) \cdot h.$$

Polynomial semirings are special cases of monoid semirings. For example, the ring $\mathbf{Z}[X_1, \ldots, X_n]$ of integer polynomials in n indeterminates is the monoid semiring $\mathbf{Z}\langle FAM_n \rangle$ where FAM_n denotes the free abelian monoid in n generators.

As mentioned in the introduction, unification in commutative/monoidal theories can be reduced to solving systems of linear equations in semirings. Similarly to unification in abelian monoids [LS75], problems without constants correspond to systems of homogeneous equations. For problems with constants one has to solve in addition systems of inhomogeneous equations.

Modules over semirings are a generalization of vector spaces over fields. Since $(\mathcal{S}, \cdot, 1)$ need not be commutative, we have to distinguish between left and right $\mathcal{S}$-modules. Solutions of homogeneous systems form right $\mathcal{S}$-modules. The unification type of a theory will depend on whether these modules are finitely generated or not. The n-fold cartesian product $\mathcal{S}^n$ is a left (right) $\mathcal{S}$-module if semiring elements are multiplied to vectors by from the left (right). A subset M of $\mathcal{S}^n$ is a *finitely generated* right $\mathcal{S}$-module if there exist finitely many $x_1, \ldots, x_k \in \mathcal{S}^n$ such that $M = \{x_1 s_1 + \cdots + x_k s_k \mid s_1, \ldots, s_k \in \mathcal{S}\}$.

3 Commutative and Monoidal Theories

In this section we shall give the definitions of commutative and monoidal theories, and show in what sense these two notions are equivalent.

3.1 Definitions and Examples

Originally commutative theories have been defined in the framework of a categorical reformulation of unification. It has been shown that the categorical definition is equivalent to an algebraic one [Ba89a], which will be given below.

Let $\mathcal{E} = (\Sigma, E)$ be an equational theory. A constant symbol e of the signature Σ is called *idempotent in* $\mathcal{E}$ if for all symbols $f \in \Sigma$ we have $f(e, \ldots, e) =_{\mathcal{E}} e$. Note that for nullary f this means $f =_{\mathcal{E}} e$. A family $o = (o_X)_X$ of mappings $o_X: \mathcal{F}_{\mathcal{E}}(X)^n \rightarrow \mathcal{F}_{\mathcal{E}}(X)$, where X ranges over all sets of variables, is an *n-ary implicit operation* if the following holds: there exist a Σ-term t and a sequence $x_1, \ldots, x_n$ of distinct variables which contains all the variables occurring in t such that $o_X(s_1, \ldots, s_n) = t[x_1/s_1, \ldots, x_n/s_n]$ for all $s_1, \ldots, s_n \in \mathcal{F}_{\mathcal{E}}(X)$. For example, if $\mathcal{E}$ is the theory of groups where the signature contains a binary symbol "$\cdot$" and a unary symbol "$^{-1}$", then the binary implicit operation given by $(x \cdot y^{-1}; x, y)$ expresses division in a group.

We are now ready to give an algebraic definition of commutative theories. An equational theory $\mathcal{E} = (\Sigma, E)$ is called *commutative* if the following holds:

1. the signature Σ contains a constant symbol e which is idempotent in $\mathcal{E}$
2. there is a binary implicit operation "$*$" in $\mathcal{F}(\mathcal{E})$ such that the constant e is a neutral element for "$*$" in any algebra $\mathcal{F}_{\mathcal{E}}(X)$, and for any n-ary function symbol $f \in \Sigma$, any algebra $\mathcal{F}_{\mathcal{E}}(X)$, and any $s_1, \ldots, s_n, t_1, \ldots, t_n \in \mathcal{F}_{\mathcal{E}}(X)$ we have $f(s_1 * t_1, \ldots, s_n * t_n) = f(s_1, \ldots, s_n) * f(t_1, \ldots, t_n)$.

Though it is not explicitly required by the definition, the implicit operation "$*$" turns out to be associative and commutative [Ba89a]. This justifies the name "commutative theory."

Well-known examples of commutative theories are the theory AM of abelian monoids, the theory AIM of idempotent abelian monoids (sometimes called AC1 in the literature), and the theory AG of abelian groups (see [Ba89a]). In these theories, the implicit operation "$*$" is given by the explicit binary operation in the signature. An example for a commutative theory where "$*$" is really implicit can also be found in [Ba89a] (Example 5.1). We shall now consider examples of commutative theories where the signature contains some additional function symbols (see [Ba90, Nu90] for more examples).

Examples 3.1 We consider the following signatures: $\Sigma := \{+, 0, h\}$, where "+" is binary, 0 is nullary, and h is unary; $\Delta := \{+, 0, f\}$, where "+" is binary, 0 is nullary, and f is binary; and $\Omega := \{+, 0, -, i\}$, where "+" is binary, 0 is nullary, and $-$ and i are unary.

AMH = (Σ, E_{AMH}), the theory of *abelian monoids with a homomorphism.* E_{AMH} consists of the identities which state that "+" is associative, commutative with neutral element "0", and the identities which state that h is a homomorphism, i.e., the identities $h(x+y) \doteq h(x)+h(y)$, $h(0) \doteq 0$.

AMIn = $(\Sigma, E_{\text{AMIn}})$, the theory of *abelian monoids with an involution.* E_{AMIn} consists of the identities of E_{AMH}, and the additional identity $h(h(x)) \doteq x$, which states that h is an involution.

COM = (Δ, E_{COM}). E_{COM} consists of the identities which state that "+" is associative, commutative with neutral element 0, and the identities $f(x+x', y+y') \doteq f(x,y)+f(x',y')$ and $f(0,0) \doteq 0$ which ensure that COM is really commutative.

GAUSS = $(\Omega, E_{\text{GAUSS}})$. E_{GAUSS} consists of the identities which state that "+" is the binary operation of an abelian group with neutral element 0 and inverse $-$, and the additional identity $x + i(i(x)) \doteq 0$.

With the exception of the third example, the additional function symbols—i.e., the function symbols apart from the binary symbol yielding the implicit operation, and the idempotent constant symbol—are all unary symbols. This motivates the definition of monoidal theories. An equational theory $\mathcal{E} = (\Sigma, E)$ is *monoidal* if

1. Σ contains a constant symbol 0, a binary function symbol "+", and all the other symbols in Σ are unary

2. "+" is associative and commutative, 0 is the neutral element for "+", i.e. $0+x =_{\mathcal{E}} x+0 =_{\mathcal{E}} x$, and every unary symbol h is a homomorphism for "+" and 0, i.e. $h(x+y) =_{\mathcal{E}} h(x)+h(y)$ and $h(0) =_{\mathcal{E}} 0$.

It is easy to see that monoidal theories are always commutative theories. Obviously, the theories AM, AIM, AG, AMH, AMIn, and GAUSS are monoidal. The theory COM is not monoidal, since its signature contains an additional *binary* function symbol. However, we shall see in the next subsection that COM may also be regarded as monoidal theory if the signature is translated appropriately.

3.2 Adding Monoids of Homomorphisms

There is an interesting difference between the theory GAUSS on the one hand, and the theories AMH and AMIn on the other hand. The additional identity $x+i(i(x)) \doteq 0$ in the theory GAUSS establishes a closer connection between the unary symbol i and the binary symbol "+" than just the fact that i is a homomorphism for "+". This is not the case for the additional identity $h(h(x)) \doteq h(x)$ in AMIn which says something about h alone. This observation will now be put into a more general setting.

Let $\mathcal{E} = (\Sigma, E)$ be a monoidal theory, and let H be a monoid generated by the finitely many elements $h_1, \ldots, h_n$. We define the augmented theory $\mathcal{E}\langle H\rangle = (\Sigma', E')$ as follows: the signature Σ' extends Σ by the unary function symbols $h_1, \ldots, h_n$; the set of identities E' extends E with the identities which state that $h_1, \ldots, h_n$ are homomorphisms, and the identities $\{h_{i_1}(\ldots h_{i_k}(x)\ldots) \doteq h_{j_1}(\ldots h_{j_l}(x)\ldots) \mid h_{i_1}\ldots h_{i_k} = h_{j_1}\ldots h_{j_l}$ holds in $H\}$. In Sections 5 and 6 we shall study unification in theories of the form $\mathcal{E}\langle H\rangle$.

The theory AMH is AM$\langle h^*\rangle$ where h^* stands for the free monoid in one generator, and AMIn is AM$\langle Z_2\rangle$ where Z_2 stands for the cyclic group of order 2, i.e., Z_2 consists of two elements e and h, and the multiplication in Z_2 is defined as $e\cdot e = e$, $h\cdot e = e\cdot h = h$, and $h\cdot h = e$. On the other hand, one can prove that GAUSS cannot be represented in the form AG$\langle H\rangle$ because of the interaction between i and "+" stated by $x + i(i(x)) \doteq 0$.

3.3 Equivalence of Commutative and Monoidal Theories

Next we show that by means of a signature transformation every commutative theory can be turned into a monoidal theory that, from the viewpoint of unification, is equivalent.

Let Σ and Σ' be signatures. A *signature transformation from Σ' to Σ* is a mapping θ that associates to every Σ'-term a Σ-term such that

1. $x\theta = x$ for every variable x
2. $f(t_1, \ldots, t_n)\theta = (f(x_1, \ldots, x_n)\theta)[x_1/t_1\theta, \ldots, x_n/t_n\theta]$ if f is an n-ary symbol and $x_1, \ldots, x_n$ are n distinct variables.

It follows from the definition that θ is completely defined by the images of the flat terms $f(x_1, \ldots, x_n)$ where f ranges over Σ'. Intuitively, θ interprets every Σ'-symbol by a Σ-term, and then extends this interpretation consistently to arbitrary Σ'-terms.

To every commutative theory $\mathcal{E} = (\Sigma, E)$ we associate a theory $\widehat{\mathcal{E}} = (\widehat{\Sigma}, \widehat{E})$ and a signature transformation θ from $\widehat{\Sigma}$ to Σ as follows. The signature $\widehat{\Sigma}$ consists of a constant 0, a binary symbol "+", and unary symbols $f_1, \ldots, f_n$ for every n-ary symbol $f \in \Sigma$, where $n \geq 1$. To define the set of identities $\widehat{E}$ we need the transformation θ. Let e be the idempotent constant in $\mathcal{E}$ and let $(t_*; x, y)$ be the pair corresponding to the implicit operation "$*$" in $\mathcal{E}$. We define θ by $0\theta := e$, $(x + y)\theta := t_*$, and $f_i(x)\theta := f(e, \ldots, x, \ldots, e)$, where $f(e, \ldots, x, \ldots, e)$ has the variable x in the i-th argument position and the constant e in the other positions. Now, with the help of this signature transformation we define $\widehat{E}$ as $\widehat{E} := \{\hat{s} \doteq \hat{t} \mid \hat{s}\theta =_{\mathcal{E}} \hat{t}\theta\}$. That is, $\widehat{E}$ is the preimage of $=_{\mathcal{E}}$ under θ.

Proposition 3.2 *Let $\mathcal{E} = (\Sigma, E)$ be a commutative theory with associated theory $\widehat{\mathcal{E}} = (\widehat{\Sigma}, \widehat{E})$ and signature transformation θ. Then:*

1. *$\widehat{\mathcal{E}}$ is a monoidal theory*
2. *$\hat{s} =_{\widehat{\mathcal{E}}} \hat{t}$ implies $\hat{s}\theta =_{\mathcal{E}} \hat{t}\theta$ for all $\widehat{\Sigma}$-terms $\hat{s}$, $\hat{t}$.*

Let $\mathcal{E} = (\Sigma, E)$ and $\mathcal{E}' = (\Sigma', E')$ be equational theories. We say that $\mathcal{E}$ and $\mathcal{E}'$ are *equivalent* if there exist signature transformations θ' from Σ to Σ' and θ from Σ' to Σ such that

1. $s =_{\mathcal{E}} t$ implies $s\theta' =_{\mathcal{E}'} t\theta'$ for all Σ-terms s and t and $s' =_{\mathcal{E}'} t'$ implies $s'\theta =_{\mathcal{E}} t'\theta$ for all Σ'-terms s' and t'
2. $s\theta'\theta =_{\mathcal{E}} s$ for all Σ-terms s, and $s'\theta\theta' =_{\mathcal{E}'} s'$ for all Σ'-terms s'.

The first condition means that θ and θ' can be seen as mappings on equivalence classes of terms. The second says that θ and θ' are inverses of each other modulo the equational theories.

One of the most prominent examples of equivalent theories are boolean rings and boolean algebras. If two theories are equivalent they describe essentially the same structures. Therefore equivalent theories share the same unification properties. The next theorem says that, in particular, from the viewpoint of unification there is no difference between commutative and monoidal theories.

Theorem 3.3 *Let $\mathcal{E} = (\Sigma, E)$ be a commutative theory with associated theory $\widehat{\mathcal{E}} = (\widehat{\Sigma}, \widehat{E})$. Then $\mathcal{E}$ and $\widehat{\mathcal{E}}$ are equivalent.*

Proof. Let θ be the signature transformation from $\widehat{\Sigma}$ to Σ. To show the equivalence of $\mathcal{E}$ and $\widehat{\mathcal{E}}$ we exhibit a signature transformation $\hat{\theta}$ from Σ to $\widehat{\Sigma}$. Define $\hat{\theta}$ by $e\hat{\theta} = 0$, and $f(x_1, \ldots, x_n)\hat{\theta} = f_1(x_1) + \cdots + f_n(x_n)$ for every n-ary symbol f in Σ. Then it is straightforward to see that θ and $\hat{\theta}$ have the required properties. □

4 Unification in Commutative/Monoidal Theories

We first show how commutative/monoidal theories give rise to semirings. Suppose $\mathcal{E} = (\Sigma, E)$ is a commutative/monoidal theory with idempotent constant 0 and implicit binary operation "+". Let x be a variable and $\mathcal{S}_\mathcal{E}$ be the set of Σ-homomorphisms from $\mathcal{F}_\mathcal{E}(x)$ to $\mathcal{F}_\mathcal{E}(x)$. Note that every such homomorphism σ is uniquely determined by the term $x\sigma$. We define an addition on $\mathcal{S}_\mathcal{E}$ by $x(\sigma + \tau) := x\sigma + x\tau$ for all $\sigma, \tau \in \mathcal{S}_\mathcal{E}$. This addition is associative and commutative, and has the constant mapping 0 as neutral element. The composition "$\cdot$" of homomorphisms is an associative multiplication on $\mathcal{S}_\mathcal{E}$ for which the identity mapping *id* is a neutral element. Moreover, the laws of distributivity and the zero laws hold.

Proposition 4.1 *If $\mathcal{E}$ is a commutative/monoidal theory, then $(\mathcal{S}_\mathcal{E}, +, 0, \cdot, id)$ is a semiring.*

As an example, consider an arbitrary Σ-homomorphism $\sigma: \mathcal{F}_{\mathrm{AMH}}(x) \rightarrow \mathcal{F}_{\mathrm{AMH}}(x)$. Then there exist $a_0, \ldots, a_k \in \mathbf{N}$ such that $x\sigma =_{\mathrm{AMH}} a_0 x + a_1 h(x) + \ldots + a_k h^k(x)$. We associate with the morphism σ the polynomial $a_0 + a_1 X + \ldots + a_k X^k$, which is an element of the semiring $\mathbf{N}[X]$ of polynomials in one indeterminate X with nonnegative integer coefficients.

Examples 4.2 The semirings corresponding to the theories AM and AG are $\mathbf{N}$ and $\mathbf{Z}$, respectively. The theories of Example 3.1 yield the following semirings (see [Nu90, Ba90]):

$\mathcal{S}_{\mathrm{AMH}}$, the semiring corresponding to the theory AMH of abelian monoids with a homomorphism, is isomorphic to $\mathbf{N}[X]$, the semiring of polynomials in one indeterminate X with nonnegative integer coefficients.

$\mathcal{S}_{\mathrm{AMIn}}$, which corresponds to the theory of abelian monoids with an involution, is the monoid semiring $\mathbf{N}\langle Z_2\rangle$, where Z_2 denotes the cyclic group of order 2.

$\mathcal{S}_{\mathrm{COM}}$, the semiring corresponding to the theory COM, is isomorphic to $\mathbf{N}\langle X, Y\rangle$, the semiring of polynomials in two *noncommuting* indeterminates X, Y with nonnegative integer coefficients. Note that $\mathbf{N}\langle X, Y\rangle$ is the monoid semiring $\mathbf{N}\langle\{X, Y\}^*\rangle$, where $\{X, Y\}^*$ denotes the free monoid in two generators X, Y.

$\mathcal{S}_{\mathrm{GAUSS}}$ is isomorphic to the ring of Gaussian numbers $\mathbf{Z} \oplus i\mathbf{Z}$, consisting of the complex numbers $m + in$, where $m, n \in \mathbf{Z}$.

The first two examples suggest that there is a close connection between augmenting a commutative/monoidal theory by a monoid (as defined at the end of Subsection 3.2) and adjoining a monoid to the corresponding semiring (as defined in Subsection 2.2). For AMIn = AM$\langle Z_2\rangle$, for instance, one can verify that the semirings $\mathcal{S}_{\mathrm{AM}\langle Z_2\rangle}$ and $\mathcal{S}_{\mathrm{AM}}\langle Z_2\rangle$ are isomorphic. It is easy to see that this kind of connection holds in general.

Theorem 4.3 *Let $\mathcal{E}$ be a commutative/monoidal theory, and let H be a finitely generated monoid. Then $\mathcal{S}_{\mathcal{E}\langle H\rangle}$, the semiring corresponding to $\mathcal{E}$ augmented by H, and the monoid semiring $\mathcal{S}_\mathcal{E}\langle H\rangle$ are isomorphic.*

The isomorphism between $\mathcal{S}_{\mathcal{E}\langle H\rangle}$ and $\mathcal{S}_\mathcal{E}\langle H\rangle$ will be used in the next two sections to study the unification problem for commutative/monoidal theories of the form $\mathcal{E}\langle H\rangle$ in an algebraic setting.

It is well known that unification problems in AM correspond to systems of linear equations over the nonnegative integers $\mathbf{N}$ [LS75]. As seen above, $\mathbf{N}$ is isomorphic to the semiring $\mathcal{S}_{\mathrm{AM}}$. We shall illustrate with an example from AMH how this correspondence can be generalized to arbitrary commutative/monoidal theories. Consider the AMH-unification problem

$$\Gamma = \langle y_1 + h(y_2) + h^2(y_3) \doteq y_1 + h^2(y_1),\;\; h(y_1) + h(y_1) \doteq h(y_1) + y_2 + h(y_3)\rangle.$$

Note that every term occurring in Γ has the form $s = \sum_{j=1}^{m} s_j(y_j)$, where $s_j(y_j)$ contains no variable other than y_j. Such a representation exists for every term in a commutative/monoidal theory. We associate to Γ matrices $M' = (\mu_{ij})$ and $N' = (\nu_{ij})$ over $\mathcal{S}_{\text{AMH}}$ as follows: if $\sum_{j=1}^{m} s_{ij}(y_j) \doteq \sum_{j=1}^{m} t_{ij}(y_j)$ is the i-th equation in Γ, then we define the Σ-homomorphisms μ_{ij}, ν_{ij} by $\mu_{ij} := [x/s_{ij}(x)]$ and $\nu_{ij} := [x/t_{ij}(x)]$. Using the fact that $\mathcal{S}_{\text{AMH}}$ and $\mathbf{N}[X]$ are isomorphic semirings, we transform M', N' into matrices M, N over $\mathbf{N}[X]$ by replacing the elements of $\mathcal{S}_{\text{AMH}}$ by the corresponding elements of $\mathbf{N}[X]$. In th example, we thus obtain

$$M = \begin{pmatrix} 1 & X & X^2 \\ 2X & 0 & 0 \end{pmatrix} \quad \text{and} \quad N = \begin{pmatrix} 1+X^2 & 0 & 0 \\ X & 1 & X \end{pmatrix}.$$

Solving Γ modulo AMH is now equivalent to solving the system of linear equations $My = Ny$ over the semiring $\mathbf{N}[X]$. Every solution of $My = Ny$ gives rise to a *basic unifier* of Γ, i.e., a unifier that introduces exactly one variable. For instance, the solutions $y^{(1)} = (1, 0, 1)$ and $y^{(2)} = (1, X, 0)$ yield unifiers $\delta^{(1)}$ and $\delta^{(2)}$, which introduce variables z_1, z_2, respectively. The term substituted for y_i by $\delta^{(k)}$ is $z_j y_i^{(k)}$, where $y_i^{(k)}$ is conceived as a Σ-homomorphism from $\mathcal{F}_{\text{AMH}}(z_j)$ to $\mathcal{F}_{\text{AMH}}(z_j)$. Thus, $\delta^{(1)} = [y_1/z_1, y_2/0, y_3/z_1]$ and $\delta^{(2)} = [y_1/z_2, y_2/h(z_2), y_3/0]$. For basic unifiers $\delta^{(1)}, \ldots, \delta^{(n)}$ that introduce distinct variables the sum $\delta^{(1)} + \cdots + \delta^{(n)}$ is again a unifier, which is more general than each $\delta^{(k)}$. For instance, $\delta = \delta^{(1)} + \delta^{(2)} = [y_1/z_1 + z_2, y_2/h(z_2), y_3/z_1]$ is more general than each of $\delta^{(1)}$ and $\delta^{(2)}$. One can show that a sum $\delta^{(1)} + \cdots + \delta^{(n)}$ of basic unifiers is a most general unifier if and only if the corresponding solution vectors $y^{(1)}, \ldots, y^{(n)}$ generate all solutions of the system of linear equations $My = Ny$, i.e., every solution is a linear combination of $y^{(1)}, \ldots, y^{(n)}$. This fact implies the following theorem [Nu88, Nu90, Ba90]:

Theorem 4.4 *A commutative/monoidal theory $\mathcal{E}$ is unitary w.r.t. unification without constants if and only if $\mathcal{S}_\mathcal{E}$ satisfies the following condition: for any pair M, N of $m \times n$-matrices over $\mathcal{S}_\mathcal{E}$ the set*

$$\mathcal{U}(M, N) := \{x \in \mathcal{S}_\mathcal{E}{}^n \mid Mx = Nx\}$$

is a finitely generated right $\mathcal{S}_\mathcal{E}$-module.

Since constant-free unification problems in commutative/monoidal theories are either unitary or of type zero [Nu88, Ba89a, Nu90], the theorem yields that the theory $\mathcal{E}$ is of type zero iff there exist matrices M, N over $\mathcal{S}_\mathcal{E}$ such that the right $\mathcal{S}_\mathcal{E}$-module $\mathcal{U}(M, M_\tau)$ is not finitely generated. Using this characterization, it can be shown that the theories AMH and COM are of type zero (see [Ba89a, Ba90]). The theories AMIn and GAUSS are unitary w.r.t. unification without constants (see [Ba89a] for the first, and [Nu90] for the second result).

5 A Sufficient Condition for Unification Type Zero

In this section we shall generalize the "type zero" result for the theory AMH to a whole class of commutative/monoidal theories. This class will be defined by properties of the corresponding semiring. Before we can do that, we need one more notation.

Let $\mathcal{S}$ be a semiring which is not a ring. That means that the abelian monoid $(\mathcal{S}, +, 0)$ is not a group, i.e., there exists an element $p \in \mathcal{S}$ such that, for all $q \in \mathcal{S}$, we have $p + q \neq 0$. We shall call such an element p of $\mathcal{S}$ *non-invertible*. An element $s \in \mathcal{S}$ which has an inverse w.r.t. "+" is called *invertible*. For the semiring $\mathbf{N}$, all elements different from 0 are non-invertible. For the direct product $\mathbf{N} \times \mathbf{Z}$, an element (n, z) is invertible iff $n = 0$. Here are some trivial facts about invertible and non-invertible elements.

1. The elements $s_1, \ldots, s_k$ of $\mathcal{S}$ are invertible if and only if their sum $s_1 + \cdots + s_k$ is invertible.

2. The element $\sum_{h\in H} s_h \cdot h$ of the monoid semiring $\mathcal{S}\langle H\rangle$ is non-invertible if and only if there exists $h \in H$ such that s_h is non-invertible in $\mathcal{S}$. Thus, if $\mathcal{S}$ is not a ring, then $\mathcal{S}\langle H\rangle$ is not a ring for any monoid H.

Recall that the theory AMH corresponds to the semiring $\mathbf{N}[X]$ of polynomials in one indeterminate X with nonnegative integer coefficients. That means that we have a monoid semiring $\mathcal{S}\langle H\rangle$ where *all the nonzero elements* of $\mathcal{S}$ are non-invertible, and where the monoid H is the free monoid X^* in one generator. The "type zero" result for AMH can now be generalized to the case where $\mathcal{S}$ contains *at least one* non-invertible element.

Theorem 5.1 *Let $\mathcal{E}$ be a commutative/monoidal theory such that the corresponding semiring $\mathcal{S}_\mathcal{E}$ is isomorphic to a monoid semiring $\mathcal{S}\langle X^*\rangle$. If $\mathcal{S}$ is not a ring, i.e., if $\mathcal{S}$ contains at least one non-invertible element, then $\mathcal{E}$ is of unification type zero.*

As mentioned before the monoid semiring $\mathcal{S}\langle X^*\rangle$ is just the polynomial semiring $\mathcal{S}[X]$. The theorem is proved if we can find matrices M, N over $\mathcal{S}[X]$ such that the right $\mathcal{S}[X]$-module $\mathcal{U}(M,N)$ is not finitely generated.

In the following we shall show that the 1×3-matrices $M := (X,X,0)$ and $N := (0,1,X^2)$ have the required property. Thus we consider the homogeneous linear equation

$$X\cdot x_1 + X\cdot x_2 \;=\; x_2 + X^2\cdot x_3 \tag{1}$$

which has to be solved by a vector $L \in \mathcal{S}[X]^3$. If L is such a vector, we denote its components by $L^{(1)}$, $L^{(2)}$, $L^{(3)}$.

Let p be a non-invertible element in $\mathcal{S}$. Obviously, for any $n \geq 1$, the vector L_n which consists of the components $L_n^{(1)} := p$, $L_n^{(2)} := pX + \cdots + pX^{n+1}$, $L_n^{(3)} := pX^n$ is a solution of (1).

Now assume that $\mathcal{U}(M,N)$ is finitely generated, i.e., there exist finitely many solutions $G_1,\cdots,G_m$ of (1) which generate all the solutions of (1). Let $n \geq 1$ be arbitrary but fixed. Since L_n is a solution of (1) there exist $l_1,\cdots,l_m \in \mathcal{S}[X]$ such that

$$L_n = \sum_{i=1}^{m} G_i l_i. \tag{2}$$

If we consider (2) in the first component, we get $p = \sum_{i=1}^m G_i^{(1)} l_i$. For $i = 1,\ldots,m$, let $p_i \in \mathcal{S}$ be the constant coefficient of the polynomial $G_i^{(1)}$, and $h_i \in \mathcal{S}$ be the constant coefficient of l_i. The last equation implies that $p = \sum_{i=1}^m p_i h_i$. Since p is non-invertible, there exists some j with $1 \leq j \leq m$ such that $p_j h_j$ is non-invertible.

Lemma 5.2 *The polynomial $G_j^{(3)}$ is of degree at least n.*

Proof. Assume that the degree of $G_j^{(3)}$ is less than n. Since G_j is a solution of (1), we know that $G_j h_j$ is also a solution, that is,

$$X\cdot G_j^{(1)}h_j + X\cdot G_j^{(2)}h_j = G_j^{(2)}h_j + X^2\cdot G_j^{(3)}h_j. \tag{3}$$

The components of the solution $G_j h_j$ satisfy the following properties:

- The constant coefficient of the polynomial $G_j^{(1)}h_j$ is $e_1 := p_j h_j$. Thus we know by the choice of j that e_1 is non-invertible.
- The polynomial $G_j^{(2)}h_j$ has constant coefficient 0. This is an immediate consequence of the equation (3).

- All the coefficients of $G_j^{(3)}h_j$ are invertible. This can be seen by considering equation (2) in the third component, which yields $pX^n = \sum_{i=1}^{m} G_i^{(3)}l_i$. Since $G_j^{(3)}h_j$ contains only monomials of degree less than n, all these monomials vanish during the summation. Consequently, all the coefficients of these monomials have to be invertible.

From the fact that the coefficient of X in $X \cdot G_j^{(1)}h_j$ is e_1 and in $X \cdot G_j^{(2)}h_j$ is 0 we get by (3) that the coefficient of X in $G_j^{(2)}h_j + X^2 \cdot G_j^{(3)}h_j$ is also e_1. Hence, the coefficient of X in $G_j^{(2)}h_j$ is e_1.

Starting with the fact the coefficient e_1 of X in $G_j^{(2)}h_j$ is non-invertible, we shall now deduce that the coefficient of X^2 in $G_j^{(2)}h_j$ is also non-invertible. Since the coefficient of X in $G_j^{(2)}h_j$ is e_1, the coefficient of X^2 in $X \cdot G_j^{(2)}h_j$ is also e_1. Thus the coefficient of X^2 on the left hand side of (3) is $e' := e_1 + e$ for some e. The coefficient e' is non-invertible because otherwise e_1 could not be non-invertible. By (3), the coefficient of X^2 in $G_j^{(2)}h_j + X^2 \cdot G_j^{(3)}h_j$ is also e'. Since all the coefficients of $X^2 \cdot G_j^{(3)}h_j$ are invertible, this finally shows that the coefficient e_2 of X^2 in $G_j^{(2)}h_j$ is non-invertible.

This argument can be iterated to show that, for all $k \geq 1$, the coefficient e_k of X^k in $G_j^{(2)}h_j$ is non-invertible. This is a contradiction to the fact that the polynomial $G_j^{(2)}h_j$ has only finitely many nonzero coefficients. □

We have just shown that, for any $n \geq 1$, there exists a j such that $G_j^{(3)}$ is of degree at least n. This is a contradiction to our assumption that there are finitely many generators G_j of all solutions of (1). This completes the proof of the theorem.

6 Adding Finite Monoids of Homomorphisms

In this section we investigate commutative/monoidal theories that are augmented with finite monoids of homomorphisms. In contrast to the case of free monoids, that was treated in the previous section, we can derive the positive result that adding finite monoids does not change the unification type and that algorithms for the original theory can be used to solve problems in the augmented theory. An example for such a theory is AMIn, the theory of abelian monoids with an involution. Recall that AMIn can be written as $\mathrm{AM}\langle Z_2\rangle$, and that the corresponding semiring is $\mathrm{N}\langle Z_2\rangle$.

Throughout this section, we suppose that $\mathcal{E}$ is a commutative/monoidal theory and H is a finite monoid. Since unification problems in $\mathcal{E}\langle H\rangle$ are equivalent to systems of linear equations over $\mathcal{S}_\mathcal{E}\langle H\rangle$, our basic technique will be to reduce such systems to systems of linear equations over $\mathcal{S}_\mathcal{E}$. As a first step we shall establish a one-to-one correspondence between vectors.

Every vector $x \in \mathcal{S}_\mathcal{E}\langle H\rangle^n$ has a unique representation as $x = \sum_{h\in H} x_h \cdot h$ where $a_h \in \mathcal{S}_\mathcal{E}{}^n$. As an example the vector

$$x = (1 + 2h, h) \in \mathrm{N}\langle Z_2\rangle$$

can be written as

$$x = (1\cdot e + 2\cdot h, 0\cdot e + 1\cdot h) = (1,0)\cdot e + (2,1)\cdot h.$$

We can formally justify this notation if we consider $\mathcal{S}_\mathcal{E}$ and H as subsets of $\mathcal{S}_\mathcal{E}\langle H\rangle$. This can be done by identifying every element $s \in \mathcal{S}_\mathcal{E}$ with $s\cdot e \in \mathcal{S}_\mathcal{E}\langle H\rangle$, where e is the unit in H, and every element $h \in H$ with $1\cdot h \in \mathcal{S}_\mathcal{E}\langle H\rangle$.

Suppose the elements of H are numbered as $h_1, \ldots, h_{|H|}$. If $x \in \mathcal{S}_\mathcal{E}\langle H\rangle^n$ has a representation as $x = x_{h_1}\cdot h_1 + \cdots + x_{h_{|H|}}\cdot h_{|H|}$, we define

$$\widehat{x} = (x_{h_1}, \ldots, x_{h_{|H|}})$$

as the vector obtained from x by writing the vectors x_h one below another. Continuing our example from above we have

$$\widehat{x} = (1,0,2,1).$$

We thus obtain a bijection between $\mathcal{S}_\mathcal{E}\langle H\rangle^n$ and $\mathcal{S}_\mathcal{E}{}^{n|H|}$. In particular, every vector in $\mathcal{S}_\mathcal{E}{}^{n|H|}$ has a representation as $\widehat{x}$ for some $x \in \mathcal{S}_\mathcal{E}\langle H\rangle^n$. Obviously, for all x, $y \in \mathcal{S}_\mathcal{E}{}^{n|H|}$ and all $s \in \mathcal{S}_\mathcal{E}$ we have

$$\widehat{x+y} = \widehat{x} + \widehat{y} \qquad \text{and} \qquad \widehat{x}\cdot s = \widehat{x\cdot s}. \tag{4}$$

In algebraic terms we can rephrase these equalities by saying that the mapping "$\widehat{\cdot}$" is a right $\mathcal{S}_\mathcal{E}$-module isomorphism.

Next we will associate to every $m{\times}n$-matrix M with entries in $\mathcal{S}_\mathcal{E}\langle H\rangle$ an $m|H|{\times}n|H|$-matrix $\widehat{M}$ with entries in $\mathcal{S}_\mathcal{E}$, such that $\widehat{Mx} = \widehat{M}\widehat{x}$ holds for every $x \in \mathcal{S}_\mathcal{E}\langle H\rangle^n$. To derive an appropriate definition of $\widehat{M}$, observe that, similar to a vector, the matrix M has a unique representation $M = \sum_{h\in H} M_h \cdot h$, where the M_h are matrices with entries in $\mathcal{S}_\mathcal{E}$. Applying M to a vector x yields $Mx = (\sum_{f\in H} M_f \cdot f)(\sum_{g\in H} x_g \cdot g) = \sum_{f,g\in H} M_f x_g \cdot f \cdot g = \sum_{h\in H}(\sum_{h=fg} M_f x_g)\cdot h = \sum_{h\in H}(\sum_{g\in H}(\sum_{h=f\cdot g} M_f)x_g)\cdot h$. This series of equalities says that the component of the vector $\widehat{Mx}$ corresponding to the element h is obtained by summing over all g the products $(\sum_{h=f\cdot g} M_f)x_g$. This shows that we have to define $\widehat{M}$ as the $m|H|{\times}n|H|$-matrix consisting of the submatrices

$$\widehat{M}_{i,j} = \sum_{\substack{h\in H \\ h_i = h\cdot h_j}} M_h,$$

where a sum over an empty set of indices is to be understood as the zero matrix. With this definition we obtain

$$\widehat{Ma} = \widehat{M}\widehat{a}. \tag{5}$$

Returning to our example theory AMIn, consider a matrix M over $\mathrm{N}\langle Z_2\rangle$. If $M = M_e\cdot e + M_h\cdot h$, then the associated matrix is

$$\widehat{M} = \begin{pmatrix} M_e & M_h \\ M_h & M_e \end{pmatrix}.$$

Thus, our general approach gives us the same representation of unification problems in AMIn as the one derived in [Ba89a].

Next we apply our transformation technique to unification problems without constants.

Proposition 6.1 *Let M, N be $m{\times}n$-matrices over $\mathcal{S}_\mathcal{E}\langle H\rangle$, and $x \in \mathcal{S}_\mathcal{E}\langle H\rangle^n$. Then:*

1. *$x \in \mathcal{U}(M,N)$ if and only if $\widehat{x} \in \mathcal{U}(\widehat{M},\widehat{N})$*
2. *$\mathcal{U}(M,N)$ is generated by $x_1,\ldots,x_k$ if $\mathcal{U}(\widehat{M},\widehat{N})$ is generated by $\widehat{x}_1,\ldots,\widehat{x}_k$.*

Proof. 1. Let $x \in \mathcal{S}_\mathcal{E}\langle H\rangle^n$. Then we have $x \in \mathcal{U}(M,N)$ if and only if $Mx = Nx$ if and only if $\widehat{Mx} = \widehat{Nx}$ if and only if $\widehat{M}\widehat{x} = \widehat{N}\widehat{x}$ if and only if $\widehat{x} \in \mathcal{U}(\widehat{M},\widehat{N})$.

2. It suffices to show that every $x \in \mathcal{U}(M,N)$ is a linear combination of $x_1,\ldots,x_k$. If $x \in \mathcal{U}(M,N)$, then $\widehat{x} \in \mathcal{U}(\widehat{M},\widehat{N})$ by part (1). Hence, $\widehat{x} = \widehat{x}_1\cdot s_1 + \cdots + \widehat{x}_k\cdot s_k$. Using equalities (4), we conclude that $x = x_1\cdot s_1 + \cdots + x_k\cdot s_k$. Thus, x is a linear combination of $x_1,\ldots,x_k$. □

If $\mathcal{E}$ is unitary w.r.t. unification without constants, then for all matrices M, N with entries from $\mathcal{S}_\mathcal{E}\langle H\rangle$ the right $\mathcal{S}_\mathcal{E}$-module $\mathcal{U}(\widehat{M},\widehat{N})$ is finitely generated, and by the preceding proposition, $\mathcal{U}(M,N)$ is finitely generated. Together with Theorem 4.4 this proves our next theorem.

Theorem 6.2 *If $\mathcal{E}$ is unitary w.r.t. unification without constants, then $\mathcal{E}\langle H\rangle$ is unitary w.r.t. unification without constants.*

Proposition 6.1 tells us how we can use an algorithm for $\mathcal{E}$ to solve problems in $\mathcal{E}\langle H\rangle$. An $\mathcal{E}\langle H\rangle$-unification problem without constants is given by $m\times n$-matrices M, N with entries in $\mathcal{S}_{\mathcal{E}\langle H\rangle} \simeq \mathcal{S}_{\mathcal{E}}\langle H\rangle$. We compute the transforms $\widehat{M}$ and $\widehat{N}$ and solve the equation $\widehat{M}y = \widehat{N}y$ over $\mathcal{S}_{\mathcal{E}}$, which we can do with the algorithm for $\mathcal{E}$. If the set of solutions of the matrix equation over $\mathcal{S}_{\mathcal{E}}$ is generated by vectors $y_1, \ldots, y_k \in {\mathcal{S}_{\mathcal{E}}}^{n|H|}$, we compute $x_1, \ldots, x_k \in \mathcal{S}_{\mathcal{E}}\langle H\rangle^n$ such that $\widehat{x}_i = y_i$. Then the set of solutions of the original equation is generated by $x_1, \ldots, x_k$ and, as shown in Section 4, a straightforward transformation yields a most general unifier of the given problem.

7 Conclusion

With this paper we pursue an algebraic approach to unification. We have combined techniques for commutative and monoidal theories that had been developed independently. We have shown that both classes of theories are essentially the same in that every monoidal theory is commutative, and every commutative theory can be turned into a monoidal theory by a signature transformation.

One of the major topics of research in unification in recent years was to construct algorithms for the combination of equational theories. This problem has been solved—at least in principle—for theories with disjoint signatures [SS89]. Of course, the case where signatures are not disjoint is too difficult to be treated in full generality. We concentrated on a special case, namely the combination of a commutative/monoidal theory with a monoid of homomorphisms. By exploiting the algebraic structure of the canonical semiring associated to such a theory, we have found combinations that are of unification type zero, and others that are of type unitary or finitary. For the latter case we have pointed out how a unification algorithm can be derived.

References

[Ba89a] F. Baader, "Unification in Commutative Theories," *JSC* 8, 1989.

[Ba89b] F. Baader, "Unification Properties of Commutative Theories: A Categorical Treatment," *Proc. CTCS*, LNCS 389, 1989.

[Ba90] F. Baader, *Unification in Commutative Theories, Hilbert's Basis Theorem, and Gröbner Bases*, SEKI-Report SR-90-01, Universität Kaiserslautern, Germany, 1990.

[BN90] F. Baader, W. Nutt, *Adding Homomorphisms to Commutative/Monoidal Theories or How Algebra Can Help in Equational Unification*, Research Report RR-16-90, German Research Center for Artificial Intelligence (DFKI), Kaiserslautern, Germany, 1990.

[KK90] C. Kirchner, F. Klay, "Syntactic Theories and Unification," *Proc. LICS 90*, 1990.

[LS75] M. Livesey, J. Siekmann, *Unification of Bags and Sets*, Report MEMO SEKI-76-II, Institut für Informatik I, Universität Karlsruhe, Germany, 1976.

[Nu88] W. Nutt, "Unification in Monoidal Theories," presented at UNIF'88, Val d'Ajol, France, 1988.

[Nu90] W. Nutt, "Unification in Monoidal Theories," *Proc. 10th CADE*, LNCS 449, 1990.

[SS89] M. Schmidt-Schauß, "Combination of Unification Algorithms," *JSC* 8, 1989.

[Si89] J. H. Siekmann, "Unification Theory: A Survey," *JSC* 7, 1989.

Undecidable Properties of Syntactic Theories *

Francis Klay

INRIA Lorraine
615 rue du Jardin Botanique - BP 101
F-54600 Villers les Nancy
E-mail: klay@loria.crin.fr

Abstract

Since we are looking for unification algorithms for a large enough class of equational theories, we are interested in syntactic theories because they have a nice decomposition property which provides a very simple unification procedure. A presentation is said resolvent if any equational theorem can be proved using at most one equality step at the top position. A theory which has a finite and resolvent presentation is called syntactic. In this paper we give decidability results about open problems in syntactic theories: unifiability in syntactic theories is not decidable, resolventness of a presentation and syntacticness of a theory are even not semi-decidable. Therefore we claim that the condition of syntacticness is too weak to get unification algorithms directly.

1 Introduction

Unification, the problem of solving equations in abstract algebras is at the heart of mechanizing mathematics and is thus a major component of theorem provers as well as logic programming language interpretors. The problem of solving equations in a theory E presented by a finite number of equational axioms is undecidable in general so that it is crucial to find subclasses of theories where the problem becomes decidable. For that purpose the concept of syntactic theories was introduced by C. Kirchner in [7, 8] in order to compute efficient unification procedure directly from a finite presentation, called resolvent, of the theory. A presentation is resolvent when a proof between two arbitrary equal terms can always be performed with at most one equality applied at the top of the terms in the equational proof. In this kind of theories, [7, 8] gives sufficient conditions in order to compute a resolvent presentation from which a unification procedure can be deduced. This has been implemented and enhanced by J. Christian [1] in order to dynamically

*This research has been partially supported by ESPRIT Basic Research Action COMPASS #3264, by the GRECO de Programmation of CNRS and by the Fonds National Suisse pour la Recherche Scientifique.

compute a unification algorithm in the completion algorithm modulo a set of axioms [6]. In [12], D. Rémy gives a sufficient condition for a presentation to be resolvent in order to built a typing algorithm for a record oriented version of the programming language ML. Recently T. Nipkow also found a sufficient condition for resolventness and showd that the inference rules for E-equality can be reduced if the presentation is resolvent [11]. In the same paper he gives a matching algorithm in syntactic theories with finite equivalence classes, this algorithm being directly computed from a resolvent presentation. In [9] we investigate the relationship between unifiability of some kind of equations of the form $f(v_1,\ldots,v_n) =^? g(v_{n+1},\ldots,v_m)$ that we call general, and syntacticness of the theory. We state that a theory is syntactic if and only if every general equation has a finite complete set of E-solutions. All these works used the syntacticness property but the following problems remained open:

- Is the unifiability in syntactic theory decidable ?
- Is the resolventness decidable ?
- Is the syntacticness decidable ?

This paper gives answers to these questions. We give the precise definition of syntactic theories and their properties in Section 2. In Section 3 we show that unifiability in syntactic theories is not decidable by a reduction to the Post's Correspondence Problem. Finally we prove in Sections 4 and 5 that the resolventness and the syntacticness are even not semi-decidable. So we are able to conclude that the syntacticness is a too weak condition to make unifiability decidable. Missing proofs can be found in [10]. In the following we are using the standard notions in equational logic [3, 4] and unification [7, 13, 5] and we are using mainly the notations and the approach of [5, 9].

2 Syntactic theories

In this section we present a subclass of equational theories called the subclass of syntactic theories. These theories have some nice properties which are likely to be interesting to design unification algorithm. A presentation A is **resolvent** [8] iff for any terms t and t' such that $t =_A t'$ there exists a A-proof

$$t = s_0 \longleftrightarrow_A^{\alpha_1} s_1 \longleftrightarrow_A \ldots \longleftrightarrow_A^{\alpha_k} s_k = t'$$

with at most one of the α_j equal to ε for $j \in [1..k]$. Obviously all theories have a resolvent presentation: it is the equational theory itself. An equational theory E is said **syntactic** (the definition of syntactic theory has been extended to almost syntactic in [2]) if there exists a finite and resolvent presentation of E.

Example 1 *The theory generated by associativity: $A = \{x + (y + z) \doteq (x + y) + z\}$ is a syntactic theory since A is a finite and resolvent presentation. Thus for any A-equal* *terms there exists an A-proof with at most one equality step at ε. For example this A-proof between t and t':*

$$\begin{aligned} t = \; & ((a+(b+c))+d)+(e+f) \longleftrightarrow_A^{\varepsilon} (a+(b+c))+(d+(e+f)) \longleftrightarrow_A^{1} \\ & ((a+b)+c)+(d+(e+f)) \longleftrightarrow_A^{\varepsilon} (a+b)+(c+(d+(e+f))) = t' \end{aligned}$$

has two equality steps at ε. Since A is resolvent this means that there exists an A-proof between t and t' with at most one equality step at ε:

$$\begin{aligned} t = \; & ((a+(b+c))+d)+(e+f) \longleftrightarrow_A^{11} (((a+b)+c)+d)+(e+f) \longleftrightarrow_A^{1} \\ & ((a+b)+(c+d))+(e+f) \longleftrightarrow_A^{\varepsilon} (a+b)+((c+d)+(e+f)) \longleftrightarrow_A^{2} \\ & (a+b)+(c+(d+(e+f))) = t' \end{aligned}$$

From now on we suppose that the theories E considered are **collapse free.** The resolventness of a presentation is equivalent to a property between any two A-equal terms and their decompositions:

Proposition 1 *[8] A set A of axioms is resolvent iff for all terms $t = f(\vec{t})$ and $t' = f'(\vec{t'})$ we have:*

$$t =_A t' \quad \Leftrightarrow \quad \left(\begin{array}{c} \exists f(\vec{u}) \doteq f'(\vec{u'}) \in A \cup A^{-1}, \ \exists \rho, \ \text{such that} \\ \vec{t} =_A \rho\vec{u} \text{ and } \vec{t'} =_A \rho\vec{u'} \end{array} \right) \text{ or } (f = f' \text{ and } \vec{t} =_A \vec{t'})$$

From this proposition we can deduce a transformation rule on unificands which is sound and complete.

Proposition 2 *[8] Let $e = (f(\vec{t}) =^? f'(\vec{t'}))$ be an equation. The following transformation rule* Mutate *is sound and complete in the theory $=_A$ if the presentation A is resolvent:*

$$\textsf{Mutate} \qquad \frac{f(\vec{t}) \stackrel{?}{=} f'(\vec{t'})}{\bigvee_{f(\vec{u}) \doteq f'(\vec{u'}) \in A \cup A^{-1}} (\vec{t} \stackrel{?}{=} \vec{u} \wedge \vec{t'} \stackrel{?}{=} \vec{u'}) \quad \bigvee_{if\ f=f'} (\vec{t} \stackrel{?}{=} \vec{t'})}$$

Note that if the theory E is syntactic, then there exists a presentation A of E which is resolvent and finite. One among the interests of syntactic theories is that with this presentation, the disjunctions of systems deduced with the previous transformation rule are always finite.

Example 2 *For commutativity the transformation rule* Mutate *is:*

$$\textsf{Mutate} \qquad \frac{t_1 + t_2 \stackrel{?}{=} t_3 + t_4}{(t_1 \stackrel{?}{=} t_4 \wedge t_2 \stackrel{?}{=} t_3) \vee (t_1 \stackrel{?}{=} t_3 \wedge t_2 \stackrel{?}{=} t_4)}$$

We characterize now the syntacticness of a collapse free theory using a unification property of the theory.

Definition 1 *Let E be a theory, $V = \{v_1, \ldots, v_m\}$ a set of distinct variables, n and $m-n$ the arities of the symbols f and g in F. The equation $f(v_1, \ldots, v_n) =^? g(v_{n+1}, \ldots, v_m)$ is called* **general equation** *on f and g. These kind of equations is the most general one that can be built on two symbols f and g.*

Theorem 1 *[9] Let E be a collapse free theory generated by A. E is syntactic iff all general equations are finitary E-unifying.*

Example 3 *The theory AC generated by $A = \{x + y \doteq y + x,\ (x + y) + z \doteq x + (y + z)\}$ is syntactic since every equation is finitary unifying in AC.*

This theorem has been recently generalized to collapse theories by J.P. Jouannaud. Syntactic theory can also be characterized by a property of the inference rules system for equational logic. In fact the inference rules can be restricted in such a way that the inference rules system remain complete.

Theorem 2 *[11] If A is a resolvent presentation then the following set of inference rules is sound and complete for $=_A$.*

Reflexivity		$\Longrightarrow$	$x = x$	
Congruence	$\vec{s} = \vec{t}$	$\Longrightarrow$	$f(\vec{s}) = f(\vec{t})$	$(\forall f \in F)$
Mutate	$\vec{x} = \vec{s}$ and $\vec{y} = \vec{t}$	$\Longrightarrow$	$f(\vec{x}) = g(\vec{y})$	$(\forall f(\vec{s}) = g(\vec{t}) \in A \cup A^{-1})$

The important point of this theorem is that transitivity can be removed from the inference rules system if A is a resolvent presentation.

3 Unifiability in syntactic theories

In this section we prove that unifiability is not decidable in syntactic theories. The idea is to associate an equational theory for each instance of the **Post's Correspondence Problem** (for short PCP). Then we show that this theory is syntactic and finally we prove that a unificand has a solution if and only if the instance of PCP has a solution.

Definition 2 *Let $V = \{c_1, \ldots, c_m\}$ be a finite alphabet and $L = [(w_1, w'_1), \ldots, (w_n, w'_n)]$ a list of ordered pairs of words in V^+. We say that this instance of PCP has a solution if there exists a sequence of integer $(i_1, i_2, \ldots, i_k)$ with $k \geq 1$ such that:*

$$w_{i1} w_{i2} \ldots w_{ik} = w'_{i1} w'_{i2} \ldots w'_{ik}$$

It is well-known that the solvability of an instance of PCP is semi-decidable but not decidable.

Example 4 *Let* $V = \{0,1\}$ *and* $L = [(1,111),(10111,10),(10,0)]$ *this instance of PCP has a solution because the sequence of integer is such that:* $w_2w_1w_1w_3 = w'_2w'_1w'_1w'_3 =$ 101111110.

In order to built the particular theory associated with the PCP instance, we define a signature F as $F_1 \cup F_2 \cup F_3$ where:

- $F_0 = \{\bot\}$.
- $F_1 = \{c_1,\ldots,c_m\} \cup \{a_1,\ldots,a_n\} \cup \{s_1,s_2\}$ where there is one monadic symbol c_i for each element of V, a symbol a_i for each pair of words in L and two symbols s_1, s_2.
- $F_3 = \{f,h\}$ and $F = F_0 \cup F_1 \cup F_3$.

Notation: In the following, the sequences of monadic symbols are written as elements of F_1^*.

The presentation A of the theory associated with the PCP instance is composed of three axioms groups. In each group there is one axiom for each pair of words:

$$A = \begin{cases} A_1 = \bigcup\limits_{\forall i \in [1..n]} & f(a_ix,y,z) = s_1s_2f(x,a_iy,w_iz) \\ A_2 = \bigcup\limits_{\forall i \in [1..n]} & h(\bot,a_iy,z) = s_2f(\bot,a_iy,z) \\ A_3 = \bigcup\limits_{\forall i \in [1..n]} & h(a_ix,y,z) = s_2s_1h(x,a_iy,w'_iz) \end{cases}$$

Example 5 *Consider the instance of PCP of example 4 and its associated theory* $=_A$. *We will prove that there exists a substitution* σ *such that* $\sigma(s_2f(x,\bot,\bot)) =_A \sigma(h(x,\bot,\bot))$ *iff the instance of PCP has a solution. For our instance of PCP a possible substitution* σ *is* $\sigma = \{x \mapsto a_3a_1a_1a_2\bot\}$ *and the A-proof has the following form:*

$$\begin{array}{llll} s_2f(a_3a_1a_1a_2\bot,\bot,\bot) & \longleftrightarrow_{A_1} & s_2f(a_1a_1a_2\bot,a_3\bot,10\bot) & \overset{3}{\longleftrightarrow}_{A_1} \\ s_2f(\bot,a_2a_1a_1a_3\bot,101111110\bot) & \longleftrightarrow_{A_2} & h(\bot,a_2a_1a_1a_3\bot,101111110\bot) & \overset{3}{\longleftrightarrow}_{A_3} \\ h(a_1a_1a_2\bot,a_3\bot,0\bot) & \longleftrightarrow_{A_3} & h(a_3a_1a_1a_2\bot,\bot,\bot) & \end{array}$$

3.1 A is a resolvent presentation

To prove the syntacticness of $=_A$ we will show in fact the resolventness of A. We first orient the presentation into a canonical rewriting system R. Then we study the innermost R-proofs in order to show that for all A-equals terms there alway exists an A-proof with at most one step at ε. Another but more complicated way of proving the syntacticness of $=_A$ is to use the theorem 1: by narrowing we compute the complete sets of A-solutions of general equations and since these sets are finite we conclude to the syntacticness of $=_A$. A can be turned into the following canonical rewriting system for $=_A$:

$$R = \begin{cases} R_1 = \bigcup_{\forall i \in [1..n]} & f(a_i x, y, z) \longrightarrow s_1 s_2 f(x, a_i y, w_i z) \\ R_2 = \bigcup_{\forall i \in [1..n]} & h(\perp, a_i y, z) \longrightarrow s_2 f(\perp, a_i y, z) \\ R_3 = \bigcup_{\forall i \in [1..n]} & h(a_i x, y, z) \longrightarrow s_2 s_1 h(x, a_i y, w'_i z) \end{cases}$$

Definition 3 *Let $\mathcal{R}$ be a canonical rewriting system. An $\mathcal{R}$-derivation $t_0 \xrightarrow{n}_{\mathcal{R}} t_n$ is said* **innermost** *iff for all rewrite step $t_i \longrightarrow_{\mathcal{R}}^{\alpha_i, \sigma_i} t_{i+1}$, the reducible position (for short* **redpos***) α_i is such that for all position $\beta > \alpha_i$ in $Dom(t_i)$ we have $t_{i|\beta} = t_{i|\beta}\downarrow_{\mathcal{R}}$.*

Remark 1 *The following straightforward remarks about R can be done:*

a. *If there is one step $t_1 \longrightarrow_{R_2}^{\varepsilon, \sigma} t_2$ in an innermost R-derivation then t_2 is in R-normal form: t_2 is in R-normal form because no rule can be applied to $t_2 = \sigma(s_2 f(\perp, a_k y, z))$ since $\sigma(a_k y)$, and $\sigma(z)$ are in R-normal form.*

b. *In every innermost R-derivation $t_1 \longrightarrow_R^{\varepsilon} t_2 \xrightarrow{*}_R t_3$ there is no rewrite step at ε or 1 between t_2 and t_3: If $t_1 \longrightarrow_{R_1}^{\varepsilon, \sigma} t_2 = \sigma(s_1 s_2(u))$ or if $t_1 \longrightarrow_{R_3}^{\varepsilon, \sigma} t_2 = \sigma(s_2 s_1(u))$ then there is no rewrite step at ε or 1 between t_2 and t_3 because for all $l \longrightarrow r \in R$ $l(\varepsilon) \notin \{s_1, s_2\}$. If $t_1 \longrightarrow_{R_2}^{\varepsilon} t_2$ then t_2 is in R-normal form.*

Lemma 1 *If $t_0 \longrightarrow_{l_1 \longrightarrow r_1 \in R_1}^{\varepsilon, \sigma_1} t_1 \longrightarrow \ldots t_{n-1} \longrightarrow_{l_n \longrightarrow r_n \in R}^{\sigma_n} t_n$ is an innermost R-derivation then:*

- *For all $i \in [0..n]$, t_i is of the form $(s_1 s_2)^i f(\vec{u})$.*
- *For all $i \in [0..n]$, if t_i is R-reducible then t_i has a unique R-redpos at 1^{2i}.*
- *All rewrite step between t_0 and t_n use rules of R_1.*

Proof: For the proof we proceed by induction on n.

$n = 0$ $t_0 = \sigma_1(l_1) = \sigma_1(f(a_k x, y, z))$ has a unique R-redpos at $\varepsilon = 1^{2n}$ since we consider an innermost R-derivation.

$n > 0$ By hypothesis $t_{n-1} = (s_1 s_2)^{n-1} f(\vec{u})$ has a unique R-redpos $\alpha = 1^{2(n-1)}$. Thus we can only apply a rule $l_n \longrightarrow r_n$ in R_1 at α. Hence $t_n = \sigma_n((s_1 s_2)^{n-1} r_n) = \sigma_n((s_1 s_2)^n f(x, a_k y, z))$. Since we consider an innermost R-derivation $\sigma_n(a_k x)$, $\sigma_n(y)$,and $\sigma_n(z)$, are in R-normal form. It means that if t_n is not in R-normal form then $\alpha = 1^{2n}$ is the unique R-redpos of t_n.

□

Lemma 2 *In every innermost R-proof of the form* $t_0 \longrightarrow^{\varepsilon}_{R_1} t_1 \xrightarrow{*}_R t \xleftarrow{*}_R t'_1 \longleftarrow^{\varepsilon}_{R_1} t'_0$, *$t_0$ is equal to t'_0.*

Proof: By the lemma 1, $t = (s_1 s_2)^n f(\vec{u})$ and we can write the R-proof as:

$$t_0 \longrightarrow^{\varepsilon}_{R_1} t_1 \xrightarrow{n-1 \neq \varepsilon}_{R_1} t_n = t = t'_n \xleftarrow{n-1 \neq \varepsilon}_{R_1} t'_1 \longleftarrow^{\varepsilon}_{R_1} t'_0$$

where $\forall i \in [0..n-1]$ there is a unique R-redpos at 1^{2i} in t_i and t'_i. Thus $\forall i \in [0..n]$, $t_i = t'_i$ since there is no superpositions between right members of rules of R_1. □

Lemma 3 *In every innermost R-proof of the form:* $t_0 \longrightarrow^{\varepsilon}_{R_3} t_1 \xrightarrow{*}_R t \xleftarrow{*}_R t'_1 \longleftarrow^{\varepsilon}_{R_3} t'_0$, *$t_0$ is equal to t'_0. (The proof of this lemma is similar to lemma 2).*

Lemma 4 *In every innermost R-proof of the form:* $t_0 \longrightarrow^{\varepsilon}_{R_2} t_1 \xrightarrow{*}_R t \xleftarrow{*}_R t'_1 \longleftarrow^{\varepsilon}_{R_2} t'_0$, *$t_0$ is equal to t'_0 and t_1 and t'_1 are in R-normal form.*

Proof: By the remark 1a and since there is no superpositions between right members of rules of R_2. □

Proposition 3 *A is a resolvent presentation.*

Proof: Given t_1, t'_1 any two A-equals terms we study one among their innermost R-proofs. If there is more than one rewrite step at ε in the R-proof then we show that there exists an A-proof with at most one equality step at ε between these two terms. If there is more than one rewrite step at ε between t_1 and t'_1 then by the remark 1b the innermost R-proof has the following form:

$$t_1 \xrightarrow{* \neq \varepsilon}_R t_2 \longrightarrow^{\varepsilon}_{l \longrightarrow r} t_3 \xrightarrow{* \neq \varepsilon}_R t_1\downarrow = t = t'_1\downarrow \xleftarrow{* \neq \varepsilon}_R t'_3 \longleftarrow^{\varepsilon}_{l' \longrightarrow r'} t'_2 \xleftarrow{* \neq \varepsilon}_R t'_1$$

where $t_3(\varepsilon) = t(\varepsilon) = t'_3(\varepsilon)$ since there is no step at ε between t_3 and t'_3. Thus $l \longrightarrow r$, $l' \longrightarrow r' \in R_1$ or $l \longrightarrow r$, $l' \longrightarrow r' \in R_2 \cup R_3$.

- If $l \longrightarrow r, l' \longrightarrow r' \in R_1$ then by the lemma 2 $t_2 = t'_2$, thus there exists a A-proof of $t_1 =_A t'_1$ with at most one step at ε: $t_1 \xleftrightarrow{* \neq \varepsilon}_A t_2 = t'_2 \xleftrightarrow{* \neq \varepsilon}_A t'_1$
- If $l \longrightarrow r, l' \longrightarrow r' \in R_2$ then the proof is similar by the lemma 4.
- If $l \longrightarrow r, l' \longrightarrow r' \in R_3$ then proof is similar by the lemma 3.
- If $l \longrightarrow r \in R_2$ and $l' \longrightarrow r' \in R_3$ then $t_1\downarrow = t_3 = s_2 f(\vec{u})$ by the remark 1a, and $t'_1\downarrow = s_2 s_1(s')$ since $t'_3 = s_2 s_1 f(\vec{u'})$ and since there is no rewrite set at ε or 1 between t'_3 and $t'_1\downarrow$ by the remark 1b. Thus this case is not possible because $t_1\downarrow \neq t'_1\downarrow$.

□

3.2 Unifiability is not decidable in syntactic theories

Now we may prove that unifiability is not decidable in syntactic theories. The idea is to study the unifiability of $U = (s_2 f(x,\bot,\bot) =^? h(y,\bot,\bot))$ in the theory A. In order to test the unifiability of this unificand we can use narrowing as a complete unification procedure since R is a canonical rewriting system. Hence we prove that the narrowing leads to a A-solution of U iff the instance of PCP associated with $=_A$ has a solution.

Lemma 5 *There exist a narrowing derivation:*

$$U = s_2 f(x_0,\bot,\bot) \stackrel{?}{=} h(y_0,\bot,\bot) \rightsquigarrow_R^{\rho_1} \cdots \rightsquigarrow_R^{\rho_p} l \stackrel{?}{=} r$$

iff there exist $k, m \in N$ such that:

$$\begin{array}{llll} Form1l: \ l = & (s_2 s_1)^k & s_2 f(x_k, a_{i_k} \ldots a_{i_1}\bot, w_{i_k} \ldots w_{i_1}\bot) & \sigma(x_0) = a_{i_1} \ldots a_{i_k} x_k \\ Form1r: \ r = & (s_2 s_1)^m & h(y_m, a_{j_m} \ldots a_{j_1}\bot, w'_{j_m} \ldots w'_{j_1}\bot) & \sigma(y_0) = a_{j_1} \ldots a_{j_m} y_m \quad or \\ Form2r: \ r = & (s_2 s_1)^{m+1} & s_2 f(\bot, a_{j_{m+1}} \ldots a_{j_1}\bot, w'_{j_{m+1}} \ldots w'_{j_1}\bot) & \sigma(y_0) = a_{j_1} \ldots a_{j_{m+1}}\bot \end{array}$$

where $\sigma = \rho_n \ldots \rho_1$, and x_k, y_m are new variables.

Proof: The narrowing derivations of the two members of U are independent since all variables are distinct in this unificand. Hence we prove this lemma by an induction on k and m. k is the number of narrowing step issued from the left member of U, and m is the number of narrowing step issued from its right member. Let n be the number of pairs of words in the instance of PCP associated with $=_A$.

If $k = 0$ then l_0 is in $Form1l$ and $\sigma_0(x_0) = x_0$ since l_0 is the left member of the initial unificand U.

If $k > 0$ then by hypothesis l_{k-1} is in $Form1l$ and $\sigma_{k-1}(x_0) = a_{i_1} \ldots a_{i_{k-1}} x_{k-1}$. Thus there exist $n = Card(R_1)$ possibilities to continue the derivation:

$$\begin{array}{lll} l_{k-1} & = & (s_2 s_1)^{k-1} \quad s_2 f(x_{k-1}, a_{i_{k-1}} \cdots a_{i_1}\bot, w_{i_{k-1}} \cdots w_{i_1}\bot) \rightsquigarrow_{R_1}^{\{x_{k-1} \mapsto a_{i_k} x_k\}} \\ l_k & = & (s_2 s_1)^k \quad s_2 f(x_k, a_{i_k} \ldots a_{i_1}\bot, w_{i_k} \ldots w_{i_1}\bot) \end{array}$$

where l_k is in $Form1l$ and:

$$\sigma_k(x_0) = \{x_{k-1} \mapsto a_{i_k} x_k\}(a_{i_1} \ldots a_{i_{k-1}} x_{k-1}) = a_{i_1} \ldots a_{i_k} x_k$$

Conversely, for every term l_k in $Form1l$ there exists a term l_{k-1} in $Form1l$.

The induction on m is similar to the induction on k. □

Lemma 6 *Let $U = (s_2 f(x,\bot,\bot) =^? h(y,\bot,\bot))$ be a unificand. U has an A-solution iff the instance of PCP associated with $=_A$ has a solution.*

Proof: In order to test the unifiability of U we can use narrowing as a complete unification procedure since R is a canonical rewriting system. U has an A-solution iff from U we can deduce by narrowing a unificand $U' = (l =^? r)$ which can be unified in the empty theory. By the lemma 5 we know the form of these unificands. Hence $l =^? r$ can be unified in the empty theory iff:

l in *Form1l*:	$(s_2 s_1)^k$	$s_2 f(x_k, a_{i_k} \ldots a_{i_1}\bot, w_{i_k} \ldots w_{i_1}\bot)$	and
r in *Form2r*:	$(s_2 s_1)^{m+1}$	$s_2 f(\bot, a_{j_{m+1}} \ldots a_{j_1}\bot, w'_{j_{m+1}} \ldots w'_{j_1}\bot)$	and
$k = m + 1$	and		
$a_{ik} \ldots a_{i1}\bot = a_{jk} \ldots a_{j1}\bot$	and		
$w_{ik} \ldots w_{i1}\bot = w'_{jk} \ldots w'_{j1}\bot$			

because the other cases lead to a clash of symbols. If n is the number of pairs of words in the instance of PCP associated with $=_A$ then by the lemma 5 we know that for all $k \geq 1$ and all $i_1, \ldots, i_k \in [1..n]$ there exists a unificand:

$$(s_2 s_1)^k s_2 f(x_k, a_{ik} \ldots a_{i1}\bot, w_{ik} \ldots w_{i1}\bot) \stackrel{?}{=} (s_2 s_1)^k s_2 f(\bot, a_{ik} \ldots a_{i1}\bot, w'_{ik} \ldots w'_{i1}\bot)$$

which is deduced by narrowing from U. Therefore U has a A-solution iff there exist $i_1, \ldots, i_k$ such that: $w_{ik} \ldots w_{i1}\bot = w'_{ik} \ldots w'_{i1}\bot$ thus iff the instance of PCP associated with the theory $=_A$ has a solution: $w_{i1} \ldots w_{ik} = w'_{i1} \ldots w'_{ik}$. □

Theorem 3 *The following problem is not decidable:*

INPUT:
- *$l =^? r$ a linear equation such that $Var(l) \cap Var(r) = \emptyset$.*
- *E a collapse free and syntactic theory which has a regular, linear, and resolvent presentation A.*

QUESTION: - *Does it exist a A-solution of $l =^? r$?*

Proof: It is an immediate consequence of lemma 6 since the PCP is not decidable. □

Corollary 1 *Unifiability in syntactic theories is not decidable.*

4 Test of resolventness

In this section we prove that the resolventness of a presentation is even not semi-decidable. To show that, we complete the resolvent presentation A of the previous section in such a way that the new presentation A' is resolvent iff the instance of PCP associated with $=_{A'}$ has no solution. Therefore we deduce that the resolventness is even not semi-decidable.

We complete the signature presented in the previous section by three new symbols:

- $F'_0 = F_0$
- $F'_1 = F_1 \cup \{g_1, g_2, g_3\}$
- $F'_3 = F_3$
- $F' = F'_0 \cup F'_1 \cup F'_3$

For the presentation of E' we add to the presentation A of E two new axioms:

$$A' = A \cup \left\{ \begin{array}{lll} A_4 = & \{g_1(x) & = & g_2(s_2(f(x,\bot,\bot)))\ \} \\ A_5 = & \{g_3(x) & = & g_2(h(x,\bot,\bot))\ \} \end{array} \right.$$

Example 6 *Consider the instance of PCP of example 4 and its associated theory $=_{A'}$. We will prove that there exists t and t' such that all A'-proof between t and t' have more than one step at ε iff the instance of PCP has a solution. For our instance of PCP, a possibility is $t = g_1(a_3a_1a_1a_{2\perp})$ and $t' = g_3(a_3a_1a_1a_{2\perp})$. An A'-proof with two steps at ε is for example:*

$$
\begin{array}{llll}
g_1(a_3a_1a_1a_{2\perp}) & \longleftrightarrow^{\varepsilon}_{A_4} & & \\
g_2s_2f(a_3a_1a_1a_{2\perp},\perp,\perp) & \longleftrightarrow^{\neq\varepsilon}_{A_1} & g_2s_2f(a_1a_1a_{2\perp},a_{3\perp},10_\perp) & \overset{3}{\longleftrightarrow}^{\neq\varepsilon}_{A_1} \\
g_2s_2f(\perp,a_2a_1a_1a_{3\perp},101111110_\perp) & \longleftrightarrow^{\neq\varepsilon}_{A_2} & g_2h(\perp,a_2a_1a_1a_{3\perp},101111110_\perp) & \overset{3}{\longleftrightarrow}^{\neq\varepsilon}_{A_3} \\
g_2h(a_1a_1a_{2\perp},a_{3\perp},0_\perp) & \longleftrightarrow^{\neq\varepsilon}_{A_3} & g_2h(a_3a_1a_1a_{2\perp},\perp,\perp) & \longleftrightarrow^{\varepsilon}_{A_5} \\
g_3(a_3a_1a_1a_{2\perp}) & & &
\end{array}
$$

A' can be turned into the following canonical rewriting system R' where R is the rewriting system of the previous section:

$$
R' = R \cup \left\{ \begin{array}{lllll} R_4 = & \{g_1(x) & \longrightarrow & g_2(s_2(f(x,\perp,\perp))) & \} \\ R_5 = & \{g_3(x) & \longrightarrow & g_2(h(x,\perp,\perp)) & \} \end{array} \right.
$$

Remark 2 *The following remarks about R' can be done:*

a. *Every innermost R'-derivation $t_0 \longrightarrow^{\varepsilon}_{R'} t_1 \xrightarrow{*}_{R'} t_n$ has in fact the following form $t_0 \longrightarrow^{\varepsilon}_{R'} t_1 \xrightarrow{*}^{\neq\varepsilon}_{R} t_n$: if the first step uses a rule in $R'\backslash R$ then we can only apply a rule in R after it since we consider an innermost proof. Hence no rule in $R'\backslash R$ can be applied later since the rules of R does not introduce symbols in $\{g_1 \cup g_3\}$.*

b. *Every innermost R'-proof $t_1 \longrightarrow^{\varepsilon}_{r\in R'} t_2 \xrightarrow{*}_{R'} t \xleftarrow{*}_{R'} t'_2 \longleftarrow^{\varepsilon}_{r'\in R'} t'_1$ is such that $r, r' \in R$ or $r, r' \in R'\backslash R$: it is a direct consequence of the previous remark.*

Lemma 7 *For every innermost R'-proof $t_1 \longrightarrow^{\varepsilon}_{R} t_2 \xrightarrow{*}_{R'} t \xleftarrow{*}_{R'} t'_2 \longleftarrow^{\varepsilon}_{R} t'_1$ there exists an A'-proof of $t_1 =_{A'} t'_1$ with at most one step at ε.*

Proof: By the remark 2a this R'-proof is the R-proof $t_1 \longrightarrow^{\varepsilon}_{R} t_2 \xrightarrow{*}^{\neq\varepsilon}_{R} t \xleftarrow{*}^{\neq\varepsilon}_{R} t'_2 \longleftarrow^{\varepsilon}_{R} t'_1$ thus there exists an A-proof of $t_1 =_{A'} t'_1$ with at most one step at ε since A is a resolvent presentation. □

Lemma 8 *For every innermost R'-proof $t_1 \longrightarrow^{\varepsilon,\sigma}_{R_4} t_2 \xrightarrow{*}_{R'} t \xleftarrow{*}_{R'} t'_2 \longleftarrow^{\varepsilon,\sigma'}_{R_4} t'_1$ there exists an A'-proof of $t_1 =_{A'} t'_1$ with at most one step at ε.*

Proof: By the remark 2a we can write this R'-proof as:

$$
t_1 \longrightarrow^{\varepsilon,\sigma}_{R_4} t_2 = \sigma(g_2s2(f(x,\perp,\perp))) \xrightarrow{*}^{\neq\varepsilon}_{R} t \xleftarrow{*}^{\neq\varepsilon}_{R} \sigma'(g_2s2(f(y,\perp,\perp))) = t'_2 \longleftarrow^{\varepsilon,\sigma'}_{R_4} t'_1
$$

Since $\{l(\varepsilon) \mid l \longrightarrow r \in R\} \cap \{g_2, s_2\} = \emptyset$ there exists an innermost R-proof:

$$
u_2 = \sigma(f(x,\perp,\perp)) \xrightarrow{k}_{R} u \xleftarrow{l}_{R} \sigma'(f(y,\perp,\perp)) = u'_2
$$

If $k = l = 0$ then $t_1 = t'_1$ else there exists a unique R_1-redpos at ε in u_2 or u'_2 since σ and σ' are R'-normalized. But by the lemma 1 this means that $k > 0$, $l > 0$, and $k = l$. Therefore $u_2 = u'_2$ by the lemma 2 and $t_1 = t'_1$. □

Lemma 9 *For every innermost R'-proof $t_1 \xrightarrow{\varepsilon,\,\sigma}_{R_5} t_2 \xrightarrow{*}_{R'} t \xleftarrow{*}_{R'} t'_2 \xleftarrow{\varepsilon,\,\sigma'}_{R_5} t'_1$ there exists an A'-proof of $t_1 =_{A'} t'_1$ with at most one step at ε. (The proof of this lemma is similar to the one of lemma 8).*

Lemma 10 *There exists an innermost R'-proof $t_1 \xrightarrow{\varepsilon,\,\sigma}_{R_4} t_2 \xrightarrow{*}_{R'} t \xleftarrow{*}_{R'} t'_2 \xleftarrow{\varepsilon,\,\sigma'}_{R_5} t'_1$ iff the instance of PCP associated to $=_{A'}$ has a solution. Furthermore if such a R'-proof exists then there is no A'-proof of $t_1 =_{A'} t'_1$ with at most one step at ε.*

Proof: By the remark 2a we can write the R'-proof as:

$$t_1 \xrightarrow{\varepsilon,\,\sigma}_{R_4} t_2 = \sigma(g_2 s2(f(x,\bot,\bot))) \xrightarrow{*\;\neq\varepsilon}_{R} t \xleftarrow{*\;\neq\varepsilon}_{R} \sigma'(g_2(h(y,\bot,\bot))) = t'_2 \xleftarrow{\varepsilon,\,\sigma'}_{R_5} t'_1$$

Since there is no rewrite step at ε between t_2 and t'_2 such a R-proof exists iff there exists an innermost R-proof:

$$\sigma(s_2(f(x,\bot,\bot))) \xrightarrow{*}_{R} t \xleftarrow{*}_{R} \sigma'(h(y,\bot,\bot))$$

Thus this R-proof exits iff $\sigma \cup \sigma'$ is a A-solution of $s_2(f(x,\bot,\bot)) =^? (h(y,\bot,\bot))$ and by the lemma 6 iff the instance of PCP associated with $=_{A'}$ has a solution. Obviously if there exists a R'-proof of $t_1 =_{A'} t'_1$ then there is no A'-proof of $t_1 =_{A'} t'_1$ with at most one step at ε because there is no axioms $l \longrightarrow r$ in $A' \cup A'^{-1}$ such that $l(\varepsilon) = g_1$ and $r(\varepsilon) = g_3$. □

As a direct consequence of the lemmas 7, 8, 9, 10, we get:

Lemma 11 *A' is resolvent iff the instance of PCP associated with $=_{A'}$ has no solution.*

Theorem 4 *The following problem is even not semi-decidable:*

INPUT: - *A a regular, linear and collapse free presentation.*

QUESTION: - *Is A a resolvent presentation ?*

Proof: Let us consider a presentation A' associated with an instance of PCP. If we suppose that this question is semi-decidable then by the lemma 11 It means that it is semi-decidable to know if an instance of PCP has no solution. But it is well-known that it is not the case. □

5 Test of syntacticness

In this section we prove that the syntacticness of a theory is even not semi-decidable. To show that, we consider the presentation A' of the previous section. The idea to check the syntacticness of the theory generated by A' is to use the theorem 1. Thus we study the general equation $g_1(x) =^? g_3(y)$ in order to show that this equation has a finite and complete set of A'-solutions iff the instance of PCP has no solution.

Remark 3 *Here are some simple primary results:*

a. *If* $e = (l_0 =^? r_0)$ *is an equation such that* $Symb(l) \cup Symb(r) \subseteq F \cup X$ *then the narrowing R-derivations and the narrowing R'-derivations issued from e are the same.*

b. *Let g be any symbol in F. If there exists no rule* $l \longrightarrow r$ *in R' such that* $l(\varepsilon) = g$ *then the complete sets of A'-solutions* $CSU(A', t =^? t')$ *and* $CSU(A', g(t) =^? g(t'))$ *are the same.*

c. *If* $t_1 =_{A'} t_1'$ *and* $t_2 =_{A'} t_2'$ *then the complete sets of A'-solutions of* $t_1 =^? t_2$ *and* $t_1' =^? t_2'$ *are the same.*

Lemma 12 *All complete sets of A'-solutions of the general equation* $g_1(x) =^? g_3(y)$ *are infinite iff the instance of PCP associated with* $=_{A'}$ *has a solution.*

Proof: Let be e the general equation $g_1(x) =^? g_3(y)$. We can rewrite e into e_1:

$$g_2(s_2(f(x, \bot, \bot))) \stackrel{?}{=} g_2(h(y, \bot, \bot))$$

This new equation has the same sets of A'-solutions than e by the remark 3c. The equation $e_2 : s_2(f(x, \bot, \bot)) =^? h(y, \bot, \bot)$ has the same sets of A'-solutions than e_1 by the remark 3b. We have proved in lemma 6 that e_2 is A-unifiable iff the instance of PCP has a solution thus:

- If the instance of PCP has no solution then the complete set of A'-solution of e is empty.
- In the lemma 6 we show that the instance of PCP has a solution $(i_1, \ldots, i_k)$ iff there exists a narrowing R-derivation:

$$e_2 \overset{\sigma_1}{\leadsto}_R \cdots \overset{\sigma_n}{\leadsto}_R e_3$$

such that e_3 can be unified in the empty theory and e_3 is of the form:

$$(s_2 s_1)^k s_2 f(x_k, a_{ik} \ldots a_{i1}\bot, w_{ik} \ldots w_{i1}\bot) \stackrel{?}{=} (s_2 s_1)^k s_2 f(\bot, a_{ik} \ldots a_{i1}\bot, w'_{ik} \ldots w'_{i1}\bot)$$

By the lemma 5 we know that:

$$\text{if} \quad \sigma = \sigma_n \ldots \sigma_1 \quad \text{then} \quad \sigma_{|x,y} = \begin{cases} x & \mapsto & a_{i1} \ldots a_{ik} x_k \\ y & \mapsto & a_{i1} \ldots a_{ik}\bot \end{cases}$$

Let $(i_1, \ldots, i_m)$ be one among the shortest solutions of the instance of the PCP. If we write out as W the word $a_{im} \ldots a_{i1}$ then the following infinite set is a subset of any complete set of A-solution of e_3:

$$\bigcup_{j \in N^+} \begin{cases} x & \mapsto & W^j\bot \\ y & \mapsto & W^j\bot \end{cases}$$

It means by the remark 3a that this infinite set is a subset of any complete set of A'-solution of e_3. Thus if the instance of PCP has a solution then all complete sets of A'-solution of e are infinite.

□

Lemma 13 *$=_{A'}$ is a syntactic theory iff the instance of PCP associated with $=_{A'}$ has no solution.*

Proof: If the instance of PCP has no solution then by the lemma 11 A' is a resolvent presentation thus $=_{A'}$ is a syntactic theory. If the instance of PCP has a solution then by the lemma 12 the general equation $g_1(x) =^? g_3(y)$ has no finite complete set of A'-solutions thus by the theorem 1 $=_{A'}$ is not a syntactic theory. □

As a direct consequence we get:

Theorem 5 *The following problem is even not semi-decidable:*

INPUT: - *A a regular, linear and collapse free presentation.*

QUESTION: - *Is the theory generated by A syntactic ?*

6 Conclusion

Syntactic theories have been introduced in unification theory in order to help to build automatically unification algorithms. This work proves that the syntacticness condition alone is too weak to reach this goal. An interesting point is that this result is proved for linear theory and unificand with distinct variables. Hence to built automatically unification algorithms is a very difficult goal since furthermore we prove that syntacticness and resolventness are even not semi-decidable. We think that syntacticness is a necessary condition to built automatically unification algorithm because without this condition we cannot know when we can decompose a unificand. Thus the main open problem is: which kind of conditions must be added to syntacticness to get the decidability of the unifiability. These conditions probably exist, since for example, syntacticness is even not semi-decidable but some people gave powerful sufficient conditions. Thanks to T.Nipkow's work, our results hold in equational theories where the inference rule of transitivity can be removed from the inference rules system without losing the completeness.

Acknowledgements: I would like to thank C. Kirchner, E. Domenjoud, P. Marchand, J.P. Jouannaud, H. Comon and T. Nipkow for fruitful discussions.

References

[1] J. Christian. High performance permutative completion. Technical report ACT-AI-303-89, MCC, 1989. PhD thesis.

[2] H. Comon. Unification et disunification. Théories et applications. Thèse d'Université de l'Institut Polytechnique de Grenoble, 1988.

[3] N. Dershowitz and J.-P. Jouannaud. *Handbook of Theoretical Computer Science*, volume B, chapter 15: Rewrite systems. North-Holland, 1990. Also as: Research report 478, LRI.

[4] G. Huet and D. Oppen. Equations and rewrite rules: A survey. In R. Book, editor, *Formal Language Theory: Perspectives and Open Problems*, pages 349–405. Academic Press, New York, 1980.

[5] J.-P. Jouannaud and C. Kirchner. Solving equations in abstract algebras: A rule-based survey of unification. Research report, CRIN, 1990. To appear in *Festschrift for Robinson*, J.-L. Lassez and G. Plotkin Editors, MIT Press.

[6] J.-P. Jouannaud and H. Kirchner. Completion of a set of rules modulo a set of equations. *SIAM Journal of Computing*, 15:1155–1194, 1986. Preliminary version in Proceedings 11th ACM Symposium on Principles of Programming Languages, Salt Lake City, 1984.

[7] C. Kirchner. Méthodes et outils de conception systématique d'algorithmes d'unification dans les théories équationnelles. Thèse d'état, Université de Nancy I, 1985.

[8] C. Kirchner. Computing unification algorithms. In *Proceeding of the First Symposium on Logic In Computer Science, Boston (USA)*, pages 206–216, 1986.

[9] C. Kirchner and F. Klay. Syntactic theories and unification. In *Proceedings 5th IEEE Symposium on Logic in Computer Science, Philadelphia (Pennsylvania, USA)*, pages 270–277, 1990.

[10] F. Klay. Undecidable properties of syntactic theories. Rapport interne crin, Centre de Recherche en Informatique de Nancy, 1990.

[11] T. Nipkow. Proof transformations for equational theories. In *Proceedings 5th IEEE Symposium on Logic in Computer Science, Philadelphia (Pennsylvania, USA)*, pages 278–288, 1990.

[12] D. Rémy. Algèbres touffues. Application au typage polymorphe des objets enregistrements dans les langages fonctionnels. Thèse de l'Université de Paris 7, 1990.

[13] J. Siekmann. Unification theory. *Journal of Symbolic Computation*, 7:207–274, 1989. Special issue on unification. Part one.

GOAL DIRECTED STRATEGIES FOR PARAMODULATION

Wayne Snyder and Christopher Lynch
Boston University
Department of Computer Science
111 Cummington St.
Boston, MA 02215
snyder@cs.bu.edu

It is well-known that the *set of support* strategy is incomplete in paramodulation theorem provers if paramodulation into variables is forbidden. In this paper, we present a paramodulation calculus for which the combination of these two restrictions is complete, based on a lazy form of the paramodulation rule which delays parts of the unification step. The refutational completeness of this method is proved by transforming proofs given by other paramodulation strategies into set of support proofs using this new inference rule. Finally, we consider the completeness of various refinements of the method, and conclude by discussing related work and future directions.

1 Introduction

A central problem in making theorem provers more efficient is restricting the inference rules to eliminate irrelevant inferences without sacrificing completeness. In resolution theorem provers [17], for example, a wide variety of refinements has been developed (see [12,5,19]), the most successful of which has perhaps been the *set of support* strategy [24,25], which seeks a refutation for an unsatisfiable set of clauses S by isolating a set $T \subseteq S$ such that $S - T$ is satisfiable (for example, $S - T$ might be the set of hypotheses, and T the negation of the theorem to be proved) and forbidding resolution inferences which take place only among the clauses in $S - T$; hence it is a form of backward reasoning (for a good discussion of the central importance of this, see [16]). In paramodulation theorem provers [18], the crucial refinement was showing that neither explicit equality axioms, nor functional reflexivity axioms and paramodulation into variable positions is necessary for completeness [4,15,11,2]; this is crucially important, since *any* equation can paramodulate into a variable, which generates so many clauses that only the simplest theorems can be proved without the restriction. (Another important refinement in both resolution and paramodulation is the use of orderings, see [12,5,11,2,26,3].)

Unfortunately, these two restrictions can *not* be used together in a paramodulation theorem prover without sacrificing completeness. For example,

$$\{f(a,b) \doteq a,\ a \doteq b\} \models \exists x.\, f(x,x) \doteq x,$$

This research was partially supported by NSF Grant No. CCR-8910268.

so that the set of clauses

$$\{\ \{f(a,b) \doteq a\},\ \{a \doteq b\},\ \{\neg f(x,x) \doteq x\}\ \}$$

is unsatisfiable, but if we pick the obvious set of support, namely the third clause, then we can only obtain a refutation by paramodulating into a variable.[1] This problem has been mentioned in various places [1,23,25], but has curiously received little attention per se. In general it seems to be important to understand precisely what the crux of the problem is, and to find out under exactly what circumstances we *can* combine these two restrictions, and whether the resultant method has any advantages over current approaches to automated theorem proving in first order logic with equality.

In this paper we present a paramodulation calculus which solves this problem by using a *lazy* form of paramodulation, which basically delays the essential parts of the unification step. In the previous example, we would delay the attempt to unify $f(a,b)$ and $f(x,x)$ in the process of paramodulating the first clause into the third (the set of support), and save some parts of the unification problem in the paramodulant, producing

$$\{\neg a \doteq x, \neg b \doteq x, \neg a \doteq x\},$$

which can then be factored into

$$\{\neg a \doteq x, \neg b \doteq x\}$$

(we assume clauses are multisets). Paramodulating the second clause from the original set into this produces

$$\{\neg b \doteq x, \neg b \doteq x\}$$

from which the empty clause can be derived using factoring and equality resolution (i.e., reflection). The use of lazy paramodulation as the basis for a complete inference system for general E-unification was introduced by Jean Gallier and the first author [9,10,21], and a refinement of this technique, called *relaxed* paramodulation, was developed later by Dougherty and Johann [7]. The current paper is an extension of this work from E-unification to the general refutational setting.

In the rest of this paper, we shall develop this method in detail, and sketch the completeness proof from the full paper by showing how to transform paramodulation proofs without set of support into proofs which use this restriction (and where some paramodulation inferences are replaced by relaxed paramodulations); this kind of proof technique (see for example [2]) is technically rather involved, but, as with Herbrand's and with Gentzen's results in proof theory, gives us a good deal of information about the proofs we obtain, and will allow us in our setting to compare the two strategies in detail. For example, proofs with set of support seem to be longer than those without this restriction, but of course their structure is easier to uncover, since it is somehow derived in a backward direction from the theorem being proved. In addition, this proof technique allows the adaptation of other restrictions (such as ordering restrictions on supported clauses) on paramodulation to provide further restrictions of this new method,

1 Technically, of course, we could define the set of support as the whole set, but this is uninteresting, since then there is no restriction.

by observing what structure is preserved when transforming such a proof into a set of support proof.

The outline of this paper is as follows. After a short section of preliminaries, we present the relaxed paramodulation calculus and sketch its refutational completeness; this is the major result of the paper. Next we discuss several possible restrictions, one of which whose completeness is a corollary of our main result, one which is incomplete, and others for which completeness is still open. Finally, we discuss the relationship of our method with previous work, and conclude with our plans for the immediate future of this project.

2 Preliminaries

In this section we present the basic concepts of paramodulation, using the elegant notation of [3] to present paramodulation in a Gentzen sequent calculus. For further information about first-order logic and Gentzen systems, see [8].

We assume the reader is familiar with the basic notions of first–order logic in clausal form. If t is a term, we use $t[s]_\alpha$ indicate that t has a subterm s at address α, and then by $t[u]_\alpha$ we denote the result of replacing s by u at α in t; if α is not significant we omit it. An equation is an atomic formula $s \doteq t$; we write $s \approx t$ to refer (ambiguously) to either $s \doteq t$ or $t \doteq s$. We consider a clause to be a *multiset*[2] of literals, represented as a sequent $A_1, \ldots, A_n \rightarrow B_1, \ldots, B_m$, which is to be interpreted as $\neg A_1 \vee \ldots \vee \neg A_n \vee B_1 \vee \ldots \vee B_m$. The empty clause is then simply $\rightarrow$. We shall use Γ, Δ, Λ, and Θ for multsets of literals, and Γ, Δ is to be understood as $\Gamma \cup \Delta$, and Γ, A as $\Gamma \cup \{A\}$, etc.

As in [3] and [22], for notational simplicity we shall consider only equational languages, i.e., where all atomic formulae are equations; for our purposes this is without loss of generality, since we can always consider a language of two sorts, add a new constant $\mathbf{T}$, and represent an atom A by the equation $A \doteq \mathbf{T}$ (for details, see [8,22]).

The *Herbrand Base* is the set of all ground atomic formula (including equations), and a *Herbrand Structure* is a subset of the Herbrand Base. A Herbrand structure H satisfies a ground clause $\Delta \rightarrow \Gamma$ iff $\Delta \subseteq H$ implies $\Gamma \cap H \neq \emptyset$; H satisfies a clause iff it satisfies all of its ground instances; and H satisfies a set of clauses iff it satisfies every member of the set. A *Herbrand E–Structure* is a Herbrand structure which satisfies the standard set of equational axioms (we henceforth assume that all structures are E–structures). It is well–known that a set of clauses has a model iff it has a Herbrand model.

A *substitution* is any function from variables to terms which is almost everywhere equal to the identity. The *support* of a substitution σ is $D(\sigma) = \{ x \,|\, \sigma(x) \neq x \}$. The set of variables *introduced by* σ is $I(\sigma) = \bigcup_{x \in D(\sigma)} Var(\sigma(x))$. A substitution ρ is a *renaming substitution away from* W if $\rho(x)$ is a variable for every $x \in D(\rho)$, $I(\rho) \cap W = \emptyset$, and for every $x, y \in D(\theta)$, $\rho(x) = \rho(y)$ implies $x = y$. If W is unimportant, then ρ is simply called a *renaming*. In the remainder of this paper, before any clause C is used, we assume it is *renamed away* from all variables in the current context. Thus, when we refer to a *variant* of a clause C, we mean a clause

2 Recall that a multiset is an unordered collection with possible duplicate elements.

$\rho(C)$ with ρ a renaming substitution such that $I(\rho)$ consists of fresh variables that have not appeared before in the current construction, proof, etc.

The *composition* of σ and θ is the substitution denoted by $\sigma\theta$ such that for every variable x we have $\sigma\theta(x) = \widehat{\theta}(\sigma(x))$. A substitution σ is *idempotent* if $\sigma\sigma = \sigma$. We assume the reader is familiar with the notion of an *mgu* of two terms, and we use $mgu(s,t)$ to (ambiguously) refer to an arbitrary *mgu* σ of s and t, and we will assume that all *mgus* are such as produced by the Herbrand–Martelli–Montanari set of transformations [13,10], so that they are idempotent and introduce no new variables.

We now introduce the standard rules for paramodulation, following [3].

Definition 2.1 The *paramodulation calculus* **PC** consists of the following inference figures.

Equality Resolution:

$$\frac{\Gamma, s \approx t \to \Delta}{\sigma(\Gamma \to \Delta)}$$

where $\sigma = mgu(s,t)$. This rule is sometimes called *reflection*.

Factoring:

$$left: \frac{\Gamma, A, A' \to \Delta}{\sigma(\Gamma, A \to \Delta)} \qquad right: \frac{\Gamma \to \Delta, A, A'}{\sigma(\Gamma \to \Delta, A)}$$

where $\sigma = mgu(A, A')$. This is often referred to as *binary* factoring.

Paramodulation:

$$left: \frac{\Gamma \to \Delta, s \approx t \qquad \Lambda, A[s'] \to \Theta}{\sigma(\Gamma, \Lambda, A[t] \to \Delta, \Theta)}$$

$$right: \frac{\Gamma \to \Delta, s \approx t \qquad \Lambda \to A[s'], \Theta}{\sigma(\Gamma, \Lambda \to A[t], \Delta, \Theta)},$$

where $\sigma = mgu(s, s')$ and s' is *not* a variable.

The clauses on the top of the figures are called *premises*, and those below, *conclusions*.

A *proof* is a member of the smallest set of unordered trees whose nodes are clauses, and such that (i) if C is a variant of a clause, then C is a proof; (ii) if Π is a proof with root C, and $\frac{C}{D}$ is an instance of either Equation Resolution or Factoring, then $\frac{\Pi}{D}$ is a proof; and (iii) if Π_1 is a proof with root C_1, and Π_2 a proof with root C_2, and $\frac{C_1 \quad C_2}{D}$ is an instance of Paramodulation, then $\frac{\Pi_1 \quad \Pi_2}{D}$ is a proof. The clause at the root of a proof or subproof is called the *conclusion* of the proof, and the clauses at the leaves are called *axioms*. A proof all of whose axioms are variants of clauses from a set S is called a *proof from* S, and a proof from S with the empty clause as conclusion is called a *refutation of* S.

Note that the variant assumption and the fact that the unification procedure never introduces new variables assures us that proofs are variable pure in the sense that clauses at independant nodes (i.e., not on the same path) have disjoint sets of variables. The major theoretical result concerning this calculus is the following.

Theorem 2.2 (Soundness and Completeness of **PC**) A set of clauses S is unsatisfiable iff there exists a refutation of S.

This theorem is still true if we restrict the rules in various ways, for example by incorporating reduction orderings, or by eliminating either left or right Factoring, but not both (see [4,15,11,19,12,5,3,26]). The strategy we are most concerned with, due to [24], is as follows.

Definition 2.3 Let $T \subseteq S$. A node C in a proof from S has *T-support* iff at least one of the axioms in the subproof rooted at C is a variant of a clause from T. A proof has *T-support* iff every one of its interior (i.e., non-leaf) nodes has T–support.[3]

The intent of the set of support strategy is to find a small T such that $S - T$ is satisfiable; this T is then used to direct the search for a proof by constructing a refutation proof backward from T, which is the reason that S is unsatisfiable. As shown by example in the introduction, this goal directed strategy, although sound, is not complete in general for **PC**. It is complete only in special cases, such as in the non-equational case, or when S is ground [24], or when the set $S - T$ consists of a set of Horn clauses which induces a canonical rewrite relation. It can also be shown that this strategy is complete when the set of equational axioms is added to the set S, since then "$\doteq$" is like any other predicate, and also it can be proved that if functional reflexivity axioms are added, and the restriction "s' is *not* a variable" is removed from Paramodulation, this strategy is complete [25]. As discussed in the introduction, these approaches have basically been rejected as too inefficient, and in large measure negate the advantages of the set of support refinement. We now proceed to discuss our solution to this problem.

3 The Relaxed Paramodulation Calculus

In this section we present the calculus **RPC**, which is based on the notion of *top unification* developed by [7] to refine the method for general E-unification presented by the first author and Jean Gallier in [10].[4]

Definition 3.1 Given an equation $s \approx t$, $TopUnif(s \approx t)$ is a set of equations defined recursively as follows:

$$TopUnif(f(\ldots) \approx g(\ldots)) \text{ is } \textit{undefined} \text{ if } f \neq g;$$
$$TopUnif(x \approx t) = \{x \approx t\};$$
$$TopUnif(f(s_1, \ldots, s_n) \approx f(t_1, \ldots, t_n)) = \bigcup_{1 \leq i \leq n} TopUnif(s_i \approx t_i),$$

3 Note that this is slightly different from the definition in [24], where factoring is permitted on clauses in $S - T$. This is certainly sensible in practice; we are here trying for conceptual clarity.

4 Our original formulation of **RPC** used the form of lazy paramodulation from [10], but clearly the refinement to top unification is superior, since it restricts the number of inferences, and conceptually it clarifies the dividing line between unification and completely lazy unification.

for $n \geq 0$, provided these last are defined; otherwise the result is undefined.

The idea is, informally, to decompose an equation down completely into variable–term or term–variable equations; note that this does not imply that such an equation is unifiable, since in the second case, perhaps $x \in Var(t)$. For example

$$TopUnif\big(f(c,x,g(x,a)) \approx f(c,h(x),g(y,a))\big) \quad = \quad \{x \approx h(x), x \approx y\},$$

but $TopUnif\big(f(c,x,g(x,a)) \approx f(d,h(x),g(y,a))\big)$ is undefined. Two terms s and t are said to *top unify* if $TopUnif(s \approx t)$ is defined. Our new method is as follows.

Definition 3.2 The *relaxed paramodulation calculus* **RPC** consists of the inference figures from **PC** plus:
Relaxed Paramodulation:

$$\textit{left:}\ \frac{\Gamma \to \Delta, s \approx t \qquad \Lambda, A[s'] \to \Theta}{\Gamma, \Lambda, \Psi, A[t] \to \Delta, \Theta}$$

$$\textit{right:}\ \frac{\Gamma \to \Delta, s \approx t \qquad \Lambda \to A[s'], \Theta}{\Gamma, \Lambda, \Psi \to A[t], \Delta, \Theta},$$

where $\Psi = TopUnif(s, s')$ is defined and s' is *not* a variable.

Note that, strictly speaking, the Paramodulation rule is now superfluous, since it can be simulated by one application of Relaxed Paramodulation plus some number of applications of Equality Resolution; however in order to consider the relationship of the two rules in detail it seems clearer to retain it.

Example 3.3 A refutation of the example from the introduction can thus be given in **RPC** as:

$$\dfrac{\dfrac{\dfrac{\to a \doteq b \qquad \dfrac{\dfrac{\to f(a,b) \doteq a \qquad f(x,x) \doteq x \to}{a \doteq x, b \doteq x, a \doteq x \to}\,RP}{a \doteq x, b \doteq x \to}\,ER}{b \doteq x, b \doteq x \to}\,P_L}{b \doteq b \to}\,ER}{\to}\,ER$$

This calculus is an extension of the one presented for general E-unification in [10] and refined (by introducing top unification) in [7]; in [22] we presented a lazy paramodulation method for the Horn clause case for conditional E-unification and for logic programming with equality, based on similar ideas; the proofs in these papers are not simply refutational in nature, but show that these systems are complete wrt the answer substitutions returned. Here extend the principal ideas from these papers to full first order logic with equality, and prove refutational completeness. Since one of our concerns is to develop an appropriate formulation of the proof theory of goal-directed computation in the presence of equality, we intend to show in the full paper in preparation how this formalism can be used to discuss all these former results in a completely rigorous and uniform way; for example, in the Horn clause case, we can show that there always exist linear proofs without factoring, and that by composing the *mgus* generated in all such proofs we can obtain a complete set of answer substitutions. Here we shall only discuss soundness and refutational completeness.

Theorem 3.4 (Soundness of **RPC**) If there exists some refutation of a set of clauses S in the calculus **RPC**, then S is unsatisfiable.

The proof proceeds along standard lines by showing that the premises of each inference figure imply the conclusion.

Theorem 3.5 (Completeness of **RPC**) If a set of clauses S is unsatisfiable, and T is a subset of S such that $S - T$ is satisfiable, then there exists a refutation of S with T–support.

For convenience, let us call a node or subproof "bad" if it is without T–support and "good" otherwise. The basic idea in proving this result is to show how to transform a bad proof (such as is given by the completeness results in [15,11,3]) into a good one. This is done by exchanging the order of inferences at critical nodes where good subproofs and bad subproofs occur together, so that good subproofs are used earlier, hence decreasing the size of the bad subproofs. Since $S - T$ is satisfiable, and **RPC** is sound, there must be at least one good subproof to get this process started, and by showing that the number of nodes in the orignal bad proof must decrease, we have our result. (In certain cases, we have to be careful to apply the transformations to critical nodes of minimal depth in order to obtain the decrease.)

The difficulty comes from the number of cases for the interaction of two inference rules (there are perhaps 32 essential cases to consider in detail) and from some technical details, such as showing that subproofs can be renamed in certain cases. To give the flavor of the proof, we simply show a few representative transformations, referring the interested reader to the full paper in preparation. Suppose there is no overlap among the applications of two equations at a critical node, e.g.,

$$\dfrac{\Gamma' \rightarrow \Delta', l \approx r \qquad \dfrac{\Gamma \rightarrow \Delta, s \approx t \qquad \Lambda, A[s'], B[l'] \rightarrow \Theta}{\sigma_1(\Gamma, \Lambda, A[t], B[l'] \rightarrow \Delta, \Theta)}\, P_L}{\sigma_2(\sigma_1(\Gamma', \Gamma, \Lambda, A[t], B[r] \rightarrow \Delta', \Delta, \Theta))}\, P_L$$

where s' and l' are not variables (note that $\sigma_1(\Gamma' \rightarrow \Delta', l \approx r)$ is just $\Gamma' \rightarrow \Delta', l \approx r$. This subproof can be transformed into the following subproof with fewer bad nodes:

$$\dfrac{\Gamma \rightarrow \Delta, s \approx t \qquad \dfrac{\Gamma' \rightarrow \Delta', l \approx r \qquad \Lambda, A[s'], B[l'] \rightarrow \Theta}{\sigma_3(\Gamma', \Lambda, A[s'], B[r] \rightarrow \Delta', \Theta)}\, P_L}{\sigma_4(\sigma_3(\Gamma', \Gamma, \Lambda, A[t], B[r] \rightarrow \Delta', \Delta, \Theta))}\, P_L$$

Note that the number of bad nodes is one less. The only thing that remains is to show that $\sigma_1\sigma_2$ is equivalent to $\sigma_3\sigma_4$ up to renaming (which is done by showing that both are *mgus* of the set of equations $\{s \doteq s', l \doteq l'\}$), and then show that we can rename the rest of the proof in accordance with this new substitution (using the variable purity of proofs and our assumptions about the unification procedure). This example shows the basic idea when there is no overlap among the application of the two rules.

This last transformation did not introduce Relaxed Paramodulation, but next suppose that there is a critical overlap among the applications of the equations $l \approx r$ and $s \approx t$, e.g., (2) If there is a critical overlap above, i.e.,

$$\dfrac{\Gamma' \to \Delta', l \approx r \qquad \dfrac{\Gamma \to \Delta, s \approx t \qquad \Lambda, A[l'[s']] \to \Theta}{\sigma_1(\Gamma, \Lambda, A[l'[t]] \to \Delta, \Theta)}\, P_L}{\sigma_2(\sigma_1(\Gamma', \Gamma, \Lambda, A[r] \to \Delta', \Delta, \Theta))}\, P_L$$

where s' is not a variable, and occurs in l' at address α. There are two cases. (2a) If l/α is a non-variable term, then there is an overlap among $l \doteq r$ and $t \doteq s$, since $\sigma_2 = mgu(l, \sigma_1(l'[t]))$, and so $\sigma_1\sigma_2$ unifies l and $l'[t]$, since $D(\sigma_1) \cap Var(l) = \emptyset$, which means that l/α and t are unifiable; since l/α is not a variable, we can transform this into

$$\dfrac{\dfrac{\Gamma \to \Delta, t \approx s \qquad \Gamma' \to \Delta', l[t']_\alpha \approx r}{\Gamma, \Gamma' \to \Delta, \Delta', l[s]_\alpha \approx r}\, P_R \qquad \Lambda, A[l'[s']] \to \Theta}{\sigma_4(\sigma_3(\Gamma', \Gamma, \Lambda, A[r] \to \Delta', \Delta, \Theta))}\, P_L$$

Now we must confirm that $\sigma_1\sigma_2$ and $\sigma_3\sigma_4$ are identical up to composition with a renaming substitution, but this is easily done by confirming that each is an *mgu* of $\{s \doteq s', l \doteq l'\}$.

Otherwise, (2b) if l/α *is* a variable or α is not an address in l, then we can do a Relaxed Paramodulation step:

$$\dfrac{\dfrac{\Gamma \to \Delta, s \approx t \qquad \dfrac{\Gamma' \to \Delta', l \approx r \qquad \Lambda, A[l'[s']] \to \Theta}{\Gamma', \Lambda, x_1 \approx u_1, \ldots, x_i \approx u_i[s'], \ldots, x_n \approx u_n, A[l'[r]] \to \Delta', \Theta}\, RP_L}{\sigma_1(\Gamma', \Gamma, \Lambda, x_1 \approx u_1, \ldots, x_i \approx u_i[t], \ldots, x_n \approx u_n, A[l'[r]] \to \Delta', \Delta, \Theta)}\, P_L}{\sigma_2^n(\ldots \sigma_2^1(\sigma_1(\Gamma', \Gamma, \Lambda, A[l'[r]] \to \Delta', \Delta, \Theta))\ldots)}\, ER^n$$

where $\sigma_1 = mgu(s, s')$ and we apply n steps of Equality Resolution to unify the equations $x_1 \approx u_1, \ldots, x_i \approx u_i[t], \ldots, x_n \approx u_n$, producing in the end $\sigma_2^1 \ldots \sigma_2^k$, which is identical up to composition with a renaming with σ_2, since both are *mgus* of $\{x_1 \approx u_1, \ldots, x_n \approx u_n\}$.

These two cases show the basic idea for the interaction of paramodulation steps; the other cases deal with the remaining interactions of various inference rules. The interaction of paramodulation with factoring is the most problematical, and involves some rather messy duplications of subproofs. But these examples suffice to show the spirit of the proof, which consists of churning through all the cases and verifying the correctness and termination of the transformations.

Another approach to proving completeness is to recast the well-known semantic proof, along the lines of [11] or [3], to incorporate the set of support restriction. But a straight–forward adaptation of the induction used in these proofs does not work, and we have been unable to find an appropriate induction measure. In fact, this is unfortunate, since then we could deal directly with the possibility of other refinements, such as the restriction to only left or only right factoring, which are problematic in the syntactic proof sketched above.

A natural next question is to what extent this calculus (and, in particular, applications of Relaxed Paramodulation) may be further restricted. We are investigating two classes of refinements, (i) those which restrict the amount of the unification computation which must be delayed, and hence refine more precisely the boundary between completeness and incompleteness for goal–directed equational inferences, and (ii) those which extend other, well–understood restrictions (e.g., as discussed in [12,5,19,11,26,3]) into the equational goal–directed setting.

To indicate some of the issues involved in the first topic, we first consider a possible refinement which turns out to be incomplete.

Definition 3.6 A proof in **RPC** *with eager paramodulation* contains no Relaxed Paramodulation inference in which s and s' are unifiable.

In other words, when searching for a refutation, we do Relaxed Paramodulation only when Paramodulation fails, which restricts the use of the former rule. Unfortunately, this refinement is not complete in this form, even in the restricted case of Horn clauses.

Example 3.7 Consider the following set of clauses:

$$\begin{aligned} &\rightarrow f(a) \doteq f(b) \\ &\rightarrow R(a) \\ &\rightarrow P(c) \\ R(x) &\rightarrow g(f(b), f(x)) \doteq c \\ P(g(y,y)) &\rightarrow \end{aligned}$$

where the last clause alone is the set of support. Then the rest of the clauses are satisfiable, but the whole set is unsatisfiable; however the reader may confirm that there is no proof under the eager paramodulation strategy, since it must first unify $g(y,y)$ and $g(f(b), f(x))$, producing the negative atom $R(b)$ which can not be resolved away. Another refinement in the same spirit is as follows.

Definition 3.8 A proof in **RPC** *with eager variable elimination* is a proof in **RPC** in which the only inference rule which is applied to a clause of the form

$$\Gamma, x \approx t \rightarrow \Delta,$$

where $x \notin Var(t)$, is Equality Resolution on the equation $x \approx t$.

In other words, whenever such an equation arises in searching for a proof, the variable x is immediately eliminated by Equality Resolution, which applies the substitution $[t/x]$ to the clause. This would in particular attempt to resolve appropriate equations produced by Relaxed Paramodulation. But the completeness of such a refinement is an open problem at this point. This strategy has a curious history in E-unification theory (which can be read about in [10]) and in the theory of Horn clause logic with equality (see [22]); various authors have (incorrectly) claimed completeness in these simpler settings, but to the authors' knowledge, no proof either way has been given. Even worse, it is not obvious that a positive answer to this question in the context of E-unification would transfer to the general setting.

As regards the second class of refinements, an immediate corollary of our completeness proof is that any refinement, e.g. an ordering restriction on paramodulation inferences, which applies to **PC** also applies to **RPC** between any two *supported* clauses which were produced without using Relaxed Paramodulation, since such fragments of the original **PC** proof remain undisturbed by the transformations. Other refinements, such as the adaptation of ordering restrictions in a more fundamental way into **RPC**, and the host of other restrictions discussed for example in [12,5,19], are currently being studied.

4 Conclusion

We have presented in this extended abstract (of a full paper in preparation) a new method for automated deduction in first order logic with equality which may be thought of as a weakening of the notion of unification in paramodulation in just the right way to allow a goal directed search for proofs. This is the most general extension in a refutational setting of the first author's thesis work on general E-unification with Jean Gallier, and of the current authors' extension of this to Horn clause logic with equality [22], and incorporates in an essential way the refinement of [7].

Among the various approaches to automated deduction in the presence of equality, the incompleteness of goal directed search when inferences into variables are forbidden was noticed very early in paramodulation systems [25], and carries over into related systems such as E-resolution [14,1], and RUE resolution [6], which attempt to remove explicit paramodulations by performing special forms of equational reasoning between opposing literals (in some ways, these are reminiscent of E-unification strategies). Since these can be simulated by a paramodulation calculus (see, for example,[1]), the incompleteness of goal directed search remains a problem. On the other hand, the *modification method* of Brand is apparently complete under set of support, a fact which is mentioned only in passing in [4]. This may be thought of as a technique for transforming a set of clauses into an equivalent form using the axioms of equality in such a way that standard resolution may be used, and historically this was the first, albeit indirect, proof of completeness of paramodulation without variable inferences. But this method, which is a rather drastic rearrangement of paramodulation calculus, does not address the central issue of resolving the conflict between goal directed computation and variable inferences in paramodulation. A related approach appears in [16], where a back–chaining (i.e., as in SLD-resolution) method for full first order logic is developed and extended to include equality; however the method requires that the equations form a canonical rewriting system.

The notion of top unification is similar in spirit to the the saving of "equational disagreements" in E-resolution and RUE resolution, except that we require that these disagreements form variable–term pairs, which eliminates far more irrelevant inferences. All of these follow the thesis (discussed in [20], but of course not original) that new forms of reasoning often require new forms of unification. But the former two methods isolate equational reasoning to the resolution of opposing literals, in the spirit of E-unification, whereas ours is a more general calculus derived in a simple and fundamental way from the basic rules of **PC**, and is intended to be a proof–theoretic foundation for understanding goal directed proofs in equational settings.

As mentioned above, our current research involves determining how various restrictions on paramodulation work in this new setting, and in attempting to constrain the application of the Relaxed Paramodulation rule. Also, since paramodulations *from* variables may be necessary in **RPC**, we are investigating a mixed backward and forward reasoning calculus, in which paramodulation from variables is restricted by allowing a limited amount of forward reasoning; hence it would combine the best of both approaches. Finally, we hope to implement these strategies and compare benchmarks with other refinements of paramodulation.

Acknowledgment The first author would like to thank Pierre Lescanne and the members of CRIN at the University of Nancy for their hospitality during a five week visit last summer, where some of the ideas presented here were developed. Many thanks also to Leo Bachmair, Dan Dougherty, Jieh Hsiang, Patty Johann, Michael Rusinowich, and, as always, Jean Gallier.

5 References

[1] Anderson, R., "Completeness Results for E-Resolution," Proceedings of the AFIPS Spring Joint Computer Conference, AFIPS Press, Reston, VA (1970) 653–656.

[2] Bachmair, L., "Proof Normalization for Resolution and Paramodulation," Proceedings of RTA89, Chapel Hill, p. 15–28.

[3] Bachmair, L. and H. Ganzinger, "On Restrictions of Ordered Paramodulation with Simplification," CADE 1990, Kaiserslautern, Germany.

[4] Brand, D., "Proving Theorems with the Modification Method," SIAM Journal of Computing **4**:4 (1975) 412–430.

[5] Chang, C., and R. Lee, *Symbolic Logic and Mechanical Theorem Proving*, Academic Press, New York (1973).

[6] Digricoli, V., and M. Harrison, "Equality-Based Binary Resolution," JACM **33**:2 (April 1986) 253–289.

[7] Dougherty, D., and P. Johann, "An Improved General E-Unification Method," Proceedings of CADE 1990, Kaiserlautern, Germany.

[8] Gallier, J., *Logic for Computer Science: Foundations of Automated Theorem Proving*, Harper and Row, New York (1986).

[9] Gallier, J. and W. Snyder, "A General Complete E-Unification Procedure," Proceedings of RTA 1987, Bordeaux, France.

[10] Gallier, J. and W. Snyder, "Complete Sets of Transformations for General E-Unification," TCS **67** (1989) 203—260.

[11] Hsiang, J., and M. Rusinowitch, "Proving Refutational Completeness of Theorem Proving Strategies: The Transfinite Semantic Tree Method," to appear in JACM (1990).

[12] Loveland, D., *Automated Theorem Proving: A Logical Basis*, North–Holland, Amsterdam (1978).

[13] Martelli, A., and U. Montanari, "An Efficient Unification Algorithm," ACM Transactions on Programming Languages and Systems **4**:2 (1982) 258–282.

[14] Morris, J., "E-Resolution: Extensions of Resolution to Include the Equality Relation," International Joint Conference on Artificial Intelligence, Washington D.C. (1969).

[15] Peterson, G., "A Technique for Establishing Completeness Results in Theorem Proving with Equality," SIAM Journal of Computing **12** (1983) 82–100.

[16] Plaisted, D., and S. Greenbaum, "Problem Representations for Back Chaining and Equality in Resolution Theorem Proving," Proceedings of the First Conference on Artificial Intelligence Applications (1984).

[17] Robinson, J., "A Machine Oriented Logic based on the Resolution Principle," JACM **12** (1965) 23–41.

[18] Robinson, G., and L. Wos, "Paramodulation and Theorem Proving in First–Order Theories with Equality," Machine Intelligence **4** (1969) 133–150.

[19] Rusinowich, M., *Démonstration Automatique: Techniques de Réécriture*, InterEditions, Paris (1989).

[20] Snyder, W., "Higher-Order E-Unification," Proceedings of CADE 1990, Kaiserlautern, Germany.

[21] Snyder, W., *Complete Sets of Transformations for General Unification*, Disseration, University of Pennsylvania (1988). To appear as a book published by Birkhäuser Boston Inc., for their series *Progress in Computer Science and Applied Logic.*

[22] Snyder, W. and C. Lynch, "Complete Inference Systems for Horn Clause Logic with Equality: A Foundation for Logic Programming with Equality," Proceedings of Second International Workshop on Conditional and Typed Rewriting Systems, Montreal, Canada (1990).

[23] Stickel, M., Tutorial on Term Rewriting, CADE 1988, Argonne, IL.

[24] Wos, L., G. Robinson, and D. Carson, "Efficiency and Completeness of the Set of Support Strategy in Theorem Proving," JACM **12**:4 (Oct. 1965) 536–541.

[25] Wos, L., and G. Robinson, "Paramodulation and Set of Support," IRIA Symposium on Automatic Demonstration, Versailles, France (1968).

[26] Zhang, H., *Reduction, Superposition, and Induction: Automated Reasoning in an Equational Logic*, Ph.D. Thesis, Rensselaer Polytechnic Institute (1988).

Minimal solutions of linear diophantine systems : bounds and algorithms

Loïc Pottier

S.A.F.I.R. Project
Institut National de Recherche en Informatique et Automatique
2004 route des Lucioles, Sophia Antipolis, 06565 Valbonne CEDEX, FRANCE
Email : `pottier@mirsa.inria.fr`

Abstract : We give new bounds and algorithms for minimal solutions of linear diophantine systems. These bounds are simply exponential, while previous known bounds were, at least until recently, doubly exponential.

1 Introduction

A linear diophantine system $Ax \leq b$ is a set of inequations with integer coefficients (A is a matrix of integers with m rows and n columns, b a vector of $Z^m$and x a vector of n indeterminates), whose we search integer solutions.
Recall that to decide if such a system has at least one integer solution is NP-complete (it is the NP-completeness of integer linear programming).
We are interested here in describing and computing the set of solutions. Remark that these systems arise in pattern matching compilation theory.
We will reduce our problem to the study of an equivalent problem, which is solving in non-negative integers the systems $Ax = 0$ (Frobenius problem). These systems arise in several sub-fields of equational rewriting theory, for instance in associative-commutative unification, or in Makanin algorithm. Remark that solving systems $Ax = b, x \geq 0$, occurring in AC-unification, reduces also to Frobenius problem by adding a new variable to x.
The non-negative integer solutions of $Ax = 0$ form a sub-monoid M of N^n, generated by its non zero minimal elements for the partial order $(x_1, \ldots, x_n) \preceq (y_1, \ldots, y_n) \iff \forall i, 1 \leq i \leq n, x_i \leq y_i$. They form a finite set. We will call this set "the Hilbert basis of M" (after [F.Giles and W.R.Pulleyblank 79]), and denote it by $\mathcal{H}(M)$.
In the two next sections we will bound and compute the elements of $\mathcal{H}(M)$. The last section applies the previous results to the initial problem, i.e. the resolution of systems $Ax \leq b$.

2 Bounds of $\mathcal{H}(M)$

It is known since [J.Von zur Gathen and M.Sieveking 78] that if $\mathcal{H}(M)$ is non-empty, it contains an element with norm (for example $\|x\|_\infty$, where $\|x\|_\infty = sup_i| x_i |$) at most simply exponential in the size of A (for example $n.m.(log\|A\|_\infty + 2)$).
But we are interested here to uniformly bound the norms of the elements of $\mathcal{H}(M)$.
Let

$$\|M\|_\infty = sup_{x\in\mathcal{H}(M)}\|x\|_\infty$$

and

$$\|M\|_1 = sup_{x\in\mathcal{H}(M)}\|x\|_1$$

(with $\|x\|_1 = \sum_i | x_i |$).
[I.Borosh and L.B.Treybig 76] have upper bounded $\|M\|_\infty$ with an expression which is doubly exponential in the size of A.
As far as we know, two simply exponential bounds exist to bound $\|M\|_\infty$ or $\|M\|_1$.
We have given the first one in [L.Pottier 90].
The second one can be deduced from an rather unknown result of [J.L.Lambert 87], and has been found independently by [E.Domenjoud 90] in a better form.
These two bounds are essentially different, because of their expressions and their proofs.
We give here two new finest bounds, the first one inspired from [L.Pottier 90], the second one from [J.L.Lambert 87] and [A.Koscielski and L.Pacholski 90],.
Recall that in the case of one equation ($m = 1$), [G.Huet 78] and [J.L.Lambert 87] have given bounds only depending of $\|A\|_\infty$. In the case of two equations, [J.F.Romeuf 89] gave a bound which is quadratic in the size of A.

2.1 First bound

This bound is inspirated by [L.Pottier 90].
Let $\|A\|_{1,\infty} = sup_i\{\sum_j | a_{ij} |\}$, and let r be the rank of A.

Theorem 1

$$\|M\|_1 \leq (1 + \|A\|_{1,\infty})^r = B_0$$

Proof :

We can without restriction choose r independent equations of $Ax = 0$. Let $x = (x_1, \ldots, x_n)$ be a non zero element of M, $p = \|x\|_1$, and $\{e_1, \ldots, e_n\}$ be the canonical basis of R^n. For every y in R^n, we note C_y the cube of volume 1 defined by

$$z \in C_y \Leftrightarrow z = y + \sum_{i=1}^{n} \lambda_i e_i, \ \forall i \in [1, n], \ \lambda_i \in [0, 1]$$

We will recursively define a sequence $y^0, \ldots, y^p$ of N^n and a sequence $z^0, \ldots, z^p$ of R^n verifying :

$y^0 = 0 \prec y^1 \prec \ldots \prec y^p = x.$
$\forall k \in [0, p-1],\ \exists j, y^{k+1} = y^k + e_j$
$\forall k \in [0, p], z^k \in C_{y^k} \cap [0, x].$

$y^0 = z^0 = 0$ are clearly convenient .

Suppose we have built y^k, and $0 \leq k \leq p-1$. $[0, x] \cap C_{y^k}$ is the set of all z which writes $\lambda \sum_i x_i e_i$ with $0 \leq \lambda \leq 1$ and

$$\forall i \in [1, n],\ y_i^k \leq \lambda x_i \leq y_i^k + 1$$

It is by hypothesis non empty, it is a segment. Take $z^{k+1} = \lambda_k \sum_i x_i e_i$ its bound where λ_k is maximum. x is non zero, then there exists a j such that

$$\lambda_k = \frac{y_j^k + 1}{x_j} = inf_{i | x_i \neq 0}\{\frac{y_i^k + 1}{x_i}\}$$

Now, let $y^{k+1} = y^k + ej$. We have now

$$\forall i \in [1, n],\ y_i^{k+1} \leq \lambda_k x_i \leq y_i^{k+1} + 1$$

and z^{k+1} belongs then to the cube $C_{y^{k+1}}$.

The points z^{k+1} and y^{k+1} are then correctly built. Finally, if $k = p-1$, then $y^p = x$, because $y^p \preceq x$ and $\|y^p\|_1 = p = \|x\|_1$, and we take $z^p = x$.

Now, let $y'^k = z^k - y^k$. We have now :

$$\forall i, 0 \leq y_i'^k \leq 1$$

then, if $(Ay^k)_i$ is the ith coordinate of Ay^k :

$$|(Ay^k)_i| = |(Az^k)_i - (Ay'^k)_i| = |(Ay'^k)_i|$$

As $0 \leq y_i'^k \leq 1$, there is then at most $\sum_j |a_{ij}| + 1$ possible values for $(Ay^k)_i$ and then at most B_0 distinct vectors Ay^k.

Now, suppose $p > B_0$. By the pigeon holes principle, it exists then i and j, $i > j > 0$ with $Ay^i = Ay^j$. Let $z = y^i - y^j$. We have now $Az = 0$. More we have $0 \prec z \prec x$, and $z \in M$. Then $x \notin \mathcal{H}(M)$.

□

2.2 Second bound

Let a_{ij} be the term of row i and of column j of the matrix A, and $\|A\|_1 = \sum_{i,j} |a_{ij}|$. Let D be the largest absolute value of the minors of A. [J.L.Lambert 87] gives the following result :

Theorem 2 (Lambert)

$$\|M\|_\infty \leq nD$$

From the proof of this result, and also from those of [A.Koscielski and L.Pacholski 90], also inspirated by [J.L.Lambert 87], we obtain the following improvements :

Theorem 3 *Let D_r be the largest absolute value of the minors of order r of A .*

$$\|M\|_\infty \leq (n-r)D_r = B_1$$

and then

$$\|M\|_\infty \leq (n-r)\left(\frac{\|A\|_1}{r}\right)^r = B_2$$

Remark : the first bound is the same as the bound of [E.Domenjoud 90], found independently.
Proof :

Let C be the cone of R^n of non-negative real solutions of $Ax = 0$. Let C_j be its intersection with the hyperplane of equation $x_j = 0$. It is clear that C is the convex hull of the union of the cones C_j. We can recursively apply this decomposition of C to the C_j, while the dimension of built cones is largest than 1. C is then the convex hull of the union of these cones of dimension 1, called "edges" of C.

Every of these edges is then the set of non-negative solutions of a system of equations obtained by choosing r independent equations of $Ax = 0$, and by adding to them $n-r-1$ equations of type $x_j = 0$ in order to keep the system of maximum rank, i.e. $n-1$.

We can then obtain director vectors (with non-negative integer coefficients) of edges by computing the n minors of order $n-1$ for every the preceding systems, that reduces to compute minors of order r of A.

Let $g_1, \ldots, g_k$ be these vectors, which have then their coordinates upper bounded in absolute value by D_r.

M is then included in the non-negative cone that they generate (the linear combinations with non-negative real coefficients), which is exactly C, with dimension at most $n-r$. We have then, with the theorem of Carathéodory :

$$Ax = 0, x \geq 0 \implies \exists j_1, \ldots, j_{n-r}, \exists \alpha_1, \ldots, \alpha_{n-r} \geq 0, x = \sum_{l=1}^{n-r} \alpha_i g_{j_i}$$

If now x is minimal, it is clear that the α_i are strictely smaller than 1. We obtain then the first part of the theorem.

The second part is a simple upper bound of the determinant of a square submatrix A' of order r of A :

$$| det(A') | \leq \prod_j \sum_i | a'_{ij} | \leq \left(\frac{\sum_{i,j} | a'_{ij} |}{r}\right)^r \leq \left(\frac{\|A\|_1}{r}\right)^r$$

(The second inequality is an upper bound of geometric average with arithmetic average).

□

The bound B_1 can be optimal , as we will see on examples, but it is not reasonably computable in practice: is it better to compute *all* the principal minors of A than to directly compute $\mathcal{H}(M)$, for example with the algorithm of [E.Contejean and H.Devie 89] which does not use a bound of $\mathcal{H}(M)$?

2.3 Comparison of B_0, B_1 and B_2

It is clear that these three bounds are simply exponential in the size of A. The following examples show that we can not compare in general the first and the last, the second being sometimes optimal, but being not computable in practice.
We have the following inequalities :

$$\|x\|_1 \leq B_0$$

$$\|x\|_1 \leq nB_2$$

$$\|x\|_\infty \leq B_0$$

$$\|x\|_\infty \leq B_2$$

So we will study the behaviours of the ratios $\frac{B_2}{B_0}$ (bounds of $\|x\|_\infty$) and $\frac{nB_2}{B_0}$ (bounds of $\|x\|_1$), when n or $\|A\|_\infty$ increases to infinity.

2.3.1 Example 1

Let a be an integer greater or equal to 3 and A the matrix

$$\begin{pmatrix} a & 1-a & & \\ & \ddots & \ddots & \\ & & a & 1-a \end{pmatrix}$$

where the non written coefficients are zero.
We have $r = m = n-1$ and $\mathcal{H}(M)$ has only one element : $((a-1)^{n-1}, a(a-1)^{n-2}, \ldots, a^{n-1})$.
Then :

$$\|M\|_\infty = B_1 = a^{n-1}, \|M\|_1 = a^n - (a-1)^n, B_2 = (2a-1)^{n-1}, B_0 = (2a)^{n-1}$$

B_1 is then optimal, B_2 and B_0 being very close.
Asymptoticaly, we have finally :

$$\lim_{n\to\infty} \frac{B_2}{B_0} = 0, \lim_{n\to\infty} \frac{nB_2}{B_0} = 0, \lim_{a\to\infty} \frac{B_2}{B_0} = 1, \lim_{a\to\infty} \frac{nB_2}{B_0} = n$$

2.3.2 Example 2 : magic square matrices

A square matrix is called magic if the sums of its coefficients of a row and of a column are all equal. The magic square matrices of order k with non-negative coefficients are then the non-negative solutions of the system $Ax = 0$, where $n = k^2 + 1, m = 2k$ and :

$$A = \begin{pmatrix} 1 & \dots & 1 & & & & & -1 \\ & & & 1 & \dots & 1 & & -1 \\ & & & & & & \ddots & \vdots \\ 1 & & & 1 & & & & -1 \\ & \ddots & & & \ddots & & \dots & \vdots \\ & & 1 & & & 1 & & -1 \end{pmatrix}$$

where the non written coefficients are zero.
We have $r = 2k - 1$. The Hilbert basis of M is the set of matrices of permutations of order k (cf [R.P.Stanley 83]). Then :

$$\|M\|_\infty = 1, \|M\|_1 = k + 1, B_1 \geq k^2 - 2k + 2$$

$$B_2 = (k^2 - 2k + 2)\left(\frac{2k(k+1)}{2k-1}\right)^{2k-1}, B_0 = (k+2)^{2k-1}$$

and :

$$\lim_{n\to\infty} \frac{B_2}{B_0} = \infty, \lim_{n\to\infty} \frac{nB_2}{B_0} = \infty$$

which gives the inverse behaviour of the preceding example.

3 Algorithms

The subject of this section is the computation of all the elements of $\mathcal{H}(M)$.
The first algorithms are based on the bounds of [G.Huet 78] and [J.L.Lambert 87] relative to one equation, extended to a system of equations, but giving then doubly exponential bounds. They are the followings :

Property 1 *Let $x = (x_1, \dots, x_p, y_1, \dots, y_q)$ be an element of the Hilbert basis of the equation*

$$a_1x_1 + \dots + a_px_p + b_1y_1 + \dots + b_qy_q = 0.$$

where the a_i are non-negative and the b_j are negative. Then :

$$\forall i, | x_i | \leq sup_j | b_j |$$

(Huet)

$$\sum_i x_i \leq sup_j | b_j |$$

(Lambert).
(the part concerning the y_j is symmetric).

$\mathcal{H}(M)$ is then obtained by enumeration under the bound, and in the case of more than one equations, we iterate the method in injecting the solutions of treated equations in the followings (after eventualy having triangularized the matrix A).
In the case of two equations [J.F.Romeuf 89] gives an original method for building a finite automaton enumerating $\mathcal{H}(M)$, and a quadratic bound in this case.

3.1 Algorithm of Contejean-Devie

[E.Contejean and H.Devie 89] have found a elegant algorithm which does not need any bound of $\mathcal{H}(M)$. The principle is the following. Let us order N^n by the order $\preceq$ defined before, and obtain a DAG (directed acyclic graph) of root 0. The algorithm enumerates a part of this DAG with the following principle :
begin with 0, and if the current vertex is a non zero vector x such that for no one among its ancestors y we have $A(x-y)=0$, visite its sons $x+e_j$ verifying $Ax.Ae_j \leq 0$ (the . denoting the scalar product of R^n).
This algorithm suprizingly terminates and is complete. If we do not visite twice a vertex of the DAG, and keep only minimal solutions for $\preceq$, we then obtain $\mathcal{H}(M)$.
Different refined versions of this algorithm exist, which eliminate early in the process some unusefull parts of the DAG.
The only result of complexity about this algorithm is, to our knowledge, a consequence of [L.Baratchart and L.Pottier 89], which gives a doubly exponential bound on the number of visited vertices.
This algorithm has good behaviour in practice, but is expensive if the elements of $\mathcal{H}(M)$ have large norms.

3.2 Algorithm of Domenjoud

In [E.Domenjoud 90] is described an algorithm which only builds solutions of $Ax = 0$ to compute minimal solutions (as the second algorithm that we present does). This recent algorithm would be interesting in pratice.

3.3 An algorithm inspirated by theorem 3

The analysis of the proof of the theorem 1 allows to modify the method of the algorithm of [E.Contejean and H.Devie 89] in only increment x by the e_l such that for every i, the i-th coordinate of $A(x+e_l)$ is between $-\sum_j a_{ij}^+$ and $\sum_j a_{ij}^-$.
The generators are then all obtained as points of the sequences strictly increasing built similarly to the preceding algorithm.

3.4 Use of standard basis

We give here a new algorithm using the preceding bounds on $\|M\|_\infty$ and $\|M\|_1$, based on the theory of standard basis (or Gröbner basis).
Let us recall basic notions of standard basis.
For a polynomial P of the ring $K[X_1,\dots,X_n]$, we note $in(f)$ the maximum monomial of f w.r.t a choosen admissible ordering on monomials (i.e. a total ordering stable by

multiplication of monomials, and with the monomial 1 as minimal element). Then a familly F of polynomials of an ideal $\mathcal{I}$ is called a *standard basis* of $\mathcal{I}$ if and only if $\{in(f), f \in F\}$ generates the ideal $\{in(f), f \in \mathcal{I}\}$.
A standard basis can be computed by completion algorithms (see [B.Buchberger 83], [A.Galligo 85]).
In our problem, the idea is to see the columns of A as the exponents of monomials in m variables, and the solutions of $Ax = 0$ in Z^nas sysygies relative to these monomials. This idea has been introduced by [F.Ollivier 90] for computation of standard basis of sub-algebras. Then a computation of an appropriate standard basis gives a canonical rewriting system whose the inverse enumerates M by increasing norm. Finaly it suffices to only keep the minimal solutions for $\preceq$ and of norm smaller than $inf\{nB_2, B_0\}$.
Let $T, X_1, \ldots, X_m, Y_1, \ldots, Y_n$ be $n+m+1$ variables, and k be an arbitrary field.
We note a_j for the jth column of A.
For all $\alpha \in Z^m$ and $\beta \in Z^n$, we note X^α and Y^β the monomials $X_1^{\alpha_1} \ldots X_m^{\alpha_m}$ and $Y_1^{\beta_1} \ldots Y_n^{\beta_n}$.
α^+ is the maximum of α and zero (for the partial order $\preceq$), and α^- is the maximum of $-\alpha$ and zero. Then $\alpha = \alpha^+ - \alpha^-$.
For every $j \in [1,n]$, we define a polynomial P_j in the ring $R = k[T, X_1, \ldots, X_m, Y_1, \ldots, Y_n]$:

$$P_j = X^{a_j^+} - Y_j X^{a_j^-}$$

Let $\mathcal{I}$ be the ideal of R generated by the P_j and the polynomial $P_0 = TY_1 \ldots Y_n - 1$, and $\mathcal{J}$ its trace (i.e. its intersection) on the ring $R' = k[Y_1, \ldots, Y_n]$.
Now, let $\mathcal{B}_\mathcal{I}$ be the reduced standard basis of $\mathcal{I}$ for the following ordering on the monomials of R :
we compare first lexicographically the X_i, and in case of equality we use the degree order, and finally the lexicographic order.
Let $\mathcal{B}_\mathcal{J}$ be the set of polynomials of $\mathcal{B}_\mathcal{I}$ where the X_i's and T do not appear.
$\mathcal{B}_\mathcal{J}$ is then a standard basis of the ideal $\mathcal{J}$ for the degree order (from a remark of D.Bayer and M.Stillman). More, its elements are differences of monomials (because those of $\mathcal{B}_\mathcal{I}$ are).
Then let $Y^{\alpha_k} - Y^{\beta_k}$ be the elements of $\mathcal{B}_\mathcal{J}$, $k \in [1,p]$ and Y^{α_k} being the leading monomials.
Now, note $\longrightarrow$ the rewriting relation corresponding to the division of polynomials by the standard basis $\mathcal{B}_\mathcal{J}$, and $\xrightarrow{*}$ its transitive reflexive closure.
We write $m1 \downarrow m2$ when two monomials $m1$ and $m2$ rewrite in the same monomial, or equivalently when $m1 - m2 \xrightarrow{*} 0$.
Then :

Property 2

$$\forall x \in Z^n \text{ , } Ax = 0 \iff Y^{x^+} - Y^{x^-} \in \mathcal{I} \iff Y^{x^+} \downarrow Y^{x^-}$$

Proof :

The first equivalence is easy to show by equational reasoning on the equations $Y_j = X^{a_j^+} X^{-a_j^-}$ derived from polynomials P_j and with the equation $P_0 = 0$ which only allows to eliminate monomials in factor in polynomials of $\mathcal{I}$.

The second assertion is just the fact that a Gröbner basis is a canonical rewriting system equivalent to the relation $P = Q \Leftrightarrow P - Q \in \mathcal{J}$. □

As a consequence :

Property 3

$$\forall x \in N^n \ , \ x \in M \iff Y^x \xrightarrow{*} 1$$

This last property allows to test if M is non reduced to $\{0\}$:

Theorem 4 *The system $Ax = 0$ has a positive solution if and only if it exists in $\mathcal{B}_{\mathcal{I}}$ a polynomial of the form $Y^\alpha - 1$.*

More, we have an effective representation of M with of rewriting rules :
Let SR_M the system of rewriting rules on monomials obtained in reversing the polynomials of $\mathcal{B}_{\mathcal{J}}$:

$$SR_M = \{Y^{\beta_1} \longrightarrow Y^{\alpha_1}, \dots, Y^{\beta_p} \longrightarrow Y^{\alpha_p}\}$$

Note $\longrightarrow_i$ its rewriting relation (it is the symmetric of $\longrightarrow$, and it is not noetherian).
Then

$$x \in M \iff 1 \xrightarrow{*}_i Y^x$$

We can then generate all the elements of M by exploration of the tree of rewritings of 1 by $\longrightarrow_i$, and obtain $\mathcal{H}(M)$ in only keeping the minimal elements of degree smaller than the bounds nB_2 and B_0 (This method is complete because $\longrightarrow_i$ increases the degrees of monomials, and then the norms $\|.\|_1$ of the solutions).
More precisely :

Theorem 5 *The following algorithm stops and returns $\mathcal{H}(M)$:*

1. $E := \{1\}$
2. **While** $\exists x \in E, y \notin E$, *with* $x \longrightarrow_i y$, *and* $deg(y) \leq inf\{nB_2, B_0\}$

   Do $E := E \cup \{y\}$
3. **Return** $\mathcal{H}(M)$:= *minimal elements for* $\preceq$ *of vectors of exponents of monomials of* $E - \{1\}$.

4 Application to $Ax \leq b$

Now come back to the initial problem, i.e. the resolution of a system $Ax \leq b$. Let C be the set of its solutions in Z^n. Then :

Corollary 1 *It exists two finite parts C_1 and C_2 of Z^n such that :*

$$x \in C \Leftrightarrow x = x_1 + x_2 + \ldots + x_k, \text{ with } x_1 \in C_1, \text{ and } x_2, \ldots, x_k \in C_2$$

and

$$\forall x \in C_1 \cup C_2, \ \|x\|_1 \leq (2 + \|A\|_{1,\infty} + \|b\|_\infty)^m$$

Proof :

We will reduce the problem to solve in N a system of homogeneous equations.

Let ψ be an endomorphism of R^n which only change the signs of some coordinates of its argument, and $\psi(A)$ the obtained matrix when changing the signs of the corresponding columns of A.

Let $y = (y_1, \ldots, y_m)$ be a vector of m new variables, z a last variable, t the vector obtained in catenating x, y,and z, and let ϕ be the projection mapping t in x.

Let A' be the matrix obtained in catenating $\psi(A)$, the identity of order m and the opposite of b.

We have now clearly the equivalence :

$$Ax \leq b \ \psi(x) \in N^n \iff \exists t \in N^{n+m+1}, A't = 0, z = 1, x = \psi(\phi(t))$$

More $rank(A') = m$, and $\|A'\|_{1,\infty} \leq \|A\|_{1,\infty} + 1 + \|b\|_\infty$.

Let $\mathcal{H}$ the Hilbert basis of $A't = 0$, and C_1^ψ (resp. C_2^ψ) the image by ϕ of the elements of $\mathcal{H}$ such that $z = 1$ (resp. $z = 0$).

We take then C_1 (resp. C_2) equal to the union of the C_1^ψ (resp. C_2^ψ) for the 2^n possible choices of ψ.

As $\|\psi(x)\|_1 = \|x\|_1$ and $\|\phi(t)\|_1 \leq \|t\|_1$, we obtain the second part of the result. □

5 Acknowledgements

We would like to thank C.Traverso and A.Galligo for usefull discussions about first versions of algorithm 3.3., J.F.Romeuf who indicated us theorem 2, and the referees for their remarks on the first version of this paper.

References

[L.Baratchart and L.Pottier 89] "Un résultat sur les systèmes d'addition de vecteurs", manuscript, INRIA Sophia Antipolis, France, feb. 1989.

[I.Borosh and L.B.Treybig 76] "Bounds of non-negative integral solutions of linear diophantine equations", Proc. AMS v.55, n.2, march 1976.

[B.Buchberger 83] "Gröbner basis: an algorithmic method in polynomial ideal theory" Camp. Publ. Nr. 83-29. 0, nov. 1983.

[E.Contejean and H.Devie 89] "Solving systems of linear diophantine equations", UNIF'89, proc. of the third international Workshop on unification, Lambrecht, RFA 89.

[E.Domenjoud 90] "Solving Systems fo Linear Diophantine Equations : An Algebraic Approch", UNIF'90, International Workshop on Unification, Leeds UK, july 1990.

[A.Galligo 85] "Algorithmes de calcul de bases standard", Preprint Université de Nice, France, 1985.

[F.Giles and W.R.Pulleyblank 79] "Total dual integrality and integer polyedra" Linear algebra and its applications, 25, pp191-196, 1979.

[G.Huet 78] "An algorithm to generate the basis of solutions to homogeneous linear diophantine equations", Information Processing Letters, vol.3, No.7, 1978.

[A.Koscielski and L.Pacholski 90] "Exponent of periodicity of minimal solutions of word equations", manuscript, University of Wroclaw, Poland, june 1990.

[J.L.Lambert 87] "Une borne pour les générateurs des solutions entières positives d'une équation diophantienne linéaire." Comptes Rendus de l'Académie des Sciences de Paris, t.305, Série I, pp39-40, 1987.

[J.L.Lambert 87] "Un problème d'accessibilité dans les réseaux de Petri" Phd thesis, theorem I.5., p 18, University of Paris-Sud, Orsay, France, 1987.

[F.Ollivier 90] "Le problème de l'identifiabilité structurelle globale : approche théorique, méthodes effectives et bornes de complexité", Phd Thesis, Ecole Polytechnique, France, june 1990.

[L.Pottier 90] "Bornes et algorithmes de calcul des générateurs des solutions de systèmes diophantiens linéaires", internal report, INRIA, feb. 90, Comptes Rendus de l'Académie des Sciences de Paris, t.311, Série I, p813-816,1990.

[J.F.Romeuf 89] "Solutions of a linear diophantine system", UNIF'89, proc. of the third international Workshop on unification, Lambrecht, RFA 89.

[R.P.Stanley 83] "Combinatorics and commutative algebra", Progress in Mathematics, Birkäuser ed., 1983.

[L.B.Treybig 75] "Bounds in piecewise linear topology", Trans.AMS, v.201, 1975.

[J.Von zur Gathen and M.Sieveking 78] "A bound on solutions of linear integer equalities and inequalities" Proc. AMS 72, pp155-158, 1978.

Proofs in Parameterized Specifications

Hélène Kirchner
CRIN & INRIA-Lorraine BP 239
54506 Vandœuvre-lès-Nancy Cedex (France)
e-mail: hkirchner@loria.crin.fr

Abstract

Theorem proving in parameterized specifications has strong connections with inductive theorem proving. An equational theorem holds in the generic theory of the parameterized specification if and only if it holds in the so-called generic algebra. Provided persistency, for any specification morphism, the translated equality holds in the initial algebra of the instantiated specification. Using a notion of generic ground reducibility, a persistency proof can be reduced to a proof of a protected enrichment. Effective tools for these proofs are studied in this paper.

1 Introduction

Parameterization is a generic way for building families of specifications and for reusing specifications. An important concern is to make use of parameterization at the proof level and to develop a generic proof method. As argued in [7], this approach has several advantages: First, it allows performing proofs in a structured way that reflects the program structure. So generic proofs are performed for parameterized equational specifications. Second, a generic proof in a parameterized specification must be given only once and can be reused for each instantiation of the parameter.

Rather than promoting completely new ideas on the subject of parameterization, this paper is aimed to clarify some points, to gather different ideas and combine several results presented in [2, 7, 8, 9, 14, 16, 23]. More precisely, the goals here are the following ones:
- to emphasize the connection between (protected) enrichment and (persistent) parameterized specifications.
- to show that a persistency proof can be reduced to a proof of a protected enrichment, using an adequate notion of generic ground reducibility.
- to precise the use of persistency for theorem proving: an equational theorem holds in the generic theory of the parameterized specification if and only if it is an inductive theorem in a particular initial algebra, called the generic algebra. Provided persistency, such a theorem is generic: for any specification morphism m, the translated equality using m holds in the initial algebra of the instantiated specification.
- to provide effective tools to prove that a parameterized specification is persistent, to prove generic theorems and to prove generic ground reducibility.

Sections 2 and 3 recall the necessary definitions and results about enrichment and parameterization. Section 4 defines the central notion of generic ground reducibility. Section 5 is devoted to providing tools for proving persistency, while Section 6 is concerned with generic theorem proving. For lack of space, most proofs are omitted. They can be found in an extended version of this paper.

2 Enrichments

All notations are compatible with [13]. Given a many-sorted signature Σ, the set of terms built on Σ and a denumerable set of sorted variables $\mathcal{X}$ is denoted by $\mathcal{T}(\Sigma \cup \mathcal{X})$. $\mathcal{V}(t)$ denotes the set of variables occurring in the term t.

Definition 1 *A* specification, *denoted $SP = (\Sigma, E)$ is given by a signature Σ, composed of a set of* sort symbols *S and a set of* function symbols *$\mathcal{F}$ with rank declarations, and a set E of universally quantified equalities $(\forall X, t = t')$ where $\mathcal{V}(t) \cup \mathcal{V}(t') \subseteq X$. (The quantification may be omitted when $X = \mathcal{V}(t) \cup \mathcal{V}(t')$).*

A specification $SP = (\Sigma, E)$ actually describes a class of algebras, namely the class of Σ-algebras satisfying the equalities E, denoted $ALG(SP)$. $ALG(SP)$ with SP-homomorphisms is a category also denoted by $ALG(SP)$.

Componentwise inclusion of specifications corresponds to enrichments.

Definition 2 *An* enrichment *of a specification $SP = (\Sigma, E)$ is a specification $SP' = (\Sigma', E')$ such that $\Sigma \subseteq \Sigma'$ and $E \subseteq E'$.*

SP is often referred to as the *primitive* or *basic* specification, while SP' is called the *enriched* specification. A forgetful functor is associated to an enrichment.

Definition 3 *Assume that $SP \subseteq SP'$. Then the* forgetful functor *$\mathcal{V}$ from $ALG(SP')$ to $ALG(SP)$ is defined as follows:*
$\forall A' \in ALG(SP')$, $A = \mathcal{V}(A')$ is the SP-algebra such that $\forall s \in \Sigma, A_s = A'_s$ and $\forall f \in \Sigma, f_A = f_{A'}$.
$\forall h'$ SP'-morphism, $h = \mathcal{V}(h')$ is the SP-homomorphism such that $h_s = h'_s$ for any $s \in \Sigma$.

Let $\overset{*}{\longleftrightarrow}_E$ denote the replacement of equals by equals on $\mathcal{T}(\Sigma \cup \mathcal{X})$, which is correct and complete for deduction in $ALG(SP)$: $t \overset{*}{\longleftrightarrow}_E t'$ iff $\forall \mathcal{A} \in ALG(SP), \mathcal{A} \models (\forall X, t = t')$.

Enrichments are classified according to their effect on the initial algebra of the enriched specification. Mainly, enrichments can produce *junks*, that is new terms that are not equivalent to an already existing term, or *confusions*, that is new equivalences between terms originally distincts.

Definition 4 *Let $SP = (\Sigma, E) \subseteq SP' = (\Sigma', E')$ be an enrichment. The enrichment is* consistent *if for any sort $s \in \Sigma$, any ground terms t and t' of sort s in $\mathcal{T}(\Sigma)$, $t \overset{*}{\longleftrightarrow}_E t'$ iff $t \overset{*}{\longleftrightarrow}_{E'} t'$. The enrichment is* sufficiently complete *if for any sort $s \in \Sigma$, any ground term t' of sort s in $\mathcal{T}(\Sigma')$, there exists a term t of sort s in $\mathcal{T}(\Sigma)$ such that $t \overset{*}{\longleftrightarrow}_{E'} t'$. An enrichment which is both consistent and sufficiently complete is said* protected.

In the case where the theories are presented by ground convergent rewrite systems (i.e. confluent and terminating on ground terms), a proof by consistency method has been developped by [2] to prove theorems in initial algebras. The method is based on the notion of ground reducibility. Given a ground convergent rewrite system R, a term t is *ground reducible* with R if all its ground instances are R-reducible. An equality $(\forall X, t = t')$ is *ground reducible* with R if for any ground substitution σ such that $\sigma(t) \neq \sigma(t')$, either $\sigma(t)$ or $\sigma(t')$ is R-reducible. The property of ground reducibility is decidable for finite rewrite systems [16, 24]. Algorithms for deciding ground reducibility in the case of left-linear rules have been given, for instance in [14, 16, 4, 19, 21]. The general case is considered in [6, 5, 17]. Ground reducibility for a class rewrite system R/E is undecidable when E is a set of associative and commutative axioms [16] but is decidable when R is left-linear [15]. The sufficient completeness is in general undecidable [16] but in some cases, it is equivalent to ground reducibility. The goal of this paper is to extend these proofs techniques to parameterized specifications.

3 Parameterization

3.1 Parameterized specifications

Definition 5 *A parameterized specification PSP is a pair (SP, SP') of specifications where $SP = (\Sigma, E)$ is called the* formal parameter, *$SP' = SP + (\Sigma'', E'')$ is called the* target specification, *and (Σ'', E'') is called the* body *of the parameterized specification. Terms built on the signature Σ and a denumerable set X_{SP} of* parameter variables, *i.e. variables of sort $s \in \Sigma$, are called* parameter terms.

Example 1 Let us axiomatize the list structure on any kind of elements. Classically "*nil*" is the empty list and "*cons*" the constructor for lists. Concatenation of lists is denoted by the function "*append*". In order to define a product operation on lists that computes the product of its elements, it is needed to constrain the elements to be in a monoid with an identity element. Then the formal parameter is the following specification MONOID, in which equalities have been oriented into rewrite rules:

$$\begin{array}{rcl} & sort\ Elem & \\ id: & \mapsto & Elem \\ \pi: Elem, Elem & \mapsto & Elem \\ \forall e: Elem,\ \pi(e, id) & \rightarrow & e \\ \forall e: Elem,\ \pi(id, e) & \rightarrow & e \\ \forall e_1, e_2, e_3: Elem,\ \pi(\pi(e_1, e_2), e_3) & \rightarrow & \pi(e_1, \pi(e_2, e_3)). \end{array}$$

Let us consider the following parameterized specification with the formal parameter MONOID and the body:

$$\begin{array}{rcl} & sort\ List & \\ cons: Elem, List & \mapsto & List \\ nil: & \mapsto & List \\ append: List, List & \mapsto & List \\ prod: List & \mapsto & Elem \\ \forall l: List,\ append(nil, l) & \rightarrow & l \\ \forall e: Elem, l, l': List,\ append(cons(e, l), l') & \rightarrow & cons(e, append(l, l')) \\ \forall l: List,\ prod(nil) & \rightarrow & id \\ \forall e: Elem, l: List,\ prod(cons(e, l)) & \rightarrow & \pi(e, prod(l)). \end{array}$$

3.2 Semantics

Semantics for parameterized specifications have been widely studied, for instance in the many-sorted case in [8, 9, 23] and in the order-sorted case in [25, 12, 11]. The case of Horn clauses parameterized specifications is considered for example in [10, 20].

To give a semantics to a parameterized specification consists in associating to SP the class $ALG(SP)$ with its SP-homomorphisms, to SP' the class $ALG(SP')$ with its SP'-homomorphisms and to PSP a functor from the category $ALG(SP)$ to $ALG(SP')$. Let $\mathcal{V}$ be the forgetful functor associated to the enrichment $SP \subseteq SP'$. From a given SP-algebra $\mathcal{A} \in ALG(SP)$, a SP'-algebra denoted $\mathcal{T}_{SP'}(\mathcal{A})$ can be built in the following way:
$Const(\mathcal{A}) = \{a :\mapsto s \mid a \in \mathcal{A}_s, s \in \Sigma\}$
$Eqns(\mathcal{A}) = \{f(a_1, ..., a_n) = f_{\mathcal{A}}(a_1, ..., a_n) \mid \forall a_i \in \mathcal{A}, \forall f \in \Sigma\}$
$SP'(\mathcal{A}) = (\Sigma' \cup Const(\mathcal{A}), E' \cup Eqns(\mathcal{A}))$
$\mathcal{T}_{SP'}(\mathcal{A}) = \mathcal{V}_{\mathcal{A}}(\mathcal{T}_{SP'(\mathcal{A})})$, where $\mathcal{V}_{\mathcal{A}}$ is the forgetful functor from the category of the $SP'(\mathcal{A})$-algebras to the category of the SP'-algebras.

$\mathcal{T}_{SP'}(\mathcal{A})$ is *the free construction* on $\mathcal{A}$ w.r.t. $\mathcal{V}$. The concepts of free construction and free functor are precisely defined for instance in [8].

Theorem 1 *[8] Let $\mathcal{V}$ be the forgetful functor from $ALG(SP')$ to $ALG(SP)$. For any $\mathcal{A} \in ALG(SP)$, let $\mathcal{F}(\mathcal{A})$ be defined as $\mathcal{T}_{SP'}(\mathcal{A})$. Then $\mathcal{F}$ extends to a free functor from $ALG(SP)$ to $ALG(SP')$ called the* free functor *w.r.t. $\mathcal{V}$.*

Definition 6 *$\mathcal{F}$, the free functor w.r.t. $\mathcal{V}$, is the* semantics of the parameterized specification PSP.

Note that a first kind of genericity is obtained with the free functor: from a class $ALG(SP)$ of SP-algebras, the free functor $\mathcal{F}$ generates the class of algebras $\mathcal{F}(ALG(SP)) = \{\mathcal{F}(\mathcal{A}) \mid \mathcal{A} \in ALG(SP)\}$. A second kind of genericity obtained from a parameterized specification, is to generate specifications and, for this purpose, the notion of parameter passing is necessary.

3.3 Parameter passing

Parameter passing is intended to formalize the instantiation of the formal parameter specification SP into an actual specification SP_1.

Definition 7 *A* signature morphism m *from* Σ *to* Σ_1 *is a function* $m : \Sigma \mapsto \Sigma_1$ *such that:*
$\forall f : s_1, ...s_n \mapsto s \in \Sigma, m(f) : m(s_1), ..., m(s_n) \mapsto m(s) \in \Sigma_1$.

Given a signature morphism m from Σ to Σ_1 and a Σ-axiom $e = (\forall X, t = t')$, the *translated axiom* denoted $m^*(e) = (\forall m^*(X), m^*(t) = m^*(t'))$ is inductively defined by
$m^*(x : s) = x : m(s)$ and $m^*(f(t_1, ..., t_n)) = m(f)(m^*(t_1), ..., m^*(t_n))$.

Definition 8 *A* specification morphism m *from* $SP = (\Sigma, E)$ *to* $SP_1 = (\Sigma_1, E_1)$ *is a signature morphism from* Σ *to* Σ_1 *such that for any axiom* $e \in E$*, the translated axiom* $m^*(e)$ *is valid in the initial* SP_1*-algebra.*

Given a specification morphism from the formal parameter specification to the actual parameter specification, an instantiated specification can be built.

Definition 9 *Given a parameterized specification* $PSP = (SP, SP')$ *and a specification morphism* m *from* SP *to* SP_1*, a* parameter passing *is given by:*
- the specification morphism m' *defined by*

$$\begin{aligned} \forall s' \in \Sigma', m'(s') = \ & \text{if } s' \in S \text{ then } m(s') \text{ else } s' \\ \forall f' : s'_1, ..., s'_n \mapsto s' \in \Sigma', m'(f') = \ & \text{if } f' \in \Sigma \\ & \text{then } m(f') : m(s'_1), ..., m(s'_n) \mapsto m(s') \\ & \text{else } f' : m'(s'_1), ..., m'(s'_n) \mapsto m'(s'). \end{aligned}$$

- the instantiated specification $SP'_1 = SP_1 + (m'(\Sigma' - \Sigma), m'^*(E' - E))$.
- the specification morphisms $p : SP \mapsto SP'$ *and* $p_1 : SP_1 \mapsto SP'_1$ *which are inclusions.*

$\mathcal{V}_m$, $\mathcal{V}_{m'}$, $\mathcal{V}_p$ and $\mathcal{V}_{p_1}$ denote the forgetful functors [8] respectively associated to morphisms m, m', p, p_1.

Example 2 Let us consider the parameterized specification of Example 1 and the actual parameter NAT:

$$\begin{aligned} &\textit{sort } Nat \\ 0 : &\mapsto Nat \\ s : Nat &\mapsto Nat \\ + : Nat, Nat &\mapsto Nat \\ \forall n : Nat,\ n + 0 &\rightarrow n \\ \forall n, m : Nat,\ n + s(m) &\rightarrow s(n + m). \end{aligned}$$

Let m be the specification morphism defined by: $m(Elem) = Nat, m(id) = 0, m(\pi) = +$. The instantiation of MONOID by NAT needs to prove that the equalities

$$\begin{aligned} \forall n : Nat,\ n + 0 &= n \\ \forall n : Nat,\ 0 + n &= n \\ \forall n_1, n_2, n_3 : Nat,\ ((n_1 + n_2) + n_3 &= n_1 + (n_2 + n_3)). \end{aligned}$$

hold in the initial algebra of NAT. The construction of the instantiated specification is left to the reader.

The question is now: is the syntactic construction of SP'_1 compatible with the semantics respectively chosen for PSP, SP_1 and SP'_1? Here the semantics given to specifications SP_1 and SP'_1 are the initial algebras of these specifications denoted respectively by $\mathcal{T}_{SP_1}$ and $\mathcal{T}_{SP'_1}$.[1] The answer is yes, provided *correctness.*

Definition 10 *The parameter passing is* correct *for a parameterized specification PSP and a specification morphism m from SP to SP_1 if*
1. $\mathcal{V}_{p_1}(\mathcal{T}_{SP'_1}) = \mathcal{T}_{SP_1}$, *property called* protection of actual parameter,
2. $\mathcal{V}_{m'}(\mathcal{T}_{SP'_1}) = \mathcal{F} \circ \mathcal{V}_m(\mathcal{T}_{SP_1})$, *property called* compatibility of parameter passing.

The first property says that SP'_1 is a protected enrichment of SP_1. The second property expresses the fact that the semantics $\mathcal{F}$ of PSP agrees with the semantics of the instantiated specification SP'_1.
This definition of correctness is relative to one specification morphism. In order to get a notion of correctness that holds for any specification morphism, a stronger property on functors is needed.

3.4 Persistency

The correctness of parameter passing for every specification morphism requires that the functor $\mathcal{F}$ be persistent. Intuitively, persistency means that for any SP-algebra $\mathcal{A} \in ALG(SP)$, $\mathcal{A}$ is protected in $\mathcal{F}(\mathcal{A})$.

Definition 11 *[8] Given a parameterized specification PSP and the forgetful functor $\mathcal{V}_p : ALG(SP') \mapsto ALG(SP)$, the free functor $\mathcal{F} : ALG(SP) \mapsto ALG(SP')$ is said* persistent *if $\mathcal{V}_p \circ \mathcal{F} = \mathcal{I}$, where $\mathcal{I}$ is the identity functor on $ALG(SP)$, up to a natural isomorphism. The parameterized specification PSP is also said persistent.*

Proposition 1 *[8] Given a parameterized specification PSP with a persistent functor $\mathcal{F}$ and a specification morphism m from SP to SP_1, there exists a persistent functor $\mathcal{F}_1 : ALG(SP_1) \mapsto ALG(SP'_1)$, called* extension *of $\mathcal{F}$ according to m. Moreover $\mathcal{F}_1$ is uniquely defined by $\mathcal{V}_{m'} \circ \mathcal{F}_1 = \mathcal{F} \circ \mathcal{V}_m$ and $\mathcal{V}_{p_1} \circ \mathcal{F}_1 = \mathcal{I}_1$, where $\mathcal{I}_1$ is the identity functor on $ALG(SP_1)$.*

If $\mathcal{F}$ is persistent, then for any specification morphism m, the functor $\mathcal{F}_1$ exists and is persistent. This is exactly what is needed for correctness of parameter passing, for each specification morphism.

Theorem 2 *[8] Given a parameterized specification PSP, the parameter passing is correct for PSP and any specification morphism m iff PSP is persistent in $ALG(SP)$.*

Example 3 A example of a non-persistent parameterized specification [8] is given by $PSP = (SP, SP')$ where $SP = \{\{s\}, \emptyset, \emptyset\}$ and $SP' = \{\{s\}, \{e\}, \emptyset\}$. Let NAT be the usual specification of natural numbers, NAT $= \{\{Nat\}, \{0 :\mapsto Nat, succ : Nat \mapsto Nat\}, \emptyset\}$, consider now the actual parameter $SP_1 =$ NAT $+\{succ(succ(x)) = x\}$, and the specification morphism m defined by $m(s) = Nat$. Then the respective domains of sort Nat of $\mathcal{T}_{SP_1}$, $\mathcal{F} \circ \mathcal{V}_m(\mathcal{T}_{SP_1})$, and $\mathcal{T}_{SP'_1}$ are respectively $\{0, succ(0)\}$, $\{0, succ(0), e\}$ and $\{0, succ(0), e, succ(e)\}$. Neither the compatibility of parameter passing nor the protection of actual parameter are satisfied.

3.5 Generic algebra

We now consider different questions: how to prove correctness of parameter passing, for any specification morphism m? How to prove a generic assertion, that is, how to prove that for any specification morphism m, the translated assertion (using m) is valid in the initial algebra of the instantiated specification? Both questions have answers that need the introduction of a generic algebra.

[1] It is implicitely assumed that SP_1 and SP'_1 have no empty sorts

Let $SP = (\Sigma, E)$ be a specification and $\mathcal{X}$ a denumerable set of variables whose sorts are in Σ. Let $\mathcal{T}_{SP}(\mathcal{X})$ be the initial term algebra associated to the specification $(\Sigma \cup \mathcal{X}, E)$. $\mathcal{T}_{SP}(\mathcal{X})$ is a Σ-algebra whose carrier is the quotient $\mathcal{T}(\Sigma \cup \mathcal{X})/E$ of the set of terms $\mathcal{T}(\Sigma \cup \mathcal{X})$ by the congruence $\overset{*}{\longleftrightarrow}_E$. Note that if $\mathcal{X}$ is any set of variables with sorts in Σ, $\mathcal{T}_{SP}(\mathcal{X})$ is the free SP-algebra generated by $\mathcal{X}$. $\mathcal{T}_{SP}(\emptyset)$ is the initial SP-algebra. The *PSP-generic algebra* is obtained for a third choice of $\mathcal{X}$:

Definition 12 *Let $PSP = (SP, SP')$ be a parameterized specification and X_{SP} a set of variables of parameter sorts. The* PSP-generic algebra *is the Σ'-algebra $\mathcal{T}_{SP'}(X_{SP})$, whose carrier $\mathcal{T}(\Sigma' \cup X_{SP})/E'$ is the quotient by $\overset{*}{\longleftrightarrow}_{E'}$ of the set $\mathcal{T}(\Sigma' \cup X_{SP})$ of* PSP-generic terms.

3.6 Generic theory of a parameterized specification

The set of theorems valid in the class of algebras $\mathcal{F}(ALG(SP))$ associated to a parameterized specification, defines the *generic theory* of the parameterized specification.

Definition 13 *The* generic theory *of the parameterized specification PSP denoted $Th(PSP)$ is the set $\{(\forall X, t = t') | \forall A \in ALG(SP), \mathcal{F}(A) \models (\forall X, t = t')\}$.*

The generic theory is also called equational theory of PSP in [23]. Note that in the degenerated case where SP is the empty specification, then $Th(PSP)$ is the inductive theory of SP'.

Definition 14 *Any substitution $\sigma : X \mapsto \mathcal{T}(\Sigma' \cup X_{SP})$ is called a* PSP-generic substitution.

As a consequence of previous definitions, an equality $(\forall X, t = t')$ holds in the PSP-generic algebra, denoted $\mathcal{T}_{SP'}(X_{SP}) \models (\forall X, t = t')$, if for any PSP-generic substitution $\sigma : X \mapsto \mathcal{T}(\Sigma' \cup X_{SP})$, $\sigma(t) \overset{*}{\longleftrightarrow}_{E'} \sigma(t')$.

Example 4 Let us consider again the parameterized specification of Example 1.
The term *append(cons(e, nil), nil)* with *e* a variable of sort *Elem*, is a PSP-generic term. Substitutions $(l \mapsto nil)$, $(l \mapsto cons(e, nil))$, $(l \mapsto append(cons(e, nil), nil))$ are PSP-generic substitutions. The equality $(\forall e : Elem, l : List,\ append(cons(e, nil), l) = cons(e, l))$ holds in the PSP-generic algebra, just because it holds in the whole class of SP'-algebras.

The next theorem states that the generic theory is exactly the set of theorems valid in the PSP-generic algebra.

Theorem 3 *[23] Let $PSP = (SP, SP')$ be a parameterized specification, X_{SP} a set of variables of parameter sorts and $\mathcal{T}_{SP'}(X_{SP})$ the PSP-generic algebra.*

$$\mathcal{T}_{SP'}(X_{SP}) \models (\forall X, t = t') \text{ iff } (\forall X, t = t') \in Th(PSP).$$

The following result explains in which sense a theorem in $Th(PSP)$ is generic: validity of an equality in the PSP-generic algebra means validity of the translated equality in any instantiation of the parameterized specification, provided that the parameterized specification is persistent. A similar result is given in [7].

Theorem 4 *Let $PSP = (SP, SP')$ be a persistent parameterized specification and X_{SP} a set of variables of parameter sorts. Then the following properties are equivalent:*
1. $\mathcal{T}_{SP'}(X_{SP}) \models (\forall X, t = t')$
2. for any specification morphism m, $\mathcal{T}_{SP'_1} \models (\forall m'^*(X), m'^*(t) = m'^*(t'))$.

The next theorem relates the notion of persistency with proof theoretical properties.

Theorem 5 *[9] Let $PSP = (SP, SP')$ be a parameterized specification and X_{SP} a set of variables of parameter sorts. PSP is persistent iff the following two properties are satisfied:*
1. $PSP = (SP, SP')$ is generic sufficiently complete*, i.e.: $\forall t \in \mathcal{T}(\Sigma' \cup X_{SP})$, t of parameter sort, $\exists t_0 \in \mathcal{T}(\Sigma \cup X_{SP})$ such that $t \xleftrightarrow{*}_{E'} t_0$.*
2. $PSP = (SP, SP')$ is generic consistent*, i.e.: $\forall t, t' \in \mathcal{T}(\Sigma \cup X_{SP})$ of parameter sorts, $t \xleftrightarrow{*}_{E'} t'$ iff $t \xleftrightarrow{*}_{E} t'$.*

The properties of generic sufficient completeness and generic consistency express in other words that the enrichment $(\Sigma \cup X_{SP}, E) \subseteq (\Sigma' \cup X_{SP}, E')$ is protected.

In order to go further and design effective tools for parameterized proofs, we now focus on equational theories described by rewrite systems.

4 Generic ground reducibility

In order to check persistency of a parameterized specification and validity in the PSP-generic algebra, the notion of PSP-generic ground reducibility is needed. Let $PSP = (SP, SP')$ be a parameterized specification with $SP = (\Sigma, R)$ and $SP' = (\Sigma', R')$ where R and R' are rewrite systems. Let X_{SP} be an infinite set of variables of parameter sorts.

Definition 15 *Given a rewrite system R', confluent and terminating on $\mathcal{T}(\Sigma' \cup X_{SP})$, a term $t \in \mathcal{T}(\Sigma' \cup X)$ is* PSP-generic ground reducible *with R' if for any PSP-generic substitution $\sigma : X \mapsto \mathcal{T}(\Sigma' \cup X_{SP})$, $\sigma(t)$ is reducible using R'. An equality $(\forall X, t = t')$ is* PSP-generic ground reducible *with R' if for any PSP-generic substitution $\sigma : X \mapsto \mathcal{T}(\Sigma' \cup X_{SP})$, such that $\sigma(t) \neq \sigma(t')$, either $\sigma(t)$ or $\sigma(t')$ is reducible using R'.*

Algorithms for checking ground reducibility can be extended to check PSP-generic ground reducibility. Here, the test for ground reducibility in the case of left-linear rules given in [14] is generalized to a test for generic ground reducibility. The goal is to exhibit a finite set of substitutions S such that a term is generic ground reducible iff all its instances by substitutions in S are reducible. In the case of left-linear rules in R', the idea to construct S is that left-hand sides of rules have a finite depth which bounds the number of substitutions to be tested. Some preliminary definitions are needed [14].

The length of a term t is the maximal size of positions ω in the term t. Let $d = depth(R')$ be the maximal depth of left-hand sides of rules in R'. The top of a term t at depth i is a term defined by:
$top(t, i) = t$ if $depth(t) < i$
$top(f(t_1, ..., t_n), 0) = f(x_1, ..., x_n)$ where x_i are new variables
$top(f(t_1, ..., t_n), i) = f(top(t_1, i-1), ..., top(t_n, i-1))$, for any symbol $f \in \mathcal{F}$.

Let $\mathcal{S}(R') = \{\ top(t_0, d) \mid t_0$ is an R'-irreducible PSP-generic term $\}$ be called the *PSP-generic test set*. For practical reasons, variables in terms of $\mathcal{S}(R')$ are assumed distinct, which is always possible by renaming them: $\forall t, t' \in \mathcal{S}(R')$, $\mathcal{V}(t) \cap \mathcal{V}(t') = \emptyset$. The set $\mathcal{S}(R')$ is computed as the limit of a stationary sequence of sets $\mathcal{S}_i$ defined as follows:
$\mathcal{S}_i = \{top(t_0, d) \mid t_0$ is an R'-irreducible PSP-generic term such that $depth(t_0) \leq i\}$.
Then $\mathcal{S}(R') = \mathcal{S}_k$ as soon as $\mathcal{S}_k = \mathcal{S}_{k+1}$ for some k.

Theorem 6 *A term t is PSP-generic ground reducible by a left-linear rewrite system R' iff all its instances $\{\sigma(t) \mid \sigma : \mathcal{V}(t) \mapsto \mathcal{S}(R')\}$, obtained by substituting variables of t by terms in $\mathcal{S}(R')$, are reducible by R'.*

Proof: The proof is similar to the proof in [15]. □

Example 5 Consider the parameterized specification of Example 1. In order to check the PSP-generic ground reducibility of $prod(append(l, l'))$, the following generic test set needs to be considered: $\{nil, cons(e_0, nil), cons(e_1, cons(e_2, nil))\}$, where e_0, e_1, e_2 are new variables of sort *Elem* in X_{SP}. For each deduced substitution α, $\alpha(t)$ is reducible using R'.

5 Proof of persistency

5.1 Proof of generic sufficient completeness

We now extend the proof of sufficient completeness for convergent rewrite systems of Kapur, Narendran and Zhang in [16]. Their result states the equivalence between sufficient completeness and a check for ground reducibility, provided that terms built on the imported signature are preserved.

Definition 16 *Let $PSP = (SP, SP')$ be a parameterized specification such that $SP = (\Sigma, R)$ and $SP' = (\Sigma', R')$. R'* preserves parameters *if $\forall(l \to r) \in R'$, whenever l has a parameter sort, r has a parameter sort, and whenever l is a parameter term, r is a parameter term.*

Proposition 2 *Let $PSP = (SP, SP')$ be a parameterized specification such that $SP = (\Sigma, R) \subseteq SP' = (\Sigma', R')$, X_{SP} is a denumerable set of variables of parameter sorts, and R' is a convergent rewrite system on $\mathcal{T}(\Sigma' \cup X_{SP})$ preserving parameters. Then the following propositions are equivalent:*
1. $\forall f \in \Sigma' - \Sigma$, whose range is a sort $s \in \Sigma$, $f(x_1, ..., x_n)$ is PSP-generic ground reducible with R',
2. $\forall t' \in \mathcal{T}(\Sigma' \cup X_{SP})$ of sort $s \in \Sigma$, $\exists t \in \mathcal{T}(\Sigma \cup X_{SP})$ such that $t \overset{*}{\longleftrightarrow}_{R'} t'$.

Proof: The proof is an extension of [16]. □

Note that the property of preserving parameters is very simple to achieve in most cases. It is satisfied for instance when every rule in $R' - R$ contains in its left-hand side at least a function symbol of $\Sigma' - \Sigma$. In a structured programming methodology and especially in parameterized specifications, this hypothesis does not appear as a restriction.

5.2 Proof of generic consistency

In the case where the basic specification is given with a ground convergent rewrite systems, the completion process appears as an interesting tool to both prove consistency of an enrichment and produce simultaneously a ground convergent rewrite system for the enriched specification. Given an enrichment $SP = (\Sigma, R) \subseteq SP' = (\Sigma', E')$ with R a ground convergent rewrite system on $\mathcal{T}(\Sigma)$, the general idea is to complete E' into a ground convergent system R' on $\mathcal{T}(\Sigma')$ and to check that whenever a rewrite rule, whose left and right-hand sides both belong to $\mathcal{T}(\Sigma)$, is added, then this rule is an inductive consequence of R. We design a similar consistency proof procedure in the slightly more general framework of a parameterized specification PSP with $SP = (\Sigma, R) \subseteq SP' = (\Sigma', E')$ with R a convergent rewrite system on $\mathcal{T}(\Sigma \cup X_{SP})$. (This can be checked by completion). PSP is generic consistent if whenever an equality, whose left and right-hand sides both belong to $\mathcal{T}(\Sigma \cup X_{SP})$, is added, then this is a theorem valid in $ALG(SP)$, that is $t \overset{*}{\longleftrightarrow}_R t'$. In order to detect inconsistencies, we need the following definition:

Definition 17 *Let us consider a parameterized specification PSP with $SP = (\Sigma, R) \subseteq SP' = (\Sigma', E')$ with R a convergent rewrite system. A set of equalities C is* provably inconsistent *with $Th(SP)$ if it contains an equality $(\forall X, t = t')$ such that t and t' are in $\mathcal{T}(\Sigma \cup X_{SP})$ and are not $\overset{*}{\longleftrightarrow}_R$-equivalent.*

If SP is itself a parameterized specification, say $SP = (SP_0, SP_1)$, then $Th(SP)$ is its generic theory and the previous definition must be refined by replacing $\overset{*}{\longleftrightarrow}_R$-equivalence by validity in the SP_0-generic algebra.

In the completion process described below, it is convenient to split the set of equalities $E' - R$ into two parts: one, called C, which contains only parameter equalities (i.e. built on terms of $\mathcal{T}(\Sigma \cup X_{SP})$), and the other, called P, that contains non-parameter ones. $OCP(P \cup R, P)$ denote the set of ordered critical pairs [1, 3] obtained by superposition of $P \cup R$ on P. There is no need to superpose rules in R with themselves. There is also no need to try superpositions of P on R because by construction, terms involved in P contain at least one function symbol in $\Sigma' - \Sigma$. $CP(R,C)$ as usual denote the set of critical pairs obtained by superposition of R on C and conversely. Note that whenever an equality contains only parameter terms and can be superposed with R, then the equality is dropped in C and such superpositions are taken into account in $CP(R,C)$.

Let P be a set of equalities (quantified pairs of terms), C a set of conjectures (parameter equalities), R the underlying rewrite system on parameter terms, terminating and confluent on $\mathcal{T}(\Sigma \cup X_{SP})$, and $>$ a reduction ordering that contains R and can be extended to a reduction ordering on $\mathcal{T}(\Sigma' \cup X_{SP})$ total on E'-equivalence classes. The generic consistency proof procedure is expressed by the following set $\mathcal{GC}$ of inference rules:

1. Deduce
$$\frac{P, C}{P \cup \{p = q\}, C} \quad \text{if } (p,q) \in OCP(P \cup R, P)$$

2. Deflation
$$\frac{P \cup \{p = q\}, C}{P, C \cup \{p = q\}} \quad \text{if } p \textit{ and } q \textit{ parameter terms}$$

3. Delete
$$\frac{P \cup \{p = p\}, C}{P, C}$$

4. Collapse
$$\frac{P \cup \{p = q\}, C}{P \cup \{p' = q\}, C} \quad \text{if } (p \rightarrow^{l \rightarrow r}_{R \cup P>} p' \;\&\; p \sqsupset l) \textit{ or } (p \rightarrow_{R \cup P>} p' \;\&\; q > p')$$

5. Deduce conjecture
$$\frac{P, C}{P, C \cup \{p = q\}} \quad \text{if } (p,q) \in CP(R,C)$$

6. Delete conjecture
$$\frac{P, C \cup \{p = q\}}{P, C} \quad \text{if } (p = q) \in Th(SP)$$

7. Simplify
$$\frac{P, C \cup \{p = q\}}{P, C \cup \{p' = q\}} \quad \text{if } p > p' \;\&\; p \overset{+}{\longleftrightarrow}_R p'$$

8. Compose
$$\frac{P, C \cup \{p = q\}}{P, C \cup \{p' = q\}} \quad \text{if } p > p' \;\&\; p \longleftrightarrow^{g=d}_{C} p' \;\&\; p \sqsupset g > d$$

9. Generic consistency Disproof
$$\frac{P, C \cup (p = q)}{Disproof} \quad \text{if } (p = q) \textit{ provably inconsistent with } Th(SP)$$

In these inference rules, $\sqsupset$ denotes the strict encompassment ordering defined by $t \sqsupset t'$ if $t_{|\omega} = \sigma(t')$, for some position ω of t and some substitution σ, with $\omega \neq \epsilon$ or $\sigma \neq Id$.

Note that this set of inference rules is more general than the one for unfailing completion [1], obtained as a subset when $C = \emptyset$, and the one for proof by consistency [1], obtained for $P = \emptyset$. The *Deflation* inference rule is used first to split the set of equalities $E' - R$ added in the enrichment, into P and C such that C contains only parameter equalities and P_0 contains all other equalities.

Theorem 7 *Let us consider a parameterized specification PSP with $SP = (\Sigma, E) \subseteq SP' = (\Sigma', E')$ and R a convergent rewrite system on $\mathcal{T}(\Sigma \cup X_{SP})$ presenting E. Let $>$ be a reduction ordering that contains R and can be extended to a reduction ordering on $\mathcal{T}(\Sigma' \cup X_{SP})$, total on E'-equivalence classes, and such that no parameter term is greater than a non-parameter one.*
Let $(P_0, C_0) = (E', \emptyset) \vdash (P_1, C_1) \vdash \ldots$ be a derivation using $\mathcal{GC}$, $P_* = \bigcup_{i\geq 0} P_i$, $C_* = \bigcup_{i\geq 0} C_i$, $P_\infty = \bigcup_{i\geq 0} \bigcap_{j>i} P_j$, $C_\infty = \bigcup_{i\geq 0} \bigcap_{j>i} C_j$. Assume that $OCP(P_\infty \cup R, P_\infty) \cup CP(R, C_\infty)$ is a subset of $P_* \cup C_*$.
If C_∞ is empty, $(P_\infty \cup R)$ is Church-Rosser with respect to $>$ on $\mathcal{T}(\Sigma' \cup X_{SP})$ and the parameterized specification is generic consistent.
If a provable inconsistency with $Th(SP)$ has been detected, the parameterized specification is not generic consistent.

Proof: First the soundness of $\mathcal{GC}$ is proved by looking at the different inference rules: If $(P, C) \vdash (P', C')$, then the congruences $\overset{*}{\longleftrightarrow}_{P \cup C \cup R}$ and $\overset{*}{\longleftrightarrow}_{P' \cup C' \cup R}$ coincide on $\mathcal{T}(\Sigma' \cup X_{SP})$.

Then the proof reduction relation $\Longrightarrow$ that reflects the inference rule system $\mathcal{GC}$, is defined. If $>$ is a reduction ordering that can be extended to a reduction ordering on $\mathcal{T}(\Sigma' \cup X_{SP})$, total on E'-equivalence classes, then the proof reduction relation is noetherian. This is done by finding a complexity measure c on proofs and a noetherian ordering $>_c$, such that: for any proof reduction rule $L \Longrightarrow R$ defining the proof reduction relation, $c(L) >_c c(R)$.

More precisely, let us define the complexity measure of elementary proof steps by:

$$\begin{array}{rcll}
c(s \overset{g=d}{\longleftrightarrow}_P t) & = & (s, g, t, 2) & if \quad s > t \\
c(s \overset{g=d}{\longleftrightarrow}_P t) & = & (t, d, s, 2) & if \quad t > s \\
c(s \overset{g=d}{\longleftrightarrow}_C t) & = & (s, g, t, 1) & if \quad s > t \\
c(s \overset{g=d}{\longleftrightarrow}_C t) & = & (t, d, s, 1) & if \quad t > s \\
c(s \rightarrow_R^{g\rightarrow d} t) & = & (s, g, t, 0) &
\end{array}$$

where $s, t \in \mathcal{T}(\Sigma' \cup X_{SP})$. Let $> \cup \rhd$ denote the union of the reduction ordering $>$ and the strict subterm ordering $\rhd$. Complexities of elementary proof steps are compared using the lexicographic combination, denoted $>_{ec}$ of $> \cup \rhd$, $\sqsupset$, $> \cup \rhd$ and the standard ordering on natural numbers. The complexity of a non-elementary proof is the multiset of the complexities of elementary proof steps that it contains. Complexities of non-elementary proofs are compared using the multiset extension $>_c$ of $>_{ec}$.

Assume that C_∞ is empty. Let us prove by induction on $\Longrightarrow$ that for any $i \geq 0$, for any proof $t \overset{*}{\longleftrightarrow}_{P_i \cup C_i \cup R} t'$ where $t, t' \in \mathcal{T}(\Sigma' \cup X_{SP})$, there exists a proof $t \overset{*}{\longrightarrow}_{R \cup P_\infty^{\geq}} t'' \overset{*}{\longleftarrow}_{R \cup P_\infty^{\geq}} t'$. Since C_∞ is empty, this means that any C_i-equality step can be replaced by R-equality steps. So if the proof $t \overset{*}{\longleftrightarrow}_{P_i \cup C_i \cup R} t'$ is not already of the desired form, then either it contains a peak of $R \cup P_\infty$, or a non-persisting equality in P. In both cases, the proof is reducible by $\Longrightarrow$ into $t \overset{*}{\longleftrightarrow}_{P_j \cup C_j \cup R} t'$ and the induction hypothesis gives the result.

Given two terms, t and t' of $\mathcal{T}(\Sigma \cup X_{SP})$ of parameter sorts, such that $t \overset{*}{\longleftrightarrow}_{P_0 \cup C_0 \cup R} t'$, there exists a proof $t \overset{*}{\longrightarrow}_{R \cup P_\infty^{\geq}} t'' \overset{*}{\longleftarrow}_{R \cup P_\infty^{\geq}} t'$. But since t and t' are parameter terms and greater then any other term occurring in the proof, all these terms must be parameter terms. So no equality in P_∞ can apply since P_∞ contains non-parameter terms. So we get $t \overset{*}{\longleftrightarrow}_R t'$.

If a provable inconsistency with $Th(SP)$ has been detected, this means that there exist $g, d \in \mathcal{T}(\Sigma \cup X_{SP})$ such that $g \longleftrightarrow_{C_k} d$, for some k, but do not satisfy $g \overset{*}{\longleftrightarrow}_R d$. So g and d are $\overset{*}{\longleftrightarrow}_{P_0 \cup C_0 \cup R}$-equivalent but not $\overset{*}{\longleftrightarrow}_R$-equivalent, which proves that the parameterized specification is not generic consistent. □

Example 6 The parameterized specification of Example 1 is persistent.
The first step is to prove that the enrichment $(\Sigma \cup X_{SP}, R) \subseteq (\Sigma' \cup X_{SP}, R')$ is generic consistent. For that, the consistency proof procedure is applied. Since there is no critical pair, the enrichment

is obviously consistent.
Second, in order to prove that the enrichment $(\Sigma \cup X_{SP}, R) \subseteq (\Sigma' \cup X_{SP}, R')$ is generic sufficiently complete, we check that R' preserves parameters.
Then we have to check that $prod(l)$ is PSP-generic ground reducible with R'. The instantiations to be checked are: $(l \mapsto nil)$, $(l \mapsto cons(e_0, nil))$, $(l \mapsto cons(e_1, cons(e_2, l)))$, where $e_0, e_1, e_2 \in X_{SP}$ and l is a variable of sort *List*. For these three instantiations σ, the term $\sigma(prod(l))$ is clearly reducible.

6 Proof of an equational theorem in the theory of a parameterized specification

In the context of proving and disproving theorems in a parameterized specification, we need a slightly different notion of provable inconsistency, that is directly inspired by the one used in [2]. We assume in this section that PSP is a parameterized specification defined by the formal parameter $SP = (\Sigma, R)$ and $SP' = (\Sigma', R')$, where R and R' are rewrite systems.

Definition 18 *Let us consider a parameterized specification PSP with $SP = (\Sigma, R) \subseteq SP' = (\Sigma', R')$, where R' is a terminating rewrite system and $>$ a reduction ordering that contains R'. A set of equalities C is* provably inconsistent with $Th(PSP)$ *if it contains an equality $(\forall X, t = t')$ which satisfies either $t > t'$ and t is not PSP-generic ground reducible, or $(\forall X, t = t')$ is not PSP-generic ground reducible.*

Replacing the notion of provable inconsistency by the notion of provable inconsistency with $Th(PSP)$ allows applying the proof by consistency method to the proof of theorems in $Th(PSP)$.

Let C be a set of conjectures, R' the underlying rewrite system of the parameterized specification, terminating and confluent on $\mathcal{T}(\Sigma' \cup X_{SP})$, and $>$ a reduction ordering that contains R'. The generic proof procedure is expressed by the following set $\mathcal{GI}$ of inference rules, that is a subset of $\mathcal{GC}$, obtained with $P = \emptyset$.

1. Deduce conjecture
$$\frac{C}{C \cup \{p = q\}} \quad \text{if } (p, q) \in CP(R', C)$$

2. Delete conjecture
$$\frac{C \cup \{p = q\}}{C} \quad \text{if } (p = q) \in Th(PSP)$$

3. Simplify
$$\frac{C \cup \{p = q\}}{C \cup \{p' = q\}} \quad \text{if } p > p' \ \&\ p \overset{+}{\longleftrightarrow}_{R'} p'$$

4. Compose
$$\frac{C \cup \{p = q\}}{C \cup \{p' = q\}} \quad \text{if } p > p' \ \&\ p \longleftrightarrow^{g=d}_{C} p' \ \&\ p \sqsupset g > d$$

5. Generic consistency disproof
$$\frac{C \cup (p = q)}{Disproof} \quad \text{if } (p = q) \text{ provably inconsistent with } Th(PSP)$$

Theorem 8 *Let $PSP = (SP, SP')$ be a parameterized specification and X_{SP} a set of variables of parameter sorts. Let R' be a terminating and confluent rewrite system on $\mathcal{T}(\Sigma' \cup X_{SP})$ presenting E', C_0 be the set of PSP-generic conjectures to be proved, and $>$ be a reduction ordering that contains R'. Let $C_0 \vdash C_1 \vdash ....$ be a derivation using $\mathcal{GI}$ such that $CP(C_\infty, R)$ is a subset of C_*. If no provable inconsistency with $Th(PSP)$ has been detected, then C_0 is included in $Th(PSP)$.*

Proof: It is a consequence of the proof by consistency method in $\mathcal{T}_{SP'}(X_{SP})$ [1, 2]. □

Example 7 In the parameterized specification of Example 1, let us prove the validity of the PSP-generic conjecture $\forall l, l' : List,\ prod(append(l, l')) = \pi(prod(l), prod(l'))$.

This conjecture is PSP-generic ground reducible. By choosing a precedence such that $prod > \pi$, the left-hand side is bigger than the right-hand side. Superpositions on the bigger term are enough. So two superpositions have to be computed:
- With the rule $\forall l : List,\ append(nil, l) \rightarrow l$, we get $prod(l') = \pi(prod(nil), prod(l'))$. The term $\pi(prod(nil), prod(l'))$ reduces to $prod(l')$ and the conjecture becomes a trivial equality.
- With the rule $\forall e : Elem, l, l' : List,\ append(cons(e, l), l') \rightarrow cons(e, append(l, l'))$, we get the equality $prod(cons(e, append(l_1, l_2))) = \pi(prod(cons(e, l_1)), prod(l_2))$. After reduction on both sides, we get $\pi(e, prod(append(l_1, l_2))) = \pi(e, \pi(prod(l_1), prod(l_2)))$ that can be simplified using the initial conjecture.

7 Conclusion

Proofs in parameterized specifications have been experimented in several systems like CEC [10] or Reveur4 [27]. To some extent, this paper may be understood as a clarification of the concepts underlying these provers.

Now several research directions can be outlined for improving this work. Persistency is a very strong property that sometimes one may want to drop. So a first question that arises is the following: which results remain true if persistency is not assumed? Actually persistency is assumed in Theorem 4, and only the compatibility of parameter passing is used in its proof. Theorem 4 could be weakened as follows: if $\mathcal{T}_{SP'}(X_{SP}) \models (\forall X, t = t')$, then for any specification morphism m satisfying the compatibility of parameter passing, $\mathcal{T}_{SP'_1} \models (\forall m'^*(X), m'^*(t) = m'^*(t'))$. However the problem of checking the compatibility of parameter passing, even for a specific specification morphism, is not solved. Moreover assuming compatibility of parameter passing for every specification morphism has been proved equivalent to persistency in [22].

This work focussed on equational parameterized specifications and has to be extended to conditional specifications: mainly, this can be done along the lines of Navarro and Orejas [20], but an adequate notion of persistency needs the introduction of so-called LOG-algebras, that have an initial boolean domain, together with a property of persistency with respect to booleans. Note that Theorem 5 has been proved by Ganzinger [9] in the case where the considered class of algebras is the class $ALG(SP)$ of all algebras satisfying SP, and by Orejas and Navarro [20] in the case where the class of algebras is restricted to LOG-algebras.

We have only considered here the techniques of proofs by consistency. Everone agrees now that they are sometimes inefficient and more direct inductive proof methods have been developed, for instance by U.Reddy [26] in the equational case, or by E.Kounalis and M.Rusinowitch [18] in the Horn clause case. Extending these methods for proving theorems in parameterized specifications does not seem too difficult, and would lead to interesting applications.

Acknowledgements: I sincerely thank Jean-Luc Rémy for his contribution to a first part of this work presented at the COMPASS Workshop in Bremen in 1989, and his useful comments on this version. My thanks also to the anonymous referees for their constructive remarks.

References

[1] L. Bachmair. *Proof methods for equational theories.* PhD thesis, University of Illinois, Urbana-Champaign, 1987. Revised version, August 1988.

[2] L. Bachmair. Proof by consistency in equational theories. In *3rd Symp. Logic in Computer Science*, pages 228–233, Edinburgh, Scotland, 1988. IEEE.

[3] L. Bachmair, N. Dershowitz, and D. Plaisted. Completion without failure. In H. Ait-Kaci and M. Nivat, editors, *Resolution of Equations in Algebraic Structures, Volume 2: Rewriting Techniques*, pages 1–30. Academic Press, 1989.

[4] R. Bundgen. Design, implementation and application of an extended ground reducibility test. Technical Report 88-05, University of Delaware USA, December 1987.

[5] H. Comon. An effective method for handling initial algebras. In P. Lescanne J. Grabowski and W. Wechler, editors, *Proc. Int. Workshop on Algebraic and Logic Programming*, pages 108–118. Akademie Verlag, 1988. Also in Springer-Verlag Lecture Notes in Computer Science, volume 343.

[6] N. Dershowitz. Computing with rewrite systems. *Information and Control*, 65(2/3):122–157, 1985.

[7] L. Duponcheel, L. Jadoul, and W. Van Puymbroeck. Generic proofs by consistency. Technical report, Bell telephone Mfg. Co., Francis Wellesplein 1, B-2018 Antwerp, Belgium, 1989.

[8] H. Ehrig and B. Mahr. *Fundamentals of Algebraic Specification 1. Equations and initial semantics*, volume 6 of *EATCS Monographs on Theorical Computer Science.* Springer-Verlag, 1985.

[9] H. Ganzinger. Parameterized specifications: parameter passing and implementation with respect to observability. *ACM Transactions on Programming Languages and Systems*, 5(3):318–354, 1983.

[10] H. Ganzinger. Ground term confluence in parametric conditional equational specifications. In F. Brandenburg, G. Vidal-Naquet, and M. Wirsing, editors, *Proceedings STACS 87*, volume 247 of *Lecture Notes in Computer Science*, pages 286–298. Springer-Verlag, 1987.

[11] J. Goguen. Parameterized programming. *Transactions on Software Engineering*, SE-10(5):528–543, September 1984.

[12] J. Goguen, J. Meseguer, and D. Plaisted. Programming with parameterized abstract objects in OBJ. In *Theory and Practice of Software Technology*, pages 163–193. North-Holland, 1983.

[13] G. Huet and D. Oppen. Equations and rewrite rules: A survey. In R. Book, editor, *Formal Language Theory: Perspectives and Open Problems*, pages 349–405. Academic Press, New York, 1980.

[14] J.-P. Jouannaud and E. Kounalis. Proof by induction in equational theories without constructors. In *Proceedings 1st Symp. on Logic In Computer Science*, pages 358–366, Boston (USA), 1986.

[15] J.-P. Jouannaud and E. Kounalis. Automatic proofs by induction in theories without constructors. *Information and Computation*, 82:1–33, 1989.

[16] D. Kapur, P. Narendran, and H. Zhang. On sufficient completeness and related properties of term rewriting systems. *Acta Informatica*, 24:395–415, 1987.

[17] E. Kounalis. Testing for inductive (co)-reducibility. In A. Arnold, editor, *Proceedings 15th CAAP, Copenhagen (Denmark)*, volume 431 of *Lecture Notes in Computer Science*, pages 221–238. Springer-Verlag, May 1990.

[18] E. Kounalis and M. Rusinowitch. Mechanizing inductive reasoning. In *Proceedings of the AAAI Conference*, pages 240–245, Boston, 1990. AAAI Press and the MIT Press.

[19] G. A. Kucherov. A new quasi-reducibility testing algorithm and its application to proofs by induction. In P. Lescanne J. Grabowski and W. Wechler, editors, *Proc. workshop on Algebraic and Logic Programming*, pages 204–213. Akademie Verlag, 1988. Also in Springer-Verlag Lecture Notes in Computer Science, volume 343.

[20] M. Navarro and F. Orejas. Parameterized Horn clause specifications: proof theory and correctness. In *Proceedings TAPSOFT Conference*, volume 249 of *Lecture Notes in Computer Science*. Springer-Verlag, 1987.

[21] T. Nipkow and G. Weikum. A decidability result about sufficient completeness of axiomatically specified abstract data types. In *6th GI Conference*, volume 145 of *Lecture Notes in Computer Science*, pages 257–268. Springer-Verlag, 1983.

[22] F. Orejas. A characterization of passing compatibility for parameterized specifications. *Theoretical Computer Science*, pages 205–214, 1987.

[23] P. Padawitz. The equational theory of parameterized specifications. *Information and Computation*, 76:121–137, 1988.

[24] D. Plaisted. Semantic confluence and completion method. *Information and Control*, 65:182–215, 1985.

[25] A. Poigné. Parameterisation for order-sorted algebraic specifications. Technical report, Dept. of Computing, Imperial College, London, 1986.

[26] U. Reddy. Term rewriting induction. In M. Stickel, editor, *Proceedings 10th International Conference on Automated Deduction, Kaiserslautern (Germany)*, volume 449 of *Lecture Notes in Computer Science*, pages 162–177. Springer-Verlag, 1990.

[27] J.L. Rémy and H. Zhang. Reveur4 : a system for validating conditional algebraic specifications of abstract data types. In T. O'Shea, editor, *Proceedings of the 5th European Conference on Artificial Intelligence*, Pisa, Italy, 1984. ECAI, Elsevier Science Publishers (North Holland).

Completeness of Combinations of Constructor Systems†

Aart Middeldorp‡

Centre for Mathematics and Computer Science,
Kruislaan 413, 1098 SJ Amsterdam, The Netherlands.
email: ami@cwi.nl

Yoshihito Toyama

Centre for Mathematics and Computer Science,
Kruislaan 413, 1098 SJ Amsterdam, The Netherlands;
NTT Basic Research Laboratories
3-9-11 Midori-cho, Musashino-shi, Tokyo 180, Japan.
email: toyama@cwi.nl

ABSTRACT

A term rewriting system is called complete if it is both confluent and strongly normalizing. Barendregt and Klop showed that the disjoint union of complete term rewriting systems does not need to be complete. In other words, completeness is not a modular property of term rewriting systems. Toyama, Klop and Barendregt showed that completeness is a modular property of left-linear TRS's. In this paper we show that it is sufficient to impose the constructor discipline for obtaining the modularity of completeness. This result is a simple consequence of a quite powerful divide and conquer technique for establishing completeness of such constructor systems. Our approach is not limited to systems which are composed of disjoint parts. The importance of our method is that we may decompose a given constructor system into parts which possibly share function symbols and rewrite rules in order to infer completeness. We obtain a similar technique for semi-completeness, i.e. the combination of confluence and weak normalization.

Introduction

A property of term rewriting systems is *modular* if it is preserved under disjoint union. Starting with Toyama [19], several authors studied modular aspects of term rewriting systems. Toyama [19] showed that confluence is a modular property. In [20] Toyama refuted the modularity of strong normalization by means of the following term rewriting systems:

$$\mathcal{R}_1 = \{\, F(0, 1, x) \rightarrow F(x, x, x) \,\}$$

$$\mathcal{R}_2 = \begin{cases} g(x, y) & \rightarrow \; x \\ g(x, y) & \rightarrow \; y. \end{cases}$$

Both systems are terminating, but their union admits the following cyclic reduction:

---

† This paper is an abbreviated version of CWI report CS-R9059.

‡ Partially supported by ESPRIT Basic Research Action 3020, INTEGRATION.

$$
\begin{aligned}
F(g(0,1), g(0,1), g(0,1)) &\to F(0, g(0,1), g(0,1)) \\
&\to F(0, 1, g(0,1)) \\
&\to F(g(0,1), g(0,1), g(0,1)).
\end{aligned}
$$

His counterexample inspired Rusinowitch [18] to the formulation of sufficient conditions for the strong normalization of the disjoint union of strongly normalizing term rewriting systems $\mathcal{R}_1$ and $\mathcal{R}_2$ in terms of the distribution of collapsing and duplicating rules among $\mathcal{R}_1$ and $\mathcal{R}_2$. Rusinowitch's results were extended by Middeldorp [11]. Barendregt and Klop gave an example showing that completeness (i.e. the combination of confluence and strong normalization) is not a modular property, see Toyama [20]. Independently, Drosten [3] gave the following simpler counterexample:

$$
\mathcal{R}_1 = \begin{cases} F(0,1,x) & \to \quad F(x,x,x) \\ F(x,y,z) & \to \quad 2 \\ 0 & \to \quad 2 \\ 1 & \to \quad 2 \end{cases}
$$

$$
\mathcal{R}_2 = \begin{cases} g(x,y,y) & \to \quad x \\ g(y,y,x) & \to \quad x. \end{cases}
$$

Both systems are easily shown to be complete. However, because both $g(0,1,1) \twoheadrightarrow 0$ and $g(0,1,1) \twoheadrightarrow 1$, the term $F(g(0,1,1), g(0,1,1), g(0,1,1))$ has a cyclic reduction akin to the one in the previous counterexample. Toyama, Klop and Barendregt [22] showed that the restriction to left-linear term rewriting systems is sufficient for obtaining the modularity of completeness. Middeldorp [10] showed that the property of having unique normal forms is modular for general term rewriting systems. An interesting alternative approach to modularity is explored in Kurihara and Kaji [7]. Middeldorp [12, 13, 14] extended the above results to conditional term rewriting systems. Kurihara and Ohuchi [8] showed that strong normalization is a modular property of term rewriting systems whose strong normalization can be shown by a simplification ordering. They extended this result in [9] to term rewriting systems which share constructors. Constructors are function symbols which do not occur at the leftmost position in left-hand sides of rewrite rules. Dershowitz [1], Geser [4] and Toyama [21] give further results on combinations of term rewriting systems with common function symbols. A comprehensive survey of combinations of (conditional) term rewriting systems can be found in Middeldorp [15].

The starting point of the present paper is the refutation of the modularity of completeness. We show that instead of requiring left-linearity it is also possible to impose the so-called *constructor discipline* for obtaining the modularity of completeness. In a constructor system (a term rewriting system which obeys the constructor discipline) all function symbols occurring at non-leftmost positions in left-hand sides of rewrite rules are constructors. Many term rewriting systems that occur in practice follow this discipline, see e.g. O'Donnell [17]. Actually we prove a much stronger result. We show that a constructor system is complete if it can be *decomposed* into complete constructor systems. The important observation is that our notion of decomposition does not imply disjointness. Consider for example the constructor system

$$\mathcal{R} = \begin{cases} 0+x & \to x \\ S(x)+y & \to S(x+y) \\ 0\times x & \to 0 \\ S(x)\times y & \to x\times y+y \\ f(0) & \to 0 \\ f(S(x)) & \to f(x)+S(x). \end{cases}$$

We can decompose $\mathcal{R}$ into

$$\mathcal{R}_1 = \begin{cases} 0+x & \to x \\ S(x)+y & \to S(x+y) \\ 0\times x & \to 0 \\ S(x)\times y & \to x\times y+y \end{cases} \quad \text{and} \quad \mathcal{R}_2 = \begin{cases} 0+x & \to x \\ S(x)+y & \to S(x+y) \\ f(0) & \to 0 \\ f(S(x)) & \to f(x)+S(x). \end{cases}$$

Both systems are easily shown to be complete and our decomposition result yields the completeness of $\mathcal{R}$. Neither the result of Kurihara and Ohuchi [9] (because $\mathcal{R}_1$ and $\mathcal{R}_2$ share the non-constructor symbol +) nor the result of Dershowitz [1] (because $\mathcal{R}_1$ and $\mathcal{R}_2$ are not right-linear) applies.

In the next section we give a concise introduction to term rewriting. Extensive surveys are Dershowitz and Jouannaud [2] and Klop [6]. In Section 2 we introduce the concept of marked reduction which plays a crucial role in the proof of our main results. Section 3 contains our main results. We define a notion of decomposability and we show that completeness is a decomposable property of constructor systems. To appreciate the non-triviality of our result, it may be contrasted with the fact that neither confluence nor strong normalization is decomposable. We further show that semi-completeness (i.e the combination of confluence and weak normalization) is a decomposable property of constructor systems.

1. Preliminaries

Let $\mathcal{V}$ be a countably infinite set of *variables*. A *term rewriting system* (TRS for short) is a pair $(\mathcal{F}, \mathcal{R})$. The set $\mathcal{F}$ consists of *function symbols*; associated to every $F \in \mathcal{F}$ is a natural number denoting its arity. Function symbols of arity 0 are called *constants*. The set $\mathcal{T}(\mathcal{F}, \mathcal{V})$ of *terms* built from $\mathcal{F}$ and $\mathcal{V}$ is the smallest set such that $\mathcal{V} \subset \mathcal{T}(\mathcal{F}, \mathcal{V})$ and if $F \in \mathcal{F}$ has arity n and $t_1, \dots, t_n \in \mathcal{T}(\mathcal{F}, \mathcal{V})$ then $F(t_1, \dots, t_n) \in \mathcal{T}(\mathcal{F}, \mathcal{V})$. Identity of terms is denoted by $\equiv$. The *root symbol* of a term t is defined as follows: $root(t) = F$ if $t \equiv F(t_1, \dots, t_n)$ and $root(t) = t$ if $t \in \mathcal{V}$. The set $\mathcal{R}$ consists of pairs (l, r) with $l, r \in \mathcal{T}(\mathcal{F}, \mathcal{V})$ subject to the following two constraints:

(1) the left-hand side l is not a variable,

(2) the variables which occur in the right-hand side r also occur in l.

Pairs (l, r) are called *rewrite rules* and will henceforth be written as $l \to r$. A rewrite rule $l \to r$ is *left-linear* if l does not contain multiple occurrences of the same variable. A *left-linear* TRS only contains left-linear rewrite rules.

A *substitution* σ is a mapping from $\mathcal{V}$ to $\mathcal{T}(\mathcal{F}, \mathcal{V})$ such that its *domain* $\{x \in \mathcal{V} \mid \sigma(x) \not\equiv x\}$ is finite. Substitutions are extended to morphisms from $\mathcal{T}(\mathcal{F}, \mathcal{V})$ to $\mathcal{T}(\mathcal{F}, \mathcal{V})$, i.e. $\sigma(F(t_1, \dots, t_n)) \equiv F(\sigma(t_1), \dots, \sigma(t_n))$ for every n-ary function symbol F and terms $t_1, \dots, t_n$. We call $\sigma(t)$ an *instance* of t. We write t^σ instead of $\sigma(t)$. An instance of a left-hand side of a rewrite rule is a *redex* (reducible expression). Let $\square$ be a special constant symbol. A *context* $C[\,,\ldots,\,]$ is a term in $\mathcal{T}(\mathcal{F}\cup\{\square\},\mathcal{V})$. If $C[\,,\ldots,\,]$ is a context with n occurrences of $\square$ and $t_1,\ldots,t_n$ are terms then $C[t_1,\ldots,t_n]$ is the result of replacing from left to right the occurrences of $\square$ by $t_1,\ldots,t_n$. A context containing precisely one occurrence of $\square$ is denoted by $C[\,]$. A term s is a *subterm* of a term t if there exists a context $C[\,]$ such that $t \equiv C[s]$. If $C[\,] \not\equiv \square$ then s is a *proper* subterm of t. We write $s \sqsubseteq t$ to indicate that s is a subterm of t.

The *rewrite relation* $\rightarrow_{\mathcal{R}}$ is defined as follows: $s \rightarrow_{\mathcal{R}} t$ if there exists a rewrite rule $l \rightarrow r$ in $\mathcal{R}$, a substitution σ and a context $C[\,]$ such that $s \equiv C[l^\sigma]$ and $t \equiv C[r^\sigma]$. The transitive-reflexive closure of $\rightarrow_{\mathcal{R}}$ is denoted by $\twoheadrightarrow_{\mathcal{R}}$; if $s \twoheadrightarrow_{\mathcal{R}} t$ we say that s *reduces* to t. We write $s \leftarrow_{\mathcal{R}} t$ if $t \rightarrow_{\mathcal{R}} s$; likewise for $s \twoheadleftarrow_{\mathcal{R}} t$. The transitive closure of $\rightarrow_{\mathcal{R}}$ is denoted by $\rightarrow_{\mathcal{R}}^{+}$ and $\leftrightarrow_{\mathcal{R}}$ denotes the symmetric closure of $\rightarrow_{\mathcal{R}}$ (so $\leftrightarrow_{\mathcal{R}} = \rightarrow_{\mathcal{R}} \cup \leftarrow_{\mathcal{R}}$). The transitive-reflexive closure of $\leftrightarrow_{\mathcal{R}}$ is called *conversion* and denoted by $=_{\mathcal{R}}$. If $s =_{\mathcal{R}} t$ then s and t are *convertible*. Two terms t_1, t_2 are *joinable*, notation $t_1 \downarrow_{\mathcal{R}} t_2$, if there exists a term t_3 such that $t_1 \twoheadrightarrow_{\mathcal{R}} t_3 \twoheadleftarrow_{\mathcal{R}} t_2$. Such a term t_3 is called a *common reduct* of t_1 and t_2. We often omit the subscript $\mathcal{R}$.

A term s is a *normal form* if there is no term t with $s \rightarrow t$. A TRS is *weakly normalizing* if every term reduces to a normal form. A TRS is *strongly normalizing* if there are no infinite reduction sequences $t_1 \rightarrow t_2 \rightarrow t_3 \rightarrow \ldots$. In other words, every reduction sequence eventually ends in a normal form. A TRS is *confluent* or has the *Church-Rosser* property if for all terms s, t_1, t_2 with $t_1 \twoheadleftarrow s \twoheadrightarrow t_2$ we have $t_1 \downarrow t_2$. A well-known equivalent formulation of confluence is that every pair of convertible terms is joinable ($t_1 = t_2 \Rightarrow t_1 \downarrow t_2$). A TRS is *locally confluent* if for all terms s, t_1, t_2 with $t_1 \leftarrow s \rightarrow t_2$ we have $t_1 \downarrow t_2$. A *complete* TRS is confluent and strongly normalizing. A *semi-complete* TRS is confluent and weakly normalizing. These properties of TRS's specialize to terms in the obvious way. If a term t has a unique normal form then we denote this normal form by $t\!\downarrow$.

The following well-known result is due to Newman [16].

NEWMAN'S LEMMA. *Every strongly normalizing and locally confluent* TRS *is confluent.* $\square$

Let $l_1 \rightarrow r_1$ and $l_2 \rightarrow r_2$ be renamed versions of rewrite rules of a TRS $\mathcal{R}$ such that they have no variables in common. Suppose $l_1 \equiv C[t]$ with $t \notin \mathcal{V}$ such that t and l_2 are unifiable, i.e. $t^\sigma \equiv l_2^\sigma$ for a most general unifier σ. The term $l_1^\sigma \equiv C[l_2]^\sigma$ is subject to the reduction steps $l_1^\sigma \rightarrow r_1^\sigma$ and $l_1^\sigma \rightarrow C[r_2]^\sigma$. The pair of reducts $\langle C[r_2]^\sigma, r_1^\sigma\rangle$ is a *critical pair* of $\mathcal{R}$. If $l_1 \rightarrow r_1$ and $l_2 \rightarrow r_2$ are renamed versions of the same rewrite rule, we do not consider the case $C[\,] \equiv \square$. A critical pair $\langle s, t\rangle$ of a TRS $\mathcal{R}$ is *convergent* if $s \downarrow_{\mathcal{R}} t$. The following lemma of Huet [5] expresses the significance of critical pairs.

CRITICAL PAIR LEMMA. *A* TRS $\mathcal{R}$ *is locally confluent if and only if all its critical pairs are convergent.* $\square$

A *constructor system* (CS for short) is a TRS $(\mathcal{F}, \mathcal{R})$ with the property that $\mathcal{F}$ can be partitioned into disjoint sets $\mathcal{D}$ and $\mathcal{C}$ such that every left-hand side $F(t_1,\ldots,t_n)$ of a rewrite rule of $\mathcal{R}$ satisfies $F \in \mathcal{D}$ and $t_1,\ldots,t_n \in \mathcal{T}(\mathcal{C},\mathcal{V})$. Function symbols in $\mathcal{D}$ are called *defined symbols* and those in $\mathcal{C}$ *constructors*. To emphasize the partition of $\mathcal{F}$ into $\mathcal{D}$ and $\mathcal{C}$ we write $(\mathcal{D}, \mathcal{C}, \mathcal{R})$ instead of $(\mathcal{F}, \mathcal{R})$ and $\mathcal{T}(\mathcal{F},\mathcal{V})$ is denoted by $\mathcal{T}(\mathcal{D},\mathcal{C},\mathcal{V})$.

Since the behaviour of a Turing machine can be simulated by a CS (see Klop [6] for details), CS's have universal computing power. The restriction on the left-hand sides of rewrite rules of CS's enables a considerable simplification of many concepts and proofs. For instance, if $\langle s, t\rangle$ is a critical pair of a CS $(\mathcal{D}, \mathcal{C}, \mathcal{R})$ then there exist different rewrite rules $l_1 \rightarrow r_1, l_2 \rightarrow r_2 \in \mathcal{R}$ (with variables suitably renamed) and a most general unifier σ of l_1 and l_2 such that $s \equiv r_1^\sigma$ and $t \equiv r_2^\sigma$.

2. Marked Reduction

In this section we introduce a new rewrite relation which plays an essential role in the proofs of our decomposition results. Due to lack of space no proofs are presented in this section. They can be found in the full version of the paper. Throughout this section we will be dealing with an arbitrary CS $(\mathcal{D}, \mathcal{C}, \mathcal{R})$.

DEFINITION 2.1.
(1) The set $\mathcal{D}^* = \{F^* \mid F \in \mathcal{D}\}$ consists of *marked* defined symbols. Terms in $\mathcal{T}(\mathcal{D}^* \cup \mathcal{D}, \mathcal{C}, \mathcal{V})$ are called *marked terms.* An *unmarked term* belongs to $\mathcal{T}(\mathcal{D}, \mathcal{C}, \mathcal{V})$.
(2) If t is a marked term then $e(t) \in \mathcal{T}(\mathcal{D}, \mathcal{C}, \mathcal{V})$ denotes the term obtained from t by erasing all marks and t^* denotes the term obtained from t by marking every unmarked defined symbol in t.
(3) Two marked terms s and t are *similar*, notation $s \approx t$, if $e(s) \equiv e(t)$. If s and t are similar then their *intersection* is the unique term $s \wedge t$ such that $s \wedge t \approx s \approx t$ and a defined symbol occurrence in $s \wedge t$ is marked if and only if the corresponding symbols in s and t are marked.
(4) The set $\mathcal{R}^*$ of *marked rewrite rules* is defined as $\{l^* \rightarrow r^* \mid l \rightarrow r \in \mathcal{R}\}$.

EXAMPLE 2.2. Consider the CS $(\mathcal{D}_1, \mathcal{C}_1, \mathcal{R}_1)$ with $\mathcal{D}_1 = \{F, G\}$, $\mathcal{C}_1 = \{S, 0\}$,

$$\mathcal{R}_1 = \begin{cases} F(S(x), y) & \rightarrow G(x) \\ G(x) & \rightarrow S(0) \end{cases}$$

and the reduction sequence

$$t \equiv F(S(G(0)), G(0)) \rightarrow F(S(G(0)), S(0)) \rightarrow G(G(0)) \rightarrow S(0).$$

If we mark some defined symbols in t then we can easily mimic this sequence by a reduction sequence in $\mathcal{R}_1 \cup \mathcal{R}_1^*$, for instance

$$F^*(S(G^*(0)), G(0)) \rightarrow_{\mathcal{R}_1} F^*(S(G^*(0)), S(0)) \rightarrow_{\mathcal{R}_1^*} G^*(G^*(0)) \rightarrow_{\mathcal{R}_1^*} S(0).$$

This correspondence does not hold for non-left-linear CS's. Consider the CS $(\mathcal{D}_2, \mathcal{C}_2, \mathcal{R}_2)$ with $\mathcal{D}_2 = \{F\}$, $\mathcal{C}_2 = \{S\}$, $\mathcal{R}_2 = \{F(x, x) \rightarrow S(x)\}$ and the reduction step

$$F(F(0, S(0)), F(0, S(0))) \rightarrow S(F(0, S(0))).$$

The marked term $F^*(F^*(0, S(0)), F(0, S(0)))$ cannot be reduced in $\mathcal{R}_2 \cup \mathcal{R}_2^*$.

By modifying the rewrite relation associated to $\mathcal{R} \cup \mathcal{R}^*$ we are able to mimic every unmarked reduction sequence, irrespective of the marking attached to the starting term.

DEFINITION 2.3. We write $s \rightarrow_m t$ if there exists a context $C[\,]$, a rewrite rule

$$C_1[x_1, \ldots, x_n] \rightarrow C_2[y_1, \ldots, y_m]$$

in $\mathcal{R} \cup \mathcal{R}^*$ (with all variables displayed) and terms $s_1, \ldots, s_n, t_1, \ldots, t_m$ such that the

following three conditions are satisfied:
(1) $s \equiv C[C_1[s_1, \ldots, s_n]]$ and $t \equiv C[C_2[t_1, \ldots, t_m]]$,
(2) $s_i \approx s_j$ whenever $x_i \equiv x_j$ for $1 \leq i < j \leq n$,
(3) $t_i \equiv \wedge\{s_j \mid x_j \equiv y_i\}$ for $i = 1, \ldots, m$.
We call $C_1[s_1, \ldots, s_n]$ a *marked redex* and the relation $\to_m$ is called *marked reduction.*

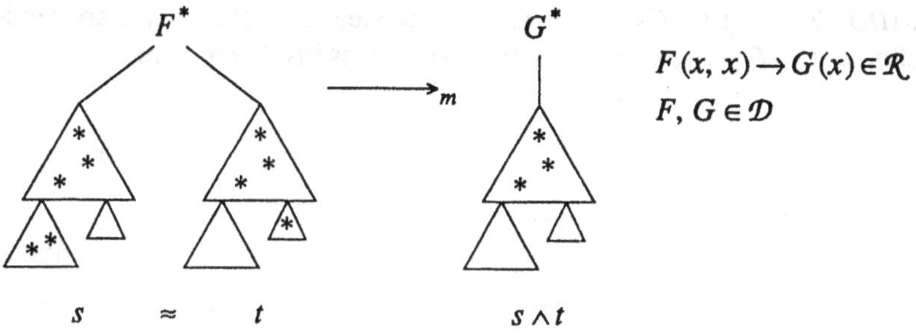

FIGURE 1.

Notice that $\to_m$ coincides with $\to_{\mathcal{R} \cup \mathcal{R}^*}$ whenever $\mathcal{R}$ is left-linear.

EXAMPLE 2.4. Consider the CS $(\mathcal{D}, \mathcal{C}, \mathcal{R})$ with $\mathcal{D} = \{F, G\}$, $\mathcal{C} = \{S, 0\}$,

$$\mathcal{R} = \begin{cases} F(x, x) & \to & S(x) \\ G(x) & \to & 0 \end{cases}$$

and the reduction sequence $F(G(F(0, S(0))), G(F(0, S(0)))) \to S(G(F(0, S(0)))) \to S(0)$. We have $F^*(G^*(F^*(0, S(0))), G^*(F(0, S(0)))) \to_m S(G^*(F(0, S(0)))) \to_m S(0)$.

The next proposition relates marked reduction to ordinary reduction. In part (2) it is essential that we restrict ourselves to CS's.

PROPOSITION 2.5.
(1) *If* $s \to_m t$ *then* $e(s) \to e(t)$.
(2) *If* $s \to t$ *and* $e(s') \equiv s$ *then there exists a term* t' *such that* $s' \to_m t'$ *and* $e(t') \equiv t$.
□

DEFINITION 2.6. If $t \equiv C[t_1, \ldots, t_n]$ such that all defined symbols in $C[\,, \ldots, \,]$ are marked and every t_i $(i = 1, \ldots, n)$ is unmarked then we call t a *capped* term. Furthermore, if $root(t_i) \in \mathcal{D}$ for $i = 1, \ldots, n$ then we write $t \equiv C{*}[t_1, \ldots, t_n]{*}$.

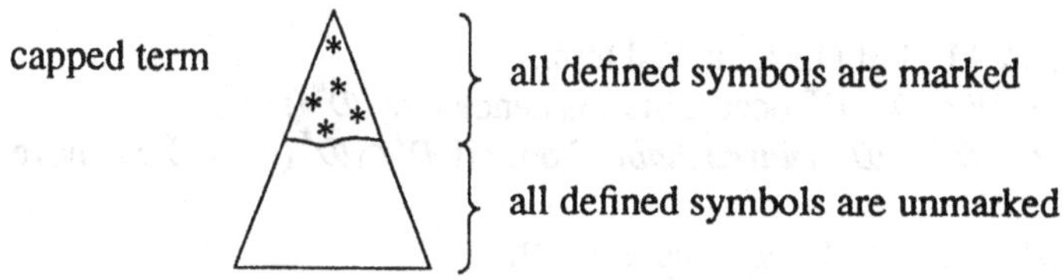

FIGURE 2.

DEFINITION 2.7. Let $s \equiv C*[s_1, \ldots, s_n]*$ be a capped term.

(1) Suppose $s \rightarrow_m t$ by contraction of the marked redex Δ. We write $s \rightarrow_m^i t$ if Δ occurs in one of $s_1, \ldots, s_n$ and we write $s \rightarrow_m^o t$ otherwise. The relation $\rightarrow_m^i$ is called *inner* marked reduction and $\rightarrow_m^o$ is called *outer* marked reduction.

(2) We call s *inside normalized* if it is a normal form with respect to $\rightarrow_m^i$.

DEFINITION 2.8. Let $t \equiv C*[t_1, \ldots, t_n]*$ be a capped term. If $t_1, \ldots, t_n$ are semi-complete then we define $\psi(t) \equiv C[t_1\downarrow, \ldots, t_n\downarrow]$. Notice that $\psi(t)$ is inside normalized.

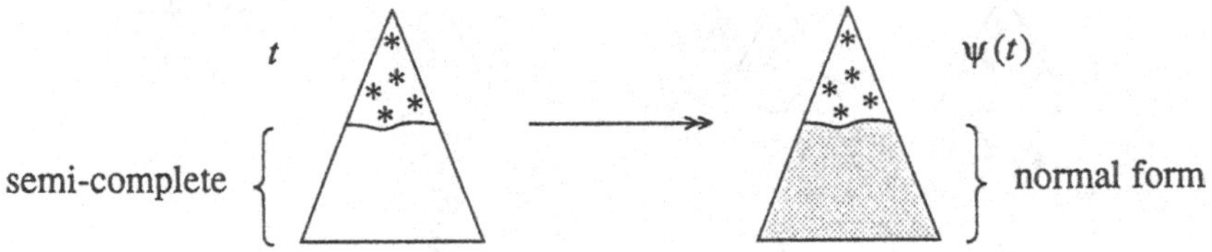

FIGURE 3.

LEMMA 2.9. *Let s be a capped term such that $\psi(s)$ is defined.*

(1) *If $s \rightarrow_m^o t$ then $\psi(t)$ is defined and $\psi(s) \rightarrow_m^{o+} \psi(t)$.*

(2) *If $s \rightarrow_m^i t$ then $\psi(t)$ is defined and $\psi(s) \equiv \psi(t)$.*

(3) *If s is a normal form then $\psi(s) \equiv s$.*

□

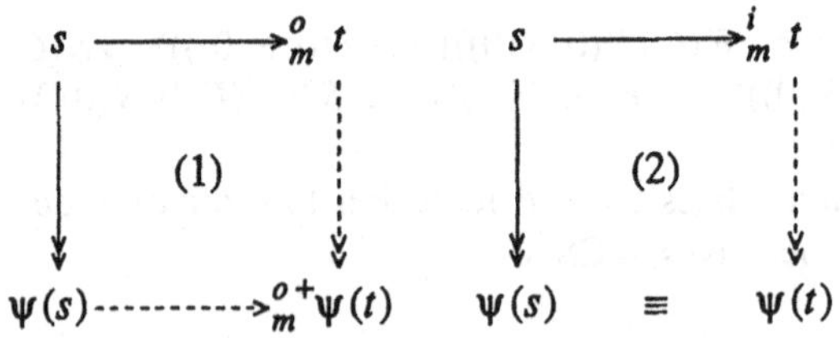

FIGURE 4.

In the remainder of this section we give some further properties of marked reduction which are needed in the next section.

LEMMA 2.10. *Let s be a capped term and suppose $s \twoheadrightarrow_m t$. For every subterm u of t with $root(u) \in \mathcal{D}$ we can find terms $s' \subseteq s$ and $t' \subseteq t$ such that $root(s') \in \mathcal{D}$, $s' \twoheadrightarrow t'$ and $u \subseteq t'$.* □

DEFINITION 2.11. Let t be a marked term.

(1) The set $\{F \in \mathcal{D} \mid F^* \text{ occurs in } t\}$ is denoted by $\mathcal{D}^*(t)$.

(2) A subset $\mathcal{D}'$ of $\mathcal{D}$ is *unreachable* from t if $\mathcal{D}' \cap \mathcal{D}^*(t') = \varnothing$ whenever $t \twoheadrightarrow_m t'$.

DEFINITION 2.12. Let $\mathcal{D}'$ be a subset of $\mathcal{D}$.

(1) A set of pairs $\phi = \{\langle s_1, x_1\rangle, \ldots, \langle s_n, x_n\rangle\}$ is a *$\mathcal{D}'$-replacement* if $x_1, \ldots, x_n$ are mutually distinct variables and $s_1, \ldots, s_n$ are mutually distinct unmarked terms such that $root(s_i) \in \mathcal{D}'$ for $i = 1, \ldots, n$. Let $t \equiv C[t_1, \ldots, t_m]$ such that all maximal subterms of t

with root symbol in $\mathcal{D}'$ are displayed. We say that ϕ is *applicable* to t if $x_1, \ldots, x_n$ do not occur in t and $\{t_1, \ldots, t_m\} \subseteq \{s_1, \ldots, s_n\}$. In this case we may write $t \equiv C[s_{i_1}, \ldots, s_{i_m}]$ with $1 \le i_1, \ldots, i_m \le n$ and we define $\phi(t) \equiv C[x_{i_1}, \ldots, x_{i_m}]$.

(2) Let $\phi = \{\langle s_1, x_1\rangle, \ldots, \langle s_n, x_n\rangle\}$ be a $\mathcal{D}'$-replacement. Suppose $t \equiv C[x_{i_1}, \ldots, x_{i_m}]$ such that all occurrences of the variables $x_1, \ldots, x_n$ in t are displayed. The term $C[s_{i_1}, \ldots, s_{i_m}]$ is denoted by $\phi^{-1}(t)$.

PROPOSITION 2.13. *Let $\mathcal{D}'$ be a subset of $\mathcal{D}$. For every term t there exists a $\mathcal{D}'$-replacement ϕ which is applicable to t.* □

PROPOSITION 2.14. *Let ϕ be a $\mathcal{D}'$-replacement for some $\mathcal{D}' \subseteq \mathcal{D}$.*
(1) *If $s \twoheadrightarrow_m t$ then $\phi^{-1}(s) \twoheadrightarrow_m \phi^{-1}(t)$.*
(2) *If ϕ is applicable to t then $\phi^{-1}(\phi(t)) \equiv t$.*
(3) *If ϕ is applicable to a capped term t then $e(\phi(t)) \equiv \phi(e(t))$.*
□

LEMMA 2.15. *Let s be a capped term. Suppose $\mathcal{D}' \subseteq \mathcal{D}$ is unreachable from s and ϕ is a $\mathcal{D}'$-replacement applicable to s.*
(1) *If $s \to_m^o t$ then ϕ is applicable to t and $\phi(s) \to_m^o \phi(t)$.*
(2) *If s has an infinite $\to_m^o$-reduction then $\phi(s)$ has an infinite $\to_m^o$-reduction.*
□

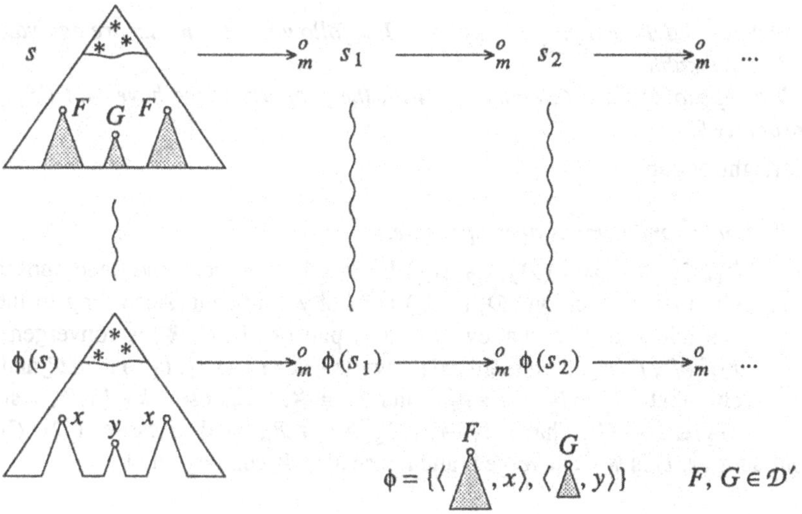

FIGURE 5.

3. Combinations of Constructor Systems

In this section we show that both completeness and semi-completeness exhibit the important compositional behaviour expressed in the next definition.

DEFINITION 3.1.
(1) Let $(\mathcal{D}, \mathcal{C}, \mathcal{R})$ be a CS and suppose $\mathcal{D}' \subseteq \mathcal{D}$. The set $\{l \to r \in \mathcal{R} \mid root(l) \in \mathcal{D}'\}$ is

denoted by $\mathcal{R} \mid \mathcal{D}'$.

(2) Two CS's $(\mathcal{D}_1, C_1, \mathcal{R}_1)$ and $(\mathcal{D}_2, C_2, \mathcal{R}_2)$ are *composable* if $\mathcal{D}_1 \cap C_2 = \mathcal{D}_2 \cap C_1 = \varnothing$ and $\mathcal{R}_1 \mid \mathcal{D}_2 = \mathcal{R}_2 \mid \mathcal{D}_1$. The second requirement is equivalent to the condition that both CS's contain all rewrite rules which 'define' a defined symbol whenever that symbol is shared. The union of pairwise composable CS's $CS_1, \ldots, CS_n$ is denoted by $CS_1 + \ldots + CS_n$ and we say that $CS_1, \ldots, CS_n$ is a *decomposition* of $CS_1 + \ldots + CS_n$.

(3) A property $\mathcal{P}$ of CS's is *decomposable* if for all pairwise composable CS's $CS_1, \ldots, CS_n$ with the property $\mathcal{P}$ we have that $CS_1 + \ldots + CS_n$ has the property $\mathcal{P}$.

The counterexample of Toyama against the modularity of strong normalization shows that strong normalization is not a decomposable property of CS's. The following example of Huet [5] shows that also confluence is not decomposable.

EXAMPLE 3.2. Consider the CS $(\mathcal{D}, C, \mathcal{R})$ with $\mathcal{D} = \{F, C\}$, $C = \{S, A, B\}$ and

$$\mathcal{R} = \begin{cases} F(x, x) & \to A \\ F(x, S(x)) & \to B \\ C & \to S(C). \end{cases}$$

Let $\mathcal{D}_1 = \{F\}$, $C_1 = C$, $\mathcal{D}_2 = \{C\}$ and $C_2 = \{S\}$. The confluent CS's $(\mathcal{D}_1, C_1, \mathcal{R}_1)$, $(\mathcal{D}_2, C_2, \mathcal{R}_2)$ constitute a decomposition of $(\mathcal{D}, C, \mathcal{R})$, but $(\mathcal{D}, C, \mathcal{R})$ is not confluent since the term $F(C, C)$ can be reduced to the different normal forms A and B.

PROPOSITION 3.3. *let $\mathcal{P}$ be a property of* CS'*s. The following statements are equivalent:*

(1) *$\mathcal{P}$ is decomposable;*

(2) *for all composable CS's CS_1 and CS_2 with the property $\mathcal{P}$ we have that $CS_1 + CS_2$ has the property $\mathcal{P}$.*

PROOF. Straightforward. □

LEMMA 3.4. *Local confluence is decomposable.*

PROOF. Let $(\mathcal{D}_1, C_1, \mathcal{R}_1)$ and $(\mathcal{D}_2, C_2, \mathcal{R}_2)$ be locally confluent and decomposable CS's. We have to show that their union $(\mathcal{D}, C, \mathcal{R})$ is locally confluent. According to the Critical Pair Lemma it sufficient to show that every critical pair of $(\mathcal{D}, C, \mathcal{R})$ is convergent. If $\langle s, t \rangle$ is a critical pair of $(\mathcal{D}, C, \mathcal{R})$ then there exist rewrite rules $l_1 \to r_1, l_2 \to r_2 \in \mathcal{R}$ and a substitution σ such that $l_1^\sigma \equiv l_2^\sigma$, $s \equiv r_1^\sigma$ and $t \equiv r_2^\sigma$. Choose $k \in \{1, 2\}$ such that $root(l_1) = root(l_2) \in \mathcal{D}_k$. We have $l_1 \to r_1, l_2 \to r_2 \in \mathcal{R}_k$ and because $(\mathcal{D}_k, C_k, \mathcal{R}_k)$ is locally confluent $\langle s, t \rangle$ is $\mathcal{R}_k$-convergent and hence also $\mathcal{R}$-convergent. □

THEOREM 3.5. *Completeness is decomposable.*

PROOF. Let $(\mathcal{D}_1, C_1, \mathcal{R}_1)$ and $(\mathcal{D}_2, C_2, \mathcal{R}_2)$ be complete and composable CS's. From Lemma 3.4 we obtain the local confluence of their union $(\mathcal{D}, C, \mathcal{R})$. According to Newman's Lemma it suffices to show the strong normalization of $(\mathcal{D}, C, \mathcal{R})$. This will be established by induction on the structure of terms $t \in \mathcal{T}(\mathcal{D}, C, \mathcal{V})$. If t is a variable or a constructor constant then t is a normal form. If t is a defined constant then t belongs to some $\mathcal{D}_k$ and because $(\mathcal{D}_k, C_k, \mathcal{R}_k)$ is strongly normalizing t cannot have an infinite reduction. For the induction step, let $t \equiv F(t_1, \ldots, t_n)$ such that $t_1, \ldots, t_n$ are strongly normalizing (and hence complete). If F is a constructor then t clearly is strongly normalizing. So assume that $F \in \mathcal{D}$. If t is not strongly normalizing then there exists an infinite reduction sequence

$$t \equiv s_1 \to s_2 \to s_3 \to \dots . \tag{1}$$

Let $t' \equiv F^*(t_1, \dots, t_n)$. According to Proposition 2.5 we can find terms s'_i with $e(s'_i) \equiv s_i$ such that

$$t' \equiv s'_1 \to_m s'_2 \to_m s'_3 \to_m \dots . \tag{2}$$

Using Lemma 2.10 and the assumption that $t_1, \dots, t_n$ are strongly normalizing, it is not difficult to show that sequence (2) contains infinitely many $\to_m^o$-steps. According to Lemma 2.9 we can transform sequence (2) into the marked reduction sequence

$$\psi(t') \equiv \psi(s'_1) \twoheadrightarrow_m^o \psi(s'_2) \twoheadrightarrow_m^o \psi(s'_3) \twoheadrightarrow_m^o \cdots \tag{3}$$

which contains infinitely many steps. Choose $k \in \{1, 2\}$ such that $F \in \mathcal{D}_k$ and let $\mathcal{D}' = \mathcal{D} - \mathcal{D}_k$. It is easy to show that $\mathcal{D}'$ is unreachable from $\psi(t')$. From Proposition 2.13 we obtain a $\mathcal{D}'$-replacement ϕ which is applicable to $\psi(t')$. By Lemma 2.15(2) the term $\phi(\psi(t'))$ has an infinite $\to_m^o$-reduction sequence

$$\phi(\psi(t')) \equiv \phi(\psi(s'_1)) \twoheadrightarrow_m^o \phi(\psi(s'_2)) \twoheadrightarrow_m^o \phi(\psi(s'_3)) \twoheadrightarrow_m^o \dots . \tag{4}$$

If we erase all markers in this sequence we obtain an infinite reduction sequence starting from the term $e(\phi(\psi(t')))$. This contradicts the strong normalization of the CS $(\mathcal{D}_k, C_k, \mathcal{R}_k)$. □

COROLLARY 3.6. *Completeness is a modular property of* CS's. □

COROLLARY 3.7. *The union of complete* CS's *which do not share defined symbols is complete.* □

We now consider a more challenging situation in which Theorem 3.5 can be applied.

EXAMPLE 3.8. Consider the CS $(\mathcal{D}, C, \mathcal{R})$ with $\mathcal{D} = \{+, \times, \mathit{fib}, <, \wedge\}$, $C = \{0, S, \mathit{true}, \mathit{false}\}$ and rewrite rules

r_1	$0+x$	$\to$	x	r_8	$x<0$	$\to$	false
r_2	$S(x)+y$	$\to$	$S(x+y)$	r_9	$0<S(x)$	$\to$	true
r_3	$0\times x$	$\to$	0	r_{10}	$S(x)<S(y)$	$\to$	$x<y$
r_4	$S(x)\times y$	$\to$	$x\times y+y$	r_{11}	$\mathit{true} \wedge \mathit{false}$	$\to$	false
r_5	$\mathit{fib}(0)$	$\to$	$S(0)$	r_{12}	$\mathit{false} \wedge \mathit{true}$	$\to$	false
r_6	$\mathit{fib}(S(0))$	$\to$	$S(0)$	r_{13}	$x \wedge x$	$\to$	x
r_7	$\mathit{fib}(S(S(x)))$	$\to$	$\mathit{fib}(S(x))+\mathit{fib}(x)$				

Consider the decomposition $(\mathcal{D}_i, C_i, \mathcal{R}_i)_{i=1}^4$ defined as follows:

i	$\mathcal{D}_i$	C_i	$\mathcal{R}_i$
1	$+ \times$	$0\ S$	$r_1\ r_2\ r_3\ r_4$
2	$+\ \mathit{fib}$	$0\ S$	$r_1\ r_2\ r_5\ r_6\ r_7$
3	$<$	$0\ S\ \mathit{true}\ \mathit{false}$	$r_8\ r_9\ r_{10}$
4	$\wedge$	$\mathit{true}\ \mathit{false}$	$r_{11}\ r_{12}\ r_{13}$

Routine arguments show that every $(\mathcal{D}_i, C_i, \mathcal{R}_i)$ is complete. Theorem 3.5 yields the completeness of $(\mathcal{D}, C, \mathcal{R})$.

The proof of the decomposability of semi-completeness is comparable to the proof of Theorem 3.5. First we show the decomposability of weak normalization.

LEMMA 3.9. *Weak normalization is decomposable.*

PROOF. Suppose $(\mathcal{D}_1, C_1, \mathcal{R}_1)$ and $(\mathcal{D}_2, C_2, \mathcal{R}_2)$ are weakly normalizing and composable CS's and let $(\mathcal{D}, C, \mathcal{R}) = (\mathcal{D}_1, C_1, \mathcal{R}_1) + (\mathcal{D}_2, C_2, \mathcal{R}_2)$. We will show by induction on the structure of t that every term $t \in \mathcal{T}(\mathcal{D}, C, \mathcal{V})$ has a normal form. The case $t \in \mathcal{D} \cup C \cup \mathcal{V}$ is easy. Suppose $t \equiv F(t_1, ..., t_n)$ and $t_1, ..., t_n$ are weakly normalizing. Let s_i be a normal form of t_i for $i = 1, ..., n$ and define $t' \equiv F(s_1, ..., s_n)$. If $F \in C$ then t' is a normal form of t. If $F \in \mathcal{D}$ then there exists a $k \in \{1, 2\}$ such that $F \in \mathcal{D}_k$. Let $\mathcal{D}' = \mathcal{D} - \mathcal{D}_k$. From Proposition 2.13 we obtain a $\mathcal{D}'$-replacement ϕ which is applicable to t'. Since $(\mathcal{D}_k, C_k, \mathcal{R}_k)$ is weakly normalizing, the term $\phi(t')$ has a normal form, say t''. Using Proposition 2.14 we obtain $t \twoheadrightarrow t' \equiv \phi^{-1}(\phi(t')) \twoheadrightarrow \phi^{-1}(t'')$. It is easy to show that $\phi^{-1}(t'')$ is a normal form. □

THEOREM 3.10. *Semi-completeness is decomposable.*

PROOF. Let $(\mathcal{D}_1, C_1, \mathcal{R}_1)$ and $(\mathcal{D}_2, C_2, \mathcal{R}_2)$ be semi-complete and composable CS's. From Lemma 3.9 we obtain the weak normalization of their union $(\mathcal{D}, C, \mathcal{R})$. Hence it is sufficient to show that every term $t \in \mathcal{T}(\mathcal{D}, C, \mathcal{V})$ has at most one normal form. We use induction on the structure of t. The case $t \in \mathcal{D} \cup C \cup \mathcal{V}$ is easy. Suppose $t \equiv F(t_1, ..., t_n)$ such that every t_i is semi-complete. If $F \in C$ then $F(t_1\downarrow, ..., t_n\downarrow)$ is the unique normal form of t. Suppose $F \in \mathcal{D}$ and let $t' \equiv F^*(t_1, ..., t_n)$. Define $\mathcal{D}_k$, $\mathcal{D}'$ and ϕ as in the proof of Theorem 3.5. First we show that if t has a normal form n then $\phi(e(\psi(t'))) \twoheadrightarrow \phi(n)$. With help of Proposition 2.5 and Lemma 2.9 we obtain a normal form n' such that $\psi(t') \twoheadrightarrow^o_m n'$ and $e(n') \equiv n$. Repeated application of Lemma 2.15(1) yields $\phi(\psi(t')) \twoheadrightarrow^o_m \phi(n')$. Erasing all markers in this sequences gives us $e(\phi(\psi(t'))) \twoheadrightarrow e(\phi(n'))$ and from Proposition 2.14(3) we obtain $e(\phi(\psi(t'))) \equiv \phi(e(\psi(t')))$ and $e(\phi(n')) \equiv \phi(e(n')) \equiv \phi(n)$. Now suppose that t has normal forms n_1 and n_2. From the above discussion we learn that $\phi(n_1) \twoheadleftarrow \phi(e(\psi(t'))) \twoheadrightarrow \phi(n_2)$. Notice that $\phi(n_1)$ and $\phi(n_2)$ are normal forms. We obtain $\phi(n_1) \equiv \phi(n_2)$ from the semi-completeness of $(\mathcal{D}_k, C_k, \mathcal{R}_k)$. Hence $n_1 \equiv \phi^{-1}(\phi(n_1)) \equiv \phi^{-1}(\phi(n_2)) \equiv n_2$ by Proposition 2.14(2). □

COROLLARY 3.11. *The union of semi-complete* CS'*s which do not share defined symbols is semi-complete.* □

References

1. N. Dershowitz, *Termination of Linear Rewriting Systems (preliminary version)*, Proceedings of the 8th International Colloquium on Automata, Languages and Programming, Acre, Lecture Notes in Computer Science **115**, pp. 448-458, 1981.
2. N. Dershowitz and J.-P. Jouannaud, *Rewrite Systems*, in: Handbook of Theoretical Computer Science, Vol. B (ed. J. van Leeuwen), North-Holland, 1990.
3. K. Drosten, *Termersetzungssysteme*, Informatik-Fachberichte **210**, Springer, 1989 (in German).
4. A. Geser, *Relative Termination*, Ph.D. thesis, University of Passau, 1990.
5. G. Huet, *Confluent Reductions: Abstract Properties and Applications to Term Rewriting Systems*, Journal of the ACM **27**(4), pp. 797-821, 1980.
6. J.W. Klop, *Term Rewriting Systems*, to appear in: Handbook of Logic in Computer

Science, Vol. I (eds. S. Abramsky, D. Gabbay and T. Maibaum), Oxford University Press, 1991.

7. M. Kurihara and I. Kaji, *Modular Term Rewriting Systems: Termination, Confluence and Strategies*, Report, Hokkaido University, Sapporo, 1988. (Abridged version: *Modular Term Rewriting Systems and the Termination*, Information Processing Letters **34**, pp. 1-4, 1990.)

8. M. Kurihara and A. Ohuchi, *Modularity of Simple Termination of Term Rewriting Systems*, Journal of IPS Japan **31**(5), pp. 633-642, 1990.

9. M. Kurihara and A. Ohuchi, *Modularity of Simple Termination of Term Rewriting Systems with Shared Constructors*, Report SF-36, Hokkaido University, Sapporo, 1990.

10. A. Middeldorp, *Modular Aspects of Properties of Term Rewriting Systems Related to Normal Forms*, Proceedings of the 3rd International Conference on Rewriting Techniques and Applications, Chapel Hill, Lecture Notes in Computer Science **355**, pp. 263-277, 1989. (Full version: Report IR-164, Vrije Universiteit, Amsterdam, 1988.)

11. A. Middeldorp, *A Sufficient Condition for the Termination of the Direct Sum of Term Rewriting Systems*, Proceedings of the 4th IEEE Symposium on Logic in Computer Science, Pacific Grove, pp. 396-401, 1989.

12. A. Middeldorp, *Confluence of the Disjoint Union of Conditional Term Rewriting Systems*, Report CS-R8944, Centre for Mathematics and Computer Science, Amsterdam, 1989. (To appear in: Proceedings of the 2nd International Workshop on Conditional and Typed Rewriting Systems, Montreal, 1990.)

13. A. Middeldorp, *Termination of Disjoint Unions of Conditional Term Rewriting Systems*, Report CS-R8959, Centre for Mathematics and Computer Science, Amsterdam, 1989.

14. A. Middeldorp, *Unique Normal Forms for Disjoint Unions of Conditional Term Rewriting Systems*, Report CS-R9003, Centre for Mathematics and Computer Science, Amsterdam, 1990.

15. A. Middeldorp, *Modular Properties of Term Rewriting Systems*, Ph.D. thesis, Vrije Universiteit, Amsterdam, 1990.

16. M.H.A. Newman, *On Theories with a Combinatorial Definition of Equivalence*, Annals of Mathematics **43**(2), pp. 223-243, 1942.

17. M.J. O'Donnell, *Equational Logic as a Programming Language*, The MIT Press, 1985.

18. M. Rusinowitch, *On Termination of the Direct Sum of Term Rewriting Systems*, Information Processing Letters **26**, pp. 65-70, 1987.

19. Y. Toyama, *On the Church-Rosser Property for the Direct Sum of Term Rewriting Systems*, Journal of the ACM **34**(1), pp. 128-143, 1987.

20. Y. Toyama, *Counterexamples to Termination for the Direct Sum of Term Rewriting Systems*, Information Processing Letters **25**, pp. 141-143, 1987.

21. Y. Toyama, *Commutativity of Term Rewriting Systems*, in: Programming of Future Generation Computer II (eds. K. Fuchi and L. Kott), North-Holland, pp. 393-407, 1988.

22. Y. Toyama, J.W. Klop and H.P. Barendregt, *Termination for the Direct Sum of Left-Linear Term Rewriting Systems*, Proceedings of the 3rd International Conference on Rewriting Techniques and Applications, Chapel Hill, Lecture Notes in Computer Science **355**, pp. 477-491, 1989.

Modular Higher-Order E-Unification

Tobias Nipkow*
University of Cambridge

Zhenyu Qian†
Universität Bremen

Abstract

The combination of higher-order and first-order unification algorithms is studied. We present algorithms to compute a complete set of unifiers of two simply typed λ-terms w.r.t. the union of α, β and η conversion and a first-order equational theory E. The algorithms are extensions of Huet's work and assume that a complete unification algorithm for E is given. Our completeness proofs require E to be at least regular.

1 Introduction

This paper is another contribution to the area of combination problems for equational theories. The overall aim of this line of research is *modularity*, the ability to construct the solutions to a problem from solutions to its subproblems. Early work on combining decision procedures was done by Nelson and Oppen [14]. A particularly active sub-area has been that of combining unification and matching algorithms for disjoint first-order equational theories [21,12,20,9,17,15,1]. More recently, the interaction of first-order and higher-order reasoning has been studied [2,3,18].

Apart from the inherent theoretical interest of this problem, there is considerable practical potential. First-order unification is used widely in high-level programming languages like OBJ [5] and in theorem provers like LP [7]. It is often crucial for the concise formulation of programs or proofs. Higher-order unification is used in extensions of logic programming like λProlog [13] and in generic theorem provers like Isabelle [16]. It is essential for an adequate representation and efficient manipulation of formal objects with bound variables, such as programs, formulae, and proofs. This work is a first step towards combining the advantages of both approaches.

Snyder [18] was the first to consider unification of two simply typed λ-terms whose constants are subject to a first-order equational theory, hereafter referred to as *higher-order E-unification*. For example, in the presence of the equation $+(x,y) = +(y,x)$, the unification problem

$$\lambda xy. + (x,y) \stackrel{?}{=} \lambda xy.F(y,x)$$

where F is a free variable, has the solution $\{F \mapsto +\}$. Snyder presents a set of inference rules that yield a highly nondeterministic algorithm enumerating a complete set of unifiers. This is in the spirit of "universal unification", extending previous inference systems for computing complete sets of unifiers for arbitrary equational theories [6], where the presentation of an equational theory itself is used to perform unification. The problem with these general algorithms is their nontermination and high nondeterminism. In contrast, we are aiming for a modular approach

*Author's address: University of Cambridge, Computer Laboratory, Pembroke Street, Cambridge CB2 3QG, England. E-mail: `Tobias.Nipkow@cl.cam.ac.uk`. Research supported by ESPRIT BRA 3245, *Logical Frameworks*.

†Author's address: Universität Bremen, FB Informatik, Postfach 330440, D-2800 Bremen 33, Germany. E-mail: `qian@informatik.uni-bremen.de`. Research partially supported by ESPRIT project 390, *PROSPECTRA*, and by ESPRIT Basic Research WG *COMPASS* 3264.

where individual unification algorithms can be combined as black boxes. Since the higher-order aspect, namely the conversion rules of the simply typed λ-calculus, is fixed, we use an extension of Huet's algorithm for higher-order unification [10,19]. This extension is parameterized by the E-unification algorithm: different first-order equational theories can be used by supplying different E-unification algorithms. Note that we confine our attention to a single E because the problem of combining different first-order equational theories is an orthogonal one and has been solved elsewhere [17,1].

Because higher-order unification alone is undecidable [8], we cannot obtain a terminating combination, even if the individual E-unification algorithms terminate. But because we can take advantage of efficient E-unification algorithms, we obtain a combination which is efficient and terminating enough for practical applications like theorem proving [16] and higher-order logic programming [13].

A review of basic definitions and notations in Section 2 is followed by two sections describing two different algorithms for higher-order equational unification. Section 3 deals with *full* higher-order E-unification, that is, the problem of enumerating a complete set of unifiers. This leads to a highly nondeterministic algorithm which is complete for all regular E. Section 4 specializes this approach to *pre*-unification: the algorithm stops as soon as a certain solved form is reached which guarantees the existence of a unifier. This idea goes back to Huet [10] and is crucial in making any kind of higher-order unification practical. Unfortunately, the completeness proof of our pre-unification algorithm requires that all E-equivalence classes are finite. Fortunately, the two most popular theories of associativity and commutativity (and their combination) meet this requirement. For lack of space, proofs are omitted or only sketched.

2 Preliminaries

We now present the basic notions we need in this paper. For further details the reader is referred to [19], [2] and [18].

2.1 Simply typed λ-calculus

Given a set $\mathcal{T}_0$ of *base types* (e.g., *int* and *nat*), the set of *(simple) types* is a set $\mathcal{T}$ inductively defined as the smallest set containing $\mathcal{T}_0$ such that $\alpha, \beta \in \mathcal{T}$ implies $\alpha \to \beta \in \mathcal{T}$. The type constructor $\to$ is assumed to associate to the right: $(\alpha_1 \to \cdots(\alpha_n \to \beta))$ is equivalent to $\alpha_1 \to \cdots \to \alpha_n \to \beta$ and, in a vector notation, to $\overline{\alpha_n} \to \beta$. In the latter case, we assume $\beta \in \mathcal{T}_0$.

For every type $\alpha \in \mathcal{T}$, we have a denumerable set $\mathcal{C}_\alpha$ of *function constants* and a denumerable set $\mathcal{V}_\alpha$ of *variables*, where all sets $\mathcal{C}_\alpha$ and $\mathcal{V}_\beta$ are pairwise disjoint. Let $\mathcal{C} = \bigcup_{\alpha\in\mathcal{T}} \mathcal{C}_\alpha$ and $\mathcal{V} = \bigcup_{\alpha\in\mathcal{T}} \mathcal{V}_\alpha$. The set $\mathcal{A}$ of *atoms* is defined as $\mathcal{A} = \mathcal{C} \cup \mathcal{V}$. Let $\mathcal{A}_\alpha = \mathcal{C}_\alpha \cup \mathcal{V}_\alpha$. The set of function constants occurring in a syntactic object O is denoted by $\mathcal{C}(O)$, the set of atoms by $\mathcal{A}(O)$.

For any type $\alpha \in \mathcal{T}$, the set $\mathcal{L}_\alpha$ of *terms* is inductively defined as the smallest set containing $\mathcal{A}_\alpha$ which is closed under the following rules of *application* and *abstraction*:

- (*Application*) If $u \in \mathcal{L}_{\alpha\to\beta}$ and $t \in \mathcal{L}_\alpha$, then $(u\ t) \in \mathcal{L}_\beta$;
- (*Abstraction*) If $t \in \mathcal{L}_\beta$ and $x \in \mathcal{V}_\alpha$, then $\lambda x.t \in \mathcal{L}_{\alpha\to\beta}$.

By analogy with first order notation, we write $(\ldots(a\ t_1)\ldots\ t_n)$ as $a(t_1,\ldots,t_n)$ or as $a(\overline{t_n})$. Similarly we abbreviate $\lambda x_1.\cdots\lambda x_k.t$ as $\lambda\overline{x_k}.t$. In the latter case, we assume t is not an abstraction.

Let $\mathcal{L} = \bigcup_{\alpha\in\mathcal{T}} \mathcal{L}_\alpha$. We denote the type of a term $t \in \mathcal{L}_\alpha$ by $\tau(t) = \alpha$. Instead of $t \in \mathcal{L}_\alpha$, we also write $t : \alpha$.

We assume the usual definitions of α, β, and η-conversion. Let $=_\alpha$ denote equivalence up to α-conversion. We only compare terms modulo $=_\alpha$. Let $\longrightarrow_\beta$ and $\longrightarrow_\eta$ denote single step β and η-reduction and define $\longrightarrow_{\beta\eta} = \longrightarrow_\beta \cup \longrightarrow_\eta$. For $\mathcal{X} \in \{\beta, \eta, \beta\eta\}$ let $=_\mathcal{X}$ denote the reflexive, symmetric and transitive closures of $\longrightarrow_\mathcal{X}$ modulo $=_\alpha$.

A term is said in *β-normal form* if no β-reduction is applicable. Each term t may be β-reduced to a β-normal form, denoted by $t{\downarrow_\beta}$. Any β-normal form is of the form $\lambda\overline{x_k}.a(\overline{t_n})$, where a is an atom, called the *head* of this term, and each t_i is in β-normal from. Define $\mathcal{H}ead(\lambda\overline{x_k}.a(\overline{t_n})) = a$. A term is called *flexible* if its head is a free variable, *rigid* if not.

A *long $\beta\eta$-normal form* is a β-normal form $\lambda\overline{x_k}.a(\overline{t_n})$, where $a(\overline{t_n})$ is of base type and each t_i is a long $\beta\eta$-normal form. For a term t, we use $t{\downarrow_{l\beta\eta}}$ to denote its unique long $\beta\eta$-normal form (modulo $=_\alpha$) $\beta\eta$-equivalent to t, which can be effectively obtained by performing η-expansions on $t{\downarrow_\beta}$. Note that if t is of the type $\overline{\alpha_k} \rightarrow \beta$ with $\beta \in \mathcal{T}_0$, then $t{\downarrow_{l\beta\eta}}$ is of the form $\lambda\overline{x_k}.s$ with $\tau(x_i) = \alpha_i, i = 1, \ldots, k$, and $\tau(s) = \beta$. By α-conversion we can always assume that two long $\beta\eta$-normal forms u and v of the same type have the same outermost λ-binders, i.e. $u = \lambda\overline{x_k}.s$ and $v = \lambda\overline{x_k}.t$. For simplicity, we do not always η-expand single variables, i.e. $F{\downarrow_{l\beta\eta}}$ may still be written as F.

The relation *subterm of* is the smallest reflexive and transitive relation such that u and t are subterms of $(u\ t)$, and t is a subterm of $\lambda x.t$.

Positions are strings of digits 1 and 2, denoted by p and q. Position p is said to be *above* q, denoted by $p \leq q$, if p is a prefix of q. Positions p and q are said to be *independent*, denoted by $p \parallel q$, if $p \not\leq q$ and $q \not\leq p$. The root position is denoted by Λ, the empty string. The subterm of a term t selected by position p is defined as t/p:

$$\begin{aligned} t/\Lambda &= t \\ (\lambda x.t)/1{\cdot}p &= t/p \\ (u\ t)/1{\cdot}p &= u/p \\ (u\ t)/2{\cdot}p &= t/p \end{aligned}$$

Note that / is a partial function. The set of all p such that t/p is defined is denoted by $\mathcal{P}os(t)$.

Let $t, s_1, \ldots, s_n$ be terms and $p_1, \ldots, p_n$ pairwise independent positions with $\tau(t/p_i) = \tau(s_i), 1 \leq i \leq n$. We use $t[s_1, \ldots, s_n]_{p_1,\ldots,p_n}$, (short: $t[s_1, \ldots, s_n]$) to denote the term obtained from t by replacing t/p_i by s_i, $i = 1, \ldots, n$. As an abbreviation, we may again use the vector notation, so that $t[s_1, \ldots, s_n]_{p_1,\ldots,p_n}$ will be represented as $t[\overline{s_n}]_{\overline{p_n}}$, and more generally, $t[s_m, \ldots, s_n]_{p_m,\ldots,p_n}$ as $t[\overline{s_{m..n}}]_{\overline{p_{m..n}}}$ for $1 \leq m \leq n$.

A variable x in a term is said to *occur bound* if it occurs in some subterm of the form $\lambda x.t$; otherwise it *occurs free*. We assume that no variable is bound more than once in a term and that no variable occurs both bound and free in a term. This can always be achieved by α-conversion. The set of all bound variables in a syntactic object O is denoted by $\mathcal{BV}(O)$, that of free variables by $\mathcal{FV}(O)$.

A *substitution* is a function $\sigma : \mathcal{V} \rightarrow \mathcal{L}$ such that $\sigma(x) \neq x$ for only finitely many variables $x \in \mathcal{V}$ and $\tau(x) = \tau(\sigma(x))$ for every $x \in \mathcal{V}$. The *domain* of a substitution σ is defined as $\mathcal{D}om(\sigma) := \{x \in \mathcal{V} | \sigma(x) \neq x\}$, and the *range* as $\mathcal{R}an(\sigma) = \bigcup_{x \in \mathcal{D}om(\sigma)} \mathcal{FV}(\sigma(x))$. If $\mathcal{D}om(\sigma) = \{x_1, \cdots, x_n\}$, we may write σ as

$$\{x_1 \mapsto \sigma(x_1), \cdots, x_n \mapsto \sigma(x_n)\},$$

or as $\{\overline{x_n \mapsto \sigma(x_n)}\}$ in the vector notation.

The application of a substitution σ to a term u is the term $\sigma(u)$ obtained by simultaneously replacing all free occurrences of x in u by $\sigma(x)$. To avoid capture of free variables we assume that all bound variables in u have been α-converted such that $\mathcal{BV}(u) \cap \mathcal{R}an(\sigma) = \emptyset$. For simplicity, we also assume that $\mathcal{BV}(u) \cap \mathcal{D}om(\sigma) = \emptyset$. Hence, $\sigma(\lambda x.t) = \lambda x.\sigma(t)$ holds automatically for any substitution σ.

The *composition* $\sigma \circ \theta$ of two substitutions is defined by $(\sigma \circ \theta)(x) = \sigma(\theta(x))$ for every $x \in \mathcal{V}$. The *union* $\sigma \cup \theta$, which is only defined if $\mathcal{D}om(\sigma) \cup \mathcal{D}om(\theta) = \emptyset$, is defined by

$$(\sigma \cup \theta)(x) = \begin{cases} \sigma(x) & \text{if } x \in \mathcal{D}om(\sigma) \\ \theta(x) & \text{if } x \in \mathcal{D}om(\theta) \\ x & \text{otherwise} \end{cases}$$

Let $\mathcal{W} \subseteq \mathcal{V}$. The restriction of a substitution σ to $\mathcal{W}$, denoted by $\sigma_{|\mathcal{W}}$, is the substitution such that $\sigma_{|\mathcal{W}}(x) = \sigma(x)$ for $x \in \mathcal{W}$, $\sigma_{|\mathcal{W}}(x) = x$ otherwise. A substitution σ is said to be *away from* $\mathcal{W}$ if $\mathcal{R}an(\sigma) \cap \mathcal{W} = \emptyset$.

A substitution σ is *normalized* if $\sigma(x)$ is in long $\beta\eta$-normal form for each $x \in \mathcal{D}om(\sigma)$. Normalized substitutions θ satisfy that $\theta(t)\!\downarrow_{l\beta\eta} = \theta(t)\!\downarrow_{\beta}$ for any long $\beta\eta$-normal form t.

In the rest of this paper, unless otherwise stated, α, β and γ will stand for arbitrary types, r, s, t, u and v for arbitrary terms, f, g, h for function constants, a and b for atoms, the lower-case letters x, y and z for bound variables, the capital letters X, Y, Z, F, G and H for free variables, and σ, θ, η and ρ for substitutions.

2.2 Algebraic terms, substitutions and equations

The set of *algebraic terms* $\alpha\mathcal{L}$ is the smallest set containing $\bigcup_{\alpha \in \mathcal{T}_0} \mathcal{A}_\alpha$ and satisfying that if $f \in \mathcal{C}_{\overline{\alpha_n} \to \beta}$ with $\{\alpha_1, \ldots, \alpha_n, \beta\} \subseteq \mathcal{T}_0$, and $s_i \in \alpha\mathcal{L} \cap \mathcal{L}_{\alpha_i}$ for $1 \leq i \leq n$, then $f(\overline{s_n}) \in \alpha\mathcal{L}$. The set $\bigcup_{\{\alpha_1, \ldots, \alpha_n, \beta\} \subseteq \mathcal{T}_0} \mathcal{C}_{\overline{\alpha_n} \to \beta}$ is called the set of *algebraic function constants* and denoted by $\alpha\mathcal{C}$.

An *algebraic equation* is an unordered pair $l \simeq r$ of terms in $\alpha\mathcal{L}$ such that $\tau(l) = \tau(r)$. An *algebraic theory* is a set E of algebraic equations. The *E-equivalence*, denoted by $=_E$, is the smallest equivalence relation on $\mathcal{L}$ satisfying $u[\sigma(l)] =_E u[\sigma(r)]$ for all terms u, all substitutions σ and all algebraic equations $l \simeq r \in E$. For example, $\lambda x.f(x) =_E \lambda x.g(x)$ if $f(X) \simeq g(X) \in E$.

Let E be an algebraic theory. E is called *consistent* if $X =_E Y$ does not hold for distinct variables X and Y. E is called *regular* if each $l \simeq r \in E$ satisfies $\mathcal{FV}(l) = \mathcal{FV}(r)$. A regular theory has only consequences $u =_E v$ with $\mathcal{FV}(u) = \mathcal{FV}(v)$. A consistent theory E is called *collapse-free* if no equation $l \simeq r \in E$ satisfies $l \in \mathcal{V}$ and $r \notin \mathcal{V}$. A collapse-free theory never has a consequence $u =_E v$ with $\mathcal{H}ead(u) \notin \alpha\mathcal{C}$ and $\mathcal{H}ead(v) \in \alpha\mathcal{C}$. E is called *root-preserving* if each $l \simeq r \in E$ satisfies $\mathcal{H}ead(l) = \mathcal{H}ead(r)$. A root-preserving theory is always collapse-free and has only consequences $u =_E v$ with $\mathcal{H}ead(u) = \mathcal{H}ead(v)$.

We denote by $=_{\beta\eta E}$ the reflexive, symmetric and transitive closure of $=_E \cup \longrightarrow_{\beta\eta}$. Breazu-Tannen [2] has shown:

Lemma 2.1 *For any two terms u and v, $u =_{\beta\eta E} v$ if and only if $u\!\downarrow_{l\beta\eta} =_E v\!\downarrow_{l\beta\eta}$.*

Let $\mathcal{W} \subseteq \mathcal{V}$. We define $\sigma =_{\beta\eta E} \theta\ [\mathcal{W}]$ iff $\sigma(X) =_{\beta\eta E} \theta(X)$ for each $X \in \mathcal{W}$. Furthermore, substitutions may be ordered by $\leq_{\beta\eta E}$ such that $\sigma \leq_{\beta\eta E} \theta\ [\mathcal{W}]$ iff there is a substitution ρ such that $\rho \circ \sigma =_{\beta\eta E} \theta\ [\mathcal{W}]$. We write $\sigma \doteq_{\beta\eta E} \theta\ [\mathcal{W}]$ iff $\sigma \leq_{\beta\eta E} \theta\ [\mathcal{W}]$ and $\theta \leq_{\beta\eta E} \sigma\ [\mathcal{W}]$. The relations $=_{\beta\eta}, \leq_{\beta\eta}$ and $\doteq_{\beta\eta}$ w.r.t. $[\mathcal{W}]$ are similarly defined. $[\mathcal{W}]$ may be omitted if $\mathcal{W} = \mathcal{V}$. A substitution σ is *idempotent* if $\sigma \circ \sigma =_{\beta\eta E} \sigma$. A substitution σ is idempotent whenever $\mathcal{D}om(\sigma) \cap \mathcal{R}an(\sigma) = \emptyset$. For any substitution σ and any finite variable set $\mathcal{W} \supseteq \mathcal{D}om(\sigma)$, there exists an idempotent substitution σ' such that $\mathcal{D}om(\sigma) = \mathcal{D}om(\sigma')$ and $\sigma \doteq_{\beta\eta E} \sigma'\ [\mathcal{W}]$. So we may restrict attention only to idempotent substitutions without loss of generality.

3 Higher order E-unification

In this section, we study general higher order E-unification, i.e. higher order unification in the presence of an arbitrary algebraic theory E. A *unification pair*, denoted by $u =^? v$, is an unordered pair of long $\beta\eta$-normal forms u and v of the same type. $u =^? v$ is called *rigid-rigid* if both u and v are rigid, *flexible-rigid* if u is flexible and v rigid, *flexible-flexible* if both u and v are flexible. A *system*, often denoted by S, is a finite multiset of unification pairs. A unification pair $F =^? v \in S$ is said to be *solved in* S if F is a free variable with $F \notin \mathcal{FV}(v) \cup \mathcal{FV}(S - \{F =^? v\})$; F is called a *solved variable in* S in this case. A system is said to be *solved* if all its unification pairs are solved.

We fix a system S and an algebraic theory E.

A substitution θ is called a λE*-unifier* (or simply *unifier*) *of* S if $\theta(u) =_{\beta\eta E} \theta(v)$ for each $u =^? v \in S$. The set of all such unifiers is denoted by $\mathcal{U}_{\lambda E}(S)$. A *complete set of* λE*-unifiers of* S *away from* $\mathcal{W}$, $\mathcal{CSU}_{\lambda E}(S)[\mathcal{W}]$ for short, is a set U of substitutions such that

- (*soundness*) $U \subseteq \mathcal{U}_{\lambda E}(S)$,
- (*completeness*) $\forall \theta \in \mathcal{U}_{\lambda E}(S)\ \exists \sigma \in U.\ \sigma \leq_{\beta\eta E} \theta\ [\mathcal{FV}(S)]$, and
- (*protectiveness*) $\forall \sigma \in U.\ \mathcal{D}om(\sigma) \subseteq \mathcal{FV}(S), \mathcal{R}an(\sigma) \cap (\mathcal{D}om(\sigma) \cup \mathcal{W}) = \emptyset$, σ is normalized.

Protectiveness can be assumed without loss of generality since for any substitution θ and any finite variable set $\mathcal{W}$, there is a normalized substitution σ away from $\mathcal{D}om(\sigma) \cup \mathcal{W}$ such that $\theta \doteq_{\beta\eta E} \sigma\ [\mathcal{FV}(S)]$. In the rest of this paper, we assume that the unifiers we get from $\mathcal{U}_{\lambda E}(S)$ are away from all variables occurring in the context.

If $S = \{F_1 =^? v_1, \ldots, F_n =^? v_n\}$ is solved, we define $\vec{S} = \{F_1 \mapsto v_1, \ldots, F_n \mapsto v_n\}$. The singleton set $\{\vec{S}\}$ is a $\mathcal{CSU}_{\lambda E}(S)[\mathcal{W}]$ for any $\mathcal{W}$ such that $\mathcal{W} \cap \mathcal{FV}(S) = \emptyset$. By an appropriate renaming substitution, $\vec{S}$ can be renamed into a $\mathcal{CSU}_{\lambda E}(S)[\mathcal{W}]$ for any finite $\mathcal{W}$.

We shall give an algorithm for higher order E-unification by a set of transformation rules. Let $u =^? v$ be a unification pair with λE-unifier $\theta \in \mathcal{U}_{\lambda E}(u =^? v)$. By Lemma 2.1 we know there exist long $\beta\eta$-normal forms s and t such that $\theta(u) \longrightarrow_{\beta}^{*} s =_E t \ {}_{\beta}^{*}\!\!\longleftarrow \theta(v)$. The basic idea of our transformations is to analyze the β-reductions and E-equivalence derivations top-down, approximating θ successively by finding bindings for free variables so that the terms become E-equivalent. Outermost free variables are solved by so-called *partial bindings*, unification pairs with E-unifiable top layers are reduced to simpler unification pairs by E*-decomposition*, until a solved form is reached.

The following definition is due to Snyder and Gallier [19]. A *partial binding of type* $\overline{\alpha_n} \to \beta$ is a term of the form

$$\lambda \overline{y_n}.a(\lambda \overline{z_{p_1}^1}.H_1(\overline{y_n}, \overline{z_{p_1}^1}), \ldots, \lambda \overline{z_{p_m}^m}.H_m(\overline{y_n}, \overline{z_{p_m}^m}))$$

for some $a \in \mathcal{A}$, where

1. $\tau(y_i) = \alpha_i$ for $1 \leq i \leq n$,
2. $\tau(a) = \overline{\gamma_n} \to \beta$ with $\gamma_i = \overline{\varphi_{p_i}^i} \to \gamma_i'$, for $1 \leq i \leq m$,
3. $\tau(z_j^i) = \varphi_j^i$ for $1 \leq i \leq m$ and $1 \leq j \leq p_i$,
4. $\tau(H_i) = \alpha_1 \to \cdots \to \alpha_n \to \varphi_1^i \to \cdots \to \varphi_{p_i}^i \to \gamma_i'$ for $1 \leq i \leq m$.

A partial binding as above is called an *imitation binding* if a is a function constant or a free variable. It is called an ith *projection binding* if a is a bound variable y_i for some i, $1 \leq i \leq n$. A *variant* of such a partial binding is obtained by replacing $H_1, \ldots, H_m$ with distinct new variables. For $F \in \mathcal{V}$, a partial binding t is *appropriate to* F if $\tau(t) = \tau(F)$. For notational brevity we extend our vector style notation to represent the partial binding above as

$$\lambda\overline{y_n}.a(\overline{\lambda\overline{z^m_{p_m}}.H_m(\overline{y_n},\overline{z^m_{p_m}})}).$$

The E-decomposition process is used to unify the top layer of a unification pair w.r.t. the algebraic theory E. The top layer of a term is obtained by abstracting each maximal alien subterm by a new variable or a new function constant. Intuitively, alien subterms are the subterms that cannot be dealt with directly by E-unification. An alien subterm should be abstracted by a new variable if this alien subterm is expected to be further unified with a term whose head is an algebraic function constant. Otherwise, it should be abstracted by a new function constant.

In this section by alien subterm we mean *α-alien subterms*, which are terms whose heads are not in αC. Formally, a subterm s of $\lambda\overline{x_k}.t$ is called an *α-alien subterm of* $\lambda\overline{x_k}.t$ if $\mathcal{H}ead(s) \notin \alpha C$ where the $x_1,\ldots,x_k$ are regarded as function constants, i.e. $\mathcal{H}ead(\{\overline{x_k \mapsto c_k}\}s) \notin \alpha C$ with each c_i being an arbitrary function constant of the type of x_i. Note that a free variable is always an α-alien subterm by definition. The *positions of the maximal α-alien subterms of* $\lambda\overline{x_k}.t$, denoted by $\mathcal{PAAS}(\lambda\overline{x_k}.t)$, are the set $P \subseteq \mathcal{P}os(t)$ satisfying

- $\forall p \in P.\ t/p$ is an α-alien subterm of $\lambda\overline{x_k}.t$,
- $\forall p,q \in P.$ if $p \neq q$ then $p \parallel q$, and
- $\forall q \in \mathcal{P}os(\lambda\overline{x_k}.t).$ if t/q is an α-alien subterm, then $\exists p \in P.\ p \leq q$.

We use $\mathcal{AAS}(\lambda\overline{x_k}.t) = \{t/p \mid p \in \mathcal{PAAS}(\lambda\overline{x_k}.t)\}$ to denote the set of all maximal α-alien subterms of $\lambda\overline{x_k}.t$. Note that for notational convenience, we have defined positions in $\mathcal{PAAS}(\lambda\overline{x_k}.t)$ w.r.t. t, not w.r.t. $\lambda\overline{x_k}.t$.

Let $\lambda\overline{x_k}.u =^? \lambda\overline{x_k}.v$ be a unification pair. An *E-decomposition of* $\lambda\overline{x_k}.u =^? \lambda\overline{x_k}.v$ is a set of unification pairs ED returned by the following process:

1. Let $ED := \{\lambda\overline{x_k}.\rho(Y_1) =^? \lambda\overline{x_k}.s_1,\ldots,\lambda\overline{x_k}.\rho(Y_n) =^? \lambda\overline{x_k}.s_n\}$, where
   - $\mathcal{PAAS}(\lambda\overline{x_k}.u) = \{p_1,\ldots,p_m\}$ and $\mathcal{PAAS}(\lambda\overline{x_k}.v) = \{p_{m+1},\ldots,p_n\}$;
   - $s_i = u/p_i$ and $s_j = v/p_j$ for $i = 1,\ldots,m$, $j = m+1,\ldots,n$;
   - $Y_1,\ldots,Y_n$ are new, not necessarily distinct variables;
   - $I \subseteq \{1,\ldots,n\}$;
   - $\rho = \{Z \mapsto Z'(\overline{x_k}) \mid Z \in \mathcal{R}an(\sigma)\} \circ \sigma$ for $\sigma \in \mathcal{CSU}_E(u[\overline{Y_m}]_{\overline{p_m}}, v[\overline{Y_{m+1..n}}]_{\overline{p_{m+1..n}}})$, where $x_1,\ldots,x_k$, and the Y_i, $i \in I$, are regarded as new free function constants, and the Z' are distinct new variables;
2. Repeat doing $ED := \{Y_i \mapsto s_i\}ED$ for all $i \in I$.

Let $P = \{p_i \mid i \in I \cap \{1,\ldots,m\}\}$ and $Q = \{p_i \mid i \in I \cap \{m+1,\ldots,n\}\}$. E-decomposition is called *optimized* if

(A) $\{p \in \mathcal{PAAS}(\lambda\overline{x_k}.u) \mid u/p \text{ rigid}\} \subseteq P$ and $\{p \in \mathcal{PAAS}(\lambda\overline{x_k}.v) \mid v/p \text{ rigid}\} \subseteq Q$,

(B) $s_i = s_j$ implies $Y_i = Y_j$ for $i,j \in I$, and

(C) $s_i = s_j$ iff $Y_i = Y_j$ for $i,j \notin I$.

The sets P and Q indicate the positions where maximal α-alien subterms should be abstracted by new function constants. P and Q are not only thought to contain the positions of all rigid maximal α-alien subterms, but also the positions of some flexible maximal α-alien subterms whose heads are not instantiated in the unification.

A unification pair may have more than one E decomposition since there may be multiple E-unifiers and multiple ways of identifying different maximal α-alien subterms. Roughly speaking, a substitution is a λE-unifier of a unification pair if and only if it is a λE-unifier of some E-decomposition of this unification pair.

Since alien subterms may be replaced by new function constants, the E-unification algorithm should accept terms with arbitrary free function constants. Most interesting E-unification algorithms known so far satisfy this condition, although Bürkert [4] presented an example where unification becomes undecidable after adding free function constants to an algebraic theory with decidable unification.

The following set of transformation rules for higher-order E-unification is called $\mathcal{HEU}$.

Rule (1) removes trivial pairs.

$$\{u \stackrel{?}{=} u\} \cup S \implies S \tag{1}$$

Rules (2a) and (2b) are decomposition rules that break a unification pair into simpler ones. In (2a), the heads of both sides of the unification pair are identical. Thus the decomposition is straightforward.

$$\{\lambda\overline{x_k}.a(\overline{u_n})) \stackrel{?}{=} \lambda\overline{x_k}.a(\overline{v_n}))\} \cup S \implies \bigcup_{1\le i\le n} \{\lambda\overline{x_k}.u_i \stackrel{?}{=} \lambda\overline{x_k}.v_i\} \cup S \tag{2a}$$

where a is an arbitrary atom, i.e. a function constant, a bound variable or a free variable.

In (2b), the top layers may be influenced by algebraic equations. Therefore, E-decomposition should be used.

$$\{\lambda\overline{x_k}.u \stackrel{?}{=} \lambda\overline{x_k}.v\} \cup S \implies ED \cup S \tag{2b}$$

where $\mathcal{H}ead(\lambda\overline{x_k}.u)$ or $\mathcal{H}ead(\lambda\overline{x_k}.v) \in \alpha C$, ED is an E-decomposition of $\lambda\overline{x_k}.u =^? \lambda\overline{x_k}.v$.

Rule (3) propogates sulotions.

$$\{\lambda\overline{x_k}.F(\overline{x_k}) \stackrel{?}{=} \lambda\overline{x_k}.v\} \cup S \implies \{F \stackrel{?}{=} \lambda\overline{x_k}.v\} \cup (\{F \mapsto \lambda\overline{x_k}.v\}S)\downarrow_{l\beta\eta} \tag{3}$$

where $F \notin \mathcal{FV}(\lambda\overline{x_k}.v)$ and $F \in \mathcal{FV}(S)$

Rule (4) attempts to approximate the unifiers by finding a partial binding for a head variable.

$$\{\lambda\overline{x_k}.F(\overline{u_n}) \stackrel{?}{=} \lambda\overline{x_k}.a(\overline{v_m})\} \cup S \implies \{F \stackrel{?}{=} t, \lambda\overline{x_k}.F(\overline{u_n}) \stackrel{?}{=} \lambda\overline{x_k}.a(\overline{v_m})\} \cup S \tag{4}$$

where a is an arbitrary atom, i.e. a function constant, a bound variable or a free variable, t is a variant of an arbitrary partial (imitation or projection) binding $\lambda\overline{y_n}.b(\overline{\lambda\overline{z_{p_m}}.H_m(\overline{y_n},\overline{z_{p_m}})})$ appropriate to F such that

- if $a, b \in C - C(E)$ then $a = b$,
- if $b = y_i$ for some $1 \le i \le n$, and $\mathcal{H}ead(u_i), a \in C - C(E)$ then $\mathcal{H}ead(u_i) = a$.

In contrast to Snyder and Gallier [19], who distinguish three subcases of rule (4), we have only one rule. The reason for us to give this kind of abstract form of transformation rule is that the choice of partial bindings in the presence of equations may be in general much more arbitrary than in the case without equations. For example, if the equational theory is not collapse-free, then b may be an arbitrary function constant in $C(E)$, no matter what a is, since b may disappear by applying collapsing equations of E. For the same reason, if a is a function constant in $C(E)$, then b may be an arbitrary atom.

It is possible to give a more concrete description of rule (4) where a number of subcases are distinguished [19,18]. However, this concrete description would still be too complex for a practical implementation, while obscuring the presentation. There is no need to do this, since $\mathcal{HEU}$ should only serve as a framework for further refinements and optimizations in special cases.

3.1 Soundness

Let S and S' be systems. We use $S \underset{\mathcal{HEU}}{\Longrightarrow_i} S'$ to denote $S \Longrightarrow S'$ by rule (i) of $\mathcal{HEU}$. The soundness of rules (1), (2a), (3) and (4) are stated in the following three lemmata, which can be shown in a similar way as in [19].

Lemma 3.1 *If* $S \underset{\mathcal{HEU}}{\Longrightarrow_i} S'$, $i = 1,3$, *then* $\mathcal{U}_{\lambda E}(S) = \mathcal{U}_{\lambda E}(S')$.

Lemma 3.2 *If* $S \underset{\mathcal{HEU}}{\Longrightarrow_{2a}} S'$, *then* $\mathcal{U}_{\lambda E}(S') \subseteq \mathcal{U}_{\lambda E}(S)$. *Especially, for any substitution* θ, *if (2a) applies to* $\lambda\overline{x_k}.a(\overline{u_n}) \stackrel{?}{=} \lambda\overline{x_k}.a(\overline{v_n}) \in S$ *with* $a \notin \mathcal{D}om(\theta)$, *then* $\theta \in \mathcal{U}_{\lambda E}(S)$ *iff* $\theta \in \mathcal{U}_{\lambda E}(S')$.

Lemma 3.3 *If* $S \underset{\mathcal{HEU}}{\Longrightarrow_4} S'$, *then* $\mathcal{U}_{\lambda E}(S') \subseteq \mathcal{U}_{\lambda E}(S)$.

Now we show soundness and completeness of rule (2b), which can be proved by an exact analysis of E-decomposition process.

An E-decomposition is called *optimized w.r.t.* θ if

(A) $i \in I$ iff $\theta(s_i) \in \mathcal{AAS}(\theta(\lambda\overline{x_k}.u)) \cup \mathcal{AAS}(\theta(\lambda\overline{x_k}.v))$

(B) $Y_i = Y_j$ iff $\theta(s_i) =_{\beta\eta E} \theta(s_j)$ for $i,j \in I$,

(C) $Y_i = Y_j$ iff $s_i = s_j$ for $i,j \notin I$.

It is easy to check that an E-decomposition is optimized if it is optimized w.r.t. θ. Therefore, it suffices only to consider the E-decompositions that are optimized w.r.t. θ in the completeness proof.

Lemma 3.4 *Let* $\mathcal{W} \subset \mathcal{V}$ *be a set of variables. Let* $\lambda\overline{x_k}.u \stackrel{?}{=} \lambda\overline{x_k}.v$ *be a unification pair with* $\mathcal{H}ead(u)$ *or* $\mathcal{H}ead(v) \in \alpha\mathcal{C}$.

(i) *Then there is a* $\theta \in \mathcal{U}_{\lambda E}(\lambda\overline{x_k}.u \stackrel{?}{=} \lambda\overline{x_k}.v)$ *iff there is a* $\theta' \in \mathcal{U}_{\lambda E}(ED)$ *for some* E-*decomposition* ED *of* $\lambda\overline{x_k}.u \stackrel{?}{=} \lambda\overline{x_k}.v$ *such that* $\theta =_{\beta\eta E} \theta'$ $[\mathcal{W}]$.

(ii) *The above also holds if* E-*decomposition is optimized w.r.t.* θ.

Finally, we can prove the following theorem using these lemmata.

Theorem 3.1 *If* $S \underset{\mathcal{HEU}}{\Longrightarrow} S'$ *then* $\mathcal{U}_{\lambda E}(S') \subseteq \mathcal{U}_{\lambda E}(S)$.

As a trivial corollary we obtain the required soundness result.

Corollary 3.1 *If* $S \underset{\mathcal{HEU}}{\overset{*}{\Longrightarrow}} S'$ *and* S' *is in solved form, then* $\vec{S'}$ *is a unifier of* S.

3.2 Completeness

In this section, we consider the completeness of $\mathcal{HEU}$. At the moment, we can only prove the completeness for regular theories, although we suspect the completeness of $\mathcal{HEU}$ for arbitrary theories.

The *(non-deterministic) completeness* of the transformations $\mathcal{HEU}$ means that for any λE-unifier θ of any system S, there is some sequence of transformations from $\mathcal{HEU}$ which finds a λE-unifier σ such that $\sigma \leq_{\beta\eta E} \theta\ [\mathcal{FV}(S)]$. Similar to [19] we use a transformation relation $\mathcal{CHEU}$ on pairs $\langle\theta, S\rangle$ with $\theta \in \mathcal{U}_{\lambda E}(S)$. Starting with any system S and any of its λE-unifiers θ, there is a sequence of $\mathcal{CHEU}$-transformations leading to a solved system S' such that $\vec{S'} \leq_{\beta\eta E} \theta\ [\mathcal{FV}(S)]$.

The transformations of $\mathcal{CHEU}$ are defined as follows:

$$\langle\theta, S\rangle \Longrightarrow_1 \langle\theta, S'\rangle$$

if $S \underset{\mathcal{HEU}}{\Longrightarrow_1} S'$;

$$\langle\theta, \{\lambda\overline{x_k}.a(\overline{u_n}) \stackrel{?}{=} \lambda\overline{x_k}.a(\overline{v_n})\} \cup S\rangle \Longrightarrow_{2a} \langle\theta, S'\rangle$$

if $\{\lambda\overline{x_k}.a(\overline{u_n}) \stackrel{?}{=} \lambda\overline{x_k}.a(\overline{v_n})\} \cup S \underset{\mathcal{HEU}}{\Longrightarrow_{2a}} S'$ and $a \notin \mathcal{D}om(\theta) \cup \alpha C$;

$$\langle\theta, S\rangle \Longrightarrow_{2b} \langle\theta \cup \eta, S'\rangle$$

if $S \underset{\mathcal{HEU}}{\Longrightarrow_{2b}} S'$, if $\mathcal{D}om(\eta) \cap \mathcal{D}om(\theta) = \{\}$ and $\theta \cup \eta \in \mathcal{U}_{\lambda E}(S')$;

$$\langle\theta, S\rangle \Longrightarrow_3 \langle\theta, S'\rangle$$

if $S \underset{\mathcal{HEU}}{\Longrightarrow_3} S'$;

$$\langle\{F \mapsto s\} \cup \theta, \{\lambda\overline{x_k}.F(\overline{u_n}) \stackrel{?}{=} v\} \cup S\rangle \Longrightarrow_4 \langle\{F \mapsto s\} \cup \theta \cup \eta, \{F \stackrel{?}{=} t\} \cup S'\rangle$$

where $s = \lambda\overline{y_n}.a(\overline{s_m})$, $\{\lambda\overline{x_k}.F(\overline{u_n}) =^? v\} \cup S \underset{\mathcal{HEU}}{\Longrightarrow_4} \{F =^? t\} \cup S'$, $t = \lambda\overline{y_n}.a(\overline{\lambda\overline{z_{pm}}.H_m(\overline{y_n}, \overline{z_{pm}})})$, and $\eta = \{H_1 \mapsto \lambda\overline{y_n}.s_1, \cdots, H_m \mapsto \lambda\overline{y_n}.s_m\}$.

Lemma 3.5 *Let $\mathcal{W} \subset \mathcal{V}$ be a set of variables. If θ is a λE-unifier of S and S is not in solved form, then there exists some $\mathcal{CHEU}$-transformation*

$$\langle\theta, S\rangle \Longrightarrow_i \langle\theta', S'\rangle \qquad i \in \{1, 2a, 2b, 3, 4\}$$

such that

(i) $\theta =_{\beta\eta E} \theta'\ [\mathcal{W}]$,

(ii) $S \underset{\mathcal{HEU}}{\Longrightarrow_i} S'$, and

(iii) if $\mathcal{D}om(\theta) \cap \mathcal{R}an(\theta) = \emptyset$ then θ' is a λE-unifier of S' with $\mathcal{D}om(\theta') \cap \mathcal{R}an(\theta') = \emptyset$

The above also holds in the case $i = 2b$ if E-decomposition is optimized w.r.t. θ.

For the further discussions, we introduce some more notions. Let t be a long $\beta\eta$-normal form and $p \in \mathcal{P}os(t)$. The *list of bound variables visible at* p, denoted as $BV_p(t)$, is the list of bound variables occurring from the top of t to the position p. All bound variables in $BV_p(t)$ are distinct since we assume that no variable is bound twice in one term. Let S be a system containing t, let θ be a λE-unifier of S, and let $F \in \mathcal{D}om(\theta)$. F is said to *occur trivial at* p *in* t if

- $t/p = F(\overline{x_n})$,
- $BV_p(t) = x_1, \ldots, x_n$ and
- there are no positions $q \in \mathcal{P}os(t)$ such that $q < p$ and $\mathcal{H}ead(t/q) \in \mathcal{D}om(\theta)$.

F is said to *occur proper at p in t* if it does not occur trivial at p of t. F is called *trivial (in S)* if F occurs trivial at each position q in t with $t/q = F(\overline{u_n})$ for all terms t of S. F is called *proper (in S)* if it occurs proper somewhere in S. Note that every variable in $\mathcal{D}om(\theta)$ is either trivial or proper. We use $TriV(S)$ (and $ProV(S)$) to denote the set of all trivial (and proper) variables in S.

The distinction between proper and trivial variables is useful in proving the reduction of complexity measure for transformation (2b). The technique in [19] does not work here since E-decomposition may introduce new variables. By construction the newly introduced variables are just trivial variables.

Let the *size of a term t*, denoted by $|t|$, be defined as the number of atomic subterms in t.

We use $UnsolP(S)$ to denote the set of all unsolved unification pairs in S. Furthermore, we denote $UnsolV(S,\theta) = \mathcal{D}om(\theta) \cap \mathcal{FV}(UnsolP(S))$, which is the set of all variables still to be solved in S. Note that for a solved unification pair $F =^? v \in S$, v may contain unsolved variables.

Let u be a long $\beta\eta$-normal form. The *syntactic height* of u, denoted by $H_{syn}(u)$, is defined by

$$H_{syn}(u) = \begin{cases} max\{H_{syn}(s) \mid s \in \mathcal{AAS}(u)\} & \text{if } \mathcal{H}ead(u) \in \alpha C; \\ max\{H_{syn}(s_i) \mid 1 \le i \le n\} + 1 & \text{if } u = \lambda \overline{x_k}.a(\overline{s_n}) \text{ and } a \notin \alpha C; \end{cases}$$

where $max\{\} = 0$.

If E is a regular theory then, by Lemma 2.1, we know that for any long $\beta\eta$-normal forms u and v, $u =_{\beta\eta E} v$ implies $H_{syn}(u) = H_{syn}(v)$.

Any ordering $>$ on a set B can be extended to an ordering $>_{mul}$ on multisets over B. For any multiset M over B and $a \in B$, we use $occ(a, M)$ to denote the number of occurrences of a in M. For any two multisets M and N, $M >_{mul} N$ if and only if (i) $M \neq N$ and (ii) whenever $occ(a,N) > occ(a,M)$ for some $a \in B$ then $occ(b,M) > occ(b,N)$ for some $b \in B$ such that $b > a$. If $>$ is well-founded on B, so is $>_{mul}$ on multisets over B.

The following is the main theorem of this subsection. The regularity condition of algebraic theories is used to prove the temination of $\mathcal{CHEU}$-transformations.

Theorem 3.2 *Let E be a regular theory. For any system S, if θ is a λE-unifier of S, then there is a terminating derivation $\langle\theta, S\rangle \overset{*}{\Longrightarrow} \langle\theta', S'\rangle$ of $\mathcal{CHEU}$-transformations such that S' is solved and $\vec{S'} \le_{\beta\eta E} \theta\ [\mathcal{FV}(S)]$.*

Proof Sketch

We first prove that there is a terminating sequence of $\mathcal{CHEU}$-transformations starting with $\langle\theta, S\rangle$. We define a comlexity measure as $\langle P, H, T\rangle$, where

- $P\langle\theta, S\rangle$ is the sum of the sizes of the bindings in θ for proper unsolved variables in S, i.e. $P\langle\theta, S\rangle = \sum\{|\theta(x)| \mid x \in UnsolV(S,\theta) \cap ProV(S)\}$,
- $H\langle\theta, S\rangle$ is the multiset containing heights of unsolved unification pairs in S w.r.t. θ, i.e. $H\langle\theta, S\rangle = \{H_{syn}(\theta(u)) \mid u =^? v \in UnsolP(S)\}$;
- $T\langle\theta, S\rangle$ is the sum of the sizes of the bindings in θ for trivial unsolved variables in S, i.e. $T\langle\theta, S\rangle = \sum\{|\theta(x)| \mid x \in UnsolV(S,\theta) \cap TriV(S)\}$

We can check that if S is not in solved form, then there are always some combinations of $\mathcal{CHEU}$-transformations which strictly reduce the complexity measure of $\langle\theta, S\rangle$. More precisely, transformations (1), (2a) and (4) always strictly reduce the complexity measure. Transformations (3) strictly reduces the complexity measure in a restricted case which is enough for the completeness. Transformation (2b) strictly reduces the complexity measure if it may be followed by some steps of transformation (2a), (3) or (4).

Then there must exist a sequence of transformations

$$\langle\theta, S\rangle = \langle\theta_0, S_0\rangle \Longrightarrow \langle\theta_1, S_1\rangle \Longrightarrow \cdots \Longrightarrow \langle\theta_m, S_m\rangle$$

such that no transformation applies to $\langle\theta_m, S_m\rangle$. By Lemma 3.5, we have $\theta =_{\beta\eta E} \theta_m\ [\mathcal{FV}(S)]$ and

$$S = S_0 \Longrightarrow S_1 \Longrightarrow \cdots \Longrightarrow S_m$$

by rules of $\mathcal{HEU}$. Again by Lemma 3.5, we know S_m is in solved form. Hence, $\vec{S_m} \leq_{\beta\eta E} \theta_m =_{\beta\eta E} \theta\ [\mathcal{FV}(S)]$ □

4 Higher-order equational pre-unification

Huet's insight [10], which turned higher-order unification from a mere curiosity into computational reality, was that flexible-flexible pairs could be ignored if one was merely interested in existence of unifiers. This section presents a similar specialization of the inference rules in the preceding section. Flexible-flexible pairs are not transformed further and the result of a successful computation is a *pre-unifier*, that is a substitution together with a set of constraints in the form of unsolved flexible-flexible pairs. Further instantiations in subsequent unification steps may turn the constraints into new rigid-flexible or rigid-rigid pairs, which can again be transformed.

The purpose of pre-unification is to obtain a computationally viable algorithm. Unfortunately, we had to restrict to term finite equational theories in order to prove completeness of the algorithm. E is called *term finite* if all E-equivalence classes are finite. Note that term finite theories are necessarily collapse-free and regular, but that the reverse implication does not hold: $E = \{f(f(x)) \simeq f(x)\}$ is collapse-free and regular but not term finite.. Note also that term finiteness is additive: if E_1 and E_2 are term finite, so is $E_1 \cup E_2$. We suspect that our pre-unification algorithm is complete for all collapse-free, regular E but have not been able to show this.

A system S is in *presolved* form if every unification pair $s =^? t$ in S is either solved or both s and t are flexible. If S is in presolved form, the substitution $\vec{S} = \{F \mapsto t \mid (\lambda\overline{x_k}.F(\overline{x_k}) =^? t) \in S\}$ is a *pre-unifier* of S in the sense of [19]. Pre-unifiers extend trivially to unifiers. With every base type σ we associate some arbitrary but fixed free variable $X_\sigma : \sigma$, and with every type $\overline{\sigma_n} \to \sigma$ a term $\hat{e}_{\overline{\sigma_n}\to\sigma} = \lambda\overline{x_n}.X_\sigma$ where the $x_i : \sigma_i$ are all distinct. Defining

$$\xi_S = \bigcup\{\{F \mapsto \hat{e}_{\tau(F)}, G \mapsto \hat{e}_{\tau(G)}\} \mid (\lambda\overline{x_k}.F(\overline{s_m}) \stackrel{?}{=} \lambda\overline{x_k}.G(\overline{t_n})) \in S\}$$

we get that $\vec{S} \cup \xi_S$ is a unifier of S.

Lemma 4.1 *If S is in presolved form and σ unifies S, then $\sigma \geq_{\beta\eta E} \vec{S}\ [\mathcal{FV}(S)]$.*

The subterm abstraction process is defined so as to push as much as possible into E-unification. Instead of the α-alien subterms of Section 3 we use *non-constant* subterms:

$$\begin{aligned}
\mathcal{PNCS}(\lambda x.s) &= \{\Lambda\}\\
\mathcal{PNCS}(F(\overline{s_n})) &= \{\Lambda\}\\
\mathcal{PNCS}(x(\overline{s_n})) &= \{i{\cdot}p \mid 1 \leq i \leq n \wedge p \in \mathcal{PNCS}(s_i)\}\\
\mathcal{PNCS}(c(\overline{s_n})) &= \{i{\cdot}p \mid 1 \leq i \leq n \wedge p \in \mathcal{PNCS}(s_i)\}
\end{aligned}$$

An *E-decomposition* of $\lambda\overline{x_k}.s =^? \lambda\overline{x_k}.t$ is a set

$$\left\{ \begin{array}{c} \lambda\overline{x_k}.(\theta' X_1)\downarrow_{l\beta\eta} =^? \lambda\overline{x_k}.s/p_1, \ldots, \lambda\overline{x_k}.(\theta' X_m)\downarrow_{l\beta\eta} =^? \lambda\overline{x_k}.s/p_m, \\ \lambda\overline{x_k}.(\theta' Y_1)\downarrow_{l\beta\eta} =^? \lambda\overline{x_k}.t/q_1, \ldots, \lambda\overline{x_k}.(\theta' Y_n)\downarrow_{l\beta\eta} =^? \lambda\overline{x_k}.t/p_n \end{array} \right\}$$

such that

- $\{p_1, \ldots, p_m\} = \mathcal{PNCS}(s)$ and $\{q_1, \ldots, q_n\} = \mathcal{PNCS}(t)$,
- $X_1, \ldots, X_m, Y_1, \ldots, Y_n$ are distinct new variables,
- $\theta \in \mathcal{CSU}_E(s[\overline{X_m}]_{\overline{p_m}}, t[\overline{Y_n}]_{\overline{q_n}})$ where the $x_1, \ldots, x_k$ are considered as constants, and
- $\theta' = \{Z \mapsto Z'(\overline{x_k}) \mid Z \in \mathcal{R}an(\theta)\} \circ \theta$ where the Z' are distinct new variables.

It is possible to optimize the replacement of alien subterms by new variables slightly: equivalent subterms can be replaced by the same variable.

Note that in this definition of E-decomposition the E-unification algorithm $\mathcal{CSU}_E$ is assumed to cope with terms containing both algebraic and non-algebraic constants. These *pseudo-algebraic* terms $\beta\mathcal{L}$ are the smallest subset of $\mathcal{L}$ such that $\mathcal{V} \subseteq \beta\mathcal{L}$ and $f(\overline{s_n}) \in \beta\mathcal{L}$ if $f : \overline{\alpha_n} \to \alpha$, $s_i : \alpha_i$ and $s_i \in \beta\mathcal{L}$, $i = 1 \ldots n$. With the exception of variables, all pseudo-algebraic terms are of base type.

We assume that $\mathcal{CSU}_E$ acts correctly on pseudo-algebraic terms in the sense that if $s, t \in \beta\mathcal{L}$, then $\sigma \in \mathcal{CSU}_E(s,t)$ is

- *type correct:* $X : \alpha$ implies $\sigma(X) : \alpha$, and
- pseudo-algebraic: $\sigma(X) \in \beta\mathcal{L}$.

In practice this extension is trivial by treating all non-algebraic constants as free, and all function types as new base types. This means that σ may not be in long $\beta\eta$-normal form because higher-order variables are not applied to any arguments. Hence the final application of $\downarrow_{l\beta\eta}$ in the definition of E-decomposition.

Finally, we define a set of rules for higher-order equational pre-unification. Rules (1) and (3) are the same as in Section 3.

If both s and t are rigid, the following generalization of Huet's SIMPL procedure [11] and Snyder and Gallier's [19] rule (2′) applies:

$$\{\lambda\overline{x_k}.s \stackrel{?}{=} \lambda\overline{x_k}.t\} \cup S \implies ED \cup S \tag{2}$$

where ED is an E-decomposition of $\lambda\overline{x_k}.s =^? \lambda\overline{x_k}.t$.

Flexible-rigid pairs can be transformed by imitation or projection. In its general form, imitation is very nondeterministic:

$$\{\lambda\overline{x_k}.F(\overline{r_m}) \stackrel{?}{=} \lambda\overline{x_k}.a(\overline{s_n})\} \cup S \implies \{F \stackrel{?}{=} t,\ \lambda\overline{x_k}.F(\overline{r_m}) \stackrel{?}{=} \lambda\overline{x_k}.a(\overline{s_n})\} \cup S \tag{4a}$$

where F is not solved, t is a variant of an imitation binding $\lambda\overline{y_m}.b(\overline{\lambda\overline{z_{p_q}}.H_q(\overline{y_m}, \overline{z_{p_q}})})$ appropriate to F, and b is any constant of the right type. Fortunately, its nondeterminism can be reduced considerably, especially if E is restricted further.

1. If a is a free constant, b must be identical to a. Otherwise b may be any constant.

2. If a is a constant in E, only those b need to be considered which are *E-compatible* with a, i.e. for which the equation $a(\overline{X_m}) =^? b(\overline{Y_n})$ has a solution. If E involves only finitely many constants and has a decidable unification problem, a compatibility table can be computed in advance. In practice this table should be supplied together with the E-unification algorithm.

3. If E is *root-preserving*, i.e. every equation in E is of the form $f(\ldots) \simeq f(\ldots)$, b must be identical to a. This is of course the case for $E = \{\}$, but also for associativity and commutativity.

Projection [11], rule ($4'b$) in [19], is the same as for pure higher-order unification:

$$\{\lambda\overline{x_k}.F(\overline{r_m}) \stackrel{?}{=} \lambda\overline{x_k}.a(\overline{s_n})\} \cup S \implies \{F \stackrel{?}{=} t,\ \lambda\overline{x_k}.F(\overline{r_m}) \stackrel{?}{=} \lambda\overline{x_k}.a(\overline{s_n})\} \cup S \tag{4b}$$

where F is not solved, a is a bound variable or constant, and t is a variant of the ith projection binding $\lambda\overline{y_m}.y_i(\overline{\lambda\overline{z_{p_q}}.H_q(\overline{y_m},\overline{z_{p_q}})})$ appropriate to F.

Rules (4a) and (4b) are implicitly followed by an application of rule (3) eliminating F.

Call this set of rules $\mathcal{HEP}$. The main difference to $\mathcal{HEU}$ is the restriction of rule (4) to flexible-rigid pairs, thus arriving at pre-unification. The use of $\mathcal{PNCS}$ instead of $\mathcal{PAAS}$ in (2) permits (2a) and (2b) of $\mathcal{HEU}$ to be merged. It should also decrease nondeterminism, as the following example shows.

Let $+$ be a commutative constant of type $\alpha \to \alpha \to \alpha$, and let c, d, f, and g be distinct constants of type $(\beta \to \gamma) \to \alpha$. The unification pair $c(r) + d(s) =^? f(u) + g(v)$, where r, s, u, and v are arbitrary terms of type $\beta \to \gamma$, does obviously not have a solution. Using $\mathcal{PAAS}$ one is left with the E-unification problem $X_1 + X_2 =^? X_3 + X_4$, which has two solutions, none of which lead anywhere. Using $\mathcal{PNCS}$ yields the E-unification problem $c(Y_1) + d(Y_2) =^? f(Y_3) + g(Y_4)$ which immediately fails.

The following soundness theorem is proved along the same lines as Corollary 3.1.

Theorem 4.1 *If $S \stackrel{*}{\underset{\mathcal{HEP}}{\Longrightarrow}} S'$ and S' is in presolved form, then $\vec{S'}$ is a pre-unifier of S.*

4.1 Completeness

Rather than go through the completeness proof in detail, we present its main ideas. Completeness is proved as in [19] using a rewrite relation $\mathcal{CHEP}$ between pairs $\langle\sigma, S\rangle$ of substitutions and equation systems. It turns out that for completeness, rules (1) and (3) can be ignored — they should be considered (sound) optimizations. Rule (2) induces

$$\langle\sigma, S\rangle \implies \langle\sigma \cup \delta', S'\rangle \tag{2}$$

iff $S \Longrightarrow S'$ via rule (2) of $\mathcal{HEP}$ and there is a substitution δ' such that $\delta'\theta' X_i =_{\beta\eta E} \sigma(s/p_i)$, $i = 1 \ldots m$, $\delta'\theta' Y_j =_{\beta\eta E} \sigma(t/q_j)$, $j = 1 \ldots n$, and $\mathcal{D}om(\delta') \subseteq \mathcal{R}an(\theta')$. The two rules (4a) and (4b) are packaged into one:

$$\langle\sigma, S\rangle \implies \langle\sigma \cup \delta, S'\rangle \tag{4}$$

iff $S \Longrightarrow S'$ via either (4a) or (4b) of $\mathcal{HEP}$, $\sigma(F)\downarrow_{l\beta\eta} = \lambda\overline{y_m}.a(\overline{u_q})$, and

$$\delta = \{H_j \mapsto \lambda\overline{y_m}.u_j \mid j = 1 \ldots q\}$$

The most difficult part of the completeness proof is to show termination of $\mathcal{CHEP}$, i.e., to prove that eventually a pre-solved system is reached. We have not been able to show termination, or at least normalization, under the weaker assumption of regularity, as in Section 3. Nevertheless we strongly suspect that collapse-freeness and regularity suffice.

Let $|t|$ denote the *size* of a term, i.e., the number of subterms and define

$$|t|_E = max\{|t'| \mid t =_E t'\}.$$

Note that $|.|_E$ is well-defined only if E is term finite. The crucial property of $|.|_E$ is its monotonicity w.r.t. the subterm ordering:

Lemma 4.2 *If E is term finite and s is a proper subterm of t, then $|s|_E < |t|_E$.*

Lemma 4.3 *If σ unifies S, then any $\mathcal{CHEP}$ reduction starting from $\langle\sigma, S\rangle$ terminates.*

Proof sketch. We give a complexity measure which decreases with every $\mathcal{CHEP}$ reduction.

The size of an equation $|s =^? t|_E$ is defined as $|s|_E$, which only makes sense if $s =_E t$. The size of a multiset of equations $|S|_E$ is the *multiset* $\{|e|_E \mid e \in S\}$. Let

$$\begin{aligned} H(\sigma,S) &= |\sigma(Uns(S))\downarrow_{l\beta\eta}|_E \\ M(\sigma,S) &= \{|(\sigma X)\downarrow_{l\beta\eta}| \mid X \in \mathcal{D}om(\sigma) - Sol(S)\} \end{aligned}$$

where $Sol(S)$ are the solved variables in S and $Uns(S)$ are the unsolved equations in S, i.e., equations not of the form $\lambda\overline{x_k}.F(\overline{x_k}) =^? t$ with $F \in Sol(S)$. The complexity of $\langle\sigma, S\rangle$ is the pair $\langle H(\sigma,S), M(\sigma,S)\rangle$. A careful analysis shows that $\langle\sigma,S\rangle \Longrightarrow_i \langle\sigma',S'\rangle$ implies $H(\sigma,S) > H(\sigma',S')$ if $i = 2$, and $H(\sigma,S) \geq H(\sigma',S')$ and $M(\sigma,S) > M(\sigma',S')$ if $i = 4$. □

Theorem 4.2 *Let E be term finite. If σ is a unifier of S, then there exists an S' in presolved form such that $S \overset{*}{\underset{\mathcal{HEP}}{\Longrightarrow}} S'$, $\sigma \geq_{\beta\eta E} \vec{S'}$ $[\mathcal{FV}(S)]$, and σ unifies all flexible-flexible pairs in S'.*

5 Conclusion

This paper is clearly only a first step towards a complete solution to the problem of integrating first-order and higher-order unification algorithms. There are some obvious directions for further research.

The strong requirements we have for E need to be relaxed. For full unification this looks a problem of finding the right proofs which do not need regularity of E. However, pre-unification, as described in Section 4, definitely requires at least collapse-freeness. What we need there is an extension of the algorithm which copes with collapsing E without sacrificing efficiency.

It is also planned to implement our pre-unification algorithm in Isabelle, extending its higher-order unification engine by commutativity and associativity-commutativity. This should give us some idea about the efficiency of our approach.

On a more theoretical level, one might also want to abstract from Huet's unification algorithm. This would result in a completely modular approach where both E-unification and unification of typed λ-terms come as black boxes.

References

[1] Boudet, A.: Unification in a combination of equational theories: an efficient algorithm. *Proc. 10th Int. Conf. Automated Deduction*, LNCS 449 (1990), 292–307.

[2] Breazu-Tannen, V.: Combing algebra and higher-order types. *Proc. 3rd IEEE Symp. Logic in Computer Science* (1988), 82–90.

[3] Breazu-Tannen, V. and Gallier, J.: Polymorphic rewriting conserves algebraic strong normalization and confluence. *Proc. ICALP*, LNCS 372 (1989), 137–150.

[4] Bürckert, H.-J.: Matching - a special case of unification? *J. Symbolic Computation* 8 (1989), 523–536.

[5] Futatsugi, K., Goguen, J.A., Jouannaud, J.-P., Meseguer, J.: Principles of OBJ2. *Proc. 12th ACM Symp. Principles of Programming Languages* (1985), 52–66.

[6] Gallier, J. and Snyder, W.: Complete sets of transformations for general E-unification. *Theoretical Computer Science 67* (1988), 203–260.

[7] Garland, S.J., Guttag, J.V.: An Overview of LP, The Larch Prover. *Proc. 3rd Int. Conf. Rewriting Techniques and Applications*, LNCS 355 (1989), 137–151.

[8] Goldfarb, W.: The undecidability of the second-order unification problem. *Theoretical Computer Science* 13 (1981), 225–230.

[9] Herold, A.: Combination of Unification Algorithms. *Proc. 8th Int. Conf. Automated Deduction*, LNCS 230 (1986), 450–469.

[10] Huet, G.: A Unification Algorithm for Typed λ-Calculus. *Theoretical Computer Science* 1 (1975), 27–57.

[11] Huet, G.: *Résolution d'equations dans les languages d'ordre 1,2,...,ω*. Thése d'Etat, Université de Paris VII (1976).

[12] Kirchner, C.: *Méthodes et outils de conception systématique d'algorithmes d'unification dans les théories équationnelles*, Thèse d'état de l'Université de Nancy I (1985).

[13] Nadathur, G. and Miller, D.: An overview of λProlog. *Proc. 5th Int. Conf. Logic Programming*, eds. R.A.Kowalski and K.A. Bowen, MIT Press (1988), 810–827.

[14] Nelson, G. and Oppen, D.: Simplification by cooperating decision procedures. *ACM TOPLAS* 1 (1979), 245–257.

[15] Nipkow, T.: Combining matching algorithms: the regular case. *Proc. 3rd Int. Conf. Rewriting Techniques and Applications*, LNCS 355 (1989), 343–358.

[16] Paulson, L.C.: Isabelle: The Next 700 Theorem Provers. In P. Odifreddi (editor), *Logic and Computer Science*, Academic Press (1990), 361–385.

[17] Schmidt-Schauß, M.: Unification in a combination of arbitrary disjoint equational theories. *J. Symbolic Computation* 8 (1989), 51–99.

[18] Snyder, W.: Higher-order E-unification. *Proc. 10th Int. Conf. Automated Deduction*, LNCS 449 (1990), 573–587.

[19] Snyder, W and Gallier, J.: Higher-order unification revisited: complete sets of transformations. *J. Symbolic Computation* 8 (1989), 101–140.

[20] Tidén, E.: Unification in Combinations of Collapse-Free Theories with Disjoint Sets of Function Symbols. *Proc. 8th Int. Conf. Automated Deduction*, LNCS 230 (1986), 431–449.

[21] Yelick, K.A.: Unification in combinations of collapse-free regular theories. *J. Symbolic Computation* 3 (1987), 153–182.

On Confluence for Weakly Normalizing Systems

Pierre-Louis Curien
LIENS (CNRS), 45 rue d'Ulm, 75230 Paris Cedex 05.

Giorgio Ghelli[1]
Dipartimento di Informatica, Università di Pisa, Corso Italia 40

Abstract: We present a general, abstract method to show confluence of weakly normalizing systems. The technique consists in constructing an interpretation of the source system into a target system which is already confluent. If the interpretation satisfies certain simple conditions, then the source system is confluent. The method has been used implicitly in a number of applications, but does not seem to have been presented so far in its generality. We present, as digressions, two other methods for proving confluence.

1. A general method

When a reduction system (see e.g. [HueOp] for a general introduction to rewrite systems) is both locally confluent and strongly terminating, then it is confluent. This basic property, known as Newman lemma, yields a general method to establish confluence. But what about reduction systems which are only (or only known as) weakly terminating? In this note we propose a general method to establish the confluence of such a system, relying on an interpretation into another reduction system which is already known to be confluent.

Recall that in the most general setting, a reduction system is just a binary relation $\to$ on a set A, and that the object of study is its reflexive and transitive closure $\to^*$. The reflexive, symmetric and transitive closure of $\to$ is called interconvertibility and will be written here ~. We reserve = to mean coincidence. NF is the set of normal forms, i.e. those b for which there is no b' s.t. b$\to$b'. The main basic properties of interest are:

- *confluence*: if $a \to^* a'$, $a \to^* a''$, then $a' \to^* b$, $a'' \to^* b$ for some b (equivalently, if a~a', then $a \to^* b$, $a' \to^* b$ for some b)
- *local confluence*: if $a \to a'$, $a \to a''$, then $a' \to^* b$, $a'' \to^* b$ for some b
- *strong termination* (or normalization): there is no infinite chain $a_1 \to a_2 \to \ldots$
- *weak termination*: every point a has a normal form b, i.e. there exist $a_1, \ldots, a_n$ s.t.

[1] This work was carried on with the partial support of E.E.C., Esprit Basic Research Action 3070 FIDE and of Italian C.N.R., P.F.I. "Sistemi informatici e calcolo parallelo".

$a \equiv a_1 \to \ldots \to a_n \equiv b$ and b is a normal form.

- *unique normal form*: if a~b and a, b are normal forms, then a=b.

The interest of establishing confluence of weakly normalizing systems lies in the following property: if a system is confluent and has an effective (weak) normalization strategy, then it is decidable. Indeed take a and b, normalize both to a', b' respectively. Then a~b iff a'~b' iff a'=b'. Indeed, when confluence holds, if a~b and b is a normal form, then $a \to^* b$, and if moreover a is a normal form, then a=b. In other words confluence implies unique normal form property.

What we mean by an interpretation of a reduction system $(A,\to)$ into another reduction system $(B,\to')$ is just a mapping G from A to B. Of course we expect that G is stable w.r.t. the relations in some sense. We focus here on the following properties:

(C1) $(B,\to')$ is confluent
(C2) $(A,\to)$ is weakly normalizing
(C3) If $a \to a'$ in A, then G(a)~'G(a') (equivalently: if a~a', then G(a)~'G(a'))
(C4) G translates $(A,\to)$-normal forms into $(B,\to')$-normal forms
(C5) G is injective on A-normal forms.

Proposition 1: If B, A and G are as above, then A is confluent.

Proof: Let $a \to^* b$ and $a \to^* c$:
by (C2): $\exists b'\ b' \in NF_A$ and $b \to^* b'$, $\exists c'\ c' \in NF_A$ and $c \to^* c'$
by (C3): $a \to^* b \to^* b'$ and $a \to^* c \to^* c' \Rightarrow$ G(b')~'G(c')
by (C4): $b' \in NF_A$ and $c' \in NF_A \Rightarrow G(b') \in NF_B$ and $G(c') \in NF_B$
by (C1): G(b')=G(c')
by (C5): b'=c'.
Thus both b and c can be reduced in A to a common reduct b'=c'.

The proof is illustrated by the following picture:

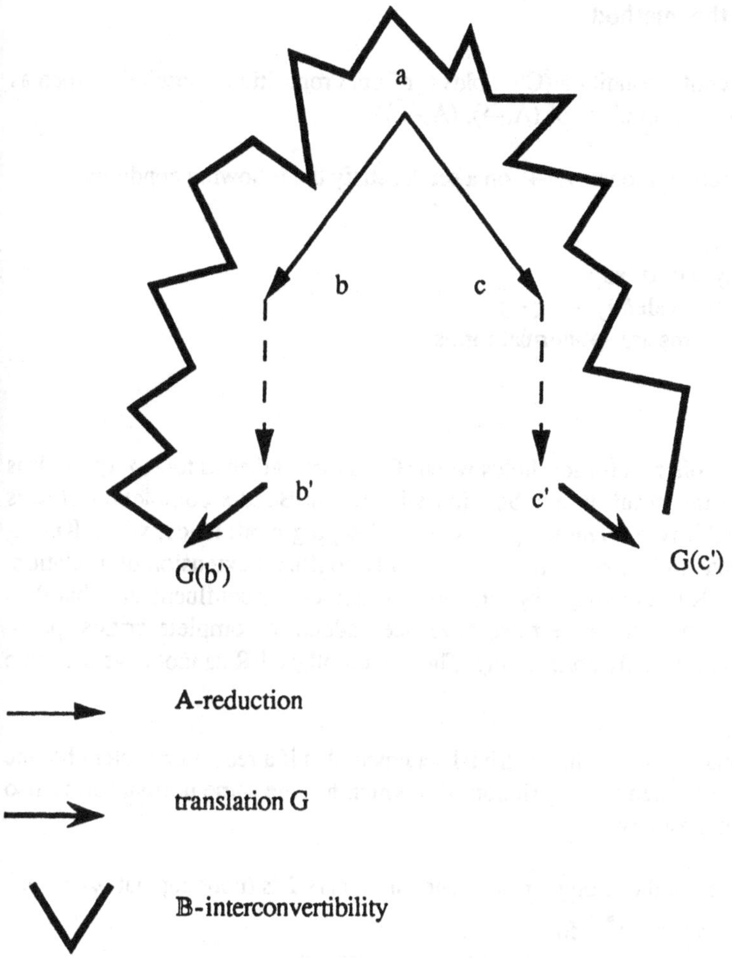

Remark 1: If only (C3), (C4) and (C5) hold, and (C1) is weakened into:

- B has the unique normal form property

then we get that A has the unique normal form property.

Remark 2: Actually the same kind of argument yields a more powerful result. Define a relation $\sim_G$ on A by:

- $a \sim_G b \Leftrightarrow_\Delta G(a) \sim' G(b)$

Then, under the assumptions of proposition 1, one has:

- If $a \sim_G b$, then $a \rightarrow^* c$ and $b \rightarrow^* c$ for some c.

This property has two easy corollaries:

- proposition 1 (notice that (C3) can be rephrased as $\sim \subseteq \sim_G$)
- $\sim = \sim_G$.

In section 4 we shall give examples of applications of Proposition 1, with a non trivial interpretation G. We shall first concentrate on the specialization of the method to the case where A=B and G is the identity.

2. A specialization of the method

If A=B and G is the identity, condition (C5) holds for free. Proposition 1 specializes then as follows (we write freely $\rightarrow$, $\rightarrow'$ in place of $(A,\rightarrow)$, $(A,\rightarrow')$).

Corollary 1: If two reduction relations $\rightarrow$, $\rightarrow'$ on a set A satisfy the following conditions:

(C1) $\rightarrow'$ is confluent
(C2) $\rightarrow$ is weakly normalizing
(C3) $\rightarrow \subseteq \sim'$ (equivalently: $\sim \subseteq \sim'$)
(C4) $\rightarrow$-normal forms are $\rightarrow'$-normal forms

then $\rightarrow$ is confluent.

Remark 3: Of course the corollary a fortiori holds when (C3) is strengthened to: $\rightarrow \subseteq \rightarrow'$. It is then interesting to look at this result in combination with Knuth-Bendix completion: it says something about the possibility of *removing* rules while keeping confluence, while Knuth-Bendix completion proceeds to *add* rules to obtain a locally confluent extension of a relation. As a consequence, suppose R is a rewriting system which is not locally confluent, and that R' is a locally confluent extension of it where rules have been added to complete critical pairs. Suppose furthermore that R is weakly normalizing. Then by corollary 1 R cannot have the same normal forms as R'.

Remark 4: Remark 1 specializes to the almost trivial statement that if a reduction system has the unique normal form property, then any restriction of it which has the same normal forms also has the unique normal form property

Remark 5: The specialization of the stronger result stated in remark 2 is (noticing that $\sim_G = \sim'$):

- If $a \sim' b$, then $a \rightarrow^* c$ and $b \rightarrow^* c$ for some c.

In particular $\sim = \sim_G$. (This was also observed by Y. Toyama [Toy]).

Remark 6: If (C1) is further specialized to: $\rightarrow'$ is strongly normalizing and locally confluent, and (C3) to: (C3) $\rightarrow \subseteq \rightarrow'$, then (C2) holds for free, so that one just has to concentrate on (C4).

The specialization quoted in remark 4 has been used by J.-W. Klop [Klo], and later by Klop and R. de Vrijer [KloVri] to show the unique normal form property for extensions of untyped λ-calculus with non linear rules. Specifically, consider the extension $\lambda\beta\delta$ of the $\lambda\beta$-calculus with a constant δ and the rule

(δ) $\delta xx \rightarrow x$.

As Klop showed, this extension fails to be confluent. But he also proved that it enjoys nevertheless the unique normal form property. One defines an extension of the rule $\delta xx \rightarrow x$, namely

(δ') $\delta xy \rightarrow x$ if $\lambda\beta\delta \vdash x=y$

(the point is to get around the non-linearity of δ). Now $\lambda\beta\delta'$ is confluent and has the same normal forms as $\lambda\beta\delta$. The unique normal form property was shown more recently for $\lambda\beta\pi$ (the $\lambda\beta$-calculus with surjective pairing, which Klop had also shown to be non confluent) [KloVri], with almost the same technique. Klop and de Vrijer consider a linear extension $\lambda\beta\pi'$ of $\lambda\beta\pi$ (with rules of the kind of δ' to replace surjective pairing). De Vrijer had shown that $\lambda\beta\pi'$ is confluent modulo an equivalence relation which does not prevent the unique normal form property to hold [Vri].

We shall illustrate the use of corollary 1 with the $\lambda\sigma$-calculus [ACCL], an extension of λ-calculus allowing for an explicit calculus of substitutions. In this calculus (as well as in the calculus of categorical combinators [CuMon,Har]), β is split into a rule Beta, and a set of rules σ (Subst in [Har]) which carry out the substitutions explicitly. The (two-sorted) syntax and the rules are described below:

Terms $a ::= 1 \mid ab \mid \lambda a \mid a[s]$
Substitutions $s ::= id \mid a.s \mid s \circ t \mid \uparrow$

Rules:

Beta

$(\lambda a)b \rightarrow a[b.id]$

σ

$(\lambda a)[s] \rightarrow \lambda(a[1.(s \circ \uparrow)])$
$(ab)[s] \rightarrow (a[s])(b[s])$
$1[a.s] \rightarrow a$
$1[id] \rightarrow 1$
$a[s][t] \rightarrow a[s \circ t]$

$\uparrow \circ (a.s) \rightarrow s$
$\uparrow \circ id \rightarrow \uparrow$
$id \circ s \rightarrow s$
$a[s][t] \rightarrow a[s \circ t]$
$(a.s) \circ t \rightarrow a[t].(s \circ t)$
$(s \circ t) \circ u \rightarrow s \circ (t \circ u)$

One of the key results in $\lambda\sigma$-calculus is that σ is strongly normalizing (a non trivial result due to T. Hardin and A. Laville)[2] and locally confluent. Corollary 1 (and specifically remark 6) allows to deduce from this that various restrictions of σ are still confluent (on ground terms). For instance, the system obtained by removing $a[s][t] \rightarrow a[s \circ t]$ is confluent. We refer to [ACCL] for details.

3. A digression

Corollary 1 is helpful for situations where an extension of the system under study is confluent. It is natural to ask whether something can be said when a *restriction* is known to be confluent. The following almost trivial, but useful property has been coined by (at least) de Vrijer:

---

[2] This result can be seen as a form of theorem of finite developments, if one thinks of performing Beta as "marking redexes".

Proposition 2: If a reduction system $(A,\rightarrow)$ has a confluent restriction $(A,\rightarrow')$ generating the same equivalence as $(A,\rightarrow)$, then $(A,\rightarrow)$ is confluent.

Proof: Suppose $a\sim b$. Then also $a\sim' b$ since $\sim=\sim'$. Then, by confluence of $\sim'$, $a \rightarrow'^{*} c$ and $b \rightarrow'^{*} c$ for some c. Hence also $a \rightarrow^{*} c$ and $b \rightarrow^{*} c$, since $\rightarrow' \subseteq \rightarrow$.

Remark 6: Corollary 1 says us that if in the end $(A,\rightarrow)$ is confluent, then *any* choice of a restriction $\rightarrow'$ would have done well in proposition 2, provided it is weakly normalizing and reaches the normal forms of $\rightarrow$.

As an illustration of proposition 2, we provide a proof of confluence of simply typed $\lambda\beta$. The idea of the proof is due G. Pottinger [Pot] (the presentation given here is a further, minor simplification of the proof in [Pot]). This proof does not rely on strong normalization, and does not either use parallelization à la Martin-Löf. It goes as follows. It is rather easy to prove strong termination of the following restriction of β-reduction: $(\lambda x.M)N \rightarrow M[x\leftarrow N]$ only when N is in normal form. One associates with every term P the multiset of the degrees (i.e. the size of the type $A\rightarrow B$ of $\lambda x.M$) of the redexes $(\lambda x.M)N$ of P. Then the restricted reduction relation strictly decreases this measure (the fundamental fact, see e.g. [Gir, chapter 4], being that created redexes have smaller degrees than the creating redex). Now take β as $\rightarrow$ and restricted β as $\rightarrow'$. Local confluence of $\rightarrow'$ is established by checking that the restricted diagrams of local confluence can be completed using restricted β-reductions only. Finally we check that $\rightarrow$ is included in $\sim'$, so that $\sim=\sim'$. Suppose that $(\lambda x.M)N$ is an unrestricted redex, reducing to $M[x\leftarrow N]$. Let P be the (restricted) normal form of N. Then the following sequence of equalities is provable in the restricted system: $(\lambda x.M)N \sim' (\lambda x.M)P \sim' M[x\leftarrow P] \sim' M[x\leftarrow N]$.

One easily checks that normal forms of $\rightarrow'$ are also normal forms for $\rightarrow$, which entails that β is weakly normalizing. But these observations are not needed for the proof of confluence.

4. Applications of the general method

In this section we illustrate the general method with two examples in which G is a non trivial mapping. Both examples concern $\beta\eta$-reduction, in two different extensions of simply typed λ-calculus. Consider first the simply typed $\lambda\beta\eta$-calculus. It is known that $\beta\eta$-confluence holds there, but not on the raw (i.e. possibly not typable) terms. Consider the following example:

$$\lambda x{:}T.(\lambda y{:}U.y)x \rightarrow_\eta \lambda y{:}U.y \quad , \quad \lambda x{:}T.(\lambda y{:}U.y)x \rightarrow_\beta \lambda x{:}T.x$$

The typing rules of the simply typed or second-order λ-calculus force T and U to coincide. The situation is less simple in the higher-order λ-calculus $F\omega$, where T and U may be interconvertible. But in $F\omega$ the type conversion system is separate from the term conversion system: we can by a separate argument get that T,U both converge to say V, so that both $\lambda x{:}T.x$ and $\lambda y{:}U.y$ converge to $\lambda y{:}V.y$.

The completion of the critical pair becomes much more complicated in the two following settings:

- (first order) *dependent types* (as in the Automath languages [Daal] language of Logical Frameworks (LF for short) [HHP]): terms may appear as subexpressions of type expressions (while the syntax of terms is the same as in the simply typed λ-calculus). The problem then is that the typing only guarantees that T and U are interconvertible, so that we need to use confluence to complete the diagram to some λy:V.y, where V is a common reduct of T, U. There is thus some circularity to turn around.

- *subtyping* [CW]: there are subtyping relations T≤U, and as soon as a:T, one may infer a:U. In this setting λx:T.(λy:U.y)x is well typed as soon as T≤U, and we are forced to add λy:U.y ~ λx:T.x (suitably oriented) to the system.

We give now some specific hints on the use of Proposition 1 to settle both confluence problems. They were solved in [Daal] for Automath languages (and adapted to LF in [Sal]) and in [CG2] for subtyping, to which we refer for more details.

Dependent types: The proof given in [Daal, section VI.3] (and even more the proof in [Sal]) can be structured as an application of Proposition 1. The basic idea of Van Daalen is to make use of an equivalence relation ~ defined as the congruence closure of:

$$\frac{T \sim_{\beta\eta} U}{\lambda x{:}T.a \sim \lambda x{:}U.a}$$

and to establish first confluence modulo ~. Formally one takes (LF,βη) as A, (LF/~,βη/~) as B, and the quotient map as G. Now, briefly, properties (C1)-(C5) are argued as follows:

(C1) Checking the confluence of B is a matter of following the usual techniques (local confluence + strong normalization [Daal], or parallelization of β + commutation of β and η [Sal]). The point is that the above critical pair is solved thanks to ~, which has of course been defined on purpose to identify both reducts of λx:T.(λy:U.y)x.

(C2) Even strong normalization is known [Daal,HHP]

(C3) Obvious

(C4) Obvious

(C5) One has to show: a ~ b, a and b in βη normal form ⇒ a=b. The proof is by induction on the sum of the sizes of a and b. One goes by cases on the proof of a~b. The critical case is

$$\frac{T \sim_{\beta\eta} U}{\lambda x{:}T.a \sim \lambda x{:}U.a}$$

We can use (C1),(C3) and (C4) to infer T ~ U from $T \sim_{\beta\eta} U$, and then we can apply induction.

Subtyping: We need some preliminaries. The equation λy:U.y ~ λx:T.x (for T≤U) is not easy to handle. First one has to be more precise, and write:

λy:U.y ~ λx:T.x: T→U (T≤U).

Indeed the consequence has been established via the term λx:T.(λy:U.y)x, whose type is T→U, which is a supertype (and in general a strict supertype) of T→T and U→U (we should mention to the unfamiliar reader that the structural subtyping for arrows goes contravariantly in the first argument: T'≤T, U≤U' ⇒ T→U ≤ T'→U'). Thus one cannot apply the rule, in either direction, to a subexpression of a given typed expression a, just by looking if it matches one hand side: we also have to check that there is a proof of a:V, for some V, which goes through assigning the supertype $T \to U$ to the redex subexpression. The legal applications of the rule can be specified by giving *typed* reduction rules. But then it is not very clear how one can adapt known techniques like Knuth-Bendix in such a setting where typing and reduction are so tightly related.

Another way to handle this subtle issue is to introduce an explicit syntax for proofs of subtyping and proofs of typing. Instead of writing a:U when we have a:T and $T \leq U$, we write $c<a>:U$, to keep track of the fact that we have first established a:T (c, which is a proof of $T \leq U$, is called a *coercion*). With this new syntax (which was introduced in [CG1]) the equation becomes:

$$c'<\lambda y{:}U.y> \sim \lambda x{:}T.c<x>$$

where c, c' prove $T \leq U$, $U \to U \leq T \to U$ respectively (notice that now there is no choice for the type $T \to U$ of both hand sides). In order to preserve the expressivity of η, we have to reformulate it in the explicit calculus: indeed the explicit version of $\lambda x{:}T.(\lambda y{:}U.y)x$ is $\lambda x{:}T.(\lambda y{:}U.y)(c<x>)$, on which we cannot apply η as it is. Here is the version of η that we need in the explicit calculus :

$$(c\eta) \quad \lambda x{:}T.d<a(c<x>)> = e<a>$$

where c, d, e prove respectively $T \leq T'$, $U \leq U'$ and $T' \to U \leq T \to U'$.

In the explicit syntax, we can safely replace subexpressions which match redexes, and thus play the normal game of completion of critical pairs.

In [CG2] we use Proposition 1 twice, to show successively that :

1) the Knuth-Bendix completion of the explicit $\beta\eta$ theory is confluent

2) the corresponding typed reduction system in the original calculus (without coercions) is confluent.

We give brief hints on both uses of Proposition 1.

1) We establish confluence of the explicit version of $\beta\eta$ (actually a maximum type Top comes also into play in [CG2], but it would take too long to enter into that here) via a translation to a second-order λ-calculus (with a terminal type). The translation is a variant of the translation proposed in [BCGS]. The idea is to associate with every proof of a subtyping $T \leq U$ a λ-term of type U which has a unique occurrence of a unique free variable (written []) of type T. For example:

- the axiom $T \leq T$ is translated by the single variable []:T
- an assumption $t \leq T$, where t is a type variable, is translated into $t^{\#}[]$ ($t^{\#}$ is a new variable of type $t \to T$, and the $^{\#}$ on the variable name is a meaningful information to be properly handled when proving (C5), see below)
- if e is a proof of $T' \to U \leq T \to U'$ built from proofs c, d of $T \leq T'$, $U \leq U'$, if c and d are translated to a and b, then the translation of e is $\lambda z{:}T.a[[]b[z]]$, where by a[a'] we mean $a[[] \leftarrow a']$
- if c is translated to a and b is translated to b', then $c<a>$ is translated to a[b'].

One takes thus as A the explicit calculus, as B the (second-order) λ-calculus (extended with a terminal type), as G the translation sketched above. (C1) (confluence of B) is proved in [CDC]. The verification of (C2)-(C4) is technical, but does not involve special insights. We only sketch here how we argue for (C5). We exhibit a left inverse for the restriction of G to normal forms. The basic idea is to take profit from the difference between η and cη. Taking a normal form a in A, we know from (C4) that G(a) is a normal form. But considered as a term in A (indeed B can be considered as a sublanguage of A where no use is made of the subtyping facility), G(a) is not a normal form. And the repeated use of cη on G(a) has the very nice effect of reconstructing the coercions that G had interpreted. Of course, for this to work, one needs to be careful in designing the exact translation G, and the exact embedding of B as a sublanguage of A. We hope that the following example will give the flavour of what is going on. Suppose that we start with c<x>, where x:T→U, and c proves T→U ≤ t→U, where t is a type variable, and where t≤T is assumed in the environment. c<x> is a normal form of the explicit calculus, its translation is $\lambda z{:}t.x(t^{\#}z)$. When embedding this expression back to B, we turn the special variables $t^{\#}$ back to coercions, and get λz:t.x(d<z>) where d is a name for the assumption t≤T. Finally we apply cη and obtain back c<x>.

Summarizing, we define on the set of translations of normal forms of A a mapping G' by composing the embedding of B into A with cη normalization, and we prove that, for any a in normal form in A, we have G'(G(a))=a.

2) To transfer the confluence from the explicit calculus to the original calculus, we take as A the original calculus, as B the explicit calculus, and as G a mapping choosing for each well-typed term one of its typing proofs. (C5) (injectivity of G on normal forms) is obvious since G has coercion erasure as inverse (by removing coercions from an explicit term, one recovers the underlying original term). (C1) (confluence of B) is 1). We refer to [CG2] for the remaining properties.

5. A final digression

We would like to mention briefly another interesting technique for proving confluence, which was pointed out (and called interpretation method) by T. Hardin [Har]. It is also based on an interpretation mapping. The ingredients are the following. Let (A,→) be a reduction system. Suppose we have a mapping G:A→A and a reduction system (G(A),→') on the image of G which satisfy the following conditions:

(I1) (G(A),→') is confluent

(I2) $\forall a \in A\ a \rightarrow^{*} G(a)$

(I3) If $a \rightarrow a'$ then $G(a) \rightarrow'^{*} G(a')$

(I4) $\rightarrow' \subseteq \rightarrow^{*}$

It is easy to see that if →, G and →' are such that (I1)-(I4) hold, then → is confluent. Conversely, if → is confluent, if G and →' are such that (I2)-(I4) hold, and if moreover G ∘ G = G, then →' is confluent.

In practice this method can be useful when → is a term rewriting system which has a locally confluent and strongly normalizing restriction →" determined by a subset of the rules of →. Then one takes →" normalization as G (this guarantees (I2), and also G ∘ G = G).

We quote three uses of this method that we are aware of:

- It was used by T. Hardin [Har] (and more recently in [ACCL]), to investigate confluence properties of categorical combinators (a calculus of explicit substitutions). Specifically, remember from section 2 the splitting of the β reduction into a rule Beta, and a set σ of substitution rules. Take σ normalization as G. On ground terms σ normalization leads from $\lambda\sigma$-terms to λ-terms (all explicit substitutions have been carried out). Take β as $\rightarrow'$.
- It is used in [CDC] to show the confluence of the extension of various typed λ-calculi with a terminal object. The rule of terminality (which equates all terms of terminal type, and more generally all terms of a type T, where T is isomorphic to the terminal type), called (gentop), has a bad interaction with the extensionality rules (η, surjective pairing). It was helpful to use the above method with (gentop) normalization as G.
- It is used by V. Breazu-Tannen and J. Gallier [BG] to show the confluence of the extension of any confluent first-order multi-sorted term rewriting system with simply-typed or polymorphic $\lambda\beta$-calculus. They take β-normalization as G.

6. Conclusion and acknowledgements

We believe that Proposition 1, although very simple, can be useful to working "rewriters". We would like to enrich our battery of applications of Proposition 1, in particular with examples which do not come from λ-calculus.

We benefited from fruitful comments of Y. Toyama, R. de Vrijer and J.-W. Klop.

References

[ACCL] M. Abadi, L. Cardelli, P.-L. Curien, J.-J. Lévy, Explicit Substitutions, Proc. POPL 90, San Francisco, ACM (1990).

[BCGS] V. Breazu-Tannen, T. Coquand, C. Gunter, A. Scedrov, Inheritance and Explicit Coercion, Proc. Logic in Computer Science 89, Monterey, USA (1989).

[BG] V. Breazu-Tannen, J. Gallier, Polymorphic Rewriting Conserves Algebraic Confluence, Proc. ICALP 89, Stresa, Lecture Notes in Comput. Sci. 372 (1989).

[CuMon] P.-L. Curien, Categorical Combinators, Sequential Algorithms and Functional Programming, Pitman (1986).

[CDC] P.-L. Curien, Roberto Di Cosmo, Confluence in the typed λ-calculus extended with Surjective Pairing and a Terminal Type, draft (1990).

[CG1] P.-L. Curien, G. Ghelli, Coherence of Subsumption, Proc. Colloque sur les Arbres en Algèbre et en Programmation 90, Lecture Notes in Comput. Sci. 431 (1990).

[CG2] P.-L. Curien, G. Ghelli, Subtyping + Extensionality: Confluence of $\beta\eta$ reduction in $F_{\leq}$, draft (1990).

[CW] L. Cardelli, P. Wegner, On Understanding Types, Data Abstraction and Polymorphism, ACM Computing Surveys, 17 (4) (1985).

[Daal] D. van Daalen, The Language Theory of Automath, PhD Thesis, Technical University of Eindhoven, 1980.

[Gir] J.-Y. Girard, Y. Lafont, P. Taylor, Proofs and Types, Cambridge University Press (1989).

[Har] T. Hardin, Confluence Results for the Pure Strong Categorical Combinatory Logic CCL: λ-calculi as Subsystems of CCL, Theoret. Comput. Sci. 65, 291-342 (1989).

[HHP] R. Harper, F. Honsell, G. Plotkin, A Framework for Defining Logics, Proc. Logic in Computer Science 87, Ithaca (1987).

[HueOp] G. Huet, D.C. Oppen, Equations and Rewrite Rules: A Survey, in Formal Languages Theory: Perspective and Open Problems, R. Book ed., 349-393, Academic Press (1980).

[Klo] J.-W. Klop, Combinatory Reduction Systems, Mathematical Center Tracts 127 (1980).

[KloVri] J.-W. Klop, R. de Vrijer, Unique Normal Forms for Lambda-calculus with Surjective Pairing, Information and Computation (1989).

[Pot] G. Pottinger, The Church Rosser Theorem for the Typed lambda-calculus with Surjective Pairing, Notre Dame Journal of Formal Logic 22 (3), 264-68 (1981).

[Sal] A. Salvesen, The Church-Rosser Theorem for LF with βη-reduction, draft (1989).

[Toy] Y. Toyama, Term Rewriting Systems and the Church-Rosser Property, PhD Thesis, Tohoku University, 1990.

[Vri] R. de Vrijer, Surjective Pairing and Strong Normalization, Two Themes in Lambda-calculus, PhD Thesis, Universiteit van Amsterdam, 1987.

Program Transformation and Rewriting

Françoise BELLEGARDE
Western Washington University
Computer Science Department
Bellingham WA. 97226
email: bellegar@strawberry.cs.wwu.edu

Abstract

We present a basis for program transformation using term rewriting tools. A specification is expressed hierarchically by successive enrichments as a signature and a set of equations. A term can be computed by rewriting. Transformations come from applying a partial unfailing completion procedure to the original set of equations augmented by inductive theorems and a definition of a new function symbol following diverse heuristics. Moreover, the system must provide tools to prove inductive properties; to verify that enrichment produces neither junk nor confusion; and to check for ground confluence and termination. These properties are related to the correctness of the transformation.

1 Equations and program transformation

An important research topic in the area of automatic programming is transformational programming. Functional programming is not inhibited by superfluous concerns such as sequential control or storage mapping. Transformational programming offers a means to formally develop efficient programs from clear programs expressed in high level, functional languages. The program transformation paradigm is not new, but it can take its place in software design only if the transformation process is automated as much as possible. Another condition for program transformation to become a useful method for software design is that it can be used to transform large programs.

A problem with the transformation paradigm is the loss of visibility of its design. Hand transformation is a lengthy, boring and error-prone process. Transformation systems might help by making the process semi-automatic, but this is not enough. As the form of a program is changed during the transformation process, its meaning soon becomes unclear and the user gets lost.

On the other hand, the program transformation process requires knowledge about the program. Properties of the program direct its transformation, and the programmer must provide them. Sometimes, these properties are well known, and they do not need to be proved over and over again. Sometimes, the proof is easy to do by hand. This task may or may not take place during the transformation process itself. In any case, the transformation system must be able to take account of properties given by the user. Moreover, it must be able to help to prove or disprove some of the properties suggested by the programmer.

In this paper, we present a basis for program transformation using term rewriting tools. The examples has been checked using a first simple version of such a transformation system.

1.1 Equations in functional programs

Either a purely functional fragment of a language like ML or a fragment of an order-sorted language like OBJ [10] can be considered as a good candidate for a specification language. It is relatively easy to write or to translate a specification with such languages in an equational form. We will consider a *specification* given by:

- a signature Σ composed of a set of *sort symbols* and a set of *function symbols* with rank declarations.
- a set E of equations.

In this sense, a specification describes a class of algebras, namely the class of Σ-algebras satisfying the equations E. But the semantics we give to such specifications is the *initial algebra* $\Im(\Sigma, E)$.
Example 1 The following specification describes *append*:

$$
\begin{aligned}
&sort\ list[elem] \\
&nil : \mapsto list \\
&:: : elem \times list \mapsto list \\
&append\ :\ list \times list \mapsto list \\
&\forall x : list.append(nil, x)\ =\ x \\
&\forall x : elem.xs, y : list.append((x :: xs), y)\ =\ x :: append(xs, y)
\end{aligned}
$$

The possibility of describing and transforming an application by successive *enrichments* of a specification allows us to handle large programs.

Definition 1 *An enrichment of a specification $S = (\Sigma, E)$ is a specification $S' = (\Sigma', E')$ such that $\Sigma \subseteq \Sigma'$ and $E \subseteq E'$.*

Enrichments can produce either junk, that is new terms that are not equivalent to an already existing term, or confusion, that is equivalence between two terms originally distinct.
Example 2 We can enrich the specification of the example 1 by adding the inductive equations:

$$\forall x, y, z : list.append(x, append(y, z))=append(append(x, y), z) \tag{1}$$

$$\forall x : list.append(x, nil)=x \tag{2}$$

We can also enrich the specification of Example 1 by adding a new function *reverse* and equations for its definition:

$$reverse : list \mapsto list$$

$$reverse(nil)\ =\ nil \tag{3}$$

$$\forall x : elem.xs : list.reverse(x :: xs)\ =\ append(reverse(xs), x :: nil) \tag{4}$$

These two enrichments do not create junk or confusion.

For now, we consider only the particular case of a pure sorted equational language. Some extensions could be considered in the future, such as equations conditioned by premises. Our goal is to define what a transformation system based on rewriting tools can offer. Before going further, let us give basic notions and notations that are used in this paper.

1.2 Basic notions and notations

We will denote by $T(\Sigma, X)$ the set of terms built with the variables X and the functions symbols of the signature Σ. The set of ground terms or terms without variables is denoted by $T(\Sigma)$. Positions in terms are represented as a sequence of integers. t/p denotes the subterm of t at the position p. Substitutions are endomorphisms of $T(\Sigma, X)$. The replacement of the subterm t/p in t by the term u is denoted by $t[p \leftarrow u]$.

Given a binary relation, $\rightarrow$, $\rightarrow^*$ is the reflexive transitive closure of $\rightarrow$. $\leftrightarrow^*$ is its reflexive and symmetric transitive closure. A relation $\rightarrow$ is noetherian if there is no infinite sequence $t_1 \rightarrow t_2 \cdots$.

A relation $\rightarrow$ is confluent if $\leftarrow^* \circ \rightarrow^* \subseteq \rightarrow^* \circ \leftarrow^*$, where $\circ$ denotes the composition of relations. An equation is a pair of terms $s = t$. Given a set E of equations, we write $s \leftrightarrow_E t$ if $s/p = \sigma(l)$ and $t = s[p < -\sigma(r)]$ for some position p in t, substitution σ and equation $l = r$ or $r = l$ in E.

A rule is an oriented pair of terms $l \rightarrow r$. A term rewriting system is a set of rules. Given a term rewriting system R, the rewriting relation $\rightarrow_R$ is a binary relation in $T(\Sigma, X)$. $s \rightarrow_R t$ if there exists a rule $l \rightarrow r$ in R, a position p in s, a substitution σ such that $\sigma(l) = s/p$ and $t = s[p < -\sigma(r)]$. A term t is in normal form if it is irreducible.

A term rewriting system is terminating if the relation $\rightarrow_R$ is noetherian, confluent if the relation $\rightarrow_R$ is confluent, and convergent if it is both confluent and terminating. Convergence ensures existence and unicity of the normal form of every term.

Critical pairs are produced by overlaps of two redexes in a same term. A non-variable term t' and a term t overlap if there exists a non-variable position p in t such that t/p and t' are unifiable. Let $g \rightarrow d$ and $l \rightarrow r$ be two rules such that l and g overlap at the position p with the most general unifier σ. The overlapped term $\sigma(g)$ produces the critical pair (p, q) defined by $p = \sigma(g[p < -r])$ and $q = \sigma(d)$. A critical pair is convergent if p and q reduce to the same term.

The *completion procedure* [12] was introduced as a means at deriving convergent term-rewriting systems used as procedures for deciding the validity of identities (the word problem) in a given equational theory. The procedure generates new rewrite rules to resolve ambiguities resulting from existing rules that overlap. These new rules are produced by non-convergent critical pairs.

A completion procedure can fail because it is unable to orient an equation into a rule without losing the termination property of the system. However, non-orientable equations may sometimes be used for reduction anyway, because their instances can be oriented. This idea is basic to the unfailing completion procedure [2, 1]. It uses the notion of ordered rewriting which does not require that an equation always be used from left to right. An ordered rewriting system is a set of equations together with a reduction ordering $>$, i.e. a well-founded, monotonic and stable. An ordered rewriting system can be denoted $(E, >)$. When the equations in E can be oriented with $>$, we usually call them rules. The ordered rewriting relation using $(E, >)$ is the rewriting relation $\rightarrow_{E>}$ where $E >$ denotes the set of all the orientable instances of E. This allows us to extend the notion of critical pairs to ordered critical pairs and to extend the completion process to an unfailing completion process, i.e. a completion that cannot fail. The outcome of the unfailing completion procedure, when it does not loop, is either a (ground) convergent term rewriting system R when all equations are rules or a ground convergent ordered rewriting system $(E, >)$ when some equations remain unordered. By *ground convergence*, we mean termination and confluence on ground terms. Obviously, convergence implies ground convergence.

Given a ground convergent term rewriting system R, a term t is *ground (or inductively) reducible* with R if all its ground instances are R reducible.

An equation $s = t$ is an *inductive theorem (or inductive consequence)* of E if for any ground substitution σ, $\sigma(s) = \sigma(t)$.

1.3 Checking properties of enrichments

Using equational logic as a programming language was proposed by O'Donnell [17], by Gogen [10] and by Dershowitz [8]. An operational semantics can be given to functions defined by equations by using term rewriting systems.

We consider programs presented in a specification $S = (\Sigma, E)$ by a set of equations E. The specification S is constructed by successive enrichments of a specification $S_0 = (\Sigma_0, E_0)$. We consider the case when the set of functions in the signature Σ can be split into a set of constructors C and a set of defined functions D. The definition of functions of D is *sufficiently complete with respect to* C, i.e. it produces no junk, if every ground term is provably equal to a *constructor term*, which is a term built only with constructors.

When E can be partitioned into constructors and defined symbols, $E_C \cup E_D$, where E_C is the subset of equations that contain only constructors and variables. If $E_C = \emptyset$, the constructors are said to be free. The specification is consistent with respect to C, i.e. it produces no confusions, if for all constructor terms s and t, $s \longleftrightarrow^*_E t$ iff $s \longleftrightarrow^*_{E_C} t$. A good transformation system must be able to prove properties about specifications. Let us consider the principal results regarding enrichments.

Let $S = (\Sigma, E) \subseteq S' = (\Sigma, E')$ be an enrichment with only new equations: $E' = E \cup E_0$. The enrichment is consistent if every equation in E_0 is an inductive consequence of E.

When theories are presented by ground convergent term rewriting systems, the ground completion process can be used to prove consistency of an enrichment and to produce simultaneously a ground convergent term rewriting system for the enriched specification. Consider an enrichment $S = (\Sigma, R_0) \subseteq S' = (\Sigma', R_0 \cup E_0)$ with R_0 a ground convergent term rewriting system on T_Σ. The general idea is to complete first $R_0 \cup E_0$, yielding a ground convergent system R' on $T_{\Sigma'}$. Then one checks that whenever a rewrite rule, whose left and right-hand sides both belong to T_Σ, is added, then this rule is an inductive consequence of R_0. Bachmair has designed an unfailing ground completion procedure for consistency proofs in [1].

If the term rewriting system R associated with the specification is ground confluent, deciding sufficient completeness with respect to C is the same as checking that the normal form of all ground terms is a constructor term. If R preserves constructor terms, (i.e. for any rule $l \rightarrow r$ where l is a constructor term, r is also a constructor term), then it is equivalent to checking for inductive reducibility [11]. Deciding inductive reducibility can be done by using test sets. A constructive method for test sets is given by Kounalis in [13].

Ground confluence of the associated term rewriting system is required for proofs about enrichments. However, we do not always require consistency or sufficient completeness of enrichments. A specification that builds the integers modulo 2 by enriching a specification of integers is not consistent. A specification that builds integers with an infinity element by enriching a specification of integers is not sufficiently complete. Still, these kinds of construction can both be useful. Moreover, we do not really want to limit the transformation process to terminating programs. However, we are limited if we want to do automatic proofs about enrichments.

2 Program transformation

Dershowitz has shown how completion can be applied to the task of program synthesis from specifications in [7, 9]. In the same veine, the transformation process can be viewed as a *partial unfailing completion.* To recall the process, let us take the well known example of the transformation of the specification of the function *reverse* in example 2 [7]. This example provides useful comparaison between diverse systems.

Example 3 We want a more efficient implementation of *reverse*. In an attempt to find one, we enrich the specification with the definition of a new function motivated by a generalization of the right-hand side of equation 4.

$$h : list \times list \mapsto list$$
$$h(u, v) = append(reverse(u), v) \tag{5}$$

Overlaps between the right-hand side of equation 5 and the left-hand sides of equations 3 and 4 produce ordered critical pairs resulting in a direct definition of the function h:

$$h : list \times list \mapsto list$$
$$h(nil, v) = append(nil, v) \tag{6}$$
$$h(x :: xs, v) = append(append(reverse(xs), x :: nil), v) \tag{7}$$

This corresponds to applications of the instantiation law followed by an unfolding in the system of Burstall and Darlington [5]. The right-hand side of the equation 6 can be simplified using the definition of *append*:

$$h(nil, v) = v \tag{8}$$

The right-hand side of equation 7 can be simplified successively using the associativity of *append* given by equation 1, the definition of append, equation 2, and equation 5, oriented from right to left into:

$$h(x :: xs, v) = h(xs, x :: v)$$

This corresponds to applications of laws, unfoldings and finally a folding in the system of Burstall and Darlington. An overlap between the left-hand side of equation 2 and the right-hand side of equation 5 results in the equation:

$$reverse(x) = h(x, nil)$$

This overlap is another motivation for proposing equation 5. This completes the transformation of *reverse* using *append* into a tail recursive definition of *reverse* using only ::.

If we look at diverse examples, the heuristic is always the same: given a specification which defines a function f by equations, the first step consists of the introduction of a new function $h(x_1, \cdots, x_n) = e$, where e is chosen from the following heuristics:

- generalization of a subexpression e_f in the definition of f i.e. $e_f = \sigma(e)$ for some substitution σ so that e_f can be simplified into $\sigma(h(x_1, \cdots, x_n))$,
- a simple composition of functions in the definition of f and
- a tuple of subexpressions in the definition of f chosen from any of these heuristics.

Often, it happens that $f(x'_1, \cdots, x'_p)$ is a subexpression of e because the definition of f is recursive.

Overlaps between the left-hand side of the definition of h and the right-hand sides of one or more of the equations of f result in a direct definition of h by a set of equations d_h.

The second step consists in the simplification of the left-hand sides of d_h using equations of the original specification S of f and equations of an enrichment of S. Instances of e are simplified into instances of h.

If $f(x'_1, \cdots, x'_p)$ is a subexpression of e, it happens (mostly because the user has chosen e on purpose) that an instance of e can be simplified into $f(x'_1, \cdots, x'_n)$, resulting in a direct definition of f using h. In any case, because of the heuristics used to choose e, e_f can be simplified, resulting in a definition of f using h.

Let us consider another simple example to illustrate the tupling heuristic.

Example 4 The following specification (Σ, E) of integers:

$$\begin{aligned}
&sort\ Int\\
&ZERO\ :\mapsto Int\\
&S\ :\ Int \mapsto Int\\
&+\ :\ Int \times Int \mapsto Int\\
&*\ :\ Int \times Int \mapsto Int\\
&\forall x : Int.ZERO + x \ = \ x\\
&\forall x : Int.y : Int.S(x) + y \ = \ S(x + y)\\
&\forall x : Int.ZERO * x \ = \ ZERO\\
&\forall x : Int.y : Int.S(x) * y \ = \ x * y \ + \ y
\end{aligned}$$

is enriched with a definition of the function fib defining the nth fibonacci number:

$$fib\ :\ Int \mapsto Int$$

$$fib(ZERO) \ = \ ZERO \quad (9)$$

$$\forall x : Int.fib(S(ZERO) \ = \ S(ZERO) \quad (10)$$

$$\forall x : Int.fib(S(S(x))) \ = \ fib(S(x)) \ + \ fib(x) \quad (11)$$

We will now generalize $fib(S(x)) + fib(x)$ using a new function g by the tupling heuristic introducing as a new sort, pairs of integers:

$$\begin{aligned} sort &: pair[elem] \\ \langle -, - \rangle &: elem \times elem \mapsto pair \\ fst &: pair \mapsto elem \\ snd &: pair \mapsto elem \\ \forall x : elem.y : elem.fst(\langle x, y \rangle) &= x \\ \forall x : elem.y : elem.snd(\langle x, y \rangle) &= y \end{aligned}$$

We define g by:

$$\begin{aligned} g \ Int &\mapsto Int \\ \forall x : Int.g(x) &= \langle fib(S(x), fib(x) \rangle \end{aligned} \tag{12}$$

Overlaps between the left-hand side of the definition of g and the definitions of fst and snd result in:

$$fib(S(x)) = fst(g(x)) \tag{13}$$

$$fib(x) = snd(g(x)) \tag{14}$$

Equation 14 is a new definition of fib using g. Equation 11 is simplified into :

$$fib(S(S(x))) = fst(g(x)) + snd(g(x)) \tag{15}$$

Equation 12 is simplified into

$$\langle fst(g(x)), snd(g(x)) \rangle = g(x) \tag{16}$$

An overlap between 14 and 9, and an overlap between 13 and 10 results in:

$$\begin{aligned} fst(g(ZERO)) &= S(ZERO) \\ snd(g(ZERO)) &= ZERO \end{aligned}$$

instantiating 16 into:

$$g(ZERO) = \langle S(ZERO), ZERO \rangle \tag{17}$$

Overlaps between 14, 13 and 15 result in:

$$\begin{aligned} fst(g(S(x))) &= fst(g(x)) + snd(g(x)) \\ snd(g(S(x))) &= fst(g(x)) \end{aligned}$$

instantiating 16 into:

$$g(S(x)) = \langle fst(g(x)) + snd(g(x)), fst(g(x)) \rangle \tag{18}$$

Equations 14, 17, and 18 constitute a tail recursive definition of fib.

This transformation process is not restricted to simple and well known examples. The interested reader can look at the development of the Kwic example using our system, given in the appendix. Reddy gives very interesting examples in [19]. I will not address in this paper the question of the amelioration of the efficiency of a program by using this transformation process with the heuristics described above. I am only interested here in its correctness and its implementation using term-rewriting techniques.

2.1 Correctness of the transformation process

The transformation process consists primarily of that part of the unfailing completion process that I call a *partial unfailing completion.*

Definition 2 *Two specifications $S = (\Sigma, E)$ and $S' = (\Sigma, E')$ are equivalent if for any ground terms s and t, $s \longleftrightarrow^*_E t$ iff $s \longleftrightarrow^*_{E'} t$.*

In other words, S and S' have the same initial algebra. In the following, $t_1, \cdots, t_n$ are constructor terms. Recall that C is the set of constructors. Therefore, $T(C)$ is the set of ground constructor terms.

Definition 3 *Let $S_f = (\Sigma_f, E_f)$ be the specification defining the function f. The result of the transformation is a specification $S'_f = (\Sigma'_f, E'_f)$ specifying the same function f i.e. for all ground terms, $f(t_1, \cdots, t_n) \longleftrightarrow^*_{E_f} s$ if $f(t_1, \cdots, t_n) \longleftrightarrow^*_{E'_f} s$. In other words, S_f and S'_f are equivalent on the terms $T(C \cup \{f\}) \times T(C)$*

Proposition 1 *Let us call $S = (\Sigma, E)$ the enrichment of $S_f = (\Sigma_f, E_f)$ with a set of new function symbols Σ_h, their definitions E_h, and inductive consequences L of E. We have $\Sigma = \Sigma_f \cup \Sigma_h$ and $E = E_f \cup E_h \cup L$. Let $S' = (\Sigma, E')$ be the result of a partial unfailing completion of S. Then*

1. *The partial unfailing completion transforms S into an equivalent specification S'.*
2. *The transformation process transforms a specification $S_f = (\Sigma_f, E_f)$ of a function f into an equivalent specification $S'_f = (\Sigma'_f, E'_f)$ of the function f if*
   - *The E_C-equality (equality between constructors) is included into the E'_f-equality,*
   - *S is consistent with respect to the conctructors and*
   - *S'_f is a complete definition of f, i.e. for all ground terms $f(t_1, \cdots, t_n)$, there exists a constructor term s such that $f(t_1, \cdots, t_n) \longleftrightarrow^*_{E'_f} s$.*

Proof: The first result follows simply from the fact that partial unfailing completion does not modify the initial algebra. Considering the second result, the transformation process transform S_f into S'_f. First, S_f and S are equivalent because neither inductive consequences nor E_h modifies the initial algebra. Second, the partial unfailing completion does not modify the E-equality, thus $\longleftrightarrow^*_{E'_f} \subseteq \longleftrightarrow^*_E$. Let us consider a ground term $f(t_1, \cdots, t_n)$, and a constructor term s such that $f(t_1, \cdots, t_n) \longleftrightarrow^*_{E'_f} s$, then:

- $f(t_1, \cdots, t_n) \longleftrightarrow^*_E s$ by $\longleftrightarrow^*_{E'_f} \subseteq \longleftrightarrow^*_E$.
- Conversely, if $f(t_1, \cdots, t_n) \longleftrightarrow^*_{E_f} s$, then $f(t_1, \cdots, t_n) \longleftrightarrow^*_E s$ because S_f and S are equivalent. There exists a constructor term u such that $f(t_1, \cdots, t_n) \longleftrightarrow^*_{E'_f} u$. Therefore $f(t_1, \cdots, t_n) \longleftrightarrow^*_E u$ by $\longleftrightarrow^*_{E'_f} \subseteq \longleftrightarrow^*_E$. $u \longleftrightarrow^*_E s$ by transitivity of the E-equality. $u \longleftrightarrow^*_{E_C} s$ by consistency of E with respect to the conctructors. $u \longleftrightarrow^*_{E'_f} s$ because the E'_f-equality contains the E_C-equality. $f(t_1, \cdots, t_n) \longleftrightarrow^*_{E'_f} s$, by transitivity of the E-equality.

□

Proposition 2 *The transformation process preserves the consistency of a specification with respect to the conctructors.*

Proof: The partial unfailing completion does not modify the initial algebra. □

Let us consider now the operational point of view. The theory is presented by a term rewriting system R for the specification (Σ, R). Computation of a term in $T(\Sigma)$ is done by rewriting. The operationally complete definition of a function f with a specification (Σ_f, R_f) w.r.t C is when for all ground term $f(t_1, \cdots, t_n)$, there exists a constructor term s such that

$$f(t_1, \cdots, t_n) \to_R^* s$$

Operational and algebraic definitions coincide if R is ground convergent.

Proposition 3 *Let the specification (Σ_f, R_f) be an operationally complete definition of f consistent with respect to the conctructors. If it is transformed into the specification (Σ'_f, R'_f) which is an operationally complete definition of f, then the computations, i.e. the normal forms of a ground term $f(t_1, \cdots, t_n)$, by R and R' are E_C-equal.*

> **Proof:** A ground term $f(t_1, \cdots, t_n)$ has R_f-normal forms s and R'_f-normal forms u that are constructor terms. The R_f equality and the R'_f equality are contained into the $E = R_f \cup E_h \cup L$ equality where R_f is the set of rules of the new functions and L the set of the inductive consequences introduced for the transformation. Therefore, u and s are $(E, >)$-normal forms. Σ, E is consistent w.r.t C because the addition of L and E_h does not modify the consistency. $u \longleftrightarrow_{E_C}^* s$ by consistency of (Σ, E) with respect to the conctructors. □

All the above results are valid, if we replace the specification of the constructors (C, E_C) by any specification at any level of the hierarchic construction of the specification. The following result proved in [14] is interesting when we construct a specification hierarchically.

Proposition 4 *Let R_0 be a ground convergent term rewriting system, and let R_f be an enrichment of (Σ_0, R_0) with a complete definition of f w.r.t Σ_0, then R_f is ground convergent.*

As a consequence, when R_f is transformed into $R_0 \subseteq R'_f$, which gives a new complete definition of f w.r.t R_0, R'_f is ground convergent.

2.2 Implementation of the transformation process

The core of the system is the partial unfailing completion. Given a source term rewriting system R_f, we enrich the specification $S = (\Sigma_f, E_f)$ with a definition E_h of a new symbol h, like h in the example of *reverse* or like g in the example of *fibonacci*. This new definition is given in general by a unique equation $E_h : h(x_1, \cdots, x_n) = e$ where e is built following the diverse heuristics suggested above. Let us imagine how to organize the partial completion process.

1. *The system computes the ordered critical pairs between E_h and R_f*

   Let σ be the most general unifier of e with $f(t_1, \cdots, t_n)$, left-hand side of an equation $f(t_1, \cdots, t_n) = t$. Let $\sigma(e)$ be greater than the instance $\sigma(h(x_1, \cdots, x_n))$ so that the ordered critical pairs are equations $\sigma(h(x_1, \cdots, x_n)) = \sigma(t)$. If R_f contains complete definitions of every defined symbol, such equations contain a complete definition of h.

2. *The system processes simplifications*, then, the complete definition of f is simplified, and h must occur in the definition of f. The possible overlaps with E_h can give more than one possibility, as shown in the following example:

   Example 5 Let R_f be a ground convergent system for a complete definition of factorial containing a definition of + and $*$ on integers.

$$S(x) + y \to S(x + y)$$
$$ZERO + x \to x$$

$$S(x) * y \rightarrow x * y + y$$
$$ZERO * x \rightarrow ZERO$$
$$fact(S(x)) \rightarrow S(x) * fact(x) \quad (19)$$
$$fact(ZERO) \rightarrow S(ZERO) \quad (20)$$

With a definition $h(x,u) = u * fact(x)$, the process will return the equations:

$$h(x, S(u)) = h(x,u) + fact(x) \quad (21)$$
$$h(x, ZERO) = ZERO \quad (22)$$
$$h(S(x), u) = u * (h(x,x) + fact(x)) \quad (23)$$
$$h(ZERO, S(u)) = x * S(ZERO) \quad (24)$$

by overlapping $u * fact(x)$ on the definition of $*$ and on the definition of $fact$. The user may be disturbed by these two potential complete definitions of h.

3. *Additional inductive theorems can be added to help the transformation*

   Theorems are helpful

   (a) for simplifying all rules and equations

   (b) for deducting new inductive equations from the definition of the new symbol.

   Therefore, *we do not overlap the theorems with E_h and the consequences of E_h.* The system overlaps the new theorems only with E_f. Moreover, *some theorems can be used only for simplification and in this case they need not to be overlapped with E_f.* The user must indicate if the theorem must be overlapped or not. *New theorems can be provided by the user at the beginning of the whole process or during the process.* The separation of the transformation process into two steps as we illustrated here is totally artificial.

2.3 Limitations of partial completion

1. *The partial completion can loop* as every completion procedure can do. However, the user can always interrupt the process when getting a result that contains a new, hopefully better, complete definition of the function of interest.

   Example 6 Let us continue our example with the definition of factorial. First, we introduce the inductive theorem $x * S(ZERO) \rightarrow x$ for simplifying the right-hand side of equation 24 into

$$h(ZERO, S(u)) = x \quad (25)$$

   Second, we introduce the associativity of $*$ in an attempt to simplify the left-hand side $u * (h(x,x) + fact(x))$ of equation 23 and remove the occurrence of $fact(x)$. Assuming that $z*(u*fact(x))$ is greater than $(z*u)*fact(x)$, a superposition of the associativity on $h(x,u) = u * fact(x)$ generates the pair:

$$z * h(x,u) = h(x, z * u) \quad (26)$$

   Assuming that $h(x,x) + fact(x)$ is greater than $h(x, S(x))$, the right-hand side of equation 23 is simplified, resulting in:

$$h(S(x), u) = h(x, u * S(x)) \quad (27)$$

   Equations 25 and 27 give a tail recursive complete definition of h, but the other superpositions make the completion process continue indefinitely. One can notice that the process would work and be finite if superpositions were limited to being done only with the definition of $fact$ given by equations 19 and 20.

2. *The process can also fail to find the desired result, even if it exists, because of the inadequacy of the ordering.* This is the principal drawback of this method. Recursive path ordering [6] often does not work as well as polynomial interpretations [15]. Transformation orderings [3, 4] might be useful. Work remains to be done to find adequate orderings.

One way to resolve this can be to restrict the completion more severely. Let g be the function such that the overlaps between the definition of g and the definition of h must be done to find the new definition of h. Given the new function h, the function of interest f, the inductive consequences L, the means to orient equations of L, and the function g, the system or the user needs only to orient equational consequences.

The Focus system [18], which does not search for equational consequences, is even more restrictive. It superposes only g and h and simplifies by rewriting. Therefore, it does no completion at all. But this is sometimes too restrictive. For example, it generates this definition of *reverse* as

$$reverse(nil) = nil$$
$$reverse(x :: xs) = h(xs, x :: nil)$$

but it does not generate the definition $reverse(x) = h(x, nil)$. The user of Focus must provide this definition which can be proved by induction [20, 14] from the first one. The following example shows the weaknesses of the various choices.

Example 7 Let us go back to factorial. With the associativity of $*$ oriented as $x*(y*z) \to (x*y)*z$, the superposition between $u * fact(x)$ and $fact(S(x))$ is $u * fact(S(x))$ which has 3 distinct normal forms giving 3 definitions of $h(S(x), u)$ as:

1. $h(S(x), u) = u * (h(x, x) + fact(x))$
2. $h(S(x), u) = u * h(x, S(x))$
3. $h(S(x), u) = h(x, u * S(x))$

The third one gives a tail recursive definition. The partial completion will force the confluence to a unique normal form. If the ordering is well chosen (we noted above that this is the major drawback of the method), the third definition will be the normal-form. On another hand, without completion, like in Focus, you might get either the third definition with an outermost rewriting, or the first or the second ones with an innermost rewriting. The first definition is obtained by choosing to simplify first with the rule $S(x) * y \to x * y + y$, and the second definition is obtained by choosing to simplify first with the definition of $h(x, u)$.

3 Conclusion

We choose to use a partial unfailing completion process as the central part of a transformation system. For that we use the toolkit of rewriting tools provided by ORME [16]. With this simple initial implementation we have tested the well-known examples and the kwic example given in the appendix. The kwic example is interesting because it requires 3 steps of transformation and therefore shows the potential for transformation of larger specifications by composition of individual transformation steps. With abilities to perform:

1. induction proofs [14, 20],
2. check of consistency [1, 9],
3. check of complete definition and sufficient completeness [11, 13],

one could check the main properties of the specifications and prove the inductive consequences that must be added to perform the transformation. The system must also extract the specification (Σ_f, R_f) from the specification (Σ', R') resulting from a transformation step, i.e. extract a complete definition of the function of interest f and, iteratively, complete definitions of functions that occur in the definition of f. For this purpose it needs a check of a complete definition.

All this might be extended to conditional specifications, i.e. a set of conditional equations. A conditional equation is an equation or an expression $e_1 \wedge \cdots \wedge e_n \Rightarrow e$ where $e_1, \cdots, e_n$ are equations called conditions and e is an equation. Conditional equations are very useful to express specifications. The function *filter* in the appendix might rather be expressed by:

$$filter(nil) = nil$$
$$issig(x) \Rightarrow filter(cons(x, xs)) = filter(xs)$$
$$not(issig(x)) \Rightarrow filter(cons(x, xs)) = cons(x, filter(xs))$$

They are also very useful to express conditional properties that might help the transformation.

Acknowledgements: We thank U.S. Reddy for giving us a copy of the Focus System, P. Lescanne for the tools provided by ORME, R. Kieburtz for the idea of the kwic example, B. Vance, J. Hook, R. Kieburtz, C. Kirchner and P. Lescanne for their support and comments.

References

[1] L. Bachmair. Proofs methods for equational theories. PhD thesis, University of Illinois, Urbana-Champaign, 1987. Revised version, August 1988.

[2] L. Bachmair, N. Dershowitz, and D. Plaisted. Completion without failure. In *Proceedings of the colloquium on the resolution of Equations in Algebraic Structures*, 1987.

[3] F. Bellegarde and P. Lescanne. Transformation Orderings. In *12th Colloquium on Trees in Algebra and Programming*, TAPSOFT, pages 69-80, Springer Verlag, Lecture Notes in Computer Science 249, 1987.

[4] F. Bellegarde and P. Lescanne. Termination by Completion, Technical Report CRIN 90-R-028, 1990.

[5] R. M. Burstall and J. Darlington. A Transformation System For Developing Recursive Programs. *Journal of the Association for Computing Machinery*, 24, pages 44-67, 1977.

[6] N. Dershowitz. Termination. In *Proceedings of the first Conference on Rewriting Techniques and Applications*, Springer Verlag, Lecture Notes in Computer Science 202, pages 180-224, Dijon, France, 1985.

[7] N. Dershowitz. Synthesis By Completion. *Proceedings of the Ninth International Joint Conference on Artificial Intelligence*, pages 208-214, Los Angeles, 1985.

[8] N. Dershowitz. Computing with rewrite systems. *Information and Control*, 65(2/3):122-157, 1985.

[9] N. Dershowitz. Completion and its Applications. *Resolution of Equations in Algebraic Structures*, Academic Press, New York, 1988.

[10] J. Goguen and C. Kirchner and H. Kirchner and A. Megrelis and J. Meseguer and T. Winkler. An introduction to OBJ-3. In *Proceedings of the 1rst Intern. Workshop on Conditional Term Rewriting Systems*, Lecture Notes in Computer Science 308, 1988.

[11] D. Kapur, P. Narendran, and H. Zhang. On sufficient completeness and related properties of term rewriting systems. *Acta Informatica,* 24:395-415, 1987.

[12] D. E. Knuth and P. B. Bendix. Simple Word Problems in Universal Algebras. In J. Leech, editor, *Computational Problems in Abstract algebra*, pages 263-297, Pergamon Press, Oxford, U. K.,1970.

[13] E. Kounalis, Testing for Inductive (CO)-Reducibility. In *Proceedings of the 15th International Colloquium on Trees in Algebra and Programming,* Lecture Notes in Computer Science 431, pages 221-238, 1990.

[14] E. Kounalis, M. Rusinowitch. Mechanizing Inductive Reasoning. In *Proceedings of the eight National Conference on Artificial Intelligence*, AAAI-90, 1990.

[15] D.S. Lankford. On proving term rewriting systems are Noetherian, *Memo MTP-3,* Mathematic Department, Louisiana Tech. University, Ruston, LA, May 1979. (Revised October 1979).

[16] P. Lescanne, Completion Procedures as Transition Rules + Control:ORME. In *2nd Intern. Workshop Algebraic and Logic Programming,* Lecture Notes in Computer Science, 1990.

[17] M. O'Donnell. *Equational Logic as a Programming Language. Foundation of Computing,* MIT Press, 1985.

[18] U. S. Reddy. Transformational derivation of programs using the Focus system. In *Symposium Practical Software Development Environments*, pages 163-172, ACM, December 1988.

[19] U. S. Reddy, Formal methods in transformational derivation of programs. In Proceedings of the ACM Intern. Workshop on Automatic Software Design, AAAI, 1990.

[20] U. S. Reddy, Term Rewriting Induction. In *Proceedings of the Conference of Automated Deduction,* 1990.

4 Appendix: Kwic example

The problem is to produce, from a list of titles, a list of the cyclic permutations of the original titles such that we retain only those permutations that begin with a key word.

4.1 Specification

Here we show the construction of the specification by successive enrichments.

Let us represent a title by a list of words. Our input is a list of titles. We can use a specification of lists.

$$\begin{aligned}
&sort\ list[elem] \\
&nil\ : \mapsto list \\
&::\ :\ elem \times list \mapsto list \\
&append\ :\ list \times list \mapsto list \\
&\forall x : list.append(nil, x) \ = \ x \\
&\forall x : elem.xs, y : list.append((x :: xs), y) \ = \ x :: append(xs, y)
\end{aligned}$$

We enrich the specification with the definition of a function *rotations* to get an elementary cyclic permutation.

$$\begin{aligned}
\forall x : list[word].rotate(nil) \ &= \ nil \\
\forall x : word.xs : list.rotate(x :: xs) \ &= \ append(xs, x :: nil)
\end{aligned}$$

Now we want to iterate the function *rotate* to get the permutations of a title. The title itself can be used to control the iteration. The function *all* returns the complete list of cyclic permutations of a title.

$$\forall x : list[word].all(x) = repeat(x, x)$$
$$\forall x : list[word].repeat(x, nil) = nil$$
$$\forall x : word.u, xs : list[word].repeat(u, x :: xs) = u :: repeat(rotate(u), xs)$$

We can now get the permutations of all titles by a function *concall*:

$$concall(nil) = nil$$
$$\forall x : list[word].xs : list[list[word]].concall(x :: xs) = append(all(x), concall(xs))$$

The list of permutations can be filtered to extract the significant titles. The permutations whose initial word belongs to the set of insignificant words can be dropped. A predicate *issig* checks if the permutation is kept. *To specify filter, we use a ternary conditional operator cond.*

$$filter(nil) = nil$$
$$\forall x : list[word].xs : list[list[word]].$$
$$filter(x :: xs) = cond(issig(x), x :: filter(xs), filter(xs))$$

Finally, we get the desired result by:

$$\forall x : list[list[word]].sigperm(x) = filter(concall(x))$$

4.2 First transformation step

We first transform the definition of *sigperm*:

$$sigperm(nil) = nil$$
$$sigperm(x :: xs) = filter(append(repeat(x, x), concall(xs))$$

We introduce an inductive theorem:

$$filter(append(x, y)) = append(filter(x), filter(y))$$

sigperm is simplified into:

$$sigperm(x :: xs) = append(filter(repeat(x, x)), sigperm(xs))$$

The complete definition of *sigperm* is now:

$$append(nil, x) = x$$
$$append(x :: xs, y) = x :: append(xs, y)$$
$$rotate(nil) = nil$$
$$rotate(x :: xs) = append(xs, x :: nil)$$
$$repeat(x, nil) = nil$$
$$repeat(u, x :: xs) = u :: repeat(rotate(u), xs)$$
$$filter(nil) = nil$$
$$filter(x :: xs) = cond(issig(x), x :: filter(xs), filter(xs))$$
$$sigperm(nil) = nil$$
$$sigperm(x :: xs) = append(filter(repeat(x, x)), sigperm(xs))$$

4.3 Second transformation step

We are now interested in modifying the composition $filter(repeat(x, x))$ in the definition of *sigperm*. We introduce a new definition:

$$sigrot(x, y) = filter(repeat(x, y))$$

and the transformation process gives the equations:

$$sigperm(x :: xs) = append(sigrot(x, x), concfil(xs))$$
$$sigrot(x, nil) = nil$$
$$sigrot(u, x :: xs)) = cond(issig(u), u :: sigrot(rotate(u), xs), sigrot(rotate(u), xs))$$

The complete definition of *sigperm* is now:

$$append(nil, x) = x$$
$$append(x :: xs, y) = x :: append(xs, y)$$
$$rotate(nil) = nil$$
$$rotate(x :: xs) = append(xs, x :: nil)$$
$$sigperm(nil) = nil$$
$$sigperm(x :: xs) = append(sigrot(x, x), sigperm(xs))$$
$$sigrot(x, nil) = nil$$
$$sigrot(u, x :: xs)) = cond(issig(u), u :: sigrot(rotate(u), xs), sigrot(rotate(u), xs))$$

4.4 Third transformation step

Our objective is to get rid of the costly occurrences of *append* in *sigperm*. We introduce the new definition:

$$sr(x, y, u) = append(sigrot(x, y), u)$$

and the theorem:

$$append(cond(u, x, y), z) = cond(u, append(x, z), append(y, z))$$

the transformation process returns:

$$sr(x, nil, u) = u$$
$$sr(x1, x :: xs, u) = cond(issig(x), x :: sr(rotate(x1), xs, u), sr(rotate(x1), xs, u))$$
$$concfil(x :: xs) = sr(x, x, concfil(xs))$$

The complete definition of *sigperm* is now:

$$append(nil, x) = x$$
$$append(x :: xs, y) = x :: append(xs, y)$$
$$rotate(nil) = nil$$
$$rotate(x :: xs) = append(xs, x :: nil)$$
$$sigperm(nil) = nil$$
$$sigperm(x :: xs) = sr(x, x, concfil(xs))$$
$$sr(x, nil, u) = u$$
$$sr(x1, x :: xs, u) = cond(issig(x), x :: sr(rotate(x1), xs, u), sr(rotate(x1), xs, u))$$

An efficient representation of arithmetic for term rewriting

Dave Cohen

Department of Computer Science

Royal Holloway and Bedford New College

University of London

and

Phil Watson

Department of Computing Science

University of Glasgow

Abstract

We give a locally confluent set of rewrite rules for integer (positive and negative) arithmetic using the familiar system of place notation. We are unable to prove its termination at present, but we strongly conjecture that rewriting with this system terminates and give our reasons. We show that every term has a normal form and so the rewrite system is normalising.

We justify our choice of representation in terms of both space efficiency and speed of rewriting.

Finally we give several examples of the use of our system.

1 Introduction

Equational rewriting has achieved much attention as a possible paradigm for automating mathematical theorem proving. In this paper we describe a useful rewriting system which rewrites expressions involving the standard arithmetic terms into a place valued representation.

The system is locally confluent which is important if we want to use it to prove theorems. We pay particular attention to the question of termination of rewriting. Although at present we are unable to prove that rewriting with this system terminates, there is nonetheless strong circumstantial evidence that this is the case, such as the existence of normal forms for all terms.

Although our system is of interest in its own right we particularly intend that it can be used as a module in larger rewrite systems where integers are seen as atomic entities and only ground terms are required to be rewritten by our system. In this case completeness is likely to be important for ground terms, but completeness of the whole system is unnecessary. However as we have only a single sort (Int) and this is non-empty completeness and ground completeness are equivalent.

Our proof of local confluence is by machine: we used the theorem prover LP (Larch Prover) for this and our rules are given in a format acceptable to LP. LP was a natural choice as we required a theorem prover capable of handling associative and commutative operators.

2 Construction of the rewrite set

2.1 Motivation

Many examples from equational rewriting involve the use of natural numbers (or integers). It is normal to describe these in terms of the generator 0, and the successor function $s\_$ (possibly with some additional symbol $-\_$ for unary minus). This can lead to difficult long expressions involving many occurrences of the function $s\_$, e.g.

$$x * 7 \rightarrow x$$

would be a messy rewrite rule with seven occurrences of the function $s\_$. The only way around this problem is to find a different representation for integers.

We decided that it would be useful to have a system of rewrite rules which will take an arithmetic expression involving integers and the usual arithmetic functions

(+, *) and rewrite it to a concise normal form. This system could then be seen as a "module" in a larger rewriting set involving integers which enables more complicated equations to be manipulated and understood. By this we mean that the larger rewrite set would call on a sort "Int" and our rewrite set would be included as a module which exclusively rewrites terms of sort Int (to terms of sort Int).

2.2 Representation

2.2.1 Natural Numbers

The first decision that we had to make in order to construct a rewrite set was which concise normal form we would use to represent a ground term. For the moment we only consider non-negative ground terms. The problem of introducing negative integers will be tackled later.

We decided to adopt the place system of representation. This system is familiar and easy to use.

We can also justify this choice more formally as follows. We require smaller numbers to require no more symbols than larger ones, and to introduce only one new function symbol (_._) to represent the 'invisible' place operator used in everyday arithmetic, and a fixed number k of constants $(0, \ldots, k-1)$ $k > 1$. Given these requirements the number of symbols required to represent numbers in the range $k^n, \ldots, k^{n+1} - 1$ is $n + 1$, i.e. the most concise representation has space complexity $O(\log(n))$. The normal place system of arithmetic achieves this optimal bound.

Definition We will measure the length of a term t by

$$length(t) = \#\text{occurrences in t of } \_.\_$$

Theorem 1 *In any representation of numbers satisfying:*

1. $n < m \Rightarrow length(n) \leq length(m)$.
2. *We have only one function* _._ *of fixed arity two.*
3. *We have a fixed number* k *of constants,* $k > 1$.
4. *We have to represent each term in one of the two normal forms:*
   (a) *a where a is a constant.*
   (b) *t.a where a is a constant and t is a non-zero term in normal form.*

the representation of n has length at least $\lfloor \log_k(n) \rfloor$ *if* $n > 0$.

Proof 1 The proof is an easy induction argument on the number of applications of _._ and will not be given here for reasons of space.

Restricting our attention for the moment to ground terms (i.e. terms containing no variables), we can make the obvious choice of normal forms and require that each natural number has a normal form of the form above.

2.2.2 Choice of Base

At this point we made the decision to work in base 4 arithmetic. This results in a substantial saving of space over the more obvious choice of base 10, while having enough constant symbols (in our previous notation k = 4) to illustrate all the necessary points.

The other obvious choice of base, base 2, is satisfactory in itself but is markedly different from all other bases in the following way. In base 2 all terms of the form $n * p$ (where n is an integer in normal form and p is a product of variables) will rewrite (using one of the rules

```
0 * x -> 0
1 * x -> x
(y.z) * x -> (x * y).(x * z)
```

which must occur in a place arithmetic rewrite system). In a higher base, however, terms of the form $2 * p$, $3 * p$,... occur. These are in normal form but we need rules to cope with the sum of such terms, e.g. $(2 * p) + (3 * p)$, which occur in these higher bases, though obviously not in base 2. Thus by choosing a base higher than 2 we solve a more general problem. We may note that there is no corresponding difference between bases 3 and 4, or 4 and 5, etc.

Following this argument we believe that an implementation of base 10 arithmetic using the method of this paper will not run into any difficulties beyond those encountered here, except for the extra demands on the theorem prover in terms of space and time.

2.2.3 Negative Integers

The next stage in the development was to extend our range of interest to include negative integers. We introduced a new constant MIN to represent 'negative 1'.

This immediately causes a difficult problem. This is because when we introduce negative values into a place system, there are many equivalent forms for a given integer, even using only the binary operator _._ and the five constants $0, 1, 2, 3, MIN$. This applies equally to positive and negative integers and gives particular problems when terms involve variables. It becomes very hard to devise a confluent set of rewrite rules.

Again restricting our attention for the moment to ground terms, we decided upon a normal form for negative integers reminiscent of two's complement form for signed binary numbers. Normal forms of natural numbers are as above (they do not use MIN). Normal forms of negative integers are terms of the form $((\ldots(MIN.x_1).x_2)\ldots).x_n, n \geq 0$ where if $n \geq 1$ then $((\ldots(x_1.x_2)\ldots).x_n$ is the normal form of a natural number.

N.B. A term such as $MIN.1$ (the normal form of 'negative three') is not to be confused with $MIN * 1$ in the 'everyday' represention of negative integers. Recall that _._ is the place operator.

2.2.4 Non-Ground Terms

Normal forms for terms with variables are more complicated. Some such terms will have a ground normal form (e.g. the term $x + (MIN * x)$ is not a ground term but has the ground normal form 0). Terms without ground normal forms will have a normal form of the form

$$((\ldots((x_1.x_2).x_3)\ldots).x_{n-1}).x_n \text{ for } n \geq 1$$

where each of the x_i is $1, 2$, or 3 (or MIN, if none of the other x_i contains MIN) or of the form

$$\sum_{j=1}^{m_i} t_{i,j} \text{ where } m_i \geq 1$$

and for each j

$$t_{i,j} = \prod_{k=1}^{p_{i,j}} v_{i,j,k} \text{ where } p_{i,j} \geq 1$$

or

$$t_{i,j} = d * \prod_{k=1}^{p_{i,j}} v_{i,j,k} \text{ where } d \text{ is of one of the forms } 2, MIN \text{ and } p_{i,j} \geq 1$$

and the products $\prod_k v_{i,j,k}$ are all products of variables distinct up to associativity and commutativity of _*_ in that x_i.

Our one further requirement is that of course x_1 is not equal to 0 unless $n = 1$ and the whole term is identically 0.

For some examples of this normal form see section 2.3.

So we set about the task of constructing a rewrite set which would rewrite any term into a term of this form.

We chose to do our completion in LP as this allowed us to use AC-operators. We inserted rules ordered by hand in the direction that fitted with our choice of normal form. We will return to the problem of proving termination in section 4.2.

2.3 Examples of Normal Forms

We give a few examples of semantic objects and their corresponding normal forms.

'fourteen'	3.2
'forty-two'	(2.2).2
'negative twenty-two'	$((MIN.2).2).2$
$(x-2)*(x-2)$	$((MIN*x)+1).(x*x)$
$12xy+37z-1$	$(2*z).(((3*x)*y)+z).(z+MIN)$

3 Rewrite Rule Set for Place Arithmetic

The following is our rewrite system for place arithmetic. The format is that used by the Larch Theorem Prover LP. We call the system *integers*.

```
0 * x -> 0
1 * x -> x
2 * 2 -> 1 . 0
3 * x -> x . (MIN * x)
MIN * MIN -> 1
2 * MIN -> MIN . 2
(x . y) * z -> (x * z) . (y * z)
(y + z) * x -> (x * y) + (x * z)

0 + x -> x
x + x -> 2 * x
1 + 2 -> 3
```

```
1 + MIN -> 0
2 + MIN -> 1
3 + x -> 1 . (MIN + x)
(x . y) + z -> x . (y + z)
(2 * x) + x -> 3 * x
(MIN * x) + x -> 0
(2 * v1) + (MIN * v1) -> v1

MIN . 3 -> MIN
x . MIN -> (MIN + x) . 3
0 . x -> x
x . (y . z) -> (x + y) . z
```

4 Completeness

To show that *integers* is complete we have to show that it is both (locally) confluent and terminating.

4.1 Local Confluence

To prove local confluence we used Knuth-Bendix completion [KB70] in the theorem prover LP, as already mentioned.

4.2 Termination

To show termination is harder as we have not used any particular termination ordering to orient our rewrite rules, but simply oriented them according to our choice of normal forms.

In fact we can see very quickly that neither of the two most commonly used termination orderings can orient our rules as we wish:

1. The Knuth-Bendix Ordering is unable to accept the rule

   ```
   x.MIN -> (MIN + x).3
   ```

   oriented in the direction we have chosen.

2. The Recursive Path Ordering is unable to accept the above rule and also the rule

```
x.(y.z) -> (x+y).z
```

oriented in the direction we have chosen.

Each of these orderings also fails on other rules apart from the ones mentioned.

Despite this we strongly conjecture that rewriting with *integers* always terminates. Our reasons for this are as follows:

1. Every term has a (unique up to AC) normal form;
2. Despite considerable effort we have been unable to find a counter-example.

4.3 Normalisation

We can show that every term rewrites using *integers* to one of the normal forms in 2.2.

Theorem 2 *Every term can be rewritten using integers to a term in normal form.*

Proof 2 Due to lack of space we cannot include the full proof. The proof method is as follows. We give a sequence of subsets of *integers*, each of which is easily proved to be normalising (in most cases, terminating), with the intention that in normalising a term, these subsets are applied in the order given. The class of normal forms reached after the application of each subset is given. After the application of the last subset the possible normal forms are exactly those given in 2.2.

4.4 Efficiency

We claim that rewriting of mathematical expressions is (time) efficient in *integers* compared to the standard representations, and so the use of this system will improve the efficiency of rewrite sets. We have not proved this formally. Our claim is based on the following argument.

Depending of course on the implementation, one of the major costs (and possibly the single greatest cost) of a term rewriting system is the time spent by the machine in *matching*, i.e. attempting to apply a rule to a term at some given position. Our representation of integers has been proved to be space-efficient. Compared to the common 'successor' notation our terms are much smaller, i.e. have fewer positions, so matching may be expected to take less time. We may also expect that this will more than compensate for the slightly larger number of rules required by our system compared to a system with 'successor' notation.

Practical tests are needed to compare the two notations.

Also, as we have proved, the space efficiency of our system is asymptotically optimal. Naturally this is of great convenience to the user compared with 'successor' notation.

5 Use of Place Arithmetic Rewrite Set

We will give a series of examples of the use of our system. We are particularly interested in examples of the rewriting of ground terms as we expect that for most purposes integer arithmetic will be more appropriate than multinomial arithmetic. We also give a very interesting example to illustrate our intention that *integers* can be used as a module in a larger system. In this example we add a set of rewrite rules for the factorial function (normalising on all integer values greater than 0), and then normalise a term representing 'factorial ten'. This is pleasingly quick to normalise.

5.1 Some examples of rewrites and normal forms

All of the following examples were run using the Larch theorem prover (LP) on a SUN 3/60. We will give the LP output in each case, unedited except where space demands and to remove line numbers, which take the form 1.,2., etc. and so may be confused with our notation for the place operator. The command "sh no" means normalise and display normal form of the given term.

Example 1 We begin by normalising a representation of the integer 'fifty-five'.

```
-> sh no (3.2).MIN


The sequence of reductions leading to the normal
form of the term is:
   (3 . 2) . MIN
   ((3 . 2) + MIN) . 3
```

```
(3 . (2 + MIN)) . 3
(3 . 1) . 3
```

Example 2 Next we do a simple example involving MIN.

```
-> sh no (MIN * x * y) * (MIN * z)


The sequence of reductions leading to the normal
form of the term is:
 ((MIN * x) * y) * (MIN * z)
 1 * x * y * z
 x * y * z
```

Example 3 We now demonstrate the use of *integers* as a module in a larger system. We add two extra rules to our system to handle a new unary function FACT(_), intended to represent factorial.

```
FACT(1) -> 1
FACT(x) -> x * FACT(MIN + x)
```

It is easily seen that this system of two rules, in conjunction with *integers*, is normalising on integer ground terms greater than 0. Within a reasonable time we can normalise quite large terms, for example 'factorial ten':

```
-> sh no fact(2.2)


The sequence of reductions leading to the normal
form of the term is:
 FACT(2.2)
 (2.2)*FACT(2.1)
 (2.1)*(2.2)*FACT(2.0)
 (2.0)*(2.1)*(2.2)*FACT((2+MIN).3)
 (2.0)*(2.1)*(2.2)*FACT(1.3)
 (1.3)*(2.0)*(2.1)*(2.2)*FACT(1.2)
 (1.2)*(1.3)*(2.0)*(2.1)*(2.2)*FACT(1.1)
 (1.1)*(1.2)*(1.3)*(2.0)*(2.1)*(2.2)*FACT(1.0)
```

```
(1.0)*(1.1)*(1.2)*(1.3)*(2.0)*(2.1)*(2.2)*FACT((1+MIN).3)
(1.0)*(1.1)*(1.2)*(1.3)*(2.0)*(2.1)*(2.2)*FACT(0.3)
(1.0)*(1.1)*(1.2)*(1.3)*(2.0)*(2.1)*(2.2)*FACT(3)
(1.0)*(1.1)*(1.2)*(1.3)*(2.0)*(2.1)*(2.2)*3*FACT(2)
(1.0)*(1.1)*(1.2)*(1.3)*(2.0)*(2.1)*(2.2)*2*3
((1.0).((1.0)*MIN))*(1.1)*(1.2)*(1.3)*(2.0)*(2.1)*(2.2)*2
                        .
                        .
                        .

(three hundred and eleven rewrite steps omitted)

((((((((3.1).3).1).1).3).(1+2)).0).(0*2)).(0*2)).(0*2)
((((((((3.1).3).1).1).3).3).0).(0*2)).(0*2)).(0*2)
((((((((3.1).3).1).1).3).3).0).0).(0*2)).(0*2)
((((((((3.1).3).1).1).3).3).0).0).0).(0*2)
((((((((3.1).3).1).1).3).3).0).0).0).0
```

In total three hundred and twenty-nine rewrites suffice to reduce FACT(2.2), representing 'factorial ten', to its normal form. We may be sure that a system using 'successor notation' would take many more steps.

6 Further Research

We intend to further investigate the efficiency of *integers* in two ways. The first will be a 'benchmarking' process, running our system against a typical set of rules for integer arithmetic in 'successor' notation. The second will be a more theoretical approach bringing in complexity theory, and defining the efficiency of a term rewriting system in terms of all the costs of rewriting, including matching, and not simply counting the number of rewrites performed.

Also we will attempt to prove the termination of *integers* as a matter of priority.

7 Conclusion

We have presented a locally confluent and normalising set of rewrite rules (*integers*) which performs place arithmetic (with variables) in base 4. We have shown that normal forms under *integers* are asymptotically optimal in terms of space efficiency, and that in some natural examples many fewer rewrites are required using *integers* than would be required by a system using 'successor' notation.

Acknowledgements

The authors would like to thank all who contributed to this paper, particularly Deryck Brown of the University of Glasgow.

References

[De87] N. Dershowitz, Termination of rewriting, J. Symbolic Computation, 3, 69-116 (1987)

[KB70] D.E. Knuth, P. B. Bendix, Simple word problems, In: J. Leech (ed.), *Computational Problems in Abstract Algebra*, 263-297, Pergamon Press (1970)

Query optimization using rewrite rules

Sieger van Denneheuvel (UVA)
Karen Kwast (UVA)
Gerard R. Renardel de Lavalette (RUU)
Edith Spaan (UVA)

Dept. of Mathematics and Computer Science, Univ. of Amsterdam (UVA) &
Applied Logic Group, Dept. of Philosophy, Univ. of Utrecht (RUU)

Abstract

In literature on query optimization the normal form for relational algebra expressions consisting of Projection, Selection and Join, is well known. In this paper we extend this normal form with Calculation and Union and define a corresponding language UPCSJL. In addition we show how the normal form can be used for query optimization.

1 Introduction

PSJ expressions are relational algebra expressions containing only project, select and join operators. This restricted class of expressions, called **PSJL** in the sequel, is commonly used in relational databases. **PSJL** expressions are studied in [YAN87] and [LAR85], where it is mentioned without proof (which is not very difficult) that they can be reduced into a normal form where first the join operators are applied, then selection and finally projection. Such normalization procedures play an important role in query optimization: see [ULL89] and [YAN87]. Standard optimization techniques can be used to further optimize **PSJL** normal form expressions: e.g. in special circumstances the 'selection before join' heuristic can be applied to push selection down to the relational database tables ([ULL89]).

In this paper we add the relational operators *calculate* and *union* to the above mentioned relational operators, thus obtaining the language of **UPCSJL** expressions. The question arises whether **UPCSJL** expressions also can be reduced into a normal form. We prove the existence of a normal form, where the joins are followed by selection, calculation, projection and finally union; moreover, our proof yields a direct construction. This **UPCSJL** normal form already performs some optimization, but the normal form procedure can serve as the starting point for further optimization (just as for **PSJL** normal forms).

There are several reasons to apply normalization before optimization. Firstly, normalization reduces the number of relational operators in a relational expression. As a consequence, optimization after normalization can be more efficient since the number of reducible subexpressions (redexes) on which optimization rules are applied is also reduced. Moreover, the optimization rules can benefit from the fixed structure of a normal form.

Secondly, there is a functional difference between rules used for normalization and rules used for optimization: the former are unconditional, whereas the latter are conditional. A normalization rule should always be applicable on a subexpression of the proper syntactical format or else a normal form could not be obtained. On the other hand optimization rules only rewrite if in addition to a proper format also a condition involving the subexpression is satisfied. Therefore optimization is in general more expensive than normalization, since applications of optimization rules may fail.

Our interest for PCSJ expressions lies in its role in the integration of relational databases and constraint solving. This integration is one of the aims of the declarative Rule Language **RL**. **RL** was defined by Peter van Emde Boas in [VEMD86a], where a relational semantic model is given to interpret **RL** (see also [VEMD86b], [VEMD86c]).

2 The data language DL

We begin with the definition of the *data language* **DL**: it will act as a parameter for the language **UPCSJL**. **DL** is a many-sorted language containing variables (denoted by the metavariables $x, y, \ldots$, also called *attributes*), constants ($c, d, \ldots$, also called *values*), functions ($f, g, \ldots$), $=$ (the equality predicate), predicates, propositional connectives ($\neg, \wedge, \vee, \rightarrow$) and the propositional constant $\mathtt{true}$. Terms ($s, t, \ldots$) and assertions ($A, B, \ldots$, also called *constraints* or *conditions*) are defined as usual.

If E is any of the items defined above (or a collection of these), then $var(E)$ is the set of variables occurring in E. Furthermore, we assume some evaluation mechanism $eval(\_)$ for **DL** to be given, which evaluates closed terms (terms without variables) to constants and closed assertions to truth values. We give an example language for **DL**, defined by the following sorts, constants, functions and predicates:

sorts:	**NUM** (natural numbers), **STR** (strings of characters)
constants:	0,1,2,...in **NUM**, all finite strings in **STR**
functions:	$*, + :$ **NUM** $\times$ **NUM** $\rightarrow$ **NUM**
	cat: **STR** $\times$ **STR** $\rightarrow$ **STR** (concatenation)
	length: **STR** $\rightarrow$ **NUM** (length of a string)
	digits: **NUM** $\rightarrow$ **STR** (converts a number to its string representation)
predicates:	$<, >, \leq, \geq, \neq$ (binary predicates, both on **NUM** and on **STR**)

3 Functions of the language UPCSJL

Before we define the sorts of the language **UPCSJL**, we introduce the following.

Definition 1 *A solution is an expression of the form $x = t$ with $x \notin var(t)$.*

Definition 2 *A solution set is a finite set $\{x_1 = t_1, \ldots, x_n = t_n\}$, satisfying:*
1. $\|\{x_1, \ldots, x_n\}\| = n$, *(the variables are distinct)*
2. $\{x_1, \ldots, x_n\} \cap var(\{t_1, \ldots, t_n\}) = \emptyset$

A *tuple* (denoted by $\phi, \psi, \ldots$) is a solution set of the form $\{x_1 = c_1, \ldots, x_n = c_n\}$. For tuples ϕ, we often write $attr(\phi)$ instead of $var(\phi)$. Tuples are called *similar* if they have the same attributes. A *base relation* is a finite set of similar tuples.

Assume that an instance of **DL** is given, i.e. some language with sorts, variables, constants, etc. We now present the functions used in the definition of the language **UPCSJL**. **UPCSJL** is a four-sorted language with expressions (thus named to distinguish them from **DL**-terms) and equations. The sorts are:

- $\mathcal{V}$ (finite sets of **DL**-variables denoted by $X, Y, Z, \ldots$)
- $\mathcal{C}$ (constraints, i.e. **DL**-assertions denoted by $A, B, C, \ldots$)
- $\mathcal{S}$ (solutions sets denoted by $\Phi, \Psi, \ldots$)
- $\mathcal{R}$ (relations denoted by $R, S, T, \ldots$)

An expression of sort $\mathcal{R}$ is also called a *relational expression*. A base relation may be followed by a bracketed list denoting all its attributes. In the sequel we use the notation E^X for renaming an expression E with respect to a variable set X:

Definition 3
$E^X = E'$ *with* E' *obtained by renaming all variables* $var(E) - X$ *uniquely.*

To allow a brief notation, a renaming $(\_)^X$ can be applied at several places in an expression. In that case all occurrences of $(\_)^X$ denote the *same* renaming:

Example 1 *Renaming variables in expressions with* $X = \{x\}$, $Y = \{y\}$
$\sigma((r(x,y))^X \bowtie s(y,z)^Y, (x > y)^X \wedge (y > z)^Y) \rightarrow \sigma(r(x,u_1) \bowtie s(y,u_2), x > u_1 \wedge y > u_2)$

Next we present the functions of **UPCSJL**. They are grouped according to their range.

3.1 Functions with range $\mathcal{V}$

Besides the usual set operations $\cup, \cap$ and $-$, we have:

Definition 4 *(attributes of a relational expression)*
$attr(\_) : \mathcal{R} \rightarrow \mathcal{V}$

Definition 5 *(head and tail variables of a solution set)*
$hvar(\_) : \mathcal{S} \rightarrow \mathcal{V}$, $hvar(\{x_1 = t_1, \ldots, x_n = t_n\}) = \{x_1, \ldots, x_n\}$
$tvar(\_) : \mathcal{S} \rightarrow \mathcal{V}$, $tvar(\{x_1 = t_1, \ldots, x_n = t_n\}) = var(\{t_1, \ldots, t_n\})$

3.2 Functions with range $\mathcal{C}$

Besides $\wedge$ (conjunction), we have the *merge* of two solution sets, yielding a constraint. The merge function is defined as follows:

Definition 6 $\oplus : \mathcal{S} \times \mathcal{S} \rightarrow \mathcal{C}$, $\Phi \oplus \Psi = \{s = t \mid x = s \in \Phi, x = t \in \Psi\}$

Example 2 $\{x = \text{'bob'}\} \oplus \{x = y \; cat \; z\} = \{\text{'bob'} = y \; cat \; z\}$
$\{x = u + 2, y = 3\} \oplus \{x = v + 2\} = \{u + 2 = v + 2\}$
$\{x = y + 2\} \oplus \{z = x, y = 2 * x\} = \emptyset$

Solution sets $\Phi = \{x_1 = t_1, \ldots, x_n = t_n\}$ can be interpreted as substitutions $[x_1 := t_1, \ldots, x_n := t_n]$ which can be applied to (collections of) items. So we have an operation *apply*:

Definition 7 $\_(\_) : \mathcal{S} \times \mathcal{C} \rightarrow \mathcal{C},\ \Phi(A) = A[x_1 := t_1, \ldots, x_n := t_n]$
with $\Phi = \{x_1 = t_1, \ldots, x_n = t_n\}$

Example 3 $\{x = u + 2\}(\{x = u + 1\}) = \{u + 2 = u + 1\}$
$\{x = u + 2, y = v + 2\}(\{x > y\}) = \{u + 2 > v + 2\}$

3.3 Functions with range $\mathcal{S}$

Here, too, we have the usual set operations $\cup, \cap$ and $-$; besides, we introduce the *restrict* and *delete* functions:

Definition 8 *(restricting and deleting solutions)*
1. $\_[\_] : \mathcal{S} \times \mathcal{V} \rightarrow \mathcal{S},\ \Phi[X] = \{x = t \in \Phi \mid x \in X\}$
2. $\_\langle\_\rangle : \mathcal{S} \times \mathcal{V} \rightarrow \mathcal{S},\ \Phi\langle X\rangle = \{x = t \in \Phi \mid x \notin X\}$

Further we also have substitution on solutions:

Definition 9 $\_(\!(\_)\!) : \mathcal{S} \times \mathcal{S} \rightarrow \mathcal{S},\ \Phi(\!(\Psi)\!) = \{x = eval(\Phi(t)) \mid x = t \in \Psi\}$
if $hvar(\Psi) \cap tvar(\Phi) = \emptyset$

3.4 Functions with range $\mathcal{R}$

Here we find the usual operators on relations, together with the calculate operator. The definitions are:

Definition 10 *(primitive relational operators)*
1. $\pi : \mathcal{R} \times \mathcal{V} \rightarrow \mathcal{R}$
$\pi(R, X) = \{\phi[X] \mid \phi \in R\}$ *if* $X \subset attr(R)$
2. $\sigma : \mathcal{R} \times \mathcal{C} \rightarrow \mathcal{R}$
$\sigma(R, A) = \{\phi \in R \mid eval(\phi(A)) = \mathbf{true}\}$ *if* $var(A) \subset attr(R)$
3. $\kappa : \mathcal{R} \times \mathcal{S} \rightarrow \mathcal{R}$
$\kappa(R, \Phi) = \{\psi \cup \psi(\!(\Phi)\!) \mid \psi \in R\}$ *if* $tvar(\Phi) \subset attr(R)$ *and* $hvar(\Phi) \cap attr(R) = \emptyset$
4. $\_ \bowtie \_ : \mathcal{R} \times \mathcal{R} \rightarrow \mathcal{R}$
$R \bowtie S = \{\phi \cup \psi \mid \phi \in R, \psi \in S, \forall x \in attr(\phi) \cap attr(\psi)\ (\phi[x] = \psi[x])\}$
5. $\_ \cup \_ : \mathcal{R} \times \mathcal{R} \rightarrow \mathcal{R}$
$R \cup S = \{\phi \mid \phi \in R \vee \phi \in S\}$ *if* $attr(R) = attr(S)$

One readily observes that the project, select, calculate and union operators are *partial* since they are only defined when certain conditions on the arguments are met. These conditions are referred to as *wellformedness* conditions. They are quite reasonable: the wellformedness condition for projection ensures that a relation is not projected on attributes that are not part of the relation; the wellformedness condition for selection takes care that the constraint A can indeed be evaluated to **true** or **false**; the first part of the wellformedness condition for the calculate operator ensures that the tails of solutions in Φ can be evaluated, the second part rules out the possibility that the head of a solution is also determined directly by an attribute of the relation R. The constraint set A in $\sigma(R, A)$ is the *condition* of the select operator and the solution set Φ in $\kappa(R, \Phi)$ is the *instruction* of the calculate operator.

Example 4 $r(x, y) := \{\{x = 1, y = 2\}\}$
$\kappa(r(x, y), \{u = x + y, v = \text{`bob'}\}) = \{\{x = 1, y = 2, u = 3, v = \text{`bob'}\}\}$

4 Relational rewrite rules

In this section we describe a large number of rules handling the properties of the primitive relational operators. Since these operators involve wellformedness conditions we introduce a special notation to emphasize in what direction a rule can be applied:

Definition 11 $R \rightarrow S$ *iff* *R is wellformed* $\Rightarrow$ *S is wellformed* & $R = S$

For some rewrite rules $R \rightarrow S$, wellformedness of R does not imply wellformedness of all subexpressions of S and in this case the wellformedness conditions on the violating subexpressions are represented in the rewrite rule as a condition.

4.1 Projection, selection and calculation

The projection, selection and calculate operators have in common that they are applied on a single relational argument and therefore we discuss them together in this section. In the sequel we cluster rules that are similar in one proposition if possible.

Proposition 12 *(cascade rules)*
1. $\pi(\pi(R, X), Y) \rightarrow \pi(R, Y)$ .... $[\pi\pi]$
2. $\sigma(\sigma(R, A), B) \leftrightarrow \sigma(R, A \wedge B)$ .... $[\sigma\sigma]$
3. $\sigma(\sigma(R, A), B) \leftrightarrow \sigma(\sigma(R, B), A)$ .... $[\sigma\sigma*]$
4. $\kappa(\kappa(R, \Phi), \Psi) \rightarrow \kappa(R, \Phi \cup \Phi(\!(\Psi)\!))$ .... $[\kappa\kappa]$
5. $\kappa(\kappa(R, \Phi), \Psi) \leftrightarrow \kappa(R, \Phi \cup \Psi)$ *if* $hvar(\Phi) \cap tvar(\Psi) = \emptyset$ .... $[\kappa\kappa*]$
6. $\pi(R, attr(R)) \leftrightarrow R$ .... $[\pi attr]$

Proposition 13 *(selection and projection)*
1. $\sigma(\pi(R, X), A) \rightarrow \pi(\sigma(R, A), X)$ .... $[\sigma\pi]$
2. $\pi(\sigma(R, A), X) \rightarrow \pi(\sigma(\pi(R, attr(R) \cap (var(A) \cup X)), A), X)$ .... $[\pi\sigma]$
3. $\pi(\sigma(R, A), X) \rightarrow \sigma(\pi(R, X), A)$ *if* $var(A) \subset X$ .... $[\pi\sigma*]$

Proposition 14 *(selection and calculation)*
1. $\sigma(\kappa(R, \Phi), A) \rightarrow \kappa(\sigma(R, \Phi(A)), \Phi)$ .... $[\sigma\kappa]$
2. $\kappa(\sigma(R, A), \Phi) \rightarrow \sigma(\kappa(R, \Phi), A)$ .... $[\kappa\sigma]$

Proposition 15 *(calculation and projection)*
1. $\kappa(\pi(R, X), \Phi) \rightarrow \pi(\kappa(R^X, \Phi), hvar(\Phi) \cup X)$ .... $[\kappa\pi]$
2. $\kappa(\pi(R, X), \Phi) \rightarrow \pi(\kappa(R, \Phi), hvar(\Phi) \cup X)$ *if* $attr(R) \cap hvar(\Phi) = \emptyset$ ...... $[\kappa\pi*]$
3. $\pi(\kappa(R, \Phi), X) \rightarrow \pi(\kappa(R, \Phi[X]), X)$ .... $[\pi\kappa]$
4. $\pi(\kappa(R, \Phi), X) \rightarrow \pi(\kappa(\pi(R, tvar(\Phi) \cup (X \cap attr(R))), \Phi), X)$ .... $[\pi\kappa*]$

Note that in rule $(\kappa\pi)$ by the renaming in R^X and the wellformedness of $\kappa(\pi(R, X), \Phi)$ all variables $hvar(\Phi)$ are different from the variables $attr(R^X)$, as required.

4.2 Join and union

In this section we add the join and union operators in our list of rules. As before some rules have both conditional and unconditional versions.

Proposition 16 *(symmetry and associativity of $\bowtie$ and $\cup$)*
1. $R \bowtie S \leftrightarrow S \bowtie R$ .... [SYMo $\bowtie$]
2. $(R \bowtie S) \bowtie T \leftrightarrow R \bowtie (S \bowtie T)$ .... [ASSo $\bowtie$]
3. $R \cup S \leftrightarrow S \cup R$ .... [SYM o $\cup$]
4. $(R \cup S) \cup T \leftrightarrow R \cup (S \cup T)$ .... [ASS o$\cup$]

Proposition 17 *(join and calculation)*
1. $R \bowtie \kappa(S, \Phi) \rightarrow \kappa(\sigma(R \bowtie S, \Phi[attr(R)]), \Phi\langle attr(R)\rangle)$ .... [$\bowtie\kappa$]
2. $R \bowtie \kappa(S, \Phi) \rightarrow \kappa(R \bowtie S, \Phi)$ *if* $hvar(\Phi) \cap attr(R) = \emptyset$ .... [$\bowtie\kappa*$]
3. $R \bowtie \kappa(S, \Phi) \rightarrow \sigma(R \bowtie S, \Phi)$ *if* $hvar(\Phi) \subset attr(R)$ .... [$\bowtie\kappa+$]
4. $\kappa(R \bowtie S, \Phi) \rightarrow R \bowtie \kappa(S, \Phi)$ *if* $tvar(\Phi) \subset attr(S)$ .... [$\kappa\bowtie$]

Proposition 18 *(join and projection)*
1. $R \bowtie \pi(S, X) \rightarrow \pi(R \bowtie S^X, attr(R) \cup X)$ .... [$\bowtie\pi$]
2. $R \bowtie \pi(S, X) \rightarrow \pi(R \bowtie S, attr(R) \cup X)$ *if* $attr(R) \cap attr(S) \subset X$ .... [$\bowtie\pi*$]
3. $\pi(R \bowtie S, X) \rightarrow \pi(R \bowtie \pi(S, attr(S) \cap (attr(R) \cup X)), X)$ .... [$\pi\bowtie$]
4. $\pi(R \bowtie S, X) \rightarrow \pi(R \bowtie \pi(S, attr(S) \cap X), X)$ *if* $attr(R) \cap attr(S) \subset X$ .... [$\pi\bowtie*$]

Proposition 19 *(join and selection)*
1. $R \bowtie \sigma(S, A) \rightarrow \sigma(R \bowtie S, A)$ .... [$\bowtie\sigma$]
2. $\sigma(R \bowtie S, A) \rightarrow R \bowtie \sigma(S, A)$ *if* $var(A) \subset attr(R)$ .... [$\sigma\bowtie$]

Proposition 20 *(distribution rules for union)*
1. $\pi(R \cup S, X) \rightarrow \pi(R, X) \cup \pi(S, X)$ .... [$\pi\cup$]
2. $\sigma(R \cup S, A) \rightarrow \sigma(R, A) \cup \sigma(S, A)$ .... [$\sigma\cup$]
3. $\kappa(R \cup S, \Phi) \rightarrow \kappa(R, \Phi) \cup \kappa(S, \Phi)$ .... [$\kappa\cup$]
4. $R \bowtie (S \cup T) \rightarrow R \bowtie S \cup R \bowtie T$ .... [$\bowtie\cup$]

Distribution rules for the join operator are described in the next section.

4.3 Generalized relational rules

Before we can proceed further we need rules which straightforwardly generalize rules of the previous section. However, the last generalized rule ($\kappa\bowtie\kappa$) is quite involved and crucial for the construction of a normal form that includes calculation.

Proposition 21 *(derived from ($\bowtie\pi$) and ($\bowtie\pi*$))*
1. $\pi(R, X) \bowtie \pi(S, Y) \rightarrow \pi(R^X \bowtie S^Y, X \cup Y)$ .... [$\pi\bowtie\pi$]
2. $\pi(R, X) \bowtie \pi(S, Y) \rightarrow \pi(R \bowtie S, X \cup Y)$ *if* $attr(R) \cap attr(S) \subset X \cap Y$ .... [$\pi\bowtie\pi*$]

Proposition 22 *(derived from ($\pi\bowtie$) and ($\pi\bowtie*$))*
1. $\pi(R \bowtie S, X) \rightarrow$
$\pi(\pi(R, attr(R) \cap (attr(S) \cup X)) \bowtie \pi(S, attr(S) \cap (attr(R) \cup X)), X)$ .... [$\pi\bowtie+$]
2. $\pi(R \bowtie S, X) \rightarrow$
$\pi(R, attr(R) \cap X) \bowtie \pi(S, attr(S) \cap X)$ *if* $attr(R) \cap attr(S) \subset X$ .... [$\pi\bowtie\#$]

Proposition 23 *(derived from* $(\bowtie\sigma)$*)* .... $[\sigma\bowtie\sigma]$
$\sigma(R, A) \bowtie \sigma(S, B) \rightarrow \sigma(R \bowtie S, A \wedge B)$

The next proposition we need for derivation of $(\kappa\bowtie\kappa)$. In general if both Φ and Ψ are solution sets, the expression $\Phi \cup \Psi$ might not be a solution set. One possible reason is that heads from Φ also occur as heads from Ψ. In this case the $\oplus$ operator can be used to merge the tails in Φ and the tails in Ψ together:

Proposition 24 *(introduction of the merge operator)* .... $[\kappa\bowtie\kappa*]$
$\kappa(R, \Phi) \bowtie \kappa(S, \Psi) \rightarrow \kappa(\sigma(R \bowtie S, \Phi \oplus \Psi), \Phi) \quad \textit{if} \quad hvar(\Phi) = hvar(\Psi)$

Proposition 25 *(generalization of* $(\bowtie\kappa)$*)* .... $[\kappa\bowtie\kappa]$
$\kappa(R, \Phi) \bowtie \kappa(S, \Psi) \rightarrow$
$\kappa(\sigma(R \bowtie S,$
$\quad \Phi \oplus \Psi \wedge \Phi[attr(S)] \wedge \Psi\langle hvar(\Phi)\rangle[attr(R)]),$
$\quad \Phi\langle attr(S)\rangle \cup \Psi\langle hvar(\Phi)\rangle\langle attr(R)\rangle)$

Proof: Informally we first explain the construction. In the righthand side of $(\kappa\bowtie\kappa)$ the merge operator $\oplus$ handles the case that there is a solution $x = t \in \Phi$ and a solution $y = s \in \Psi$ with $x = y$, in analogy to rule $(\kappa\bowtie\kappa*)$. Also another case needs to be checked. Suppose there is a solution $x = t \in \Phi$ such that $x \in attr(S)$. If this solution were put in the instruction of the calculate operator, then the resulting expression would be unwellformed. The problem can be handled by recognizing that $x = t$ now satisfies the wellformedness conditions of the select operator, viz. $var(x = t) \subset attr(R) \cup attr(S)$. So the restriction operator inserts the solution $x = t$ in the condition of the select operator and the delete operator deletes it from the calculate instruction. The symmetric case that there is a solution $x = t \in \Psi$ such that $x \in attr(R)$, is handled in the same way.
It should be noted that in the construction the expression

$$\Psi\langle hvar(\Phi)\rangle[attr(R)]$$

in the select condition can be replaced by the more simple expression $\Psi[attr(R)]$ since the following holds:

$$\Phi \oplus \Psi \wedge \Phi[attr(S)] \wedge \Psi\langle hvar(\Phi)\rangle[attr(R)] = \Phi \oplus \Psi \wedge \Phi[attr(S)] \wedge \Psi[attr(R)] \quad \ldots\ldots [**]$$

However this could lead to duplicate use of solutions from Ψ in the select condition and since the rule is to be used for query optimization we want to avoid this duplication.

For a formal derivation of $(\kappa\bowtie\kappa)$ we first observe:

$$\Phi \oplus \Psi = \Phi[hvar(\Psi)] \oplus \Psi[hvar(\Phi)] \quad \ldots\ldots\ldots\ldots [*]$$

Now put:

$$\begin{array}{ll} \Phi_1 = \Phi[attr(S)] & \Psi_1 = \Psi[attr(R)] \\ \Phi_2 = \Phi[hvar(\Psi)] & \Psi_2 = \Psi[hvar(\Phi)] \\ \Phi_3 = \Phi\langle attr(S) \cup hvar(\Psi)\rangle & \Psi_3 = \Psi\langle attr(R) \cup hvar(\Phi)\rangle \end{array}$$

Both Φ and Ψ can be obtained as mutually disjoint unions of the above solution sets:

$$\Phi = \Phi_1 \cup \Phi_2 \cup \Phi_3, \Psi = \Psi_1 \cup \Psi_2 \cup \Psi_3$$

We have:

$\kappa(R,\Phi) \bowtie \kappa(S,\Psi) \rightarrow \kappa(R,\Phi_1 \cup \Phi_2 \cup \Phi_3) \bowtie \kappa(S,\Psi_1 \cup \Psi_2 \cup \Psi_3)$

$\rightarrow \kappa(\kappa(\kappa(R,\Phi_1),\Phi_2),\Phi_3) \bowtie \kappa(\kappa(\kappa(S,\Psi_1),\Psi_2),\Psi_3)$ .... $(\kappa\kappa*)$

$\rightarrow \kappa(\kappa(\kappa(\kappa(R,\Phi_1),\Phi_2) \bowtie \kappa(\kappa(S,\Psi_1),\Psi_2),\Psi_3),\Phi_3)$ .... $(\bowtie\kappa*,\bowtie\kappa*)$

$\rightarrow \kappa(\kappa(\kappa(R,\Phi_1),\Phi_2) \bowtie \kappa(\kappa(S,\Psi_1),\Psi_2),\Phi_3 \cup \Psi_3)$ .... $(\kappa\kappa*)$

$\rightarrow \kappa(\kappa(\sigma(\kappa(R,\Phi_1) \bowtie \kappa(S,\Psi_1),\Phi_2 \oplus \Psi_2),\Phi_2),\Phi_3 \cup \Psi_3)$ .... $(\kappa\bowtie\kappa*)$

$\rightarrow \kappa(\kappa(\sigma(\kappa(R,\Phi_1) \bowtie \kappa(S,\Psi_1),\Phi \oplus \Psi),\Phi_2),\Phi_3 \cup \Psi_3)$ .... $(*)$

$\rightarrow \kappa(\kappa(\sigma(\sigma(\sigma(R \bowtie S,\Psi_1),\Phi_1),\Phi \oplus \Psi),\Phi_2),\Phi_3 \cup \Psi_3)$ .... $(\bowtie\kappa+,\bowtie\kappa+)$

$\rightarrow \kappa(\sigma(R \bowtie S,\Phi_1 \wedge \Psi_1 \wedge \Phi \oplus \Psi),\Phi_2 \cup \Phi_3 \cup \Psi_3)$ .... $(\sigma\sigma,\sigma\sigma,\kappa\kappa*)$

The last expression yields $(\kappa\bowtie\kappa)$ by backsubstitution of Φ_1, Φ_2, Φ_3, Ψ_1 and Ψ_3 and application of $(**)$. ■

5 The language PCSJL

In this section we define a sublanguage **PCSJL** of **UPCSJL** together with a normal form for **PCSJL**. This normal form will be used in the next section for the construction of the **UPCSJL** normal form.

Definition 26 *The language* **PCSJL** *consists of expressions constructed from the functions:* $\bowtie,\pi,\kappa,\sigma$

Definition 27 *A* **PCSJL** *normal form is an expression of the form* $\pi(\kappa(\sigma(R_1 \bowtie \ldots \bowtie R_n, A),\Phi),X)$ *where* $R_1,\ldots,R_n$ *are base relations.*

Another feasible normal form (i.e. format to which all expressions are reducible) exchanges the positions of selection and calculation: $\pi(\sigma(\kappa(R_1 \bowtie \ldots \bowtie R_n,\Phi),A),X)$. However the disadvantage of this normal form is that unnecessary computations are performed for tuples for which the select condition evaluates to false. Derivation of our normal form below is achieved by applying four unconditional rules of the form $R \rightarrow S$ for each of the primitive four relational operators.

Proposition 28 *The expression* $\pi(\kappa(\sigma(R,A),\Phi),X)$ *is wellformed iff:*

1. $var(A) \subset attr(R)$
2. $tvar(\Phi) \subset attr(R)$
3. $hvar(\Phi) \cap attr(R) = \emptyset$
4. $X \subset attr(R) \cup hvar(\Phi)$

Proposition 29 *(projection rule derived from* $(\pi\pi)$*)* .... $[\pi\pi\kappa\sigma]$

$\pi(\pi(\kappa(\sigma(R,A),\Phi),X),Y) \rightarrow \pi(\kappa(\sigma(R,A),\Phi),Y)$

Proposition 30 *(selection rule derived from* $(\sigma\pi)$, $(\sigma\kappa)$ *and* $(\sigma\sigma)$*)* .... $[\sigma\pi\kappa\sigma]$

$\sigma(\pi(\kappa(\sigma(R,A),\Phi),X),B) \rightarrow \pi(\kappa(\sigma(R,A \wedge \Phi(B)),\Phi),X)$

Proposition 31 *(calculation rule derived from* $(\kappa\pi)$ *and* $(\kappa\kappa)$*)* .... $[\kappa\pi\kappa\sigma]$

$\kappa(\pi(\kappa(\sigma(R,A),\Phi),X),\Psi) \rightarrow \pi(\kappa(\sigma(R^X,A^X),\Phi^X \cup \Phi^X((\Psi))),hvar(\Psi) \cup X)$

Proposition 32 *(join rule derived from* $(\pi\bowtie\pi)$, $(\kappa\bowtie\kappa)$ *and* $(\sigma\bowtie\sigma)$*)* .... $[\pi\kappa\sigma\bowtie\pi\kappa\sigma]$

$\pi(\kappa(\sigma(R,A),\Phi),X) \bowtie \pi(\kappa(\sigma(S,B),\Psi),Y) \rightarrow$

$$\pi(\kappa(\sigma(R^X \bowtie S^Y,$$
$$A^X \wedge B^Y \wedge \Phi^X \oplus \Psi^Y \wedge \Phi^X[attr(S^Y)] \wedge \Psi^Y\langle hvar(\Phi^X)\rangle[attr(R^X)]),$$
$$\Phi^X\langle attr(S^Y)\rangle \cup \Psi^Y\langle hvar(\Phi^X)\rangle\langle attr(R^X)\rangle\ ),$$
$$X \cup Y)$$

Proposition 33 *Every wellformed relational expression in* **PCSJL** *can be transformed into an equivalent wellformed* **PCSJL** *normal form.*

Proof: Basis: $R \rightarrow \pi(\kappa(\sigma(R, \mathbf{true}), \emptyset), attr(R))$.
Induction: rules $(\pi\pi\kappa\sigma)$, $(\sigma\pi\kappa\sigma)$, $(\kappa\pi\kappa\sigma)$ and $(\pi\kappa\sigma\bowtie\pi\kappa\sigma)$.
■

Note that the normal form is not unique. Especially the renaming of clashing variables is a rich source of equivalent expressions. We conclude this section with some examples of normalization.

Example 5
$\kappa(\sigma(\pi(\kappa(\sigma(r(v,w), v > w), \{x = v + w\}), \{w, x\}), x > 0), \{v = x + 2\})$
$\rightarrow \pi(\kappa(\sigma(r(u_1, w), u_1 + w > 0 \wedge u_1 > w), \{v = u_1 + w + 2\}), \{v, w, x\})$

$\pi(\kappa(\sigma(r(w,y,v), w > y), \{x = w + y\}), \{w, x\})$
$\quad \bowtie \pi(\kappa(\sigma(s(x,z,v), x > z), \{y = x + z\}), \{x, y, z\})$
$\rightarrow \pi(\kappa(\sigma(r(w, u_1, u_2) \bowtie s(x, z, u_3), w > u_1 \wedge x > z \wedge x = w + u_1), \{y = x + z\}), \{w, x, y, z\})$

6 The language UPCSJL

The next step is the addition of the union operator, yielding the language **UPCSJL**. The normal form procedure presented in this section heavily relies on the normal form construction for **PCSJL** expressions as discussed before. By the availability of this construction, derivation of a **UPCSJL** normal form is more or less straightforward.

Definition 34 *If* $R_1, \ldots, R_n$ *are* **PCSJL** *normal forms then the following is a* **UPCSJL** *normal form:*

$$R_1 \cup \ldots \cup R_n$$

Proposition 35 *Every wellformed relational expression in* **UPCSJL** *can be transformed into an equivalent wellformed* **UPCSJL** *normal form.*

Proof: It suffices to distinguish the following six cases:
1. Basis: $R \rightarrow \pi(\kappa(\sigma(R, \mathbf{true}), \emptyset), attr(R))$.
Induction: in the remaining cases, using the construction of Proposition 33, it is assumed that R_i and S_i are reduced to **PCSJL** normal form. The expressions T_i are the resulting **PCSJL** normal forms:
2. $\pi(R_1 \cup \ldots \cup R_n, X) \rightarrow \pi(R_1, X) \cup \ldots \cup \pi(R_n, X) \rightarrow T_1 \cup \ldots \cup T_n$ ..... $(\pi\cup, \pi\pi\kappa\sigma)$
3. $\sigma(R_1 \cup \ldots \cup R_n, X) \rightarrow \sigma(R_1, X) \cup \ldots \cup \sigma(R_n, X) \rightarrow T_1 \cup \ldots \cup T_n$ ..... $(\sigma\cup, \sigma\pi\kappa\sigma)$
4. $\kappa(R_1 \cup \ldots \cup R_n, X) \rightarrow \kappa(R_1, X) \cup \ldots \cup \kappa(R_n, X) \rightarrow T_1 \cup \ldots \cup T_n$ ......$(\kappa\cup, \kappa\pi\kappa\sigma)$
5. $(R_1 \cup \ldots \cup R_n) \bowtie (S_1 \cup \ldots \cup S_m) \rightarrow R_1 \bowtie S_1 \cup \ldots \cup R_n \bowtie S_m$
$\rightarrow T_1 \cup \ldots \cup T_{n*m}$ .... $(\bowtie\cup, \pi\kappa\sigma\bowtie\pi\kappa\sigma)$
6. $(R_1 \cup \ldots \cup R_n) \cup (S_1 \cup \ldots \cup S_m) \rightarrow R_1 \cup \ldots \cup R_n \cup S_1 \cup \ldots \cup S_m$ ....(ASS $\circ\cup$)
Note that the rules $(\pi\cup)$, $(\sigma\cup)$, $(\kappa\cup)$, $(\bowtie\cup)$ and (ASS $\circ\cup$) were given in Section 4.2. ■

7 Query optimization of UPCSJL normal forms

Optimization of **PSJL** expressions (i.e. expressions involving projection, selection and join) is rather well understood (see the literature on query transformations, e.g. [YAN87], [ULL89]). The heuristic of performing selections and projections before joins is effective because the number and size of tuples to be joined can be reduced. If also calculations are applied before joins duplicate computations are avoided. For instance in the next example the variable v is calculated twice instead of four times if the indicated rewriting takes place:

Example 6 $r(x,y) := \{\{x = 100, y = 3\}, \{x = 200, y = 3\}\}$
$s(y,z) := \{\{y = 3, z = 10\}, \{y = 3, z = 20\}\}$
$\kappa(r(x,y) \bowtie s(y,z), v = x + 1) \rightarrow \kappa(r(x,y), v = x + 1) \bowtie s(y,z)$

A procedure for direct optimization of relational expressions, excluding the calculate operator, consists of the following steps (adapted from [ULL89]):

Algorithm 36 *(direct optimization)*

1. *Apply* $(\sigma\sigma)$ *to separate all compound select conditions:*

$$\sigma(R, A_1 \wedge \ldots \wedge A_n) \rightarrow \sigma(\ldots\sigma(R, A_1)\ldots, A_n)$$

2. *Apply* $(\sigma\sigma*)$, $(\sigma\pi)$, $(\sigma\bowtie)$ *and* $(\sigma\cup)$ *to move selection down.*

3. *Apply* $(\pi\pi)$, $(\pi\bowtie+)$, $(\pi\cup)$, $(\pi\sigma)$ *and* $(\pi attr)$ *to move projection down. Rules* $(\pi\pi)$ *and* $(\pi attr)$ *cause some projections to disappear, while rule* $(\pi\sigma)$ *splits a projection into two projections, one of which can be migrated downwards if possible.*

4. *Apply* $(\sigma\sigma)$, $(\pi\pi)$ *and* $(\sigma\pi)$ *to combine cascades of selections and projections into a single selection, a single projection or a single projection followed by a single projection.*

A disadvantage of the above algorithm is that the optimization rules are applied on the entire relational expression. An appealing prospect would therefore be first to reduce the number of relational operators, by bringing the expression in **UPCSJL** normal form. Subsequently it should be possible to apply standard optimization techniques for **PSJL** expressions on the normal form. Our aim is to give an optimization algorithm for **UPCSJL** expressions using the heuristic rules of Algorithm 36.

A necessary rewrite rule to accomplish this goal transforms a **PCSJL** normal form to an expression that contains a 'maximal' **PSJL** normal form:

Proposition 37 *(derived from* $(\pi\kappa*)$ *and* $(\pi\kappa)$*)* .... $[\pi\kappa\sigma]$
$\pi(\kappa(\sigma(R,A),\Phi),X) \rightarrow \pi(\kappa(\pi(\sigma(R,A), tvar(\Phi) \cup (attr(R) \cap X)), \Phi[X]), X)$

In $(\pi\kappa\sigma)$ a projection operator is inserted between calculation and selection such that: 1) the computation Φ can still be performed, 2) attributes of R that are in X remain available for the outermost projection.

After application of $(\pi\kappa\sigma)$ we need to optimize the newly created **PSJL** normal form. Algorithm 36 contains the rules $(\pi\sigma)$ and $(\pi\bowtie)$ to move projection down in the relational expression. To avoid that projection is first pushed down over selection and subsequently distributed over the join operator we combine these two steps into a single rule:

Proposition 38 *(derived from ($\pi\sigma$) and ($\pi\bowtie$))* .... [$\pi\sigma\bowtie$]
$\pi(\sigma(R \bowtie S, A), X) \rightarrow \pi(\sigma(R \bowtie \pi(S, attr(S) \cap (attr(R) \cup var(A) \cup X)), A), X)$

In the above rule ($\pi\sigma\bowtie$) the expression S is projected on the union of: 1) $attr(R)$ to make sure that the joined attributes remain after projection, 2) $var(A)$ so that the condition A can be evaluated, 3) X to enforce that necessary attributes of S that are not in A remain after projection.

Once the **PSJL** expression has been processed also calculation should take part in the optimization. Calculation can be pushed down over projection, selection and join with a single rule:

Proposition 39 *(derived from ($\kappa\pi$), ($\kappa\sigma$) and ($\kappa\bowtie$))* .... [$\kappa\pi\sigma\bowtie$]
$\kappa(\pi(\sigma(R \bowtie S, A), X), \Phi) \rightarrow \pi(\sigma(R^X \bowtie \kappa(S^X, \Phi), A^X), hvar(\Phi) \cup X)$
if $tvar(\Phi) \subset attr(S)$

Efficiency can be gained by generalizing ($\sigma\bowtie$), ($\pi\sigma\bowtie$) and ($\kappa\pi\sigma\bowtie$) to an arbitrary number of joins, so that processing of nested joins is avoided:

Proposition 40 *(generalization of $\sigma\bowtie$)* .... [$\sigma\bowtie\bowtie$]
$\sigma(R_1 \bowtie \ldots \bowtie R_i \bowtie \ldots \bowtie R_n, A_1 \wedge \ldots \wedge A_j \wedge \ldots \wedge A_p) \rightarrow$
$\sigma(R_1 \bowtie \ldots \bowtie \sigma(R_i, A_j) \bowtie \ldots \bowtie R_n, A_1 \wedge \ldots \wedge A_{j-1} \wedge A_{j+1} \wedge \ldots \wedge A_p)$
if $var(A_j) \subset attr(R_i)$

Proposition 41 *(generalization of $\pi\sigma\bowtie$)* .... [$\pi\sigma\bowtie\bowtie$]
$\pi(\sigma(R_1 \bowtie \ldots \bowtie R_i \bowtie \ldots \bowtie R_n, A), X) \rightarrow$
$RREWR\pi(\sigma(R_1 \bowtie \ldots \bowtie$
$\pi(R_i, attr(R_i) \cap (attr(R_1 \bowtie \ldots \bowtie R_{i-1} \bowtie R_{i+1} \bowtie \ldots \bowtie R_n) \cup var(A) \cup X))$
$\bowtie \ldots \bowtie R_n, A), X)$

Proposition 42 *(generalization of $\kappa\pi\sigma\bowtie$)* .... [$\kappa\pi\sigma\bowtie\bowtie$]
$\kappa(\pi(\sigma(R_1 \bowtie \ldots \bowtie R_i \bowtie \ldots \bowtie R_n, A), X), \Phi_1 \cup \ldots \cup \Phi_j \cup \ldots \cup \Phi_q) \rightarrow$
$\kappa(\pi(\sigma(R_1^X \bowtie \ldots \bowtie \kappa(R_i^X, \Phi_j) \bowtie \ldots \bowtie R_n^X, A^X), hvar(\Phi_j) \cup X),$
$\Phi_1 \cup \ldots \cup \Phi_{j-1} \cup \Phi_{j+1} \cup \ldots \cup \Phi_q)$ *if* $tvar(\Phi_j) \subset attr(R_i)$

Now we combine the above generalized rules into an optimization algorithm for **UPCSJL** (i.e. **PCSJL** expressions extended with union) that uses normalization:

Algorithm 43 *(normalization before optimization)*

1. *Bring the expression into* **UPCSJL** *normal form:* $S_1 \cup \ldots \cup S_m$
2. *Apply the following steps (a),(b),(c) on each* S_j *for* $j = 1, \ldots m$. S_j *has the form* $\pi(\kappa(\sigma(R_1 \bowtie \ldots \bowtie R_n, A), \Phi), X)$:
   (a) *Apply* ($\pi\kappa\sigma$) *yielding:* $\pi(\kappa(\pi(\sigma(R_1 \bowtie \ldots \bowtie R_n, A), Y), \Phi), X)$
   (b) *Optimize the subexpression* $\pi(\sigma(R_1 \bowtie \ldots \bowtie R_n, A), Y)$:
      i. *Apply* ($\sigma\sigma$) *to separate select conditions yielding:*

$$\pi(\sigma(R_1 \bowtie \ldots \bowtie R_n, A_1 \wedge \ldots \wedge A_p), Y)$$

ii. Apply $(\sigma\bowtie\bowtie)$ *at most p times to move selection down.*

iii. Apply $(\pi\sigma\bowtie\bowtie)$ *exactly n times to move projection down.*

iv. Apply $(\sigma\sigma)$ *to combine cascades of selections and* $(\pi attr)$ *to eliminate superfluous projections.*

(c) Optimize the created subexpression $\kappa(\pi(\sigma(R'_1 \bowtie \ldots \bowtie R'_n, A'), Y), \Phi)$*:*

i. Apply $(\kappa\kappa*)$ *to separate calculate computations yielding:*

$$\kappa(\pi(\sigma(R'_1 \bowtie \ldots \bowtie R'_n, A'), Y), \Phi_1 \cup \ldots \cup \Phi_q)$$

ii. Apply $(\kappa\pi\sigma\bowtie\bowtie)$ *at most q times to move calculation down.*

iii. Apply $(\kappa\kappa*)$ *to combine cascades of calculations.*

8 Conclusions

In this paper we have defined **UPCSJL**, a language with expressions built up using the operations union, projection, selection, join and calculate. We have shown that a Normal Form Theorem for this language exists, using *unconditional* rules to reduce arbitrary relational expressions into normal form. In addition we showed that after normalization *conditional* optimization rules can benefit from the fixed structure of the normal form.

References

[DEN90] van Denneheuvel, S., Kwast K. & Renardel de Lavalette G. R., *A normal form for PCSJ expressions,* Computing Science in the Netherlands, Nov 1990, 109-120, (1990).

[LAR85] Larson, P.A., Yang, H.Z., *Computing Queries from Derived Relations,* Proc. of the 11th Intl. Conf. on VLDB, 259-269, (1985).

[ULL89] Ullman, J.D., *Principles of Data and Knowledge - Base Systems,* Volume II: The New Technologies, Computer Science Press, (1989).

[VEMD86a] van Emde Boas, P., *RL, a Language for Enhanced Rule Bases Database Processing,* Working Document, Rep IBM Research, RJ 4869 (51299), (1986).

[VEMD86b] van Emde Boas, P., *A semantical model for the integration and modularization of rules,* Proceedings MFCS **12**, Bratislava, LNCS **233**, 78-92, (1986).

[VEMD86c] van Emde Boas, H. & van Emde Boas, P., *Storing and Evaluating Horn-Clause Rules in a Relational Database,* IBM J. Res. Develop. **30** (1), 80-92, (1986).

[YAN87] Yang, H. Z., Larson, P. A., *Query Transformations for PSJ-queries,* Proc. of the 13th Int. Conf. on VLDB, Brighton, 245-254, (1987).

Boolean Algebra Admits No Convergent Term Rewriting System

Rolf Socher-Ambrosius
Fachbereich Informatik, Universität Kaiserslautern
Postfach 3049, D-6750 Kaiserslautern, W.- Germany

Abstract: Although there exists a normal form for the theory of Boolean Algebra w.r.t. associativity and commutativity, the so called set of prime implicants, there does not exist a convergent equational term rewriting system for the theory of boolean algebra modulo AC. The result seems well-known, but no formal proof exists as yet. In this paper a formal proof of this fact is given.

Keywords: Boolean Algebra, Term Rewriting, Automated Theorem Proving.

1 Introduction

The existence (and computability) of a normal form is a very pleasant property of mathematical structures. It guarantees the decidability of the word problem and it also overcomes the difficulty with choosing the simplicity criterion for simplification procedures. The existence of a normal form usually comes along with the existence of a convergent (i.e. terminating and confluent) term rewriting system. Knuth & Bendix (1970) introduced a method to construct convergent term rewriting systems from given equational theories. Their approach, however, fails, for instance, for commutativity axioms, since equations like $fxy=fyx$ cannot be employed as reductions without violating the finite terminating condition. Equational term rewriting systems have been introduced by Lankford & Ballantyne (1977) and Peterson & Stickel (1981), in order to overcome the difficulties with such equations. Equational term rewriting systems are composed of a rewriting system R and an equational system E, which contains those axioms that cannot be used as rules. Equational rewriting can be seen as rewriting on the equivalence classes of terms modulo an equational system. Such equational rewriting systems modulo associativity and commutativity exist, for instance, for abelian groups or boolean rings. It is not, however, for the theory of boolean algebra (BA). Since the 1950s it is known that boolean algebra admits a normal form, which is called the *set of prime implicants*. However, no equational term rewriting system that rewrites a given BA-term into this normal form has ben found as yet. The set of prime implicants can only be produced on an algorithmic way. It was Quine (1952) and (1959), who first developed such an algorithm, and others (Slagle, Chang & Lee 1970, Tison

1969) followed. The non-existence of a convergent system for boolean algebra is well-known. It seems, however, that there does not exist any formal proof of this fact as yet. Sometimes it is argued that the minimal set of prime implicants is not unique for boolean algebra terms. This argument only provides some intuition, it lacks, however, the formal proof that clausal form could be the only possible normal form for boolean algebra terms. It is, for instance, not obvious that the boolean algebra equation

$$\neg(x \vee y) = \neg x \wedge \neg y$$

should be directed from left to right, which is required for the clausal form transformation.

Hullot (1980) and Peterson & Stickel (1981) report attempts to find a convergent system for BA using the Knuth-Bendix (1970) completion procedure, which failed to terminate in all experiments. Why then does there exist an algorithm for transforming BA-terms into normal form, but not a term rewriting system? The deeper reason seems to be the essential role that resolution plays in the algorithm. A resolution rewrite rule had to look like

$$(x \vee y) \wedge (\neg x \vee z) \rightarrow (x \vee y) \wedge (\neg x \vee z) \wedge (y \vee z)$$

Such a rule, however, obviously violates the condition of being noetherian.
In the following we will give a formal proof that a convergent term rewriting system for boolean algebra cannot exist.

2 Boolean Algebra and Term Rewriting Systems

In the following we assume a term set $\mathcal{T} = \mathcal{T}(\mathcal{F},\mathcal{V})$ over a signature $\mathcal{F}$ and a set $\mathcal{V}$ of variables. For any object o, let $\mathcal{V}(o)$ denote the set of all variables occurring in o.

2.1 Definition (Equational System):

An **equational** system E is a set of termpairs s=t. This system generates an equality relation $=_E$ in the following way: We define a relation $=_{1,E}$ by $s =_{1,E} t$, iff there exists an occurrence u in s, an equation s'=t' or t'=s' in E, and a substitution σ, such that $s/u = s'\sigma$ and $t = s[u \leftarrow t'\sigma]$. The relation $=_E$ is defined as the transitive, reflexive closure of $=_{1,E}$. It is clear that $=_E$ is an equivalence relation. The equivalence class of t modulo E will be denoted as $[t]_E$.

2.2 Definition (Equational Term Rewriting System):

A **term rewriting system** R (over $\mathcal{T}$) is a set of termpairs $l \rightarrow r$ (the so called **rules**), such that $\mathcal{V}(r) \subseteq \mathcal{V}(l)$ (and $l,r \in \mathcal{T}$). A term t_1 **R-reduces** to a term t_2, written $t_1 \Rightarrow_R t_2$, iff there exists an occurrence u in t_1, a rule $l \rightarrow r$ in R, and a substitution σ, such that $t_1/u = l\sigma$ and $t_2 = t_1[u \leftarrow r\sigma]$.
A term t_1 **E,R-reduces** to t_2, written $t_1 \Rightarrow_{R/E} t_2$, iff there exist $t'_1 \in [t_1]$, $t'_2 \in [t_2]$ with $t'_1 \Rightarrow_R t'_2$.
$\Rightarrow_{R/E}^+$ denotes the transitive, $\Rightarrow_{R/E}^*$ denotes the reflexive transitive closure of $\Rightarrow_{R/E}$ and $=_{R/E}$ denotes the reflexive, symmetric, and transitive closure of $\Rightarrow_{R/E}$.

The pair (E,R) is called an equational term rewriting system (ETRS). It can be understood also as a rewriting system for $\mathcal{T}/{=_E} = \{[t] \mid t \in \mathcal{T}\}$.
(E,R) is **noetherian,** iff there is no infinite sequence of E,R-reductions from any term.
(E,R) is **confluent,** iff $t \Rightarrow_{R/E}^* t_1$ and $t \Rightarrow_{R/E}^* t_2$ implies the existence of a term t_3 with $t_1 \Rightarrow_{R/E}^* t_3$ and $t_2 \Rightarrow_{R/E}^* t_3$.
A noetherian and confluent system is called **convergent.**
A term t_1 is called **(E,R-)irreducible,** iff there is no term t_2 with $t_1 \Rightarrow_{R/E} t_2$, and (E,R-)reducible otherwise.
An irreducible term t is called a **normal form** for t_1, iff $t_1 \Rightarrow_{R/E}^* t$.
If (E,R) is convergent, then each term t has a normal form $t\!\downarrow$, and $s\!\downarrow =_E t\!\downarrow$ holds for each term s with $s =_{R/E} t$.

2.3 Definition:

Let $\mathfrak{R}$ be a convergent ETRS over $\mathcal{T}$. Then the noetherian partial ordering $>_{\mathfrak{R}}$ on $\mathcal{T}$ generated by $\mathfrak{R}$ is defined by $s >_{\mathfrak{R}} t$ iff $s \Rightarrow_{\mathfrak{R}}^+ t$. In the following we shall usually drop the index $\mathfrak{R}$.

2.4 Lemma:

Let $\mathfrak{R}$ be a convergent system on $\mathcal{T}$ and let E be the congruence generated by $\mathfrak{R}$.

a) The ordering > generated by $\mathfrak{R}$ is compatible with substitutions, that is, $s>t$ implies $s\sigma > t\sigma$ for any $s,t \in \mathcal{T}$ and any substitution σ.
b) Let $s,t \in \mathcal{T}$. If $s =_E t$ and t is $\mathfrak{R}$-irreducible, then $s=t$ or $s>t$ holds.

Proof: Obvious.

In the following let AC be the equational system
$AC = \{x \vee y = y \vee x,\ x \wedge y = y \wedge x,\ x \vee (y \vee z) = (x \vee y) \vee z,\ x \wedge (y \wedge z) = (x \wedge y) \wedge z\}$.
and ACD the system $AC \cup \{x \vee (y \wedge z) = (x \vee y) \wedge (x \vee z),\ x \wedge (y \vee z) = (x \wedge y) \vee (x \wedge z)\}$.

2.5 Definition (Boolean Algebra):

A **boolean algebra** is an algebra $(B, \wedge, \vee, \neg, 0, 1)$ with the binary operators $\wedge, \vee$ and the unary operator $\neg$, which satisfies:

a) $(B, \wedge, \vee)$ is a distributive lattice, that is for all $a,b \in B$:

$a \vee b = b \vee a$	$a \wedge b = b \wedge a$
$a \vee (b \vee c) = (a \vee b) \vee c$	$a \wedge (b \wedge c) = (a \wedge b) \wedge c$
$(a \vee b) \wedge b = b$	$(a \wedge b) \vee b = b$
$a \wedge (b \vee c) = (a \wedge b) \vee (a \wedge c)$	$a \vee (b \wedge c) = (a \vee b) \wedge (a \vee c)$
$0 \vee b = b$	$1 \wedge b = b$

b) $(a \wedge \neg a) = 0$ $\qquad$ $(a \vee \neg a) = 1$

The axioms of boolean algebra imply the following well-known properties of the operators $\vee, \wedge$, and $\neg$:

2.6 Lemma:

Let $(B,\wedge,\vee,\neg)$ be a boolean algebra. Then for all $a,b\in B$:

$1\vee a = 1$ $\qquad$ $0\wedge a = 0$

$a\vee a = a$ $\qquad$ $a\wedge a = a$

$\neg(a\vee b) = \neg a\wedge\neg b$ $\qquad$ $\neg(a\wedge b) = \neg a\vee\neg b$

$\neg\neg a = a$

2.7 Lemma

Let $(B,\wedge,\vee,\neg,0,1)$ be a boolean algebra, and let $x_1,\ldots,x_n\in B$ with $x_1\wedge\ldots\wedge x_n = 1$. Then $x_i=1$ holds for all $i\in\{1,\ldots,n\}$.

Proof: Let $x_1\wedge\ldots\wedge x_n = 1$. Then $x_i = (x_1\wedge\ldots\wedge x_n)\vee x_i = 1\vee x_i = 1$ holds for each $i\in\{1,\ldots,n\}$.

In the following we shall consider exclusively the term set $\mathcal{T}=\mathcal{T}(F_B,\mathcal{V})$, where F_B is the signature $(\wedge,\vee,\neg,0,1)$ of boolean algebra.

In the following equality will tacitly be understood to be equality modulo AC. Equality modulo BA will be denoted by $\equiv$, and terms which are equal under BA, will also be called *equivalent*. We will use the customary notion of *literals, clauses* and a *conjunctive normal form* (CNF). A term t is called a literal, iff it is either of the form a, or of the form $\neg a$, with a being a constant or a variable. The term t is a *clause*, if $t = s_1\vee\ldots\vee s_n$, with pairwise distinct literals s_i. A term t is called a *CNF-term*, if $t = s_1\wedge\ldots\wedge s_n$, where the s_i are pairwise distinct clauses. A term with topsymbol $\vee$ is also called a *disjunction*, a term with topsymbol $\wedge$ a *conjunction*, and a term with topsymbol $\neg$ a *negation*.

2.8 Lemma:

There is no convergent system (ACD,R) such that $=_{ACD,R}$ coincides with $\equiv$.

Proof: Let $\mathfrak{R}$=(ACD,R) be a convergent system with $=_{\mathfrak{R}}\ =\ \equiv$, and let $>$ be the partial order generated by $\mathfrak{R}$. First we remark that from $x\wedge x\equiv x$, and $x\vee(x\wedge y)\equiv x$ follows $x\wedge x > x$, and $x\vee(x\wedge y) > x$ for any $x,y\in\mathcal{V}$, since the term x is irreducible.
Consider the term $t = x\vee(y\wedge z)$. We have

$$t =_{ACD} (x\vee y)\wedge(x\vee z) =_{ACD} (x\wedge x)\vee(x\wedge z)\vee(y\wedge x)\vee(y\wedge z) > x\vee(y\wedge z) = t,$$

which is a contradiction.

2.9 Theorem:

There exists no convergent ETRS (AC,R) such that $=_{AC,R}$ coincides with $\equiv$.

Note that we deal exclusively with term rewriting systems over the fixed signature F_B. There exists, for instance, a convergent system over the extended signature $(\wedge,\vee,\neg,+,*,0,1)$, see Hsiang (1985).

In order to prove the theorem above, we first provide some lemmata. For the remainder of the paper, we shall assume that there exists a convergent system $\mathfrak{R} = (AC,R)$ for BA. Let $>$ be the noetherian ordering associated with $\mathfrak{R}$.

2.10 Lemma:

The following relations hold:

$(x \vee y) \wedge y > y$

$\neg x \vee x > 1$

$x \vee x > x$

$x \vee 0 > x$

$x \vee 1 > 1$

$\neg\neg x > x$

$(x \vee y) \wedge (\neg x \vee y) > y$

Proof: For each line, the two terms are equivalent according to definition 2.5 and lemma 2.6. Furthermore, each right hand side is obviously irreducible, hence the assertion follows from lemma 2.4.b.

The proof of our main theorem proceeds essentially by considering a particular term t, and proving that all terms $t' \cong t$ are reducible. The following lemmata will provide two important techniques to prove a term t reducible, which are used heavily in the sequel. The first states that the normal form of a symmetric term must be symmetric.

If t is a term and p,q are distinct variables or constants occurring in t, and $t(p,q) = t(q,p)$, then the term t is called *symmetric* in (p,q). t is called *semi-symmetric* in (p,q), iff $t(p,q) \cong t(q,p)$.

2.11 Lemma (Symmetry Lemma):

Let $x,y \in \mathcal{V}$ with $x \neq y$, and let $t=t(y,x)$ be irreducible. If t is semi-symmetric in (x,y), then t is even symmetric in (x,y).

Proof: Assume $t(x,y) \neq t(y,x)$. Then we have $t(x,y) > t(y,x)$, since the latter is irreducible. But then, according to 2.4.a also $t(x,y)\sigma > t(y,x)\sigma$ for $\sigma = \{x \rightarrow y; y \rightarrow x\}$, which implies $t(y,x) > t(x,y)$, a contradiction.

The symmetry lemma can also be stated as follows: If the term t is symmetric in (x,y), then $t\downarrow$ is also symmetric in (x,y).

The next "subterm lemma" shows that a term t is reducible, if a subterm of t can be replaced by a shorter term, without changing the original term's value.

2.12 Lemma (Subterm Lemma):

Let $t = s_1 \wedge \ldots \wedge s_n$, with $n \geq 1$, and let $\sigma = \{x \rightarrow t_0\}$ be a substitution with $x \in \mathcal{V}(t)$ and $x \notin \mathcal{V}(t_0)$. If $s_1\sigma \cong s_1$ does not hold, and $s_1\sigma \wedge s_2 \wedge \ldots \wedge s_n \cong t$, then t is reducible.

Proof: Assume that t is irreducible. Let $s_1' = (s_1\sigma)\downarrow$, and let $t' = s_1' \wedge s_2 \wedge \ldots \wedge s_n$. Then, since $s_1\sigma \cong s_1$ does not hold, and $t' \cong t$, we have $t' > t$. In particular, we have

$t'\sigma > t\sigma,$

which implies

$s_1'\sigma \wedge s_2\sigma \wedge \ldots \wedge s_n\sigma > s_1\sigma \wedge s_2\sigma \wedge \ldots \wedge s_n\sigma,$

and, since $s_1\sigma \geq s_1' = s_1'\sigma$, we have

$s_1'\sigma \wedge s_2\sigma \wedge \ldots \wedge s_n\sigma > s_1'\sigma \wedge s_2\sigma \wedge \ldots \wedge s_n\sigma,$

which is a contradiction.

It should be noted that the assertion of the subterm lemma also holds for a disjunction $t = s_1 \vee \ldots \vee s_n$.

2.13 Example:

Let $t = (x \vee y) \wedge \neg x$. We show that t is reducible. Let $\sigma = \{x \to 0\}$. First it is easy to see that $t \cong y \wedge \neg x$, and $y = y\sigma$, and $y\sigma \cong (x \vee y)$ does not hold. If t were irreducible, then we had

$y \wedge \neg x > (x \vee y) \wedge \neg x$

hence

$y \wedge \neg 0 = (y \wedge \neg x)\sigma > ((x \vee y) \wedge \neg x)\sigma = (0 \vee y) \wedge \neg 0 > y \wedge \neg 0$

which is a contradiction.

2.14 Lemma:

Let t be a term with $\mathcal{V}(t) = \{x_1, \ldots, x_n\}$. Then there is a unique CNF-term $t^{\tilde{}} = c^{\tilde{}}_1 \wedge \ldots \wedge c^{\tilde{}}_m$, where each $c^{\tilde{}}_i$ is a clause containing all x_j's, and $t^{\tilde{}} \cong t$. The term $t^{\tilde{}}$ is called the *standardized CNF* of t. Each $c^{\tilde{}}_i$ is called a *standard clause* of t. The notion of a *standardized DNF* is defined analogously.

Proof: See, for instance, Rudeanu (1974).

2.15 Example:

Let $t = (\neg x \vee y) \wedge (\neg x \vee \neg z)$. Then $t^{\tilde{}} = (\neg x \vee y \vee z) \wedge (\neg x \vee y \vee \neg z) \wedge (\neg x \vee \neg y \vee \neg z)$ is the standardized CNF of t.

2.16 Lemma:

If $t = t_1 \wedge \ldots \wedge t_n$, then for each $i \in \{1, \ldots, n\}$, there are standard clauses $c^{\tilde{}}_{i1}, \ldots, c^{\tilde{}}_{ik_i}$, with

$t_i \cong c^{\tilde{}}_{i1} \wedge \ldots \wedge c^{\tilde{}}_{ik_i}.$

Moreover,

$\bigcup_{i=1..n} \bigcup_{j=1..n_j} c^{\tilde{}}_{ij} = \{c^{\tilde{}}_1, \ldots, c^{\tilde{}}_n\}.$

2.17 Lemma:

Let $t = x \vee y$. Then either $t\downarrow = t$, or $t\downarrow = \neg(\neg x \wedge \neg y)$.

Proof: Obvious.

2.18 Lemma:

Let $t = (x \vee y) \wedge (y \vee z) \wedge (z \vee x)$. Then $t\downarrow \in \{t_1, \ldots, t_8\}$, where

$t_1 = (x\wedge y) \vee (y\wedge z) \vee (z\wedge x)$,
$t_2 = \neg(\neg y\vee\neg z) \vee \neg(\neg x\vee\neg z) \vee \neg(\neg y\vee\neg x)$,
$t_3 = (x\vee y) \wedge (y\vee z) \wedge (z\vee x)$,
$t_4 = \neg(\neg y\wedge\neg z) \wedge \neg(\neg x\wedge\neg z) \wedge \neg(\neg y\wedge\neg x)$
$t_5 = \neg[\neg(y\vee z) \vee \neg(x\vee z) \vee \neg(y\vee x)]$,
$t_6 = \neg[(\neg y\wedge\neg z) \vee (\neg x\wedge\neg z) \vee (\neg y\wedge\neg x)]$,
$t_7 = \neg[(\neg y\vee\neg z) \wedge (\neg x\vee\neg z) \wedge (\neg y\vee\neg x)]$,
$t_8 = \neg[\neg(y\wedge z) \wedge \neg(x\wedge z) \wedge \neg(y\wedge x)]$.

Proof:
a) Let $t\!\downarrow = s_1\vee\ldots\vee s_n$, and let $t^{\sim}$ be the standardized DNF of t. Then $t^{\sim} = d_1\vee d_2\vee d_3\vee d_4$, with

$d_1 = x\wedge y\wedge z$, $d_2 = \neg x\wedge y\wedge z$, $d_3 = x\wedge\neg y\wedge z$, $d_4 = x\wedge y\wedge\neg z$.

According to 2.16, each s_i is equivalent to a disjunction of d_j's. Moreover, $t\!\downarrow$ must be symmetric in (x,y), in (y,z), and in (x,z), and thus there are only the following cases: Either $t\!\downarrow = s_1\vee s_2$, with $s_1 \equiv d_1$, and $s_2 \equiv d_2\vee d_3\vee d_4$, or $t\!\downarrow = s_1\vee s_2\vee s_3$, with the following possibilities:

$s_1 \equiv d_1\vee d_2$, $s_2 \equiv d_1\vee d_3$, $s_3 \equiv d_1\vee d_4$,
$s_1 \equiv d_1\vee d_2\vee d_3$, $s_2 \equiv d_1\vee d_3\vee d_4$, $s_3 \equiv d_1\vee d_2\vee d_4$.

Let $t\!\downarrow = s_1\vee s_2$ with $s_1 \equiv d_1$, and $s_2 \equiv d_2\vee d_3\vee d_4$, and let $\sigma=\{z\to 1\}$. Then $s_1\sigma$ is not equivalent to s_1. We show that $s_1\sigma\vee s_2 \equiv s_1\vee s_2$: We have

$s_1\sigma\vee s_2 \equiv (x\wedge y\wedge z) \vee (x\wedge y\wedge\neg z) \vee d_2 \vee d_3 \equiv$
$(x\wedge y) \vee d_2 \vee d_3 \equiv (x\wedge y) \vee (x\wedge y\wedge\neg z) \vee d_2 \vee d_3 \equiv s_1\vee s_2$

Hence the subterm lemma implies that $s_1\vee s_2$ is reducible.
Let $t\!\downarrow = s_1\vee s_2\vee s_3$. If $s_1 \equiv d_1\vee d_2 \equiv y\wedge z$, $s_2 \equiv d_1\vee d_3 \equiv x\wedge z$, $s_3 \equiv d_1\vee d_4 \equiv y\wedge x$, then we have either $s_1 = y\wedge z$, $s_2 = x\wedge z$, $s_3 = x\wedge y$, and $t\!\downarrow= t_1$, or $s_1 = \neg(\neg y\vee\neg z)$, $s_2 = \neg(\neg x\vee\neg z)$, $s_3 = \neg(\neg y\vee\neg x)$, and $t\!\downarrow = t_2$.
If $s_1 \equiv d_1\vee d_2\vee d_3 \equiv (x\vee y)\wedge z$, $s_2 \equiv d_1\vee d_3\vee d_4 \equiv x\wedge(y\vee z)$, $s_3 \equiv d_1\vee d_2\vee d_4 \equiv y\wedge(x\vee z)$, then let $\tau=\{x\to 0\}$. It is easy to see that

$s_1\tau\vee s_2\vee s_3 \equiv s_1\vee s_2\vee s_3$,

and $s_1\tau$ is not equivalent to s_1. Hence the subterm lemma implies that $s_1\vee s_2 \vee s_3$ is reducible.
b) Let $t\!\downarrow = s_1\wedge\ldots\wedge s_n$. Analogously to a) it can be shown that $t\!\downarrow\in\{t_3,t_4\}$ in this case.
c) Let $t\!\downarrow = \neg t'$, with $t' = s_1\vee\ldots\vee s_n$. Then $t\!\downarrow \equiv \neg s_1\wedge\ldots\wedge\neg s_n$. Let $t^{\sim}$ be the standardized CNF of t. Then $t^{\sim} = c_1\wedge c_2\wedge c_3\wedge c_4$, with

$c_1 = x\vee y\vee z$, $z_2 = \neg x\vee y\vee z$, $z_3 = x\vee\neg y\vee z$, $z_4 = x\vee y\vee\neg z$.

Then each $\neg s_i$ is equivalent to a conjunction of c_j's, and analogously to part a) it can be shown that either $t\!\downarrow$ is reducible according to the subterm lemma, or $t\!\downarrow\in\{t_5,t_6\}$. The case where $t' = s_1\wedge\ldots\wedge s_n$ is treated analogously.

2.19 Lemma:

If the terms $x\vee(y\wedge z)$ and $x\wedge(y\vee z)$ are both irreducible, then $\mathfrak{R}$ is not convergent.

Proof: The assumption of the lemma implies $(x\vee y)\wedge(x\vee z) > x\vee(y\wedge z)$, $(x\wedge y)\vee(x\wedge y) > x\wedge(y\vee z)$, and, in particular, since both $y\wedge z$ and $y\vee z$ are irreducible, $\neg(\neg y\wedge\neg z) > y\vee z$, and

$\neg(\neg y \vee \neg z) > y \wedge z$. This proves all terms $t_1, \ldots, t_8$ of the previous lemma to be reducible, hence $\mathfrak{R}$ cannot be confluent.

Hence it will be assumed in the following that one of the terms $x \vee (y \wedge z)$ and $x \wedge (y \vee z)$ is reducible. It is sufficient to assume the term $x \vee (y \wedge z)$ to be reducible, the alternative case admitting an analogical proof. In particular, this assumption implies that each disjunct s_i of an irreducible term $t = s_1 \vee \ldots \vee s_n$ is either a negation or an atom.

2.20 Lemma:

Either the term $x \vee y$ or the term $x \wedge y$ is reducible.

Proof: We consider the term $t = (\neg x \vee y) \wedge (\neg y \vee x) \wedge (x \vee z)$. Since t is semi-symmetric in (x,y), but not symmetric, t must be reducible.

a) Let $t{\downarrow} = s_1 \wedge \ldots \wedge s_n$, where the s_i are not conjunctions.
If $n \geq 3$, let a be an arbitrary constant and let $\sigma = \{x \to a, y \to a, z \to \neg a\}$. We have $t > t{\downarrow}$, and in particular $t\sigma > t{\downarrow}\sigma$, where $t\sigma = (\neg a \vee a) \wedge (\neg a \vee a) \wedge (a \vee \neg a)$, and $t{\downarrow}\sigma = s_1\sigma \wedge \ldots \wedge s_n\sigma$. From $t\sigma \cong 1$ follows $t{\downarrow}\sigma \cong 1$, and hence $s_i\sigma \cong 1$, for each $i \in \{1, \ldots, n\}$. Hence $s_i\sigma > 1$, and, since $s_i\sigma$ is composed solely of the literals a and $\neg a$, the last step of this derivation must be of the form $a \vee \neg a \Rightarrow 1$. Thus we have the reduction $(\neg a \vee a) \wedge (\neg a \vee a) \wedge (a \vee \neg a) \Rightarrow^+_{\mathfrak{R}} (\neg a \vee a) \wedge \ldots \wedge (a \vee \neg a)$, where the second term has $n \geq 3$ conjuncts, which obviously contradicts the finite termination property of $\mathfrak{R}$.
Now let $n=2$, that is $t{\downarrow} = s_1 \wedge s_2$. Let $\tilde{t}$ be the standardized CNF of t. Then $\tilde{t} = c_1 \wedge \ldots \wedge c_5$, with

$c_1 = \neg x \vee y \vee z, c_2 = x \vee \neg y \vee z, c_3 = \neg x \vee y \vee \neg z, c_4 = x \vee \neg y \vee \neg z, c_5 = x \vee y \vee z.$
We distinguish two cases:

Case 1: s_1 is symmetric in (x,y). Then s_2 is also symmetric in (x,y), since $t{\downarrow}$ is. From lemma 2.16 follows that s_1 and s_2 are equivalent to conjunctions of the c_i. Taking into account the symmetry property, there remain the following possibilities:

$s_1 \cong c_1 \wedge c_2, s_2 \cong c_3 \wedge c_4 \wedge c_5,$
$s_1 \cong c_3 \wedge c_4,$ or $s_1 \cong c_3 \wedge c_4 \wedge c_5,$ and $s_2 \cong c_1 \wedge c_2 \wedge c_5,$
$s_1 \cong c_1 \wedge c_2 \wedge c_3 \wedge c_4, s_2 \cong c_5, s_2 \cong c_1 \wedge c_2 \wedge c_5,$ or $s_2 \cong c_3 \wedge c_4 \wedge c_5.$

In the first line, let $\sigma = \{z \to 0\}$. We have $s_1\sigma \wedge s_2 \cong t$, and s_1 is not equivalent to $s_1\sigma$. From the subterm lemma follows that $s_1 \wedge s_2$ is reducible.
In the second line, let $\tau = \{z \to 1\}$. We have $s_1\tau \wedge s_2 \cong t$, and s_1 is not equivalent to $s_1\sigma$. From the subterm lemma follows that $s_1 \wedge s_2$ is reducible.
In the third line, let $\varphi = \{x \to y\}$. We obtain in all three cases $s_1 \wedge s_2\varphi \cong t$, and s_2 is not equivalent to $s_2\varphi$, and from the subterm lemma follows that $s_1 \wedge s_2$ is reducible.

Case 2: s_1 is not symmetric in (x,y). Then $s_1 = s_2\{x \to y; y \to x\}$, and for each c_i occurring in s_1, $c_i\{x \to y; y \to x\}$ must occur in s_2. Hence both s_1 and s_2 must consist of at least 3 c_i's, and both contain c_5. We have the following possibilities:

$s_1 \cong c_1 \wedge c_3 \wedge c_5, s_2 \cong c_2 \wedge c_4 \wedge c_5,$
$s_1 \cong c_1 \wedge c_4 \wedge c_5, s_2 \cong c_2 \wedge c_3 \wedge c_5,$
$s_1 \cong c_1 \wedge c_2 \wedge c_3 \wedge c_5, s_2 \cong c_1 \wedge c_2 \wedge c_4 \wedge c_5,$
$s_1 \cong c_2 \wedge c_3 \wedge c_4 \wedge c_5, s_2 \cong c_1 \wedge c_3 \wedge c_4 \wedge c_5.$

In the first, third, and fourth line, let $\sigma=\{z\to 1\}$. In either case, we have $s_1\sigma\wedge s_2 \equiv t$, and s_1 is not equivalent to $s_1\sigma$, hence $s_1\wedge s_2$ must be reducible according to the subterm lemma. In the second line, we have $s_1 \equiv (y\vee z)\wedge(x\vee\neg y\vee\neg z)$, and $s_2 \equiv (x\vee z)\wedge(\neg x\vee y\vee\neg z)$. Let $\tau=\{z\to\neg x\}$. Since $s_1\tau\wedge s_2 \equiv t$, and s_1 is not equivalent to $s_1\tau$, $s_1\wedge s_2$ must be reducible according to the subterm lemma.

b) Let $t\downarrow = s_1\vee\ldots\vee s_n$. Let $t^{\sim}$ be the standardized DNF of t. Then $t^{\sim} = c_1\vee c_2\vee c_3$, with

$d_1 = \neg x\wedge\neg y\wedge z$, $d_2 = x\wedge y\wedge z$, $d_3 = x\wedge y\wedge\neg z$.

Obviously, $n\leq 3$, since otherwise one s_i, say s_n, would be redundant, that is $t\downarrow \equiv s_1\vee\ldots\vee s_{n-1}$, which obviously contradicts the irreducibility of $t\downarrow$. If n=3, then $t\downarrow = s_1\vee s_2\vee s_3$, with $s_i \equiv d_i$. But then $s_2\vee s_3 \equiv x\wedge y \equiv \neg(\neg x\vee\neg y)$, hence $s_2\vee s_3$ is reducible.
Thus we have $t\downarrow = s_1\vee s_2$, where both s_1 and s_2 are negations, with the following possibilities:

$s_1 \equiv d_1$, $s_1 \equiv d_1\vee d_3$, or $s_1 \equiv d_1\vee d_2$, and $s_2 \equiv d_2\vee d_3$,
$s_1 \equiv d_3$, or $s_1 \equiv d_1\vee d_3$, and $s_2 \equiv d_1\vee d_2$,
$s_1 \equiv d_2$, $s_2 \equiv d_1\vee d_3$,

In the first line, $s_2 \equiv d_2\vee d_3 \equiv x\wedge y \equiv \neg(\neg x\vee\neg y)$ holds. One of the last two terms is irreducible, hence $s_2 = x\wedge y$, or $s_2 = \neg(\neg x\vee\neg y)$. But s_2 is a negation, hence $t\downarrow = s_1\vee\neg(\neg x\vee\neg y)$, from which follows that $\neg(\neg x\vee\neg y)$ is irreducible and thus $x\wedge y$ is reducible. In both the second and the third line, let $\sigma=\{z\to 1\}$. Then $s_1\sigma \vee s_2 \equiv t$, and from the subterm lemma follows that $s_1\wedge s_2$ is reducible.

c) Let $t\downarrow = \neg s$. Then either $t\downarrow = \neg(s_1\vee \ldots \vee s_n)$, which can be treated analogously to a), or $t\downarrow = \neg(s_1\wedge \ldots \wedge s_n)$. In this case we obtain, similarly to b), $t\downarrow = \neg(s_1' \wedge s_2')$, with $s_1' \equiv d_1'$, or $s_1' \equiv d_1'\wedge d_2'$, or $s_1' \equiv d_1'\wedge d_3'$ and $s_2' \equiv d_2'\wedge d_3'$, where

$d_1' = x\vee y\vee\neg z$, $d_2' = \neg x\vee\neg y\vee\neg z$, $d_3' = \neg x\vee\neg y\vee z$.

First of all, $t\downarrow = \neg(s_1' \wedge s_2')$ implies that $\neg(x\wedge y)$ is irreducible, hence $\neg x\vee\neg y$ is reducible. We have $s_2' \equiv d_2'\wedge d_3' \equiv \neg x\vee\neg y$, and since s_2' is irreducible, $s_2' = \neg(x\wedge y)$. Now $t\downarrow = \neg(s_1' \wedge \neg(x\wedge y))$ implies that $\neg(x\wedge\neg y)$ is irreducible, hence $\neg x\vee y$ is reducible. Assume that s_1' is a disjunction, say $s_1' = u_1\vee\ldots\vee u_m$. Then each u_j must be an atom, since both $x\vee(y\wedge z)$ and $x\vee\neg y$ are reducible. But it is easy to see that there is no disjunction of the atoms x, y, and z can be equivalent to one of the terms d_1', $d_1'\wedge d_2'$, or $d_1'\wedge d_3'$. Hence s_1' must be of the form $s_1' = \neg u$, which implies that $t\downarrow = \neg(\neg u \wedge \neg(x\wedge y))$ is irreducible. Hence also $\neg(\neg x \wedge \neg y)$ is irreducible, which implies that $x\vee y$ is reducible.

<u>2.21 Lemma:</u>

Either the terms $x\vee y$ and $\neg(x\wedge y)\wedge\neg(x\wedge z)$ are both reducible, or the terms $x\wedge y$ and $\neg(x\vee y)\vee\neg(x\vee z)$ are both reducible.

Proof: According to the previous lemma, either $x\vee y$ or $x\wedge y$ is reducible.
Case 1: $x\vee y$ is reducible. Consider the term $t = (\neg x\vee y)\wedge(\neg y\vee x)\wedge(\neg x\vee\neg z)$. Since t is semi-symmetric in (x,y), but not symmetric, t must be reducible. Since $x\vee y$ is reducible, $t\downarrow$ cannot be a disjunction. Hence we have either $t\downarrow = s_1\wedge\ldots\wedge s_n$ or $t\downarrow = \neg s$. The first case is treated analogously to case a) of the previous lemma. In the case, where $t\downarrow = \neg s$, we

have $t\downarrow = \neg(s_1' \wedge s_2')$, with $s_1' \cong d_1'$, or $s_1' \cong d_1' \wedge d_2'$, or $s_1' \cong d_1' \wedge d_3'$ and $s_2' \cong d_2' \wedge d_3'$, where

$$d_1' = \neg x \vee \neg y \vee z, \quad d_2' = x \vee y \vee z, \quad d_3' = x \vee y \vee \neg z.$$

Analogouly to case c) of the previous lemma, we obtain $s_2' = \neg(\neg x \wedge \neg y)$, hence from $t\downarrow = \neg(s_1' \wedge s_2')$ follows that the term $t_0 := \neg(x \wedge \neg(\neg x \wedge \neg y))$ is irreducible, which in turn implies that $t_1 := \neg(x \wedge y) \wedge \neg(x \wedge z)$, which is equivalent to t_0, is reducible.

Case 2: $x \wedge y$ is reducible. Consider the term $t = (x \vee y \vee z) \wedge (\neg x \vee \neg y)$. Since $x \wedge y$ is reducible, t is also reducible, and, moreover, $t\downarrow$ cannot be a conjunction. Hence we have either $t\downarrow = s_1 \vee \dots \vee s_n$ or $t\downarrow = \neg s$. The first case is treated analogously to case a) of the previous lemma. In the case, where $t\downarrow = \neg s$, we have $t\downarrow = \neg(s_1' \vee s_2')$, with $s_1' \cong d_1'$, or $s_1' \cong d_1' \vee d_2'$, or $s_1' \cong d_1' \vee d_3'$ and $s_2' \cong d_2' \vee d_3'$, where

$$d_1' = \neg x \wedge \neg y \wedge \neg z, \quad d_2' = x \wedge y \wedge z, \quad d_3' = x \wedge y \wedge \neg z.$$

Analogouly to case c) of the previous lemma, we obtain $s_2' = \neg(\neg x \vee \neg y)$, hence from $t\downarrow = \neg(s_1' \wedge s_2')$ follows that the term $t_0 := \neg(x \vee \neg(\neg y \vee \neg z))$ is irreducible, which in turn implies that $t_1 := \neg(x \vee y) \vee \neg(x \vee z)$, which is equivalent to t_0, is reducible.

2.22 Corollary:

$\mathfrak{R}$ is not confluent.

Proof: We consider again the term $t = (x \vee y) \wedge (y \vee z) \wedge (z \vee x) \cong (x \wedge y) \vee (y \wedge z) \vee (z \wedge x)$ of lemma 2.16.

Case 1: The terms $x \vee y$ and $\neg(x \wedge y) \wedge \neg(x \wedge z)$ are both reducible. The reducibility of $x \vee y$ excludes t_1, t_2, t_3, t_5, t_6, and t_7 of lemma 2.16 from being irreducible, and the reducibility of $\neg(x \wedge y) \wedge \neg(x \wedge z)$ excludes both t_4 and t_8 from being irreducible.

Case 2: The terms $x \wedge y$ and $\neg(x \vee y) \vee \neg(x \vee z)$ are both reducible. The reducibility of $x \wedge y$ excludes t_1, t_3, t_4, t_6, t_7, and t_8 of lemma 2.16 from being irreducible, and the reducibility of $\neg(x \vee y) \vee \neg(x \vee z)$ excludes both t_2 and t_5 from being irreducible.

This corollary provides the proof of our main theorem 2.6.

Acknowledgement

I would like to thank J. Avenhaus for helpful suggestions and comments.

References

Hullot, J.M. (1980). A Catalogue of Canonical Term Rewriting Systems. Report CSL-113, SRI International, Menlo Park, CA.

Hsiang, J. (1985). Refutational Theorem Proving using Term-rewriting Systems. *Artificial Intelligence* 25, 255 - 300.

Knuth, D.E.; Bendix, P.B. (1970). Simple Word Problems in Universal Algebra, in: J. Leech (Ed.): *Computational Problems in Universal Algebra*, Pergamon Press.

Lankford, D.S. & Ballantyne, A.M. (1977). Decision procedures for simple equational theories with a commutative axiom: Complete sets of commutative reductions. Tech. Report, Mathematics Dept., University of Texas, Austin, Texas.

Peterson, G.E. & Stickel, M.E. (1981). Complete Sets of Reductions for Some Equational Theories. *Journal of the ACM*, **28**, 233 - 264.

Quine, W.V. (1952). The Problem of Simplifying Truth Functions. *American Math. Monthly*, **59**, 521 - 531.

Quine, W.V. (1959). On Cores and Prime Implicants of Truth Functions. *Am. Math. Monthly*, **66**, 755 - 760.

Rudeanu, S. (1974). *Boolean Functions and Equations*. North-Holland, Amsterdam.

Slagle, J.R.; Chang, C.L. & Lee, R.C.T. (1970). A New Algorithm for Generating Prime Implicants. *IEEE Trans. on Comp.* **19/4**, 304 - 310.

Tison, P. (1967). Generalized Consensus Theory and Application to the Minimization of Boolean Functions. *IEEE Trans. on Comp.* **16/4**, 446 - 456.

Decidability of confluence and termination of monadic term rewriting systems

Kai Salomaa[1]
Department of Mathematics
University of Turku, SF-20500 Turku
Finland

Abstract. *Term rewriting systems where the right-hand sides of rewrite rules have height at most one are said to be monadic. These systems are a generalization of the well known monadic Thue systems. We show that termination is decidable for right-linear monadic systems but undecidable if the rules are only assumed to be left-linear. Using the Peterson-Stickel algorithm we show that confluence is decidable for right-linear monadic term rewriting systems. It is known that ground confluence is undecidable for both left-linear and right-linear monadic systems. We consider partial results for deciding ground confluence of linear monadic systems.*

1. Introduction

A term rewriting system is said to be confluent if any two terms that are reduced from a common term are joinable. A system is terminating if it allows no infinite reductions. It is well known that both the termination and confluence properties are in general undecidable, cf. [De1,DJ,HL,Ja,Kl]. Recently it has been shown in [Da] that termination is undecidable already for systems with only one rewrite rule. For ground term rewriting systems where the rules are not allowed to contain variables confluence and termination are decidable, cf. [DTHL,HL,Oy].

For some applications instead of confluence it is sufficient to consider the weaker property of ground confluence, i.e., confluence on the set of ground terms, see [Gö,KNO,Ku]. For instance, it is possible that a restricted Knuth-Bendix type completion procedure yields a ground confluent system even if the general Knuth-Bendix procedure loops. However, as observed in [KNO] it is usually more difficult to test for ground confluence than for the stronger confluence property. Ground confluence is not even semi-decidable even for terminating systems, cf. [Ku].

Here we investigate the decidability of confluence, ground confluence, and termination properties of monadic term rewriting systems. Monadic term rewriting systems have been first considered in [GB] as an extension of monadic Thue systems, cf. [Bo1,Bo2,BJW,Ja]. The right-hand sides of monadic systems are restricted to be terms of height one. It is shown in [GB,Sa] that the congruence classes corresponding to canonical monadic term rewriting systems can be recognized by deterministic tree pushdown automata.

We show that termination is decidable for right-linear monadic systems but undecidable in the left-linear case. In [KNO] it is proved that the ground confluence property is undecidable for left-linear and right-linear monadic systems. Here we

[1]This research has been supported by the Academy of Finland.

contrast this result by showing that one can effectively decide whether a right-linear monadic system is confluent. The above question is reduced to checking confluence of an equational term rewriting system that can be made to be terminating and to satisfy the assumptions needed for the Peterson-Stickel local confluence test.

The results of [KNO] leave open the question whether ground confluence is decidable for linear monadic systems, i.e., systems that are both left- and right-linear. Using the decidability of equivalence of deterministic multitape finite automata, cf. [HK], we have obtained partial decidability results for ground confluence of linear monadic systems. On the basis of these results we conjecture that ground confluence of linear monadic systems would indeed be decidable.

2. Definitions

The reader is assumed to be familiar with terms and term rewriting systems. Here we only fix notations and briefly recall some definitions. For a more detailed exposition cf. e.g. [BD,De1,De2,DJ,GS,Hu,HO,JK,Kl,KB,PS].

The symbol Σ denotes always a finite ranked alphabet and Σ_m is the set of m-ary symbols, $m \geq 0$. The rank of an element $\sigma \in \Sigma$ is denoted by rank(σ). It is always assumed that $\Sigma_0 \neq \varnothing$. We denote by X a fixed enumerable set of variables $\{x_1, x_2, ...\}$. The set of words (resp. nonempty words) over a set A is denoted A^* (resp. A^+) and the empty word by λ. The cardinality of a finite set A is denoted #A. The set of positive integers is denoted by N_+.

The set of **Σ-terms** over the variables X (or ΣX-terms) is denoted $F_\Sigma(X)$ and the set of **ground terms** (i.e., terms without variables) by F_Σ. The **domain** of a term t $\in F_\Sigma(X)$, dom(t) $\subseteq N_+^*$ is defined in the usual way. The label ($\in \Sigma \cup X$) of a node u $\in$ dom(t) is denoted by t(u). The set of nodes u $\in$ dom(t) such that t(u) $\in \Sigma$ is denoted by **sdom(t)**. A node u is a **predecessor** of v, $u \leq v$, if u is a prefix of v. Nodes u and v are **independent**, $u \| v$, if neither one is a predecessor of the other.

Furthermore, we assume that the concepts: the height, a subterm, the root and a leaf of a term are well known. The subterm of t at node u is denoted t/u and the height of t is denoted by hg(t). The set of variables occurring in a term t is denoted **var(t)**. A term t is said to be **linear** (in variables of X) if t contains at most one occurrence of any variable.

A **substitution** is a mapping $\varphi: X \to F_\Sigma(X)$ such that the domain of φ, $D(\varphi) = \{x \in X \mid x\varphi \neq x\}$ is finite. A substitution φ has a unique homomorphic extension to a mapping $F_\Sigma(X) \to F_\Sigma(X)$ that is also denoted by φ. If $t' = t\varphi$ for some substitution φ, we say that t' is an **instance** of t. In our notation substitutions operate from the right whereas otherwise functions operate from the left.

Let $m \geq 1$, $t, t_1, ..., t_m \in F_\Sigma(X)$ and $x_1, ..., x_m \in$ var(t), $u_1, ..., u_m \in$ dom(t). Then $t(x_1 \leftarrow t_1, ..., x_m \leftarrow t_m)$ denotes the term that is obtained from t by replacing every occurrence of a variable x_i by t_i, $i = 1, ..., m$, and $t(u_1 \leftarrow t_1, ..., u_m \leftarrow t_m)$ denotes the term that is obtained from t by replacing the subterm t/u_i by t_i, $i = 1, ..., m$. Furthermore, we say that terms $t_1', ..., t_m'$ have been obtained from $t_1, ..., t_m$ by **standardizing the**

variables apart if t_i' is obtained from t_i by renaming some variables and $var(t_i')\cap var(t_j') = \emptyset$ when $i \neq j$.

Let Σ and Ω be ranked alphabets and let for every $m \geq 0$, $Y_m = \{y_1,...,y_m\}$ be a set of variables disjoint with X. Suppose that for every m such that $\Sigma_m \neq \emptyset$ there is given a mapping $h_m : \Sigma_m \rightarrow F_\Omega(Y_m)$. The mappings h_m, $m \geq 0$, determine a **tree homomorphism** (cf. [GS]) $h: F_\Sigma(X) \rightarrow F_\Omega(X)$ as follows.

(i) For $x \in X$, $h(x) = x$.

(ii) Let $t = \sigma(t_1,...,t_m)$, $m \geq 0$, $\sigma \in \Sigma_m$. Then $h(t) = h_m(\sigma)(y_1 \leftarrow h(t_1),...,y_m \leftarrow h(t_m))$.

In the following we give some basic definitions concerning term rewriting systems for general reduction systems. This is done so that we do not have to repeat the definitions later when considering equational reduction systems.

A **reduction system** is a pair (A,S) where A is the carrier set and S determines a binary reduction relation $\rightarrow_S$ on A. The transitive and the reflexive, transitive closures of $\rightarrow_S$ are denoted respectively by $\rightarrow_S^+$ and $\rightarrow_S^*$. Also we denote $\leftrightarrow_S = \rightarrow_S \cup \rightarrow_S^{-1}$ and then $\leftrightarrow_S^*$ is the reflexive, symmetric and transitive closure of $\rightarrow_S$. The pair (A,S) is usually denoted simply by S.

An element $b \in A$ is said to be **irreducible** (modulo S) if there does not exist $c \in A$ such that $b \rightarrow_S c$. The set of irreducible elements is denoted IRR(S). Elements b_1 and b_2 are said to be (S-)**joinable** if there exists $c \in A$ such that $b_i \rightarrow_S^* c$, $i = 1,2$. The reduction system S (or $\rightarrow_S$) is **confluent** if for all $b,c_1,c_2 \in A$ the condition $b \rightarrow_S^* c_i$, $i = 1,2$, implies that c_1 and c_2 are joinable. The system S is **locally confluent** if the condition $b \rightarrow_S c_i$, $i = 1,2$, implies that c_1 and c_2 are joinable. The system S is **terminating** (or well-founded) if it does not admit infinite reductions $b_1 \rightarrow_S b_2 \rightarrow_S \cdots$. Finally, a system that is both terminating and confluent is said to be **canonical**. The following well known result is originally from [Ne], see also [Hu,KB].

Lemma 2.1. (Newman's lemma). A terminating reduction system S is locally confluent iff it is confluent.

A **term rewriting system** (trs) over the ranked alphabet Σ is defined as a reduction system $(F_\Sigma(X),S)$ where S is a finite set of rules of the form $r \rightarrow s$, $r,s \in F_\Sigma(X)$ and $var(s) \subseteq var(r)$. The set of term rewriting systems over Σ is denoted TRS(Σ). The reduction relation $\rightarrow_S \subseteq F_\Sigma(X) \times F_\Sigma(X)$ associated with the trs S is defined as follows.

Let $t_1,t_2 \in F_\Sigma(X)$. Then

(2.1) $\quad t_1 \rightarrow_S t_2$ iff $t_1 = t(u \leftarrow r\varphi)$ and $t_2 = t(u \leftarrow s\varphi)$

where $t \in F_\Sigma(X)$, $u \in dom(t)$, $(r \rightarrow s) \in S$ and φ is a substitution such that $D(\varphi) \subseteq var(r)$. If in (2.1) we want to specify the rewrite rule $d{:}r \rightarrow s$ of S that is used in the reduction step, or also the node u where it is applied, this is denoted by $t_1 \rightarrow_S\langle d\rangle\ t_2$, or $t_1 \rightarrow_S\langle d,u\rangle\ t_2$.

A trs S is said to be **left- (right-) linear** if all the left-hand (respectively right-hand) sides of rules of S are linear, and S is **linear** if it is both left- and right-linear.

The restriction of S to the set of ground terms is the reduction system $(F_\Sigma, S(g))$. Here $\rightarrow_{S(g)}$ is the restriction of $\rightarrow_S$ to $F_\Sigma \times F_\Sigma$. The system S is said to be **ground confluent** if S(g) is confluent. Note that the definition of S(g) should not be confused with a ground term rewriting system, cf. [DTHL,HL,Oy].

3. Monadic term rewriting systems

Monadic term rewriting systems have been introduced in [GB] as an extension of monadic Thue systems, see also [KNO,Sa].

A trs S is said to be **monadic** if $hg(s) \leq 1$ and $hg(r) \geq hg(s)$ for every rule $r \rightarrow s \in S$. The set of monadic term rewriting systems over Σ is denoted MON(Σ). Similarly, LINMON(Σ), LLINMON(Σ) and RLINMON(Σ) denote the subclasses of MON(Σ) consisting respectively of the linear, left- and right-linear systems. If the ranked alphabet Σ is arbitrary, we use the notations MON, LINMON, LLINMON and RLINMON.

In the following we assume that $S \in$ MON does not contain rules of the form $x \rightarrow x$ or $x \rightarrow \sigma$, where $x \in X$, $\sigma \in \Sigma_0$. It is easy to see that when considering the decidability of the termination and confluence properties, this can be done without loss of generality. We will show that without restriction one can assume also that a monadic system is normalized as defined in the following.

We say that $S \in$ MON is **normalized** if every right-hand side of a rule of S is either a variable or of the form $\sigma(x_1,...,x_m)$, $m \geq 0$, $\sigma \in \Sigma_m$, $x_1,...,x_m \in X$. Here the variables $x_1,...,x_m$ are not necessarily distinct. Hence a normalized monadic system allows in the right-hand sides of rules elements of Σ only at the root. In the following we construct corresponding to every $S \in$ MON a normalized monadic system NORM(S) such that S and NORM(S) are "essentially equivalent".

Let Σ be a ranked alphabet. Corresponding to Σ we define the ranked alphabet $\Omega(\Sigma)$ as follows. $\Omega(\Sigma)$ consists of Σ and elements of the form

(3.1) $\qquad \sigma[z_1,...,z_m]$,

where $m \geq 1$, $\sigma \in \Sigma_m$, $z_1,...,z_m \in \Sigma_0 \cup \{e\}$. Here e is a new symbol, $z_{i_1} = ... = z_{i_p} = e$, $p \geq 0$, $i_1,...,i_p \in \{1,...,m\}$, and $z_j \in \Sigma_0$ when $j \notin \{i_1,...,i_p\}$.

In $\Omega(\Sigma)$ the rank of an element $\sigma \in \Sigma_m$ is m, and the rank of an element as in (3.1) is p (= the number of symbols e in the sequence $z_1,...,z_m$). We define the tree homomorphism h: $F_{\Omega(\Sigma)}(X) \rightarrow F_\Sigma(X)$ as follows.

(i) $h_n(\sigma) = \sigma(y_1,...,y_n)$ if $\sigma \in \Sigma_n$, $n \geq 0$.

(ii) Let $\sigma[z_1,...,z_m]$ be as in (3.1). Then $h_p(\sigma[z_1,...,z_m]) = \sigma(z_1',...,z_m')$, where $z_i' = z_i$ if $i \notin \{i_1,...,i_p\}$, and $z_i' = y_j$ if $i = i_j$, $1 \leq j \leq p$.

We define $R \in$ LINMON($\Omega(\Sigma)$) as follows. Let $\sigma[z_1,...,z_m]$ be as in (3.1) and $j \in \{i_1,...,i_p\}$ (i.e., $z_j = e$). Let $\gamma \in \Sigma_0$. Then R has the rule

$\sigma[z_1,...,z_j,...,z_m](x_1,...,x_{j-1},\gamma,x_{j+1},...,x_p) \rightarrow \sigma[z_1,...,\gamma,...,z_m](x_1,...,x_{j-1},x_{j+1},...,x_p)$.

Clearly if $t_1 \rightarrow_R t_2$, then $h(t_1) = h(t_2)$. It is easy to see that for every $t \in F_{\Omega(\Sigma)}(X)$ there exists a unique term norm(t) $\in$ IRR(R) such that $t \rightarrow_R^*$ norm(t) and h(t) = h(norm(t)). The normal form norm(t) is obtained from t simply by attaching all leaves labeled by elements of Σ_0 to their immediate predecessor.

Now let $S \in MON(\Sigma)$ be arbitrary. We define $NORM(S) \in MON(\Omega(\Sigma))$ by $NORM(S) = R \cup S'$ where S' is defined as follows:

(N1) Let $(r \to s) \in S$ and let φ be a substitution such that $D(\varphi) \subseteq var(r)$ and $(\forall x \in D(\varphi))\ x\varphi \in \Sigma_0$. Then

$$norm(r\varphi) \to norm(s\varphi) \in S'.$$

(N2) Suppose that $\sigma_1 \to \sigma_2 \in S$ where $\sigma_1, \sigma_2 \in \Sigma_0$. Let $\sigma[z_1,...,z_i,...,z_m]$ be as in (3.1), where $z_i = \sigma_1$, $i \in \{1,...,m\}$. Then

$$\sigma[z_1,...,\sigma_1,...,z_m](x_1,...,x_p) \to \sigma[z_1,...,\sigma_2,...,z_m](x_1,...,x_p) \in S'.$$

Above σ_1 and σ_2 correspond to the ith argument.

Clearly NORM(S) is finite and normalized, and we have the following result.

Lemma 3.1. Let $t_1, t_2 \in F_{\Omega(\Sigma)}(X)$.
(i) If $h(t_1) \to_S h(t_2)$ then $norm(t_1) \to_{NORM(S)^+} norm(t_2)$.
(ii) If $t_1 \to_{S'} t_2$ then $h(t_1) \to_S h(t_2)$.

Now using Lemma 3.1 one can easily prove:

Lemma 3.2. Let $S \in MON(\Sigma)$.
(i) S is terminating iff NORM(S) is terminating.
(ii) S is confluent iff NORM(S) is confluent.
(iii) S is ground confluent iff NORM(S) is ground confluent.

Theorem 3.3. Let W be any of the families MON, LLINMON, RLINMON or LINMON. Let Z stand for the property of termination, confluence, or ground confluence. Suppose that the following problem is decidable:

Instance: $S \in W$ such that S is normalized.

Question: Does S have the property Z?

Then also the corresponding question without the assumption that S is normalized is decidable.

Proof. Let $NORM(S) = S' \cup R$ be as defined above. Since the rules of R are linear, it follows from the definition of S' that $S \in W$ iff $NORM(S) \in W$ for all families W. Also clearly NORM(S) is effectively constructible. Hence the result follows from Lemma 3.2. Q.E.D.

In the following we always assume that MON, LLINMON, RLINMON and LINMON stand for the corresponding classes of normalized monadic systems. This makes the constructions in the next sections, in particular the proof of Theorem 5.5, essentially easier. By Theorem 3.3 this restriction can be done without loss of generality.

4. Termination

In this section we consider the decidability of the termination property of monadic term rewriting systems. Suppose that S is monadic and let $t \in F_\Sigma(X)$ be fixed. If $t \rightarrow_S^* t'$, then $hg(t') \le hg(t)$. Thus for a given t it is trivially decidable whether there exist infinite reductions starting from t. However, in the following it is seen that generally termination of monadic systems is decidable only in the right-linear case.

Definition 4.1. Let $S \in$ RLINMON(Σ) and $t \rightarrow_S^* t'$. We define a partial function $c(t,t')$: dom(t) $\rightarrow$ dom(t') as follows.

Suppose that $t \rightarrow_{S\langle d,u\rangle} t'$ where d is a rule $r \rightarrow s \in S$. Let $v \in$ dom(t). Then $c(t,t')(v)$ is defined by (i) - (iii).

(i) If $v \le u$ or $v \,||\, u$, $c(t,t')(v) = v$.

(ii) If $v = uu_1$, $u_1 \in$ dom(r), $u_1 \ne \lambda$, and $r(u_1) \notin X$, then $c(t,t')(v)$ is undefined.

(iii) Suppose that $v = uu_1u_2$, where $r(u_1) = x \in X$. If $x \notin$ var(s), then $c(t,t')(v)$ is undefined. In the case $x \in$ var(s) we have two possibilities:

(a) If s is of the form $\sigma(x_1,\ldots,x_m)$, then $c(t,t')(v) = uiu_2$, where $x = x_i$, $1 \le i \le m$. (Since S is right-linear, the index i is uniquely defined.)

(b) If $s = x$, then $c(t,t')(v) = uu_2$.

The above definition of $c(t,t')$ is extended to all pairs of trees t and t' such that $t \rightarrow_S^* t'$ by setting $c(t,t)$ as the identity function on dom(t) and if $t \rightarrow_S^* t_1 \rightarrow_S t'$ then $c(t,t')(v) = c(t_1,t')[c(t,t_1)(v)]$.

Intuitively for $v \in$ dom(t), $c(t,t')(v)$ is the address in t' into which v is taken in the reduction $t \rightarrow_S^* t'$. To be completely precise, the function $c(t,t')$ depends on the specific reduction from t to t'. If the node v has been deleted in the reduction $t \rightarrow_S^* t'$, then $c(t,t')(v)$ is undefined. We say that the node $c(t,t')(v)$ of t' **corresponds** to the node $v \in$ dom(t).

Lemma 4.2. Suppose that $S \in$ LINMON(Σ) contains only rules of the form

(4.2.1) $\quad \sigma(x_1,\ldots,x_m) \rightarrow \tau(x_{p(1)},\ldots,x_{p(m)})$,

where p is a permutation of $\{1,\ldots,m\}$. Suppose that $t \rightarrow_{S\langle d,v\rangle} t_1 \rightarrow_S^* t_2$ and let $u \in$ dom(t). If $t(u) = \rho_1$, where rank$(\rho_1) = n$ and $t_2(c(t,t_2)(u)) = \rho_2$, then rank$(\rho_2) = n$ and

(4.2.2) $\quad \rho_1(x_1,\ldots,x_n) \rightarrow_S^* \rho_2(x_{q(1)},\ldots,x_{q(n)})$

for some permutation q of the set $\{1,\ldots,n\}$. Furthermore, if $u = v$, then the first rule used in the reduction (4.2.2) may be assumed to be d. (The last part will be needed in the proof of Theorem 5.5.)

Proof. The proof is an easy induction on the length of the reduction $t \rightarrow_S^* t_2$. Q.E.D.

Using the fact from the previous section that monadic systems can be assumed to be normalized we now obtain easily the decidability of the termination property for right-linear systems.

Theorem 4.3. The following problem is decidable.

Instance: $S \in$ RLINMON.

Question: Is S terminating?

Proof. Suppose that there exists an infinite reduction $t_0 \rightarrow_{S\langle d_0,u_0\rangle} t_1 \rightarrow_{S\langle d_1,u_1\rangle} \cdots$. If $r_1 \rightarrow_{S\langle d\rangle} r_2$, $(d \in S)$, where the left-hand side of d is not linear or contains at least two symbols of Σ or the right-hand side is a variable, then $\#dom(r_2) < \#dom(r_1)$, (S is right-linear normalized). So there exists an integer M such that for all $n \geq M$ the rule d_n is of the form (4.2.1). For $n \geq M$, $dom(t_n) = \{c(t_M,t_n)(v) \mid v \in dom(t_M)\}$. Thus there exists $v \in dom(t_M)$ such that the intersection of the sets $\{(c(t_M,t_n)(v), n) \mid n \geq M\}$ and $\{(u_M, M), (u_{M+1}, M+1), \ldots\}$ is infinite, let us denote it $\{(u_{n_1}, n_1), (u_{n_2}, n_2), \ldots\}$, $n_1, n_2, \ldots \geq M$. In the above notation (u, j), $u \in dom(t_j)$, $j \geq M$, the second components are used just to distinguish between identical elements of N_+^* belonging to domains of different terms t_j. There exist $i < j$ such that $d_{n_i} = d_{n_j}$. If d_{n_i} is as in (4.2.1), then by Lemma 4.2 ($\rho_1 = \rho_2 = \sigma$) we have

(4.3.1) $\quad \sigma(x_1,\ldots,x_m) \rightarrow_S^* \sigma(x_1,\ldots,x_m)$.

On the other hand, the existence of $\sigma \in \Sigma_m$ such that (4.3.1) holds is clearly also a sufficient condition for the nontermination of S. Condition (4.3.1) can of course be checked effectively. Q.E.D.

In the following we show that termination is undecidable for left-linear monadic systems. This is done by reducing the well known Post Correspondence Problem, PCP, to the question of termination of left-linear systems. An instance of PCP is a four-tuple $\mathbf{A} = (A,n,\mathbf{u},\mathbf{v})$ consisting of a finite alphabet A, an integer $n \geq 1$ and n-tuples $\mathbf{u} = (u_1,\ldots,u_n)$ and $\mathbf{v} = (v_1,\ldots,v_n)$ of nonempty words over A. A sequence $i_1,\ldots,i_k$ $(i_j \in \{1,\ldots,n\})$ is said to be a solution of the PCP $\mathbf{A}$ if $u_{i_1}\ldots u_{i_k} = v_{i_1}\ldots v_{i_k}$. It is well known that it is undecidable whether a given PCP has a solution.

Theorem 4.4. The following problem is undecidable.

Instance: $S \in$ LLINMON.

Question: Is S terminating?

Proof. Let $\mathbf{A} = (A,n,\mathbf{u},\mathbf{v})$, $\mathbf{u} = (u_1,\ldots,u_n)$, $\mathbf{v} = (v_1,\ldots,v_n)$, $n \geq 1$, $u_i, v_i \in A^+$, $i = 1,\ldots,n$, be an arbitrary instance of PCP. Define the ranked alphabet $\Sigma = \Sigma_0 \cup \Sigma_1 \cup \Sigma_5$ as follows: $\Sigma_0 = \{\tau\}$, $\Sigma_5 = \{\omega_1,\omega_2\}$, $\Sigma_1 = A \cup \{1,\ldots,n\}$. We construct $S \in$ LLINMON(Σ) such that S is nonterminating iff the PCP $\mathbf{A}$ has a solution.

The set S consists of the rules

(4.4.1) $\quad \omega_1(\tau,\tau,\tau,x,y) \rightarrow \omega_2(x,x,y,x,y)$, and

(4.4.2) $\quad \omega_2(a_1\cdots a_p(x_1),b_1\cdots b_q(x_2),c(x_3),x,y) \rightarrow \omega_1(x_1,x_2,x_3,x,y)$,

(4.4.3) $\quad \omega_1(a_1\cdots a_p(x_1),b_1\cdots b_q(x_2),c(x_3),x,y) \rightarrow \omega_1(x_1,x_2,x_3,x,y)$,

where $c \in \{1,\ldots,n\}$, $a_i, b_j \in A$, $i = 1,\ldots,p$, $j = 1,\ldots,q$, and $u_c = a_1\cdots a_p$, $v_c = b_1\cdots b_q$.

Suppose that the PCP $\mathbf{A}$ has a solution $i_1,\ldots,i_k$. Denote $u_{i_1}\cdots u_{i_k} = w$ $(= v_{i_1}\cdots v_{i_k})$. Then

$$\begin{aligned}
&\omega_2(w(\tau),w(\tau),i_1\cdots i_k(\tau),w(\tau),i_1\cdots i_k(\tau))\\
&\rightarrow_S \omega_1(u_{i_2}\cdots u_{i_k}(\tau),v_{i_2}\cdots v_{i_k}(\tau),i_2\cdots i_k(\tau),w(\tau),i_1\cdots i_k(\tau))\\
&\rightarrow_S \cdots \rightarrow_S \omega_1(\tau,\tau,\tau,w(\tau),i_1\cdots i_k(\tau))\\
&\rightarrow_S \omega_2(w(\tau),w(\tau),i_1\cdots i_k(\tau),w(\tau),i_1\cdots i_k(\tau)),
\end{aligned}$$

and thus $\rightarrow_S$ does not terminate.

Conversely it can be shown that an infinite reduction of S necessarily contains a subreduction essentially as above. The details are omitted here. Thus the nontermination of S implies that the PCP A has a solution. Q.E.D.

Note that in the above proof S is even locally confluent. The rules of S can be slightly simplified if one removes this requirement.

It is easy to see that $S \in TRS(\Sigma)$ is terminating iff its restriction to the set of ground terms S(g) is terminating (assuming that $\Sigma_0 \neq \varnothing$). However, the confluence and ground confluence conditions are not equivalent and in the following we will see that for certain monadic systems it is even possible that confluence is decidable whereas ground confluence is undecidable.

5. Confluence

Here we consider the decidability of the (ground) confluence property of monadic systems. We reduce the question of confluence of a right-linear monadic system to testing for local confluence of an equational term rewriting system. For this reason we recall some definitions concerning equational systems.

By considering the rules of a trs S as two-directional equations, S defines the equational theory $\leftrightarrow_S^*$. In the following, term rewriting systems that are seen as (one-directional) reduction systems are generally denoted by the letters R and S, and rewriting systems that are considered as two-directional equations are denoted by E. The congruence class of $\leftrightarrow_E^*$ corresponding to $t \in F_\Sigma(X)$ is denoted by $[t]_E$ (or simply $[t]$) and $F_\Sigma(X)/E = \{ [t]_E \mid t \in F_\Sigma(X) \}$.

Equational term rewriting systems are discussed e.g. in [BD,DJ,Hu,HO,JK,PS]. Here we use the terminology of [PS] where the reductions operate directly on the congruence classes of the equational theory.

Let $R, E \in TRS(\Sigma)$. Then the **reduction system R modulo E** (or equational reduction system) $(F_\Sigma(X)/E, R/E)$ is defined as follows. Let $[r_1]_E, [r_2]_E \in F_\Sigma(X)/E$. Then

(5.1) $\qquad [r_1]_E \rightarrow_{R/E} [r_2]_E$

iff there exist $t_i \leftrightarrow_E^* r_i$, $i = 1,2$, such that $t_1 \rightarrow_R t_2$. We will consider only equational theories with finite congruence classes. Hence it is clear that for arbitrary $r_1, r_2 \in F_\Sigma(X)$ one can decide whether (5.1) holds by generating all elements of $[r_i]_E$, $i = 1,2$.

In the following, let $R \in TRS(\Sigma)$ and E be an equational theory over Σ. Let $t_1, t_2 \in F_\Sigma(X)$. We assume that the concepts of an **E-unifier** of t_1 and t_2 and a **complete set of E-unifiers** of t_1 and t_2 are known, cf. e.g. [FH,He,Hu,HO,JK,PS]. Unless otherwise mentioned, we always assume that all terms t_1 and t_2 have a finite complete set of E-unifiers that can be constructed effectively, i.e., there exists a finite complete E-unification algorithm.

We say that R is **E-compatible** (cf. [PS]) if for all $t_1, t_2 \in F_\Sigma(X)$ such that $[t_1]_E \rightarrow_{R/E} [t_2]_E$ the following condition holds. There exists $u \in dom(t_1)$, a rule $(r \rightarrow s) \in R$ and a substitution φ with $D(\varphi) \subseteq var(r)$ such that

$$t_1/u \leftrightarrow_E^* r\varphi \text{ and } [t_2]_E \rightarrow_{R/E}^* [t_1(u \leftarrow s\varphi)]_E.$$

Next we define the set of **critical pairs** CP(R/E) of the equational reduction system R/E as in [PS]. Suppose that $(r_1 \to s_1), (r_2 \to s_2) \in R$, where the variables of r_1 and r_2 have been standardized apart. Let $u \in sdom(r_1)$ be such that r_1/u and r_2 are E-unifiable. Let Γ be a finite complete set of E-unifiers of r_1/u and r_2. Then for every $\varphi \in \Gamma$: $([r_1(u \leftarrow s_2)\varphi]_E, [s_1\varphi]_E) \in CP(R/E)$.

By choosing the complete sets of unifiers Γ differently one may obtain different sets of critical pairs CP(R/E). However, in the following we only need to know that CP(R/E) is finite and that it can be constructed effectively, i.e., the sets Γ may be chosen arbitrarily. It is known (cf. [FH]) that if finite complete sets of unifiers exist then one can actually always choose a minimal complete set of unifiers. If $E = \emptyset$, the critical pairs defined above are the standard critical pairs of R, cf. [Hu,KB].

The following fundamental result is proved in [PS]. Assuming that R/E is terminating it gives with Newman's lemma a decision algorithm for testing confluence of the reduction system R/E.

Theorem 5.1. Suppose that R is E-compatible and E has finite complete sets of unifiers. Then the equational reduction system R/E is locally confluent iff for every critical pair $(A_1,A_2) \in CP(R/E)$, A_1 and A_2 are R/E-joinable.

A theory E is said to be **variable-permuting** if all rules of E are of the form $\sigma_1(x_1,...,x_m) \to \sigma_2(x_{p(1)},...,x_{p(m)})$, where $m \geq 0$, $\sigma_1,\sigma_2 \in \Sigma_m$, and p is a permutation of $\{1,...,m\}$. Note that this definition is not equivalent to the definitions of permutative and variable-permutative theories as given for instance in [Sc1,Sc2]. These theories do not always have finite complete sets of unifiers.

The following lemma is proved easily using Theorem 9.5 of [PS].

Lemma 5.2. Let $R \in TRS(\Sigma)$ and suppose that E is a variable-permuting theory over Σ. Then R is E-compatible.

Lemma 5.3. Suppose that E is variable-permuting. Then E has a finite complete unification algorithm.

Proof. A complete E-unification algorithm can be constructed essentially as for commutative theories, see [He,Si,Sl]. The details are left for the reader. Q.E.D.

An equational theory E is said to be **cyclic** if $t \to_E^* t'$ for every t and t' such that $t \leftrightarrow_E^* t'$ (i.e., $\to_E^* = \leftrightarrow_E^*$.) The easy proof of Lemma 5.4 is again omitted.

Lemma 5.4. Let $S \in TRS(\Sigma)$ be written as a disjoint union $S = R \cup E$. Suppose that the corresponding equational theory E is cyclic. Then S is confluent iff R/E is confluent.

Now we can state the main result of this section.

Theorem 5.5. The following question is decidable.

Instance: $S \in$ RLINMON.

Question: Is S confluent?

Proof. Let E be the subset of S consisting of all rules $r \to s$ such that $s \to_S^* r$, and denote $R = S - E$. As in the proof of Theorem 4.3 it is easy to see that all the rules of E are variable-permuting, and furthermore it can be shown that E is cyclic.

Now by Lemma 5.2, R is E-compatible. We show that the rewrite relation R/E terminates. From the fact that E is variable-permuting it follows that if $[t] = [r]$, then $\#dom(t) = \#dom(r)$. Thus we can define the size of the equivalence class $[t]$ by $\#[t] = \#dom(t)$. Since S is right-linear and normalized, if $t \to_S^* r$, then $\#dom(t) \geq \#dom(r)$ and hence also

$$\#[t] \geq \#[r] \text{ for all } [t] \to_{R/E}^* [r].$$

Let $[t_0] \to_{R/E} [t_1] \to_{R/E} [t_2] \to_{R/E} \dots \to_{R/E} [t_n]$, where $n = (\#\Sigma)(\#dom(t_0))+1$. We claim that $\#[t_n] < \#[t_0]$, from this it then follows that R/E is terminating. Assume to the contrary that $\#[t_0] = \#[t_1] = \dots = \#[t_n]$. Since E is cyclic, there exists a reduction of S

(5.5.1) $$t_0 \to_E^* r_1 \to_{R\langle d_1,u_1\rangle} s_1 \to_E^* t_1 \to_E^* \dots$$
$$\to_E^* t_{n-1} \to_E^* r_n \to_{R\langle d_n,u_n\rangle} s_n \to_E^* t_n.$$

By the choice of n we can choose $u \in dom(r_1)$ such that the intersection of the sets $\{ (c(r_1,r_i)(u), i) \mid i = 1,\dots,n \}$ and $\{ (u_1,1),\dots,(u_n,n) \}$ contains at least $\#\Sigma+1$ elements (u_{i_j},i_j), $j = 1,\dots,\#\Sigma+1$. Here the function $c(r_1,r_i)$ is from Definition 4.1. As in the proof of Theorem 4.3, the notation (v,i) denotes the node v of r_i. This means that in the nodes corresponding to u in the reduction, a rule of R is applied at least $\#\Sigma+1$ times. Choose $a, b \in \{1,\dots,\#\Sigma+1\}$, $a \neq b$ such that $r_{i_a}(u_{i_a}) = r_{i_b}(u_{i_b})$. Denote $r_{i_a}(u_{i_a}) = \sigma \in \Sigma_m$. Since $\#dom(t_0) = \#dom(t_n)$, it follows that also the rules $d_1,\dots,d_n$ are variable-permuting, i.e., all rules in the reduction (5.5.1) are as in (4.2.1). Thus by Lemma 4.2,

(5.5.2) $$\sigma(x_1,\dots,x_m) \to_{R\langle d_{i_a}\rangle} z \to_S^* \sigma(x_{p(1)},\dots,x_{p(m)})$$

where z is the right-hand side of d_{i_a} and p is a permutation. From (5.5.2) it follows that also $z \to_S^* \sigma(x_1,\dots,x_m)$. Hence, by the choice of the equations E we have $d_{i_a} \in E$ which is impossible since $d_{i_a} \in R$. This completes the proof of the termination of R/E.

Now by Theorem 5.1, Lemma 5.3, and Lemma 2.1, R/E is confluent if and only if for every critical pair (A_1,A_2) of R/E there exists C such that

(5.5.3) $$A_1 \to_{R/E}^* C \text{ and } A_2 \to_{R/E}^* C.$$

By Lemma 5.3 we can effectively generate all finitely many critical pairs of R/E and thus are able to check whether condition (5.5.3) holds. Since E is cyclic, R/E is confluent iff S is confluent by Lemma 5.4. Q.E.D.

The decidability of confluence of left-linear monadic term rewriting systems remains open. The standard approach to prove the confluence property decidable is to reduce it to checking local confluence and thus joinability of critical pairs. However, in this test one seems to need to assume the termination of the system (at least in the form of a suitable equational theory as in Theorem 5.5). In view of Theorem 4.4 this is probably very difficult for left-linear monadic systems.

In [KNO] it is shown that ground confluence is undecidable for both RLINMON and LLINMON. We have been able to obtain partial decidability results for ground confluence of LINMON. Let $n \geq 1$, $q \geq 0$ and $t \in F_\Sigma(X)$. Denote by $f_q(t)$ the number of (occurrences of) subterms of t of height exactly q. Then t is said to be (n,q)-bounded if $f_q(t) \leq n$ and we denote $B(\Sigma,n,q) = \{ t \in F_\Sigma(X) \mid f_q(t) \leq n \}$. Let $S \in$ LINMON(Σ). The restriction of S to the set of (n,q)-bounded terms is the reduction system ($B(\Sigma,n,q)$, S[n,q]) where $\rightarrow_{S[n,q]}$ is the restriction of $\rightarrow_S$ to $B(\Sigma,n,q) \times B(\Sigma,n,q)$.

We have been able to show that for all $n \geq 1$, $q \geq 0$ and $S \in$ LINMON one can effectively decide whether S[n,q] is ground confluent. (This will be presented in a forthcoming paper.) Essentially as in Theorem 5.5 one first constructs a terminating equational system R/E such that S[n,q] is ground confluent iff all (n,q)-bounded ground instances of every critical pair of R/E are joinable. For each critical pair (A_1,A_2) one can then construct deterministic multitape finite automata M_1 and M_2 such that the above condition holds for (A_1,A_2) iff M_1 and M_2 are equivalent. The main difficulty in the construction is that one has to be able to make the automata deterministic. Equivalence of deterministic multitape finite automata is decidable, cf. [HK].

The above result can be taken as evidence for assuming that ground confluence of LINMON would be decidable. This is because all natural undecidable properties of term rewriting systems seem to remain undecidable when restricted to bounded terms (for sufficiently large n and q.) For instance, the undecidability results for monadic systems considered here and in [KNO] are valid already for (n,0)-bounded terms for small values of n.

6. Summary

The figure below lists the main decidability results obtained here and in [KNO].

	MON	LLINMON	RLINMON	LINMON
Termination	U	U	D	D
Confluence	?	?	D	D
Ground confluence	U(1)	U(1)	U(1)	?(2)
Canonicity	U	U	D	D

Here the symbols "D", "U" and "?" stand respectively for "decidable", "undecidable" and "open". The results (1) are from [KNO]. The question (2) is decidable for bounded terms. The decidability of canonicity for RLINMON follows from Theorems 4.3 and 5.5 and the undecidability of canonicity for LLINMON is seen using a modification of the proof of Theorem 4.4.

References

[BD] L.Bachmair and N.Dershowitz, Completion for rewriting modulo a congruence, *Theoret. Comput. Sci.* **67** (1989) 173-201.

[Bo1] R.V.Book, Confluent and other types of Thue systems, *J.Assoc.Comput.Mach.* **29** (1982) 171-182.

[Bo2] R.V.Book, Thue systems as rewriting systems, *J. Symbolic Computation* **3** (1987) 39-68.

[BJW] R.V.Book, M.Jantzen and C.Wrathall, Monadic Thue systems, *Theoret. Comput. Sci.* **19** (1982) 231-251.

[Da] M.Dauchet, Simulation of Turing machines by a left-linear rewrite-rule, Proc. of 3rd RTA, Lect. Notes Comput. Sci. **355** (1989) 109-120.

[DTHL] M.Dauchet, S.Tison, T.Heuillard and P.Lescanne, Decidability of the confluence of ground term rewriting systems, Proc. of the 2nd LICS, 1987, 353-359.

[De1] N.Dershowitz, Termination of rewriting, *J. Symbolic Computation* **3** (1987) 69-116.

[De2] N.Dershowitz, Completion and its applications, *in:* H.Aït-Kaci and M.Nivat, eds., Resolution of equations in algebraic structures, Vol. 2, Academic Press (1989) 31-85.

[DJ] N.Dershowitz and J.-P.Jouannaud, Rewrite systems, *in:* J. van Leeuwen, ed., Handbook of Theoretical Computer Science, Vol. B, Elsevier (1990) 243-320.

[FH] F.Fages and G.Huet, Complete sets of unifiers and matchers in equational theories, *Theoret. Comput. Sci.* **43** (1986) 189-200.

[GB] J.H.Gallier and R.V.Book, Reductions in tree replacement systems, *Theoret. Comput. Sci.* **37** (1985) 123-150.

[GS] F.Gécseg and M.Steinby, Tree automata, Akadémiai Kiadó, 1984.

[Gö] R.Göbel, Ground confluence, Proc. of the 2nd RTA, Lect. Notes Comput. Sci. **256** (1987) 156-167.

[HK] T.Harju and J.Karhumäki, Decidability of the multiplicity equivalence of multitape finite automata, Proceedings of the 22nd STOC (1990) 477-481.

[He] A.Herold, Combination of unification algorithms in equational theories, Ph.D. thesis, University of Kaiserslautern, 1987.

[Hu] G.Huet, Confluent reductions: Abstract properties and applications to term rewriting systems, *J. Assoc. Comput. Mach.* **27** (1980) 797-821.

[HL] G.Huet and D.S.Lankford, On the uniform halting problem for term rewriting systems, Rapport Laboria 283, INRIA, Le Chesnay, France, 1978.

[HO] G.Huet and D.C.Oppen, Equations and rewrite rules, *in:* R.V.Book, ed., Formal language theory, Perspectives and open problems, Academic Press (1980) 349-393.

[Ja] M.Jantzen, Confluent string rewriting, EATCS Monographs on Theoretical Computer Science **14**, Springer-Verlag, 1988.

[JK] J.-P.Jouannaud and H.Kirchner, Completion of a set of rules modulo a set of equations, *SIAM J. Comput.* **15** (1986) 1155-1194.

[KNO] D.Kapur, P.Narendran and F.Otto, On ground confluence of term rewriting systems, *Information and computation* **86** (1990) 14-31.

[Kl] J.W.Klop, Term rewriting systems: From Church-Rosser to Knuth-Bendix and beyond, Proc. of 17th ICALP, Lect. Notes Comput. Sci. **443** (1990) 350-369.

[KB] D.Knuth and P.Bendix, Simple word problems in universal algebras, *in:* J.Leech, ed., Computational problems in abstract algebra, Pergamon Press (1970) 263-297.

[Ku] G.A.Kucherov, A new quasi-reducibility testing algorithm and its application to proofs by induction, Proc. of Algebraic and Logic Programming '88, Lect. Notes Comput. Sci. **343** (1988) 204-213.

[Ne] M.H.A.Newman, On theories with a combinatorial definition of "equivalence", *Ann. Math.* **43** (1942) 223-243.

[Oy] M.Oyamaguchi, The Church-Rosser property for ground term-rewriting systems is decidable, *Theoret. Comput. Sci.* **49** (1987) 43-79.

[PS] G.Peterson and M.Stickel, Complete sets of reductions for some equational theories, *J. Assoc. Comput. Mach.* **28** (1981) 233-264.

[Sa] K.Salomaa, Deterministic tree pushdown automata and monadic tree rewriting systems, *J. Comput. System Sci.* **37** (1988) 367-394.

[Sc1] M.Schmidt-Schauss, Solution to problem P140 and P141, Bull. of EATCS **34** (1988) 274-275.

[Sc2] M.Schmidt-Schauss, Unification in permutative equational theories is undecidable, *J. Symbolic Computation* **8** (1989) 415-421.

[Si] J.Siekmann, Matching under commutativity, Proc. of EUROSAM'79, Lect. Notes Comput. Sci. **72** (1979) 531-545.

[Sl] L.Slagle, Automated theorem proving for theories with simplifiers, commutativity and associativity, *J. Assoc. Comput. Mach.* **21** (1974) 622-642.

BOTTOM-UP TREE PUSHDOWN AUTOMATA and REWRITE SYSTEMS[1]

J-L. Coquidé(1), M. Dauchet(1), R. Gilleron(2), S. Vàgvölgyi(3)
(1),(2): {Coquide,Dauchet,Gilleron}@lifl.lifl.fr
(3): h1002vag@ella.uucp
(1)LIFL, URA 369 CNRS, IEEA Université de Lille I, 59655 Villeneuve d'Ascq Cedex France
(2)LIFL, URA 369 CNRS, IUT A Université de Lille I, 59653 Villeneuve d'Ascq Cedex France
(3)Research Group on Theory of Automata, Hungarian Academy of Sciences, Somogyi u.7, H-6720 Szeged, Hungary

Abstract. Studying connections between term rewrite systems and bottom-up tree pushdown automata (tpda), we complete and generalize results of Gallier, Book and K. Salomaa. We define the notion of *tail reduction free* rewrite systems (trf rewrite systems). Using the decidability of inductive reducibility (Plaisted), we prove the decidability of the trf property. Monadic rewrite systems of Book, Gallier and K. Salomaa become an obvious particular case of trf rewrite systems. We define also *semi-monadic* rewrite systems which generalize monadic systems but keep their fair properties. We discuss different notions of bottom-up tree pushdown automata, that can be seen as the algorithmic aspect of classes of problems specified by trf rewrite systems. Especially, we associate a deterministic tpda with any left-linear trf rewrite system.

Key words. Rewrite systems (= term rewrite systems). Church-Rosser (= confluent) systems.Noetherian systems. Convergent (= noetherian + confluent) systems. (Deterministic) tree automaton with pushdown store. (Recognizable, Context-free) tree languages.

1. INTRODUCTION

Equations and rewrite systems have been extensively used to specify programs and data types (see Goguen, Thatcher, Wagner and Wright [12] for one of the seminal papers, Huet and Oppen [16], Dershowitz and Jouannaud [8] for overviews). The paradigm of Rewrite Systems modelizes evaluation in Logic Programming as well as interpreters in Functional Programming; in this case they usually get the confluence (i.e. the Church-Rosser) property. Obviously, most interesting properties are undecidable, but works are devoted to study partial algorithms or special kinds of rewrite systems. The most popular result is the Knuth-Bendix algorithm, which permits in many cases to compute normal forms of data types. So, researchers get decidable properties of fragments of theories or of subclasses of rewrite systems to supply tools for software engineering.

At the same time, stacks are a basic data structure in computer science, for example in syntactical analysis or for recursive procedures calls. In the string case, the equivalence between context-free grammars and pushdown automata is a well-known fact, and connections between string rewrite systems and pushdown automata have been studied too (see Book [2] for a survey).

Following works of Book and Gallier [3] and K. Salomaa [18], the goal of this paper is to study connections between tree pushdown automata and rewrite systems. Tree pushdown automata can be seen as the algorithmic aspect of problems specified by rewrite systems.

[1]This research was performed while S. Vàgvölgyi was visiting the department of computer science (L.I.F.L, URA 369 CNRS, I.E.E.A), University of Lille Flandres-Artois. This work was supported in part by the "PRC Mathématiques et Informatique" and ESPRIT2 Working Group ASMICS.

The main originality of our approach is that we point out the importance of a new concept, that we call *tail reduction free property* - trf property for the simplicity. Generalizing in an obvious way the common sense, the *tail* of a term is the sequence of subterms that we get by erasing the root; we say (see Definition in Section 4.2) that a trs S is *tail reduction free* iff for every right handside r of S and any irreducible (for S) ground substitution σ, σ(tail(r)) is irreducible too. (This means that if σ is irreducible, σ(r) can only be reduced by matching at the root of r). We prove that this property is decidable, tracing it back to the decidability of inductive reducibility (Plaisted [17]) (see Theorem in Section 4.4).

Consider the two following rewrite systems.

Example1 (Peano rules). Consider the ranked alphabet {+, s, 0} where + is of rank 2 and s is of rank 1 and 0 is of rank 0. The rewrite system S1 is defined by the following two rules: $+(s(x),y) \rightarrow +(x,s(y))$, $+(0,x) \rightarrow x$.

Example 2. Consider the ranked alphabet {a, b, §} where a,b are of rank 1 and § is of rank 0. S2 contains only the rule $b(a(x)) \rightarrow a(b(x))$.

As no + occurs in a ground irreducible term, the rewrite system S1 is trf, but S2 is not trf, considering the ground irreducible substitution σ(x) = a§. We show how the trf property is related to tree bottom-up pushdown automata (see Section 4.5).This notion both generalizes preceeding studies and illuminates the deep connection between pushdown automata and rewrite systems.

More precisely, Book, Gallier and K. Salomaa associated tree pushdown automata only with what they call *monadic* rewrite systems.

A rewrite system S on a ranked alphabet Σ is monadic if, for each rule $l \rightarrow r$, $depth(l) \geq 1$and $depth(r) \leq 1$. In this paper, we distinguish two purposes which are mixed in the preceeding papers:

- How to assign a tree pushdown automaton to a rewrite system.
- Find rewrite systems classes with "good" properties (decision and complexity results, relation between context-free and recognizable tree languages...).

For the first aim, we introduce the notion of being tail reduction free. Monadic rewrite systems are obviously trf (the tail of a right-hand side contains variables and nullary function symbols!). Morever, we show how to simulate a Turing machine by trf rewrite systems (see Section 4.3).

For the second aim, we introduce the class of *semi-monadic* rewrite systems (see Sections 4.3 and 5.1). This class keeps the same good properties as the monadic one, and embeds the monadic class and the class of ground rewrite systems, the theory of which is decidable [7].

The paper is organized as follows. In part 2, we recall some useful definitions. In part 3, we compare different definitions of bottom-up tree pushdown automata (tpda for short). *Determinism* is always an important notion; here we define this notion only in the left-linear case (see Section 3.2). (Let us recall that a rule is *left-linear* if no variable occurs more than once in the left-hand side.) The reason of this restriction is that in this case we can associate a simple syntactical criterion to a realistic notion of determinism; more precisely, we say that a tpda is deterministic iff the set of rules has no critical pair (i.e. there are no overlappings between left-hand sides), which ensures the confluence only in the left-linear case (Huet [14], Huet & Levy [15]).

In part 4, we introduce the notion of tail reduction free rewrite system and assign to any trf rewrite system a tpda by a natural construction (prop.1 in section 4.5). Note that it could be possible to assign to any rewrite system a tpda (or a trf rewrite system). By this way, we could generalize to the tree case the simulation of a Turing machine (see Section 4.3). We do not present this construction here because the construction is somewhat artificial and unefficient. We also make precise the connection between the *linearity* of rewrite systems and the needless of *control* in a tpda. Indeed, in order to compute S-normal forms of ground terms, we assign a deterministic tpda to a left-linear convergent trf rewrite system S and a tpda with control to a convergent trf rewrite system S (see Section 4.5).

The formal language aspect is sketched in Section 5. Extending a result of K.Salomaa [18], we prove that recognizability ([11]) is preserved by linear semi-monadic rewrite systems (see Section 5.1).

2. PRELIMINARIES

The purpose of Section 2 is to make the paper self-contained, but usual notions are only sketched.

2.1. Terms,substitutions

Let Σ be a finite ranked alphabet and T_Σ be the set of terms (trees) over Σ. Let X be a denumerable set of variables, $X_m=\{x_1,...,x_m\}$, $T_\Sigma(X)$ and $T_\Sigma(X_m)$ be the set of terms over $\Sigma \cup X$ and $\Sigma \cup X_m$ respectively. We denote the set of variables occurring in t by *V(t)*. A term t is *linear* if no variable occurs twice in t. A term t is *ground* if no variable occurs in t. A *context* c is a term in $T_\Sigma(X_m)$ such that each variable occurs exactly once in c. The result of the substitution of each x_i by a term t_i is denoted $c(t_1,...,t_m)$.

We define the set of *occurences* (or positions) of a term t as usual. Let $t_{/o}$ denote the subterm of t rooted at occurence o. By *depth(t)* we mean the depth of the tree t, *path(t)* denotes the set of all strings of symbols in Σ that occur as labels of a path from the root to a leaf of t and *fr(t)* denotes the frontier of a tree t. We also define a *substitution* as usual. $D(\sigma)$ denotes the domain of a substitution σ and we use the same notation for the substitution and its extension. The tree t *matches* l if there exists a substitution σ such that $\sigma(l)=t$. The trees t and u are *unifiable* if there exists a substitution σ such that $\sigma(t)=\sigma(u)$.

A *k-normal tree* t is a linear term of $T_\Sigma(X_m)$ for some m such that for every occurence o in O(t), either ($|o|=k$ & $t_{/o} \in X$), or ($|o|<k$ & $root(t_{/o}) \in \Sigma$), and $fr(t) \in \Sigma^* x_1 \Sigma^* ... x_m \Sigma^*$. Two distinct k-normal trees are not unifiable and for every term u in T_Σ, there exists one (and only one) k-normal tree t such that u matches t. Let $T_\Sigma(X,k)$ be the set of k-normal trees.

2.2. Rewrite systems

A *rewrite system* $S=\{l \rightarrow r \,/\, l, r \in T_\Sigma(X),\ V(r) \subset V(l)\}$ on T_Σ is a finite set of pairs of terms in $T_\Sigma(X)$. $\rightarrow_S$ is the rewrite relation induced by S and $\xrightarrow{*}_S$ is the reflexive and transitive closure of $\rightarrow_S$. A rewrite system is *ground,* (resp. *linear, left-linear, right-linear*) if, for each rule $l \rightarrow r$, the terms l, r are ground (resp. l, r are linear, l is linear, r is linear). We denote the set of irreducible terms for S by *IRR(S)*. A substitution σ is *ground irreducible* for S if for every variable x in $D(\sigma)$, $\sigma(x)$ is ground and irreducible.

Let S be a rewrite system and < be a partial order on S, we define the *innermost rewrite relation induced by S with respect to <* , denoted by $\rightarrow_{i,S,<}$, as follows: $t \rightarrow_{i,S,<} t'$ iff $t \rightarrow_S t'$ by a rule that is applied at an occurence o such that no other rule is applicable below o and there is no greater rule applicable at this occurence o. If there is no order, then $\rightarrow_{i,S}$ is the innermost rewrite relation induced by S and is the innermost strategy of rewriting. Note we can also consider a total order on S.

For more developments see ([8] and [16]).

2.3. Tree languages

A *bottom-up tree automaton* is a quadruple $A=(\Sigma,Q,Q_f,R)$ where Σ is a finite ranked alphabet, Q is a finite set of states of arity 0, Q_f, the set of final states, is a subset of Q, R is a finite set of rules of the following type: $f(q_1,\ldots,q_n) \rightarrow q$ with $n \geq 0$, $f \in \Sigma_n$, $q_1,\ldots,q_n, q \in Q$. Note we can consider R as a ground rewrite system on $T_{\Sigma \cup Q}$.

The tree language recognized by A is $L(A)=\{t \in T_\Sigma \,/\, t \xrightarrow{*}_A q,\ q \in Q_f\}$, where $\rightarrow_A$ is the rewrite relation $\rightarrow_R$. Let A be a bottom-up tree automaton, then there exists a deterministic bottom-up tree automaton B such that L(B)=L(A) (we recall that B is deterministic if B has no two rules with the same left-hand side). A tree language F is *recognizable* if there exists a bottom-up tree automaton A such that L(A)=F. For more developments see [11].

3. BOTTOM-UP TREE PUSHDOWN AUTOMATA

This section is devoted to a discussion of different notions of bottom-up tree pushdown automata. Two "normalization" problems arise: the first one is about the size of the popped terms, and the second one is about the arity of the states. These problems are not too deep, but it is useful to carefully study the situation. In the word case, a transition rule of a pda is of the form $(q, a, \alpha) \rightarrow (q', \beta)$ where q and q' are states, a is a letter of the input alphabet or the empty word ε , α is a letter of the stack alphabet and β is a string over the stack alphabet. During the corresponding move, a is read, α is popped and β is pushed. It is easy to check that if we generalize the definition, allowing a to be a string, then we do not modify the power of the pda; the reason is that we can pop the letters of the pda step by step, memorizing it in the state.

The situation is quite different in the tree case for the reason illustrated by the following example: consider the following ε-rule of a tpda: $q(b(b(x,y),z)) \rightarrow q'(c(x,y,z))$ where q and q' are states, b and c are stack symbols, and x,y,z are variables; if we impose rank one for the states, the rule cannot be simulated by a sequence of rules popping only one letter, because when we pop the first b, we delete one of the subtrees at its two sons. A way to overcome this problem is accepting any rank for states; so, using an intermediate state q" of rank 2, we simulate the above rule by the rules $q(b(x,z)) \to q''(x,z)$ and $q''(b(x,y),z) \to q'(c(x,y,z))$ where q,q' and q" are states, b and c are stack symbols, and x,y,z are variables. We see that this construction is quite natural, and it is not surprising that the following three classes of bottom-up tree pushdown automata are equivalent: tpda with any rank for states and any depth for popped terms; tpda with any rank for states and popping terms of depth 1; tpda with rank 1 for states and any size for popped terms.

An other option concerning the definitions of rules is the following: either use rules that only *read* or *reduce* as Book, Gallier and K. Salomaa or use rules which *read* and *reduce* at the same time as in the string case (reduce rules, as defined by Book and Gallier, are ε-rules, the expression "reduce" does not mean here to reduce the depth of the stack but refers to the associated rewriting).

We easily prove that the two points of view are equivalent, as it is illustrated by the following example: the rule $a(q_1(u_1),\ldots q_n(u_n)) \to q(u)$ ($u_1,\ldots,u_n$ are stack supertrees), which reads a in the input and reduces at the same time, can be simulated by the read rule $a(q_1(x_1),\ldots q_n(x_n)) \to q''(a(x_1,\ldots, x_n))$ (q" is a new state and $x_1,\ldots,x_n$ denote variables, as usual) and the reduce rule $q''(a(u_1, \ldots ,u_n)) \to q(u)$.

3.1. General definitions

Definition1. A *bottom-up tree pushdown automaton* (tpda for short) is a quintuple $T=(\Sigma,\Gamma,Q,Q_f,R)$, where Σ is a finite ranked alphabet of input symbols, Γ is a finite ranked alphabet of stack symbols, Q is a finite set of ranked states, Q_f (included in Q_1, the set of states of rank 1 in Q) is the set of final states, R is a finite set of rewrite rules over $T_{\Sigma\cup\Gamma\cup Q}$ of the following two types:
standard rules: $u= \alpha(q_1(u_{11},\ldots,u_{1n_1}),\ldots,q_m(u_{m1},\ldots,u_{mn_m})) \to q(u_1,\ldots,u_n)$ where $\alpha\in\Sigma_m$, $q_i\in Q$, n_i is the rank of q_i, n is the rank of q, u_{ij}, $u_k \in T_\Gamma(X)$, $j=1,\ldots,n_i$, $i=1,\ldots,m$, $k=1,\ldots,n$, the sets of variables $V(u_{ij})$ are disjoint.
ε-rules: $q(u_1,\ldots,u_n) \to q'(u'_1,\ldots,u'_{n'})$ where q and $q'\in Q$ with rank n and n', respectively, $u_i,u'_j\in T_\Gamma(X)$, $i=1,\ldots,n$, $j=1,\ldots,n'$.

Let $T=(\Sigma,\Gamma,Q,Q_f,R)$ be a tree pushdown automaton.
A *configuration* c of T is a term of $T_{\Sigma\cup\Gamma\cup Q}$ such that path(c) is included in $\Sigma^*Q\Gamma^+\cup\Sigma^*$. The *move relation* $\to_T$ is the rewrite relation $\to_R$ on the set of configurations of T.
The *computation relation* $\overset{*}{\to}_T$ is the reflexive and transitive closure of $\to_T$.
An *initial configuration* is $c = t\in T_\Sigma$. If $c = t_0(q_1(u_{11},\ldots,u_{1n_1}),\ldots,q_m(u_{m1},\ldots,u_{mn_m}))$ and $t=t_0(t_1,\ldots,t_m)$, then $t_1,\ldots,t_m$ have been scanned and the ith read head is in state q_i with associated tree stacks $u_{i1},\ldots,u_{in_i}$.
A *final configuration* is $c = q(v)$ for some final state q and tree stack v.

The *tree transformation induced by T* is the set: $\tau(T)=\{(u,v)\in T_\Sigma x T_\Gamma / u \overset{*}{\to}_T q(v), q\in Q_f\}$.

T and T' are *equivalent* ($T \equiv T'$) if and only if $\tau(T) = \tau(T')$.

The maximum rank of the states of T, denoted by *rank(T)*, is called the rank of T and the maximum depth of the stack supertrees appearing in the left-hand sides of the rules, denoted by *depth(T)*, is called the depth of T. Let $TPDA_{**}$, $TPDA_{n*}$, $TPDA_{*k}$, $TPDA_{nk}$ denote respectively the classes of all tree pushdown automata of any rank and any depth, of rank n and any depth, of any rank and depth k, of rank n and depth k, respectively.

Proposition 1.
For each T in $TPDA_{**}$, there exists a T' in $TPDA_{1*}$ such that $T\equiv T'$.
For each T in $TPDA_{*1}$, there exists a T' in $TPDA_{12}$ such that $T\equiv T'$.
For each T in $TPDA_{1*}$, there exists a T' in $TPDA_{*1}$ such that $T\equiv T'$.

The technical proofs are omitted here.(see[4]).

Thus we prove that the classes $TPDA_{**}$, $TPDA_{1*}$, $TPDA_{*1}$ and $TPDA_{12}$ are equivalent to each other. The most usual class is $TPDA_{1*}$. From [3] we adopt another class of bottom-up tree pushdown automata, the class of bottom-up tree pushdown automata with read rules and reduce rules (reduce rules are called tree stack update rules in [3]).

Definition2. A *bottom-up tree pushdown automaton with read rules and reduce rules* ($tpda^{rr}$ for short) is a quintuple $A=(\Sigma,\Gamma,Q,Q_f,R)$, where Σ is a finite ranked alphabet of input symbols, Γ is a finite ranked alphabet of stack symbols, Q is a finite set of states of arity 1, Q_f (included in Q) is the set of final states, R is a finite set of rewrite rules over $T_{\Sigma\cup\Gamma\cup Q}$ of the following two types:
(a) *Read rules.*
$\alpha(q_1(x_1),...,q_k(x_k)) \to q(\beta(x_1,...,x_k))$ where $\alpha\in\Sigma_k$, $k\geq 0$, $q\in Q$, $q_i\in Q$, $x_i\in X$, $i=1,...,k$, $\beta\in\Gamma_k$;
(b) *Reduce rules* (tree stack update rules).
$q(l) \to q'(r)$ where $q,q'\in Q$, $l,r\in T_\Gamma(X)$.

Let us denote the class of bottom-up tree pushdown automata with read rules and reduce rules by $TPDA^{rr}$. Note that $TPDA^{rr}$ is a subclass of $TPDA_{1*}$, hence for each $tpda^{rr}$ A, we define the set of configurations, the initial configurations, the final configurations, the move relation, the computation relation, the tree transformation induced by A in the same way as for a bottom-up tree pushdown automaton.

Proposition 2.
For each T in $TPDA_{1*}$, there exists an A in $TPDA^{rr}$ such that $T\equiv A$.

Proof. Let $T=(\Sigma,\Gamma,Q,Q_f,R)$ be a bottom-up tree pushdown automaton in $TPDA_{1*}$. We construct the tpda^{rr} $A=(\Sigma,\Gamma',Q',Q_f,R')$ as follows.
For each standard rule r: $\alpha(q_1(u_1),...,q_m(u_m)) \rightarrow q(u)$ of T, we introduce the state q_r, the symbol $\bar{\alpha}$ in Γ' and the following new rules in R':
$\alpha(q_1(x_1),...,q_m(x_m)) \rightarrow q_r(\bar{\alpha}(x_1,...,x_m))$ and $q_r(\bar{\alpha}(u_1,...,u_m)) \rightarrow q(u)$.
It should be clear that $\tau(T)=\tau(A)$.❑

Note: Let < be an order on R, then we define the move relation $\rightarrow_{<A}$ as the innermost relation induced by R with respect to < (see Section 2.2) on the set of configurations of T. We need this definition in Section 4.5 to compute with the strategy "reduce before read". We then say that A is a *tpda^{rr} with priority*.

3.2. Deterministic Tree pushdown automata

Definition. A bottom-up tree pusdown automaton is *deterministic* if the set of rewrite rules is left-linear and is without critical pairs.

Note that we restrict our definition to left-linear tree pushdown automata, indeed, since the set R of rewrite rules is without critical pairs and is left linear, $\rightarrow_R$ is confluent [14].

Let us denote the deterministic subclasses of bottom-up tree pushdown automata classes of the previous Section by $DTPDA_{**}$, $DTPDA_{n*}$, $DTPDA_{*k}$, $DTPDA_{nk}$ and $DTPDA^{rr}$, respectively . We will prove the deterministic version of the main result of the previous section that is, we show that $DTPDA_{**}$, $DTPDA_{1*}$, $DTPDA_{*1}$ and $DTPDA^{rr}$ are equivalent to each other.

Proposition.
For each T in $DTPDA_{**}$, there exists a T' in $DTPDA_{1*}$ such that $T \equiv T'$.
For each T in $DTPDA_{*1}$, there exists a T' in $DTPDA_{12}$ such that $T \equiv T'$.
For each T in $DTPDA_{1*}$, there exists a $T' \in DTPDA_{*1}$ such that $T \equiv T'$.
For each T in $DTPDA_{1*}$, there exists a T' in $DTPDA^{rr}$ such that $T \equiv T'$.

The technical proofs are omitted here.(see[4]).

4. PUSHDOWN AUTOMATA AND REWRITE SYSTEMS

4.1. Introduction

In Section 4.2, we define the main notion of this paper, the class of *tail reduction free* rewrite systems (trf rewrite systems). Tracing the trf property back to inductive reducibility (Plaisted, [17]), we prove the decidability of the trf property (Th.4.4). We give a lot of examples which illustrate the fact that this class is fairly powerful. The *monadic* rewrite systems of Book, Gallier and K. Salomaa are an obvious particular case of trf rewrite systems.

We define also (Section 4.3) *semi-monadic* rewrite systems which generalize monadic rewrite systems but keep their fair properties: a rule is semi-monadic if in its right-hand side with depth greater than 0, sons of the root are variables or ground terms. If a semi-monadic rewrite system is convergent, it is easy to construct an equivalent semi-monadic trf rewrite system (see Section 4.3).

Theorem 1 in Section 4.5 associates a tpda *with reduce priority* which computes normal forms of terms for any convergent trf rewrite system . A tpda with reduce priority is a tpda which applies reduce rules before read rules (i.e. it applies a read rule rewriting the subtree $\alpha(q_1(u_1),\ldots,q_n(u_n))$ in the tree $q(u)$ only when no reduce rule is applicable for the subtrees $q_i(u_i), 1 \leq i \leq n$).

In the left-linear case, the notion of priority is needless and Theorem 2 in Section 4.5 associates a deterministic tpda computing normal forms of ground terms with any left-linear convergent trf rewrite system..

We introduce the notion of tail reduction free rewrite systems with the next two examples.

Example1. Let $\Sigma=\{a,b,c,\$\}$ where a, b, c are of rank 1 and \$ is of rank 0. Let $S = \{ acbx \rightarrow cx \}$ be a rewrite system. We define a tpda $A=(\Sigma,\Sigma,\{q\},\{q\},R)$ with $R=\{\$\rightarrow q\$;\ aqx\rightarrow qax;\ bqx\rightarrow qbx;\ cqx\rightarrow qcx;\ qacbx\rightarrow qcx\}$. If we consider, for example u=aaacbb\$, we have $u \overset{*}{\rightarrow}_A qaaacbb\$$ or $u \overset{*}{\rightarrow}_A qaacb\$$ or $u \overset{*}{\rightarrow}_A qac\$$. Moreover, if we consider $<$ such that the read rules (the four first rules) are less than the reduce rule (the fifth rule) w.r.t $<$, we get $u \overset{*}{\rightarrow}_{<A} qac\$$. Thus, in this case, using a tpda with reduce priority, we can compute the S-normal form of u.

Example 2. Let $S=\{ abx \rightarrow bax \}$ be a rewrite system. We take the tpda $A=(\Sigma,\Sigma,\{q\},\{q\},R)$ with $R=\{\$\rightarrow q\$;\ aqx\rightarrow qax;\ bqx\rightarrow qbx;\ qabx\rightarrow qbax\}$. Consider the tpda with reduce priority as in the previous example, we cannot compute the normal form of each term u. Indeed, for u=aabb\$, we have $u \overset{*}{\rightarrow}_{<A} qbaab\$$ and baab\$ is not the S-normal form of u.

4.2. Tail reduction free rewrite systems

Definition. Let S be a rewrite system, then S has the *tail reduction free* (trf) property iff for each rule $l \rightarrow r$ in S with $r = b(r_1,\ldots,r_n)$ where b is in Σ_n, for each ground irreducible substitution σ, and for each i, $1 \leq i \leq n$, $\sigma(r_i)$ is irreducible.

Note: We consider rewrite systems such that there exists at least one ground irreducible substitution. For example, we avoid, as usual, that the left-hand side of a rule is a variable.

Example: The rewrite system S of Example 1 of the previous section obviously has the trf property and the rewrite system of Example 2 obviously has not (consider the ground substitution $\sigma(x)=b\$$).

4.3. Examples of trf rewrite systems

Example 1: Monadic rewrite systems.

Definition. a rewrite system is said to be *monadic* if, for every rule $l \to r$ in S, depth(l)≥1 and depth(r)≤1.

Proposition. *Each monadic rewrite system has the trf property.*
The proof is obvious.❑

Example 2: Semi-monadic rewrite systems

Definition. A rewrite system is said to be *semi-monadic* if, for every rule $l \to r$ in S, depth(l)≥1 and either r is a variable ($r \in X$) or $r=b(y_1,...,y_k)$ where b is in Σ_k, y_i is either a variable ($y_i \in X$) or a ground term ($y_i \in T_\Sigma$), i=1,...,k.

We note that monadic and ground rewrite systems are particular cases of semi-monadic rewrite systems (a rewrite system is ground if no variable occurs in the rewrite rules).

Proposition. *Let S be a semi-monadic and convergent rewrite system. Then there exists a semi-monadic and convergent rewrite system S' equivalent to S such that S' has the trf property.*
Proof. For each ground term in a right-hand side, we substitute its normal form for S.❑

Example 3: Simulation of a Turing machine.

Theorem. *We can simulate any Turing machine by a trf rewrite system associated to a* $tpda_{12}$.
The technical construction can be found in [4].

4.4 Decidability of the trf property

Theorem. *The trf property is decidable.*

Proof. We prove this result (see [4]) tracing it back to the decidability of inductive reducibility (Plaisted [17]).

4.5. trf rewrite systems and tpda

Proposition 1. *For every trf rewrite system S, there exists a bottom-up tree pushdown automaton* $A=(\Sigma,\Gamma,Q,Q_f,R)$ *and an order < on R such that:*
$(t \xrightarrow{*}_{i,S} t'$ *and* $t' \in IRR(S)) \Leftrightarrow (t \xrightarrow{*}_{<A} q(t')$ *and* $q \in Q_f)$ *holds for every* $t, t' \in T_\Sigma$.

Proof. Let $A=(\Sigma,\Sigma,Q=\{q,q'\},Q_f=\{q'\},R)$ with R the set of rules of the following three types: (i) $q(l) \rightarrow q(r)$, for each rule $l \rightarrow r$ in S.

(ii) $q(x) \rightarrow q'(x)$.

(iii) $b(q'(x_1),...,q'(x_k)) \rightarrow q(b(x_1,...,x_k))$, for each b in Σ_k.

Let < be the partial order defined by: $(q(x) \rightarrow q'(x)) < (q(l) \rightarrow q(r))$, for each rule $q(l) \rightarrow q(r)$ in R. Roughly speaking, the computation relation $\xrightarrow{*}_{<A}$ is defined such that A applies a read rule only when all the corresponding tree stacks are irreducible for the rules (i) and (ii), and then the tree stacks are irreducible for S.

The proof of the equivalence is in [4].❑

Proposition 2. *For every left-linear trf rewrite system S and for each total order < on S, there exists a deterministic bottom-up tree pushdown automaton B such that:*

$(t \xrightarrow{*}_{i,S,<} t'$ and $t' \in IRR(S)) \Leftrightarrow (t \xrightarrow{*}_B q(t')$ and $q \in Q_f)$ holds for every $t, t' \in T_\Sigma$.

Proof. Let S be a left-linear trf rewrite system, $k = \max\{depth(l) \,/\, l \rightarrow r \in S\}+1$ and $T_\Sigma(X,k)$ be the set of k-normal trees. Let us first define $B=(\Sigma,\Sigma,Q,Q_f,R)$ with $Q = \{q_t \,/\, t \in T_\Sigma(X,k)\} \cup \{q\}$, $Q_f = \{q_t \,/\, t$ is not unifiable with a left-hand-side of a rule in S$\}$ and R set of rules defined as follows:

Read rules: (i) $\forall b \in \Sigma, \forall q_i \in Q_f$, $b(q_1(x_1),...,q_n(x_n)) \rightarrow q(b(x_1,...,x_n))$

Reduce rules: (ii) $q(t) \rightarrow q_t(t)$, for each $t \in T_\Sigma(X,k)$

(iii) $q_t(l) \rightarrow q(r)$, for each $q_t \notin (Q_f \cup \{q\})$, where $l \rightarrow r$ is the greatest rule for < such that t matches l.

It is obvious that B is deterministic. The proof of the equivalence is in [4].❑

Note: In the non left-linear case, we cannot use a finite set of states (or terms of depth less than k) to express that a term is irreducible as we have to check equality at an unbounded depth.

For convergent (i.e noetherian and confluent) trf rewrite systems, we have:

Theorem 1. *For every convergent trf rewrite system S, there exists a bottom-up tree pushdown automaton A and an order < on R such that:for every $t, t' \in T_\Sigma$ $(t \xrightarrow{*}_S t'$ and $t' \in IRR(S)) \Leftrightarrow (t \xrightarrow{*}_{<A} q(t')$ and $q \in Q_f)$.*

Theorem 2. *For every convergent left-linear trf rewrite system S, there exists a deterministic bottom-up tree pushdown automaton B such that:for every $t, t' \in T_\Sigma$, $(t \xrightarrow{*}_S t'$ and $t' \in IRR(S)) \Leftrightarrow (t \xrightarrow{*}_B q(t')$ and $q \in Q_f)$.*

The proofs of these two theorems are obvious, using the two previous propositions and the hypothesis S convergent.❑

5. LANGUAGES, PUSHDOWN AUTOMATA AND REWRITE SYSTEMS

5.1. Semi-monadic rewrite systems

We prove in this section that recognizability is preserved for semi-monadic rewrite systems. This result embeds the same result of K. Salomaa ([18]) for monadic rewrite systems and the same known result for ground rewrite systems([5],[6]).

Let S be a rewrite system and L be a recognizable tree language, we denote by S*(L) the set of descendants of L modulo S, i.e $S^*(L)=\{ t' / t \xrightarrow{*}_S t' \;\&\; t \in L\}$.

Theorem. *Let S be a linear semi-monadic rewrite system and L be a recognizable tree language , then S*(L) is recognizable.*

The technical proof can be found in [9].

Note: This result is obviously false if S is non right-linear (consider the rewrite system reduced to the rule $f(x)\rightarrow g(x,x)$ and the recognizable tree language L= fa*$), however, we presume that the same result holds if S is a right-linear semi-monadic rewrite system.

5.2 Languages and pushdown automata

It is easy to define (by empty stacks or by final states) the tree language defined by a tpda. For connections between context-free tree languages and tree pusdown automata, see [10] for the bottom-up case and [13] for the top-down case.

We saw that we can simulate any Turing machine with a bottom-up tree pushdown automaton in the class $TPDA_{12}$; it is easy to modify the construction to obtain the same result in the class $TPDA_{21}$. But what about the class $TPDA_{11}$? We conjecture that the associated class of tree languages is a subclass of context free tree languages, and that it can be analized in real time.

Conclusion

Developments of the present work, from the language and the complexity point of view, study of the connection between rewriting and top-down pushdown automata, could be further works for a better knowledge of particular kinds of rewrite systems.

References

[1] R.V. Book, M. Jantzen and C. Wrathall, Monadic Thue Systems, Theoret. Comput. Sci. **19** (3)(1982) 231-252.

[2] R.V. Book, Thue systems as Rewriting Systems, J. of Symbolic Computation, 3, (1987).

[3] R.V. Book and J.H. Gallier, Reductions in Tree Replacement Systems, Theoretical Computer Science, 37, (1985).

[4] J.L.Coquidé, M.Dauchet, R.Gilleron and S.Vàgvölgyi, Bottom-up tree pushdown automata and rewrite systems, Technical Report, I.T 190, L.I.F.L, Université de Lille.

[5] J.L.Coquidé and R.Gilleron, Proofs and Reachability Problem for Rewrite Systems, IMYCS, to appear in Lec. Notes Comp. Sci, (1989).

[6] M. Dauchet and S.Tison, Decidability of Confluence in Ground Term Rewriting Systems, Fondations of Computation Theory, Cottbus, Lec. Notes Comp. Sci., 199, (1985)

[7] M.Dauchet and S. Tison, The theory of Ground Rewrite System is Decidable, IEEE Symposium on Logic in Computer Science, Philadelphie, to appear, (1990)

[8] N.Dershowitz and J.P. Jouannaud, Rewrite systems, Handbook of Theoretical Computer Science, J.V.Leeuwen editor, North-Holland, to appear.(1989).

[9] Z. Fülöp and S. Vàgvölgyi, Ground Term Rewriting rules for the Word Problem of Ground Term Equations, submitted paper, (1989).

[10] J.H. Gallier and K.M. Schimpf, Parsing Tree Languages using Bottom-up Tree Automata with Tree Pushdown Stores, J. Computer Science, (1985).

[11] F. Gecseg and M. Steinby, Tree automata, Akademiai Kiado, (1984).

[12] J. Goguen, J.W. Thatcher, E. Wagner and E. Wright, An initial algebra approach to the specification correctness, and implementation of abstract data types, in: *Current Trends in Programming Methodology Vol. 4* (Prentice-Hall, Englewood Cliffs, NJ, 1977) 80-149.

[13] I. Guessarian, Pushdown Tree Automata, Math. Systems Theory, 16, (1983).

[14] G. Huet, Confluent Reductions: Abstract properties and applications to Term Rewriting Systems, J.A.C.M., 27, (1980).

[15] G.Huet and J-J. Levy, Call by need Computation in non-ambiguous linear Term Rewriting Systems, TR359, I.N.R.I.A., Le Chesnay, France, (1979).

[16] G.Huet and D.C. Oppen, Equations and Rewrite Rules: A survey, in R.V.Book, ed., New York, Academic Press, Formal Language Theory: Perspectives and Open Problems, (1980).

[17] D.A. Plaisted, Semantic Confluence Tests and Completion Methods, Information and Control, 65, (1985).

[18] K. Salomaa, Deterministic Tree Pushdown Automata and Monadic Tree Rewriting Systems, Journal of Comput. and Syst. Sci., 37, (1988).

[19] K.M. Schimpf, A Parsing Method for Context-free Tree Languages, Ph.D., Univ. of Pennsylvania, (1982).

On relationship between term rewriting systems and regular tree languages

G.A.Kucherov
Institute of Informatics Systems
Siberian Division, USSR Academy of Sciences
Novosibirsk, 630090, USSR

The paper presents a new result on the relationship between term rewriting systems (TRSs) and regular tree languages. Important consequences (concerning, in particular, a problem of ground-reducibility) are discussed.

1. Introduction.

It is well known that in many applications of term rewriting techniques non-left-linear TRSs (i.e. those with possibly multiple occurences of a variable in the left-hand side of a rule) are essentially more difficult to handle than left-linear ones (see, e.g., [Der81,GKM83,Huet80]). According to general mathematical intuition this is not very surprising, since in most fields of mathematics non-linear objects (e.g., differential equations, algebraic equations, finite difference schemes, etc.) are essentially more complex. But a formal comparative study of the power of linear and non-linear objects as well as the problem of "linear approximation" is always of great interest.

Obviously, a comparison between left-linear and non-left-linear TRSs may have different purposes and, accordingly, may be stated in different ways. For example, in [DC87] the expressive power of different ways of "generating" sets of ground terms is compared (here "generating" means reduction of a set of ground terms by a given TRS). Our paper takes another approach where the powers of left-linear and non-left-linear TRSs are compared with respect to the corresponding sets of reducible ground terms. Within this approach the following question may be propounded: if $\mathcal{R}$ is a non-left-linear TRS and $\mathcal{S}$ is the set of $\mathcal{R}$-reducible ground terms then we ask whether a "weaker" left-linear TRS $\mathcal{L}$ exists such that every term in $\mathcal{S}$ is $\mathcal{L}$-reducible.

Let us give a toy example which illustrates also a more pragmatic motivation of the problem. Suppose we want to specify an equality predicate *eq* over the natural numbers defined by the free constructors *0* and *s*. The axiom

$$eq(x,x) \rightarrow true \qquad (1)$$

naturally comes into mind, but it does not form a complete axiomatization of eq since it remains undefined when its arguments are different. To complete the specification we have to add, e.g., the following three axioms.

$$eq(0,s(x)) \rightarrow false \qquad (2)$$
$$eq(s(x),0) \rightarrow false \qquad (3)$$
$$eq(s(x),s(y)) \rightarrow eq(x,y) \qquad (4)$$

Apparently, (1)-(4) form a complete specification of eq, i.e. all terms of the set $S=\{eq(M_1,M_2) \mid M_1,M_2 \in T_F\}$ are reducible. Now we note that axiom (1) and no other axiom is applicable only when both arguments of eq are 0. Thus, preserving the completeness of the specification, axiom (1) can be transformed into $eq(0,0) \rightarrow true$ so that the resulting specification is left-linear. (The resulting specification may be less effective but in this paper we are concerned only about matters of principle.)

In fact, the example above illustrates also a main result of the paper which states that such a "preserving" transformation is possible to perform whenever S is a *regular tree language*. This result has a number of important consequences, two of them are discussed in the paper. The first one characterizes a class of regular tree languages defined by TRSs. It is known ([GB85]) that left-linear TRSs define regular languages of reducible ground terms, but the question whether this "TRS-definable" subclass of regular tree languages enlarges if non-left-linear TRSs are admitted is still to be answered. In this work we give a negative answer to this question. We show that if some (non-left-linear) TRS define a regular tree language, than there exists a left-linear TRS defining the same language.

Another consequence concerns a problem of ground-reducibility of a term for a given TRS [KZ85,JK86]. A term t is $\mathcal{R}$-ground-reducible if all ground instances of t are $\mathcal{R}$-reducible. Ground-reducibility is very important in reasoning about abstract data types [JK86, Com88, Kuc88, Fr86, KNZ87]. As it was shown in [Pla85,KNZ87], ground-reducibility is decidable for any term t and any TRS $\mathcal{R}$. However, while practical and transparent ground-reducibility testing algorithms are known for left-linear systems (see [Kuc88]), the problem is much more complicated in the non-left-linear case. Several new solutions have been presented recently [Com88,Kou90], but in our opinion the design of a practicable algorithm for the general case is still on the agenda. Besides, the complexity of the problem also has to be studied.

Some complexity measures for related problems have been presented in [KNRZ87].

In our work we show that if a linear term is $\mathcal{R}$-ground-reducible then it is $\mathcal{R}^*$-ground-reducible, where $\mathcal{R}^*$ is a left-linear system derived from $\mathcal{R}$ as in the above example. For example, if we have to specify completely a function f, i.e. we should make the term $f(x_1,\ldots,x_n)$ ground-reducible, then we can always do without non-left-linear rules. Thus, non-left-linearity may only make the computation of f more effective, and is never inevitable for defining its values. In other words, non-left-linear TRSs are not more expressive than left-linear ones in defining linear terms completely. Since ground instances of a linear term form a regular set, this gives another illustration of the leit-motif of the work: *in specifying a regular set of ground terms non-left-linear rules can always be avoided.*

2. Preliminaries

2.1. Term rewriting systems

For our presentation we will need only basic definitions from term rewriting systems theory (for more details see [HO80]). Let T_F denote the set of ground terms over a given signature F. $T_F(\mathcal{X})$ stands for the set of terms over a set of generators $\mathcal{X}$. In particular, $T_F(X)$ denotes the set of terms with variables, and $var(t) \subseteq X$ is the set of variables in $t \in T_F(X)$. We define the set of occurences $O(t)$ in the usual way as sequences of natural numbers (including the empty sequence corresponding to the root occurence). If $t \in T_F(\mathcal{X})$, then $\hat{O}(t) \subseteq O(t)$ denotes the subset of occurences of generators in t (for instance, if $t \in T_F(X)$, then $\hat{O}(t)$ is the set of variable occurences in t). For two occurences $u,v \in O(t)$ define $u \leqslant v$ iff u is a prefix of v and $u<v$ iff $u \leqslant v$ and $v \neq u$.

Given a TRS $\mathcal{R}$, a term $t \in T_F(X)$ is ($\mathcal{R}$-)reducible iff a subterm of t is matched by the left-hand side of a rule in $\mathcal{R}$. A term $t \in T_F(X)$ is ($\mathcal{R}$-)ground-reducible iff for every ground substitution δ $\delta(t)$ is ($\mathcal{R}$-)reducible. Throughout the paper we consider only reducibility (ground- reducibility) of terms, but not the results of reductions. Therefore we identify a TRS $\mathcal{R}$ with the set of its left-hand sides and we treat $\mathcal{R}$ as an ordinary set of terms. $Gr(\mathcal{R})$ stands for the set of ground instances of the terms of $\mathcal{R}$ and $Red(\mathcal{R})$ stands for the set of $\mathcal{R}$-reducible ground terms.

For a term, $t \in T_F(X)$ $\sigma(t)$ is the result of superposition (the superposant) of a term s into t at $u \in O(t) \setminus \hat{O}(t)$, if σ is a most general unifier of s and t/u (σ is called a superposition substitution).

2.2 Regular tree languages

Regular tree languages [GS84] are a natural generalization of "classical" regular languages (of words over a finite alphabet). Most basic results about regular languages have their counterparts in the tree case. In particular, there are three main presentations of regular tree languages:

- tree automata;
- regular tree grammars;
- regular expressions.

In what follows, we use the first two of these. We will always consider ground terms with the natural tree structure as language objects. In the rest of the section we recall the definitions of tree automata and regular grammars for ground terms languages.

A *bottom-up tree automaton* is a quadruple $\mathcal{A}=(Q,F,Q_{fin},P)$, where Q is a finite set of states, F a signature, $Q_{fin} \subseteq Q$, and P is a finite set of transition rules

$f(q_1,\ldots,q_n) \to q_0$, where $f \in F_n$, $n \geq 1$, $\{q_0,\ldots,q_n\} \subseteq Q$, or

$a \to q_0$, where $a \in F_0$, $q_0 \in Q$.

Informally, given a ground term M the automaton attempts to visit all the nodes of M moving from the leaves towards the root. A node can be visited only if all its immediate descendants have been visited.

An automaton $\mathcal{A}$ defines a (multi-valued, partial) function $\psi_{\mathcal{A}}: T_F \to Q$ defined recursively as follows.

- if $M=a \in F_0$ and P contains a rule $a \to q_0$, then $\psi_{\mathcal{A}}(M)=q_0$,
- if $M=f(M_1,\ldots,M_n)$, $f \in F_n$, for every $i \in (1,n)$ $\psi_{\mathcal{A}}(M_i)=q_i$, and P contains a rule $f(q_1,\ldots,q_n) \to q_0$, then $\psi_{\mathcal{A}}(M)=q_0$.

The automaton $\mathcal{A}$ is said to accept a term M, if $\psi(M) \in Q_{fin}$. An automaton $\mathcal{A}$ defines a regular language of terms accepted by $\mathcal{A}$. The automaton $\mathcal{A}$ is called

- *deterministic*, if there are no two transition rules in P with the same left-hand sides,
- *completely defined*, if for every $f \in F_n$ and every $q_1,\ldots,q_n \in Q$ there exists a transition rule with the left-hand side $f(q_1,\ldots,q_n)$,
- *equivalent to an automaton* $\mathcal{B}$, if $\mathcal{A}$ and $\mathcal{B}$ define the same language,

It is known [GS84] that every bottom-up tree automaton $\mathcal{A}$ has an equivalent completely defined deterministic automaton $\mathcal{B}$. Obviously, for such an automaton B a function ψ_B is single-valued and total.

A *regular tree grammar* is a quadruple $\Phi=(N,F,r,R)$, where N is a finite set of nonterminal symbols, F a signature, $r\in N$ an axiom, and R a finite set of derivation rules

$$p\to t,\ \text{where } p\in N,\ t\in T_F(N).$$

Terms of $T_F(N)$ (note that $T_F\subseteq T_F(N)$) are called *sentential forms*. A sentential form s_2 is derived (in one step) from a sentential form s_1, if there exists a derivation rule $p\to t$ and $u\in\hat{O}(s_1)$ such that $s_1/u=p\in N$ and $s_2=s_1[u\leftarrow t]$. A language generated by a grammar $\Phi=(N,F,r,R)$ is the set of ground terms derivable from r in an arbitrary number of steps.

Regular grammars and tree automata are two different ways of defining the class of all regular tree languages.

3. TRSs and regular tree languages

For the sake of convenience we will write "a linear set" as a synonym for "a set of linear terms".

A fundamental result about the relationship between linear TRSs and regular tree languages is due to [GB85].

Theorem 3.1.[GB85] If $\mathcal{R}$ is a linear set then $Red(\mathcal{R})$ is a regular tree language.

Obviously, the theorem can be extended to any non-linear set $\mathcal{R}$ for which $Gr(\mathcal{R})=Gr(\mathcal{L})$ for a linear TRS $\mathcal{L}$. This property has been extensively studied in [LM87].

On the other hand, it is known that there are regular languages of ground terms, which cannot be represented as $Red(\mathcal{R})$ for any TRS $\mathcal{R}$ [FV88]. (A trivial example [Com89a] is the language of even numbers $\{s^{2n}(0)\,|\,n\geqslant 0\}$. The term 0 belongs to the language, but if 0 is reducible then any term $s^k(0)$ is reducible.)

However, the following question arises. Do left-linear TRSs generate all regular tree languages representable as $Red(\mathcal{R})$? In other words, if for a non-linear set $\mathcal{R}$ $Red(\mathcal{R})$ is a regular tree language, does a linear set $\mathcal{L}$ always exist such that $Red(\mathcal{L})=Red(\mathcal{R})$? In this section we answer this question positively.

The following definition formalizes a notion of "transformation into a linear set" which we illustrated in the introduction.

Definition 3.1. A set $\mathcal{R}^*$ is called an *instantiation* of a set $\mathcal{R}$ iff every term in $\mathcal{R}^*$ is an instance of a term in $\mathcal{R}$. $\mathcal{R}^*$ is called a *linear instantiation* if it is linear in addition.

Clearly, if $\mathcal{R}^*$ is a linear instantiation of $\mathcal{R}$, then for every $t\in\mathcal{R}^*$ there exists $s\in\mathcal{R}$ s.t. $\sigma(s)=t$ and σ assigns a ground term to every non-linear variable of s.

Suppose $Lan\subseteq T_F$ is a regular ground term language, $\mathcal{R}$ a TRS, and $Lan\subseteq Red(\mathcal{R})$, i.e. every term $M\in Lan$ is $\mathcal{R}$-reducible. Let us split Lan into two sublanguages as follows. $Lan=Lan'\uplus Lan''$, where Lan' consists of all $M\in Lan$, reducible only at root (and irreducible at any $u\neq\varepsilon$), and Lan'' consists of all $M\in Lan$ reducible at some $u\neq\varepsilon$. Obviously, $Lan'\subseteq Gr(\mathcal{R})$. The following key lemma says that Lan' can be "approximated" with a linear instantiation of $\mathcal{R}$.

Lemma 3.1. Let $\mathcal{R}$ and $Lan=Lan'\uplus Lan''$ be defined as above (in particular, $Lan'\subseteq Gr(\mathcal{R})$). Then there exists a linear instantiation $\mathcal{R}^*$ of $\mathcal{R}$ s.t. $Lan'\subseteq Gr(\mathcal{R}^*)$.

Proof. The proof of the lemma itself requires technical lemmas and is too long to be given here. It can be found in the full version of the paper [Kuc90]. Here we only explicate a scenario of the proof to illustrate the main idea.

If $\mathcal{R}$ is linear, then $\mathcal{R}^*=\mathcal{R}$. Otherwise consider a non-linear term $p\in\mathcal{R}$ and suppose $y\in var(p)$ occurs twice or more in p. Consider all instances $\delta_1(p),\delta_2(p),\ldots$ s.t.

(a) $\forall i\ \delta_i(p)\in Lan'$, and

(b) $\forall i\ \delta_i(p)\notin Gr(\mathcal{R}\setminus\{p\})$ (i.e. $\delta_i(p)$ is an instance of no other $t\in\mathcal{R}$).

If there are no such instances then we delete p from $\mathcal{R}$ preserving the condition $Lan'\subseteq Gr(\mathcal{R})$ and reapply recursively the procedure. If such instances exist, consider the set $\Delta=\{\delta_1(y),\delta_2(y),\ldots\}$ of ground terms assigned to y. If Δ is finite, then we instantiate p replacing y by the terms of Δ. Thus, we remove a non-linearity (still keeping up $Lan'\subseteq Gr(\mathcal{R})$), and we reapply again the procedure. Finally, assume that Δ is infinite and show that this entails a contradiction with the lemma.

Suppose Lan is defined by a minimal completely specified deterministic bottom-up tree automaton $\mathcal{A}$ with a set of states Q. We construct a term $\hat{M}$ which is accepted by $\mathcal{A}$ but is irreducible, thus coming to contradiction. Now let us give an idea of constructing $\hat{M}$.

First note that if $M_1,M_2\in T_F$, $M_1\in Lan$, $u\in O(M_1)$ and $\Phi_{\mathcal{A}}(M_1/u)=\Phi_{\mathcal{A}}(M_2)$, then $M_1[u\Leftarrow M_2]\in Lan$. Since Δ is infinite, it contains infinitely many terms corresponding to some state $q_0\in Q$. Therefore without loss of generality we assume $\Phi_{\mathcal{A}}(\delta_i(y))=q_0$ for all $\delta_i(y)\in\Delta$. We also assume that all terms of Δ are different.

Suppose $u_1, u_2 \in \hat{O}(p)$ are two different occurences of y in p. We assume $\hat{M}$ to have one of the forms $\hat{M}=\delta_i(p)[u_1 \Leftarrow \delta_j(y)]$ or $\hat{M}=\delta_i(p)[u_2 \Leftarrow \delta_j(y)]$ for $i \neq j$. On the one hand, $\hat{M} \in Lan$ (according to the above remark). On the other hand, $\hat{M}$ is not an instance of p, since different terms are assigned to y. Finally, $\hat{M}$ is $\mathcal{R}$-irreducible at any $u \in \hat{O}(p)$, since the subterm $\hat{M}/u$ coincides with $\delta_i(p)/u$ (or $\delta_j(p)/u$). Thus, to complete the proof we show that there exist i,j such that a term $\hat{M}$ cannot be reduced at any $u \in O(p) \setminus \hat{O}(p)$.■

Lemma 3.1 allows us to state as its consequence the main theorem of this section.

Theorem 3.2. Let $\mathcal{R}$ be an arbitrary TRS. If $Red(\mathcal{R})$ is a regular tree language, then there exists a linear instantiation $\mathcal{R}^*$ s.t. $Red(\mathcal{R}^*)=Red(\mathcal{R})$.

Proof. Suppose $Lan=Red(\mathcal{R})$. Consider the subset $Lan' \subseteq Lan$ consisting of the terms, reducible only at root. By lemma 3.1 there exists a linear instantiation $\mathcal{R}^*$ of $\mathcal{R}$ s.t. $Lan' \subseteq Gr(\mathcal{R}^*)$. Since every term of Lan has a subterm belonging to Lan' (every reducible term has a reducible subterm which has no proper reducible subterm), then every term of Lan is $\mathcal{R}^*$-reducible and, hence, $Lan \subseteq Red(\mathcal{R}^*)$. On the other hand, $\mathcal{R}^*$ is an instantiation of $\mathcal{R}$ and therefore $Red(\mathcal{R}^*) \subseteq Lan$. Thus, $Red(\mathcal{R}^*)=Lan$.■

To sum up, theorem 3.2 shows that the class of regular ground term languages defined by arbitrary TRSs coincides with that defined by left-linear TRSs.

4 Main result

Using lemma 3.1 we prove in this section the main result of the paper. In contrast to the previous section we use here a regular grammar representation for regular tree languages. The reason is partly explained by the following lemma whose counterpart for tree automata does not hold.

Lemma 4.1. Suppose Lan is a regular ground term language defined by a regular grammar $\Phi=(N,F,r,P)$, and suppose $\mathcal{L}$ is a linear set. Then the regular language $Lan \cap Gr(\mathcal{L})$ can be defined by a grammar $\Phi'=(N,F,r,P')$ (i.e. a grammar with the same set of nonterminals).

Proof is omitted because of space limitations.

Now we are in position to prove the main result.

Theorem 4.1. Suppose *Lan* is a regular ground term language, and $Lan \subseteq Red(\mathcal{R})$. Then there exists a linear instantiation $\mathcal{R}^*$ of $\mathcal{R}$ s.t. $Lan \subseteq Red(\mathcal{R}^*)$.

Proof. Let $\Phi=(N,F,r,P)$ be a regular tree grammar defining *Lan*. Instead of considering *Lan* we shall prove the theorem for the regular sublanguage $Lan_0 \subseteq Lan$ defined by the grammar $\Phi_0=(N,F,r,P_0)$, where Φ_0 is derived from Φ by deleting every derivation rule with r occuring in its right-hand side. This restriction is correct since $Lan \subseteq Red(\mathcal{R})$ is equivalent to $Lan_0 \subseteq Red(\mathcal{R})$.

We proceed by induction over the number of nonterminals in N.

Base case. If $N=\{r\}$, then Lan_0 is finite, for it consists of all right-hand sides of P which have no occurences of r (=are ground terms). For a finite language the theorem trivially holds.

Induction step. Suppose the theorem holds for any language for which the number of nonterminals is less than k, and suppose a N contains k elements. Without loss of generality we assume P_0 to have a single rule having r in its left-hand side. A generalization to the case of finite number of such rules is straightforward.

For every $p \in N$ denote by $Lan_0(p)$ a set of ground terms derivable from p in the grammar Φ_0 (thus, $Lan_0=Lan_0(r)$). Since r does not occur in the right-hand sides of rules of P_0 then the rule $r \to t$ can be applied only at the first step of the derivations of terms of Lan_0 and cannot be applied otherwise. In particular, the rule $r \to t$ is not applicable in deriving terms of $Lan_0(p)$ for any $p \in N \setminus \{r\}$. Hence, $Lan_0(p)$ is defined by a grammar with the number of nonterminals smaller than k.

Now consider the rule $r \to t$. Let $\hat{O}(t)=\{u_1,\ldots,u_m\}$ and $p_1,\ldots,p_n \in N$ be the nonterminals occuring in t. Substitute at $u_1,\ldots,u_m$ different variables $x_1,\ldots,x_m$ (so that every x_l corresponds to a nonterminal $p_{j_l} \in \{p_1,\ldots,p_n\}$) and let $\tilde{t}$ be the resulting linear term. Every $M \in Lan_0$ is an instance of $\tilde{t}$ and is uniquely associated with the tuple $\langle M/u_1,\ldots,M/u_m\rangle$, where M/u_l is an element of the corresponding language $Lan_0(p_{j_l})$. By the theorem condition M is $\mathcal{R}$-reducible, and there are two alternatives:

(a) M is reducible at some $u \in O(\tilde{t}) \setminus \hat{O}(\tilde{t})$, but is not at any $v \geqslant u_l$ for any $u_l \in \hat{O}(\tilde{t})$, that is, every component of the tuple $\langle M/u_1,\ldots,M/u_m\rangle$ is irreducible,

(b) M is reducible at some $v \geqslant u_l$ for some $u_l \in \hat{O}(\tilde{t})$, that is, there is a reducible component in $\langle M/u_1,\dots,M/u_m\rangle$.

By using lemma 3.1 (here we skip some technical details that can be found in [Kuc90]) there exists a linear instantiation $\mathcal{R}^*$ such that every $M\in Lan_0$ satisfying (a) is $\mathcal{R}^*$-reducible at some $u\in O(\tilde{t})\setminus\hat{O}(\tilde{t})$, i.e. M is an instance of $\{\delta_1(\tilde{t}),\dots,\delta_K(\tilde{t})\}$, where $\{\delta_1(\tilde{t}),\dots,\delta_K(\tilde{t})\}$ is the linear set of superposants of $\mathcal{R}^*$ into $\tilde{t}$. So the language $Lan_0\cap Gr(\{\delta_1(\tilde{t}),\dots,\delta_K(\tilde{t})\})$ is reducible by a linear instantiation of $\mathcal{R}$.

Now we show that the language $Lan_0\setminus Gr(\{\delta_1(\tilde{t}),\dots,\delta_K(\tilde{t})\})$ is also reducible by a linear instantiation of $\mathcal{R}$. Using the results of [LM87], for a linear set $\{\delta_1(\tilde{t}),\dots,\delta_K(\tilde{t})\}$ we compute a linear set $\{\eta_1(\tilde{t}),\dots,\eta_J(\tilde{t})\}$ such that $Gr(\{\eta_1(\tilde{t}),\dots,\eta_J(\tilde{t})\})=Gr(\{\tilde{t}\})\setminus Gr(\{\delta_1(\tilde{t}),\dots,\delta_K(\tilde{t})\})$. Since $Lan_0\cap Gr(\{\delta_1(\tilde{t}),\dots,\delta_K(\tilde{t})\})$ contains all the terms satisfying (a), then $Lan_0\cap Gr(\{\eta_1(\tilde{t}),\dots,\eta_J(\tilde{t})\})$ contains only the terms satisfying (b). Thus, for every $M\in Lan_0\cap Gr(\{\eta_1(\tilde{t}),\dots,\eta_J(\tilde{t})\})$ the tuple $\langle M/u_1,\dots,M/u_m\rangle$ is reducible. Denote

$$H_1=\langle\eta_1(x_1),\dots,\eta_1(x_m)\rangle,\dots,H_J=\langle\eta_J(x_1),\dots,\eta_J(x_m)\rangle$$

and assume without loss of generality that $H_1,\dots,H_J$ are pairwise non-unifiable. If for $M\in Lan_0$ the tuple $\langle M/u_1,\dots,M/u_m\rangle$ is an instance of $H_1,\dots,H_J$, then there exists a reducible component M/u_l. Then for every $H_k=\langle\eta_k(x_1),\dots,\eta_k(x_m)\rangle$ there exists a component $\eta_k(x_l)$ s.t. every term $N\in Lan_0(p_{j_l})\cap Gr(\eta_k(x_l))$ is $\mathcal{R}$-reducible. (Otherwise there exists $M\in Lan_0\cap Gr(\{\eta_1(\tilde{t}),\dots,\eta_J(\tilde{t})\})$ for which the tuple $\langle M/u_1,\dots,M/u_m\rangle$ is not reducible). Consider the language $Lan_0(p_{j_l})\cap Gr(\eta_k(x_l))$. Since $Lan_0(p_{j_l})$ is defined by a grammar with the number of nonterminals less than k, then by lemma 4.1 the language $Lan_0(p_{j_l})\cap Gr(\eta_k(x_l))$ can be defined by a grammar with the same number of nonterminals. By induction hypothesis every term of the language $Lan_0(p_{j_l})\cap Gr(\eta_k(x_l))$ is reducible by a linear instantiation $\mathcal{R}_k^*$ of $\mathcal{R}$. Thus, all tuples $\langle M/u_1,\dots,M/u_m\rangle$ which are instances of $H_1,\dots,H_J$ are reducible by a linear instantiation $\mathcal{R}_1^*\cup\dots\cup\mathcal{R}_L^*$, i.e. the language $M\in Lan_0\cap Gr(\{\eta_1(\tilde{t}),\dots,\eta_J(\tilde{t})\})$ is also reducible by a linear instantiation of $\mathcal{R}$. This completes the proof.∎

Thus, we have proved the main result which states that when specifying a regular language of ground terms (so that every term of the language is reducible) non-left-linearity of rules is always possible to avoid.

If $Lan=Red(\mathcal{R})$, we obtain immediately theorem 3.2 of the previous section. If $\mathit{Lan}=\mathit{Red}(\mathcal{L})$ for a linear set $\mathcal{L}$ we come up with another important theorem.

Theorem 4.2. Suppose $\mathcal{L},\mathcal{R}$ are finite sets, $\mathcal{L}$ is linear, and $\mathit{Red}(\mathcal{L})\subseteq\mathit{Red}(\mathcal{R})$. Then there exists a linear instantiation $\mathcal{R}^*$ of $\mathcal{R}$ s.t. $\mathit{Red}(\mathcal{L})\subseteq\mathit{Red}(\mathcal{R}^*)$.

In the next section we show how theorem 4.2 is applied to the problem of ground-reducibility.

To end this section we point out a relation between theorem 4.2 and the results of [LM87] which we have already used in the proof of theorem 4.1. Given a set of terms $\mathcal{R}$ a set $\bar{\mathcal{R}}$ is said to be a complement of $\mathcal{R}$ [Kuc88] if $\mathit{Gr}(\bar{\mathcal{R}})=T_F\backslash\mathit{Gr}(\mathcal{R})$. We can briefly summarize [LM87] as follows. For a finite set $\mathcal{R}$ a finite complement (in terms of [LM87] "disjunction of terms") exists iff a linear instantiation $\mathcal{R}^*$ exists s.t. $\mathit{Gr}(\mathcal{R}^*)=\mathit{Gr}(\mathcal{R})$. An algorithm has been proposed for testing this condition and computing $\bar{\mathcal{R}}$ when possible. Using the former statement we can easily deduce the following proposition.

Proposition 4.1. Suppose $\mathcal{R}_1,\mathcal{R}_2$ are finite sets , $\mathcal{R}_1$ is linear, and $\mathit{Gr}(\mathcal{R}_1)\subseteq\mathit{Gr}(\mathcal{R}_2)$. Then there exists a linear instantiation $\mathcal{R}_2^*$ of $\mathcal{R}_2$ s.t. $\mathit{Gr}(\mathcal{R}_1)\subseteq\mathit{Gr}(\mathcal{R}_2^*)$.

This proposition is exactly our theorem 4.2 where *Red* is replaced by *Gr*. Clearly, using theorem 4.1 we could further generalize proposition 4.1 replacing $\mathit{Gr}(\mathcal{R}_1)$ by an arbitrary regular language *Lan*. Thus, our results can be interpreted as "lifting" some results of [LM87] from the level of matching to that of reducibility (=matching at an arbitrary occurence).

Unfortunately, it is impossible to generalize the notion of complement to our case. As simple examples show, even if $\mathcal{R}$ is a linear set, a finite set $\hat{\mathcal{R}}$ s.t. $\mathit{Red}(\hat{\mathcal{R}})=T_F\backslash\mathit{Red}(\mathcal{R})$ does not generally exist.

5. Application to ground-reducibility

In the introductory example we have built a linear instantiation of a given TRS, preserving ground-reducibility of the term $eq(x,y)$. In this section we answer the question whether such a transformation is always possible to perform. In other words, given a TRS $\mathcal{R}$ and a $\mathcal{R}$-reducible term t does a linear instantiation $\mathcal{R}^*$ always exist s.t. t is $\mathcal{R}^*$-ground-reducible?

Lemma 5.1. Suppose $\mathcal{L},\mathcal{R}$ are TRSs. The following conditions are equivalent.

1. $Red(\mathcal{L}) \subseteq Red(\mathcal{R})$.

2. for every $t \in \mathcal{L}$ t is $\mathcal{R}$-ground-reducible.

Proof. Obvious.■

Lemma 5.1 together with theorem 4.2 give a positive answer to the question above provided that t is a linear term.

Lemma 5.2. If $\mathcal{R}$ is a TRS, and a term t is linear and $\mathcal{R}$-ground-reducible, then there exists a linear instantiation $\mathcal{R}^*$ of $\mathcal{R}$ s.t. t is $\mathcal{R}^*$-ground-reducible.

Proof. Follows directly from lemma 5.1 and theorem 4.2.■

Now consider the case when t is non-linear. Suppose $t \in T_F(X)$ and $var(t)=\{x_1,\dots,x_n\}$. Here we use the same technique as in the proof of theorem 4.1. Compute all superpositions of the terms of $\mathcal{R}$ into t and suppose $\{\sigma_1(t),\dots,\sigma_m(t)\}$ is the set of corresponding superposants. Now introduce a new n-ary function symbol h and consider a term $h(x_1,\dots,x_n)$. Denote $\sigma_l(h(x_1,\dots,x_n))=h(\sigma_l(x_1),\dots,\sigma_l(x_n))= h(s_{l1},\dots,s_{ln})$. Clearly, t is $\mathcal{R}$-ground-reducible iff $h(x_1,\dots,x_n)$ is ground-reducible by $\mathcal{R}\cup\{h(s_{11},\dots,s_{1n}),\dots,h(s_{m1},\dots,s_{mn})\}$. According to lemma 5.2 we conclude that $h(x_1,\dots,x_n)$ is ground-reducible by a linear instantiation $\mathcal{R}^*\cup\{h(s_{11},\dots,s_{1n}),\dots,h(s_{k1},\dots,s_{kn})\}^*= \mathcal{R}^*\cup\{h(s^*_{11},\dots,s^*_{1n}),\dots,h(s^*_{k1},\dots,s^*_{kn})\}$. Turning back to t we can see that if $h(s_{j1},\dots,s_{jn})$ corresponds to a superposition of some $l \in \mathcal{R}$ into t, then $h(s^*_{j1},\dots,s^*_{jn})$ corresponds to a superposition of an instance of l into t s.t. the superposition substitution is non-restricting, i.e. assignes to variables of t linear terms with pairwise disjoint variable sets (see [Fr86]). Thus, we have proved the following result.

Theorem 5.1. Given a TRS $\mathcal{R}$ suppose t is $\mathcal{R}$-ground-reducible. Then there exists an instantiation $\mathcal{R}^*$ s.t. t is $\mathcal{R}^*$-ground-reducible, and every $l \in \mathcal{R}^*$ is linear or else l superposes into t yielding a non-restricting superposition substitution.

Theorem 5.1 is a generalization of lemma 5.2 to the non-linear case. If we are to specify completely a non-linear term, then we can avoid those non-linearities in left-hand sides which generate restricting superposition substitutions. For example, specifying a term $f(x,x,y)$, the left-hand-sides $f(s(x),s(x),s(y))$, $f(0,x,x)$ are non-linear, but do not produce restricting superposition substitutions and cannot generally be instantiated. On the contrary, a left-hand-side $f(s(x),s(y),s(y))$ presents a non-linearity which can be always avoided.

6. Concluding remarks

New results on relationship between term rewriting systems and regular tree languages have been presented. A central result is theorem 4.1 which is fundamental and has several important consequences. One of them (theorem 3.2) states that left-linear TRSs define all possible regular sets of reducible ground terms. Together with the remark after theorem 3.1 this can be viewed as a characterization of the class of term rewriting systems defining a regular set of reducible ground terms. Another corollary (theorem 4.2) has important applications to the problem of ground-reducibility. We have proved (lemma 5.2), for example, that ground-reducibility of a linear term can be always preserved after an instantiation of the rewriting system into a left-linear one.

A possible further development of the work may primarily concern algorithmic issues. The presented proofs do not allow us to extract any reasonable algorithms from them. Meanwhile, a problem of designing such algorithms is very important. E.g., if we had an algorithm for constructing a linear instantiation for lemma 4.2, we could always reduce a ground-reducibility checking problem to the left-linear case for which practicable decision algorithms are known (see introduction). An algorithm for testing regularity of $Red(\mathcal{R})$ for a given TRS $\mathcal{R}$ is also of great importance. A problem of designing such an algorithm was raised in [Com89b] where an assumption of regularity of $Red(\mathcal{R})$ played a principal role. However, this algorithm is deeply related to a general ground-reducibility testing algorithm and therefore seems to be of similar complexity. Design and complexity analysis of the abovementioned algorithms may be a subject for the future work.

Most of the presented results were obtained during the author's visit to Computer Science Research Center of Nancy. Many thanks must be given to Hubert Comon and Pierre Lescanne for stimulating discussions and practical help. I wish also to express my gratitude to Magnus Steinby who made numerous helpful comments on an earlier version of the paper.

References

[Com88] Comon H., An effective method for handling initial algebras.-Lect. Notes Comput. Sci., vol.343, 1989, p.108-118.

[Com89a] Comon H., Private communication, October 1989, Nancy.

[Com89b] Comon H., Inductive proofs by specifications transformation.- Lect. Notes Comput. Sci., 1989, vol.355, p.76-91.

[DC87] Dauchet M., De Comite, A gap between linear and non linear term-rewriting systems.- Lect. Notes Comput. Sci., 1987, vol.256, p.95-104.

[Der81] Dershowitz N., Termination of linear term rewriting systems.-Lect. Notes Comput. Sci., 1981, vol.115, p.448-458.

[Fr86] Fribourg L., A strong restriction of the inductive completion procedure, Lect. Notes Comput. Sci., 1986, vol. 226, p.105-115.

[FV88] Fulop Z., Vagyvolgyi S., A characterization of irreducible sets modulo left-linear term-rewriting systems by tree automata.-Lect. Notes Comput. Sci., vol.343, 1989, p.157.

[GB85] Gallier J.H., Book R.V., Reductions in tree replacement systems.- Theoretical Computer Science, 1985, vol.37, p.123-150.

[GKM83] Guttag J.V., Kapur D., Masser D.R., On proving uniform termination and restricted termination of term rewriting systems.-SIAM Journ. Comput., 1983, vol.12, n.1, p.189-214.

[GS84] Geoseg F., Steinby M., Tree automata.-Akademiai Kiado, Budapest, 1984.

[Huet80] Huet G., Confluent reductions: abstract properties and applications to term rewriting systems.- Journ. ACM, 1980,vol.27, n.4, p.797-821.

[HO80] Huet G., Oppen D.C., Equations and rewrite rules: a survey.- Formal Languages Theory: Perspectives and Open Problems, Academic Press, New York, 1980, p.349-405.

[JK86] Jouannaud J.-P., Kounalis E., Automatic proofs by induction in equational theories without constructors.- Proc. Symp. Logic in Comput. Sci., Cambridge, Mass., June 16-18, 1986, p.358-366.

[KNZ87] Kapur D., Narendran P., Zhang H., On sufficient-completeness and related properties of term rewriting systems.- Acta Informatica, 1987, 24, p.395-415.

[KNRZ87] Kapur D., Narendran P.,Rosenkrantz D.J., Zhang H., Sufficient-Completeness, Quasi-Reducibility and Their Complexity. -TR 87-26, Comp. Sci. Dep., State Univ. of New York at Albany, 1987.

[KZ85] Kounalis E., Zhang H., A general completeness test for equational specifications.- Proc. 4th Hungarian Comput. Sci. Conf., 1985, p.185-200.

[Kou90] Kounalis E., Testing for inductive (co)-reducibility.- Lect. Notes Comput. Sci., vol.431, 1990, p.221-238.

[Kuc88] Kucherov G.A. A new quasi-reducibility testing algorithm and its application to proofs by induction.-Lect. Notes Comput. Sci., vol.343, 1989, p.204-213.

[Kuc90] Kucherov G.A. On relationship between term rewriting systems and regular tree languages.- INRIA, Rapport de Recherche, No.1273, August 1990.

[LM87] Lassez J.-L., Marriott K., Explicit representation of terms defined by counter examples, Journ. Automated Reasoning, 1987, 3, p.301-318.

[Pla85] Plaisted D.A., Semantic confluence tests and completion methods.- Inf. and Contr., 1985, 65, p.182-215.

The Equivalence of Boundary and Confluent Graph Grammars on Graph Languages of Bounded Degree

Franz J. Brandenburg

Lehrstuhl für Informatik, University of Passau
Innstr. 33, D 8390 Passau, Germany

Abstract

Let B-edNCE and C-edNCE denote the families of graph languages generated by boundary and by confluent edNCE graph grammars, respectively. Boundary means that two nonterminals are never adjacent, and confluent means that rewriting steps are order independent. By definition, boundary graph grammars are confluent, so that B-edNCE $\subseteq$ C-edNCE. Engelfriet et. al. [8] have shown that this inclusion is proper, in general, using certain graph languages of unbounded degree as a witness. We prove that equality holds on graph languages of bounded degree, i.e., $\text{B-edNCE}_{deg} = \text{C-edNCE}_{deg}$, where the subscript "deg" refers to graph languages of bounded degree. Thus, for bounded degree, boundary graph grammars are the operator normal form of confluent graph grammars and e.g., the characterization results obtained independently for B-edNCE and C-edNCE can be merged. Our result confirms boundary and confluent graph grammars as notions for context-free graph grammars.

Keywords

graph grammars, boundary and confluent graph grammars, operator normal form, graph languages of bounded degree

---

This research was supported in part by the Deutsche Forschungsgemeinschaft under Br 835/1.

1. Introduction

The theory of graph grammars is an area of active research within Theoretical Computer Science. It deals with the rewriting of graphs and hypergraphs, and so it generalizes the theories of string, term and tree rewriting. Graph grammars provide an attractive way of formalizing the notion of recursively defined sets of graphs and introduce new methods for the description of sets of graphs. These are syntax based and hierarchical and are complementary to classical methods from graph theory. Unfortunately, there are many distinct types of graph grammars depending on the kind of target objects, e.g., graphs or hypergraphs, directed or undirected graphs and with or without edge labels, on the type of the replaced left-hand sides, e.g., a vertex, an edge, a hyperedge, or a hyperedge with ports, on the embedding mechanism, and on the general framework, e.g., the algorithmic, the logic/axiomatic, or the algebraic approach. See, e.g. [3-6].

The crucial notion is that of a "context-free graph grammar". Many types of graph grammars look context-free, but they are not. This stems from the embedding mechanism, which specifies how the right-hand sides of the productions are connected with the neighborhood of the replaced object. Nowadays there is a broad agreement on the notion of context-free graph grammars. Courcelle et al. [4] state that C-edNCE graph grammars are the "largest known class of context-free vertex rewriting graph grammars". These grammars operate with vertex and edge labeled, directed graphs and use a neighborhood controlled embedding mechanism. The context-freeness comes from the confluence. Confluence is a dynamic property on the derivations. It means that the rewriting is order independent, and induces a proper hierarchical structure on the derivations which can be described by derivation trees. There are many deep results on C-edNCE graph languages, including characterizations, e.g., through normalized C-edNCE graph grammars, leftmost edNCE graph grammars, separated handle-rewriting hypergraph grammars, regular tree embeddings, and algebraic term-rewriting grammars, see, e.g., [4, 6]. Further results deal with complexity results, decision and closure properties and various applications.

Confluence is decidable for edNCE graph grammars. However, the design of a confluent edNCE graph grammar for a particular language may be a tedious task. For such practical purpose, static properties on graph grammars are much easier to work with. Static means a structural property of the productions which guarantees confluence. The separation of nonterminals by terminals is such a property. For context-free string grammars this leads to the operator normal form, which is common for the infix notation of expressions. In this case, separation does not reduce the generative power. However, for arbitrary string grammars, separation may lead to terminal bounded grammars, which generate only context-free languages. This means a big loss in power. See [10, chapter 10.2].

For graph grammars, the situation is similar. In [4], Courcelle et al. introduced separated handle hypergraph grammars which precisely generate the C-edNCE graph languages. Here the separation is for hypergraphs with ports, which can be seen as "big" left-hand sides. A local separation leads to boundary graph grammars, introduced first by Rozenberg and Welzl for undirected graphs without edge labels [12]. This pioneering paper has strongly influenced the theory of graph grammars. Boundary graph grammars are the operator normal form of vertex replacement graph grammars. They are easy to handle, as demonstrated by examples for many important sets of graphs., e.g., outerplanar graphs, graphs with tree-width, cutwidth or bandwidth $\leq k$, flow graphs of well-structured programs, etc. See, e.g. [12]. Many constructions and results have been generalized from context-free string grammars to boundary graph grammars. These are based on special normal forms, including the context consistent and neighborhood preserving normal forms, providing nonterminals with a control over the environment. These investigations have led to a rich theory of boundary graph grammars and their graph languages, which includes, e.g. Chomsky and Greibach normal forms, closure properties, decidability and complexity results and a characterization in terms of regular tree languages. See e.g. [6-9, 12]. However, the boundary restriction reduces the generative power of context-free graph grammars, at least - and as we shall show only - for graph languages with unbounded degree. This has been shown by Engelfriet et al. [8] using the set of edge complements of binary trees as a witness.

The degree of a graph is one of its most important features. There are many important sets of graphs with bounded degree, e.g., binary trees, grids, and boolean networks or VLSI circuits with bounded fan. Thus, the restriction to sets of graphs of bounded degree is natural and well-motivated. It is known that boundary and confluent graph grammars are invariant w.r.t bounded degree, i.e. their classes of languages are closed under intersection with graph languages of bounded degree. Here, we go a step further and show as our main result that the set of graphs of degree at most k generated by a confluent edNCE graph grammar is a boundary edNCE graph language. Hence, on graph languages of bounded degree, boundary and confluent edNCE graph grammars are equivalent, and boundary graph grammars are the operator normal form of confluent graph grammars. Together these classes of graph grammars are nicer than the other promising types of context-free graph grammars. On graph languages of bounded degree, the "good" properties of confluent and boundary graph grammars and graph languages merge, including the characterizations in terms of various descriptions, normal forms, complexity results and closure properties.

Let us state these results in set theoretic terms: Let B-edNCE and C-edNCE denote the classes of graph languages generated by boundary and confluent edNCE graph grammars and let B-edNCE_{deg} and C-edNCE_{deg} denote the corresponding classes of graph languages of bounded degree. Then $\text{B-edNCE} \subset \text{C-edNCE}$ and $\text{B-edNCE}_{deg} = \text{C-edNCE}_{deg}$.
Our proof of $\text{B-edNCE}_{deg} = \text{C-edNCE}_{deg}$ is direct. It uses graph grammars with edge labels. However, the direction of the edges is unimportant, so that the result also holds for classes like B-eNCE_{deg} and C-eNCE_{deg}. The proof makes use of some deep observations on the structure of confluent graph grammars. First, confluence permits the restriction to leftmost derivations. Secondly, if g is a generated graph and g' is a subgraph of g generated from a single vertex, then there are at most K edges connecting g' with the rest-graph $g-g'$, where K depends on the degree of g and on the graph grammar. These edges are called bridges and their endpoints in g' are called ports. Hence, the local environment of the subderivation of g' is bounded. The key of our construction is a look-ahead from a nonterminal vertex A to the terminal subgraph generated from A. We guess the ports generated from A. These ports separate A from its environment and make the grammar a boundary graph grammar. Clearly, also the bridges must be guessed. The correctness of each guess is verified by the use of appropriate vertex and edge labels.

2. Graphs and Graph Grammars

We assume that the reader is familiar with the basic notions from graph theory and with graph rewriting systems. Our objects are sets of vertex and edge labeled graphs, which are generated by some graph grammar from the family of edNCE vertex replacement systems.

Definition

A graph $g = (V, E, m)$ consists of finite sets of vertices V, a set of directed, labeled edges $E \subseteq V \times \Delta \times V$ and a vertex labeling function $m: V \to (N \cup T)$. N, T and Δ are the alphabets of nonterminal and terminal vertex labels and of terminal edge labels. m associates a nonterminal or a terminal symbol with each vertex v, and accordingly, v is called a nonterminal or a terminal vertex. Each $e = (u, x, v)$ in E is an edge with label x from vertex u to vertex v. A set of graphs L (over the terminal alphabets T and Δ) is called a graph language.

Let $g = (V, E, m)$ be a graph. A vertex u is a (B, x, d)-neighbor of a vertex v, if u has the label B and there is an edge e with edge label x in direction d from v to u, i.e., $e = (v, x, u)$, if d = out and $e = (u, x, v)$, if d = in. u is a neighbor of v, if u is a (B, x, d)-neighbor for some $(B, x, d) \in (N \cup T) \times \Delta \times \{in, out\}$. The degree of a graph g, deg(g), is the maximal number of neighbors of any vertex. The degree of a graph language L is $deg(L) = max\{deg(g) \mid g \in L\}$. L is of bounded degree, if $deg(L) \leq k$ for some k. Let $GRAPH_{deg}$ denote the class of graph languages of bounded degree.

For a graph $g = (V, E, m)$, a subset of the vertices $V' \subseteq V$ induces the subgraph $g' = (V', E', m')$ with $E' = E \cap V' \times \Delta \times V'$ and $m'(v) = m(v)$ for each $v \in V'$. Define ports and bridges by port(g', g) $= \{v' \in V' \mid v'$ is a neighbor of some vertex $u \in V-V'\}$ and bridge(g', g) $= \{(v, a, w), (w, a, v) \mid v \in V', w \in V-V'\}$. Thus port(g', g) consists of all vertices in g' with a neigbor in the vertex-complement graph g–g' and bridge(g', g) consists of all edges connecting g' with g–g'.

Definition

An edNCE graph grammar is a system $GG = (N, T, \Delta, P, S)$, where N, T and Δ are the alphabets of nonterminal vertex labels, terminal vertex labels and terminal edge labels. $S \in N$ is the axiom. P is a finite set of productions of the form $p = (A, R, C)$, where $A \in N$ is the label of a replaced vertex, R is the right-hand side graph of p, and C is the connection relation. C is a finite set of tuples (B, x, d, u, x', d'), where $B \in N \cup T$ is a vertex label, $x, x' \in \Delta$ are edge labels, $\{d, d'\} \in \{in, out\}$ specify edge directions and u is a vertex in R.

A direct derivation step means the application of a production $p = (A, R, C)$ to some vertex v. It defines a binary relation on graphs, $g \Rightarrow_{(v, p)} g'$. p is applicable to v, if v has the label A. Then v is replaced by a new isomorphic copy of R, which is vertex disjoint with g. Finally establish connections between the neighbors of v and the target vertices of (the copy of) R as specified by the connection relation C. A tuple (B, x, d, u, x', d') selects all (B, x, d)-neighbors w of v in the graph g. Each such w is connected to the specified vertex u of the right-hand side R by an edge with label x' in direction d'. Thus, the (B, x, d)-neighbors of v become (B, x', d')-neighbors of u. The vertices of R are called descendants of v.

A derivation $D : g \Rightarrow^* h$ from a graph g to a graph h consists of a sequence of direct derivation steps. It is defined up to isomorphism, and we suppose that vertices from the right-hand sides are new. The direct descendants induce a tree structure on the vertices of a derivation. Each vertex is recorded once according to its first occurrence. A vertex v occurring in some graph g' of D induces a subderivation D(v). D(v) consists of the derivations steps of D applied to v and its descendants, and generates the subgraph g(v) of h. D(v) corresponds to a subtree of the derivation tree of D with root v. The rest graph g'-{v} and its rewriting is ignored. Notice that D(v) and g(v) are uniquely determined by v due to the isomorphism and uniqueness assumptions.

The language generated by a graph grammar GG consists of all terminal graphs generated from the axiom, $L(GG) = \{g \mid S \Rightarrow^* g, g = (V, E, m) \text{ and } m(v) \in T \text{ for every } v \in V\}$.

An edNCE graph grammar looks context-free, because of its productions and the tree structure of its derivations. However, the embedding induces some context-dependence. This is illustrated, e.g., by the set of all graphs (over $T = \Delta = \{a\}$) which can be generated by the following strategy: for a graph $g = (V, E, m)$ first generate a complete graph with $|V|$ vertices labeled by a nonterminal A. Then apply productions, which only rename vertex labels, i.e., $A \rightarrow B$, $B \rightarrow A$ for a nonterminal B. The embedding is such that it preserves all edges except when both endpoints are labeled by B, i.e., $(B, a, d, u, a, d) \notin C$, if $p = (A, \cdot_B, C)$. So each edge $(u, a, v) \notin E$ is cut from the complete graph by relabeling u and v by B. Finally, a production $A \rightarrow a$ relabels vertices to terminal vertices. Clearly, the order of relabelings of a vertex $A \rightarrow B$, $B \rightarrow A$ is significant for the deletion of the edges. Such a situation is avoided by order independence, called confluence in graph grammar theory.

Definition

An edNCE graph grammar GG is confluent or a C-edNCE graph grammar, if derivation steps on distinct vertices can be done in any order. Thus, if v and v' are vertices of some graph g with $v \neq v'$ then $g \Rightarrow_{(v, p)} h \Rightarrow_{(v', p')} k$ and $g \Rightarrow_{(v', p')} h' \Rightarrow_{(v, p)} k'$ imply $k = k'$ (up to isomorphism). Let C-edNCE denote the class of graph languages generated by C-edNCE graph grammars.

In C-edNCE graph grammars, derivations have a proper tree structure, just as in the case of context-free string grammars. There is a one-to-one correspondence between derivation trees and leftmost derivation. A leftmost derivation is obtained by inducing an order on the vertices of the graphs of the right-hand sides of the productions, and translating this order to the derivation steps using the Bubble Sort Algorithm.
Notice that confluence is decidable for edNCE graph grammars.

The subclass of confluent graph grammars is considered the most powerful type of context-free graph grammars. This is based on recent characterizations of the class C-edNCE in terms of distinct types of graph grammars, recursive systems of equations, regular tree embeddings and second order monadic logic. See [4, 6]. Clearly, confluence reduces the generative power of graph grammars, unfortunately very severely. E.g., the set of all graphs cannot be generated by C-edNCE graph grammars. Hence, in context-free graph grammar theory, there is no analogue to the regular set of all strings Σ^*.

3. Results

We now come to our first technical result which is due to Schuster [14]. It stems from the fact that the embedding mechanism of an edNCE graph grammar GG = (N, T, Δ, P, S) is based on (B, x, d)-neighbors. Hence, GG distinguishes at most $|N \cup T| \cdot 2|\Delta|$ classes. This induces an upper bound for the number of ports and bridges of induced subgraphs in terms of the degree of the derived graphs.

Lemma 1

Let GG = (N, T, Δ, P, S) be a C-edNCE graph grammar and $c = 2^{2|N \cup T|\,|\Delta|}$.

Let $D: S \Rightarrow^* g$ be a derivation, v a vertex occurring in some graph of D, and g(v) the induced subgraph generated by D(v). Let k = deg(g) be the degree of g.

Then $|port(g(v), g)| \leq c \cdot k^2$ and $|bridge(g(v), g)| \leq c \cdot k^2$.

Next, we turn to boundary graph grammars, for which confluence is a static property. The concept of boundary graph grammars has been introduced by Rozenberg and Welzl [12]. In a boundary graph grammar a nonterminal vertex has only terminal neighbors. Clearly, this property suffices for the right hand sides of the productions. Thus, nonterminal vertices cannot interfere with each other in derivation steps, which immediately implies confluence. More formally, an edNCE graph grammar GG = (N, T, Δ, P, S) is boundary, or a B-edNCE graph grammar, if for every production (A, R, C) no nonterminal vertex of R has a nonterminal neighbor.
Let B-edNCE denote the class of graph languages generated by boundary edNCE graph grammars.

From the definition it is clear that B-edNCE ⊆ C-edNCE. In fact, this inclusion is proper, i.e., B-edNCE ⊂ C-edNCE. This has been shown by Engelfriet et al. [8] using the set of graphs which are the edge complements of the binary trees as a witness. In fact, Engelfriet et. al. consider undirected graphs and eNCE graph grammars. The extension to directed graphs is straightforward.

Before we state our new result let us explain the relevant notions in terms of strings. A string $a_1a_2...a_n$ can be regarded a linear graph. Each symbol a_i (except at the ends) has two neighbors, a_{i-1} and a_{i+1}. Then each context-free string grammar can be seen as a confluent graph grammar with a simple embedding mechanism. Now, the boundary restriction means the separation of any two nonterminals by at least one terminal. This is the well-known operator normal form of context-free string grammars. It comes from productions for expressions in infix notation, e.g., E → E+E. It is well-known, that every context-free string language can be generated by some context-free string grammar in operator normal form. See, e.g. [10]. Thus, boundary and confluent are equivalent for context-free string grammars.

For C-edNCE graph grammars the degree plays a crucial role. Let B-edNCE_{deg} and C-edNCE_{deg} denote the classes of languages of bounded degree generated by boundary and confluent edNCE graph grammars. Boundary and confluent graph grammars are consistent with bounds on the degree. First, it is decidable whether an edNCE graph grammar GG generates a language of bounded degree, and if so, the degree is effectively computable. Secondly, B-edNCE and C-edNCE are closed under intersection with graph languages of degree k, and finally, if L(GG) is of bounded degree, then there is an equivalent graph grammar GG' of the same type such that every generated graph has bounded degree. These facts follow directly by a generalization of results of Rozenberg and Welzl [13] and from results of Courcelle [3].

Lemma 2
It is decidable whether an edNCE graph grammar GG has bounded degree.
The degree deg(L(GG)) can effectively be computed.

Lemma 3
For every C-edNCE (B-edNCE) graph language L and every constant k there is a C-edNCE (B-edNCE) graph grammar GG such that $L(GG) = \{g \in L \mid deg(g) \leq k\}$ and GG has bounded degree, i.e., there is some constant k' such $deg(g) \leq k'$ for every graph g generated by GG.

Our main result sharpens Lemma 3. It establishes the close connections between boundary and confluent graph grammars. Notice that the direction of the edges is not important, so the results holds accordingly for eNCE graph grammars.

THEOREM
For every C-edNCE graph grammar GG and every constant k there is a B-edNCE graph grammar GG' such that $L(GG') = \{g \in L(GG) \mid deg(g) \leq k\}$.
Proof (sketch).
Our proof is a generalization of the direct construction of an operator normal form for context-free string grammars in [10, Thm. 4.8.1]. However, the neighborhood is much more involved.
For convenience, suppose that $GG = (N, T, \Delta, P, S)$ is a C-edNCE graph grammar in binary normal form and is of bounded degree. Then GG has nonterminal productions of the form $p = (A, R, C)$ with $R = (\{v', v''\}, E, m)$ or $R = (\{v'\}, \emptyset, m)$ and terminal productions of the form $(A, \bullet_a, C)$.
Let $K = c \cdot k^2$ be the bound on the size of the sets of ports and bridges of induced subgraphs. Each such set of ports and bridges can be described by a graph with at most K real vertices for the ports and at most K bridges, which are edges from ports to new, virtual vertices. Let Π denote the set of all such graphs. Then $|\Pi| \leq r$ for some (large!) constant r.
Consider a nonterminal production $p = (A, R, C)$, which is not a boundary production. For an illustration, let $p = (A, A'\bullet\!\!-\!\!\bullet A'', C)$, where the vertices v' and v'' are identified with the labels A' and A'' and are connected by some edges. In GG', p shall be replaced by a finite set of boundary productions of the form $p(\pi) = (A, A'\bullet\!\!-\!\!\bullet_{\pi}\!\!-\!\!\bullet A'', C')$. $\pi \in \Pi$ consists of at most K ports of a subgraph g(A'). $p(\pi)$ has edges from the ports to the left vertex A', between the ports and from the ports and to the right vertex A'', the latter being bridges of the induced subgraph g(A'). Hence, $p(\pi)$ is a boundary production, since the "old" edges between A' and A'' are broken at the ports.
Where does π come from? Consider a leftmost derivation $D : S \Rightarrow^* h$ with $h \in L(GG)$ and the application of p to some vertex v with label A in some graph g'. Then v is replaced by the graph $R = A'\bullet\!\!-\!\!\bullet A''$, which is embedded into $g-\{v\}$ as described by the embedding relation C. Suppose that A' is left of A''. Let D(A') be the subderivation of D rewriting A' into the terminal graph g(A'). Consider D when D(A') is done. Then the graph $h' = g(A') \otimes A''$ has been generated.

h' consists of the subgraph g(A') and of the single vertex A" together with the bridges connecting g(A') with A".
The boundary graph grammar GG' simulates the derivation $v \Rightarrow^* h$ in a different order. GG' takes a look-ahead to the ports and bridges of the graph g(A'). In the first step, using the production p(π), GG' generates a graph of the form A'•—•π—•A". The vertices of π are the ports of induced subgraph g(A') of h and the edges between the vertices of π and A" are the bridges of g(A'). Then, GG' completes the generation of g(A'). Since the ports and bridges a generated aforehead, GG' must omit their generation when completing g(A') and must control their construction. This is done by augmented vertex labels. Terminal productions $(A, \cdot_a, C)$ of GG are simulated by erasing productions of the form (Å, Ø,Ø), where Å is an augmented label of the form Å = (A, η, γ), where the components η and γ match if the simulation is correct; otherwise, if there is a mis-match, the augmented nonterminal cannot be replaced.

In more detail, construct a boundary edNCE graph grammar GG' = (N', T, Δ', P', S') as follows: The nonterminals are triples $(A, \eta, \gamma) \in N'$, where $A \in N$ is the base label, $\eta = \{(B_1, x_1, d_1),\ldots, (B_q, x_q, d_q)\}$, each $B_i \in N \cup T$, $x_i \in \Delta$, $d_i \in \{in, out\}$, is the exact specification of $q \leq K$ edges, and γ is a description of $m \leq K$ ports in a subgraph induced by A, together with their incident edges. Both, η and γ consider multiplicities of edge specifications (B, x, d). η plays the role of the context describing function from [12] and records the incident edges of a nonterminal vertex v just before v is rewritten. $\gamma = (\gamma_1,\ldots, \gamma_m)$ with each γ_j of the form γ_j = (j, m(j), N-bridges(j), T-bridges(j), port-bridges(j)). The first component of each γ_j is the name or number of a port, say $1 \leq j \leq K$. These components are distinct for γ. $m(j) \in T$ is the terminal vertex label of port j. N-bridges, T-bridges and port-bridges are multisets of edge specifications of the form (z, x, d) with $z \in N \cup T \cup \{1,\ldots,K\}$. x is an edge label and d specifies the edge direction relative to the vertex j. Each edge is recorded individually by a triple (z, x, d). Thus, if there are m edges incident with j, then N-bridges contains m tuples (z, x, d), some of which may be the same. This holds accordingly for the T-bridges. port-bridges are distinguished by the port numbers.
The edge labels consist of the old edge labels or of the new label "*", i.e., $\Delta' = \Delta \cup \{*\}$.
Let S' = (S, Ø,Ø).
For every production p = (A, R, C) of GG let GG' contain a finite set of productions of the form p(π) = (Å, R', C') with Å = (A, η, γ), where η describes the context and γ records each port and bridge derived apriori from A.

If R = ({v', v"}, E, m) with m(v') = A', m(v") = A", and E ≠ Ø, i.e. R = A'•—•A"
then R' = (V', E', m'), where
V' = {v', v"} ∪ {w ∈ new-port(v'), $1 \leq w \leq K$}
E' = {(v', m'(v'), w) | w ∈ new-port(v')}
∪ {N-bridges(w), T-bridges(w), port-bridges(w) | w ∈ new-port(v')}

new-port(v') consists of a set of new vertices, which are numbered by w ∈ {1,...,K} and stored in the γ-component of the label of v'. The numbers are chosen such that the first components of the γ-components are distinct. These vertices are ports of a subgraph g(v') induced by v' in some leftmost derivation and are used to separate g(v') from v". They occur in this step. There may be some "old ports" generated in earlier steps for the separation of g(v') and v".

The new ports and their incident edges are recorded in the γ' component of the label of v'. The old ports and their edges have already been recorded at the augmented label of the replaced vertex v. Their description is transferred one-to-one to the labels of v' and v".

Let $m'(v') = (A', \eta', \gamma')$, $m'(v'') = (A'', \eta'', \gamma'')$ and $m'(w) \in T$ for each $w \in$ new-port(v'). The augmented vertex labels are computed as follows. Notice that each item is recorded twice. One entry represents sufficiency, the other necessity. Recall that elements are computed according to their multiplicities. Partition γ into $\gamma = \hat{\gamma} \cup \gamma''$ with $\hat{\gamma} \cap \gamma'' = \emptyset$ and define $\gamma' = \hat{\gamma} \cup \{(w, m'(w)$, N-bridges(w), T-bridges(w), port-bridges(w) $\mid w \in$ new-port(v')$\}$. Here, w is chosen such that $1 \leq w \leq K$ and each w is unique in γ'. $m'(w) \in T$. N-bridges(w) consists of triples of the form (A", x, d), which are recorded one-by-one in η'' as described below. A triple (a, x", d") is in T-bridges(w) and (a, x, d, w, x", d") $\in$ C' iff there is some (a, x, d) $\in \eta$ with (a, x, d, v', x', d') $\in$ C and (a, x', d') $\in \eta'$. Finally, (j, x, d) $\in$ port-bridges(w) iff $j \neq w$ and (m'(w), x, –d) $\in \eta''$. Here –d means the opposite direction such that –in = out and –out = in. Hence, γ is distributed over the labels of v' and v" and is augmented by information from the new ports. η' describes the neighborhood of v' and is computed from the right-hand side R and from an update of the neighborhood η of A under the embedding relation C. (A", x, out) $\in \eta'$ iff (v', x, v") $\in$ E, (A", x, in) $\in \eta'$ iff (v", x, v') $\in$ E and (B, y, d') $\in \eta'$ iff (B, x, d) $\in \eta$ and (B, x, d, v', y, d') $\in$ C. Similarly, η'' describes the neighborhood of v" and is computed from η and from the new ports. (B, y, d') $\in \eta''$ iff (B, x, d) $\in \eta$ and (B, x, d, v", y, d') $\in$ C. For every new port w added to γ' and every (A", x, d) $\in$ N-bridges(w) add (m'(w), x, -d) to η'' and simultaneously, record (A", x, d) in N-bridges(w).

The embedding relation C' is adapted to the new nonterminal vertex labels and the new production p(π). For a terminal vertex label a and old edge labels $x, x' \in \Delta$, (a, x, d, v', x', d') $\in$ C iff (a, x, d, v', x', d') $\in$ C and similarly for v". Edges with new edge labels are transferred one-to-one from v to v' and v". For the new terminal ports w, C' is adapted as described above.

If p is a nontermial production with $E = \emptyset$, then there are no new ports. Then the productions of p(π) are computed as above with an update of the augmented labels, i.e. of the η and γ components.

If p = (A, R, C) is a terminal production of GG, i.e. R = ({v'}, $\emptyset$, m) with m(v') = a, or R = $\bullet_a$ then P' contains a "copy" ((A,$\emptyset$,$\emptyset$), R, C) and deletions (A', $\emptyset$, $\emptyset$) with A' = (A, η, γ), provided that η and γ match. Otherwise, the "illegal" vertex (A, η, γ) never disappears.

(η, γ) match, if γ = (w, a, N-bridges(w), T-bridges(w), $\emptyset$) stores a single port w and there is a one-to-one correspondence modulo the embedding relation between the N-and T-bridges and the neighborhood specification. I.e. for $B \in N$, (B, x, d) $\in$ N-bridges(w) iff (B, x', d', v', x, d) $\in$ C and (B, x', d') $\in \gamma$ and for $a \in T$, (a, x, d) $\in$ T-bridges(w) iff (B, x', d', v', x, d) $\in$ C and (B, x', d') $\in \gamma$. Each occurrence of a triple counts.

The correctness of this construction can be shown by a lengthly induction on leftmost derivations.

For an illustration consider an example. The productions of GG and GG' are self-explaining from the derivation. The vertices are identified with their labels $N = \{A, A', A'', B', B''\}$ and $T = \{b', b''\}$. Let $\Delta = \{x, x', x'', y', y''\}$. The direction d of the edges is ignored.

The productions are $A \rightarrow A'\underline{\quad X \quad}A''$, $A' \rightarrow B'\underline{\quad Y \quad}B''$, $B' \rightarrow b'$ and $B'' \rightarrow b''$. The connection relation relabels edges by adding primes.

$$
\begin{array}{cccccccc}
& & & & B' & & b' & & b' \\
& & & & |y & & |y' & & |y'' \\
A & \Rightarrow & A'\underline{\ X\ }A'' & \Rightarrow & B''\underline{\ x'\ }A'' & \Rightarrow & B''\underline{\ x'\ }A'' & \Rightarrow & b''\underline{\ x''\ }A''
\end{array}
$$

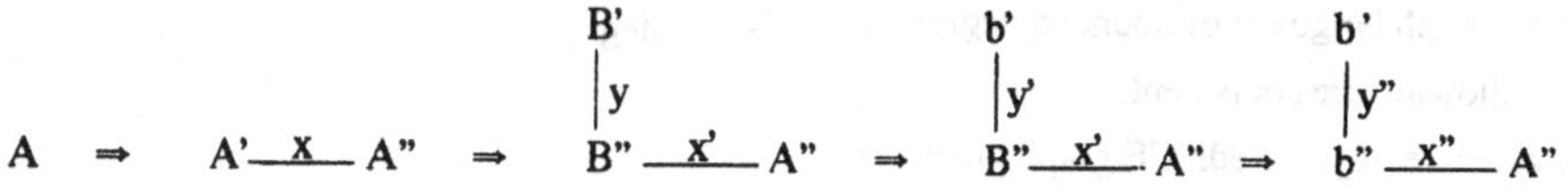

In GG', the vertices u, u' and u'' correspond to A, A' and A'' and v' and v'' correspond to B' and B''. Vertex b'' is a port of g(A') and g(B''). It is a new port for A' and an old port for B''. Vertex b' is a new port of g(B'). Brackets "{, }" are omitted from augmented labels.

The oabove derivation is simulated in GG' as follows:

u $m(u) = (A, \emptyset, \emptyset)$

$\Rightarrow$ $u'\underline{\ *\ }b''\underline{\ x''\ }u''$ $m(u') = (A', (A'',x,d), (b'',b'',(A'',x',d),\emptyset,\emptyset))$

                         $m(u'') = (A'', (b'',x'',d), \emptyset, \emptyset, \emptyset)$

$\Rightarrow$ v'

        |*

        b'

        |* \y'' $m(v') = (B', (B'', y, d), (b',b',(B'', y', d), \emptyset, (b'', y'', d)))$

      $v''\underline{\ *\ }b''\underline{\ x''\ }u''$ $m(v'') = (B'',\{(A'',x',d), (b',y',d)\}, (b'',b'',(A'',x'',d),\emptyset, (b',y'',d)$

$\Rightarrow$ b'

        |* \y'' $(m(v'), \emptyset, \emptyset)$ is a production in GG',

      $v''\underline{\ *\ }b''\underline{\ x''\ }u''$ since there is a match by $(B'', y, d, v', y', d) \in C$

$\Rightarrow$ b'

         \y'' $(m(v''), \emptyset, \emptyset)$ is a production in GG', since there is a match by

        $b''\underline{\ x''\ }u''$ $(A'', x', d, v'', x'', d) \in C$ and $(b', y', d, v'', y'', d) \in C$

Corollary

$\text{C-edNCE} \cap \text{GRAPH}_{deg} = \text{B-edNCE} \cap \text{GRAPH}_{deg} = \text{B-edNCE}_{deg} \subseteq \text{B-edNCE}$

and

$\text{B-edNCE}_{deg} = \text{C-edNCE}_{deg}$.

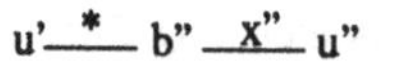

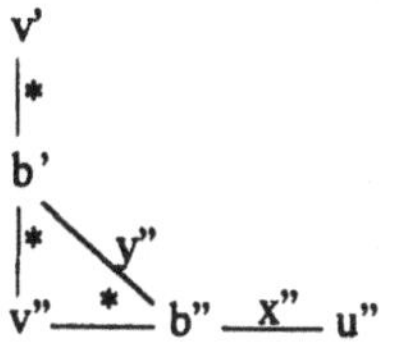

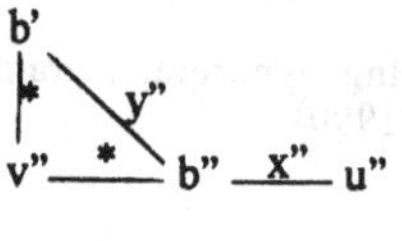

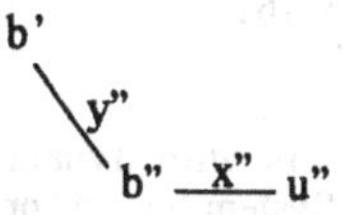

Corollary

For graph languages with bounded degree, boundary edNCE graph grammars are the operator normal form of confluent edNCE graph grammars (and similarly for eNCE).

Corollary

Let L be a graph language of bounded degree, L ∈ $GRAPH_{deg}$.

Then the following are equivalent:

L is generated by a C-edNCE graph grammar.

L is leftmost generated by a C-edNCE graph grammar, L ∈ L-edNCE (see [4, 6].

L is generated by a separated hyper-handle grammar (S-HH, see [4]).

L is generated by a hypergraph rewriting system (see [9])

L is generated by a regular tree embedding (see [6]).

L is regular path defined (see [8]).

L is defined by a polynomial system of equations (see [4]).

Similarly, for graph languages of bounded degree, all results known for boundary graph grammars carry over to confluent graph grammars, such as further normal forms including Chomsky and Greibach normal form,the context-consistent and neighborhood preserving normal forms, closure properties and decidability using through second order formulas, and strong complexity results, etc. [2, 3, 4].

References

[1] F.J. Brandenburg, "On partially ordered graph grammars", Lecture Notes in Computer Science 291 (1987), 99-111.

[2] F.J. Brandenburg, "On polynomial time graph grammars" Lecture Notes in Computer Science 294 (1988), 227-236

[3] B. Courcelle, "An axiomatic definition of context-free rewriting and its applications" Theoret. Comput. Sci. 55 (1987), 141-181.

[4] B. Courcelle, J. Engelfriet and G. Rozenberg, "Handle-rewriting hypergraph grammars", Techn. Report, University of Leiden and University of Bordeaux (1990)

[5] H. Ehrig, M. Nagl, G. Rozenberg and A. Rosenfeld (eds.) Graph Grammars and Their Application to Computer Science, Lecture Notes in Computer Science 291 (1987)

[6] J. Engelfriet, "Context-free NCE graph grammars", Proc. FCT 1989, Lecture Notes in Computer Science 380 (1989), 148-161

[7] J. Engelfriet and G. Leih, "Complexity of boundary graph grammars" RAIRO Theoret. Inform. Appl.24, (1990), 267-274

[8] J. Engelfriet, G. Leih, and E. Welzl, "Boundary graph grammars with dynamic edge relabeling", Techn. Report, University of Leiden (1988), J. Comput. System Sci. to appear.

[9] J. Engelfriet and G. Rozenberg, "A comparison of boundary graph grammars and context-free hypergraph grammars, Inform. Comput. 84 (1990), 163 -206.

[10] M. Harrison, "Introduction to Formal Language Theory", Addison Wesley, 1978

[11] D. Janssens, G. Rozenberg and E. Welzl, "The bounded degree problem for NLC grammars is decidable", J. Comput. System Sci. 33 (1986), 415-422.

[12] G. Rozenberg and E. Welzl, "Boundary NLC graph grammars - basic definitions, normal forms and complexity", Inform. Control 69 (1986), 131-167.

[13] G. Rozenberg and E. Welzl, "Combinatorial properties of boundary NLC graph languages", Discrete Appl. Math. 16 (1987), 58-73

[14] R. Schuster, "Graphgrammatiken und Grapheinbettungen: Algorithmen und Komplexität" Ph.D. Thesis and MIP-8711 Report, Universität Passau (1987).

Left-to-Right Tree Pattern Matching

Albert Gräf
FB Mathematik, Arbeitsgruppe Informatik
Johannes Gutenberg-Universität Mainz
Email: Graef@dmzrzu71.bitnet

Abstract

We propose a new technique to construct left-to-right matching automata for trees. Our method is based on the novel concept of *prefix unification* which is used to compute a certain *closure* of the pattern set. From the closure a kind of deterministic matching automaton can be derived immediately. We also point out how to perform the construction *incrementally* which makes our approach suitable for applications in which pattern sets change dynamically, such as in the Knuth-Bendix completion algorithm.

Our method, like most others, is restricted to *linear* patterns (the case of non-linear matching can be handled as usual by checking the consistency of variable bindings in a separate pass following the matching phase).

1 Introduction

Terms are usually interpreted as trees labelled with symbols from a ranked alphabet $\Sigma = \bigcup_{n \in N_0} \Sigma_n$ s.t. each vertex labelled with a symbol α of arity n (i.e. $\alpha \in \Sigma_n$) has exactly n sons. With certain symbols from Σ_0 being distinguished as *variable symbols*, an *instance* of a Σ-tree t can be obtained by replacing leaves labelled with variable symbols by corresponding subtrees.

In this paper we focus on a restricted form of the *tree pattern matching problem* which can be stated as follows:

(*) Given a finite set $P = \{p_1, \ldots, p_n\}$ of Σ-trees (the *patterns*) and another Σ-tree q (the *subject*), determine whether q is an instance of any of the $p_i \in P$.

We generally assume that patterns are *linear* (i.e., double occurrences of variables are not allowed). Non-linear patterns can be handled as usual (do linear matching first, then check for the consistency of variable bindings).

The problem (*) arises in a variety of applications, such as the implementation of equational and functional programming languages. For most applications, the straightforward algorithm to solve (*) (successively check q against all p_i, $i = 1, \ldots, n$) is unacceptable since its running time depends on the number of patterns n.

The usual approach to solve problems like (*) efficiently is to preprocess patterns to produce some sort of *matching automaton*. The tree matching algorithms of Hoffmann and O'Donnell (see [7], and also [4,6] for a discussion of regular tree languages and bottom-up tree acceptors) and the discrimination net approach (see e.g. [3]) are of this kind. Similar work has been done in the field of functional programming language implementation [1,9,10,11]. All these techniques are based on the idea of somehow *factorizing* patterns to speed up the matching process.

We will propose a new method to construct left-to-right matching automata for trees which is easy to implement and is comparable in efficiency with the technique discussed in [11]. That is, matching time is independent of the number of patterns, and the sizes of matching automata are quite reasonable (usually much better than the exponential bounds obtained by other techniques).

A left-to-right tree matching algorithm must be able to handle ambiguities that arise when prefixes of patterns overlap. For instance, consider the patterns $p_1 = f(x, a)$ and $p_2 = f(g(x), b)$ (where x is a variable). When scanning the subject term $q = f(g(y), a)$ from left to right, we are not able to determine that q matches p_1, and does not match p_2, until we have seen the last symbol a of q. Hence the matching algorithm, after having seen the symbols f and g of q, must remember the fact that q at this point could still match the instance $p_1' = f(g(x), a)$ of p_1.

To solve this problem, we introduce the novel concept of *prefix unification* which generalizes ordinary term unification to arbitrary term prefixes. Prefix unification is used to determine overlaps between pattern prefixes, like the overlap between p_1 and p_2 at the position of x in p_1 in the example above. By simultaneously unifying all possible combinations of pattern prefixes we may then compute a (finite) *closure* of the pattern set from which a kind of deterministic matching automaton can be derived immediately.

The formal properties of prefix unification and the closure operation make it easy to systematically derive efficient algorithms for the construction of pattern set closures. In particular, we will also point out how closures can be computed *incrementally*, which is interesting in applications in which pattern sets change dynamically, such as in the Knuth-Bendix completion algorithm [8,12].

The paper is organized as follows. Section 2 introduces basic terminology. In Section 3 we develop a formal model for left-to-right tree matching automata. Section 4 covers prefix unification and the computation of closures. Section 5 states our main result concerning the construction of left-to-right matching automata. In Section 6 we briefly discuss some results related to the complexity of our matching technique. Section 7 summarizes results and discusses open problems.

2 Preliminaries

To make precise the notions of Σ-trees, terms, instances of terms and left-to-right tree matching automata mentioned in the previous section, let us first introduce some terminology.

Given a finite, nonempty alphabet Σ, by Σ^* we denote the set of all strings over Σ. $|\nu|$ is the length of a string $\nu \in \Sigma^*$, and ε is the empty string (the string of length 0). By $\mathit{Pref}(\nu)$ we denote the set of all prefixes of $\nu \in \Sigma^*$, including the empty string and ν itself. We also write $\mathit{Pref}(M)$ for the union of all $\mathit{Pref}(\nu)$, $\nu \in M \subseteq \Sigma^*$.

We consider a finite *ranked* alphabet Σ which is the disjoint union $\Sigma = \biguplus_{n \in \mathbb{N}_0} \Sigma_n$ of alphabets Σ_n, $n \in \mathbb{N}_0$. If $\alpha \in \Sigma_n$ we say that α has *arity*, or *rank*, n, and write $\#\alpha = n$. We generally assume that there is a distinguished nullary symbol $\omega \in \Sigma_0$ which plays the role of an *anonymous variable symbol*.

The set T_Σ of Σ-*terms* is inductively defined as the least subset of Σ^* containing all $\alpha t_1 \cdots t_n$ for which $\alpha \in \Sigma_n$, $n \in \mathbb{N}_0$, and $t_1, \ldots, t_n \in T_\Sigma$. Observe that, in particular, $\alpha \in T_\Sigma$ for each $\alpha \in \Sigma_0$, and each term $\alpha t_1 \cdots t_n$ naturally represents the labelled tree with root label α and subtrees $t_1, \ldots, t_n$ (if $n = 0$, then $\alpha t_1 \cdots t_n = \alpha$ is a leave). Throughout the rest of this paper, we will use the terms "Σ-term" and "Σ-tree" synonymously.

The length $|t|$ of $t \in T_\Sigma$ as a string is also called the *size* of t, and by $h(t)$ we denote the *height* of t as a tree.

We say that $q \in T_\Sigma$ *matches*, or is an *instance* of, $p \in T_\Sigma$, written $p \leq q$, iff q can be obtained by replacing the ω's in p by Σ-terms. $\leq$ is called the *subsumption ordering* on terms. Note that $p \leq q$ implies that $|p| \leq |q|$ and $h(p) \leq h(q)$, and $p \leq q, q \leq p$ implies that $p = q$. In a slight abuse of the usual terminology we also extend the notion of subsumption to arbitrary strings over Σ, i.e. for $\nu, \mu \in \Sigma^*$ we write $\nu \leq \mu$ iff μ can be obtained by replacing ω's in ν with terms accordingly.

3 Left-to-Right Tree Matching Automata

To specify the notion of a left-to-right matching automaton on Σ-trees, we use finite state systems which in each move may either read one symbol or an entire subterm from the input tape, and change states accordingly.

More precisely, we call a *term acceptor*, or TA, for short, a finite automaton A with *state set* S, a *start state* $s_0 \in S$ and a set of *final states* $F \subseteq S$. The possible moves of A are specified by a *state transition relation* δ which consists of pairs $s\alpha \to s'$, $s, s' \in S$, $\alpha \in \Sigma$. *Configurations* of a TA A are denoted by strings $s\mu$, $s \in S$, $\mu \in \Sigma^*$, and A *accepts* a term $q \in T_\Sigma$ iff $s_0 q \xrightarrow{*} s$ s.t. $s \in F$, where $\xrightarrow{*}$ is the reflexive/transitive

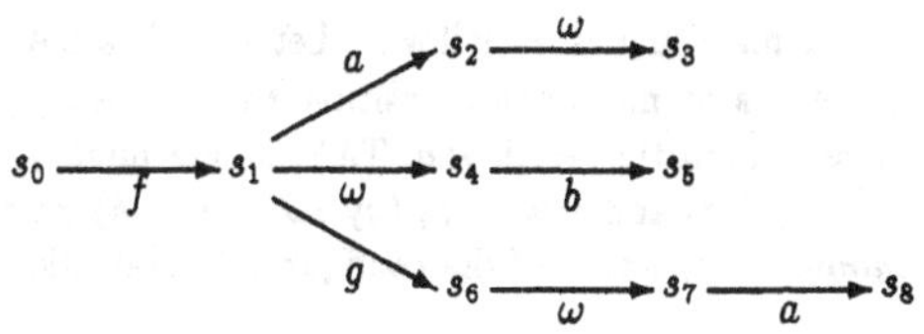

Figure 1: Transition diagram.

closure of the *configuration transition relation* defined by:

$$s\alpha\mu \to s'\mu \text{ iff } s\alpha \to s' \in \delta,\ \alpha \in \Sigma \setminus \{\omega\}$$
$$su\mu \to s'\mu \text{ iff } s\omega \to s' \in \delta,\ u \in T_\Sigma.$$

That is, a TA A essentially is a *finite* (string) *automaton* (FA, for short) with the additional capability to skip over arbitrary subterms in the input. For a TA A, we will also consider the corresponding FA which performs the same moves as A except that in a transition on ω, the FA may only read the literal symbol ω, and not an arbitrary subterm.

3.1 Lemma. A, as a TA, accepts an input term q if and only if A, as an FA, accepts some term p s.t. $p \leq q$. □

If the transition diagram of A is *acyclic*[1], then A, as an FA, accepts only finitely many terms. Hence:

3.2 Theorem. The following are equivalent:

a) $L \subseteq T_\Sigma$ is the set of all instances of some finite pattern set $P \subseteq T_\Sigma$.

b) $L \subseteq T_\Sigma$ is accepted by an acyclic TA A. □

Theorem 3.2 justifies our choice of acyclic TA's as devices to solve the restricted tree matching problem (*) stated at the beginning of this paper. In the following, we generally assume that TA's are acyclic.

3.3 Example. A TA A accepting the term language L of all instances of some finite pattern set $P \subseteq T_\Sigma$, which even is *deterministic* as an FA, can easily be found by carrying out the following "prefix construction" (also known as the construction of a "trie" or a "discrimination net" for P): $S = Pref(P)$, $s_0 = \varepsilon$, $F = P$ and $\delta = \{s\alpha \to s' \mid s' = s\alpha\}$. A TA obtained from $P = \{fa\omega, f\omega b, fg\omega a\}$ in this manner ($\#f = 2$, $\#g = 1$, $\#a = \#b = 0$) is depicted in fig. 1. □

The TA of fig. 1, although deterministic as an FA, is *not* deterministic as a TA, i.e. there are inputs μ s.t. $s' \overset{*}{\leftarrow} s\mu \overset{*}{\rightarrow} s''$, but $s' \neq s''$. E.g., for $\mu = fg\omega$ we have that $s_0\mu \overset{*}{\rightarrow} s_4$, but also $s_0\mu \overset{*}{\rightarrow} s_7$. This is the kind of ambiguity in left-to-right matching of trees pointed out in Section 1.

One solution to this problem is to employ some *backtracking strategy* to find an accepting transition chain if there is one. Such a scheme has been used, e.g., in [1] and [3], but it may be quite inefficient because the subject term may have to be scanned as often as the number of patterns (asymptotically).

The second solution is to construct *deterministic TA's* for matching. Unfortunately, it turns out that for many term languages accepted by TA's there are no deterministic TA's accepting the same language, see [5] for details.

Instead, we will now propose another type of TA's, *canonical TA's*, which accept all languages acceptable by (acyclic) TA's, while still admitting a *deterministic matching strategy*.

[1]The *transition diagram* of a TA A, as usual, is obtained by connecting each pair of states s, s' with those edges labelled α for which $s\alpha \to s' \in \delta$. See fig. 1 for an example.

The idea behind canonical TA matching is as follows. Let us call a TA *weakly deterministic* (*w.d.*) iff for any pair $s \in S$, $\alpha \in \Sigma$ there is at most one transition $s\alpha \to s' \in \delta$. That is, a TA is w.d. iff its underlying string automaton is deterministic. With w.d. TA's, indeterminacy is restricted to situations in which $s\alpha\nu \to s_1\nu$ (by some $s\alpha \to s_1 \in \delta$) and $s\alpha\nu \to s_2$ (by $s\omega \to s_2 \in \delta$) for some $\alpha \in \Sigma \setminus \{\omega\}$, $\alpha\nu \in T_\Sigma$. Let us define the following *canonical* restriction of the configuration transition relation:

$$s\alpha\nu\mu \stackrel{c}{\to} \begin{cases} s_1\nu\mu & \text{if } s\alpha \to s_1 \in \delta,\ \alpha \neq \omega \\ s_2\mu & \text{otherwise, if } s\omega \to s_2 \in \delta,\ \alpha\nu \in T_\Sigma \end{cases}$$

That is, an ω-transition is taken only when no applicable transition on some $\alpha \neq \omega$ exists. If A is w.d., then $\stackrel{c}{\to}$ is a partial function by definition, and hence for any input μ and state s there is at most one canonical configuration transition chain $s\mu \stackrel{c*}{\to} s'$ ($s' \in S$). However, some q accepted by A may not always also be accepted through a canonical configuration transition chain. We say that a w.d. TA A is *canonical* if it meets this condition, i.e. if for any $q \in T_\Sigma$ accepted by A we also have that $s_0q \stackrel{c*}{\to} s$ for some $s \in F$.

Observe that the TA in fig. 1 is w.d., but not canonical, since $q = fgab \geq f\omega b$, but $s_0q \stackrel{c*}{\to} s_7b$ which does not lead to a final state. However, this could be fixed by introducing a new "closure pattern" $fg\omega b$ which is determined by "overlapping" the pattern $f\omega b$ and the prefix fg of the pattern $fg\omega a$. In fact, the w.d. TA constructed from the new pattern set P' obtained by adding the closure pattern $fg\omega b$ to $P = \{fa\omega, f\omega b, fg\omega a\}$ yields a canonical TA accepting the same language (note that $fg\omega b \geq f\omega b \in P$ and hence the set of instances of P' is the same as that of P).

In the following section we will develop the tools to construct pattern set closures systematically, by introducing the concept of prefix unification. Section 5 then proposes our canonical TA construction method and proves it correct.

4 Prefix Unification

While ordinary term unification is based on determining least upper bounds w.r.t. the subsumption ordering on terms, we analogously define prefix unification in terms of the following *prefix subsumption ordering*:[2]

4.1 Definition. Let $\nu, \mu \in \Sigma^*$. We write $\nu \trianglelefteq \mu$ iff some prefix of μ is an instance of ν, i.e. there is some $\mu' \in Pref(\mu)$ s.t. $\nu \leq \mu'$. $\trianglelefteq$ is called the *prefix subsumption ordering* on Σ^*. Note that $\trianglelefteq$ is a well-founded ordering on Σ^* with $\varepsilon \trianglelefteq \nu\ \forall\, \nu \in \Sigma^*$, $\nu \trianglelefteq \mu \Rightarrow |\nu| \leq |\mu|$, and $\nu \trianglelefteq \mu, \mu \trianglelefteq \nu \Rightarrow \nu = \mu$ for all $\nu, \mu \in \Sigma^*$. We also remark that if $p, q \in T_\Sigma$, then $p \trianglelefteq q \iff p \leq q$. □

For instance, we have that $\nu = f\omega \trianglelefteq fg\omega a = \mu$ (f, g and a as in Example 3.3), since the prefix $fg\omega$ of μ is an instance of ν.

It is easy to see that for any pair $\nu, \mu \in \Sigma^*$ there is a unique *greatest lower bound* $\nu \wedge \mu \in \Sigma^*$ w.r.t. $\trianglelefteq$ (i.e. $\nu \wedge \mu \trianglelefteq \nu, \mu$, and for any $\lambda \in \Sigma^*$ with $\lambda \trianglelefteq \nu, \mu$ we have that $\lambda \trianglelefteq \nu \wedge \mu$), which can be obtained as follows:

- $\nu \wedge \mu = \alpha(\nu' \wedge \mu')$ if $\nu = \alpha\nu'$, $\mu = \alpha\mu'$, $\alpha \in \Sigma$.
- $\nu \wedge \mu = \omega(\nu' \wedge \mu')$ if ν and μ have different head symbols and $\nu = u\nu'$, $\mu = v\mu'$, $u, v \in T_\Sigma$.
- $\nu \wedge \mu = \varepsilon$ in all other cases.

It then follows from lattice theory [2] that for any pair $\nu, \mu \in \Sigma^*$ which have a common upper bound w.r.t. $\trianglelefteq$, there is a unique least upper bound for ν and μ, denoted $\nu \vee \mu$. As usual, we may complete Σ^* with a new "top" element $\top \notin \Sigma^*$ assumed to be maximal w.r.t. $\trianglelefteq$, and set $\nu \vee \mu = \top$ for all $\nu, \mu \in \Sigma^*$ which do not have a common upper bound w.r.t. $\trianglelefteq$.

4.2 Proposition. $(\Sigma^*_\top, \trianglelefteq) := (\Sigma^* \cup \{\top\}, \trianglelefteq)$ is a complete lattice. □

[2] In fact, we will define prefix unification on arbitrary Σ-strings, but what we are actually interested in is the unification of pattern prefixes, hence the name "prefix unification."

The binary operations $\wedge$ and $\vee$ on $\Sigma_\top^*$ are commonly referred to as *meet* and *join*, respectively. Note that $\wedge$ and $\vee$ are associative, commutative, idempotent and distribute over each other. Also observe that $\vee$ applied to *terms* is really nothing but the ordinary unification of (linear) terms.

The following lemma summarizes some useful rules for the computation of joins in $\Sigma_\top^*$. We assume that $\nu\top = \top\nu = \top \; \forall \, \nu \in \Sigma_\top^*$ (extending the concatenation operation on Σ^*).

4.3 Lemma. Let $\nu, \mu \in \Sigma_\top^*$.

a) $\nu \vee \mu = \top$ if either $\nu = \top$, $\mu = \top$, or ν and μ have different head symbols $\neq \omega$.

b) $\nu \vee \varepsilon = \varepsilon \vee \nu = \nu$.

c) $\alpha\nu \vee \alpha\mu = \alpha(\nu \vee \mu) \; \forall \, \alpha \in \Sigma$.

d) $\omega\nu \vee \alpha\mu = \alpha(\omega^n\nu \vee \mu) \; \forall \, \alpha \in \Sigma_n \setminus \{\omega\}$. □

For instance, we have that $f\omega b \vee fg = f(\omega b \vee g) = fg(\omega b \vee \varepsilon) = fg\omega b$.

We now turn to the definition of the closure operation. It is useful to define the closure operation not only on pattern sets $P \subseteq T_\Sigma$, but also on any set $M \subseteq \Sigma^*$ of term suffixes for the same prefix $\lambda \in \Sigma^*$. As we shall see, this enables us to compute closures by a simple recursive algorithm (Lemma 4.6).

4.4 Definition. $M \subseteq \Sigma^*$ is called a *suffix set* if there is some $\lambda \in \Sigma^*$ s.t. $\lambda M \subseteq T_\Sigma$ (i.e. $\lambda\mu \in T_\Sigma \; \forall \, \mu \in M$). Note that, in particular, $\emptyset$ and $\{\varepsilon\}$ are suffix sets, and if M is a suffix set containing ε, then $M = \{\varepsilon\}$. Furthermore, any subset of a suffix set is again a suffix set.

We say that a suffix set M is *closed* iff for each $\mu \in M$, $\nu \in \mathit{Pref}(M)$ s.t. $\mu \vee \nu \neq \top$ we have that $\mu \vee \nu \in M$. □

In the following we generally assume that all suffix sets are finite.

4.5 Definition. For any (finite) suffix set M let the *closure* $\overline{M}$ of M be defined by $\overline{M} = \{\mu_0 \vee \nu_1 \vee \cdots \vee \nu_n \neq \top \mid \mu_0 \in M, \nu_i \in \mathit{Pref}(M), n \geq 0\}$. □

It can be shown that $\overline{M}$ is actually the *unique minimal closed suffix set* containing M. We also remark that for any $\overline{\mu} \in \overline{M}$ there are $\mu_0 \in M$, $\nu_\mu \in \mathit{Pref}(\mu)$ for each $\mu \in M$, $\mu \neq \mu_0$, s.t. $\overline{\mu}$ may be written as $\overline{\mu} = \mu_0 \vee \bigvee_{\mu \in M \setminus \{\mu_0\}} \nu_\mu$. This is because for any $\nu, \mu \in \Sigma^*$ with $\nu \trianglelefteq \mu$ (or $\nu \in \mathit{Pref}(\mu)$, in particular) we have that $\nu \vee \mu = \mu$. Hence $\overline{M}$ must be finite (on the basis of our general assumption that M is finite). Also observe that $M \subseteq \overline{M}$ and that any $\overline{\mu} \in \overline{M}$ is an instance of some $\mu_0 \in M$.

The closure operation has the usual properties justifying its name: *extensionality* ($M \subseteq \overline{M}$), *monotonicity* ($N \subseteq M \Rightarrow \overline{N} \subseteq \overline{M}$) and *idempotence* ($\overline{\overline{M}} = \overline{M}$).

The following lemma provides us with a useful recursive algorithm to compute closures. For a suffix set M and $\alpha \in \Sigma$ we let $M/\alpha = \{\mu \in \Sigma^* \mid \alpha\mu \in M\}$ be the suffix set obtained by "factoring out" the head symbol α.[3]

4.6 Lemma. Let $M \subseteq \Sigma^*$ be a suffix set.

a) $M \subseteq \{\varepsilon\} \Rightarrow \overline{M} = M$.

b) $M \not\subseteq \{\varepsilon\} \Rightarrow \overline{M} = \bigcup_{\alpha \in \Sigma} \alpha\overline{M_\alpha}$ where

$$M_\omega = M/\omega$$

$$M_\alpha = \begin{cases} M/\alpha \cup \omega^{\#\alpha} M/\omega & M/\alpha \neq \emptyset \\ \emptyset & \text{otherwise} \end{cases}$$

$\forall \, \alpha \neq \omega$.

[3] For completeness, let us note that we assume the following precedence of operations: first /, then concatenation, then $\wedge$ and $\vee$, and finally the set operations $\cup$ and $\setminus$.

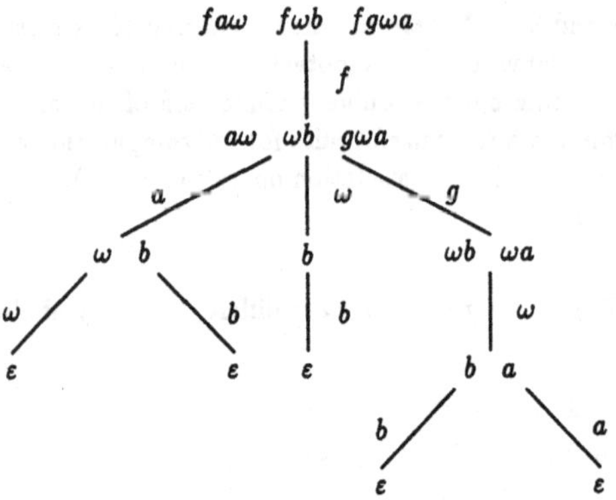

Figure 2: Computation of closed pattern set.

Proof: a) is clear from Definition 4.5. For b), we can show that $\overline{M_\alpha} = \overline{M}/\alpha \;\forall\, \alpha \in \Sigma$ using Lemma 4.3. □

Note that a computation of $\overline{M}$ following Lemma 4.6 always terminates (given a finite suffix set $M \subseteq \Sigma^*$) since each step corresponding to 4.6b) reduces the maximum size $k_M = \max_{\mu\in\overline{M}} |\mu|$ of the closure element suffixes under computation.

4.7 Example. Consider again the pattern set $P = \{fa\omega, f\omega b, fg\omega a\}$ of Example 3.3. The computation of $\overline{P}$ is depicted in fig. 2. We have that

$$\overline{P} = f\overline{P/f} = f\overline{\{a\omega, \omega b, g\omega a\}} =: f\overline{M},$$

where

$$\begin{aligned}\overline{M} &= a\overline{M_a} \cup \omega\overline{M_\omega} \cup g\overline{M_g}\\ &= a\overline{\{\omega, b\}} \cup \omega\overline{\{b\}} \cup g\overline{\{\omega b, \omega a\}}\\ &= \{a\omega, ab, \omega b, g\omega b, g\omega a\}.\end{aligned}$$

□

To compute closures incrementally by combining existing closures, we can make use of the extensionality, monotonicity and idempotence properties stated above. Suppose we want to compute $\overline{M \cup N}$ for some suffix set $M \cup N$, and have already constructed $\overline{M}$ and $\overline{N}$. We then have that $\overline{M \cup N} \subseteq \overline{\overline{M} \cup \overline{N}} \subseteq \overline{\overline{M \cup N}} = \overline{M \cup N}$ and hence $\overline{M \cup N} = \overline{\overline{M} \cup \overline{N}}$. Because $\overline{M}$ and $\overline{N}$ are already closed, to construct $\overline{\overline{M} \cup \overline{N}}$ we need only consider "mixed" closure elements $\mu \vee \nu$ s.t. $\mu \in Pref(\overline{M})$, $\nu \in Pref(\overline{N})$, as specified in the following lemma.

4.8 Lemma. Let $M \cup N$ be a suffix set s.t. M and N are closed, i.e. $\overline{M} = M$, $\overline{N} = N$. Then:

a) If $M = \emptyset$, $N = \emptyset$, or $M = N = \{\varepsilon\}$, then $\overline{M \cup N} = M \cup N$.

b) If $M \cup N \not\subseteq \{\varepsilon\}$, then $\overline{M \cup N} = \bigcup_{\alpha\in\Sigma} \alpha(\overline{M_\alpha \cup N_\alpha})$ where $M_\omega = M/\omega$, $N_\omega = N/\omega$,

$$M_\alpha = \begin{cases} M/\alpha & M/\alpha \neq \emptyset \\ \omega^{\#\alpha} M/\omega & M/\alpha = \emptyset \neq M/\omega, N/\alpha \neq \emptyset \\ \emptyset & \text{otherwise} \end{cases}$$

$$N_\alpha = \begin{cases} N/\alpha & N/\alpha \neq \emptyset \\ \omega^{\#\alpha} N/\omega & N/\alpha = \emptyset \neq N/\omega, M/\alpha \neq \emptyset \\ \emptyset & \text{otherwise} \end{cases}$$

$\forall\, \alpha \neq \omega$.

Proof: Apply Lemma 4.6. □

Note that if M is closed then M/α, $\alpha \in \Sigma$ is so, too. This makes it possible to apply 4.8 recursively. Also note that by 4.8a), we may stop applying 4.8b) as soon as either M or N becomes empty.

4.9 Example. In particular, Lemma 4.8 can be used to *add* new patterns to an existing closure as follows. For (closed) suffix sets M and N let $\nabla(M,N) = \overline{M \cup N}$. Given $\overline{P}$ and $Q \subseteq T_\Sigma$, compute $\overline{Q}$ and then $\nabla(\overline{P},\overline{Q})$. Note that if Q is a singleton pattern set $Q = \{q\}$ then $\overline{Q} = Q$ and hence $\overline{P \cup Q} = \nabla(\overline{P}, Q)$.

For instance, consider $P = \{fa\omega, f\omega b\}$ ($\Rightarrow \overline{P} = \{fa\omega, f\omega b, fab\}$) and $Q = \{fg\omega a\}$. Then

$$\nabla(\{fa\omega, f\omega b, fab\}, \{fg\omega a\}) = f\nabla(\{a\omega, \omega b, ab\}, \{g\omega a\}) =: fM$$

where

$$\begin{aligned} M &= a\nabla(\{\omega, b\}, \emptyset) \cup \omega\nabla(\{b\}, \emptyset) \cup g\nabla(\{\omega b\}, \{\omega a\}) \\ &= \{a\omega, ab, \omega b\} \cup g\omega\nabla(\{b\}, \{a\}) \\ &= \{a\omega, ab, \omega b\} \cup g\omega b\nabla(\{\varepsilon\}, \emptyset) \cup g\omega a\nabla(\emptyset, \{\varepsilon\}) \\ &= \{a\omega, ab, \omega b, g\omega b, g\omega a\}. \end{aligned}$$

Hence $\overline{P \cup Q} = \{fa\omega, fab, f\omega b, fg\omega b, fg\omega a\}$. □

We conclude this section with a technique to *delete* patterns from closures. Given some suffix set M, the corresponding closure $\overline{M}$ and $N \subseteq M$, we want to compute $\overline{M \setminus N}$. Our solution is based on the idea of generating closure elements induced by $M \setminus N$ "on the fly." By this means we may identify the members of $\overline{M} \setminus (\overline{M \setminus N})$ which have to be deleted from $\overline{M}$, according to the identity

$$\overline{M \setminus N} = \overline{M} \setminus (\overline{M} \setminus (\overline{M \setminus N})).$$

4.10 Lemma. Let $N \subseteq M \subseteq \Sigma^*$ be suffix sets. Then

$$\overline{M} \setminus (\overline{M \setminus N}) = \begin{cases} \emptyset & N = \emptyset \\ \overline{M} & M \setminus N = \emptyset \\ \bigcup_{\alpha \in \Sigma} \alpha(\overline{M}/\alpha \setminus (\overline{M_\alpha \setminus N_\alpha})) & \text{otherwise} \end{cases}$$

where $M_\omega = M/\omega$, $N_\omega = N/\omega$,

$$\begin{aligned} M_\alpha &= \begin{cases} M/\alpha \cup \omega^{\#\alpha} M/\omega & \overline{M}/\alpha \neq \emptyset \\ \emptyset & \text{otherwise} \end{cases} \\ N_\alpha &= \begin{cases} N/\alpha \cup \omega^{\#\alpha} N/\omega & M/\alpha \setminus N/\alpha \neq \emptyset \\ M_\alpha & \text{otherwise} \end{cases} \end{aligned}$$

$\forall\, \alpha \neq \omega$.

Proof: Apply Lemma 4.6. □

Note that $N_\alpha \subseteq M_\alpha$ and, by Lemma 4.6b), $\overline{M}/\alpha = \overline{M_\alpha}$ $\forall\, \alpha \in \Sigma$ which makes it possible to apply 4.10 recursively.

4.11 Example. Consider $P = \{fa\omega, f\omega b, fg\omega a\}$ ($\overline{P} = \{fa\omega, f\omega b, fg\omega a, fab, fg\omega b\}$) and suppose we want to delete $f\omega b$ from the closure.

Let $\Delta(\overline{M}, M, N) = \overline{M} \setminus (\overline{M \setminus N})$. We may compute $\Delta(\overline{P}, P, \{f\omega b\})$ recursively following Lemma 4.10. The computation is depicted in fig. 3.

In the computation tree, the elements to the left of the bar denote the members of $\overline{M}$, where the elements of $\overline{M} \setminus M$ are marked with an asterisk and the members of M are unmarked; the elements to the right of the bar denote the members of N; and the sets at the leaves of the tree, under the bar, denote the final closures computed when either N or $M \setminus N$ becomes empty, in which case $\Delta(\overline{M}, M, N) = \emptyset$ or $\Delta(\overline{M}, M, N) = \overline{M}$, respectively.

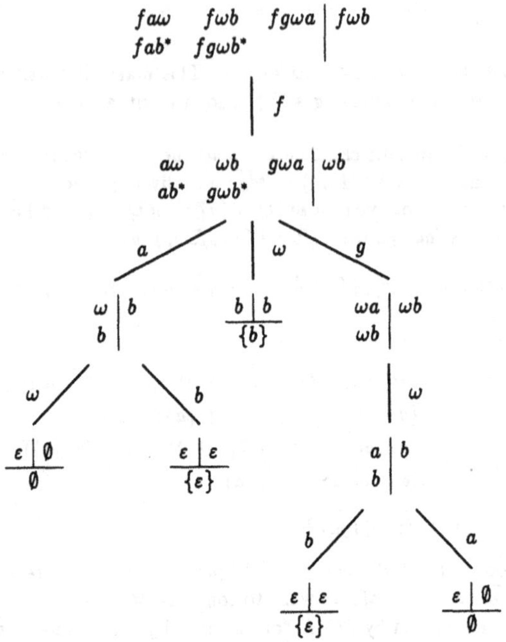

Figure 3: Pattern deletion.

For instance, we have that $\Delta(\overline{P}, P, \{f\omega b\}) = f\Delta(\overline{M}, M, N)$ where $M = \{a\omega, \omega b, g\omega a\}$, $\overline{M} = M \cup \{ab, g\omega b\}$ and $N = \{\omega b\}$. Following the edge labelled a we find that $\Delta(\overline{M}/a, M_a, N_a) = \Delta(\{\omega, b\}, \{\omega, b\}, \{b\}) = \omega\Delta(\{\varepsilon\}, \{\varepsilon\}, \emptyset) \cup b\Delta(\{\varepsilon\}, \{\varepsilon\}, \{\varepsilon\}) = \omega\emptyset \cup b\{\varepsilon\} = \{b\}$.

The final result is $\Delta(\overline{P}, P, \{f\omega b\}) = \{fab, f\omega b, fg\omega b\}$ (read the paths from the root to the leaves labelled with nonempty result sets) and hence $\overline{P \setminus \{f\omega b\}} = \overline{P} \setminus \Delta(\overline{P}, P, \{f\omega b\}) = \{fa\omega, fg\omega a\}$. □

5 Canonical TA Construction

We now are ready to state our main result and show how to effectively construct a canonical TA for any given finite pattern set P. Let $L(P) = \{q \in T_\Sigma \mid q \geq p \text{ for some } p \in P\}$ be the term language of all instances of P. The basic procedure to construct a canonical TA accepting $L(P)$ is as follows.

Construct the closure $\overline{P}$ of P, using the methods outlined in the previous section. We then have that $P \subseteq \overline{P} \subseteq L(P)$, and $\overline{P}$ is finite. Hence we may construct a deterministic FA A accepting $\overline{P}$ which also is a w.d. TA accepting $L(\overline{P}) = L(P)$. We show below that A is canonical.

It is worth noting that once the pattern set closure has been constructed, to derive the corresponding canonical TA from it is a trivial task. In fact, the closed pattern set can be stored using a tree-like representation which immediately gives the state diagram of the canonical TA (compare fig. 2 on page and fig. 5 below!); such a tree-like representation will also be helpful to implement the recursive algorithms of Lemmas 4.6, 4.8 and 4.10.

5.1 Proposition. Let $P \subseteq T_\Sigma$ be a finite pattern set and A a deterministic FA accepting $\overline{P}$. If A is *reduced* (i.e. for any non-final state s of A there is at least one final state reachable from s), then A is a canonical TA accepting $L(P)$.

Proof: A is a w.d. TA accepting $L(P)$ by construction. We have to show that A is canonical.

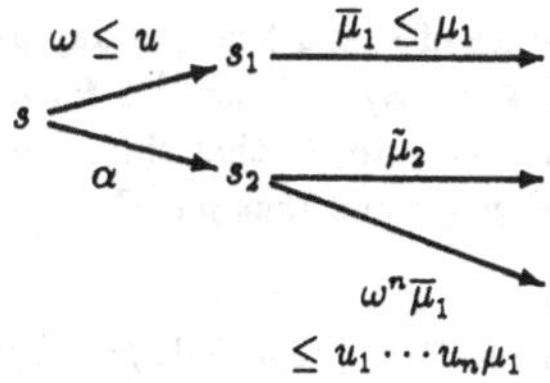

Figure 4: Proof of Proposition 5.1.

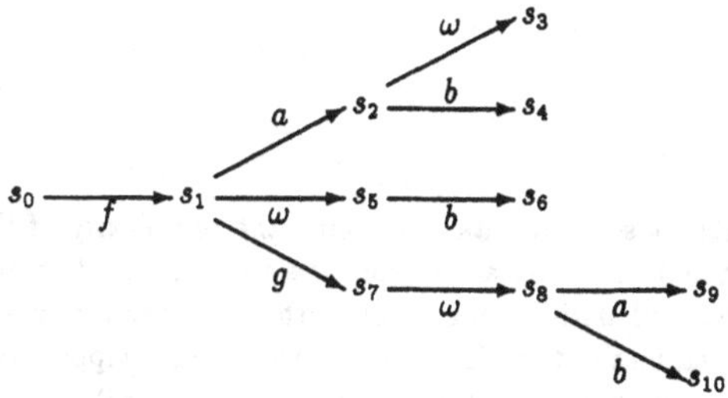

Figure 5: Canonical TA for $P = \{fa\omega, f\omega b, fg\omega a\}$.

Without loss of generality we assume that A is of the form $S = Pref(\overline{P})$, $s_0 = \varepsilon$, $F = \overline{P}$, $\delta = \{s\alpha \to s' \mid s, s' \in S, s' = s\alpha\}$ discussed in Example 3.3.[4]

Assume a state $s \in S$ and input $\mu \in \Sigma^*$ s.t. $s\mu \to s_1\mu_1 \xrightarrow{*} \overline{s}_1 \in F$, but $s\mu \not\xrightarrow{c} s_1\mu_1$. Then $\mu = u\mu_1$, $u = \alpha u_1 \cdots u_n$, $\alpha \in \Sigma_n \setminus \{\omega\}$, $u_i \in T_\Sigma$, and $s\omega \to s_1 \in \delta$ (i.e. $s_1 = s\omega$), while there is another transition $s\mu \xrightarrow{c} s_2 u_1 \cdots u_n \mu_1$ by some $s\alpha \to s_2 \in \delta$.

We have that $s_2 = s\alpha$ is the prefix of some final state $\tilde{s}_2 = s\alpha\tilde{\mu}_2 \in F$ (for some $\tilde{\mu}_2 \in \Sigma^*$). Let $\overline{s}_1 = s_1\overline{\mu}_1 = s\omega\overline{\mu}_1$ for some $\overline{\mu}_1 \in \Sigma^*$. Since $F = \overline{P}$ is closed, $\overline{s}_2 = \overline{s}_1 \vee s\alpha$ also is a final state of A. We have that $\overline{s}_2 = s\omega\overline{\mu}_1 \vee s\alpha = s(\omega\overline{\mu}_1 \vee \alpha) = s_2\omega^n\overline{\mu}_1$.

The situation is depicted in fig. 4.

Note that for any $s, s' \in S$ with $s' = s\mu'$ we have that $s\mu \xrightarrow{*} s'$ iff $\mu \ge \mu'$. Hence $s_1\mu_1 \xrightarrow{*} \overline{s}_1 \Rightarrow \mu_1 \ge \overline{\mu}_1 \Rightarrow u_1 \cdots u_n\mu_1 \ge \omega^n\overline{\mu}_1 \Rightarrow s\mu \xrightarrow{c} s_2u_1 \cdots u_n\mu_1 \xrightarrow{*} \overline{s}_2 \in F$.

We have shown that for any "critical" situation of the form

$$s_2\mu_2 \xleftarrow{c} s\mu \to s_1\mu_1 \xrightarrow{*} \overline{s}_1 \in F$$

there exists another transition chain $s_2\mu_2 \xrightarrow{*} \overline{s}_2$ leading to a final state $\overline{s}_2$. It is now easy to show, by induction on the transition relation $\to$, that A is canonical. □

5.2 Example. Consider again the pattern set $P = \{fa\omega, f\omega b, fg\omega a\}$ with $\#f = 2$, $\#g = 1$, $\#a = \#b = 0$. The closure of P is $\overline{P} = P \cup \{fab, fg\omega b\}$ (cf. Example 4.7). The corresponding canonical TA is depicted in fig. 5. □

The following lemma is useful in determining the actual patterns which have been matched in the final states of a canonical TA.

5.3 Lemma. Let A be the canonical TA constructed for some finite pattern set P, obtained through the prefix construction, and for any $s \in F = \overline{P}$ let the *match set* $P_s \subseteq P$ be defined by $P_s = \{p \in P \mid p \le s\}$.

Then $q \ge p \in P$ iff $s_0q \xrightarrow{c*} s \in F$ s.t. $p \in P_s$.

[4]Note that if the proposition holds for this automaton A, then also for any other reduced deterministic FA A' accepting $\overline{P}$, since then there always is a surjective FA homomorphism $h : A \mapsto A'$, implying that A' essentially behaves exactly like A (though A' may identify "equivalent" states of A).

Proof: If $s_0 q \xrightarrow{c*} s \in F$, then q is an instance of s, and hence of any $p \in P_s$.

For the opposite direction, let $q \geq p \in P$. By 5.1, $s_0 q \xrightarrow{c*} s$ for some $s \in F$. We have that $q \geq s$ and $q \geq p$ and hence $q \geq s \vee p \in \overline{P}$. Now it is easy to see that since $s_0 q \xrightarrow{c*} s$ and $q \geq s \vee p \geq s$ we also have that $s_0 q \xrightarrow{c*} s \vee p$. But then $s \vee p = s$, i.e. $p \leq s$ and thus $p \in P_s$. □

5.4 Example. For our sample pattern set $P = \{fa\omega, f\omega b, fg\omega a\}$, the match sets associated with final states are as follows (cf. fig. 5):

$$\begin{array}{lll} s_3 : fa\omega & s_6 : f\omega b & s_9 : fg\omega a \\ s_4 : fa\omega, f\omega b & & s_{10} : f\omega b \end{array}$$

□

6 Complexity

In this section we briefly present some results concerning the complexity of the canonical TA technique. We assume that A is the canonical TA obtained from the pattern set P as discussed in Section 5 (using the prefix construction). We will derive bounds for both the maximum matching time of A (which is $\max_{\overline{p} \in \overline{P}} |\overline{p}|$, assuming that each move of the TA – in particular, the skipping of subtrees – may be carried out in constant time) and the number of final states of A (which is $|\overline{P}|$).

We first consider the sizes of closure patterns $\overline{p} \in \overline{P}$. It is easy to show that for any $p, q \in T_\Sigma$ and $\nu \in Pref(q)$ s.t. $p \vee \nu \neq \top$, we have that $|p \vee \nu| \leq |p| + |q|$ and $h(p \vee \nu) \leq \max\{h(p), h(q)\}$. Since any $\overline{p} \in \overline{P}$ is the join of some $p \in P$ with some prefixes of other patterns in P, it follows that

$$|\overline{p}| \leq \sum_{p \in P} |p| \;\forall\, \overline{p} \in \overline{P}. \tag{1}$$

and

$$h(\overline{p}) \leq \max_{p \in P} h(p) \;\forall\, \overline{p} \in \overline{P}, \tag{2}$$

i.e. $|\overline{p}| = O(r^h)$ where r is the maximum arity of the symbols in Σ and h is the maximum pattern height.

(1) is essentially the same result as that for the matching technique proposed in [11]. (2) tells us, in particular, that the maximum matching time of A is independent of the number of different patterns.

Bounds for $|\overline{P}|$ can be derived by counting the maximum number of different joins of pattern prefixes. Since any $\overline{p} \in \overline{P}$ may be written as $\overline{p} = \bigvee_{p \in P} \nu_p$ s.t. $\nu_p \in Pref(p)$, $p \in P$, we have that

$$|\overline{P}| \leq \prod_{p \in P} (|p| + 1). \tag{3}$$

Again, this parallels the corresponding result in [11]. Tighter bounds for $|\overline{P}|$ can be obtained by further restricting the number of different combinations ν_p, $p \in P$ to consider, depending on the structure of the pattern set. For any pair $p, q \in P$ let $n_{p,q}$ be the number of different joins $p \vee \nu \notin \{p, \top\}$ for $\nu \in Pref(q)$. Let k_p be the number of $n_{p,q} > 0$, $q \in P$, and let n_p be the average of all $n_{p,q} > 0$ (i.e. $n_p = 0$ if $k_p = 0$, and $n_p = \frac{1}{k_p} \sum_{n_{p,q} > 0} n_{p,q}$ otherwise). For each $p \in P$ we then have that the number of different joins $p \vee \bigvee_{q \neq p} \nu_q \in \overline{P}$ is $\leq \prod_{q \in P} (n_{p,q} + 1) \leq (n_p + 1)^{k_p}$. Hence:

$$|\overline{P}| \leq \sum_{p \in P} (n_p + 1)^{k_p}. \tag{4}$$

For instance, for the pattern set $P = \{fa\omega, f\omega b, fg\omega a\}$ used as a running example throughout this paper, we have that $\prod_{p \in P}(|p|+1) = 80$, but $(n_{fa\omega}+1)^{k_{fa\omega}} + (n_{f\omega b}+1)^{k_{f\omega b}} + (n_{fg\omega a}+1)^{k_{fg\omega a}} = 2^1 + 2^2 + 1^0 = 7$. In fact, (4) indicates that if the k_p's are bounded (i.e., there is a bound on the maximum number of patterns with nontrivial overlapping prefixes), the size of $\overline{P}$ is linear in $|P|$ and polynomial w.r.t. the n_p's, which in turn are always bounded by the size of the longest pattern.

All bounds derived in this section are sharp (asymptotically), see [5] for details. This means that there are always pathological examples of pattern sets which produce excessively large canonical TA's, and that the bound (4), in general, is no better than (3). Up to now, we have not been able to find a more specific measure for the actual size of closures which can be determined from the given pattern set P easily. We think, however, that the k_p parameters of (4) might be useful in characterizing certain classes of "good-natured" pattern sets in terms of syntactic criteria.

Some final remarks about the closure construction algorithms sketched out in Section 4 are in order. For simplicity, let us assume that all basic operations performed in these algorithms (such as test for emptiness, computation of M/α, etc.) may be carried out in constant time.

We then have that the recursive closure computation following Lemma 4.6 needs time proportional to $|\mathit{Pref}(\overline{M})|$. The construction of $\nabla(M,N) = \overline{M \cup N}$ according to 4.8 may need time $O(|\mathit{Pref}(\nabla(M,N))|)$, in the worst case, and the computation of $\Delta(\overline{M},M,N) = \overline{M} \setminus \overline{M \setminus N}$ following 4.10 needs time $O(|\mathit{Pref}(\Delta(\overline{M},M,N))|)$ (assuming that in each step we only consider those $\alpha \in \Sigma$ for which $N_\alpha \neq \emptyset$).

For 4.10 we may conclude that only those positions in closure elements need to be considered that are part of at least one element of $\overline{M}$ which is to be deleted; in this sense, 4.10 is efficient. In the case of 4.8, it is not as easy to tell the effects of a change, since we may rescan large parts of M and N which in fact may fail to produce even a single new closure element. Another difficulty is that the complexity not only depends on the "sizes" of M and N but also on the overlaps between elements of M and N. Somehow restricting the possible number of these overlaps, in a manner similar to the derivation of (4) above, might prove to be a useful approach to the problem.

7 Conclusion

In this paper we presented a new technique for left-to-right tree pattern matching, which is useful in the implementation of programming languages based on term rewriting. Our technique is comparable in efficiency with existing techniques used in functional language implementations, such as the one described in [11].

The technique is based on a new approach to the construction of left-to-right tree matching automata. The theoretical background is the novel concept of *prefix unification* which extends the normal unification of terms to term prefixes. In our framework, *canonical* matching automata admitting a deterministic matching strategy are produced by computing a *closure* of the pattern set which is determined by unifying pattern prefixes.

We showed how closures can be computed recursively by simultaneously unifying pattern prefixes, and incrementally by combining existing closures and deleting patterns from a closure.

There are several directions for future research. First of all, a detailed analysis of the incremental closure construction algorithms still needs to be done. One might then consider to apply these algorithms to Knuth-Bendix completion or similar applications involving dynamically changing pattern sets. Up to now we have only used the canonical TA matching technique to implement a simple prototype of a term rewrite system compiler without incremental features.

Secondly, there remains the question of determining *minimal* canonical TA's. Given some finite pattern set P, we may always construct a "reduced" pattern set $P_0 \subseteq P$ s.t. $L(P_0) = L(P)$ and P_0 consists only of the "most general" patterns in P (i.e., if $p, q \in P_0$ s.t. $p \leq q$ then already $p = q$). There is strong evidence that the canonical TA constructed from P_0, in some sense, is a "smallest" canonical TA accepting $L(P)$. More precisely, it seems quite reasonable that *any* weakly deterministic TA accepting $L(P)$, as an FA, must accept at least $\overline{P}_0$ in order to be canonical. Together with FA state minimization techniques this could probably be turned into a theorem characterizing the minimal canonical TA's for a given pattern set P, and a corresponding procedure to effectively construct such minimal TA's.

Finally, from the theoretical point of view, prefix unification and closure construction might be useful concepts which should be explored further. In particular, it might be interesting to extend prefix unification and closure construction to nonlinear patterns and to the subterm matching problem (find *all* matches with subterms of the subject term).

References

[1] Augustsson, Lennart: Compiling pattern matching. *Proc. Functional Programming Languages and Architectures '85.* Berlin (etc.): Springer, 1985, pp. 369–381. (Lecture Notes in Computer Science 201).

[2] Birkhoff, G.: *Lattice theory.* New York: American Mathematical Society, 1948.

[3] Christian, Jim: Fast Knuth-Bendix completion: summary. *Proc. Rewriting Techniques and Applications '89.* Berlin (etc.): Springer, 1989, pp. 551–555. (Lecture Notes in Computer Science 355).

[4] Gécseg, Ferenc; Magnus Steinby: *Tree automata.* Budapest: Akadémia Kiadó, 1984.

[5] Gräf, Albert: *Efficient pattern matching for term rewriting.* Johannes Gutenberg-Universität Mainz, 1990. (Technical Report 3/90).

[6] Hemerik, C.; J.P. Katoen: Bottom-up tree acceptors. *Science of Computer Programming*, **13**, 1989, pp. 51–72.

[7] Hoffmann, C.M.; M.J. O'Donnell: Pattern matching in trees. *Journal of the ACM*, **29**, 1, 1982, pp. 68–95.

[8] Knuth, D.E.; P.B. Bendix: Simple word problems in universal algebras. *Proc. of the Conference on Computational Problems in Abstract Algebra 1967.* Oxford: Pergamon Press, 1970, pp. 263–298.

[9] Laville, A.: Implementation of lazy pattern matching algorithms. *Proc. European Symposium on Programming '88.* Berlin (etc.): Springer, 1988, pp. 298–316. (Lecture Notes in Computer Science 300).

[10] Peyton Jones, Simon L.: *The implementation of functional programming languages.* Englewood Cliffs, N.J. (etc.): Prentice Hall, 1987. (International Series in Computer Science).

[11] Schnoebelen, Ph.: Refined compilation of pattern matching for functional languages. *Science of Computer Programming*, **11**, 1988, pp. 133–159.

[12] Toyama, Yoshihito: Fast Knuth-Bendix completion with a term rewriting system compiler. *Information Processing Letters*, **32**, 1989, pp. 325–328.

Incremental Techniques for Efficient Normalization of Nonlinear Rewrite Systems

R. Ramesh
Department of Computer Science
University of Texas at Dallas
Richardson, TX 75083.

I.V. Ramakrishnan
Department of Computer Science
State University of New York at Stony Brook
Stony Brook, NY 11794.

Abstract

In [8] we described a nonlinear pattern-matching algorithm with the best known worst-case and optimal average-case time complexity. In this paper we first report on some experiments conducted on our algorithm. Based on these experiments we believe that our algorithm is useful in speeding up normalization of nonlinear rewrite systems even when it has a small number (≥ 5) of rewrite rules.

In order to find matches quickly our algorithm operates in two phases. In the first phase it scans the subject tree to collect some "information" which is then used to find matches quickly in the second phase. Scanning can become very expensive especially for large subject trees. However the normalization process is "incremental" in nature. After each reduction step the subject tree is altered and this modified tree is usually not completely different from the old tree. Hence it should be possible to avoid scanning and searching the entire tree for new matches. We describe general techniques to exploit the incremental nature of normalization to speed up each reduction step. Specifically, using these techniques we show that the search for new matches in the subject tree following a replacement can be made *independent* of the size of the subject tree.

1 Introduction

Normalization can be regarded as a process of tree replacement wherein terms are represented as trees and the input term (called the *subject*) is reduced to its normal form by repeatedly replacing its subtrees according to rules of the term-rewriting system. Each reduction step involves two phases. The first phase is *tree pattern matching* which finds a substitution called a match, that yields a subterm of the input term when applied to the left-hand-side of a rule (called the *pattern*). This is followed by the replacement phase in which this subterm is replaced by the substitution instance of the right-hand-side of the same rule.

Normalization is quite compute intensive as it involves repeated reduction steps. Hence there is a lot of interest in enhancing its efficiency. So far the matching operation in a reduction step has received a lot of attention. Several efficient pattern-matching algorithms have been reported (see [2, 4, 11] for example). But all these algorithms assume that the rewrite system is linear. Restricting patterns to be linear may appear adequate for equational programming based on term rewriting [5]. However in many other applications of term rewriting such as completion procedures [3] and theorem proving [13], patterns are typically nonlinear. Unlike linear patterns, to match nonlinear patterns we must also perform a *consistency check*, i.e., verify that the substitutions for multiple occurrences of a variable are identical. For linear patterns there is no need to compute substitutions explicitly for matching. To use linear pattern matching algorithms for nonlinear patterns a separate phase is needed to compute substitutions. This is not efficient since if n is the size of the subject and k is the number of variables in the pattern then this phase alone requires $O(nk)$ time in the worst case.

In [8] we described a new algorithm for nonlinear tree pattern matching based on explicitly computing substitutions for variables. The basic idea in our approach was to reduce tree pattern matching to string matching. In our method we first transform all the patterns into strings. Given a subject tree we transform it into a string and then perform several string matching steps involving pattern strings and subject string. The outcome of these string matching steps is then correlated to identify occurrences of pattern trees in the subject tree. For efficiency our algorithm is driven by the Aho-Corasick string matching automaton that is constructed out of pattern strings [1]. The subject string is scanned by the automaton and the state information so obtained is used to perform string matching. Our algorithm has the best known asymptotic worst-case and optimal average-case time complexity for matching nonlinear patterns. To describe our algorithm's time bound we need the following terminology.

Let $\mathcal{K}(p)$ denote the set of all those root-to-leaf path strings in pattern p that end on leaves labeled with variables. Let k be the total number of occurrences of variables in p. Then $|\mathcal{K}(p)| = k$. Let the suffix number of a string ω in $\mathcal{K}(p)$ denote the number of strings in $\mathcal{K}(p)$ which are suffixes of ω. Let k^*, the suffix index of $\mathcal{K}(p)$, be the maximum among all suffix numbers of strings in $\mathcal{K}(p)$. If k is 0 then k^* is 1. Note that the number of leaves that can be labeled with variables can range from 0 to the maximum number of leaves in the pattern and so $k^* \leq k \leq l$ where l is the number of leaves in p.

We showed that in the worst case there can be $O(nk^*)$ substitutions when matching p against a subject whose size is n. We also showed that the worst-case time complexity of our algorithm is $O(nk^*)$. It is interesting to observe that the other steps in addition to computing substitutions does not add to the asymptotic time complexity of our algorithm. We also showed that on the average our algorithm spends only $O(1)$ time per node in the subject tree.

In this paper we first report the results of some experiments conducted on our algorithm (see Section 3). We measured the time taken by our algorithm to find all matches as well as only one match. Both patterns and subject trees were generated randomly. For the same set of inputs we also measured the running time of the naive matching algorithm. We remark that only the naive algorithm is used for pattern matching in all the three well-known term-rewriting systems - Reve [10], SbReve [9] and RRL [12]. Our experiments reveal that the running time of our algorithm to find all matches as well as only one match is superior to that of the naive algorithm even when the rewrite system has only one pattern. In our algorithm we first have to transform the subject into a string and then scan it (with the automaton) before commencing the match operation. If the scanning time is added to the matching time then for finding all matches as well as only match our algorithm begins to outperform the naive algorithm when the rewrite system has five or more patterns.

Based on our experiments we believe that in practice we do not need a large term-rewriting system consisting of many patterns to achieve superior performance with our algorithm. Nevertheless it is worthwhile to explore new sources and identify new methods for improving its efficiency still further.

One important source discussed below arises from the "incremental" nature of normalization. Note that in a reduction step the match operation is followed by a replacement operation. The latter operation creates a modified subject tree on which the next reduction step is performed. Since this tree may not be completely different from the old one it may not be necessary now to scan the modified tree entirely and so the scanning time of our algorithm can be substantially reduced. Furthermore, a replacement operation may result in new matches to appear and old ones to disappear. Again it may be possible to reduce match time since it may not be necessary to search the entire tree for finding matches. Notice that if we had to search the entire tree then each reduction step is *dependent* on the subject size. An interesting problem then is to exploit the incremental changes so that each reduction step can be made *independent* of the subject size. The design of such normalization algorithms has not received much attention.

In Section 4 we describe our approach for efficiently normalizing nonlinear rewrite systems by exploiting the incremental changes caused by reduction steps. The search for matches is based on our nonlinear pattern matching algorithm. In our algorithm the pattern and subject strings were maintained in an array data structure to facilitate rapid string matching operations. However a replacement operation can cause some substrings of the subject to be replaced by other ones. So insertion and deletion now becomes an expensive operation. By suitably combining the array structure with a linked list we show that replacement can be made independent of the subject size. By exploiting an important property of the Aho-Corasick automaton we show that rescanning the subject following replacement is only dependent on the pattern size. The scanning time reported in our experiments was the time required to scan the subject tree completely. Given that the average size of a pattern is quite small compared to the subject, scanning time can now be made insignificant to matching time. Finally we also show that the search for matches can also be made independent of the subject size. The techniques developed in this paper are quite general. It can be used in conjunction with other normalization algorithms for nonlinear rewrite systems as well as in other applications (see discussion in Section 5).

The rest of this paper is organized as follows. To make this paper self-contained we begin with a review of our pattern-matching algorithm in Section 2. The details appear in [8]. Section 3 summarizes our experimental results. In Section 4 we describe incremental techniques for doing normalization of nonlinear rewrite systems. Estimates on the expected gain by use of these techniques and discussion about their generality appear in the concluding section (Section 5). For lack of space some proofs have been omitted in this paper. They can however be found in [14].

2 Review of Nonlinear Pattern Matching Algorithm

As mentioned earlier our algorithm is based on computing substitutions for variables explicitly. The basic idea is to reduce tree pattern matching to string matching. Every term is "linearized" using its Euler chain which is obtained as follows. First for every edge (v_i, v_j) create a new edge (v_j, v_i). Next list the nodes of the resulting structure as it is traversed in preorder. The *Euler string* of a term tree is obtained by replacing all the nodes in the Euler chain by their labels. For example, the Euler string of the term $f(f(a,X),X)$ is $ffafXffXf$. Note that the Euler subchain between the first and last occurrence of a node is the Euler chain of the subtree rooted at that node. We begin with a simple algorithm that captures the major ideas in our algorithm. This is followed by a brief discussion on efficiently implementing these ideas.

2.1 Simple Algorithm

Let C_s and C_p be the Euler chains of subject s and pattern p respectively. Let E_s and E_p denote their respective Euler strings and $|E_s|$ (which is proportional to subject size) and $|E_p|$ (which is proportional to pattern size) denote their respective sizes. Assume that pattern p has k variables and denote them $v_1, v_2, \ldots, v_k$. Since only leaf nodes are labeled with variables and since each leaf node appears exactly once in the Euler string E_p, each of the these variables appear exactly once in E_p.

E_s is stored in an array. Next split E_p into $k+1$ pattern strings by removing the k variables from it. We denote these strings as $\omega_1, \omega_2, \ldots, \omega_{k+1}$. For example removing both occurrences of X in the term $f(f(a,X),X)$ results in three pattern strings $\omega_1 = ffaf$, $\omega_2 = ff$, $\omega_3 = f$. We then construct boolean tables $M_1, M_2, \ldots, M_k$ where each M_i has $|E_s|$ entries. If there is a match for ω_i in E_s starting at position j then the jth entry of M_i contains 1; otherwise it is 0.

Let E'_p be the new string obtained by replacing $v_1, v_2, \ldots, v_k$ in E_p by strings $e_1, e_2, \ldots, e_k$

respectively such that each of the e_i is an Euler string of a subtree of s. Now we say that E_p matches[1] a substring of E_s at i if E_p' thus obtained is identical to the substring of E_s between positions i and $i + |E_p'| - 1$. In such a case, if $C_s(i)$ is the first occurrence of a node, say u, then it can be shown that pattern p matches the subtree of s (modulo nonlinearity) rooted at u.

Since ω_1 is a prefix of E_p it is obvious that the set of nodes where p matches s is a subset of the matches for ω_1 in E_s. We therefore look for a possible pattern match only at the non-zero entries in M_1 that correspond to the first occurrence of a node. Due to this unique feature of our algorithm we can avoid unnecessary computations on match attempts that are bound to fail. Let i be one such entry. To determine a pattern match at i we proceed as follows. $i + |\omega_1| - 1$ is the position in E_s at which the string match of ω_1 terminates. Since E_p is $\omega_1 v_1 \omega_2 v_2 \ldots \omega_k v_k \omega_{k+1}$, the substitution for v_1 must be a substring of E_s that starts at position $i + |\omega_1|$. This position must correspond to the first occurrence of a node, say x (as only subtrees can be substituted for variables) in C_s. Otherwise there cannot be a match for the pattern at i. On the other hand, if this condition is satisfied then the substitution for v_1 will be the substring of E_s between those positions that correspond to the first and last occurrence of x in C_s. Let the last occurrence of x occur at the jth position in C_s. We should therefore match ω_2 beginning from the $(j+1)^{\text{th}}$ position in E_s. Obviously $M_2[j+1]$ must be a non-zero entry. Similarly we compute substitutions for $v_2, v_3, \ldots, v_k$. Upon successful completion we conclude E_p will match E_s at position i.

2.2 Improving Efficiency

Note that in our simple algorithm we had to construct $O(k)$ tables each having $|E_s|$ entries. Therefore a direct implementation of this will require $O(nk)$ time. To improve time performance such tables should not be explicitly computed. Note that for nonlinear patterns we have to perform consistency check on the substitutions computed for variables. In [8] we show that an upper bound on the number of substitutions is $O(nk^*)$ and that our method computes at most this many substitutions.

In order to get a time bound proportional to the number of substitutions we cannot spend more than $O(1)$ time between substitutions. In our method the Euler string of the subject is stored in an array along with a pointer from the first occurrence of a node to its last occurrence. Therefore computing a substitution now merely requires verifying that it begins at the first occurrence of a node and if so skipping to its last occurrence. All this can be done in $O(1)$ time. Having thus computed the substitution for a variable v_i the steps preceding computation of the substitution for v_{i+1} involves determining whether pattern string ω_{i+1} matches a substring in E_s at the position following the substitution for v_i. Had we constructed the tables then this information can be obtained in $O(1)$ time from table M_{i+1}. In the absence of these tables the problem now is to answer the following in $O(1)$ time: given a position in E_s and a pattern string ω_i, does ω_i match the substring in E_s at that position (note that ω_i can be any one of the $k+1$ pattern strings). To answer this we do the following. We preprocess these $k+1$ pattern strings to produce an automaton that recognizes every instance of these strings. We use the Aho-Corasick algorithm [1] to construct this automaton and use it to recognize pattern strings in E_s as follows. With every position in the array containing E_s, we store the state of the automaton on reading the symbol in that position. All we need to do now, in order to conclude whether pattern string ω_j matches the substring of E_s at i, is to look at the state of the automaton in position $i + |\omega_j| - 1$ in E_s. Since E_s is stored in an array this lookup can therefore be done in $O(1)$ time. Upon obtaining the state we have to verify whether it is an accepting state for ω_j. This verification must be done in $O(1)$ time. But in

[1]When we say that string ω matches E_s at i it means that ω matches the substring in E_s between (and including) positions i to $i + |\omega| - 1$

the Aho-Corasick algorithm the output set is represented as a linked list making it unsuitable to do verification in $O(1)$ time. Hence we must copy the linked list in to an array as the last step in the construction of the automaton. In [8] we show that this step does not change the asymptotic linear time complexity of the Aho-Corasick algorithm.

Consistency Check

So far we have only identified pattern matches modulo nonlinearity. The next step is to check the consistency of the computed substitutions. To do so we proceed as follows. To each node in the subject tree we assign an integer code (varying from 1 to n) called its *signature* such that two nodes get the same signature if and only if the subtrees rooted beneath them are identical. Such an encoding can be computed in time proportional to the number of nodes in the tree using the method in [6]. The signatures are then used to do consistency check during the matching process.

Our algorithm, based on the ideas described above, finds all matches in $O(nk^*)$ time. Our algorithm can also be easily modified to find only one match. But the worst case time complexity will still remain $O(nk^*)$.

3 Experimental Results

This section summarizes the experiments conducted on our algorithm and the naive pattern-matching algorithm on a Sun 3/50. We measured the time taken by our algorithm as well as the naive algorithm to find all matches and only one match on the same set of inputs. Both patterns and subject trees were generated randomly. The maximum arity of any function symbol was limited to be at most 3 and the maximum size of the pattern was limited to be at most 6. The restrictions on these two parameters was influenced by inspection of typical examples of canonical systems. (A catalogue of such systems appear in [7].) Both the number of patterns and the size of the subject trees were varied.

Tables 1 and 2 is a summary of our experimental results for finding all matches and one match respectively. The columns under *naive* denotes the running time of the naive matching algorithm in milliseconds. Similarly the columns under *scan* and *match* denote the time in milliseconds for scanning E_s and doing matching respectively. The columns *Avg n* and *Avg m* denote the average sizes of the subject and pattern respectively. S_1 and S_2 are speedup ratios; S_1= (naive match time / our match time) whereas S_2 = (naive match time/ (scan time + our match time)).

The scanning time measured here is the time taken to scan the entire subject tree. In the next section we will show how to reduce the scan time. In particular we will make it dependent only on the pattern size. Doing so will make scan time negligible compared to matching time. Estimates on the expected reduction are discussed in the concluding section (Section 5).

Notice that increase in the number of patterns causes slight increases in the scanning time also. The reason is that during scan we also obtain information of the places in the subject's Euler chain where the first string of each pattern occurs. Match attempts are initiated only at these positions thereby avoiding unnecessary computations on some match attempts that are bound to fail.

4 Incremental Techniques

This section describes normalization of nonlinear rewrite systems based on our pattern-matching algorithm. The basic idea here is to scan the subject tree completely only once for the first reduction step. For doing subsequent reduction steps we use incremental techniques to exploit the fact that a replacement operation may not necessarily create a completely new subject tree. Hence it may

Avg n	Number of Patterns													
	1							5						
	Avg m	naive	ours		Speedups			Avg m	naive	ours		Speedups		
			scan	match	S1	S2				scan	match	S1	S2	
25.08	4.5	1.31	1.05	0.25	5.24	1.0		3.6	6.34	2.2	1.78	3.56	1.59	
98.37	4.5	4.94	3.65	0.43	11.49	1.21		4.4	26.64	5.6	2.26	11.79	3.13	
1010.47	4.5	41.49	29.5	4.89	8.48	1.21		4.2	207.28	53.8	33.82	6.13	2.37	

Avg n	Number of Patterns												
	10						25						
	Avg m	naive	ours		Speedups		Avg m	naive	ours		Speedups	
			scan	match	S1	S2			scan	match	S1	S2
25.08	3.9	12.6	1.4	2.34	5.38	3.37	3.72	31.06	3.2	4.14	7.5	4.23
98.37	4.0	48.16	3.6	2.6	18.52	7.77	3.88	121.26	12.0	11.02	11.0	5.27
1010.47	3.80	406.09	67.60	40.34	10.07	3.76	3.84	1021.49	101.6	78.12	13.08	5.68

Table 1: Experimental Results for finding All Matches

Avg n	Number of Patterns												
	1						5						
	Avg m	naive	ours		Speedups		Avg m	naive	ours		Speedups	
			scan	match	S1	S2			scan	match	S1	S2
25.08	4.5	0.86	0.95	0.12	7.48	0.81	3.6	3.38	2.2	0.52	6.5	1.24
98.37	4.5	4.35	3.55	0.25	17.26	1.14	4.4	18.14	5.6	0.84	21.6	2.82
1010.47	4.5	27.25	30.1	0.18	155.71	0.90	4.2	80.85	51.8	0.5	161.6	1.55

Avg n	Number of Patterns												
	10						25						
	Avg m	naive	ours		Speedups		Avg m	naive	ours		Speedups	
			scan	match	S1	S2			scan	match	S1	S2
25.08	3.9	9.34	1.8	1.44	6.49	2.88	3.72	25.22	2.6	1.76	14.33	5.78
98.37	4.0	38.72	5.4	0.94	41.19	6.11	3.88	89.26	10.4	5.0	17.85	5.80
1010.47	3.80	157.53	66.2	1.24	127.04	2.30	3.84	374.11	99.40	9.38	39.88	3.44

Table 2: Experimental Results for finding one match

not be necessary to scan the entire tree.

Specifically, suppose the subtree rooted at u is replaced by an instance of r ($p \rightarrow r$ is the applicable rule). Furthermore, let λ denote the length of the longest string in the Euler chain of p, h_p denote height of p, h_r denote height of r and k_r denote the number of variables in r.

- **We show that replacement takes $O(k_r\lambda)$ time.**
- **Upon replacement the states corresponding to symbols in the modified subject tree must be computed. We show that this can also be accomplished within time $O(k_r\lambda)$.**
- **Prior to recomputing matches we must also assign signatures to all the nodes in the modified subject tree. By representing these signatures in a suitable fashion we show that this can be done in time proportional to the size of r and the number of ancestors of the replaced subtree.**
- **Finally, we show that we need initiate pattern matches only at the nodes in r and at most h_p ancestors of the replaced subtree. This is somewhat surprising since nonlinearity can cause a new match to occur at any of the ancestor nodes.**

Our starting point is a term-rewriting system which has rules of the form $l \rightarrow r$. All the lhs and rhs are transformed into their Euler strings. Using the strings of lhs we construct the Aho-Corasick automaton. The subject tree is also transformed into its Euler string and scanned completely for

the first reduction step only. In the remainder of this section we describe techniques to efficiently handle subsequent reduction steps.

4.1 Replacement

We begin with a description of the issues involved in the replacement and ideas for handling it efficiently. Recall that in our algorithm the subject tree is maintained as an array. Although this enabled us to carry out pattern matching efficiently it is not suitable for performing replacement. This is because replacement involves deleting a substring of E_s and inserting another one in its place. Clearly deletion and insertion in an array cannot be performed efficiently unless the length of the deleted portion equals the length of the inserted portion. In particular, if $E_s = \alpha\beta\gamma$ and if β is to be replaced by δ then replacement requires the following steps. Suppose $|\beta| = |\delta|$ then we simply overwrite the portion of array containing β by δ. On the other hand if $|\beta| \neq |\delta|$ then δ will not fit exactly in the portion of the array that contains β. Therefore now in addition to copying δ, we must also move γ. This requires $O(|\gamma|)$ time and thus replacement will depend upon the subject size. This dependency can be eliminated had we maintained E_s as a linked list. Since we cannot do random access in a linked list we can no longer answer each string matching question in $O(1)$ time. The interesting problem now is how to combine linked list and an array so that we can not only answer each string matching question in $O(1)$ time but also do the replacement in time that is independent of the subject size.

Recall again that in order to identify occurrence of ω_i at j in E_s we need to access the state of the automaton located at the $(j+|\omega_i|-1)^{\text{th}}$ position in E_s. Suppose we maintain E_s as a *linked list of arrays* of $|\omega_i|$ records. Now we can identify occurrence of ω_i at j in $O(1)$ time since accessing the state information at $(j+|\omega_i|-1)$ requires traversing at most one link in the linked list. Furthermore if ω_i is the longest among all the pattern strings then we can also answer any string matching question involving any other pattern string in $O(1)$ time. The first idea then is to maintain E_s as a linked list of arrays where each array is of size λ. But this alone is not enough to guarantee that replacement can be done in time independent of the subject size. The number of filled entries in each array also plays a crucial role in this regard. In particular, by ensuring that at least the first $\lambda/2$ entries in each array is nonempty we have:

Lemma 1 *Let $E_s = \alpha\beta\gamma$ be a linked list of arrays of records as described above. Let δ be another linked list of arrays. Constructing $E'_s = \alpha\delta\gamma$ requires $O(\lambda)$ time.*

Proof: Omitted. Details appear in [14].

Now replacement in a reduction step requires deleting the Euler chain of the matched subtree and inserting the Euler chain of the right hand side. Both deletion and insertion together take $O(\lambda)$ time. The Euler chain of rhs is obtained by insertion of the the Euler chains of the substitutions for the variables in it. If there are k_r variables in the rhs then creating its Euler chain takes $O(k_r\lambda)$ time.

4.2 Rescanning New Subject Tree

Let E'_s denote the Euler chain of the new subject term following replacement. The task of rescanning is to identify, for each entry in E'_s, the state reached by the automaton upon reading the function symbol in it. Obviously this can be done by scanning E'_s completely from the beginning. Clearly the time required for completely rescanning is proportional to the size of the new subject.

We will show how to perform rescanning independent of the subject size. The ideas here need some basic properties of the Aho-Corasick automaton. What follows is a brief review of this automaton. The Aho-Corasick automaton consists of nodes called *states* and two types of links - *goto* and *failure*. The goto links are labeled with symbols from the alphabet of the pattern strings. These links together with the states form a trie structure whose root is the start state. Following [1] we say string ω *represents* state u if the path in the trie from the start state (the root node) to state u spells out ω. The automaton scans the input text for recognizing occurrences of pattern strings. While scanning it makes either a goto or a failure transition. Suppose the automaton is in state u after scanning the first j symbols of the input text $a_1a_2\ldots a_ja_{j+1}\ldots a_n$. If there is a goto link labeled a_{j+1} from u to w then the automaton makes a goto transition to w. Now,

Lemma 2 (Aho-Corasick) *The string represented by w is the longest suffix of $a_1a_2\ldots a_{j+1}$ that is also a prefix of some pattern string.*

On the other hand if there is no such link labeled a_{j+1} from u then it makes a failure transition. If this transition takes the automaton to a state v then:

Lemma 3 (Aho-Corasick) *The string represented by v is longest proper suffix of the string represented by u.*

If it is unable to make a goto transition from v with a_{j+1} it again makes a failure transition and continues to do so until it reaches some state from which it can make a goto transition with a_{j+1}. Note that the automaton will eventually be able to make a goto transition as there are goto transitions from the start state for every symbol.

Using the Aho-Corasick automaton we rescan E'_s as follows. Let $E_s = \alpha\beta\gamma$ and $E'_s = \alpha\delta\gamma$. Assume that δ has already been scanned. Then:

Lemma 4 *Rescanning E'_s requires $O(\lambda)$ time.*

Proof: Let $\mathcal{A} = A_1, A_2, \ldots, A_{|E_s|}$ be the sequence of states of the automaton on scanning E_s. Let $\mathcal{B} = B_1, B_2, \ldots, B_{|\delta|}$ be the sequence of states associated with δ and $\mathcal{C} = C_1, C_2, \ldots, C_{|E'_s|}$ be the sequence of states upon completely scanning E'_s. To show that $\mathcal{A}$ can be changed to $\mathcal{C}$ in $O(\lambda)$ time we proceed as follows.

Let $\mathcal{D}$ denote the sequence of states associated with $\alpha\delta\gamma$ before rescanning. So $\mathcal{D} = A_1, A_2, \ldots, A_{|\alpha|}, B_1, B_2, \ldots, B_{|\delta|}, A_{|\alpha|+|\beta|+1}, A_{|\alpha|+|\beta|+2}, \ldots, A_{|E_s|}$. Let $\mathcal{D}(i)$ and $\mathcal{C}(i)$ be the ith element in the sequences $\mathcal{D}$ and $\mathcal{C}$ respectively. Now observe that the first $|\alpha|$ function symbols in E_s and E'_s are the same. Therefore $\mathcal{C}(i) = \mathcal{D}(i)\ \forall i (1 \leq i \leq |\alpha|)$. From lemma 2 it follows that $\mathcal{C}(i)$ depends only on at most λ consecutive symbols immediately preceding the ith function symbol in E'_s. Therefore $\mathcal{C}(i) = \mathcal{D}(i)\ \forall i (|\alpha| + \lambda \leq i \leq |\alpha| + |\delta|)$ and $\forall i\ |\alpha| + |\delta| + \lambda \leq j \leq |E'_s|$. Therefore $\mathcal{D}$ differs from $\mathcal{C}$ by 2 subsequences of at most λ states. We can obtain these two subsequences by simply scanning the first λ symbols in δ and the first λ symbols in γ. This requires only $O(\lambda)$ time. □

Note that the above lemma assumes δ (the rhs) is already scanned prior to replacement. Recall that δ is created by inserting the Euler chains of the substitutions for the variables in it. The Euler chains for these substitutions are obtained from E_s which has already been scanned. Moreover the Euler chains of each rhs is already scanned prior to start of the normalization process. Therefore if k_r is the number of variables in the rhs then rescanning δ requires $O(k_r \times \lambda)$ time.

4.3 Computing New Signatures

Recall that in order to declare a match our algorithm must verify the consistency of computed substitutions. To each node in the subject tree we assign an integer code (varying from 1 to n) called its *signature* such that two nodes get the same signature if and only if the subtrees rooted beneath them are identical. We use the method in [6] to compute signatures of the nodes bottom-up. First all the leaves are radix sorted on their labels and the ranks computed are assigned as the signatures. (Note duplicates are assigned the same rank.) Now suppose the signature for all nodes upto height i have been computed. The signatures of nodes at height $i+1$ are computed as follows. Each node v at level $i+1$ is assigned a signature vector $< f, j_1, j_2, \ldots, j_n >$ where f is the label on v and j_q is the signature of the qth child of v. The signature vectors assigned to all the nodes at height $i+1$ are radix sorted. If the rank of v is α and the largest signature assigned to any node at height i is β then $\alpha + \beta$ is the signature of v. In Figure 1 s_1 is a subject tree. The node numbers are in italics whereas the signature for the nodes are in boldface.

Upon replacement the subject tree changes and therefore new signatures need to be computed for some of the nodes in the subject tree. Specifically if r is the right hand side whose instance replaces the subtree rooted at u then new signatures have to be computed for all nodes in r (but not for nodes in the substitutions) as well as all ancestors of u. The assignment of new signature to the root of a subtree is complicated by the fact that we have to assign it the same signature as any other subtree that is identical to it. This mans that we have to search the subject for identical subtrees. Obviously all the new signatures can be recomputed by the above algorithm. But this requires time proportional to size of the new subject tree.

Note that we have to compute new signatures for the nodes in r and for all the ancestors of u. This can never be done in time less than the number of new signatures to be computed. If a_u is the number of ancestors of u then the lower bound for computing new signatures will be $O(r + a_u)$. We now describe a technique for computing the new signatures within this time bound.

Computing new signatures involves searching for identical subtrees. In the bottom-up method for computing signatures the search for identical subtrees reduces to searching for identical signature vectors assigned to the roots of subtrees. The main idea in our technique is to maintain all the signature vectors as a trie. In Figure 2, t_1 is the signature trie for s_1. Each path from the root to a leaf represents a signature vector. Each leaf node is annotated by a signature corresponding to the signature vector represented by a path that ends on it and a list (within parentheses) of nodes having that signature. In t_1 the signature vector $< g, 1 >$ ends on a leaf node annotated with signature 2. Nodes 4 and 8 in parentheses are the nodes having signature 2.

Using a signature trie we can search for equal subtrees in time proportional to the length of the signatures. Specifically, suppose node v has vector $\alpha =< j_1, j_2, \ldots, j_l >$. Now identifying subtrees in the subject that are identical to that rooted at v is equivalent to checking whether there is a path in the signature trie that spells out α. This can be easily checked by regarding trie as an automaton and scanning α with it. If the scan is successful, i.e., if we reach a leaf node in the trie then there is such a subtree otherwise the subtree rooted at v is unique. Upon successful scanning we give the signature thus identified to v; otherwise a unique integer is assigned to v.

Prior to the start of normalization all the nodes in the input term are assigned signatures and a signature trie is built. During subsequent replacements new signatures are computed using this trie. The trie is also updated to include only the signature vectors of those nodes currently present in the subject. This ensures that the trie contains enough information to compute signatures during next replacement and also ensures that size of the trie does not grow arbitrarily.

The signature trie is updated as follows. Suppose the vector of node v is to be deleted from the trie. We first check to see if α is the signature of any other node in the subject. If it is not then we can remove the path of the trie that spells out α. We do so by starting from the leaf towards the root. Observe that we must stop on reaching an internal node whose arity ≥ 2. Suppose α is to be added to the trie then we simply insert into it using technique similar to that used to construct the automaton [1]. Since nodes can have the same signature we maintain a reference count at the leaf nodes of the trie that denotes the number of nodes having the signature denoted by this node.

This count is used during deletion. We illustrate the above ideas using Figures 1 and 2.

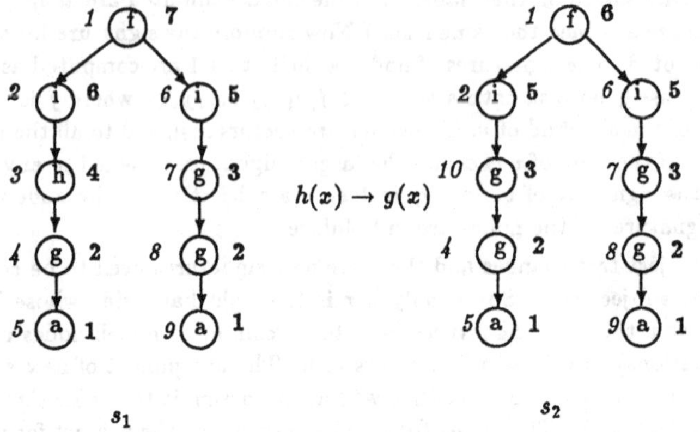

Figure 1

node numbers are in italics

node signatures are in boldface

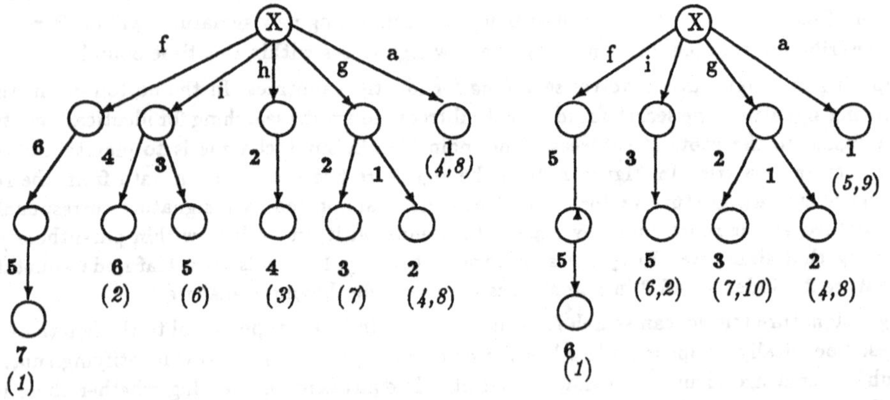

signature trie t_1 for s_1 signature trie t_2

Figure 2

Suppose s_2 is obtained after replacing the subtree $h(g(a))$ in s_1 by $g(g(a)))$ according to the rule $h(X) \rightarrow g(X)$. Observe that we need only compute signatures for nodes 10, 2 and 1. Node 10 is introduced by the rhs of the rule and nodes 2 and 1 are ancestors of 10. The vector for 10 is $< g, 2 >$. There is a path in the trie that spells out this vector. Therefore we assign 3 as the signature for node 10. Similarly we compute 5, 6 as signatures for nodes 2 and 1 respectively. Now the trie is updated. Specifically, we delete the vectors representing old signatures of nodes 2 and 1 and insert the vector corresponding to the new signature of node 1. Trie t_2 is obtained following these changes. Summarizing,

Lemma 5 *New signatures for the nodes in the subject can be computed in* $O(\mid r \mid + a_u)$ *time.*

Proof: Omitted.

Note that we assign a unique integer as the signature to the root of a subtree that is not identical to any other subtree in the new subject. Assigning a new integer is done as follows. Suppose q is the largest signature assigned so far and v has to be assigned a new signature then we assign $q+1$ as its signature. Note that in this method the size of the largest signature assigned can grow arbitrarily large and eventually we may be not be able to represent them in $O(1)$ words. To circumvent this problem we reuse the integers that are no longer assigned as signatures to any node in the subject. Specifically, we maintain a pool of integers. Whenever we identify a subtree that is unique, we pick an integer from this pool and assign it as the new signature. Similarly, whenever we identify that an integer is no longer in use as a signature for any node then we return it to this pool. We can show that the largest integer in the pool will never exceed the size of the largest subject tree during the normalization process (details omitted).

4.4 Computing New Matches

Now we are ready to compute new matches that may arise as a consequence of performing replacement. Observe again that new matches occur only at nodes of r and all the ancestors of u. In addition, because of nonlinearity a match for the pattern can occur at *any* of the ancestors of u. Since the nodes in r are not part of the old subject tree we must initiate matches at these nodes. However the interesting case is initiating matches at the ancestors of u. A naive approach is to initiate matches at all of the ancestors. However we now show that it is not necessary to do so. By using the information available from previous pattern matches we can substantially reduce the number of ancestor nodes at which a new match must be initiated. Specifically, we initiate matches over a consecutive sequence of h_p ancestors only. (h_p is height of pattern.) For all the other nodes the match information is obtained in $O(1)$ time.

Let u_i $(0 \leq i \leq s)$ denote the ith ancestor of u (u_0 is u and and u_s is root of the new subject tree).

Definition 1 *We say that u_l $(l \geq 0)$ is the critical node if l is the smallest integer such that the subtree rooted at u_l is unique.*

We remark that the subtree rooted at any ancestor of u_l is also unique. Since the subject tree is unique such an l always exists. Therefore for all u_i $(0 \leq i < l)$ there exists a subtree in the subject that is identical to the one rooted at u_i. Let v_i be the root of this identical subtree. Note that v_i cannot be a descendant of u_i. Therefore the subtree rooted at v_i is part of the subject tree prior to replacement. Now observe if there is a pattern match at u_i then there is a pattern match at v_i also. A match must have been initiated at v_i prior to replacement. The information about this match attempt can be readily used for a match at u_i also. In particular, the substitutions for match at u_i will also be identical to the substitutions computed for the match at v_i. By maintaining a link from u_i to v_i we avoid computing substitutions explicitly for u_i. Thus computing a match for any node that is a descendant of a critical node requires only $O(1)$ time. Finally, we remark that if $l > 0$ then we do not have to initiate matches at any of the nodes in r.

We now show that new matches need to be initiated at the ith ancestor of the critical node only if $i < h_p$. Let Γ denote the set of all i^{th} ancestors of u_l where $i \geq h_p$. Then,

Lemma 6 *For any $v \in \Gamma$ if there was no match at v before replacement then there will be no match after replacement.*

Proof: By contradiction. Let a match occur at v after replacement. Since v is at a distance greater than h_p from u_l it must be the case that a variable x in the pattern p falls on an ancestor w of

u_l. Recall that the subtree rooted at w is unique. Therefore x must appear only once in p for this pattern match to occur. Since the replacement only changes the subtree rooted at some descendant of w it must be the case that there is a pattern match at v prior to replacement also. □

Note that it is not necessary for every node in Γ that had a pattern match before the replacement to have a match now. In particular, suppose there was a pattern match for $v \in \Gamma$ before replacement. Assume that variable x occurs more than once in the pattern. Let w, a descendant of v, be a substitution for x. After replacement the pattern match at v will disappear as the subtree at w becomes unique.

Let us denote w as a *nonlinear* substitution for a pattern match if the subtree rooted at w is computed as a substitution for a variable x that occurs more than once in the pattern. Let N_w be the set of ancestors of w for which w is a nonlinear substitution. Let Δ denote the set of ancestors of the critical node u_l.

Lemma 7 *If w_1 is an ancestor of w_2 then $N_{w_1} \cap N_{w_2} = \emptyset$*

Proof: Suppose $v \in N_{w_1} \cap N_{w_2}$. For this to happen w_1 and w_2 must be the substitutions for two variables for the same pattern match at v. This is impossible as it would imply that nodes labeled with these variables in the pattern have ancestor-descendant relationship. □

Therefore, we can conclude : $\sum_{w \in \Delta} | N_w |=| \Delta |$.

All we need to do now is to examine N_w's and mark every node in Γ that had a pattern match before replacement. The remaining unmarked nodes in Γ that had a pattern match before must have a pattern match now also. Note that the roots of the substitutions for these pattern matches remain the same. In conclusion, we must initiate new matches only at the nodes in r and in $\Delta - \Gamma$. The match information for all the other nodes is obtained in $O(1)$ time. Now $|\Delta - \Gamma| = h_p$. Now N_w can be maintained easily and it will not increase the asymptotic time complexity of the algorithm.

5 Conclusion

An important consequence of our techniques is that we need no longer scan the subject tree completely following each replacement operation. The scanning time in Tables 1 and 3 (see Section 3) is the time taken to scan the subject completely. Based on the rescanning strategy described in Section 4 we estimate the reduction in scanning time as follows. From our experimental data we observed that the longest pattern string typically has about 4 nodes. On examining several canonical systems [7] we found that the rhs typically has about 4 nodes and 2 of these are labeled with variables. Following replacement we we need to rescan all the nodes labeled with function symbols in the Euler chain of the rhs. There are 5 such nodes. We need to scan at most λ symbols in each of the two substitutions and λ symbols immediately following the replaced subchain in the modified subject. So altogether we have to scan at most $5+(2*4)+4 = 17$ symbols. Since the time to scan the subject tree is linear on the average we can conclude for example that the time to rescan a 1000 node tree for a rewrite system with one pattern (see column 5 and row 3 of table 1) will require 0.24 msec which is a substantial reduction over 29.5 msecs! Similarly we can estimate that for rewrite systems with 5,10 and 25 patterns rescanning a 1000 node tree will requires 0.44,0.56 and 0.85 msecs respectively. Observe that the scanning time now is substantially smaller than the matching time. Therefore the speedups we get now are closer to those in column S_1. Finally, these speedups have not taken into account the reduction in matching time made possible by having to search only the replaced subtree and a few ancestors of the root of the replaced subtree. Doing so will further increase the speedup ratio.

The techniques described in this paper are quite general. For instance, the ideas underlying rescanning can also be used in any other algorithm based on string-matching automata (such as the top-down algorithm in [2]). The techniques used for computing signatures incrementally can also be used in any application that requires identifying identical subtrees. In fact we can also use it to assign signatures to all the nodes in the entire tree prior to start of the normalization process. The time complexity of our algorithm when used for computing signatures for the entire tree will be linear in the size of the tree. The Downey-Sethi-Tarjan algorithm [6] also requires linear time. Unlike their algorithm we compute signatures of the nodes one at a time. Therefore we do not require the sorting steps used in their algorithm and so it is likely that our algorithm can in fact run faster than theirs in an implementation. Finally we remark that all the techniques described in this paper can operate directly on the Euler chains. Hence we do not need to maintain separate tree representation of the subject and patterns.

Acknowledgement

The research reported in this paper has been supported in part by NSF grant CCR-8805734

References

[1] A.V. Aho and M.J. Corasick, "Efficient String Matching: An Aid to Bibliographic Search", *CACM*, Vol 18 No. 6, June 1975, pp. 333-340.

[2] C.M. Hoffmann and M.J. O'Donnell, "Pattern Matching in Trees", *JACM* 29, 1, 1982 pp. 68-95.

[3] D. E. Knuth and P. Bendix, "Simple word problems in Universal Algebras", *Computational Problems in Abstract Algebra*, J. Leech, ed., Pergammon Press, Oxford 1970, pp. 263-297.

[4] D.R. Chase, "An Improvement to Bottom-up Tree Pattern Matching", *Fourteenth Annual ACM Symposium on Principles of Programming Languages*, Jan 1987.

[5] M.J. O'Donnell, "Equational Logic as a Programming Language", Foundations of Computation Series, MIT Press 1985.

[6] P.J. Downey, R. Sethi and R.E. Tarjan, "Variations on the Common Subexpression Problem", *JACM* Vol 24 No. 4, 1980, pp. 758-771.

[7] Jean-Marie Hullot, "A Catalogue of Canonical Term Rewriting Systems", Technical Report CSL-113, SRI International, April 1980.

[8] R. Ramesh and I.V. Ramakrishnan, "Nonlinear Pattern Matching in Trees", *Proceedings of the 15th International Colloquium on Automata, Languages and Programming*, LNCS, Vol 317, Springer Verlag, pp. 473-488, July 1988. (Also to appear in JACM.)

[9] J. Hsiang and J. Mzali, "SbReve Users Guide", Technical Report, LRI 1988.

[10] P. Lescanne, "REVE: A Rewrite Rule Laboratory", *8th Int'l Conference on Automated Deduction*, LNCS 230, July 1986.

[11] J. Cai, R. Paige and R.E. Tarjan, "More Efficient Bottom-Up Tree Pattern Matching", European Symposium on Programming, 1990.

[12] D. Kapur and H. Zhang, "RRL: A Rewrite Rule Laboratory", *9th Int'l Conference on Automated Deduction*, May 1988.

[13] J. Hsiang, "Refutational Theorem Proving using Term Rewriting Systems", *Artificial Intelligence* 25, 1985, pp. 255-300.

[14] R. Ramesh and I.V. Ramakrishnan, "Incremental Techniques for Efficient Normalization of Nonlinear Rewrite Systems", Technical Report No. UTDCS-38-90, Department of Computer Science, University of Texas at Dallas, Richardson, TX 75083.

On fairness of completion-based theorem proving strategies *

Maria Paola Bonacina **Jieh Hsiang**
Department of Computer Science
SUNY at Stony Brook
Stony Brook, NY 11794-4400, USA
{bonacina,hsiang}@sbcs.sunysb.edu

Abstract

Fairness is an important concept emerged in theorem proving recently, in particular in the area of completion-based theorem proving. Fairness is a required property for the *search plan* of the given strategy. Intuitively, fairness of a search plan guarantees the generation of a successful derivation if the *inference* mechanism of the strategy indicates that there is one. Thus, the completeness of the inference rules and the fairness of the search plan form the completeness of a theorem proving strategy. A search plan which exhausts the entire search space is obviously fair, albeit grossly inefficient. Therefore, the problem is to reconciliate fairness and efficiency. This problem becomes even more intricate in the presence of *contraction inference rules* – rules that remove data from the data set.

The known definitions of fairness for completion-based methods are designed to ensure the confluence of the resulting system. Thus, a search plan which is fair according to these definitions may force the prover to perform deductions completely irrelevant to prove the intended theorem. In a theorem proving strategy, on the other hand, one is usually only interested in proving a specific theorem. Therefore the notion of fairness should be defined accordingly. In this paper we present a target-oriented definition of fairness for completion, which takes into the account the theorem to be proved and therefore does not require computing all the critical pairs. If the inference rules are complete and the search plan is fair with respect to our definition, then the strategy is complete. Our framework contains also notions of redundancy and contraction. We conclude by comparing our definition of fairness and the related concepts of redundancy and contraction with those in related works.

1 Introduction

A theorem proving strategy is composed of a set of inference rules and a search plan. Refutational completeness of a strategy involves both the inference rules and the search plan. First, it requires that for all unsatisfiable inputs, there exist successful derivations by the inference rules of the strategy. Second, it requires that whenever successful derivations exist, the search plan guarantees that the computed derivation is successful. We call these properties *completeness* of the inference

*Research supported in part by grants CCR-8805734, INT-8715231 and CCR-8901322, funded by the National Science Foundation. The first author has been also supported by Dottorato di ricerca in Informatica, Universitá degli Studi di Milano, Italy.

rules and *fairness* of the search plan, respectively. In this paper we concentrate our study on the fairness property. A search plan which exhaustively performs all possible steps is trivially fair, but tremendously inefficient. Therefore, the problem is to reconciliate fairness and efficiency. This problem is especially striking in case of Knuth-Bendix type completion procedures applied to theorem proving. Fairness for Knuth-Bendix completion appeared first in [10]. A more general definition of fairness in the proof ordering framework is given in [2]. A derivation by a completion procedure is fair according to [2] if all the critical pairs from *persisting* equations are eventually generated and all the generated equations are eventually reduced. This definition of fairness captures exactly the requirements that a derivation generating a canonical system must satisfy, but it is not intended for a derivation proving a specified theorem. Indeed, most theorem proving derivations do not satisfy this definition of fairness, since proving a theorem usually does not require generating all critical pairs.

In this paper we present a notion of fairness for completion procedures applied to theorem proving. This definition of fairness is part of a new approach to completion procedures, which was partially presented in [5]. All the fundamental concepts of the theory of completion are uniformly defined in terms of *proof reduction*. This approach allows us to regard theorem proving as the basic interpretation of a completion process, rather than as a side-effect of the generation of a confluent presentation of the theory. We define completeness of the inference rules and fairness of the search plan and we show that if the inference rules are complete and the search plan is fair according to our definitions, the procedure is complete.

Our approach to the problem of fairness is new in three respects, all relevant to theorem proving. First, our definition is target-oriented, that is it takes the theorem to be proved into account. A target-oriented definition of fairness is important for search plans to be both fair and efficient. For instance, we may have a problem where the target $s \simeq t$ is an equation on a signature F_1 and the input presentation E is the union of a set E_1 of equations on the signature F_1 and a set E_2 of equations on another signature F_2, disjoint from F_1. Such a problem can occur in definitions of abstract data types, where the signature F_1 contains the constructors and a set of defined symbols, whereas the signature F_2 is another set of defined symbols. According to our definition, a derivation where no inference from E_2 is performed is fair. On the other hand, fairness as defined in previous works on completion [2, 13, 4] would require to compute critical pairs from the equations in E_2 as well.

Second, we emphasize the distinction between fairness as a property of the search plan and completeness as a property of the inference rules. Most theorem proving strategies are simply presented by giving a set of inference rules and the task of designing a suitable search plan is left to the implementation phase. This is not satisfactory, since the actual performance of the prover depends heavily on the search plan. Therefore, we think that it is important to study search plans systematically and define their requirements formally. Our approach to the fairness problem is a first step in this direction. Finally, it is generally acknowledged that the application of contraction inference rules is mandatory for efficiency. However, the known definitions of fairness emphasize the application of expansion rules, whereas ours treats both expansion and contraction rules uniformly.

In the first two sections we outline our approach to completion procedures. We refer the reader to [5] for a more detailed treatment of this framework. The fourth section contains the definition of fairness and the theorem of completeness of strategies with fair search plan. In Section 5, we show how the classical results on Knuth-Bendix completion are covered in our framework, if uniform fairness, i.e. fairness to generate confluent systems, is assumed. Finally, in Sections 6 and 7 we discuss and compare with ours the definitions of fairness and redundancy in [13] and in [4].

2 Proof orderings for theorem proving

A basic assumption of our approach is that all the elements we deal with are partially ordered. We recall that a *simplification ordering* on terms is a monotonic and stable ordering such that a term is greater than any of its subterms. A simplification ordering is well founded [6]. A *complete simplification ordering* is also total on the set of ground terms. A *proof ordering* is a monotonic, stable and well founded ordering on proofs. Given a finite set of sentences S, we denote by $Th(S)$ the *theory* of S, $Th(S) = \{\varphi | S \models \varphi\}$, and we say that S is a *presentation* of the theory $Th(S)$.

We assume a complete simplification ordering $\succ$ on terms and literals[1] and a proof ordering $>_p$ on proofs. We denote proofs by capital Greek letters: $\Upsilon(S,\varphi)$ is a proof of φ from axioms in the presentation S. Our notion of proof ordering is different from the one introduced in [1]. According to the notion of proof ordering in [1], only two proofs $\Upsilon(S,\varphi)$ and $\Upsilon'(S',\varphi)$ of the same theorem in different presentations can be compared. We assume that two proofs $\Upsilon(S,\varphi)$ and $\Upsilon'(S',\varphi')$ of different theorems may also be comparable.

The minimum proof is the *empty proof.* We denote by *true* the theorem whose proof is empty and we assume that *true* is the bottom element in our ordering on terms and literals. Given a pair $(S;\varphi)$, we can select a minimal proof among all proofs of φ from S:

Definition 2.1 *Given a proof ordering $>_p$, we denote by $\Pi(S,\varphi)$ a minimal proof of φ from S with respect to $>_p$, i.e. a proof such that for all proofs $\Upsilon(S,\varphi)$ of φ from S, $\Upsilon(S,\varphi) \not<_p \Pi(S,\varphi)$.*

A *theorem proving problem* is a pair $(S;\varphi)$, where S is a presentation of a theory and φ the theorem to be proved, which we call the *target.* The process of proving φ from S is the process of reducing φ to *true* and $\Pi(S,\varphi)$ to the empty proof. A theorem proving derivation is a sequence of deductions

$$(S_0;\varphi_0) \vdash (S_1;\varphi_1) \vdash \ldots \vdash (S_i;\varphi_i) \vdash \ldots,$$

where at each step $(S_{i+1};\varphi_{i+1})$ replaces $(S_i;\varphi_i)$ and $\Pi(S_{i+1},\varphi_{i+1})$ replaces $\Pi(S_i,\varphi_i)$. At stage i the problem is to prove φ_i from S_i or equivalently to reduce $\Pi(S_i,\varphi_i)$. Since $S_i \neq S_{i+1}$ and $\varphi_i \neq \varphi_{i+1}$ in general, we need a proof ordering which may compare two proofs of different theorems, in order to compare $\Pi(S_i,\varphi_i)$ and $\Pi(S_{i+1},\varphi_{i+1})$. The derivation halts successfully at stage k if φ_k is *true* and $\Pi(S_k,\varphi_k)$ is empty.

3 Completion procedures

A theorem proving strategy is given by a set of *inference rules I* and a *search plan* Σ. The data are pairs $(S;\varphi)$, where S is the *presentation* and φ is the *target.* We distinguish four classes of inference rules, depending on whether a rule transforms the presentation or the target and whether it is an *expansion* inference rule or a *contraction* inference rule, as they are called in [8]. An expansion inference rule expands a set of sentences by deriving new sentences, whereas a contraction inference rule contracts a set of sentences by either deleting some sentences or replacing them by others. We assume that a target is a clause and therefore can be regarded as a set of literals:

- *Presentation inference rules:*
  - *Expansion inference rules:* f: $\dfrac{(S;\varphi)}{(S';\varphi)}$ where $S \subset S'$.

[1] A well founded, monotonic and stable ordering is sufficient.

- *Contraction inference rules*: $f\colon \frac{(S;\varphi)}{(S';\varphi)}$ where $S \not\subseteq S'$.

- *Target inference rules*:

  - *Expansion inference rules*: $f\colon \frac{(S;\varphi)}{(S;\varphi')}$ where $\varphi \subset \varphi'$.

  - *Contraction inference rules*: $f\colon \frac{(S;\varphi)}{(S;\varphi')}$ where $\varphi \not\subseteq \varphi'$.

We characterize a theorem proving derivation as a process of *proof reduction*. A target inference step modifies the target and therefore it affects the proof of the target. We require that the proof of the target is reduced:

Definition 3.1 *A target inference step* $(S;\varphi) \vdash (S;\varphi')$ *is* proof-reducing *if* $\Pi(S;\varphi) \geq_p \Pi(S;\varphi')$. *It is* strictly proof-reducing *if* $\Pi(S;\varphi) >_p \Pi(S;\varphi')$.

For a presentation inference step we allow more flexibility:

Definition 3.2 *Given two pairs* $(S;\varphi)$ *and* $(S';\varphi')$, *the relation* $(S;\varphi) \rhd_{p,\mathcal{T}} (S';\varphi')$ *holds if*

1. *either* $\Pi(S;\varphi) >_p \Pi(S';\varphi')$

2. *or*

   (a) $\Pi(S;\varphi) = \Pi(S';\varphi')$,

   (b) $\forall\psi \in \mathcal{T}$, $\Pi(S,\psi) \geq_p \Pi(S',\psi)$ *and*

   (c) $\exists\psi \in \mathcal{T}$ *such that* $\Pi(S,\psi) >_p \Pi(S',\psi)$.

Definition 3.3 *A presentation inference step* $(S;\varphi) \vdash (S';\varphi)$ *is* proof-reducing *on* $\mathcal{T}$ *if* $(S;\varphi) \rhd_{p,\mathcal{T}} (S';\varphi)$ *holds. It is* strictly proof-reducing *if* $\Pi(S;\varphi) >_p \Pi(S';\varphi)$.

The condition $(S_i;\varphi_i) \rhd_{p,\mathcal{T}} (S_{i+1};\varphi_{i+1})$ says that a step which reduces the proof of the target is proof-reducing, regardless of its effects on other theorems. However, an inference step on the presentation may not immediately decrease the proof of the target and still be necessary to decrease it eventually. Such a step is also proof-reducing, provided that it does not increase any proof and strictly decreases at least one. This notion of proof reduction applies to presentation inference steps which are either expansion steps or contraction steps which replace some sentences by others. A contraction step which deletes sentences does not modify any minimal proof. In order to characterize these steps, we introduce a notion of *redundancy*:

Definition 3.4 *A sentence* φ *is* redundant *in* S *on don ain* $\mathcal{T}$ *if* $\forall\psi \in \mathcal{T}$, $\Pi(S,\psi) = \Pi(S \cup \{\varphi\},\psi)$.

A sentence is redundant in a presentation if adding it to the presentation does not reduce any minimal proof.

Definition 3.5 *An inference step* $(S;\varphi) \vdash (S';\varphi')$ *is* reducing *on* $\mathcal{T}$ *if either it is proof-reducing or it deletes a sentence which is redundant in* S *on domain* $\mathcal{T}$.

Definition 3.6 *An inference rule* f *is* reducing *if all the inference steps* $(S;\varphi) \vdash_f (S';\varphi')$ *where* f *is applied are reducing.*

We have finally all the elements to give the definition of a completion procedure:

Definition 3.7 *A theorem proving strategy* $\mathcal{C} =< I; \Sigma >$ *is a* completion procedure *on domain* $\mathcal{T}$ *if for all pairs* $(S_0; \varphi_0)$, *where* S_0 *is a presentation of a theory and* $\varphi_0 \in \mathcal{T}$, *the derivation*

$$(S_0;\varphi_0)\vdash_{\mathcal{C}}(S_1;\varphi_1)\vdash_{\mathcal{C}}\ldots\vdash_{\mathcal{C}}(S_i;\varphi_i)\vdash_{\mathcal{C}}\ldots$$

has the following properties:

- *monotonicity:* $\forall i \geq 0,\ Th(S_{i+1}) \subseteq Th(S_i)$,
- *relevance:* $\forall i \geq 0,\ \varphi_i \in \mathcal{T}$ *and* $\varphi_{i+1} \in Th(S_{i+1})$ *if and only if* $\varphi_i \in Th(S_i)$ *and*
- *reduction:* $\forall i \geq 0$, *the step* $(S_i;\varphi_i)\vdash_{\mathcal{C}}(S_{i+1};\varphi_{i+1})$ *is reducing on* $\mathcal{T}$.

The *domain* $\mathcal{T}$ is the set of sentences where the inference rules of the completion procedure are reducing. For instance, for the Knuth-Bendix completion procedure $\mathcal{T}$ is the set of all equations. For the Unfailing Knuth-Bendix procedure, $\mathcal{T}$ is the set of all ground equations.

The *monotonicity* and *relevance* properties establish the soundness of the presentation and the target inference rules respectively. Monotonicity ensures that a presentation inference step does not create new elements which are not in the theory, while relevance ensures that a target inference step replaces the target by a new target in such a way that proving the latter is equivalent to proving the former.

Reduction is the key property which characterizes completion procedures. If the procedure is given a target, it tries to reduce the proof of the target until it is empty. If no target is given, the derivation has the form

$$(S_0;\emptyset)\vdash_{\mathcal{C}}(S_1;\emptyset)\vdash_{\mathcal{C}}\ldots\vdash_{\mathcal{C}}(S_i;\emptyset)\vdash_{\mathcal{C}}\ldots.$$

The goal is to transform the given presentation S_0 into a confluent one by reducing the proofs of all theorems. Since few theories have finite, confluent presentations, the application of completion to theorem proving is far more interesting in practice.

Note that our definition of completion procedures is different from the classical one [2]. Consequently, our idea of fairness is also different. In our view, a completion procedure is given a target and therefore fairness is fairness with respect to the goal of proving the given target.

4 Fairness

Given a theorem proving problem $(S_0;\varphi_0)$ and a set of inference rules I, the application of I to $(S_0;\varphi_0)$ defines a tree, the *I-tree rooted at* $(S_0;\varphi_0)$. The nodes of the tree are labeled by pairs $(S;\varphi)$. The root is labeled by the input pair $(S_0;\varphi_0)$. A node $(S;\varphi)$ has a child $(S';\varphi')$ if $(S';\varphi')$ can be derived from $(S;\varphi)$ in one step by an inference rule in I. The I-tree rooted at $(S_0;\varphi_0)$ represents all the possible derivations by the inference rules in I starting from $(S_0;\varphi_0)$.

Intuitively, a set I of inference rules is *refutationally complete* if whenever $\varphi_0 \in Th(S_0)$, the I-tree rooted at $(S_0;\varphi_0)$ contains successful nodes, nodes of the form $(S;\mathit{true})$. We use the term "refutational completeness" for the inference rules to differentiate it from the completeness of the theorem proving strategy. Furthermore, "refutational" emphasizes that the goal is to prove a specific theorem. The following definition is an equivalent characterization of this concept in terms of proof reduction:

Definition 4.1 *A set* $I = I_p \cup I_t$ *of inference rules is* refutationally complete *if whenever* $\varphi \in Th(S)$ *and* $\Pi(S,\varphi)$ *is not minimal, there exist derivations*

$$(S;\varphi)\vdash_I(S_1;\varphi_1)\vdash_I\ldots\vdash_I(S';\varphi')$$

such that $\Pi(S,\varphi) >_p \Pi(S',\varphi')$.

This definition says that a set of inference rules is refutationally complete if it can reduce the proof of the target whenever it is not minimal. Since a proof ordering is well founded, it follows that the I-tree rooted at $(S;\varphi)$ contains successful nodes if $\varphi \in Th(S)$. The advantage of giving the definition of completeness in terms of proof reduction is that the problem of proving completeness of I is reduced to the problem of exhibiting a suitable proof ordering [2].

Given a completion procedure $C =< I;\Sigma >$, the I-tree rooted at $(S_0;\varphi_0)$ represents the entire search space that the procedure can potentially derive from the input $(S_0;\varphi_0)$. The search plan Σ selects a path in the I-tree: the derivation from input $(S_0;\varphi_0)$ controlled by Σ is the path selected by Σ in the I-tree rooted at $(S_0;\varphi_0)$. A pair $(S_i;\varphi_i)$ reached at stage i of the derivation is a visited node in the I-tree. Each visited node $(S_i;\varphi_i)$ may have many children, but the search plan selects only one of them to be $(S_{i+1};\varphi_{i+1})$. A search plan Σ is *fair* if whenever the I-tree rooted at $(S_0;\varphi_0)$ contains successful nodes, the derivation controlled by Σ starting at $(S_0;\varphi_0)$ is guaranteed to reach a successful node. Similar to completeness, we rephrase this concept in terms of proof reduction:

Definition 4.2 *A derivation*

$$(S_0;\varphi_0)\vdash_C(S_1;\varphi_1)\vdash_C\ldots\vdash_C(S_i;\varphi_i)\vdash_C\ldots$$

controlled by a search plan Σ *is* fair *if and only if for all* $i \geq 0$, *if there exists a path*

$$(S_i;\varphi_i)\vdash_I\ldots\vdash_I(S';\varphi')$$

in the I*-tree rooted at* $(S_0;\varphi_0)$ *such that* $\Pi(S_i;\varphi_i) >_p \Pi(S';\varphi')$, *then there exists an* $(S_j;\varphi_j)$ *for some* $j > i$, *such that* $\Pi(S';\varphi') \geq_p \Pi(S_j;\varphi_j)$. *A search plan* Σ *is* fair *if all the derivations controlled by* Σ *are fair.*

In other words, if the inference rules can reduce the proof of the target at $(S_i;\varphi_i)$, a fair search plan guarantees that the proof of the target will be indeed reduced at a later stage $(S_j;\varphi_j)$. This definition is target-oriented because it only requires that the proof of the intended target is reduced. If the inference rules are refutationally complete and the search plan is fair, a completion procedure on domain $\mathcal{T}$ is complete, i.e. it is a *semidecision procedure* for $Th(S)\cap\mathcal{T}$ for all presentations S:

Theorem 4.1 *If a completion procedure* C *on domain* $\mathcal{T}$ *has refutationally complete inference rules and fair search plan, then for all derivations*

$$(S_0;\varphi_0)\vdash_C(S_1;\varphi_1)\vdash_C\ldots\vdash_C(S_i;\varphi_i)\vdash_C\ldots,$$

where $\varphi_0 \in Th(S_0)$, $\forall i \geq 0$, *if* $\Pi(S_i,\varphi_i)$ *is not minimal, then there exists an* (S_j,φ_j), *for some* $j > i$, *such that* $\Pi(S_i,\varphi_i) >_p \Pi(S_j,\varphi_j)$.

Proof: if $\Pi(S_i,\varphi_i)$ is not minimal, then by completeness of the inference rules, there exists a path $(S_i;\varphi_i)\vdash_I\ldots\vdash_I(S';\varphi')$ such that $\Pi(S_i;\varphi_i) >_p \Pi(S';\varphi')$. By fairness of the search plan, there exists an $(S_j;\varphi_j)$, for some $j > i$, such that $\Pi(S_i;\varphi_i) >_p \Pi(S';\varphi') \geq_p \Pi(S_j;\varphi_j)$. □

Corollary 4.1 *If a completion procedure* C *on domain* $\mathcal{T}$ *has refutationally complete inference rules and fair search plan, then for all inputs* $(S_0;\varphi_0)$, *if* $\varphi_0 \in Th(S_0)$, *the derivation*

$$(S_0;\varphi_0)\vdash_C(S_1;\varphi_1)\vdash_C\ldots\vdash_C(S_i;\varphi_i)\vdash_C\ldots$$

reaches a stage k, $k \geq 0$, *such that* φ_k *is the clause true.*

Proof: if $\varphi_0 \in Th(S_0)$, then by Theorem 4.1 and the well foundedness of $>_p$ the derivation reaches a stage k such that the proof $\Pi(S_k, \varphi_k)$ is minimal. Since the minimal proof is the empty proof, φ_k is the clause *true*. □

For instance, the Unfailing Knuth-Bendix procedure was proved in [9, 3] to be a semidecision procedure for equational theories.

5 Uniform fairness and saturated sets

Definition 4.2 of fairness is sufficient for theorem proving as shown by Theorem 4.1. If Definition 4.2 is applied to Knuth-Bendix completion, it is not sufficient to guarantee that a confluent rewrite system is eventually generated, since it does not guarantee that all critical pairs are eventually considered. This requires a much stronger fairness property, which we call *uniform fairness*.

The first definition of uniform fairness appeared in [10], where it is required that the search plan sorts the rewrite rules in the data set by a well founded ordering, in order to ensure that no rule is indefinitely postponed. We recall this very first notion of fairness, because it states explicitly that fairness is a property of the search plan. The later definitions of fairness are given at a much higher abstraction level, which may prevent the reader from seeing that fairness is a property of the search mechanism.

Given a derivation by completion starting from a presentation S_0, the *limit* S_∞ of the derivation is the possibly infinite set $\bigcup_{j\geq 0}\bigcap_{i\geq j} S_i$ of all the *persistent* sentences, that is the sentences which are generated at some stage and never deleted afterwards [10, 2]. The advantage of dealing with the limit S_∞ is that if the derivation halts at some stage k, $S_\infty = S_k$. Therefore, properties stated in terms of the limit S_∞ apply uniformly to both halting and non halting derivations. We denote by $I_e(S)$ the set of clauses which can be generated in one step by the presentation expansion rules of the completion procedure applied to S.

Definition 5.1 (Rusinowitch 1988) [13], (Bachmair and Ganzinger 1990) [4] *A derivation*

$$(S_0;\emptyset)\vdash_C(S_1;\emptyset)\vdash_C\ldots(S_i;\emptyset)\vdash_C\ldots$$

by a completion procedure C on domain $\mathcal{T}$ is uniformly fair *on domain $\mathcal{T}$ if $\forall\varphi\in I_e(S_\infty)$ there exists an S_j such that either $\varphi\in S_j$ or φ is redundant in S_j on domain $\mathcal{T}$.*

This definition of fairness generalizes previous definitions given in [10] and [2]. It says that every clause φ that can be persistently generated by an expansion rule during the completion process is either actually generated at some step, or replaced by other clauses which yield a smaller proof of φ and make it redundant. For instance, a Knuth-Bendix derivation such that all critical pairs from persisting equations are eventually generated or subsumed or reduced to a common term is uniformly fair according to this definition.

Uniform fairness has been studied and progressively refined in [2, 13, 4] with the purpose of solving the problem of the interaction between expansion inference rules and contraction inference rules. The intuitive meaning of uniform fairness is to be fair to the inference rules, that is to apply all the inference rules to all the data. However, this is impossible, because the inference rules include both contraction rules and expansion rules: if a clause φ is deleted by a contraction step before an expansion rule f is applied to φ, the derivation is not fair to f. The problem has been then to define fairness in such a way that the application of contraction rules is fair. This problem is solved in the definition of uniform fairness by establishing that it is fair not to perform an expansion inference step if its premises are not persistent and it is fair to replace a clause φ by clauses which make it redundant. In this way it is fair to apply contraction inference rules such as simplification and subsumption. In actuality, a uniformly fair procedure will perform exhaustively all expansion steps which are not inhibited by contraction steps.

Fairness and uniform fairness are different in several basic aspects. Fairness is *target-oriented*, whereas uniform fairness is defined for a derivation without a target. Indeed Definition 5.1 is not a definition of fairness for theorem proving. In [13, 4], Definition 5.1 is applied to refutational theorem proving, where S_0 contains the negation of the target. In this case the only persisting clause is the empty clause $\Box$ and $I_e(\bigcup_{i\geq 0}\bigcap_{j\geq i} S_j) = \{\Box\}$. Then Definition 5.1 says that the limit of the derivation is the empty clause. A notion of fairness given in terms of the limit does not represent useful information for the design of search plans because it does not say anything about how a search plan should choose the successor at any given stage of the derivation.

Our definition of fairness does not differentiate between expansion rules and contraction rules and between persisting and non-persisting clauses, because the interaction of expansion and contraction rules is no longer the issue. All inference rules are treated uniformly by considering their effect with respect to the goal of reducing the proof of the target. On the other hand, expansion rules and contraction rules are treated differently in the definition of uniform fairness. Uniform fairness emphasizes fairness with respect to the expansion inference rules, while the role of contraction inference rules is buried in the restriction to persistent clauses. The reason is that it is necessary to consider all critical pairs in order to obtain a confluent set, but it is not necessary to reduce them, although in practice completion without simplification is hopelessly inefficient.

The following example illustrates a set of conditions for an Unfailing Knuth-Bendix derivation which have been proved in [2] to be sufficient for uniform fairness. These conditions represent the most well known definition of (uniform) fairness for a completion procedure:

Example 5.1 *A derivation*
$(E_0;\emptyset)\vdash_{UKB}(E_1;\emptyset)\vdash_{UKB}\ldots\vdash_{UKB}(E_i;\emptyset)\vdash_{UKB}\ldots$
is uniformly fair if

- *for all critical pairs* $g \simeq d \in I_e(E_\infty)$, $g \simeq d \in \bigcup_{i\geq 0} E_i$ *and*
- E_∞ *is reduced.*

The first condition says that all critical pairs derivable from persisting equations are eventually generated. The second condition says that all persisting equations are eventually simplified as much as possible. As was remarked above, the application of contraction rules is allowed but not required: the first condition alone is sufficient for uniform fairness. Since at any stage of the computation it is not known which equations are going to persist and which equations are going to be simplified, the above conditions for uniform fairness prescribe in practice to apply exhaustively all the inference rules of Unfailing Knuth-Bendix completion until none applies.

The concept of uniform fairness leads to the following notion of *saturated* presentation:

Definition 5.2 (Kounalis and Rusinowitch 1988) [12], (Bachmair and Ganzinger 1990) [4] *A presentation S is* saturated *on the domain $\mathcal{T}$ of a completion procedure if and only if* $\forall\psi \in I_e(S)$, *either $\psi \in S$ or ψ is redundant in S on $\mathcal{T}$.*

In other words, a presentation is saturated if no non-trivial consequences can be added. In the equational case, as remarked in [12], a set of equations is saturated if no divergent critical pairs can be deduced, or equivalently, the set is *locally confluent.* As in the definition of uniform fairness, the application of contraction inference rules is allowed but not required: contraction inference rules may still be applicable to a saturated set. A locally confluent equational presentation is not necessarily reduced.

If a derivation is uniformly fair, S_∞ is saturated. Since uniform fairness is defined in terms of redundancy and our notion of redundancy is more general than those in [13] and [4], we give a new proof of this result:

Theorem 5.1 (Kounalis and Rusinowitch 1988) [12], (Bachmair and Ganzinger 1990) [4] *If a derivation*

$(S_0;\emptyset)\vdash_C(S_1;\emptyset)\vdash_C\ldots(S_i;\emptyset)\vdash_C\ldots$

is uniformly fair *on domain* $\mathcal{T}$, *then* S_∞ *is* saturated *on domain* $\mathcal{T}$.

Proof: we show that for all $\varphi \in I_e(S_\infty)$, either $\varphi \in S_\infty$ or φ is redundant in S_∞ on $\mathcal{T}$. By uniform fairness of the derivation, there exists an S_j, for some $j \geq 0$, such that either $\varphi \in S_j$ or φ is redundant in S_j on $\mathcal{T}$:

1. if $\varphi \in S_j$, then either φ is not deleted afterwards, that is $\varphi \in S_\infty$, or φ is deleted at some stage $i > j$. There are in turn two cases: either φ is simply deleted or it is replaced by another sentence φ':
   (a) let S_i be $S \cup \{\varphi\}$ and S_{i+1} be S. By Definition 3.7 of completion, such a step deletes a redundant sentence. Then φ is redundant in S_i on $\mathcal{T}$. Since by Definition 3.7 of completion, $\forall\psi \in \mathcal{T}$, $\Pi(S_i,\psi) \geq_p \Pi(S_\infty,\psi)$, φ is also redundant in S_∞ on $\mathcal{T}$.
   (b) if $S_i = S \cup \{\varphi\}$ and $S_{i+1} = S \cup \{\varphi'\}$, then this step is proof-reducing, that is $\forall\psi \in \mathcal{T}$, $\Pi(S_i,\psi) \geq_p \Pi(S_{i+1},\psi)$ and $\exists\psi \in \mathcal{T}$ such that $\Pi(S_i,\psi) >_p \Pi(S_{i+1},\psi)$. If φ itself represents a minimal proof of φ in S_i, then there exists a minimal proof $\Pi(S_{i+1},\varphi)$ in S_{i+1} such that $\varphi \geq_p \Pi(S_{i+1},\varphi)$. Since $\varphi \notin S_{i+1}$, $\Pi(S_{i+1},\varphi)$ is not φ itself and therefore $\varphi >_p \Pi(S_{i+1},\varphi) \geq_p \Pi(S_\infty,\varphi)$. If φ does not represent a minimal proof of φ in S_i, then there exists a minimal proof $\Pi(S_i,\varphi)$ of φ in S_i such that $\Pi(S_i,\varphi) <_p \varphi$ and therefore $\varphi >_p \Pi(S_i,\varphi) \geq_p \Pi(S_{i+1},\varphi) \geq_p \Pi(S_\infty,\varphi)$. In both cases there exists a minimal proof $\Pi(S_\infty,\varphi)$ of φ in S_∞ such that $\varphi >_p \Pi(S_\infty,\varphi)$ and by monotonicity and stability of $>_p$, $P[\varphi\sigma] >_p P[\Pi(S_\infty,\varphi)\sigma]$ for all proof contexts P and substitutions σ. In other words, φ is not involved in any minimal proof in S_∞ of a theorem of $\mathcal{T}$, since any occurrence of φ in a proof can be replaced by a proof $\Pi(S_\infty,\varphi)$ smaller than φ itself. It follows that φ is redundant in S_∞ on $\mathcal{T}$.
2. If φ is redundant in S_j on $\mathcal{T}$, since $\forall\psi \in \mathcal{T}$, $\Pi(S_j,\psi) \geq_p \Pi(S_\infty,\psi)$ by Definition 3.7 of completion, φ is redundant in S_∞ on $\mathcal{T}$. □

This theorem generalizes the following classical results:

Theorem 5.2 (Knuth and Bendix 1970) [11], (Huet 1981) [10], (Bachmair, Dershowitz and Hsiang 1986) [1] *If a derivation*

$(E_0;\emptyset)\vdash_{KB}(E_1;\emptyset)\vdash_{KB}\ldots(E_i;\emptyset)\vdash_{KB}\ldots$

by the Knuth-Bendix completion procedure does not fail and is uniformly fair *on the domain* $\mathcal{T}$ *of all equations, then* E_∞ *is a* confluent *term rewriting system.*

Knuth-Bendix completion fails if an unoriented equation persists. If a derivation by Knuth-Bendix completion does not fail, all the persistent equations are oriented into rewrite rules according to a reduction ordering and therefore E_∞ is a terminating rewrite system. By Theorem 5.1, E_∞ is saturated, i.e. locally confluent. By Newman's lemma [7], a terminating rewrite system is confluent if and only if it is locally confluent. Therefore E_∞ is confluent.

Theorem 5.3 (Hsiang and Rusinowitch 1987) [9], (Bachmair, Dershowitz and Plaisted 1989) [3] *If a derivation*

$$(E_0;\emptyset)\vdash_{UKB}(E_1;\emptyset)\vdash_{UKB}\ldots(E_i;\emptyset)\vdash_{UKB}\ldots$$

by the Unfailing Knuth-Bendix completion procedure is uniformly fair *on the domain $\mathcal{T}$ of all ground equations, then E_∞ is a* ground confluent *set of equations.*

For this second result we recall that given a set of equations E, $s \rightarrow_E t$ if $s \leftrightarrow_E t$ and $s \succ t$ for a reduction ordering $\succ$. The Unfailing Knuth-Bendix procedure assumes that $\succ$ is a complete simplification ordering. Since a complete simplification ordering is total on ground terms, $\leftrightarrow_{E_\infty} = \rightarrow_{E_\infty} \cup \leftarrow_{E_\infty}$ holds for ground terms and E_∞ is Church-Rosser on ground terms if and only if it is ground confluent. Since a complete simplification ordering is well founded, E_∞ is terminating on ground terms. The domain $\mathcal{T}$ of Unfailing Knuth-Bendix is the set of ground equations. By Theorem 5.1, E_∞ is saturated on $\mathcal{T}$, i.e. it is locally confluent on ground terms. By Newman's lemma E_∞ is ground confluent and therefore Church-Rosser on ground terms.

The Church-Rosser property on ground terms is important because $E \models \forall \bar{x} s \simeq t$ if and only if $\hat{s} \leftrightarrow_E^* \hat{t}$. If E is Church-Rosser on ground terms, $\hat{s} \leftrightarrow_E^* \hat{t}$ if and only if $\hat{s} \rightarrow_E^* \circ \leftarrow_E^* \hat{t}$ and therefore $E \models \forall \bar{x} s \simeq t$ can be decided by well founded reduction by E.

6 Redundancy

In the previous sections we introduced a notion of redundancy. The interest in redundancy of data elements in a theorem proving derivation resides in the importance of contraction inference rules. Contraction inference rules are necessary to make theorem proving feasible. However, few contraction rules are known. The purpose of studying redundancy is to get some insight about how to design new and powerful contraction rules. Therefore we conclude with a discussion on redundancy.

A notion of redundant clauses appeared in [13] and in [4], where the term "redundant" was first used. Redundant clauses according to these works are redundant in our sense. On the other hand, there are clauses which are intuitively redundant and redundant according to our definition, but not according to the definitions in [13] and [4].

Definition 6.1 (Rusinowitch 1988) [13] *A clause φ is* R-redundant *in a set S if there exists a clause $\psi \in S$ such that ψ properly subsumes φ, i.e. $\psi \gtrdot \varphi$, where $\gtrdot$ is the proper subsumption ordering on clauses.*

R-redundancy has been recently investigated in [14] in the context of proofs by resolution in first order logic. Very high numbers of R-red ındant clauses may be generated in such derivations, resulting in waste of space to hold them and in waste of time to perform the subsumption test to detect them. Two techniques to limit the generation of R-redundant clauses are proposed in [14].

The definition of redundancy in [4] assumes a well founded ordering $>^d$ total on ground clauses[2]:

Definition 6.2 (Bachmair and Ganzinger 1990) [4] *A clause φ is* B-redundant *in a set S if for all ground instances $\varphi\sigma$ of φ, there are ground instances $\psi_1 \ldots \psi_n$ of clauses in S such that $\{\psi_1 \ldots \psi_n\} \models \varphi\sigma$ and $\forall j, 1 \leq j \leq n, \varphi\sigma >^d \psi_j$.*

[2]In [4] a notion of *deletion ordering* is defined for this purpose. Well foundedness and totality on ground clauses are the only properties of a deletion ordering which are relevant to our discussion.

It is immediate to see that if the ordering $>^d$ is the proper subsumption ordering, B-redundancy specializes to R-redundancy and therefore a clause which can be subsumed is an example of a redundant clause:

Lemma 6.1 (Bachmair and Ganzinger 1990) [4] *R-redundant clauses are B-redundant.*

In our view, the intuition behind the notion of redundancy is that a clause φ is redundant in S if adding φ to S does not decrease any minimal proof in S (Definition 3.4). In fact our definition captures the meaning of Definition 6.2:

Theorem 6.1 *If a clause φ is B-redundant in S, then it is redundant on the domain of all ground clauses.*

Proof: we assume a proof ordering $>_p$ on ground proofs such that the minimal proof of a ground clause φ in S, $\Pi(S,\varphi)$, is the smallest set $\{\psi_1 \dots \psi_n\}$ of ground instances of clauses in S such that $\{\psi_1 \dots \psi_n\} \models \varphi$, according to the multiset extension $>^d{}_{mul}$ of the ordering $>^d$. Since $>^d$ is well founded and total on ground clauses, $>_p$ is well defined. Let φ be B-redundant in S. We show that $\Pi(S \cup \{\varphi\}, \psi) = \Pi(S, \psi)$ for all ground theorems ψ. Since $S \subset S \cup \{\varphi\}$, $\Pi(S \cup \{\varphi\}, \psi) \leq_p \Pi(S, \psi)$ trivially holds and therefore we simply have to show that $\Pi(S \cup \{\varphi\}, \psi) \not<_p \Pi(S, \psi)$. The proof is done by way of contradiction: if $\Pi(S \cup \{\varphi\}, \psi) <_p \Pi(S, \psi)$, then the smallest set of ground instances of clauses in $S \cup \{\varphi\}$ which logically entails ψ has the form $S' \cup \{\varphi\sigma_1 \dots \varphi\sigma_k\}$ for some set S' of ground instances of clauses in S and some ground substitutions $\sigma_1 \dots \sigma_k$. Since φ is B-redundant in S, for all $\varphi\sigma_i$, $1 \leq i \leq k$, there are ground instances $\{\psi_1^i \dots \psi_n^i\}$ of clauses in S such that $\{\psi_1^i \dots \psi_n^i\} \models \varphi\sigma_i$ and $\varphi\sigma_i >^d \psi_j^i$, $\forall j, 1 \leq j \leq n$. Therefore, $S' \cup \{\psi_1^i \dots \psi_n^i\}_{i=1}^k <^d{}_{mul} S' \cup \{\varphi\sigma_1 \dots \varphi\sigma_k\}$ and $S' \cup \{\psi_1^i \dots \psi_n^i\}_{i=1}^k \models \psi$, that is $S' \cup \{\varphi\sigma_1 \dots \varphi\sigma_k\}$ cannot be the smallest set entailing ψ. It follows that $\Pi(S \cup \{\varphi\}, \psi) = \Pi(S, \psi)$. □

On the other hand, there are cases where trivially redundant clauses are not B-redundant, whereas they are redundant according to our definition:

Example 6.1 *If $S = \{P, \neg R, R\}$, where P and R are ground atoms, P is intuitively redundant and it is redundant according to our Definition 3.4: the minimal proof of every ground theorem is given by $\{\neg R, R\}$, since $\{\neg R, R\}$ yields the empty clause and therefore any clause. However, if $R \succ P$, P is not B-redundant.*

This example shows that a notion of redundancy based on an ordering on clauses is not ideal, since different precedences on predicate symbols may be needed in order to characterize as redundant different clauses during a computation. A notion of redundancy based on a proof ordering seems to behave more satisfactorily.

7 Discussion

Intuitively, fairness of a search strategy means that every inference step which needs to be considered will eventually be considered. In completion-based methods, this usually means resolving all potential critical pairs. In theorem proving, on the other hand, one is not interested in critical pairs which may not contribute to a proof of the target theorem. Thus, in theorem proving applications fairness does not require resolving all possibile conflicts but only those which may lead to a proof. It is therefore conceivable to design a fair search strategy which ignores the majority of possible critical pairs.

One can actually even go further: since in theorem proving we are only interested in finding *one* proof, a search strategy can be considered fair as long as it does not remove the possibility of finding any proof. Thus, a search strategy may trim the search space considerably and still be fair as long as it does not trim away all the proofs. We feel that our approach, which separates the presentation from the target, provides a better framework than others for the study of such fair search strategies.

We conclude with some discussion about contraction and redundancy. A contraction inference rule either deletes sentences or replace them by others. An alternative scheme, called *deletion*, is given in [4]. The deletion scheme differs from our contraction scheme, since it only allows to infer S' from S by deleting a sentence in S. If this deletion scheme is adopted, an inference rule which replaces a sentence by other sentences has to be schematized as the composition of an expansion rule and a deletion rule. For instance, Simplification of $p \simeq q$ into $p \simeq q[r\sigma]_u$ is described as the generation of the equation $p \simeq q[r\sigma]_u$ followed by the deletion of $p \simeq q$.

This approach has the substantial drawback that it requires to consider more general inference rules than those actually used in the set of inference rules of a given strategy. For the simplification of $p \simeq q$ into $p \simeq q[r\sigma]_u$, the equation $p \simeq q[r\sigma]_u$ may not be a critical pair derivable by a superposition step. This is the case if $p \succ q$, that is the right hand side of a rewrite rule is simplified. A general paramodulation inference rule, which is not featured by the Unfailing Knuth-Bendix procedure, is then required in order to simulate simplification. We prefer our expansion/contraction schemes, since they allow us to classify directly every concrete inference rule of a given strategy as either an expansion rule or a contraction rule.

In Section 3 we have used redundancy to characterize those contraction rules which delete sentences. It is immediate to show that redundancy plays a role also for contraction rules which replace a sentence by others. If a contraction inference step $(S \cup \{\psi\}; \varphi) \vdash (S \cup \{\psi'\}; \varphi)$ is proof-reducing by Condition 2 in Definition 3.2, the deleted sentence ψ is clearly redundant in $S \cup \{\psi'\}$. However, this is not necessarily true for a contraction step which strictly reduces the proof of the target. Such a step is reducing because it reduces the proof of the target, without any provision for the proofs of the other theorems. Therefore, according to our definition of proof-reduction, a contraction inference rule may also delete non redundant sentences, provided it is sound and it strictly reduces the proof of the target.

This is a further difference between our approach and the one in [4]. In [4] redundancy is used to define the deletion scheme itself: a deletion inference rule is a rule which deletes redundant sentences. We prefer not to relate so tightly the notions of contraction and redundancy, as long as the assumed notion of redundancy is not target-oriented. According to the three definitions of redundancy we have considered, a sentence is redundant if it is useless with respect to all the theorems in the domain. Therefore, these three definitions are not target-oriented. Our definition of proof reduction instead allows in principle very strong contraction rules which may replace even non redundant sentences if they do not help in proving a specific target.

Our whole framework is target-oriented tough, and therefore it naturally leads to a target-oriented definition of redundancy:

Definition 7.1 *A sentence φ is* redundant for *ψ in S, if* $\Pi(S,\psi) = \Pi(S \cup \{\varphi\}, \psi)$.

Clearly, a clause which is redundant on a domain T is redundant for all the targets in T.

A target-oriented definition of redundancy applies for instance to the example we gave in the introduction, where the target $s \simeq t$ is an equation on a signature F_1 and the input presentation is the union of a set E_1 of equations on the signature F_1 and a set E_2 of equations on another signature F_2, disjoint from F_1. If we consider the problem in terms of the search process, it is fair to indefinitely postpone the equations in E_2. If we consider it in terms of the inference process, all the equations in E_2 are redundant for $s \simeq t$ and therefore can be eliminated. This duality of inference and search is not surprising, since most issues in theorem proving can be described both as properties of the inference mechanism and as properties of the search plan. We feel that thinking in terms of search is less categorical and therefore may give more flexibility. Further work is necessary to turn an abstract study of fairness, contraction and redundancy into concrete search plans and inference rules.

References

[1] L.Bachmair, N.Dershowitz, J.Hsiang, Orderings for Equational Proofs, in *Proceedings of the First Annual IEEE Symposium on Logic in Computer Science*, 346–357, Cambridge, MA, June 1986.

[2] L.Bachmair, Proofs Methods for Equational Theories, Ph.D. thesis, Department of Computer Science, University of Illinois, Urbana, IL.,1987.

[3] L.Bachmair, N.Dershowitz and D.A.Plaisted, Completion without failure, in H.Ait-Kaci, M.Nivat (eds.), *Resolution of Equations in Algebraic Structures*, Vol. II: Rewriting Techniques, 1–30, Academic Press, New York, 1989.

[4] L.Bachmair, H.Ganzinger, Completion of First-Order Clauses with Equality by Strict Superposition, to appear in M.Okada, S.Kaplan (eds.), *Proceedings of the Second International Workshop on Conditional and Typed Rewriting Systems*, Montreal, Canada, June 1990.

[5] M.P.Bonacina, J.Hsiang, Completion Procedures as Semidecision Procedures, to appear in M.Okada, S.Kaplan (eds.), *Proceedings of the Second International Workshop on Conditional and Typed Rewriting Systems*, Montreal, Canada, June 1990.

[6] N.Dershowitz, Orderings for term-rewriting systems, *Theoretical Computer Science*, Vol. 17, 279–301, 1982.

[7] N.Dershowitz, J.-P.Jouannaud, Rewrite Systems, Chapter 15, Volume B, *Handbook of Theoretical Computer Science*, North-Holland, 1989.

[8] N.Dershowitz, A Maximal-Literal Unit Strategy for Horn Clauses, to appear in M.Okada, S.Kaplan (eds.), *Proceedings of the Second International Workshop on Conditional and Typed Rewriting Systems*, Montreal, Canada, June 1990.

[9] J.Hsiang, M.Rusinowitch, On word problems in equational theories, in Th.Ottman (ed.), *Proceedings of the Fourteenth International Conference on Automata, Languages and Programming*, Karlsruhe, West Germany, July 1987, Springer Verlag, Lecture Notes in Computer Science 267, 54–71, 1987.

[10] G.Huet, A Complete Proof of Correctness of the Knuth-Bendix Completion Algorithm, *Journal of Computer and System Sciences*, Vol. 23, 11–21, 1981.

[11] D.E.Knuth, P.Bendix, Simple Word Problems in Universal Algebras, in J.Leech (ed.), *Proceedings of the Conference on Computational Problems in Abstract Algebras*, Oxford, England, 1967, Pergamon Press, Oxford, 263–298, 1970.

[12] E.Kounalis, M.Rusinowitch, On Word Problems in Horn Theories, in E.Lusk, R.Overbeek (eds.), *Proceedings of the Ninth International Conference on Automated Deduction*, 527–537, Argonne, Illinois, May 1988, Springer Verlag, Lecture Notes in Computer Science 310, 1988.

[13] M.Rusinowitch, Theorem-proving with resolution and superposition: an extension of Knuth and Bendix procedure as a complete set of inference rules, Thèse d'Etat, Université de Nancy, 1987.

[14] R.Socher-Ambrosius, How to Avoid the Derivation of Redundant Clauses in Reasoning Systems, to appear in *Journal of Automated Reasoning*, 1990.

Proving Equational and Inductive Theorems by Completion and Embedding Techniques

J. Avenhaus
Department of Computer Science
University of Kaiserslautern
6750 Kaiserslautern
West Germany
e-mail: avenhaus@uklirb.uucp

Abstract
The Knuth-Bendix completion procedure can be used to transform an equational system into a convergent rewrite system. This allows to prove equational and inductive theorems. The main draw back of this technique is that in many cases the completion diverges and so produces an infinite rewrite system. We discuss a method to embed the given specification into a bigger one such that the extended specification allows a finite "parameterized" description of an infinite rewrite system of the base specification. The main emphasis is in proving the correctness of the approach. Examples show that in many cases the Knuth-Bendix completion in the extended specification stops with a finite rewrite system though it diverges in the base specification. This indeed allows to prove equational and inductive theorems in the base specification.

1. Introduction

Term rewriting systems constitute an important tool to compute and reason in systems defined by equations. Given a set E of equations, the validity problem for E is to decide whether a given equation s = t holds in all models of E. We write $s =_E t$ in this case and call s = t an equational theorem of E. For abstract data types one is usually interested in the initial model $\mathfrak{I}(E)$ of E. The equation s = t holds in $\mathfrak{I}(E)$ iff $s\sigma =_E t\sigma$ for every ground substitution σ. We call s = t an inductive theorem in this case and write $s = t \in ITh(E)$. For both problems, proving equational and inductive theorems, the Knuth-Bendix completion procedure and refinements thereof have turned out to be very helpful. A major draw back of this approach is that in many cases the completion procedure diverges and so produces an infinite convergent rewrite systems. The paper discusses a method to overcome the divergence problem in such cases where the infinite rewrite system has some regularities.

There are various proposals of how to circumvent divergence of the Knuth-Bendix completion procedure, see [Her] for suggestions. In [Kir] it is proposed to describe an infinite set of rewrite rules by a finite set of meta rules containing parameters to describe the infinite set of rules. The problem is how to deal with such a parameterized system, e.g. how to test confluence. In [Kir] a rather complicated framework using order-sorted rewriting is developed. Under some strong restrictions completion of parameterized rewrite systems seems to be possible.

There is another approach based on embedding techniques and several research groups have experimented with it. Thomas/Jantke [TJa] and Lange [Lan] were the first to fix the idea of extending the given structure by natural numbers and to encode an infinite sequence of equations by a single equation using the natural numbers as parameters. In [Lan] inference rules are presented for automatically learning the encoding from the first elements of the infinite sequence. This indeed results in many cases in a finite confluent rewriting system. But the learned encoding is only assumed to be correct for the whole sequence. No attempt is made to prove the approach correct, i.e. to guarantee that reasoning in the extended structure is safe for reasoning in the base structure.

The aim of the present paper is the following: (1) We provide a sound foundation of the embedding method. To do so, we give a careful definition of "consistent enrichment" which allows to prove equational theorems. We point out the problems and give sufficient conditions to overcome them. This is motivated but not restricted to the extension by a copy of NAT. (2) We incorporate knowledge about the parameters. For instance, the method is much more successful if one uses the AC-property of + on the natural numbers. (3) We extend the method to prove inductive theorems. (4) We demonstrate the power of the extended method by giving interesting examples for proving equational and inductive theorems in cases where the classical completion procedure fails. Notice that the embedding method is very general and can be used in different contexts. For inductive theorem proving we apply it to the method of Jouannaud and Kounalis [JKo], one could easily use other methods, e.g. that of Bachmair [Bac].

We assume the reader to be familiar with rewriting and completion techniques as developed in [KBe] and [Hue]. For a survey see [AMa] and [DJo]. We use the standard notation. If R is a rewrite system then $\rightarrow_R$ denotes the rewrite relation induced by R. We call R convergent if $\rightarrow_R$ is terminating (there is no infinite sequence $t_0 \rightarrow_R t_1 \rightarrow_R t_2 \rightarrow_R \ldots$) and confluent (the relation ${}_R\overset{*}{\leftarrow} \circ \overset{*}{\rightarrow}_R$ is included in $\overset{*}{\rightarrow}_R \circ {}_R\overset{*}{\leftarrow}$). We also need rewriting modulo a congruence generated by a set A of equations. R is A-terminating if the relation $\rightarrow_{R/A} = {=_A} \circ \rightarrow_R \circ {=_A}$ is terminating. For rewriting modulo A we use the relation $\rightarrow_{R,A} \subseteq \rightarrow_{R/A}$ as defined by $s \rightarrow_{R,A} t$ iff $s \equiv s[u]$, $u =_A l\sigma$ and $t \equiv s[r\sigma]$ for some rule $l \rightarrow r$ in R and some substitution σ. We say R,A is a convergent rewrite system for the set E of equations if (i) R is A-terminating and (ii) R,A is Church-Rosser modulo A ($s =_{R \cup A} t$ implies $s \overset{*}{\rightarrow}_{R,A} s_0 =_A t_0 \, {}_{R,A}\overset{*}{\leftarrow} t$ for some terms s_0, t_0) and (iii) $=_E \, = \, =_{R \cup A}$. For completion procedures that try to transform E into R,A see e.g. [BDe], [PSt], [JKi]. For early papers on inductive theorem proving see [Mus], [KMu] and [HHu].

The paper is organized as follows. We start in section 2 with a motivating example. In section 3 the embedding technique is presented and applications to prove equational and inductive theorems are discussed. In section 4 we show how to test the conditions which are necessary for the approach to work and in section 5 we give some examples to show the power of the method.

2. An example

As a motivating example for our approach let us consider the specification NAT of the natural numbers with the gcd-function, see e.g. [Her].

$$E_1:\quad x+0 = x \qquad g(x,0) = x \qquad g(x+y,y) = g(x,y)$$
$$x+s(y) = s(x+y) \qquad g(0,y) = y \qquad g(x,y+x) = g(x,y)$$

We call $spec_1 = (\Sigma_1,F_1;E_1)$ the base specification. (For details see section 3). We would like to have a convergent rewrite system R for E, this would allow us to compute $g(s^i(o), s^j(o))$ and to prove equational theorems of E. Furthermore, we would like to prove inductive theorems of E, e.g. $g(x,x) = x$.

Completion of E using the ordering RPO (see [Der]) with precedence $+ \triangleright s$ diverges and produces terms of the form $s^n(x)$. It is natural to look for an embedding that contains a copy of the natural numbers and to express $s^n(x)$ by $S(n,x)$ where S is a new function symbol. To do so we specify the copy of natural numbers by the operators $\underline{1}$ and $\underline{+}$ and the equations (AC) and add a new operator S with defining equations (S):

$$\text{(S)}\qquad s(x) = S(\underline{1},x) \qquad\qquad S(u,S(v,x)) = S(u\underline{+}v,x)$$

$$\text{(AC)}\qquad u \underline{+} v = v \underline{+} u \qquad\qquad (u\underline{+}v) \underline{+} w = u \underline{+} (v\underline{+}w)$$

Let $spec_2 = (\Sigma_2,F_2;E_2)$ be the extended specification with E_2 consisting of E, (AC) and (S). We will prove that $spec_2$ is a consistent enrichment of $spec_1$ and so can be used to reason in $spec_1$. If we start the completion procedure with input E_2, it diverges again. But, from the infinite set of rules being produced one can see that the infinitely many equations $x + s^n(y) = s^n(x+y)$ are needed. So we add the equation

$$\text{(S1)}\qquad x + S(u,y) = S(u,x+y)$$

to E_2 and get E_3. Note that $x + S(u,y) = S(u,x+y)$ describes the infinite sequence of valid equations $x + s^n(y) = s^n(x+y)$ and so is "safe" (see section 3). Again, $spec_3 = (\Sigma_2,F_2;E_3)$ is a consistent enrichment of $spec_1$. If we now start the completion procedure with input E_3 we will get the finite rewrite system R modulo (AC):

R: $x+0 \to x$	$s(x) \to S(\underline{1},x)$
$x+S(u,y) \to S(u,x+y)$	$S(u,S(v,x)) \to S(u\underline{+}v,x)$
$g(x,0) \to x$	$g(x+y,y) \to g(x,y)$
$g(0,y) \to y$	$g(x,y+x) \to g(x,y)$
$g(S(u,x+y), S(u,y)) \to g(x,S(u,y))$	$g(S(u,x), S(u,y+x)) \to g(S(u,x),y)$
$g(S(u,x), S(u,0)) \to g(x,S(u,0))$	$g(S(u,0)), S(u,y)) \to g(S(u,0),y)$
$g(S(u\underline{+}v,x), S(u,0)) \to g(S(v,x), S(v,0))$	$g(S(u,0)), S(u\underline{+}v,x)) \to g(S(u,0), S(v,x))$

Using this finite rewrite system R we can prove the inductive theorem

$g(x,x) = x$ of E by the method "proof by consistency", see [JKo]: We complete $R \cup \{g(x,x) = x\}$ and get as result the system R and the rule $g(x,x) \to x$. Since $g(x,x)$ is inductively reducible by R,AC we have proved that $g(x,x) = x$ is indeed an inductive theorem of E.
We are going to make these ideas precise in the rest of the paper.

3. The embedding strategy

Assume we have two specifications $spec_1$ and $spec_2$ such that $spec_1$ is a subspecification of $spec_2$ (for definitions see below). In this section we study how to use $spec_2$ to compute and prove equational and inductive theorems in $spec_1$. This will need a careful definition of what it means that $spec_2$ is a consistent enrichment of $spec_1$. We will use equational reasoning in $spec_2$ but do not consider all models of $spec_2$. Intuitively, we consider all those $spec_2$-models that consist of an arbitrary $spec_1$-model extended by the initial model of the extension. This allows us to add "safe equations", and in many applications these additional equations help to get a finite convergent rewrite system in $spec_2$ to reason in the given base specification $spec_1$.

A <u>specification</u> spec = $(\Sigma,F;E)$ consists of a signature sig = (Σ,F) and a set E of defining equations. Here Σ is a set of sorts and F is a set of operators. For each sort s we assume to have a constant of sort s and a denumerable set V_s of variables such that $V_s \cap V_{s'} = \emptyset$ for $s \neq s'$. Then V is the union of all V_s and Term(F,V) is the set of terms over $F \cup V$. The terms in Term(F,V) are called spec-terms, those in $Term(F,\emptyset)$ are ground terms.

If $spec_1 = (\Sigma_1,F_1;E_1)$ and $spec_2 = (\Sigma_2,F_2;E_2) = (\Sigma_1 \cup \Sigma_0, F_1 \cup F_0; E_1 \cup E_0)$ then $spec_1$ is a subspecification of $spec_2$. Let V_1 be the set of variables of sorts $s \in \Sigma_1$ and V_0 the set of variables of sorts $s \in \Sigma_0$. Then the elements of V_0 are called <u>parameters</u>. A sort-preserving substitution ψ: $V_0 \to Term(F_2,V_1)$ is called a <u>parameter substitution</u>. We call $spec_2 = (\Sigma_2,F_2;E_2)$ a <u>consistent enrichment</u> of $spec_1 = (\Sigma_1,F_1;E_1)$ if $\Sigma_1 \subset \Sigma_2$, $F_1 \subset F_2$ and for all $spec_1$-terms t_1, t_2 we have $t_1 =_{E_2} t_2$ iff $t_1 =_{E_1} t_2$. Notice that t_1, t_2 may contain variables. We call $spec_2$ an <u>almost free</u> consistent enrichment if in addition every parameter free term of sort $s \in \Sigma_0$ is a ground term. In this case a parameter substitution ψ is ψ: $V_0 \to Term(F_2,\emptyset)$. In our applications all enrichments are almost free.

In our gcd-example of the previous section we have $spec_1 = (\{NAT\}, \{0,s,+,g\}; E_1)$. To get $spec_2$ we add the new sort NAT_+, the operators in $F_0 = \{\underline{1},\underline{+},S\}$ and the equations (AC) and (S). Notice that we are dealing with many sorted signatures so that NAT and NAT_+ are completely different sorts and + and $\underline{+}$ are different operators.

If terms $t_1,t_2 \in Term(F_1 \cup F_0, V_1 \cup V_0)$ of sort in Σ_1 contain parameters $x \in V_0$, then t_1 <u>describes</u> all the terms $t_1\psi \in Term(F_1 \cup F_0,V_1)$ with ψ a parameter substitution. So the equation $t_1 = t_2$ describes all the equations $t_1\psi = t_2\psi$. It is called <u>safe</u> with respect to $spec_2$ if $t_1\psi =_{E_2} t_2\psi$ for each parameter substitution ψ.

Fact 3.1: If $spec_2 = (\Sigma_2, F_2; E_2)$ is a consistent enrichment of $spec_1 = (\Sigma_1, F_1; E_1)$ and each equation in E is safe with respect to $spec_2$, then $spec_3 = (\Sigma_2, F_2; E_2 \cup E)$ is a consistent enrichment of $spec_1$.

Proof: Let t_1, t_2 be $spec_1$-terms and $t_1 \equiv s_1 \vdash\!\dashv s_2 \vdash\!\dashv \ldots \vdash\!\dashv s_n \equiv t_2$ be a proof of $t_1 =_{E_2 \cup E} t_2$ where $\vdash\!\dashv$ is the one-step proof relation. Let ψ be a parameter substitution and $s_i' \equiv s_i\psi$. Then $t_1 \equiv s_1' \vdash\!\dashv s_2' \vdash\!\dashv \ldots \vdash\!\dashv s_n' \equiv t_2$ is also a proof for $t_1 =_{E_2 \cup E} t_2$. If $s_i' \vdash\!\dashv s_{i+1}'$ uses an E-equation $l = r$ then it uses it in the form $l\psi' = r\psi'$ with ψ' a parameter substitution. This proof step can be replaced by a proof in E_2. This proves $t_1 =_{E_2} t_2$. Now we have $t_1 =_{E_1} t_2$ by the consistency property. ∎

Now assume $spec_2$ is a consistent enrichment of $spec_1$. We may run the completion procedure with input E_2 and an appropriate reduction ordering, thereby regarding both the "variables" in $spec_1$ and the "parameters" in $spec_2$ as variables. Then new equations are produced by equational reasoning. These equations are automatically safe, so all specifications produced by the completion procedure are consistent enrichments by Fact 3.1. In many cases one needs rewriting and completion modulo a set A of unorientable equations. This gives

Theorem 3.2:
Assume $spec_2 = (\Sigma_2, F_2; E_2)$ is a consistent enrichment of $spec_1 = (\Sigma_1, F_1; E_1)$ and E is a set of equations that are safe with respect to $spec_2$. If completion of $E_2 \cup E$ results in $R \cup A$ such that R,A is convergent then $spec_3 = (\Sigma_2, F_2; R \cup A)$ is a consistent enrichment of $spec_1$. So for s, t $\in Term(F_1, V_1)$ we have $s =_{E_1} t$ iff $\hat{s} =_A \hat{t}$ where $\hat{s}$ and $\hat{t}$ denote R,A-normal forms of s and t. ∎

Now the <u>embedding strategy</u> is as follows: Let $spec_1$ as in Theorem 3.2 be given such that the completion procedure with input E_1 diverges. From the first equations generated by the completion procedure we first construct an appropriate consistent enrichment $spec_2$ and then a set E of safe equations. If the system R,A is finite then the construction was successful. In section 4 we develop tools for proving that an enrichment is indeed consistent and equations are indeed safe. We believe that splitting the embedding process into these two steps simplifies the correctness proof.

Theorem 3.2 allows us to prove equational theorems in the base specification $spec_1$ by reasoning in a consistent enrichment of $spec_1$. We shall now use the approach of Jouannaud and Kounalis [JKo] to prove inductive theorems in the same way. The basic approach of [JKo] is: If R is a rewrite system for E and R_0 is a convergent rewrite systems for $R \cup E_0$ such that $R \subseteq R_0$ and every left-hand side of a rule in R_0 is inductively reducible by R then $E_0 \subseteq ITh(E)$. Here the condition $R \subseteq R_0$ can be dropped, then R needs only be terminating, it will automatically be ground confluent.

Notice that both terms, "inductive theorem" and "inductively reducible" depend on the underlying signature. If spec = $(\Sigma, F; E)$ is given then $s = t \in ITh(E;spec)$ means that $s\sigma =_E t\sigma$ for all spec-ground substitutions σ, analogously, t is spec-inductively reducible in R means that $t\sigma$ is reducible by R for all spec-ground substitutions σ.

Theorem 3.3:
Let $spec_2 = (\Sigma_2, F_2, R_2 \cup A)$ be a consistent enrichment of $spec_1 = (\Sigma_1, F_1, E_1)$ such that R_2/A is terminating. Let $E = E' \cup E''$ be a set of $spec_2$-equations such that E' contains $spec_1$-equations only. If R/A is a convergent rewrite system for $R_2 \cup A \cup E$ such that every left-hand side l of a rule $l \to r$ in R is $spec_2$-inductively reducible by R_2/A then $E' \subseteq ITh(E_1; spec_1)$ and $E \subseteq ITh(R_2 \cup A, spec_2)$.

Proof: Let s = t be an equation in E' and σ a $spec_1$-ground substitution, we have to prove $s\sigma =_{E_1} t\sigma$. Let s' and t' be R_2/A-normal forms of $s\sigma$ and $t\sigma$, respectively. Then $s' =_{R \cup A} t'$ since R/A is a rewrite system for $R_2 \cup A \cup E$. In addition, s' and t' are R/A-irreducible since they are R_2/A-irreducible and every left-hand side of a rule in R is $spec_2$-inductively reducible by R_2/A. Since R/A is convergent this gives $s' =_A t'$. So we have $s\sigma =_{R_2 \cup A} t\sigma$ and hence $s\sigma =_{E_1} t\sigma$ since $spec_2$ is a consistent enrichment of $spec_1$. This proves the first statement. The second statement is proved in the same way. ∎

Theorem 3.3 refers to the rewrite relation $\to_{R/A}$, but in practice one uses $\to_{R,A}$. If R,A is convergent, then (i) so is R/A and (ii) a term t is reducible by R,A iff it is reducible by R/A. In this case we may replace R_2/A and R/A in Theorem 3.3 by R_2,A and R,A, respectively.

4. Proving consistency, safety and inductive reducibility

In this section we study how to deal with the conditions that are necessary for our embedding strategy to work. The problems are: (1) Proving consistency of an enrichment, (2) Proving safeness of equations in the enrichment and (3) Proving that a term is inductively reducible by R/A. All these problems are hard in general but may be solved for our applications.

For the first problem we will give in Theorem 4.1 fairly general conditions under which $spec_2$ is a consistent enrichment of $spec_1$. In fact, this theorem does not refer to the defining equations of $spec_1$. So it leads to "uniform consistent enrichments". This result may be interesting in its own right since proving consistency is a non-trivial problem (even undecidable in general) but needed in different contexts. Theorem 4.1 allows us to construct some standard extensions to be used safely later on.

A <u>ground reduction ordering</u> over sig is a reduction ordering that is total on sig-ground terms. Many of the standard reduction orderings can be extended to be total on ground terms. For example, the lexicographical path ordering LPO of Kamin and Levy with a total precedence $\triangleright$ is a ground reduction ordering (see th survey paper [Der] for a definition of the LPO). If $f \triangleright g$ for all $f \in F_0$, $g \in F_1$ then condition (1) of the following Theorem holds.

Theorem 4.1

Let $spec_1 = (\Sigma_1, F_1; E_1)$, and $spec_2 = (\Sigma_2, F_2; R \cup E_1)$ be given with $\Sigma_2 = \Sigma_1 \cup \Sigma_0$, $F_2 = F_1 \cup F_0$, $E_2 = R \cup E_1$. Assume

(1) There is a ground reduction ordering $>$ on $sig_2 = (\Sigma_2, F_2)$ such that $s \not> t$ whenever s is F_0-free but t is not F_0-free.

(2) $l > r$ for all $l \to r$ in R.

(3) l is F_1-free for all $l \to r$ in R.

(4) R is confluent on $spec_2$-terms.

Then $spec_2$ is a consistent enrichment of $spec_1$.

We prove this theorem in the Appendix. Here we comment on the strong condition (1). One may conjecture that it is enough to use any reduction ordering. The following example, which I owe to a referee, disproves this conjecture. A ground reduction ordering is needed.

Example.

$F_1 = \{b, c, f, h, k\}$ $\qquad$ $F_2 = F_1 \cup \{a, g\}$

E_1: $\quad h(x, f(x)) \to c$ $\qquad$ R: $\quad g(a) \to fgk(a)$

$\qquad h(x, x) \to b$

$\qquad k(x) \to x$

Here E_1 is already oriented to a convergent rewrite system and R is convergent. $spec_2$ is no consistent enrichment of $spec_1$: We have $b \neq_{E_1} c$ but $b =_{E_2} c$ since

$$h(g(a), g(a)) \to_{E_1} b$$

$$h(g(a), g(a)) \to_R h(g(a), fgk(a)) \to_{E_1} h(g(a), fg(a)) \to_{E_1} c$$

Notice that $g(a) > fgk(a)$ for no ground reduction ordering $>$ since $g(a)$ is homeomorphically embedded in $fgk(a)$ (see [Der]). Notice also that $R \cup E_1$ has no critical pairs, but is not confluent and is not terminating.

We now show how Theorem 4.1 can be used to prove that $spec_2 = (\Sigma_1 \cup \Sigma_0, F_1 \cup F_0; E_1 \cup E_0)$ is a consistent enrichment of $spec_1 = (\Sigma_1, F_1; E_1)$ in the case where E_0 contains unorientable equations: Assume E_0 can be split into $E_0 = E' \cup E''$ such that (i) completion of E' results in a rewrite system R satisfying the conditions of Theorem 4.1 and (ii) E'' is a set of safe equations, then $spec_2$ is indeed a consistent enrichment of $spec_1$ by Theorem 4.1 and Fact 3.1. We will do this construction explicitely for our standard extension of $spec_1$ by a copy of NAT. (See the Appendix for a generalization of Theorem 4.1).

Lemma 4.2.

Let $spec_1 = (\Sigma_1, F_1; E_1)$ and $spec_2 = (\Sigma_1 \cup \Sigma_0, F_1 \cup F_0; E_1 \cup E_0)$ and $spec_2' = (\Sigma_1 \cup \Sigma_0, F_1 \cup F_0; E_1 \cup E_0')$ be given with $F_0 = \{\underline{1}, \underline{+}, S_1, \dots, S_n\}$

E_0: $\quad S_i(\underline{1}, x) = s_i(x) \qquad S_i(u \underline{+} v, x) = S_i(u, S_i(v, x)) \qquad s_i \in F_1 \qquad i = 1, \dots, n$

$\qquad (u \underline{+} v) \underline{+} w = u \underline{+} (v \underline{+} w)$

E_0': $\quad E_0$ and $u \underline{+} v = v \underline{+} u$

Then $spec_2$ and $spec_2'$ are almost free consistent enrichments of $spec_1$.

Proof: Let $\triangleright$ be a total precedence on $F_1 \cup F_0$ with $f \triangleright g$ for $f \in F_0$, $g \in F_1$ and status $\tau(f)$ = left-to-right for all $f \in F_1 \cup F_0$ and let $>$ be the corresponding LPO.

Then the rewrite system R resulting from E_0 by orienting the equations from left to right satisfies all the conditions of Theorem 4.1. So $spec_2$ is a consistent enrichment of $spec_1$. The equation $u \pm v = v \pm u$ is safe in $spec_2$: For any parameter substitution ψ we have $u\psi =_{E2} \underline{n} \equiv \underline{1} \pm (\underline{1} \pm \dots (\underline{1} \pm \underline{1}) \dots)$ and we have to prove $\underline{n} \pm \underline{m} =_{E2} \underline{m} \pm \underline{n}$. This can be done with the equation $(u \pm v) \pm w = u \pm (v \pm w)$. So by Fact 3.1 $spec_2'$ is a consistent enrichment of $spec_1$. ∎

Now we address to the second problem: Given a consistent enrichment $spec_2$ of $spec_1$ and a $spec_2$-equation $t_1 = t_2$, is $t_1 = t_2$ safe in $spec_2$? By definition of "safe" this is the case iff $t_1\psi =_{E2} t_2\psi$ for every parameter substitution ψ. This can be proved by Noetherian induction on the terms in $Term(F_2, V_1)$. For example, if u is a variable (parameter) of sort $s \in \Sigma_0$ occurring in t_1 and/or t_2, then $t_1[u] = t_2[u]$ is safe in $spec_2$ if one can prove $t_1[t] =_{E2} t_2[t]$ using as induction hypothesis $t_1[t'] =_{E2} t_2[t']$ for any subterm t' of t of sort s. Experience shows that this is relatively easy for equations $t_1 = t_2$ arising from applying the embedding strategy. See the examples in the next section.
If $spec_2$ is an almost free enrichment of $spec_1$, then by the next Lemma one can use any method for proving inductive theorems to prove safeness. This is due to the fact that now ψ replaces any parameter by a ground term. Let $\bar{t}$ denote the term resulting from t by replacing each variable x_j of sort in Σ_1 by a new constant $\bar{x}_j$.

Lemma 4.3.

Let $spec_2 = (\Sigma_1 \cup \Sigma_0, F_1 \cup F_0; E_2)$ be an almost free consistent enrichment of $spec_1 = (\Sigma_1, F_1, E_1)$. The $spec_2$-equation $t_1 = t_2$ is safe in $spec_2$ iff $\bar{t}_1 = \bar{t}_2 \in ITh(E_2, spec_2')$. Here $spec_2'$ results from $spec_2$ by adding the new constants $\bar{x}_j$ for x_j in t_1 or t_2.

Proof: For any $spec_2$-equation $s = t$ we have $s =_{E2} t$ iff $\bar{s} =_{E2} \bar{t}$. Notice that $\overline{t\psi} \equiv \bar{t}\psi$. Notice also that the $spec_2'$-ground instances of $\bar{t}$ are precisely the $\bar{t}\psi$ where ψ is a parameter substitution. This gives: $t_1 = t_2$ is safe iff for all parameter substitutions $t_1\psi =_{E2} t_2\psi$ iff for all ψ: $\bar{t}_1\psi =_{E2} \bar{t}_2\psi$ iff $\bar{t}_1 = \bar{t}_2 \in ITh(E_2; spec_2')$. ∎

Now we discuss the third problem: Given t,R,A; is t inductively reducible by R/A ? Notice that in our application A is a set of AC-axioms for operators of the structure by which $spec_1$ is extended. In this case the R/A-normal forms of F_0-ground terms are known. For example, if R contains $s(x) \to S(\underline{1},x)$ and $S(u,S(v,x)) \to S(u \pm v,x)$ then the R/A-normal form of any ground term containing the operators s, 0, S, $\underline{1}$, $\pm$ only is A-equal to 0 or $\underline{n}$ or $S(\underline{n},0)$. Since t is inductively reducible by R/A iff $t\sigma$ is reducible by R/A for all normalized ground substitutions σ, this may help to verify that t is indeed inductively reducible by R/A. Here σ is called normalized if for every variable x in t the ground term $x\sigma$ is in R/A-normal form.

Let R/A be a rewrite system over the signature $sig = (\Sigma,F)$. To verfiy that a term t is inductively reducible by R/A we extend a well-known method to test for R-inductive reducibility. We call $F' \subset F$ a set of constructors, if for every

F-ground term s there is a F'-ground term s' with $s \overset{*}{\rightarrow}_{R/A} s'$. If $f \notin F'$ then no R/A-normal form has top symbol f. So in a preprocessing step we try to compute a small set of constructors F'. If $f \in F'$ then some terms with top symbol f may still be reducible. So we try simultaneously to compute a set M_f such that all R/A-irreducible ground terms with top symbol f are in M_f. Let M be the union of all the M_f. Let σ be a M-substitution if $x\sigma \in M$ for all x such that $x \neq x\sigma$. Then t is inductively reducible in R/A iff tσ is reducible in R/A for all M-ground substitutions σ.

As an example let us consider the rewrite system R in the gcd-example of section 2. F' has to contain 0, $\underline{1}$, $\underline{+}$ and S. We have $M_S = \{S(\underline{n},0) \mid n \geq 1\}$. Now it is easy to see that $s,+ \notin F'$. We prove $g \notin F'$: We have to show that g(x,y) is inductively reducible by R/A. By what we already know it is enough to show that $g(t_1,t_2)$ is reducible for all ground terms t_1,t_2 of the form 0 or $S(\underline{n},0)$. If $t_1 \equiv 0$ or $t_2 \equiv 0$ then $g(t_1,t_2)$ is reducible by $g(x,0) \rightarrow x$ or $g(0,y) \rightarrow y$. For $t_1 \equiv S(\underline{n},0)$, $t_2 \equiv S(\underline{m},0)$ we have the cases n = m, n < m and m < n. There are rules in R with left-hand sides $g(S(u,x),S(u,0))$, $g(S(u,0),S(u\underline{+}v,x))$ and $g(S(u\underline{+}v,x),S(u,0))$, they can be used to reduce $g(t_1,t_2)$. So $g \notin F'$ and g(x,y) is inductively reducible. Therefore g(x,x) is inductively reducible and this had to be proved in section 2.

Notice that the arguments for R,A are the same as for R/A. Notice also that these arguments assume knowledge about the (in general well-known) structure only by which the base specification $spec_1$ is extended. No knowledge about $spec_1$ is used. So these arguments seem to be fair.

5. Examples

In this section we want to demonstrate the applicability of the embedding approach. The proposed procedure is highly interactive at the moment because the user has to provide an appropriate enrichment of the base specification and a set of safe equations in the enrichment that may describe the infinite set of equation in the base specification produced by the completion procedure. But techniques may be developed to automate the procedure to some extent, see [Lan]. The examples discussed in this paper use an enrichment of the base specification by natural numbers. Of course, one may think of enrichments by other well known structures, for example by lists.

Example Addition on the integers

E_1:	$0+y = y$	$s(x)+y = x+s(y)$	$s(p(x)) = x$
		$p(x)+y = x+p(y)$	$p(s(x)) = x$

This specification of the integers is known to produce problems to rewrite techniques if one wants to prove $x+0 = x \in ITh(E_1)$. The reason is that the completion procedure with input E_1 and $x+0 = x$ diverges and produces the terms $s^n(x)$, $p^n(x)$. To avoid this we use the extension of $spec_1$ by a copy of NAT and rewrite $s^n(x)$ to $S(\underline{n},x)$ and $p^n(x)$ to $P(\underline{n},x)$:

E_0: $s(x) = S(\underline{1},x)$ $\quad S(u,S(v,x)) = S(u\underline{+}v,x)$
$p(x) = P(\underline{1},x)$ $\quad P(u,P(v,x)) = P(u\underline{+}v,x)$

The completion procedure modulo the AC-axioms for $\underline{+}$ with input $E_1 \cup E_0$ diverges. From the sequence of rules generated one can see that the following equations are missing

E: $P(u,S(v,x)) = S(v,P(u,x))$ $\quad S(u\underline{+}v,P(v,x)) = S(u,x)$
$S(u,P(u,x)) = x$ $\quad S(u,P(u\underline{+}v,x)) = P(v,x)$

Using Lemma 4.3 it is easy to see that the equations in E are safe. Completion of $E_1 \cup E_0 \cup E$ modulo AC gives the following set R of rules. We use $S^u(x)$ as an abbreviation of $S(u,x)$. So we have $S^{1+u}(x) \equiv S(\underline{1} \underline{+} u,x)$ and $S^uP^v(x) \equiv S(u,P(v,x))$.

R:
$s(x) \rightarrow S^1(x)$ $\quad S^uS^v(x) \rightarrow S^{u+v}(x)$
$p(x) \rightarrow P^1(x)$ $\quad P^uP^v(x) \rightarrow P^{u+v}(x)$
$0+y \rightarrow y$
$S^1(x)+y \rightarrow x+S^1(y)$ $\quad P^1(x)+y \rightarrow x+P^1(y)$
$S^{1+u}(x)+y \rightarrow S^u(x)+S^1(y)$ $\quad P^{1+u}(x)+y \rightarrow P^u(x)+P^1(y)$
$S^uP^1(x)+y \rightarrow S^u(x)+P^1(y)$ $\quad S^uP^{1+v}(x)+y \rightarrow S^uP^v(x)+P^1(y)$
$P^uS^v(x) \rightarrow S^vP^u(x)$ $\quad S^uP^u(x) \rightarrow x$
$S^{u+v}P^u(x) \rightarrow S^v(x)$ $\quad S^uP^{u+v}(x) \rightarrow P^v(x)$
$S^{w+u}P^{w+v}(x) \rightarrow S^uP^v(x)$
$\underline{+}$ is AC

If we now want to prove $x+0 = x \in \mathrm{ITh}(E_1)$ then we add to R the rule $x+0 \rightarrow x$. Again the completion diverges. So we add as inductive hypothesis

R_0: $x+S^u(0) \rightarrow S^u(x)$ $\quad x+0 \rightarrow x$
$x+P^u(0) \rightarrow P^u(x)$ $\quad x+S^uP^v(0) \rightarrow S^uP^v(x)$

Now the completion stops with $R' = R \cup R_0$. By the technique described at the end of section 4 one proves easily that $F' = \{0,\underline{1},\underline{+},S,P\}$ is a set of constructors and that $M_S = \{S(\underline{n},0) \mid n \geq 1\}$ and $M_P = \{P(\underline{n},0) \mid n \geq 1\}$. So $t \equiv x+y$ is inductively reducible in R,A. This proves that every left-hand side in R_0 is inductively reducible by R,A and finally that $x+0 = x \in \mathrm{ITh}(E_1)$.

Example Binomial numbers

E_1: $0+y = y$ $\quad b(x,0) = s(0)$
$s(x)+y = s(x+y)$ $\quad b(0,s(y)) = 0$
$\quad b(s(x),s(y)) = b(x,s(y)) + b(x,y)$

We want to prove that $E' \subset \mathrm{ITh}(E_1)$ for

E': $b(x,x) = s(0)$ $\quad b(x,s(x)) = 0$

Let R_1 result from E_1 by orienting the rules from left to right. Then R_1 is convergent. Adding E', the completion procedure diverges and produces the terms $s^n(x)$. So we use the standard extension by NAT to rewrite $s^n(x)$ to $S(\underline{n},x)$ and generalize E' to E in the obvious way. This gives

E_0: $s(x) = S(\underline{1},x)$ $\quad S(u,S(v,x)) = S(u\underline{+}v,x)$
E: $b(x,x) = S(\underline{1},0)$ $\quad b(x,S(u,x)) = 0$

The completion process with input $E_1 \cup E_0$ gives R, adding E gives R_0.

$$
\begin{array}{lll}
R: & s(x) \to S^1(x) & S^uS^v(x) \to S^{u+v}(x) \\
 & 0+y \to y & S^1(x)+y \to S^1(x+y) \\
 & b(x,0) \to S^1(0) & S^{1+u}(x)+y \to S^1(S^u(x)+y) \\
 & b(0,S^1(y)) \to 0 & b(0,S^{1+u}(y)) \to 0 \\
 & b(S^1(x),S^1(y)) \to b(x,S^1(y))+b(x,y) & \\
 & b(S^{1+u}(x),S^1(y)) \to b(S^u(x),S^1(y))+b(S^u(x),y) & \\
 & b(S^1(x),S^{1+v}(y)) \to b(x,S^{1+v}(y))+b(x,S^v(y)) & \\
 & b(S^{1+u}(x),S^{1+v}(y)) \to b(S^u(x),S^{1+v}(y))+b(S^u(x),S^v(y)) & \\
R_0: & b(x,x) \to S^1(0) & \\
 & b(x,S^u(x)) \to 0 & b(S^v(x),S^{u+v}(x)) \to 0 \\
 & \underline{+} \text{ is AC} &
\end{array}
$$

The system R,A is convergent. Using the technique described at the end of section 4 one proves that $F' = \{0,\underline{1},\underline{+},S\}$ is a set of constructors for R,A. So $t \equiv b(x,y)$ is inductively reducible by R,A and $E \subset ITh(E_1)$ is proved.
Notice that the proof of $E' \subset ITh(E_1)$ by standard methods is not quite easy. Here one has only to find E in order to describe the infinite sequence of rules. No lemma is needed.

Appendix: Proof and generalization of Theorem 4.1

To prove Theorem 4.1 we need some preleminaries about unfailing completion. Let E be a set of equations over some signature sig and let > be a ground reduction ordering on sig. Let $E^> = \{l \to r \mid l = r \text{ or } r = l \text{ in } E, \ r \not> l\}$. Then (E,>) defines a Noetherian rewrite relation $\to_{E,>}$ by $s \to_{E,>} t$ iff $s \equiv s[l\sigma]$, $t \equiv s[r\sigma]$, $l \to r$ in $E^>$ and $l\sigma > r\sigma$. By $CP(E^>)$ we denote the set of critical pairs of $E^>$. The relation $\to_{E,>}$ is ground confluent if it is confluent on the sig-ground terms. A pair $\langle p,q\rangle$ of terms is $\to_{E,>}$-joinable, if for some t we have $p \overset{*}{\to}_{E,>} t \ {}_{E,>}\overset{*}{\leftarrow} q$. The following Lemma A.1 is easily proved. The Lemma A.2 is from [BDP] and [HRu]. Note that the output of the unfailing completion procedure with input (E,>) may be infinite, but it exists.

Lemma A.1

Let (E,>) be given where > is a ground reduction ordering. Then $\to_{E,>}$ is ground confluent iff all ground instances of all critical pairs in $CP(E^>)$ are $\to_{E,>}$-joinable. ▪

Lemma A.2 [BDP], [HRu]

Let (E,>) be given where > is a ground reduction ordering and let E' be the result of the unfailing completion procedure with input (E,>). Then $\to_{E',>}$ is ground confluent. ▪

Proof of Theorem 4.1: Assume t_1 and t_2 are $spec_1$-terms with $t_1 =_{E_2} t_2$, we have to prove $t_1 =_{E_1} t_2$. Let $\bar{t}_1$ denote the Skolemized form of t_1, i.e. $\bar{t}_1$ results from t_1 by replacing each variable x_j by a new constant c_j for $j = 1,\ldots,n$. We have $t_1 =_{E_i} t_2$ iff $\bar{t}_1 =_{E_i} \bar{t}_2$ for $i = 1,2$. So we have to prove that $\bar{t}_1 =_{E_2} \bar{t}_2$ implies $\bar{t}_1 =_{E_1} \bar{t}_2$.
It is easily to see that the ground reduction ordering > on $sig_2 = (\Sigma_2,F_2)$ can

be extended to a ground reduction ordering on sig' = $(\Sigma_2, F_2 \cup \{c_1,...,c_n\})$ satisfying $s \not> t$ if s,t are sig'-terms such that s is F_0-free but t is not. Now we start the unfailing completion procedure with input $(E_1,>)$ and denote the result by E. Then $=_E = =_{E_1}$, $\rightarrow_{E,>}$ is confluent on the sig'-ground terms by Lemma A.2, and $\rightarrow_{E,>}$ rewrites an F_0-free term into an F_0-free term. Note that $R \subset >$ and hence $\rightarrow := \rightarrow_{R \cup E,>} = \rightarrow_R \cup \rightarrow_{E,>}$ and $(R \cup E)^> = R \cup E^>$. We are going to show by Lemma A.1 that $\rightarrow$ is ground confluent: Note that $CP(R \cup E^>) = CP(R) \cup CP(E^>)$ by condition (3). Since both, $\rightarrow_R$ and $\rightarrow_{E,>}$ are ground confluent, all ground instances of all critical pairs in $CP((R \cup E)^>)$ are $\rightarrow$-joinable. So $\rightarrow$ is indeed ground confluent, i.e. for all sig'-ground terms s,t with $s =_{E_2} t$ we have $s \xrightarrow{*} \circ \xleftarrow{*} t$.

Note that $\bar{t}_1, \bar{t}_2$ are sig'-ground terms with $\bar{t}_1 =_{E_2} \bar{t}_2$, so $\bar{t}_1 \xrightarrow{*} \circ \xleftarrow{*} \bar{t}_2$. In this derivation no R-rule applies since $\bar{t}_1, \bar{t}_2$ are F_0-free, $\rightarrow_{E,>}$ rewrites F_0-free terms into F_0-free terms and every left-hand side of a rule in R is not F_0-free. This gives $\bar{t}_1 \xrightarrow{*}_{E,>} \circ \ {}_{E,>}\xleftarrow{*} \bar{t}_2$ and $\bar{t}_1 =_{E_1} \bar{t}_2$. This had to be shown. ∎

Theorem 4.1 can be extended to allow rewriting modulo A, where A is a set of AC-axioms. But to do so one needs A-compatible ground reduction orderings. Only recently Narendran and Rusinowitch proved in a paper contained in this volume that such orderings exist. A reduction ordering > is an <u>A-compatible ground reduction ordering</u> if (i) $s' =_A s > t = t'$ implies $s' > t'$ for all terms s,s',t,t' and (ii) $s > t$ or $t > s$ or $t =_A s$ for all ground terms s,t.

Techniques developed by Bachmair and Dershowitz in [BDe] for completion modulo A can be used to prove

Theorem A.3

Let $spec_1 = (\Sigma_1, F_1; E_1)$ and $spec_2 = (\Sigma_2, F_2; R \cup A \cup E_1)$ be given with $\Sigma_2 = \Sigma_1 \cup \Sigma_0$, $F_2 = F_1 \cup F_0$, $E_2 = R \cup A \cup E_1$. Assume

(0) A a set of AC-axioms for some operators in F_0.

(1) There is a reduction ordering > that can be extended to an A-compatible ground-reduction ordering on $Term(F_2 \cup K, V)$, where K is an infinite set of new constants.

(2) $\ell > r$ for all $l \rightarrow r$ in R.

(3) ℓ is F_1-free for all $\ell \rightarrow r$ in R.

(4) R,A is Church-Rosser modulo A.

Then $spec_2$ is a consistent enrichment of $spec_1$.

Acknowledgement: I want to thank Deepak Kapur and Leo Bachmair for discussions on Theorem 4.1 and Theorem A.3. They helped to fill a gap in a previous proof.

References

[AMa] Avenhaus, J., Madlener, K.: Term rewriting and equational reasoning, in R.B. Banerji: Formal techniques in Artificial Intelligence: A source book Elsevier, Amsterdam, 1989

[Bac] Bachmair, L.: Proof by consistency in equational theories, 3rd. LICS (1988), pp. 228-233

[BDe] Bachmair, L., Dershowitz, N.: Completion for rewriting modulo a congruence, TCS 67 (1989), pp. 173-201.

[BDP] Bachmair, L., Dershowitz, N., Plaisted, D.: Completion without Failure, Coll. on the Resolution of Equations in Algebraic Structures, Austin (1987), Academic Press, NY 1989

[Der] Dershowitz, N.: Termination, J. Symb. Comp. 3(1987), pp. 69-116

[DJo] Dershowitz, N., Jouannaud, J.P., Rewriting Systems, Handbook of Computer Science, Volume A, North-Holland, 1990.

[Her] Hermann, M.: Vademecum of divergent term rewriting systems, Research Report 88-R-082, Centre de Recherche en Informatique de Nancy, 1988

[HHu] Huet, G., Hullot, J.M.: Proofs by indcution in equational theories with constructors, 21th FOCS (1980), pp. 96-107.

[Hue] Huet, G.: Confluent reductions: Abstract properties and applications to term rewriting systems, J. ACM 27 (1980), pp. 797-821.

[HRu] Hsiang, J., Rusinowitch, M.: On word problems in equational theories, 14th ICALP (1987), LNCS 267, pp. 54-71

[JKi] Jouannaud, J.P., Kirchner, H.: Completion of a set of rules modulo a set of equations, SIAM J. Comp. 15 (1986), pp. 1155-1194

[JKo] Jouannaud, J.P., Kounalis, E.: Automatic proofs by induction in theories without constructors, Inf. and Comp. 82 (1989), pp. 1-33

[KBe] Knuth, D.E., Bendix, P.B.: Simple word problems in universal algebras, Computational problems in abstract algebra (ed.): J. Leech, Pergamon Press (1970), pp. 263-297.

[Kir] Kirchner, H.: Schematization of infinite sets of rewrite rules generated by divergent completion processes, TCS 67 (1989), pp. 303-332.

[KMu] Kapur, D., Musser, D.R.: Proof by Consistency, Proc. of an NSF Workshop on the Rewrite Rule Laboratory, Schenectady, G.E. R & D Center Report GEN84008 (1984), also in AI Journal 31 (1987), pp. 125-157.

[Mus] Musser, D.R.: On proving inductive properties of abstract data types, Proc. 7th Symp. on Principles of Prog. Languages, 1980, pp. 154-162.

[NRu] Narendran, P., Rusinowitch, M.: Any ground associative-commutative theory has a finite canonical system, this volume

[Lan] Lange, St.: Towards a set of inference rules for solving divergence in Knuth-Bendix completion, in K.P. Jantke (ed.): Analogical and inductive inference, LNCS 397 (1989), pp. 304-316

[PSt] Peterson, G.E., Stickel, M.E.: Complete sets of reduction for some equational theories, J. ACM 28 (1981), pp. 233-264

[TJa] Thomas, M., Jantke, K.P.: Inductive inference for solving divergence in Knuth-Bendix completion, in K.P. Jantke (ed.): Analogical and inductive inference, LNCS 397 (1989), pp. 288-303

Divergence Phenomena during Completion

Andrea Sattler-Klein
Universität Kaiserslautern, Fachbereich Informatik
Postfach 3049, 6750 Kaiserslautern (FRG)
email: sattler@informatik.uni-kl.de

ABSTRACT

We will show how any primitive recursive function may be encoded in a finite canonical string rewriting system. Using these encodings for every primitive recursive function f (and even for every recursively enumerable set $\mathbb{C}$) a finite string rewriting system $\mathfrak{R}$ and a noetherian ordering $>$ may be constructed such that completion of $\mathfrak{R}$ with respect to $>$ will generate a divergence sequence that encodes explicitly the input/output behaviour of f (or the set $\mathbb{C}$, respectively). Furthermore, we will show by an example that if completion of a set $\mathfrak{R}$ with respect to a noetherian ordering $>$ diverges, then there need not exist any rule that causes infinitely many other ones by overlapping.

I. INTRODUCTION

An essential problem of a Knuth-Bendix completion procedure for term rewriting systems is that in many cases it doesn't terminate, that means it generates infinitely many rules. A lot of divergence patterns have been investigated by Hermann and Privara in [9], by Hermann in [6],[7],[8] and by Mong and Purdom in [15]. Several methods to avoid divergence of completion have been proposed: One approach among others is to describe infinitely many rules in a finite way. For these purposes some new frameworks have been developed: Kirchner introduced in [10] meta-rules to describe sequences of rules that have some regularities in their structure (see also [11]). Gramlich proposed term schemes for the same purposes [5] and recently, Chen et al. introduced the notion of hyper-terms [3].

It is well known that if the word problem of a term rewriting system $\mathfrak{R}$ is decidable then there exists a finite and canonical system $\mathfrak{R}'$ that is a consistent enrichment of $\mathfrak{R}$ [14]. Thus, another approach is to use auxiliary operators in order to obtain a finite representation of an infinite set of rules. Embedding techniques have been studied in [1], [19] and in [12], [13] inference rules are described that may be used to enrich a set of rules automatically in order to rewrite infinite sequences of rules that have special structural similarities.

In both approaches the following questions are fundamental:

1) Which types of regularities may appear in the structure of rules that are generated during a nonterminating completion process ?

   In many cases these regularities are of a form that some unary function symbols are 'pumped'. Notice that all examples of this kind given in the above mentioned literature may be easily finitely described by using only variables with range $\mathbb{N}$ as exponents. So one important question is what kind of numerical functions may appear as exponents.

2) How can regularities be detected ?

   In order to apply the methods described above it is necessary to detect during a nonterminating completion process divergence sequences (infinite subsets of the canonical system to be generated) that consist of rules which have some regularities in their structure.

   It has been proposed several times (e.g. in [10], [19]) to consider the 'parents' of the rules for these purposes. Concerning this proposal the following conjecture has been made: If the completion algorithm diverges then there will be at least one rule that may be a 'parent' of infinitely many other rules.

In this paper we will focus our attention on these questions by studying infinite canonical string rewriting systems. (String rewriting systems may be viewed as a special kind of term rewriting systems (monadic term rewriting systems without constants)).
For the notions of string rewriting systems we refer for example to [2]. In the following we will use ε for the empty word and we will denote the length of a word $\mathfrak{w}$ by $|\mathfrak{w}|$.
The main tool will be an encoding of primitive recursive functions into finite canonical string rewriting systems. This is shown in lemma 1. Using this construction divergence sequences which represent explicitly the input/output behaviour of any (non-nullary) primitive recursive function can be constructed. Moreover, even any infinite recursively enumerable set can be represented by a divergence sequence. This shows once more that the restriction to string rewriting systems is not a real one, since most phenomena occur already with them. We will give a lot of concrete divergence examples in order to give an impression how difficult it will be to detect and to handle divergence sequences.
Concerning the second question we will show how regularities in the structure of the rules may be reflected in their history. In order to illustrate the relationship of the rules with respect to their origin we will use a kind of 'family-tree'. Some typical 'family-trees' will be presented and we will give an explicit counterexample to the above conjecture that if the completion algorithm diverges then there will be at least one rule that may be a 'parent' of infinitely many other rules.
These results show that any method which attempts to encode divergence sequences has to be powerfull enough to describe any recursively enumerable set (i.e. has to have full computational power) and that in order to detect divergence sequences the history of the produced rules is important but it may be also very complicated and difficult to follow.

Whether a completion process for a fixed input terminates or not may depend on the strategy used. That is, the process may generate an infinite complete system even if the canonical (i.e. complete and interreduced) system (which is uniquely determined [4]) is finite. In order to guarantee that the generated system is canonical we will assume that our completion procedure uses interreduction and generates new rules by normalizing the critical pairs. Despite of this, our examples will be of a form that no interreduction is possible.

Throughout this paper we will use a well-founded syllable ordering to prove the termination of the constructed string rewriting systems. This syllable ordering is defined in the following way:
Let $\mathfrak{u}, \mathfrak{v}$ be two words over an alphabet Σ, $>_{lex}$ be a total precedence on Σ and $\tau : \Sigma \longrightarrow \{l, r\}$ be a so-called status function that should be total, too. Then the induced syllable ordering $>$ is defined by:

$\mathfrak{u} > \mathfrak{v}$ iff $|\mathfrak{u}|_{max(\mathfrak{u},\mathfrak{v})} > |\mathfrak{v}|_{max(\mathfrak{u},\mathfrak{v})}$

or ($|\mathfrak{u}|_{max(\mathfrak{u},\mathfrak{v})} = |\mathfrak{v}|_{max(\mathfrak{u},\mathfrak{v})} = n$, $max(\mathfrak{u},\mathfrak{v}) = a$, $\mathfrak{u} = \mathfrak{u}_1 a \ldots \mathfrak{u}_n a \mathfrak{u}_{n+1}$, $\mathfrak{v} = \mathfrak{v}_1 a \ldots \mathfrak{v}_n a \mathfrak{v}_{n+1}$
$\tau(a) = l$, $\exists i \in \{1, \ldots, n+1\} : (\mathfrak{u}_j = \mathfrak{v}_j$ for all $j < i$ $(j \in \{1, \ldots, n+1\})$ and $\mathfrak{u}_i > \mathfrak{v}_i)$)

or ($|\mathfrak{u}|_{max(\mathfrak{u},\mathfrak{v})} = |\mathfrak{v}|_{max(\mathfrak{u},\mathfrak{v})} = n$, $max(\mathfrak{u},\mathfrak{v}) = a$, $\mathfrak{u} = \mathfrak{u}_1 a \ldots \mathfrak{u}_n a \mathfrak{u}_{n+1}$, $\mathfrak{v} = \mathfrak{v}_1 a \ldots \mathfrak{v}_n a \mathfrak{v}_{n+1}$
$\tau(a) = r$, $\exists i \in \{1, \ldots, n+1\} : (\mathfrak{u}_j = \mathfrak{v}_j$ for all $j > i$ $(j \in \{1, \ldots, n+1\})$ and $\mathfrak{u}_i > \mathfrak{v}_i)$)

where $max(\mathfrak{u},\mathfrak{v})$ denotes the greatest symbol (with respect to $>_{lex}$) that occurs in $\mathfrak{u}$ or $\mathfrak{v}$ and $|\mathfrak{u}|_a$ denotes the number of occurences of the symbol 'a' in $\mathfrak{u}$.
This syllable ordering is a generalization of the iterated syllable ordering that has been studied for example in [18]. We have added a status function to the usual definition. This function determines in which order (from left to right or from right to left) the 'syllables' are compared. According to the results achieved in [18] the following holds: If $\tau(x) = r$ for all symbols x then $\mathfrak{u} > \mathfrak{v}$ if and only if the corresponding monadic terms are ordered in the same way by the RPO. (In our examples we will sometimes define a partial precedence and we will assume that it is extended to a total one in some way).

II. REGULARITIES IN THE STRUCTURE OF THE RULES

The starting point of this work has been the following observation: If completion of a string rewriting system diverges then in many cases there are obvious regularities in the structure of the rules that are generated. That is, these rules may be divided into a finite number of sequences in such a way that the members of each sequence can be factored in a uniform manner such that only some exponents differ. Thus, an important question is which kind of exponents may appear in such situations and what kind of relationship between them may exist.

In this paper we will show that these exponents may represent any primitive recursive function. In order to achieve this, we first show how any primitive recursive function can be simulated by a finite and canonical string rewriting system. Doing this, the argument values as well as the function values will be represented in unary.

LEMMA 1 :

For every primitive recursive function $f : \mathbb{N}^n \longrightarrow \mathbb{N}^m$, there exists a finite canonical string rewriting system $\mathfrak{R}$ over an alphabet Σ with $\Sigma_0 = \{a,b,g,c,v,e\} \subseteq \Sigma$ and a symbol $f \in \Sigma$ such that the following holds:

i) 1) $fb^{p_1}a \dots b^{p_n}age \xRightarrow{*}_{\mathfrak{R}} b^{p_1}a \dots b^{p_n}agc^{q_1}v \dots c^{q_m}ve$ for $p_i \in \mathbb{N}$ $(i=1..n)$ and $f(p_1,\dots,p_n)=(q_1,\dots,q_m)$

2) $b^{p_1}a \dots b^{p_n}agc^{q_1}v \dots c^{q_m}ve$ is $\mathfrak{R}$-irreducible for $p_i \in \mathbb{N}$ $(i=1..n)$ and $f(p_1,\dots,p_n)=(q_1,\dots,q_m)$

ii) There exists a precedence $>_{lex} \subseteq \Sigma \times \Sigma$ satisfying $a >_{lex} b >_{lex} g >_{lex} c >_{lex} v >_{lex} e$ and a status function $\tau : \Sigma \longrightarrow \{l,r\}$ satisfying $\tau(x)=r$ for $x \in \Sigma_0$ such that for the induced syllable ordering $> \subseteq \Sigma^* \times \Sigma^*$ the following holds: $\xRightarrow{*}_{\mathfrak{R}} \subseteq >$

iii) All $(l,r) \in \mathfrak{R}$ satisfy the following conditions:

$|l| = 2$ and the first letter of l is not in Σ_0

PROOF:

The proof is done by induction on the formation of f.

BASIS: Let f be an initial primitive recursive function, i.e.: f is the zero function, the successor function or the projection function, respectively.

CASE 1: $f : \mathbb{N}^0 \longrightarrow \mathbb{N}$ is defined by: $f = 0$. Define:

$\Sigma := \Sigma_0 \cup \{f\}$

$\mathfrak{R} := \{ fg \longrightarrow gv \}$

$>_{lex}$ to be the precedence defined by $f >_{lex} a >_{lex} b >_{lex} g >_{lex} c >_{lex} v >_{lex} e$

τ to be the status function defined by $\tau(x) = r$ for all $x \in \Sigma$

CASE 2: $f : \mathbb{N} \longrightarrow \mathbb{N}$ is defined by: $f(x) = x+1$ for all $x \in \mathbb{N}$. Define:

$\Sigma := \Sigma_0 \cup \{f,d\}$

$\mathfrak{R} := \{ fb \longrightarrow bdf ,\ fa \longrightarrow af ,\ fg \longrightarrow gcv ,\ db \longrightarrow bd ,\ da \longrightarrow ad ,\ dg \longrightarrow gc \}$

$>_{lex}$ to be the precedence defined by $f >_{lex} d >_{lex} a >_{lex} b >_{lex} g >_{lex} c >_{lex} v >_{lex} e$

τ to be the status function defined by $\tau(x) = r$ for all $x \in \Sigma$

CASE 3: $f : \mathbb{N}^n \longrightarrow \mathbb{N}$ $(n \geq 1)$ is defined by: $f(x_1,x_2,\dots,x_n) = x_k$ for all $x_1,\dots,x_n \in \mathbb{N}$ where $k \in \{1,\dots,n\}$ is arbitrary but fixed. Define:

$\Sigma := \Sigma_0 \cup \{f_0, f_1, f_2, \dots, f_k, d\}$

$\mathfrak{R} := \{fb \longrightarrow (f_k b)\!\downarrow\} \cup \{f_j b \longrightarrow bf_j \,/\, j \in (\{0\} \cup \{2,\dots,k\})\} \cup \{f_j a \longrightarrow af_{j-1} \,/\, j \in \{1,\dots,k\}\}$
$\cup \{fa \longrightarrow af_{k-1} ,\ f_1 b \longrightarrow bdf_1 ,\ f_0 a \longrightarrow af_0 ,\ f_0 g \longrightarrow gv ,\ db \longrightarrow bd ,\ da \longrightarrow ad ,\ dg \longrightarrow gc\}$

where $(f_k b)\!\downarrow$ denotes the normal form of the string '$f_k b$' with respect to $\mathfrak{R}$

$>_{lex}$ to be the precedence defined by $f >_{lex} f_k >_{lex} f_{k-1} >_{lex} \dots >_{lex} f_0 >_{lex} d >_{lex} a >_{lex} b >_{lex} g >_{lex} c >_{lex} v >_{lex} e$

τ to be the status function defined by $\tau(x) = r$ for all $x \in \Sigma$

Obviously, the sets $\mathfrak{R}$ defined in these initial cases are confluent and satisfy conditions i), ii) and iii).

INDUCTION STEP: Let f be a composed function that is, it is constructed from two primitive recursive functions f_1, f_2 by means of the operation combination ($f = \langle f_1, f_2 \rangle$), composition ($f = f_1 \circ f_2$) or primitive recursion ($f = R(f_1, f_2)$), respectively. By induction hypothesis, for f_1 (f_2) there exists a finite canonical string rewriting system $\mathfrak{R}_1$ ($\mathfrak{R}_2$) over an alphabet Σ_1 (Σ_2), a symbol $f_1 \in \Sigma_1$ ($f_2 \in \Sigma_2$), a precedence $>_{lex,1}$ ($>_{lex,2}$) and a status function τ_1 (τ_2) such that the above conditions are satisfied. Let $>_1$ ($>_2$) be the syllable ordering induced by $>_{lex,1}$ and τ_1 ($>_{lex,2}$ and τ_2).
We assume w.l.o.g. that $\Sigma_1 \cap \Sigma_2 = \Sigma_0$.
To f a string rewriting system $\mathfrak{R}$ of the required form may be constructed by adding some additional rules to the union of the sets $\mathfrak{R}_1$ and $\mathfrak{R}_2$. This union obviously satisfies condition ii) too, that is it is noetherian. Since $\mathfrak{R}_1$ and $\mathfrak{R}_2$ satisfy condition iii), there are no superpositions between rules of $\mathfrak{R}_1$ and those of $\mathfrak{R}_2$. Thus, the set $\mathfrak{R}_1 \cup \mathfrak{R}_2$ is also confluent.
Here we will consider only the case of the primitive recursion. For the other cases we refer to [17].
CASE 3: $f : \mathbb{N}^{n+1} \longrightarrow \mathbb{N}^m$ is defined by: $f = R(f_1, f_2)$ i.e.

$f(0, x_1, x_2, \ldots, x_n) = f_1(x_1, x_2, \ldots, x_n)$ for all $x_1, \ldots, x_n \in \mathbb{N}$

$f(x_0+1, x_1, x_2, \ldots, x_n) = f_2(f(x_0, x_1, \ldots, x_n), x_0, x_1, \ldots, x_n)$ for all $x_0, \ldots, x_n \in \mathbb{N}$

where $f_1 : \mathbb{N}^n \longrightarrow \mathbb{N}^m$, $f_2 : \mathbb{N}^{n+m+1} \longrightarrow \mathbb{N}^m$. Define:

$\Sigma := \Sigma_1 \cup \Sigma_2 \cup \{f, y, z, A, B, C, V, r, x_1, x_2, \ldots, x_m\}$

(we assume w.l.o.g. that $f, y, z, A, B, C, V, r, x_1, x_2, \ldots, x_m \notin (\Sigma_1 \cup \Sigma_2)$)

$\mathfrak{R} := \mathfrak{R}_1 \cup \mathfrak{R}_2 \cup \mathfrak{R}_a$ where

$\mathfrak{R}_a = \{ fa \longrightarrow af_1,\ fb \longrightarrow x_m f_2 yzf,\ zb \longrightarrow Bz,\ za \longrightarrow Az,\ zg \longrightarrow z,\ zc \longrightarrow Cz,$
$zv \longrightarrow Vz,\ ze \longrightarrow e,\ AV \longrightarrow VA,\ AC \longrightarrow CA,\ Ae \longrightarrow rage,\ Ar \longrightarrow ra,$
$BV \longrightarrow VB,\ BC \longrightarrow CB,\ Br \longrightarrow rb,\ yC \longrightarrow by,\ yV \longrightarrow ay,\ yr \longrightarrow \varepsilon \}$
$\cup \{ x_i b \longrightarrow x_i \ / \ i \in \{1, \ldots, m\} \} \cup \{ x_i a \longrightarrow x_{i-1} \ / \ i \in \{2, \ldots, m\} \} \cup \{ x_1 a \longrightarrow b \}$

$>_{lex} := >_{lex,1} \cup >_{lex,2} \cup \{(f,s) / s \in \{f_1, f_2, a, b, x_m, y, z\}\} \cup \{(z,s) / s \in (\Sigma_0 \cup \{A,B,C,V\})\}$
$\cup \{(A,s) / s \in \{V,C,e,r,a,g\}\} \cup \{(B,s) / s \in \{V,C,b,r\}\} \cup \{(y,s) / s \in \{C,b,V,a,r\}\}$
$\cup \{(x_i, x_{i-1}) / i \in \{2, \ldots, m\}\} \cup \{(x_i, a) / i \in \{1, \ldots, m\}\} \cup \{(x_i, b) / i \in \{1, \ldots, m\}\}$

$\tau := \tau_1 \cup \tau_2 \cup \{(s, r) \ / \ s \in \{f, y, z, A, B, C, V, r, x_1, x_2, \ldots, x_m\}\}$

$>$ to be the syllable ordering induced by $>_{lex}$ and by τ

Using the induction hypothesis it can be proved that $>_{lex}$ and τ are well-defined. Moreover, it can be checked easily that $\overset{*}{\Longrightarrow}_{\mathfrak{R}_a} \subseteq >$. Since, in addition, $>_1$ and $>_2$ are contained in $>$ we have: $\overset{*}{\Longrightarrow}_{\mathfrak{R}} \subseteq >$. Hence, the set $\mathfrak{R}$ satisfies condition ii).
Obviously, the set $\mathfrak{R}_a$ is confluent and satisfies condition iii). Moreover, every left-hand side of these rules contains only 'new' symbols (i.e. symbols that do not occur in $\Sigma_1 \cup \Sigma_2$) or symbols of Σ_0. This fact together with the induction hypothesis implies that there is also no superposition between rules of $\mathfrak{R}_a$ and those of $\mathfrak{R}_1 \cup \mathfrak{R}_2$. Thus, $\mathfrak{R}$ is confluent too.
It remains to prove that $\mathfrak{R}$ really simulates f. Here we confine ourselves to give some intuitive explanations of this fact:
The computation of the primitive recursion is simulated as follows: If the first argument is the empty word i.e. it represents zero, then f_1 is simulated by 'f_1'. Otherwise $f(x_0, x_1, \ldots, x_n)$ is computed. In order to bring this 'old' function value in position the symbols 'a', 'b', 'c', 'v' are renamed to 'A', 'B', 'C', 'V' temporarily. After simulating f_2 the appropriate arguments are restored by 'x_i'. ∎

The main point of this lemma is given by condition i), which states that every primitive recursive function may be encoded into a finite canonical string rewriting system. But, the conditions ii) and iii) have not just been added for proof-theoretical reasons only: In the following lemma 1 will be used to construct certain divergence sequences. This will be done by adding some rules to the string rewriting systems considered in this lemma. In doing so, condition ii) and iii) will be used to ensure that the resulting string rewriting system is noetherian and that there are no superpositions between the rules added and the 'old' ones.

In the following Σ_0 denotes the set $\{a,b,g,c,v,e\}$.

The first theorem shows that to any primitive recursive function a finite string rewriting system $\mathfrak{R}$ and a noetherian ordering $>$ may be constructed such that the corresponding canonical system encodes explicitly the input/output behaviour of this function.

THEOREM 1 :

For every primitive recursive function $f : \mathbb{N}^n \longrightarrow \mathbb{N}^m$, there exists a finite string rewriting system $\mathfrak{R}$ over an alphabet Σ with $(\Sigma_0 \cup \{u,o\}) \subseteq \Sigma$ and a noetherian ordering $>$ such that the corresponding canonical system $\mathfrak{R}^*$ is of the following form :

$$\mathfrak{R}^* = \mathfrak{R} \cup \{ ub^{p_1}a \ldots b^{p_n}agc^{q_1}v \ldots c^{q_m}ve \longrightarrow o \ / \ p_i \in \mathbb{N} \ (i=1..n) \text{ and } f(p_1,\ldots,p_n) = (q_1,\ldots,q_m) \}$$

PROOF:

Let $f : \mathbb{N}^n \longrightarrow \mathbb{N}^m$ be any primitive recursive function. If $n=0$, i.e. $f = (k_1,\ldots,k_m)$ for some constants $k_1,\ldots,k_m \in \mathbb{N}$, the set $\mathfrak{R} = \{ugc^{k_1}v \ldots c^{k_m}ve \longrightarrow o\}$ is canonical and of the required form. Otherwise, if $n > 0$, then let Σ_1 , $\mathfrak{R}_1$, $>_{lex,1}$, τ_1 be those sets that belong to f according to lemma 1 and let f be the corresponding symbol. We assume w.l.o.g. that $u,o,l,w_1,w_2,\ldots,w_n \notin \Sigma_1$. Define:

$$\Sigma := \Sigma_1 \cup \{u,o,l\} \cup \{ w_i \ / \ i \in \{1,\ldots,n\}\}$$

$\mathfrak{R} := \mathfrak{R}_1 \cup \mathfrak{R}_a$ where

$$\begin{aligned} \mathfrak{R}_a = \; & \{ la \longrightarrow al, lb \longrightarrow bl, lg \longrightarrow gl, lc \longrightarrow l, lv \longrightarrow l, le \longrightarrow e, w_1a \longrightarrow ba \} \\ & \cup \{ w_iu \longrightarrow ufw_il \ / \ i \in \{1,\ldots,n\}\} \cup \{ w_ib \longrightarrow bw_i \ / \ i \in \{1,\ldots,n\}\} \\ & \cup \{ w_ia \longrightarrow aw_{i-1} \ / \ i \in \{2,\ldots,n\}\} \cup \{ w_io \longrightarrow o \ / \ i \in \{1,\ldots,n\}\} \\ & \cup \{ ua^ngc^{q_1}v \ldots c^{q_m}ve \longrightarrow o \ / \ f(0,0,\ldots,0) = (q_1,\ldots,q_m)\} \end{aligned}$$

$$\begin{aligned} >_{lex} := \; & >_{lex,1} \cup \{ (u,x) \ / \ x \in (\Sigma_0 \cup \{o,l\})\} \cup \{ (l,x) \ / \ x \in \Sigma_0 \} \\ & \cup \{ (w_i,x) \ / \ x \in \{a,b,u,l,f,o\}, i \in \{1,\ldots,n\}\} \cup \{ (w_i,w_{i-1}) \ / \ i \in \{2,\ldots,n\}\} \end{aligned}$$

$$\tau := \tau_1 \cup \{ (x,r) \ / \ x \in \{u,o,l\}\} \cup \{ (w_i,r) \ / \ i \in \{1,\ldots,n\}\}$$

$>$ to be the syllable ordering induced by $>_{lex}$ and τ

$$\hat{\mathfrak{R}} := \mathfrak{R} \cup \{ ub^{p_1}a \ldots b^{p_n}agc^{q_1}v \ldots c^{q_m}ve \longrightarrow o \ / \ p_i \in \mathbb{N} \ (i=1..n) \text{ and } f(p_1,\ldots,p_n) = (q_1,\ldots,q_m)\}$$

Using conditions ii) and iii) of lemma 1 it can be easily proved that $\hat{\mathfrak{R}}$ is noetherian and confluent. Furthermore, by induction on the vector $(p_1,\ldots,p_n)$ it can be shown that

$$ub^{p_1}a \ldots b^{p_n}agc^{q_1}v \ldots c^{q_m}ve \overset{*}{\Longleftrightarrow}_{\mathfrak{R}} o$$

for $p_i \in \mathbb{N}$ $(i=1..n)$ and for $f(p_1,\ldots,p_n) = (q_1,\ldots,q_m)$. Thus, $\hat{\mathfrak{R}}$ is equivalent to $\mathfrak{R}$.

Here, we want to give only an informal description of how the set of additional rules is generated by completing $\mathfrak{R}$: $ub^{p_1}a \ldots b^{p_n}agc^{q_1}v \ldots c^{q_m}ve \longrightarrow o$ overlaps with $w_iu \longrightarrow ufw_il$ for all $i \in \{1,\ldots,n\}$.

This will result in the critical pair $(ufw_ilb^{p_1}a \ldots b^{p_n}agc^{q_1}v \ldots c^{q_m}ve, w_io)$.

While 'w_io' reduces to 'o' , the reduction of '$ufw_ilb^{p_1}a \ldots b^{p_n}agc^{q_1}v \ldots c^{q_m}ve$' leads to '$ub^{p_1}a \ldots b^{p_{i-1}}ab^{p_i+1}ab^{p_{i+1}}a \ldots b^{p_n}agc^{r_1}v \ldots c^{r_m}ve$' , where $f(p_1,\ldots,p_{i-1},p_i+1,p_{i+1},\ldots,p_n) = (r_1,\ldots,r_m)$.

In the latter case 'l' serves for 'deleting the old function value', 'w_i' 'increases the ith argument value by one' and finally the 'f' 'computes the function value of f for the new argument vector'. ∎

The construction method of this proof can be used repeatedly in order to encode a finite number of primitive recursive functions $f_i : \mathbb{N}^{n(i)} \longrightarrow \mathbb{N}^{m(i)}$ $(i=1..k)$. The easiest way to do so is to use for each f_i new symbols 'u_i', 'l_i', '$w_{i,1}$', ..., '$w_{i,n(i)}$' instead of 'u', 'l', 'w_1', ..., 'w_n' and to extend the set of rules, the precedence, and the status function appropriately. A lot of rules may be saved if the same 'l' and the same 'w_i' 's ($w_1,\ldots,w_j$ where j is the maximum of $n(1),\ldots,n(k)$) are used to encode all functions f_i. However, the corresponding canonical system $\mathfrak{R}^*$ will be the union of the set of initial rules and the set

$$\bigcup_{j=1} \{ u_jb^{p_1}a \ldots b^{p_{n(j)}}agc^{q_1}v \ldots c^{q_{m(j)}}ve \longrightarrow o \ / \ p_i \in \mathbb{N} \ (i=1..n(j)) \text{ and } f_j(p_1,\ldots,p_{n(j)}) = (q_1,\ldots,q_{m(j)})\}.$$

But, even the same 'u' can be used to represent the input/output behaviour of several functions [17]. In that case the set of corresponding rules does not represent a real encoding since, in general, there exist infinitely many sets of functions which may be represented by it. Thus, it will be very difficult to infer a finite description of these rewrite rules, if the only known information is this infinite set of rules.

Not only primitive recursive functions may be connected with divergence sequences, but also primitive recursive sets may play an essential role as well: The part of the domain which is explicitly represented can be any primitive recursive set [17].

But, the sets involved may be more complicated: The part of the domain which is explicitly represented can be non-recursive.

THEOREM 2 :

For every recursively enumerable set $\mathbb{C} \subseteq \mathbb{N}^n$ ($\mathbb{C} \neq \emptyset$, $n \geq 1$) and for all primitive recursive functions $f_1 : \mathbb{N}^n \longrightarrow \mathbb{N}^{m_1}$, $f_2 : \mathbb{N}^n \longrightarrow \mathbb{N}^{m_2}$, there exists a finite string rewriting system $\mathfrak{R}$ over an alphabet Σ with $(\Sigma_0 \cup \{u,o\}) \subseteq \Sigma$ and a noetherian ordering $>$ such that the corresponding canonical system $\mathfrak{R}^*$ is of the following form :

$$\mathfrak{R}^* = \mathfrak{R} \cup \{ ub^{p_1}a\ldots b^{p_n}agc^{q_1}v\ldots c^{q_{m_1}}ve \longrightarrow o \ / \ p_i \in \mathbb{N}\,(i=1..n) \text{ and } f_1(p_1,\ldots,p_n)=(q_1,\ldots,q_{m_1})\}$$
$$\cup \{ ub^{p_1}a\ldots b^{p_n}agc^{q_1}v\ldots c^{q_{m_2}}ve \longrightarrow o \ / \ (p_1,\ldots,p_n) \in \mathbb{C} \text{ and } f_2(p_1,\ldots,p_n)=(q_1,\ldots,q_{m_2})\}$$

PROOF:

This proof is based on the fact that a non-empty set $A \subseteq \mathbb{N}^k$ is recursively enumerable if and only if it is the range of a primitive recursive function $f: \mathbb{N} \longrightarrow \mathbb{N}^k$.

Since $\mathbb{C}$ is recursively enumerable there exists a primitive recursive function $f : \mathbb{N} \longrightarrow \mathbb{N}^n$ which enumerates $\mathbb{C}$. This implies that also an n-ary function f_3 with the same properties exists.

Let Σ_i, $\mathfrak{R}_i$, $>_{lex,i}$, τ_i ($i \in \{1,2,3\}$) be the sets and f_i ($i \in \{1,2,3\}$) be the symbol that correspond to f_i ($i \in \{1,2,3\}$) according to lemma 1. We assume w.l.o.g. that $\Sigma_i \cap \Sigma_j = \Sigma_0$ for $i,j \in \{1,2,3\}$ with $i \neq j$ and that $(\{z,w,u,o,l\} \cup \{w_i / i \in \{1,\ldots,n\}\}) \cap (\Sigma_1 \cup \Sigma_2 \cup \Sigma_3) = \emptyset$. Furthermore, let $\mathfrak{R}_{s1}$ be the set $\mathfrak{R}_s$ of theorem 1, where 'f', 'f' and 'm' are substituted by 'f_1', 'f_1' and 'm_1', respectively and let $>_{lex,s1}$ be the corresponding precedence and τ_{s1} be the corresponding status function. Define:

$$\Sigma := \Sigma_1 \cup \Sigma_2 \cup \Sigma_3 \cup \{z,w,u,o,l\} \cup \{w_i / i \in \{1,\ldots,n\}\}$$
$$\mathfrak{R} := \mathfrak{R}_1 \cup \mathfrak{R}_2 \cup \mathfrak{R}_3 \cup \mathfrak{R}_{s1} \cup \mathfrak{R}_{s2} \quad \text{where}$$
$$\mathfrak{R}_{s2} = \{ wu \longrightarrow uf_2zf_3l,\ wo \longrightarrow o \}$$
$$\cup \{ za \longrightarrow z,\ zb \longrightarrow z,\ zg \longrightarrow z,\ zc \longrightarrow bz,\ zv \longrightarrow az,\ ze \longrightarrow ge \}$$
$$>_{lex} := >_{lex,1} \cup >_{lex,2} \cup >_{lex,3} \cup >_{lex,s1} \cup \{(z,x) / x \in \Sigma_0\} \cup \{(w,x) / x \in \{u,f_2,z,f_3,l,o\}\}$$
$$\tau := \tau_1 \cup \tau_2 \cup \tau_3 \cup \tau_{s1} \cup \{(x,r) / x \in \{w,z\}\}$$

$>$ to be the syllable ordering induced by $>_{lex}$ and τ

Then $\mathfrak{R}^* = \mathfrak{R} \cup \mathfrak{R}_{s1} \cup \mathfrak{R}_{s2}$ where

$$\mathfrak{R}_{s1} = \{ ub^{p_1}a\ldots b^{p_n}agc^{q_1}v\ldots c^{q_{m_1}}ve \longrightarrow o \ / \ p_i \in \mathbb{N}\ (i=1..n) \text{ and } f_1(p_1,\ldots,p_n)=(q_1,\ldots,q_{m_1})\}$$
$$\mathfrak{R}_{s2} = \{ ub^{p_1}a\ldots b^{p_n}agc^{q_1}v\ldots c^{q_{m_2}}ve \longrightarrow o \ / \ (p_1,\ldots,p_n) \in \mathbb{C} \text{ and } f_2(p_1,\ldots,p_n)=(q_1,\ldots,q_{m_2})\}$$

is the corresponding canonical system.

The sequence $\mathfrak{R}_{s1}$ may be generated according to the proof of theorem 1 by the rules of $\mathfrak{R}_{a1}$. It contributes to construct the sequence $\mathfrak{R}_{s2}$ by delivering all n-tuples of $\mathbb{N}^n$ as arguments. The essential overlaps are those between the rule $wu \longrightarrow uf_2zf_3l$ and those of $\mathfrak{R}_{s1}$. The corresponding critical pairs are of the form $(uf_2zf_3lb^{p_1}a\ldots b^{p_n}agc^{q_1}v\ldots c^{q_{m_1}}ve, wo)$ and can be normalized to a pair of the form $(ub^{r_1}a\ldots b^{r_n}agc^{t_1}v\ldots c^{t_{m_2}}ve, o)$ where $f_3(p_1,\ldots,p_n)=(r_1,\ldots,r_n)$ $(\in \mathbb{C})$ and $f_2(r_1,\ldots,r_n)=(t_1,\ldots,t_{m_2})$. Here, 'l' is used to 'delete the function value' and 'z' is used in order to 'delete the argument vector and to rewrite the function value to an argument value'.

Notice, that there are also overlaps between the rules of $\mathfrak{R}_{s2}$ and the rules $w_iu \longrightarrow uf_iw_il$ ($i \in \{1,\ldots,n\}$) of $\mathfrak{R}_{a1}$. But, these overlaps may cause only rules of $\mathfrak{R}_{s1}$. ∎

In this proof the sequence $\mathfrak{R}_{s1}$ is used in order to construct the sequence $\mathfrak{R}_{s2}$ by 'supplying f_3 with arguments'. Another possibility to construct a sequence of the form of $\mathfrak{R}_{s2}$ is to do it by itself in a similar way to the one mentioned in the proof of theorem 1 : For example, if f_3 is injective and its inverse function can be extended to a primitive recursive function g, then this function g could be used to compute the 'next' member of $\mathbb{C}$ according to its enumeration by f_0. Among others, another method is to use a finite number of rules $r_1, r_2, \dots, r_k$ of the sequence to be generated as a basis and to generate the rest of the sequence inductively using a finite number of primitive recursive functions $f_1, f_2, \dots, f_l$. This means, that the part of the domain which is explicitly represented will be the closure of the argument values encoded in the rules $r_1, r_2, \dots, r_k$ under $f_1, f_2, \dots, f_l$.
In the previous examples the rules of a divergence sequence have always a fixed right-hand side. The next theorem shows that the right-hand sides of a divergence sequence may also represent the function values of a primitive recursive function f where the corresponding argument values are encoded in the left-hand sides.

THEOREM 3 :
For every primitive recursive function $f: \mathbb{N}^n \longrightarrow \mathbb{N}^m$ $(n \geq 1)$, there exists a finite string rewriting system $\mathfrak{R}$ over an alphabet Σ with $(\Sigma_0 \cup \{u_1, u_2\}) \subseteq \Sigma$ and a noetherian ordering $>$ such that the corresponding canonical system $\mathfrak{R}^*$ is of the following form :

$$\mathfrak{R}^* = \mathfrak{R} \cup \{ u_1 b^{p_1} a \dots b^{p_n} age \longrightarrow b^{p_1} a \dots b^{p_n} age \;/\; p_i \in \mathbb{N}\ (i=1..n)\}$$
$$\cup \{ u_2 b^{p_1} a \dots b^{p_n} age \longrightarrow c^{q_1} v \dots c^{q_m} ve \;/\; p_i \in \mathbb{N}\ (i=1..n) \text{ and } f(p_1,\dots,p_n)=(q_1,\dots,q_m)\}$$

PROOF:
Let Σ_1, $\mathfrak{R}_1$, $>_{lex,1}$, τ_1 those sets that belong to f according to lemma 1 and let f be the corresponding symbol. We assume w.l.o.g. that $u_1, u_2, l, w_1, w_2, \dots, w_n \notin \Sigma_1$. Define:

$$\Sigma := \Sigma_1 \cup \{u_1, u_2, l\} \cup \{ w_i \;/\; i \in \{1,\dots,n\} \}$$
$$\mathfrak{R} := \mathfrak{R}_1 \cup \{ w_i u_1 \longrightarrow u_1 w_i \;/\; i \in \{1,\dots,n\}\} \cup \{ w_i b \longrightarrow b w_i \;/\; i \in \{1,\dots,n\}\}$$
$$\cup \{ w_i a \longrightarrow a w_{i-1} \;/\; i \in \{2,\dots,n\}\} \cup \{ w_1 a \longrightarrow ba \} \cup \{ u_1 a^n ge \longrightarrow a^n ge \}$$
$$\cup \{ u_2 u_1 \longrightarrow lf,\ lb \longrightarrow l,\ la \longrightarrow l,\ lg \longrightarrow \varepsilon \}$$
$$>_{lex} := >_{lex,1} \cup \{ (u_1,a), (u_1,g), (u_1,e), (u_2,u_1), (u_2,l), (u_2,f), (l,a), (l,b), (l,g)\}$$
$$\cup \{ (w_i,x) \;/\; x \in \{a,b,u_1\}, i \in \{1,\dots,n\}\} \cup \{ (w_i,w_{i-1}) \;/\; i \in \{2,\dots,n\}\}$$
$$\tau := \tau_1 \cup \{ (x,r) \;/\; x \in \{u_1,u_2,l\}\} \cup \{ (w_i,r) \;/\; i \in \{1,\dots,n\} \}$$

$>$ to be the syllable ordering induced by $>_{lex}$ and τ

Then $\mathfrak{R}^* = \mathfrak{R} \cup \mathfrak{R}_{s1} \cup \mathfrak{R}_{s2}$ where

$$\mathfrak{R}_{s1} = \{ u_1 b^{p_1} a \dots b^{p_n} age \longrightarrow b^{p_1} a \dots b^{p_n} age \;/\; p_i \in \mathbb{N}\ (i=1..n)\}$$
$$\mathfrak{R}_{s2} = \{ u_2 b^{p_1} a \dots b^{p_n} age \longrightarrow c^{q_1} v \dots c^{q_m} ve \;/\; p_i \in \mathbb{N}\ (i=1..n) \text{ and } f(p_1,\dots,p_n) = (q_1,\dots,q_m)\}$$

is the corresponding canonical system.
Completion of $\mathfrak{R}$ with respect to $>$ will generate the addional rules in the following way: The sequence $\mathfrak{R}_{s1}$ will be generated by overlapping rules of itself with those of the form $w_i u_1 \longrightarrow u_1 w_i$, while $\mathfrak{R}_{s2}$ will be generated by overlapping the rules of $\mathfrak{R}_{s1}$ with the rule $u_2 u_1 \longrightarrow lf$. ∎

Until now, we have only considered primitive recursive functions on $\mathbb{N}$. What about primitive recursive word functions ?
The results achieved here for primitive recursive functions on $\mathbb{N}$ can be carried over to primitive recursive word functions (see [17]).

We close this section with some concrete examples (encoding primitive recursive functions on $\mathbb{N}$, primitive recursive word functions and a regular set) that have been computed with our completion system COSY. Their constructions have been inspired by some additional ideas to those mentioned before. We would like to emphasize example 3 that shows that primitive recursive functions on $\mathbb{N}$ may also be encoded in a nested way. (In these examples we assume $>$ to be the syllable ordering induced by $>_{lex}$ and by τ. $\mathfrak{R}^*$ will denote the canonical system that corresponds to $\mathfrak{R}$ and $>$.)

EXAMPLE 1 : Encoding the range of a non-linear function.

Let $\mathfrak{R} = \{ od \longrightarrow o,\ bf \longrightarrow fbbb,\ af \longrightarrow ab,\ abc \longrightarrow o,\ cd \longrightarrow fc \}$,

$d >_{lex} c >_{lex} f >_{lex} b >_{lex} a >_{lex} o$, $\tau(s) = I$ for $s \in \{a,b,c,d,f,o\}$.

Completion of $\mathfrak{R}$ with respect to $>$ generates an infinite sequence $S = \{ ab^{f(i)}c \longrightarrow o \ / \ i \in \mathbb{N}\setminus\{0\} \}$ where f is a function satisfying the following conditions: $f(0) = 1$ and $f(n+1) = 3 * f(n) + 1$.

Thus, $\mathfrak{R}^* = \mathfrak{R} \cup \{ ab^nc \longrightarrow o \ / \ n \in \{ 3^m + ((3^m - 1)/2) \ / \ m \in \mathbb{N} \} \}$. □

EXAMPLE 2 : Encoding a non-linear function with only two rules.

Let $\mathfrak{R} = \{ abca \longrightarrow bcaaa,\ cb \longrightarrow bc \}$, $b >_{lex} c >_{lex} a$, $\tau(s) = I$ for $s \in \{a,b,c\}$.

Then $\mathfrak{R}^* = \{ ab^nc^na^{2^n-1} \longrightarrow b^nc^na^{2^{n+1}-1} \ / \ n \in \mathbb{N}\setminus\{0\} \} \cup \{ cb \longrightarrow bc \}$. □

EXAMPLE 3 : Nested pumping.

Let $\mathfrak{R} = \{ go \longrightarrow o,\ va \longrightarrow abv,\ vb \longrightarrow bv,\ ve \longrightarrow e,\ ua \longrightarrow ax,\ xb \longrightarrow ABx,\ xa \longrightarrow za,$
$BA \longrightarrow AB,\ Bz \longrightarrow zb,\ wa \longrightarrow aw,\ wA \longrightarrow bw,\ wz \longrightarrow a,\ gd \longrightarrow dwuv,\ dababe \longrightarrow o \}$,

$g >_{lex} d >_{lex} u >_{lex} x >_{lex} z >_{lex} a >_{lex} A >_{lex} b >_{lex} w >_{lex} B >_{lex} v >_{lex} e >_{lex} o$,

$\tau(s) = I$ for $s \in \{g,d,u,z,a,A,b,w,B,v,e,o\}$, $\tau(x) = r$.

Then $\mathfrak{R}^* = \mathfrak{R} \cup \{ d(ab^n)^{n+1}e \longrightarrow o \ / \ n \in \mathbb{N}\setminus\{0\} \}$. □

EXAMPLE 4 : Encoding the initial segments of $\mathbb{N}$.

Let $\mathfrak{R} = \{ go \longrightarrow o,\ yb \longrightarrow bxy,\ xb \longrightarrow bx,\ ga \longrightarrow abxy,\ abxd \longrightarrow o \}$,

$d >_{lex} g >_{lex} b >_{lex} y >_{lex} x >_{lex} a >_{lex} o$, $\tau(s) = I$ for $s \in \{a,b,d,g,x,y,o\}$.

Then $\mathfrak{R}^* = \mathfrak{R} \cup \{ ab^nx^nyx^{n-1}y \ldots x^2yx^1d \longrightarrow o \ / \ n \in \mathbb{N}\setminus\{0\} \}$. □

EXAMPLE 5 : Encoding the reversal function.

Let $\mathfrak{R} = \{ ad \longrightarrow da,\ bd \longrightarrow db,\ fd \longrightarrow fa,\ ae \longrightarrow ea,\ be \longrightarrow eb,\ fe \longrightarrow fb,$
$gc \longrightarrow f,\ ce \longrightarrow bc,\ cd \longrightarrow ac \}$,

$g >_{lex} d >_{lex} e >_{lex} a >_{lex} b >_{lex} c >_{lex} f$, $\tau(s) = I$ for $s \in \{a,b,c,d,e,f,g\}$.

Then $\mathfrak{R}^* = \mathfrak{R} \cup \{ g\mathfrak{w}c \longrightarrow f\rho(\mathfrak{w}) \ / \ \mathfrak{w} \in \{a,b\}^* \}$ where ρ denotes the reversal function on $\{a,b\}^*$. Using this string rewriting system $\mathfrak{R}$ a lot of interesting divergence examples can be constructed by adding some additional rules: For example, if we add the rule $aa \longrightarrow \varepsilon$ this rule will not cause any new rule by overlapping, but it will be used for reduction. The resulting canonical system will be $\mathfrak{R} \cup \{ g\mathfrak{w}c \longrightarrow f\rho(\mathfrak{w}) \ / \ \mathfrak{w} \in \{a,b\}^*$ such that 'aa' is not a substring of $\mathfrak{w}$ $\}$. We could also substitute the 'aa' by a new symbol 'h' (with $\tau(h) = I$ and $f >_{lex} h$) by adding the set of rules $\{ aa \longrightarrow h,\ hd \longrightarrow dh,\ he \longrightarrow eh,\ ha \longrightarrow ah \}$. In this case the set $\{ g\mathfrak{w}_1c \longrightarrow f\mathfrak{w}_2 \ / \ \mathfrak{w} \in \{a,b\}^*$ and $\mathfrak{w}_1, \mathfrak{w}_2$ are the normal forms of $\mathfrak{w}, \rho(\mathfrak{w})$, respectively with respect to $\{ aa \longrightarrow h,\ ha \longrightarrow ah \} \}$ is generated. Even more complicated situations will arise, if we add a rule of the form $\mathfrak{w} \longrightarrow \varepsilon$ where $\mathfrak{w}$ is not a palindrome. In this case the left-hand side of a rule of $\mathfrak{R}^*$ may be reducible by this additional rule while the right-hand side is still irreducible, or vice versa. Moreover, this new rule may produce further rules by overlapping. □

EXAMPLE 6 : Encoding a regular set.

Let $\mathfrak{R} = \{ aaba \longrightarrow baa \}$, $b >_{lex} a$, $\tau(a) = \tau(b) = I$.

Then $\mathfrak{R}^* = \mathfrak{R} \cup \{ aab\mathfrak{w}aa \longrightarrow ba\mathfrak{w}aa \ / \ \mathfrak{w} \in (\{b\}^+ \cdot (\{a\} \cdot \{b\}^+)^*) \}$ □

III. REGULARITIES IN THE HISTORY OF THE RULES

Now, we will have a closer look on how divergence sequences may be build up. The easiest way to create a divergency is to take one fixed rule that overlaps with another one, generating a new rule, that overlaps with the 'fixed' rule, too and so on.

Most of the divergence sequences that have been mentioned in the literature as well as most of those mentioned here have been constructed in this way. In these cases the rules of a sequence have a common 'parent'. Thus, the regularities in the structure of the rules is reflected by some regularities in the history of these rules.

We will use dags to illustrate the dependency of the rules that have been generated by a completion process upon each other with respect to their origin. Let $\mathfrak{R}$ be a string rewriting system, $>$ be a noetherian ordering and $\mathfrak{R}^*$ be the corresponding canonical system that has been generated using some completion procedure P. Then we associate with a set $\hat{\mathfrak{R}} \subseteq \mathfrak{R}^*$ and P a dag which is defined as follows: The set of nodes is the set $\hat{\mathfrak{R}}$. There is an edge from rule r_1 to rule r_2, if and only if r_2 has been generated by P by overlapping r_1 with another rule. Especially, there will be only one edge pointing at r_2 , if this rule has been generated by overlapping one rule with itself or with another rule that has been deleted by interreduction. The latter case will not appear in our examples.

Hence, the situations described above may be illustrated by graphs as depicted in figure 1.

figure 1:

figure 2:

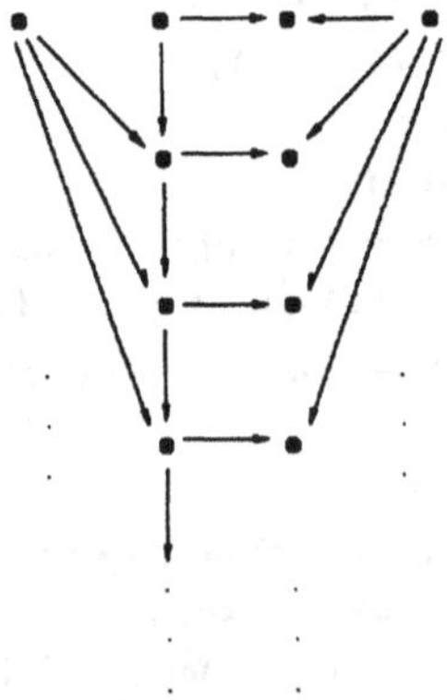

figure 3:

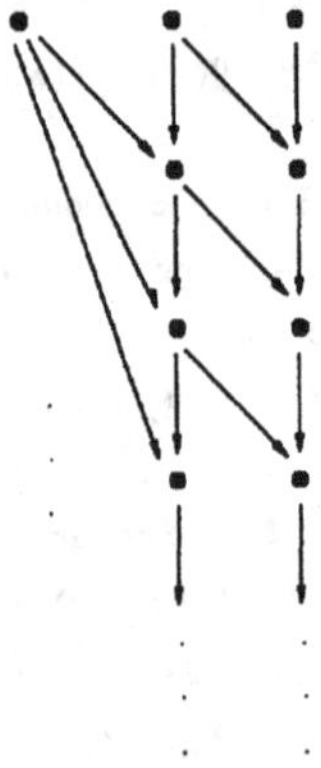

Figure 2 illustrates another typical kind of dependency that may occur. It shows that there need not be any dependencies between the rules of a divergence sequence. For example the divergence sequences mentioned in the proof of theorem 3 are constructed in this way.

The next example shows that the rules of a divergence sequence need not have a common 'parent'.

EXAMPLE 7 :

Let $\mathfrak{R}$ = { ab ⟶ ca , ddca ⟶ aeb , dcaf ⟶ aef , dae ⟶ dccg , gb ⟶ cg , gf ⟶ af }

$d >_{lex} e >_{lex} b >_{lex} c >_{lex} g >_{lex} a >_{lex} f$, $\tau(s) = 1$ for $s \in \{a,b,c,d,e,f,g\}$.

Then $\mathfrak{R}^* = \mathfrak{R} \cup \mathfrak{R}_{s1} \cup \mathfrak{R}_{s2}$ where

$\mathfrak{R}_{s1} = \{ ddc^n a \longrightarrow aeb^n \ / \ n \in \mathbb{N} \setminus \{0\} \}$, $\mathfrak{R}_{s2} = \{ dc^{n+1}af \longrightarrow aeb^n f \ / \ n \in \mathbb{N} \}$.

Figure 3 illustrates how the divergence sequences are generated. □

But, even if the rules of a divergence sequence do not depend on rules of the same sequence, then these rules may have no common 'parent' :

EXAMPLE 8 :

Let $\mathfrak{R} = \{ ab \longrightarrow ca , ddca \longrightarrow aeb , dcaf \longrightarrow ho , hdc \longrightarrow dcc \}$,

$d >_{lex} b >_{lex} a >_{lex} c >_{lex} f >_{lex} o >_{lex} e >_{lex} h$, $\tau(s) = l$ for $s \in \{a,b,c,d,e,f,h,o\}$.

Then $\mathfrak{R}^* = \mathfrak{R} \cup \mathfrak{R}_{s1} \cup \mathfrak{R}_{s2} \cup \mathfrak{R}_{s3}$ where

$\mathfrak{R}_{s1} = \{ ddc^n a \longrightarrow aeb^n \ / \ n \in \mathbb{N} \setminus \{0\} \}$, $\mathfrak{R}_{s2} = \{ dh^n o \longrightarrow aeb^n f \ / \ n \in \mathbb{N} \setminus \{0\} \}$,

$\mathfrak{R}_{s3} = \{ dc^n af \longrightarrow h^n o \ / \ n \in \mathbb{N} \setminus \{0\} \}$.

In this case, the sequences $\mathfrak{R}_{s1}$ and $\mathfrak{R}_{s3}$ are generated in the way described by figure 1 using the rule $ab \longrightarrow ca$ and $hdc \longrightarrow dcc$, respectively as the 'fixed' one. The sequence $\mathfrak{R}_{s2}$ is generated by overlapping the rules of the sequence $\mathfrak{R}_{s1}$ with those of $\mathfrak{R}_{s3}$ (see figure 4). □

Another situation that may arise is that two sequences cause each other:

EXAMPLE 9 :

Let $\mathfrak{R} = \{ Kb \longrightarrow dK , Ad \longrightarrow bA , bAZ \longrightarrow AZD \}$,

$Z >_{lex} K >_{lex} d >_{lex} b >_{lex} A >_{lex} D$, $\tau(s) = r$ for $s \in \{b,d,A,D,K\}$, $\tau(Z) = l$.

Then $\mathfrak{R}^* = \mathfrak{R} \cup \mathfrak{R}_{s1} \cup \mathfrak{R}_{s2}$ where

$\mathfrak{R}_{s1} = \{ b(AK)^n AZ \longrightarrow (AK)^n AZD \ / \ n \in \mathbb{N} \}$, $\mathfrak{R}_{s2} = \{ d(KA)^n Z \longrightarrow (KA)^n ZD \ / \ n \in \mathbb{N} \setminus \{0\} \}$

is the corresponding canonical system. In this case the sequences $\mathfrak{R}_{s1}$ and $\mathfrak{R}_{s2}$ are generated in the way described by figure 5. Here the rules $Kb \longrightarrow dK$ and $Ad \longrightarrow bA$ play the role of the 'fixed' rules. □

figure 4 :

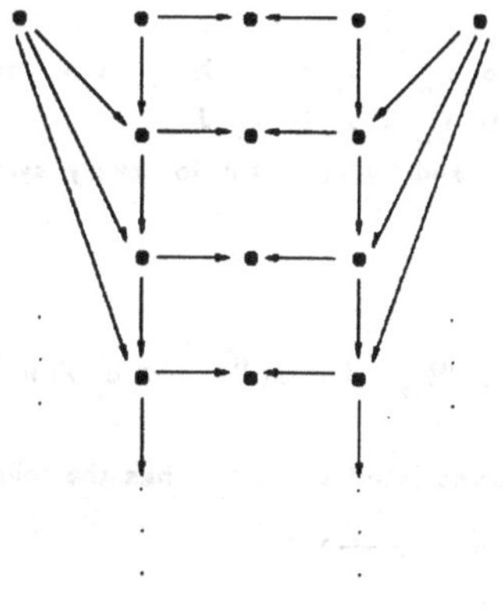

figure 5:

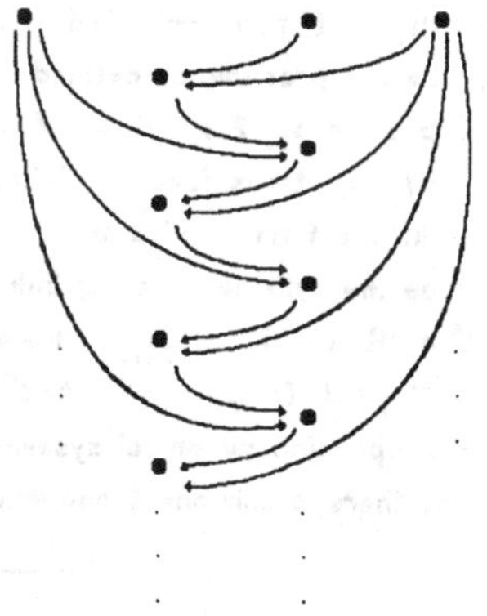

In the last three examples the graph associated with $\mathfrak{R}^*$ will be the same no matter which completion strategy is used. In general, infinitely many dependency graphs may be associated with $\mathfrak{R}^*$ since a lot of rules may cause the same rules.

We want to emphasize this fact with an example which has been used several times for other purposes in the literature (e.g. in [16]):

EXAMPLE 10 :

Let $\mathfrak{R} = \{ aba \longrightarrow ba \}$, $b >_{lex} a$, $\tau(a) = \tau(b) = l$.

Then $\mathfrak{R}^* = \{ ab^n a \longrightarrow b^n a \ / \ n \in \mathbb{N} \setminus \{0\} \}$ is the corresponding canonical system.

In this case there are the following dependencies:

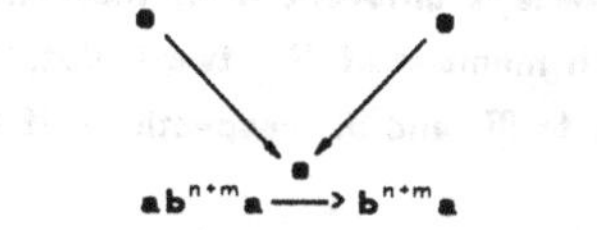

Thus, infinitely many graphs may be associated with $\mathfrak{R}^*$. Depending on the completion strategy used, the corresponding graph may be of such a form that there is no node which is adjacent to infinitely many other ones. This means, that there is no rule which has produced actually infinitely many rules. Hence, in this case the set $\mathfrak{R}^* \setminus \{aba \longrightarrow ba\}$ may not be classified in a finite number of classes each representing a set of rules that have a common 'parent'. But, even the graph mentioned by figure 1 is one of those that may be associated with $\mathfrak{R}^*$ (where the initial rule is the 'fixed' one). □

These examples may lead to the following conjecture: If the completion of a finite set $\mathfrak{R}$ with respect to a noetherian ordering diverges, then there will be at least one rule in the generated canonical system $\mathfrak{R}^*$ that may cause infinitely many other rules of $\mathfrak{R}^*$. But, the next example shows that this conjecture is not true.

EXAMPLE 11 :

Let $\mathfrak{R} = \mathfrak{R}_1 \cup \mathfrak{R}_2 \cup \mathfrak{R}_a$ where

$$\begin{array}{llllll} \mathfrak{R}_1 = \{ & co \longrightarrow od, & fo \longrightarrow fzd, & da \longrightarrow aDC, & CD \longrightarrow DC, & Cg \longrightarrow F, \\ & CF \longrightarrow F, & Cb \longrightarrow \varepsilon, & AD \longrightarrow EA, & Ag \longrightarrow Uuu, & AF \longrightarrow F, \\ & Ab \longrightarrow G, & EU \longrightarrow Uc, & EF \longrightarrow Fd, & Gb \longrightarrow G, & Gg \longrightarrow F, \\ & zU \longrightarrow c, & zF \longrightarrow zd, & za \longrightarrow zA \; \} & & \end{array}$$

$$\begin{array}{llllll} \mathfrak{R}_2 = \{ & cuo \longrightarrow J, & cJ \longrightarrow J, & fJ \longrightarrow o, & fzo \longrightarrow o, & fzY \longrightarrow uuab, \\ & fzH \longrightarrow L, & dua \longrightarrow dX, & dH \longrightarrow \varepsilon, & dY \longrightarrow L, & do \longrightarrow o, \\ & Xb \longrightarrow HIX, & Xg \longrightarrow Yg, & IH \longrightarrow HI, & IY \longrightarrow Yb, & Lb \longrightarrow L, \\ & Lg \longrightarrow o, & LY \longrightarrow L, & LH \longrightarrow L \; \} & & \end{array}$$

$$\mathfrak{R}_a = \{ \; fcuu \longrightarrow fzd, \; uuabg \longrightarrow o \; \}$$

$>_{lex}$ be the precedence defined by $g >_{lex} f >_{lex} u >_{lex} o >_{lex} z >_{lex} a >_{lex} X >_{lex} I >_{lex} H >_{lex} A >_{lex} E >_{lex} c >_{lex} F >_{lex} d >_{lex} Y >_{lex} L >_{lex} C >_{lex} D >_{lex} U >_{lex} G >_{lex} b >_{lex} J$

τ be the status function defined by $\tau(a) = \tau(o) = l$ and by $\tau(x) = r$ for every symbol x used different from 'a' and 'o'

$>$ be the syllable ordering induced by $>_{lex}$ and τ

Then $\mathfrak{R}^* = \mathfrak{R} \cup \mathfrak{R}_{s1} \cup \mathfrak{R}_{s2}$ where

$$\mathfrak{R}_{s1} = \{ \; fc^n uu \longrightarrow fzd^n \; / \; n \in \mathbb{N} \setminus \{0\} \; \}, \quad \mathfrak{R}_{s2} = \{ \; uuab^n g \longrightarrow o \; / \; n \in \mathbb{N} \setminus \{0\} \; \}$$

is the corresponding canonical system.

In this case there is only one dependency graph that may be associated with $\mathfrak{R}^*$. It has the following form:

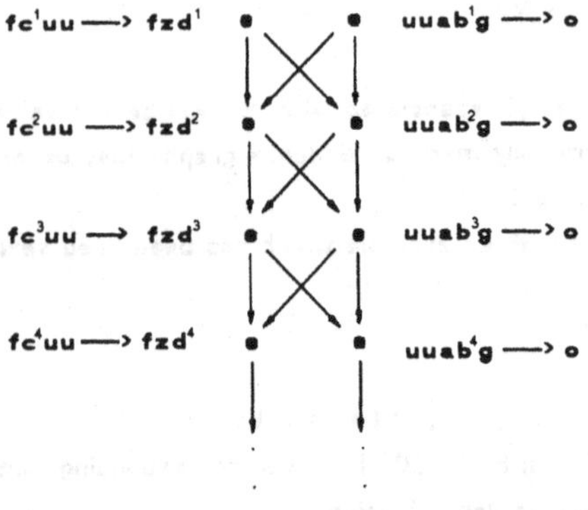

Indeed, there are infinitely many overlaps different from those mentioned in the graph: Each rule of the sequence $\mathfrak{R}_{s1}$ overlaps with each member of $\mathfrak{R}_{s2}$ twice. But, the corresponding critical pairs have a common descendant with respect to $\mathfrak{R}_1$ and $\mathfrak{R}_2$ respectively, if the number of occurrences of 'c' is different from that of 'b'. □

ACKNOWLEDGEMENT:

I would like to thank Professor Klaus Madlener for his valuable comments and suggestions.

REFERENCES

[1] Avenhaus, J.: "Transforming infinite rewrite systems into finite rewrite systems by embedding techniques", *SEKI Report SR-89-21*, Universität Kaiserslautern, 1989

[2] Book, R.V.: "Thue systems as rewriting systems", *Proc. RTA '85*, Dijon, 1985, pp. 63-94

[3] Chen, H. and Hsiang, J. and Kong, H.-C.: "On finite representations of infinite sequences of terms", Preprint, Extended abstract in: *Proc. CTRS '90*, Montreal, *1990*, pp. 37-44

[4] Dershowitz, N. , Marcus, L. and Tarlecki, A. : "Existence, uniqueness, and construction of rewrite systems. *SIAM Journal on Computation, 17(4)*, 1988, pp. 629-639

[5] Gramlich, B.: "Unification of term schemes - Theory and Applications", *SEKI Report SR-88-18*, Universität Kaiserslautern, 1988

[6] Hermann, M.: "Vademecum of divergent term rewriting systems", *Research report 88-R-082*, Centre de Recherche en Informatique de Nancy, 1988.

[7] Hermann, M.: "Chain properties of rule closures", *Proc. STACS '89*, Paderborn, 1989 , pp. 339-347

[8] Hermann, M.: "Crossed term rewriting systems", *Research report 89-R-003*, Centre de Recherche en Informatique de Nancy, 1989

[9] Hermann, M. and Privara, I.: "On nontermination of Knuth-Bendix algorithm", *Proc. ICALP '86*, Rennes, 1986, pp. 146-156

[10] Kirchner, H.: "Schematization of infinite sets of rewrite rules. Application to the divergence of completion processes", *Proc. RTA '87* , Bordeaux, 1987, pp. 180-191

[11] Kirchner, H. and Hermann, M.: "Computing meta-rules from crossed rewrite systems", Preprint, Extended abstract in: *Proc. CTRS '90*, Montreal, 1990, p. 60

[12] Lange, S.: "Towards a set of inference rules for solving divergence in Knuth-Bendix completion", *Proc. AII '89*, Reinhardsbrunn Castle, 1989, pp. 304-316

[13] Lange, S. and Jantke, K.P.: "Towards a learning calculus for solving divergence in Knuth-Bendix completion", *Communications of the Algorithmic Learning Group*, TH Leipzig, 1989, pp. 288-303

[14] Meseguer, J. and Goguen, J.A.: "Initiality, induction, and computability", in: "Algebraic methods in semantics", ed. by M. Nivat and J.C. Reynolds, Cambridge University Press, 1985, pp. 459-541

[15] Mong, C.-T. and Purdom, P.W.: "Divergence in the completion of rewriting systems", *Technical report, Dept. of Comp. Science, Indiana University*, 1987

[16] Narendran, P. and Stillman, J.: "It is undecidable whether the Knuth-Bendix completion procedure generates a crossed pair", *Proc. STACS '89*, Paderborn, 1990, pp. 348-359

[17] Sattler-Klein, A.: "Divergence phenomena during completion", *Internal report 203/90*, FB Informatik, Universität Kaiserslautern, 1990

[18] Steinbach, J.: "Comparing on Strings: Iterated syllable ordering and recursive path orderings", *SEKI Report SR-89-15*, Universität Kaiserslautern, 1989

[19] Thomas, M. and Jantke, K.P.: "Inductive inference for solving divergence in Knuth-Bendix completion", *Proc. AII '89*, Reinhardsbrunn Castle, 1989, pp. 288-303

Simulating Buchberger's Algorithm by Knuth-Bendix Completion

Reinhard Bündgen*
Wilhelm-Schickard-Institut, Universität Tübingen
D-7400 Tübingen, Fed. Rep. of Germany
⟨buendgen@informatik.uni-tuebingen.de⟩

Abstract: We present a canonical term rewriting system whose initial model is isomorphic to $GF(q)[x_1, \ldots, x_n]$. Using this set of rewrite rules and additional ground equations specifying an ideal we can simulate Buchberger's algorithm for polynomials over finite fields using Knuth-Bendix term completion modulo AC. In order to simplify our proofs we exhibit a critical pair criterion which transforms critical pairs into simpler ones.

1 Introduction

The Knuth-Bendix completion procedure [KB70] and Buchberger's algorithm [Buc65] are effective methods to compute a canonical simplifier the elements of an algebraic structure modulo an equivalence relation defined by a set of equations ([BL82]). Buchberger's algorithm computes the *standard* or *Gröbner basis* for an ideal in $F[x_1, \ldots, x_n]$ (where F is a field) presented by a set of polynomial equations. The Knuth-Bendix procedure, upon success, transforms a set of equations over first order terms into a canonical term rewriting system. Extensions of this procedure like [PS81] or [JK86] allow to handle terms modulo an equational theory.

The similarity of the two procedures is striking. Their architecture is similarly built around a critical pair computation step as key paradigm; progress in one area w.r.t. to critical pair criteria was successfully transferred to the other ([WB85]); and there are problems which can be tackled with both methods as for example theorem proving ([Hsi82], [KN85]).

Despite these similarities, so far it was not possible to describe one of the procedures in terms of the other. The term structure seems to be more general than that of polynomials, but a main obstacle to overcome with a term presentation of polynomials is that fields may not be presented purely equationally. This problem can be solved if only the initial model of a term rewriting system is considered. Then it is possible to present a confluent term rewriting system (modulo AC) whose ground term model is isomorphic to a polynomial ring over some field and we can show that Buchberger's algorithm can be simulated by term completion modulo associativity and commutativity in all cases where the coefficient domain of the polynomial ring is a finite field.

The first attempt to unify these two procedures was made in [Loo81] and the latest we know of in [KKW89]. All these approaches use term completion modulo associativity and commutativity. In and [KKW89], the authors propose a common ancestor of both completion procedures

*This work is part of the Ph.D. research of the author supervised by Prof. R. Loos.

by introducing rewriting modulo a simplification relation thus taking into account that we consider polynomials to exist only as fully simplified objects. The simplification operation may also hide the field specific operations. However, we can show that the distinction between a reduction and a simplification relation is not necessary. The completion of finitely presented rings has been studied in [LC86]. There and in [KKN85] it has been pointed out that completion in finitely presented commutatve rings can be interpreted as polynomial completion in $Z[x_1, \ldots, x_r]$. [Bün90] gives an analysis analogously to the one presented here which handles modifications of Buchberger's algorithm where the coefficient domains are commutative rings.

2 Preliminaries

We assume the reader is familiar with term rewriting systems and Buchberger's algorithm. For surveys on these topics see [Der89] and [Buc85].

Throughout this paper we denote the finite set of sorted, ranked function symbols and constants by $\mathcal{F}$ and the set of variables by $\mathcal{X}$. $T(\mathcal{F},\mathcal{X})$ denotes the set of all terms. A term t without variables is called a *ground term*. $T(\mathcal{F})$ is the set of ground terms. Let p be a position in a term t then $t|_p$ denotes the subterm of t at position p, and the result of replacing the subterm $t_1|_p$ of a term t_1 by a term t_2 is denoted by $t_1[p \leftarrow t_2]$.

A *substitution* $\sigma : \mathcal{X} \rightarrow T(\mathcal{F},\mathcal{X})$ is a mapping from the set of variables into the set of terms. If we apply a substitution σ to a term t we write $t\sigma$. $t\sigma$ is called an *instance* of t.

Let $\triangleright$ be any relation then $\triangleright^*$ is the transitive and reflexive closure of $\triangleright$. $\triangleright$ is *terminating* if there is no infinite chain $t_1 \triangleright t_2 \triangleright \ldots$. A relation $\rightarrow_A$ on terms is *stable* if $s\sigma \rightarrow_A t\sigma$ follows from $s \rightarrow_A t$ for all substitutions σ, it is *compatible* if from $s \rightarrow_A t$ follows $f(\cdots,s,\cdots) \rightarrow_A f(\cdots,t,\cdots)$ for all $f \in \mathcal{F}$. A (ground) term t is in *(ground) normal form* w.r.t. $\rightarrow_A$ if there is no t' such that $t \rightarrow_A t'$ otherwise t is *reducible*. If $t \rightarrow_A^* n$ and n is in normal form we write $t\downarrow_A = n$. $T(\mathcal{F},\mathcal{X})\downarrow_A$ is the set of terms in normal form. $\rightarrow_A$ is *confluent* if for all $s, t_1, t_2 \in T(\mathcal{F},\mathcal{X})$, where $s \rightarrow_A^* t_1$ and $s \rightarrow_A^* t_2$, follows $t_1\downarrow_A = t_2\downarrow_A$. If $\rightarrow_A$ is terminating and confluent, we say it is *canonical*.

A *term rewriting system (TRS)* $\mathcal{R}$ is a set of *rewrite rules* $l \rightarrow r$, where l and r are terms. The application of a rule defines the compatible and stable rewrite relation $\rightarrow_\mathcal{R}$. Thus $s \rightarrow_\mathcal{R} t$ if a subterm s' of s is an instance $l\sigma$ of l and t is the term s where s' is replaced by $r\sigma$.

In this paper, we consider rewrite relations modulo an equivalence relation $\mathcal{T}$. In particular, we are interested in both associativity and commutativity laws (*AC-laws*) for some operators in $\mathcal{F}$. Let $T(\mathcal{F},\mathcal{X})/\mathcal{T}$ be the set of equivalence classes of terms modulo $\mathcal{T}$. Then $\mathcal{R}/\mathcal{T}$ defines a rewrite relation on $T(\mathcal{F},\mathcal{X})$ such that $s \rightarrow_{\mathcal{R}/\mathcal{T}} t$ if there are $s' =_\mathcal{T} s$ and $t' =_\mathcal{T} t$, such that $s' \rightarrow_\mathcal{R} t'$.

If $s\sigma =_\mathcal{T} t\sigma$ for some substitution σ, then σ is called an $\mathcal{T}$-unifier of s and t. The most general unifier modulo $\mathcal{T}$ of s and t $mgu_\mathcal{T}(s,t)$ is a minimal set of substitutions, such that all unifiers of s and t are an instance of an element of $mgu_\mathcal{T}(s,t)$. If $\mathcal{T}$ consists of AC-laws an algorithm to compute the $mgu_{AC}(s,t)$ for any two terms exists, and $mgu_{AC}(s,t)$ is finite ([Sti81]).

Let $l_1 \rightarrow r_1, l_2 \rightarrow r_2 \in \mathcal{R}$, p is a non-variable position in l_2 and σ be such that $l_1\sigma =_\mathcal{T} l_2|_p\sigma$ is a *most general common instance (mgci)* of l_1 and $l_2|_p$ modulo $\mathcal{T}$ (i.e. $\sigma \in mgu_\mathcal{T}(l_1, l_2|_p)$). Then $(l_2[p \leftarrow r_1]\sigma,\ r_2\sigma)$ is called an *($\mathcal{T}$-)critical pair* of $l_1 \rightarrow r_1$ and $l_2 \rightarrow r_2$, $l_2\sigma$ is a *superposition*, and $l_2[p \leftarrow r_1]\sigma \swarrow l_2\sigma \searrow r_2\sigma$ is a *critical peak* of the two rules. A critical pair $(t_1,\ t_2)$ is *($\mathcal{T}$-)confluent* if $t_1\downarrow_{\mathcal{R}/\mathcal{T}} =_\mathcal{T} t_2\downarrow_{\mathcal{R}/\mathcal{T}}$.

Term completion procedures ([KB70]) and their extensions modulo an equational theory ([PS81],

[JK86]) take as input a set of equations $\mathcal{E}$, and a term ordering $\succ$. They compute a canonical TRS $\mathcal{R}$ which is equivalent to $\mathcal{E}$. That is, if $s =_{\mathcal{E}} t$, then s and t have a common normal form in $\mathcal{R}$ or $\mathcal{R}/\mathcal{T}$ respectively. Single steps of the completion procedures can be viewed as inferences which transform a set of equations and a set of rules into equivalent but 'simpler' sets of equations and rules: $(\mathcal{E};\mathcal{R}) \vdash (\mathcal{E}';\mathcal{R}')$. Upon success a completion procedure stops with an empty set of equations and a canonical TRS. Again see [Der89] for a survey.

Power products can be ordered in the reverse lexicographical ordering induced by a total ordering on the indeterminates; w.r.t. this ordering, we define the *leading term* of a multivariate polynomial p in distributive normal form to be the monomial with the greatest power product in p and denote it by $LT(p)$. The *reductum* of p is $RED(p) = p - LT(p)$. The *leading coefficient* is the coefficient of the leading term. The integers will be denoted by $\mathbf{Z}$ and we write $\mathbf{Z}_p$ for $\mathbf{Z}/p\mathbf{Z}$. The Galois field of order q will be denoted by $GF(q)$. In the sequel, we often use *polynomial* instead of multivariate polynomial.

3 Critical Pair Transformations

We present a critical pair transformation inference rule for term completion. Critical pair transformation can be viewed as a critical pair criterion and its correctness can be shown using the techniques of [Küc86] and [BD88]. For a survey of critical pair criteria see [BD88]. The transformation criterion seems to be weaker than most of the criteria presented in that survey but it turns out to be extremly helpful for the kind of completion we are concerned with. The proof of the following theorem for rewriting using the Peterson-Stickel rewrite relation [PS81] can be found in [Bün90].

Theorem 1 *Let $\mathcal{R} \cup \{l \rightarrow r\}$ be an $\mathcal{T}$-compatible and terminating TRS and let $\mathcal{E}$ be a set of equations such that $(\mathcal{E};\mathcal{R} \cup \{l \rightarrow r\}) \vdash^* (\mathcal{E}';\mathcal{R}' \cup \{l' \rightarrow r'\})$. Further let $\mathcal{R} \cup \{l' \rightarrow r'\}$ be $\mathcal{T}$-compatible and terminating. If there is a position q in l and a substitution τ such that $l|_q =_{\mathcal{T}} l'\tau$, then all critical pairs of $\mathcal{R} \cup \{l \rightarrow r\}$ are confluent if both $(l[q \leftarrow r'\tau], r)$ and all critical pairs of $\mathcal{R} \cup \{l' \rightarrow r'\}$ are confluent w.r.t. $\mathcal{R} \cup \{l' \rightarrow r'\}$.* □

Theorem 1 suggests the following completion strategy. Whenever a critical pair (s,t) of $l \rightarrow r$ and a rule in $\mathcal{R}$ is computed and $s \succ t$, then test whether l is reducible by $s \rightarrow t$. If this is the case, add (l,r) to $\mathcal{E}$ and continue by computing critical pairs between $s \rightarrow t$ and $\mathcal{R}$. All other critical pairs between $l \rightarrow r$ and $\mathcal{R}$, computed so far, can be deleted. So we get the following inference rule:

$$\frac{(\mathcal{E};\mathcal{R} \cup \{l \rightarrow r\})}{(\mathcal{E} \cup \{l \leftrightarrow r\};\mathcal{R} \cup \{s \rightarrow t\})} \quad \text{if} \left\{ \begin{array}{l} s \succ t,\ (s,t) \text{ is a } \mathcal{T}\text{-critical pair of} \\ \mathcal{R} \cup \{l \rightarrow r\} \text{ and } l \text{ is reducible by } s \rightarrow t. \end{array} \right.$$

The theorem may be exploited even further. It is possible to deduce any consequence of the equations and rules in $\mathcal{E}$ and $\mathcal{R}$ by computing new critical pairs between orientable equations and rules or between two orientable equations. This can be repeated until a 'convenient' rule is deduced. In other words we introduce a new inference rule for *critical pair transformations*:

$$\frac{(\mathcal{E} \cup \{l \leftrightarrow r\};\mathcal{R})}{(\mathcal{E} \cup \{l' \leftrightarrow r'\};\mathcal{R})} \quad \text{if} \left\{ \begin{array}{l} l \succ r,\ l' \succ r',\ l \rhd l'^{1},\ (l',r') \text{ is a } \mathcal{T}\text{-critical pair of} \\ l \rightarrow r \text{ and } \mathcal{R}, (l,r) \text{ is confluent by } \mathcal{R} \cup \{l' \rightarrow r'\}. \end{array} \right.$$

[1] $\rhd$ is a terminating ordering on $\mathcal{D} \times \mathcal{D}$

4 A Term Specification of $Z_p[\vec{X}]$

4.1 R_pX: A Canonical TRS for Multivariate Polynomials

We now want to specify the abstract data type *polynomial* using a many-sorted specification. There is a coefficient sort *Cœf*, a sort for indeterminate power products *Ind* and a sort *Poly* for polynomials. The set of operators together with their signatures and intended meanings is listed in Table 1. From now on we consider a, b, c to be variables of sort *Cœf*; x, y, z to be variables of sort *Ind*; and f, g, h to be variables of sort *Poly*, if they occur in terms of $T(\mathcal{F}, \mathcal{X})$. Let us introduce a convenient notation for term representations of polynomials. $\vec{X}, \vec{Y}, \vec{Z}$ or any indexed version of these vectors represents a ground term of sort *Ind*. We will use k, κ, p, q, r or any indexed version of these letters as repetition factors for coefficient terms. Thus we write ka for $\underbrace{a + \ldots + a}_{k\text{-times}}$, a^r for $\underbrace{a \cdot \ldots \cdot a}_{r\text{-times}}$, k for $k1$ and $\bigoplus_{i=1}^{n} M(k_i, \vec{X}_i)$ for $M(k_1, \vec{X}_1) \oplus \ldots \oplus M(k_n, \vec{X}_n)$.

operator	domain	range	intended meaning
$0, 1:$		$\rightarrow$ *Cœf*	zero coefficient, and coefficient one
$-:$	*Cœf*	$\rightarrow$ *Cœf*	additive inverse for coefficients
$+, \cdot:$	*Cœf* $\times$ *Cœf*	$\rightarrow$ *Cœf*	coefficient addition (AC) and multiplication (AC)
$X_i, I:$		$\rightarrow$ *Ind*	generators of type *Ind*, and the unit-element x^0 of *Ind*
$.:$	*Ind* $\times$ *Ind*	$\rightarrow$ *Ind*	commutative concatenation (AC)
$M:$	*Cœf* $\times$ *Ind*	$\rightarrow$ *Poly*	monomial constructor
$\Omega:$		$\rightarrow$ *Poly*	zero polynomial
$\oplus, \odot:$	*Poly* $\times$ *Poly*	$\rightarrow$ *Poly*	polynomial addition (AC) and multiplication (AC)

Table 1: Signature $\mathcal{F}$ of R_pX

Optional parts in a term will be put in brackets. Thus $a[+b]$ means a or $a+b$. In a rule, optional parts on both sides of the arrow belong together.

Specification 1 presents the canonical TRS for commutative rings with unit elements of characteristic p.

Specification 1 $R_p =$

$$\{\ 1: a+0 \rightarrow a,\ 2: a \cdot 0 \rightarrow 0,\ 3: a \cdot 1 \rightarrow a,\ 4: a \cdot (b+c) \rightarrow (a \cdot b) + (a \cdot c),$$
$$5: -a \rightarrow (p-1)a,\ 6: pa \rightarrow 0,\ 6x: b + pa \rightarrow b\}$$

We are only interested in the initial model of R_p which consists of the R_p-ground normal forms $\{0, 1, 1+1, \ldots, (p-1)1\}$ and is isomorphic to the field $\mathbf{Z}_p$. Although there is *no* equation defining a multiplicative inverse for every term, we can prove the following theorem which can be considered as an inductive consequence of R_p:

Theorem 2 *Let p be a prime number. Then for every ground term t which does not reduce to 0 by R_p there is a ground term t' such that $t \cdot t' \rightarrow^*_{R_p/AC} 1$.*

Proof: It is sufficient to prove the above claim only for terms in ground normal form. Let ψ be a bijective mapping from the set of ground normal forms to $\mathbf{Z}_p$. Then $t \cdot \psi^{-1}(\psi(t)^{-1}) =_{R_p/AC} 1$. For all terms t in ground normal form other than 0, $t \succ 1$ (take e.g. the term ordering $\succ_\Phi$ proposed later in this section). By compatibility of $\succ$ this holds also for a product of two ground normal forms. Thus we have $t \cdot \psi^{-1}(\psi(t)^{-1}) \rightarrow^*_{R_p/AC} 1$. □

We can extend R_p to a canonical specification of multivariate rings over $\mathbf{Z}_p$:

Specification 2

$$
\begin{array}{llll}
R_pX = R_p \cup \{ & 7: & x.I & \rightarrow x, \\
& 8: & M(0,x) & \rightarrow \Omega, \\
& 9: & M(a,x) \oplus M(b,x) & \rightarrow M(a+b,x), \\
& 9x: & f \oplus M(a,x) \oplus M(b,x) & \rightarrow f \oplus M(a+b,x), \\
& 10: & M(a,x) \odot M(b,y) & \rightarrow M(a \cdot b, x.y), \\
& 10x: & f \odot M(a,x) \odot M(b,y) & \rightarrow f \odot M(a \cdot b, x.y), \\
& 11: & f \oplus \Omega & \rightarrow f, \\
& 12: & f \odot \Omega & \rightarrow \Omega, \\
& 13: & f \odot M(1,I) & \rightarrow f, \\
& 14: & f \odot (g \oplus h) & \rightarrow (f \odot g) \oplus (f \odot h), \\
& 15: & (f \odot M(a,x)) \oplus (f \odot M(b,x)) & \rightarrow f \odot M(a+b,x), \\
& 15x: & g \oplus (f \odot M(a,x)) \oplus (f \odot M(b,x)) & \rightarrow g \oplus (f \odot M(a+b,x)), \\
& 16: & f \odot M(a+1,I) & \rightarrow f \oplus (f \odot M(a,I)), \\
& 17: & \bigoplus_{i=1}^{p} f & \rightarrow \Omega, \\
& 17x: & g \oplus \bigoplus_{i=1}^{p} f & \rightarrow g\}
\end{array}
$$

We used the ReDuX-AC term completion system based on the TC system ([Küc82]) to show that all critical pairs of R_pX are confluent. We will later give a proof for the termination of R_pX, but first we want to comment on some properties of R_pX.

The Rule $R_pX.7$ specifies that the sort *Ind* is a commutative monoid with unit element I. The ground normal forms w.r.t. R_pX of sort *Poly* have the forms $\Omega, M(k,\vec{X})$ or $\bigoplus_{i=1}^{n} M(k_i,\vec{X_i})$ where the $k, k_i \neq 0$, $\vec{X}, \vec{X_i}$ are terms in ground normal form and the $\vec{X_i}$ are disjoint within a sum of monomials for all $1 \leq i \leq n$. Thus the canonical form for polynomials produced by R_pX is the distributive form which the standard representation used with Buchberger's algorithm.

4.2 Termination of R_pX

Following [Der79,Der87], we define a simplification ordering induced by an ordering for all pairs of polynomials which results from the interpretation of the left- and right-hand sides of a rule in R_pX. We will use the interpretation function Φ to map terms to polynomials. More precisely we describe a class of functions under certain constraints:

Definition 1 *Let $\Phi : T(\mathcal{F},\mathcal{X})/AC \rightarrow \mathbf{R}[\mathcal{X}]$ be a function from the set of terms to the set of multivariate polynomials over the reals. Φ is inductively defined:*

$\Phi(0) \geq 6$, $\Phi(1) \geq 2$, $\Phi(I) \geq 6$, $\Phi(\Omega) \geq 6$, $\quad$ $\Phi(X_i) \geq 6$ *for every Ind-constant* X_i,
$\Phi(a+b) = \Phi(a) + \Phi(b) + 5$, $\quad$ $\Phi(a \cdot b) = \Phi(a)\Phi(b)$,
$\Phi(-a) = K_1(\Phi(a)+5)$, *where* $K_1 \geq p-1$, $\quad$ $\Phi(x.y) = \Phi(x)\Phi(y)$,
$\Phi(M(a,x)) = (\Phi(a)+K_2)\Phi(x)$, *where* $K_2 \geq 5$, $\quad$ $\Phi(f \oplus g) = \Phi(f) + \Phi(g) + 5$,
$\Phi(f \odot g) = \Phi(f)\Phi(g)$.

Definition 2 *Let P and Q be two polynomials in $\mathbf{R}[x_1,\ldots,x_r]$ then*

$$P >_p Q \textit{ if } \exists y \forall y_1 \ldots \forall y_r : (y_1 \geq y \wedge \ldots \wedge y_r \geq y) \supset P(y_1,\ldots,y_r) > Q(y_1,\ldots,y_r).$$

According to this definition $P >_p Q$ if $P - Q$ is eventually positive. This is a problem which is known to be decidable. The ordering $\succ_\Phi$ with $t \succ_\Phi s$ if $\Phi(t) >_p \Phi(s)$ is a simplification ordering because all polynomial interpretations are monotonic and at least linear in every indeterminate associated with a variable occurring in the interpreted term. In addition, for every ground term t, $\Phi(t) \geq 2$. Now showing that for all rules $l \rightarrow r \in R_pX$, $\Phi(l) >_p \Phi(r)$ is straight forward.

Theorem 3 *R_pX/AC is terminating.* □

4.3 Computing Gröbner Bases in $\mathbf{Z}_p[x_1,\ldots,x_r]$ with R_pX

In this section, we demonstrate the computation of Gröbner Bases in $\mathbf{Z}_p[x_1,\ldots,x_r]$ using the TRS R_pX. Every polynomial has one interpretation as a ground normal form in the term algebra specified by R_pX modulo AC. This interpretation $\psi : T(\mathcal{F})/(R_pX/AC) \rightarrow \mathbf{Z}_p[x_1,\ldots,x_r]$ is determined by the mapping of the constants of sort *Ind* to $x_1,\ldots,x_r$.

Example 1 Let $r = 2$, $\psi(X) = x_1$, $\psi(Y) = x_2$ and $I_P = \{x_1^2x_2 - 2x_2,\ x_1x_2^2 - x_1\}$. Let $\succ_u$ be an ordering on the indeterminates with $x_1 \succ x_2$. In Buchberger's algorithm I_p can then be interpreted as a set of polynomial rules $\{x_1^2x_2 \rightarrow 2x_2,\ x_1x_2^2 \rightarrow x_1\}$ which can be translated to the following set of rewrite rules:

$$I_R = \{M(1,X.X.Y) \rightarrow M(1+1,Y),\ M(1,X.Y.Y) \rightarrow M(1,X)\}$$

Completing $R_5X \cup I_R$ yields $R_5X \cup G_R$ where

$$\begin{array}{llllllll} G_R = \{ & 1: & M(a,X.Y.Y) & \rightarrow & M(a,X), & 2: & M(a,X.Y.Y.x) & \rightarrow & M(a,X.x), \\ & 3: & M(a,X.X) & \rightarrow & M(a+a,Y.Y), & 4: & M(a,X.X.x) & \rightarrow & M(a+a,Y.Y.x), \\ & 5: & M(a,Y.Y.Y) & \rightarrow & M(a,Y), & 6: & M(a,Y.Y.Y.x) & \rightarrow & M(a,Y.x)\ \} \end{array}$$

The Gröbner basis of I_P is therefore $G_P = \{x_1x_2^2 - x_1,\ x_1^2 - 2x_2^2,\ x_2^3 - x_2\}$ and the reduction relation associated with each polynomial is that described by the rules. □

Looking at this example, we can make two observations. R_pX was left unchanged during the completion process and there are always two rules associated with a single polynomial. Let us describe the input- and output rule types of a completion with R_pX.

Specification 3 (rule types)

$$\begin{array}{llllllll} p: & M(k,\vec{X}) & \rightarrow & \bigoplus_i M(k_i,\vec{X}_i), & px: & M(ka,\vec{X}.x) & \rightarrow & \bigoplus_i M(k_ia,\vec{X}_i.x) \\ c1: & M(a,\vec{X}) & \rightarrow & \bigoplus_i M(k_i'a,\vec{X}_i), & c2: & M(a,\vec{X}.x) & \rightarrow & \bigoplus_i M(k_i'a,\vec{X}_i.x) \end{array}$$

In the above specification, all $\vec{X}_i$ are smaller than $\vec{X}$ with regard to a terminating ordering on power products and no two of the $\vec{X}_i$ are equal. The right-hand side of the rules may be the zero polynomial or a single monomial instead of a sum. For all i, $k_i' = k_ik^{-1}$. We say two rules of type c1 and c2 are *associated* with a rule of type p or the respective polynomial.

Definition 3 *Rules of type p in Specification 3 are called* p-rules. *Rules of types c1 and c2 are called* c-rules. *We say a set C of c-rules associated with a polynomial is* c-complete *if all critical pairs between a rule in R_pX and a rule in C are confluent in $(R_pX \cup C)/AC$.*

We will show that only type-c rules occure in canonical rewrite systems derived from R_pX and a set of p-rules. In the context of group completion, the term *symmetrized* is used for a property similar to c-completeness. In [LC86], Le Chenadec generalizes this notion for other algebraic structures like rings and modules. For the case of polynomial completion where the coefficient domain is a commutative ring with 1 the set of c-complete rules is more complicated as shown in [Bün90].

5 Completion with R_pX

5.1 The Termination of c-Rules

To proof the termination of both R_pX and the c-rules, we look for a term ordering which orders sums of monomials similar to the reverse lexicographic orderings often used in Buchberger's algorithm. Such a reverse lexicographic ordering $\gg$ is which induced by a total ordering $\succ_u$ on the indeterminates is a total and terminating on polynomials.

Our interest is to find a terminating ordering for terms. Thus we try to extend $\gg$ to a compatible and stable terminating term ordering $\succ$. Therefore we show that instances of Φ which simulate $\gg$ on ground terms in normal form and which orient all c-rules from left to right exists. In particular the following two inequations must hold:

type-c1 rules: $(\Phi(a) + K_2)\Phi(\vec{X}) >_p \sum_i((\Phi(k_i'a) + K_2)\Phi(\vec{X}_i) + 5) - 5$
type-c2 rules: $(\Phi(a) + K_2)\Phi(\vec{X})\Phi(x) >_p \sum_i((\Phi(k_i'a) + K_2)\Phi(\vec{X}_i)\Phi(x) + 5) - 5$

For $K_2 = 5$ the first inequation implies the second one. For all ground terms t, $\Phi(t) \geq 2$ and $k_i' < p$. In order to find the interpretations of the *Ind*-constants, let n be the maximal number of monomials and r be the maximal degree of a power product occurring in a left- or right-hand side of a rule or an equation. These numbers are not known a priori, but at any given moment during the completion process they are finite and the can be determined. Now for every constant $X \in Unk\setminus\{I\}$ let $\Phi(X) > n(2p\Phi(\vec{X}') + 1)$ where $\vec{X}'$ is the greatest power product with degree less than or equal to r such that $\psi(X) \gg \psi(\vec{X}')$. In addition, the interpretation of the smallest indeterminate must be greater than n. Then for given n, r and $K_2 = 5$ we have found an instance ϕ of Φ which simulates $\gg$ and which induces a terminating term ordering for R_pX and all c-rules. Thus we get

Theorem 4 *The TRS $R_pX \cup C$ where C consists of c-rules is terminating modulo AC.* □

5.2 Computing c-Rules

Now we want to scrutinize the completion of R_pX together with a single p-rule α. Our goal is to show that the result of this completion is a canonical TRS $R_pX \cup C$ where C is the set of c-rules associated with α. For the moment we assume that the left-hand side of α contains a non-trivial power product. The following lemma states the confluence of the resulting system.

Lemma 5 *Let C be the set of c-rules associated with a p-rule $l \to r$ where $l = M(k, \vec{X})$, and $\vec{X} \neq I$. Then all critical pairs among rules in $R_pX \cup C$ are confluent modulo AC.*

Proof: by straight forward analysis of all critical pairs. □

Next we must explain that the c-rules will actually be deduced during the completion of R_pX and α. Let $\alpha : M(k, \vec{X}) \to \bigoplus_i M(k_i, \vec{X}_i)$ be in R_pX-normal form. Superposing α with the rule $R_pX.10$ we get the critical peak

$$M(k \cdot a, \vec{X}.x) \swarrow^{M(k,\vec{X}) \odot M(a,x)} \searrow (\bigoplus_i M(k_i, \vec{X}_i)) \odot M(a, x).$$

This critical pair can be normalized and oriented from left to right. If $k = 1$ the resulting rule $\alpha_2 : M(a, \vec{X}.x) \to \bigoplus_i M(k_i a, \vec{X}_i.x)$ is of type c2. Superposing α_2 with $R_pX.7$ then results in a rule α_1 of type c1. If $k > 1$ we get a rule of type px $\alpha_3 : M(ka, \vec{X}.x) \to \bigoplus_i M(k_i a, \vec{X}_i.x)$.

ka AC-unifies with the left-hand side of $R_pX.6x$. W.l.o.g. $k < p$. Since p is prime, for every $k < p$ there is a k' such that $kk' = 1 \bmod p$. Then there must exist a p' such that $kk' = pp' + 1$. Therefore the following critical peak can be created from rule $R_pX.6x$ and rule α_3:

$$M(a, \vec{X}.x) \swarrow M(kk'a, \vec{X}.x) \searrow \bigoplus_i M(k_i k'a, \vec{X}_i.x).$$

We orient the critical pair from left to right, $\alpha_2 : M(a, \vec{X}.x) \rightarrow \bigoplus_i M(k_i k'a, \vec{X}_i.x)$, and we show that α_3 is confluent w.r.t. $R_pX \cup \{\alpha_2\}$. Note that this critical pair transformation simulates a division of an equation by k and thus enables us to make the rules 'monic'. α_2 is again of type c2. The rest carries over from the case $k = 1$. Note that in both cases α_1 reduces α making it confluent in $R_pX \cup \{\alpha_1\}$ such that no other critical pairs of α need to be considered.

If the left-hand side of a p-rule is $M(k, I)$, then according to the term ordering described in Section 5.1 its right-hand side must be Ω. Thus we get a trivial TRS where all polynomials reduce to Ω because superposing $M(a, x) \rightarrow \Omega$ with rule $R_pX.13$ results in the rule $f \rightarrow \Omega$.

Theorem 6 *Let $l \rightarrow r$ be a p-rule. Then*

- $(\emptyset, R_pX \cup \{l \rightarrow r\}) \vdash^* (\emptyset, R_pX \cup C)$ *and*
- *C consists either of all c-rules associated with $l \rightarrow r$ or of the rule $f \rightarrow \Omega$ and*
- *$R_pX \cup C$ is confluent and terminating modulo AC.*

Proof: by Theorem 4, Lemma 5 and the preceding argumentation. □

Finally we show a semi-compatibility condition for reductions on terms in R_pX-normal form (see [BL82]). This resembles the semi-compatibility of reductions by polynimials used in Buchberger's algorithm.

Corollary 7 *Let S, T and P be terms of sort Poly and $\mathcal{R} = R_pX \cup C$ where C is a c-complete set of c-rules. If $l \rightarrow r \in \mathcal{R}$ such that $S \rightarrow_{\{l \rightarrow r\}/AC} T$ then $(S \oplus P)\downarrow_{R_pX}, (T \oplus P)\downarrow_{R_pX}$ and $(S \odot P)\downarrow_{R_pX}, (T \odot P)\downarrow_{R_pX}$ are confluent w.r.t. $\mathcal{R}/AC$.* □

5.3 Selection of the Leading Term

If two c-rules superpose the result must be a critical pair such that both terms are sums of monomials. In the context of Buchberger's algorithm, a critical pair can be treated as an equation over polynomials, namely, the same terms may be added to both sides such that on one side only the leading term is left. The selection of the leading term can be simulated in the term rewriting environment by applying critical pair transformations.

Let M_i, N_i be monomial terms and let $(\bigoplus_i M_i, \bigoplus_i N_i)$ be a critical pair reduced to R_pX-normal form. W.l.o.g., let M_1 be the maximal monomial in this pair. Then a superposition with $R_pX.17x$ results in

$$M_1 \swarrow M_1 \oplus \bigoplus_{j=1}^{p}(\bigoplus_{i>1} M_i) \searrow \bigoplus_i N_i \oplus \bigoplus_{j=1}^{p-1}(\bigoplus_{i>1} M_i)$$

which can be oriented from left to right. This rule makes the original critical pair confluent. Again using Theorem 1, the old critical pair can be 'transformed' into a new one.

If both sides of the original critical pair contain the maximal power product, then the monomials containing these power products must first be 'put on one side' by an additional transformation involving $R_pX.17x$.

5.4 Critical Pairs Among c-Rules

Two c-rules can only superpose at the top positions of their left-hand sides. Thus both the coefficient terms and the *Ind*-terms of the left-hand side monomials must unify modulo AC. Unifying two ground terms of type *Ind* with variable extensions results in computing 'the least common multiple' of both terms. It is actually sufficient to compute only critical pairs between two c2-rules because these pairs subsume those involving c1-rules and the associated pairs of type-c1 can be derived anyway as described in Section 5.2. We want to consider the two rules

$$\alpha : M(a, \vec{X}.x) \to \bigoplus_i M(k_i a, \vec{X}_i.x) \text{ and } \beta : M(b, \vec{Y}.y) \to \bigoplus_i M(\kappa_i b, \vec{Y}_i.y)$$

where $\vec{X} = \vec{X'}.\vec{Z'}$, $\vec{Y} = \vec{Y'}.\vec{Z'}$ and $\vec{Z} = \vec{X'}.\vec{Y'}.\vec{Z'}$ for disjoint $\vec{X'}$ and $\vec{Y'}$. Thus $mgci_{AC}(\vec{X}.x, \vec{Y}.y)$ $= \{\vec{Z},\ \vec{Z}.z\}$. If the *Ind*-terms in the left-hand sides of the rules unify we get the critical peaks:

$$\bigoplus_i M(k_i a, \vec{X}_i.\vec{Y'}.[x]) \swarrow M(a, \vec{Z}.[x]) \searrow \bigoplus_i M(\kappa_i a, \vec{Y}_i.\vec{X'}.[x]).$$

6 Arbitrary Finite Fields

The set of finite fields has been completely classified, the main results being:

- Finite fields of the same order are identical up to isomorphisms.
- If F is a finite field of order q then $q = p^r$ for some prime number p. F is then called the Galois field of order q denoted by $GF(q)$.
- $GF(q)$ is isomorphic to some field extension $\mathbf{Z}_p(\alpha)$ of $\mathbf{Z}_p$ where α is a root of an irreducible minimal polynomial $m_\alpha(x)$ of degree r. Thus $GF(q) \cong \mathbf{Z}_p[x]/(m_\alpha(x))$ holds.

Using the last relation, we can give an equational specification whose ground term model is isomorphic to $GF(q)$. From now on let $q = p^r$. For $r > 1$, we must add a new constant A of sort *Coef* to $\mathcal{F}$. Let ψ^{-1} be a mapping from $\mathbf{Z}_p(\alpha)$ to $T(\mathcal{F})$ which maps α to A. Then let

$$F_q = R_p \cup \{A^r \to \psi^{-1}(-RED(m_\alpha(\alpha))),\ a \cdot A^r \to a \cdot \psi^{-1}(-RED(m_\alpha(\alpha)))\}.$$

To prove the termination of F_q, it is sufficient to choose $\Phi(A)$ large enough. Further it is easy to see that the last two rules do not superpose with any rule in R_p. Thus F_q is confluent. For $q = p^1$ we set $F_q = R_p$.

Theorem 8 *The set G_q of ground normal forms of F_q is isomorphic to $GF(q)$.*

Proof: G_q contains terms of the form 0, t' and $\sum_i t_i$ where the t' and t_i are either 1 or products of A and no term in the sum may occur more than $p-1$ times. Thus $|G_q| = p^r = q$. For each ground term $t \neq 0$ in F_q-normal form, $\psi(t)^{q-2} = \psi(t)^{-1}$. Then $t^* = t^{(q-2)} \downarrow_{F_q}$ is the multiplicative inverse of t and $t \cdot t^* \to^*_{F_q} 1$. Thus G_q is isomorphic to $GF(q)$. □

If $q = p^1$ let $F_qX = R_pX$. Otherwise let

$$F_qX = R_pX \cup \{A^r \to \psi^{-1}(-RED(m_\alpha(\alpha)))\downarrow_{F_q},\ a \cdot A^r \to (a \cdot \psi^{-1}(-RED(m_\alpha(\alpha))))\downarrow_{F_q}\}.$$

The termination proof for F_qX carries over from F_q and R_pX. Since no new critical pairs can be created F_qX is confluent too. Computing Gröbner bases in $GF(q)[x_1, \ldots, x_n]$ with F_qX differs from their computation as described in the previous sections only in that p-rules of the form

$$M(A', \vec{X}) \ \to\ \bigoplus_i M(A_i, \vec{X}_i)$$

can occur where the A' and the A_i are terms of sort *Coef* in ground normal form which may contain the new constant A. By completeness of the completion procedure, the equation

$$M(1,\vec{X}) \leftrightarrow (\bigoplus_i M(A_i \cdot \psi(\psi^{-1}(A')^{-1}), \vec{X}_i))\downarrow_{F_qX}$$

must be derivable. Then c-rules can be derived as shown in Section 5.2. These c-rules reduce the original p-rules making them confluent such that the critical pair transformation criterion applies. All other aspects of the completion with F_qX carry over from completion using R_pX.

7 Simulating Buchberger's Algorithm

In the preceding sections, we have shown that the completion of polynomials in the polynomial ring $GF(q)[x_1,\ldots,x_r]$ can be performed using term completion modulo associativity and commutativity. As input to the AC-completion algorithm we need

F_qX, a set P of polynomials encoded as p-rules and the ordering $\succ_\phi$.

We get as result

- $F_q \cup \{RX.10\} \cup \{f \to \Omega\}$ if the Gröbner basis of P is $\{x^0\}$ or
- $F_qX \cup C$ where C is the set of c-complete rules associated with the polynomials in the Gröbner basis of P.

The termination of the procedure is ensured by the same termination arguments which apply for Buchberger's algorithm. We now can translate all features of Buchberger's algorithm for $GF(q)[x_1,\ldots,x_r]$ to the language of term rewriting:

- Multivariate polynomials in disjunctive normal form are isomorphic to the F_qX-ground normal forms of sort *Poly*.
- Lexicographical orderings on polynomials can be extended to terminating orderings on terms.
- The reduction relation associated with a polynomial maps to the rewrite relation described by its associated c-rules. These rules can be created by completing a p- or c-rule with F_qX.
- Critical pairs between polynomials correspond to critical pairs between rules of type c2.
- In Buchberger's algorithm, transforming critical pairs to rules consists of selecting the leading term of the difference of both polynomials as left-hand side. This can be simulated as described in Section 5.3.
- The division of a rule by its leading coefficient is effected by critical pair transformations.

Figure 1 presents a term completion procedure using a strategy that models Buchberger's algorithm. The inference rules mentioned in *FFPCOMPLETE* refer to those presented in [Der89]. Hence we get our final result.

Theorem 9 *For every finite field $GF(q)$ the computation of Gröbner bases in the polynomial ring $GF(q)[x_1,\ldots,x_n]$ by Buchberger's algorithm using the reverse lexicographic ordering can be simulated using term completion modulo associativity and commutativity.* □

As a consequence of our analysis Buchberger's algorithm can be considered as a special purpose completion with built-in AC-matching, -unification, R_qX-normalization and critical pair transformations. These operations have fast arithmetical realizations for the restricted rule types

$\mathcal{R}^* \leftarrow$ **FFPCOMPLETE**$(F_qX, I_E, \succ)$

[Finite field polynomial AC-term completion with 'Buchberger strategy'.
F_qX is a canonical TRS whose initial model is $GF(q)[x_1, \ldots, x_r]$. I_E is a set of ground equations of sort *Poly* and $\succ$ is a terminating term ordering which includes the reverse lexicographical ordering on multivariate polynomials described by ground terms. Then $\mathcal{R}$ is the canonical TRS derived from $F_qX \cup I_E$.]

(1) **[Initialize.]** $\mathcal{R} := F_qX$; $\mathcal{E} := I_E$; $C := \emptyset$.
(2) **[Simplify.]** **while** the *simplify*-inference rule applies **do** $(\mathcal{E}; \mathcal{R}) := Simplify((\mathcal{E}; \mathcal{R}))$.
(3) **[Delete.]** **while** the *delete*-inference rule applies **do** $(\mathcal{E}; \mathcal{R}) := Delete((\mathcal{E}; \mathcal{R}))$.
(4) **[Stop?]** **if** $\mathcal{E} = \emptyset$ **then return** $\mathcal{R}^* = \mathcal{R}$ **and stop.**
(5) **[Orient.]** Let $a \leftrightarrow b \in \mathcal{E}$; $\mathcal{E} := \mathcal{E} \setminus \{a \leftrightarrow b\}$;
 (5.1) **[Select leading term.]** Transform $a \leftrightarrow b$ into a rule $l \rightarrow r$ of type p, px or c.
 (5.2) **[Extend rule.]** $\mathcal{R} := Complete((\emptyset; F_qX \cup \{l \rightarrow r\}), \succ)$; $C := C \cup (\mathcal{R} \setminus F_qX)$.
 (5.3) **[Trivial ideal.]** **if** $f \rightarrow \Omega \in C$ **then return** $\mathcal{R}^* = \mathcal{R}$ and **stop.**
(6) **[Collapse.]** $\mathcal{R} := F_qX \cup C$; **while** the *collapse*-inference rule applies **do** $(\mathcal{E}; \mathcal{R}) := Collapse((\mathcal{E}; \mathcal{R}))$.
(7) **[Compose.]** **while** the *compose*-inference rule applies **do** $(\mathcal{E}; \mathcal{R}) := Compose((\mathcal{E}; \mathcal{R}))$.
(8) **[Deduce.]** $C := \mathcal{R} \setminus F_qX$; Compute all critical pairs P between two c-rules in C which are associated with different p-rules; $\mathcal{E} := \mathcal{E} \cup P$; **continue with step 2.** □

Figure 1: Algorithm *FFPCOMPLETE*

occuring during polynomial ideal completion. Therefore we do not suggest to use a term completion procedure for serious Gröbner basis computations. Anyway, we think that our work brings some new insight into the common structure of critical pair completion algorithms and that it may help transferring progress from one area to the other one.

8 Acknowledgements

I wish to thank Prof. Loos for initiating this work and providing an inspiring research environment. I also thank Lars Langemyr for many discussions.

References

[BD88] Leo Bachmair and Nachum Dershowitz. Critical pair criteria for completion. *Journal of Symbolic Computation*, 6:1–18, 1988.

[BL82] Bruno Buchberger and Rüdiger Loos. Algebraic simplification. In *Computer Algebra*, pages 14–43, Springer-Verlag, 1982.

[Buc65] Bruno Buchberger. *Ein Algorithmus zum Auffinden der Basiselemente des Restklassenringes nach einem nulldimensionalen Polynomideal.* PhD thesis, Universität Innsbruck, 1965.

[Buc85] Bruno Buchberger. Gröbner bases: an algorithmic method in polynomial ideal theory. In N. K. Bose, editor, *Recent Trends in Multidimensional Systems Theory*, Reidel, 1985.

[Bün90] Reinhard Bündgen. *Completion of Bases for Multivariate Polynomials over Commutative Rings with the Term Completion Procedure According to Peterson and Stickel.* Technical Report 90–8, Wilhelm-Schickard-Institut, D-7400 Tübingen, 1990.

[Der79] Nachum Dershowitz. A note on simplification orderings. *Inf. Proc. Letters*, 9:212–215, 1979.

[Der87] Nachum Dershowitz. Termination of rewriting. *Journal of Symbolic Computation*, 3:69–115, 1987.

[Der89] Nachum Dershowitz. Completion and its applications. In H. Aït-Kaci and M. Nivat, editors, *Resolution of Equations in Algebraic Structures*, Academic Press, 1989.

[Hsi82] Jieh Hsiang. *Topics in Automated Theorem Proving and Program Generation.* PhD thesis, University of Illinois at Urbana-Champaign, Urbana, Il, USA, December 1982.

[JK86] Jean-Pierre Jouannaud and Hélène Kirchner. Completion of a set of rules modulo a set of equations. *SIAM J. on Computing*, 14(4):1155–1194, 1986.

[KB70] Donald E. Knuth and Peter B. Bendix. Simple word problems in universal algebra. In J. Leech, editor, *Computational Problems in Abstract Algebra*, Pergamon Press, 1970.

[KKN85] A. Kandri-Rody, D. Kapur, and P. Narendran. An ideal-theoretic approach to word problems and unification problems over finitely presented commutative algebras. In *Rewriting Techniques and Applications (LNCS 202)*, pages 345–364, 1985.

[KKW89] A. Kandri-Rody, D. Kapur, and F. Winkler. Knuth-Bendix procedure and Buchberger algorithm – a synthesis. In *International Symposium on Symbolic and Algebraic Computation*, pages 55–67, 1989.

[KN85] Deepak Kapur and Paliath Narendran. An equational aproach to theorem proving in first order predicate calculus. In Arvind Joshi, editor, *Proceedings of the Ninth International Conference on Artificial Intelligence*, 1985.

[Küc82] Wolfgang Küchlin. *An Implementation and Investigation of the Knuth-Bendix Completion Algorithm.* Master's thesis, Informatik I, Universität Karlsruhe, D-7500 Karlsruhe, W-Germany, 1982. (Reprinted as Report 17/82.).

[Küc86] Wolfgang Küchlin. *A Generalized Knuth-Bendix Algorithm.* Technical Report 86-01, Mathematics, Swiss Federal Institute of Technology (ETH), CH-8092 Zürich, Switzerland, January 1986.

[LC86] Philippe Le Chenadec. *Canonical Forms in Finitely Presented Algebras.* Pitman, London, 1986.

[Loo81] Rüdiger Loos. Term reduction systems and algebraic algorithms. In Jörg H. Siekmann, editor, *GWAI-81 (IFB 47)*, pages 214–234, Springer-Verlag, 1981.

[PS81] G. Peterson and M. Stickel. Complete sets of reductions for some equational theories. *Journal of the ACM*, 28:223–264, 1981.

[Sti81] Mark E. Stickel. A unification algorithm for associative-commutative functions. *JACM*, 28(3):423–434, July 1981.

[WB85] Franz Winkler and Bruno Buchberger. A criterion for eliminating unnecessary reductions in the Knuth-Bendix algorithm. In *Proc. Colloquium on Algebra, Combinatorics and Logic in Computer Science*, J. Bolyai Math. Soc., 1985

On Proving Properties of Completion Strategies

Miki HERMANN
CRIN and INRIA-Lorraine
Campus Scientifique, BP 239,
54506 Vandœuvre-lès-Nancy, France
e-mail: hermann@loria.crin.fr

Abstract

We develop methods for proving the *fairness* and *correctness* properties of rule based completion strategies by means of process logic. The concepts of these properties are formulated generally within process logic and then concretized to rewrite system theory based on transition rules. We develop in parallel the notions of *success* and *failure* of a completion strategy, necessary to support the proves of the cited properties. Finally we show the necessity of another property, called *justice*, in the analysis of completion strategies.

1 Introduction

The Knuth–Bendix completion procedure [KB70] presents a key tool for completing equational theories to confluent and terminating rewrite systems. Several properties were required to be fulfilled by the completion procedure with respect to its behavior and the produced result. The first attempt in this direction was Huet's *correctness* proof of the completion procedure [Hue81]. Huet also formulated the notion of *fairness* in completion in a certain way. It should be mentioned that Huet's presentation of the completion procedure differed considerably from the original presentation in [KB70]. Bachmair, Dershowitz, and Hsiang [BDH86] have put completion in an abstract framework, based on the notion of *transition rules*. The notions of *success* and *failure*, as well as the properties *fairness*, *soundness*, and *correctness* cristalised in the work of Bachmair [Bac87] and Dershowitz. Both give formulations of *correctness* and *fairness* in terms of equations and rewrite rules processed during the application of a transition rule based completion procedure. Moreover, Bachmair [Bac87] gives a characterization of *fairness* by several lemmas.

Fairness (a key property of completion procedures), as well as other eventuality properties, were treated in a more general framework of abstract processes and abstract programs [Fra86, GPSS80, Krö87, QS83].

An obvious tool for reasoning about programs and processes are several types of modal logic. Temporal logic [Krö87] is well-suited for reasoning about properties appearing during an execution of (mostly concurrent) fixed processes, but it has problems with locating exactly the positions during the execution and also with composition of programs from smaller units. Dynamic logic [FL79] copes perfectly with the problem of program composition from smaller units. Its drawback is that it can reason only about properties occurring before or after but not during the execution of a program. These problems were resolved in process logic [HKP82], which incorporates both temporal and dynamic logics.

We want to reason about properties of completion strategies (programs), composed from transition rules (basic instructions). Thus process logic is a suitable tool for this analysis. In this article we develop methods for proving the *fairness* and *correctness* properties of rule based completion strategies. The concepts of these properties are formulated generally within process logic and then concretized to rewrite system theory based on transition rules as given in [BDH86]. We develop in parallel the notions of *success* and *failure* of a completion strategy, necessary to support the proves of cited properties. Finally we show the necessity of another property, called *justice*, in the analysis of completion strategies. Unfortunately for us, the formalism we use is not capable to prove the justice of a strategy.

The proofs and several extensions and explications, not included into this article for lack of space, can be found in the research report [Her90].

2 Basic notation and definitions

The reader is supposed to be familiar with the concepts of term rewriting theory, temporal logic, dynamic logic, and process logic. For good overviews, see [DJ90] for rewrite systems, [Krö87] for temporal logic, [FL79] for dynamic logic, and [HKP82] for process logic.

Only to recall the few notations from rewrite systems theory: $\mathcal{T}(\mathcal{F},\mathcal{X})$ denotes the set of all terms (free algebra) over variables $\mathcal{X}$ and symbols $\mathcal{F}$, $Id_{\mathcal{T}(\mathcal{F},\mathcal{X})}$ denotes the set of all identities $t \simeq t$ over all terms $\mathcal{T}(\mathcal{F},\mathcal{X})$, $s \simeq t$ denotes an equation in E, $s \rightarrow t$ denotes a rewrite rule in a rewrite system R, $\longrightarrow_R$ denotes the smallest rewriting relation containing R, $s \succ t$ denotes the reduction ordering between the terms s and t, $t|_a$ denotes the subterm of t at a position $a \in \mathcal{P}os(t)$, $\mathcal{FP}os(t)$ denotes the set of all non-variable positions of the term t, $t\sigma$ denotes the substitution instance of the term t by the substitution σ, $CP(R_1, R_2)$ denotes the set of all critical pairs between the rules of rewrite systems R_1 and R_2, $\uparrow_R$ denotes the common ancestor relation $\stackrel{*}{\longleftarrow}_R \cdot \stackrel{*}{\longrightarrow}_R$ — *meetability*, $\downarrow_R$ denotes the common descendant relation $\stackrel{*}{\longrightarrow}_R \cdot \stackrel{*}{\longleftarrow}_R$ — *joinability*.

We remain basically in the scope of the process logic defined in [HKP82], and therefore we can use the axiomatization of this logic with the support of (strict) propositional dynamic logic [HR83]. The logic can be therefore called *strict simple process logic*, denoted StSiPL. For the exact syntax and semantics of this logic see [Her90].

The formula $\langle a\rangle\, p$ means in *some* executions of a, p is true; the formula $[a]\, p$ means in *all* executions of a, p is true. The formula $\mathbf{f}\, p$ means p is true in the *first* state of a path. The formula p **suf** q means there exists a suffix x which satisfies q and all suffixes y, where $x \prec y$, satisfy p (p is true *until* q becomes true). The formula $\mathbf{n}\, p$ means *next* state in the path exists and satisfies p. The formula **some** p means there exists a suffix satisfying p, i.e. p is true *somewhere*. The formula **all** p means all suffixes satisfy p, i.e. p is true *henceforth*. The formula **last** p says there the path is of finite length and the *last* state satisfies p. The formula **fin** (**inf**) says the path is of *finite* (*infinite*) length.

3 Completion procedure

3.1 Process logic preliminaries

Define α: $S \longrightarrow \mathcal{P}(\Sigma_0)$ to be the *applicability* function, an assignment of states to sets of atomic programs Σ_0 (programs that can be applied in the given state). For an atomic program $A \in \Sigma_0$ and a state $s \in S$ we have $A \in \alpha(s)$ if and only if there exists a state t, different from s, such that $(s,t) \in \tau(A)$.

Define now the predicate *apply* on atomic programs Σ_0 in a StSiPL model M. For a path $x \in S^\omega$ and an atomic program $A \in \Sigma_0$ we have $M, x \models apply(A)$ if and only if $A \in (\alpha \circ first)(x)$, which determines if the atomic program A is applicable to the path x.

Define γ: $S \times S \longrightarrow \mathcal{P}(\Sigma_0 \cup \{skip\})$ to be the *state connectivity* function. $\gamma(s,t)$ determines the set of all atomic programs (or the *skip*) that transform the state s into the state t. For an atomic program $A \in \Sigma_0$ and states $s,t \in S$ we have $A \in \gamma(s,t)$ if and only if $(s,t) \in \tau(A)$. Moreover, for all states $s \in S$ the *skip* program is contained in $\gamma(s,s)$ according to the fact that *skip* does not do anything.

The state connectivity γ is *simple* if for all states $s,t \in S$ the set $\gamma(s,t)$ contains at most one element: $|\gamma(s,t)| \leq 1$. We consider only simple connectivity in the sequel, otherwise we would have problems with locating non-ambiguously the use of atomic programs.

Define now the predicate *use* on atomic programs Σ_0 in a StSiPL model M. For a path $x \in S^\omega$ and an atomic program $A \in \Sigma_0$ we have $M, x \models use(A)$ if and only if $A \in \gamma(first(x), second(x))$, where $second = first \circ next$. The predicate *use* determines whether A was used in the path x as the first applied atomic program to arrive at $next(x)$. It is sometimes necessary to know the fact that an atomic program was used the last time in a path x. This is expressed by a derived predicate $lastuse(A) = use(A) \wedge \mathbf{n}\ \mathbf{all}\ \neg use(A)$. It is clear that for all atomic programs A we have $M, x \models lastuse(A) \supset use(A)$ and $M, x \models use(A) \supset apply(A)$.

We need also the (polymorphic) predicate *empty* operating on sets, preferably on the set of equations E.

3.2 Transition rules

A *transition system* [QS83] is a triple $\mathcal{S} = (S, \Sigma_0, \vdash_{\Sigma_0})$, where S is a countable set of transition *states*, Σ_0 is a *finite* set of *transition rules*, and $\vdash_{\Sigma_0}$ is a set of binary relations on S in bijection with the transition rules Σ_0.

In the sequel we observe the transition system $\mathcal{KB} = (S_{KB}, KB, \vdash_{KB})$, where states S_{KB} are formed by pairs $(E;R)$ of equations E and rewrite rules R. The Knuth–Bendix completion procedure is based on the following set KB of six transition rules:

Delete:	$(E \cup \{s \simeq s\}; R) \vdash (E; R)$
Compose:	$(E; R \cup \{s \to t\}) \vdash (E; R \cup \{s \to u\})$ if $t \longrightarrow_R u$
Simplify:	$(E \cup \{s \simeq t\}; R) \vdash (E \cup \{u \simeq t\}; R)$ if $s \longrightarrow_R u$
Orient:	$(E \cup \{s \simeq t\}; R) \vdash (E; R \cup \{s \to t\})$ if $s \succ t$
Collapse:	$(E; R \cup \{s \to t\}) \vdash (E \cup \{u \simeq t\}; R)$ if $s \longrightarrow_R u$ by $l \to r \in R$ with $s \blacktriangleright l$
Deduce:	$(E; R) \vdash (E \cup \{s \simeq t\}; R)$ if $s \simeq t \in CP(R,R) - E$

where $\blacktriangleright$ denotes a proper encompassment ordering. We write $(E;R) \vdash_{KB} (E';R')$ if the latter may be obtained from the former by one application of a rule in KB.

The transition rule *Deduce* is highly nondeterministic and therefore it would pose too much problems to use it for deducing only one critical pair at a time. The whole set of critical pairs $CP(R, R)$ is always generated at once. Thus the transition rule *Deduce* can be replaced in *KB* by "*Deduction:* $(E; R) \vdash (E \cup CP(R, R); R)$" with the operational equivalence *Deduction* = **while** *apply*(*Deduce*) **do** *Deduce* **od**.

The StSiPL model, corresponding to the transition system $\mathcal{KB}$, will be denoted by M_{KB}. A strict regular program based on the set *KB* of transition rules as atomic programs is called a *completion strategy* (or simply *strategy*). A path x corresponding to an execution of a completion strategy a is called a *completion path*.

3.3 Observed strategy

Taking advantage of the defined predicate *apply* (and *empty*), we can easily write the *KBc* completion strategy as a strict regular program based on the transition rules *KB*. First, to structure well the completion strategy we use the subprograms *rR* and *rE* to describe the reduction of all rules and equations, respectively.

```
program rR is
begin
  while apply(Compose) do Compose od;
  while apply(Collapse) do Collapse od
end
```

```
program rE is
begin
  while apply(Simplify) do Simplify od;
  while apply(Delete) do Delete od
end
```

The observed completion strategy has the form

```
program KBc(E) is
begin
  while apply(Delete) do Delete od;
  while ¬empty(E) do
        if apply(Orient) then while apply(Orient) do Orient; rR; rE od
                         else fail
        fi;
        Deduction; rE
  od
end
```

To make the proofs concerning the strategy more convenient, we describe parts of the *KBc* strategy by subprograms. These are the orient loop and the loop body:

```
program ol is
begin
  while apply(Orient) do
        Orient; rR; rE
  od
end
```

and

```
program lb is
begin
  if apply(Orient) then ol
                   else fail
  fi;
  Deduction; rE
end
```

Then the main loop can be written as

program *ml* **is**
begin
  while $\neg empty(E)$ **do** *lb* **od**
end

3.4 Term rewriting theory within process logic

Classical (finite) rewrite systems can be investigated under process logic and transition system formalism, too. In this case the rewrite rules R become the atomic programs Σ_0. A pair (of terms) (s,t) is contained in the interpretation of an atomic program (rewrite rule) $l \to r \in R = \Sigma_0$ if and only if there exists a nonvariable position $a \in \mathcal{FP}os(s)$ and a substitution σ, such that $s|_a = l\sigma$ and $t = s[r\sigma]_a$. The predicate *apply* is then equivalent to the presence of a redex in the first term (state) of a path.

The basic applied strategy (if we can speak of a certain strategy at all) is a nondeterministic choice of rewrite rules from R, denoted just by the symbol R. Thereafter the computation of a normal form (normalization) could be expressed as

$$norm(R) \quad = \quad \textbf{while } \exists p((p \in R) \land apply(p)) \textbf{ do } R \textbf{ od}$$

The fact that R is terminating is expressed just as $[norm(R)]\,\textbf{fin}$, which follows naturally from the intuitive meaning that there are no infinite rewritings. The *diamond lemma* saying "*A terminating rewrite system R is confluent iff it is locally confluent*" is then expressed by the following process logic expression

$$[norm(R)]\,\textbf{fin} \supset (LConf(R) \;\equiv\; Conf(R)) \qquad (1)$$

using the predicates $Conf(R) = \uparrow_R \subseteq \downarrow_R$ and $LConf(R) = CP(R,R) \subseteq \downarrow_R$ for confluence and local confluence respectively. The set $CP(R,R)$ is, in principle, interpreted as the relation $\longleftarrow_R \cdot \longrightarrow_R$.

4 Properties of the completion strategy

4.1 Success of completion

Following the intuitive meaning, a path x is *unfailing* (*successful*) if during the computation, expressed by x, no *fail* instruction was used. The appropriate PL expression formalizing this fact is $M, x \models unfailing \equiv \textbf{all}\, \neg use(fail)$.

Now, if the *fail* instruction was used somewhere during the completion path x of the strategy KBc, we have

$$M_{KB}, x \models \textbf{some}\, use(fail) \equiv \textbf{fin} \land \textbf{last}\, (\neg empty(E) \land \neg apply(Orient)) \qquad (2)$$

from which we deduce $M_{KB}, x \models \textbf{some}\, use(fail) \supset \textbf{fin} \land \textbf{last}\, \neg empty(E)$, which is equivalent to

$$M_{KB}, x \models (\textbf{fin} \supset \textbf{last}\, empty(E)) \supset \textbf{all}\, \neg use(fail) \qquad (3)$$

From the structure of the KBc strategy we deduce

$$[KBc]\,(\textbf{all}\, \neg use(fail) \supset (\textbf{fin} \supset \textbf{last}\, empty(E))) \qquad (4)$$

Comparing (3) and (4) we get $M_{KB}, x \models \textbf{all}\,\neg use(fail) \equiv \textbf{fin} \supset \textbf{last}\; empty(E)$. Thus, in the case of KBc we can write

$$unfailing \;=\; \textbf{fin} \supset \textbf{last}\; empty(E) \tag{5}$$

From (5), we derive immediately the following theorem:

Theorem 4.1 *A finite completion path x of the completion strategy KBc is* **successful** *(or* unfailing*) if and only if in the last state of x the set of equations E is empty.*

4.2 Correctness

The intuitive meaning of *correctness* is a predicate coupled to the notion of success. A strategy a is correct with respect to the predicate $P(a)$, if the validity of this predicate is implied by each successful and finite computation. In the PL formalism:

$$[a]\,(\textbf{fin} \wedge unfailing \supset \textbf{last}\, P(a)) \tag{6}$$

Within a completion strategy a the predicate $P(a)$ is expressed by $Conf(R)$, meaning that a completing strategy is correct with respect to the confluence of the produced rewrite rules R. That justifies the following theorem.

Theorem 4.2 *A completion strategy b is* **correct** *if and only if b produces a confluent rewrite system R whenever b finishes successfully.*

Applying the equality (5) to (6), we get the expression

$$M_{KB} \models [KBc]\,(\textbf{fin} \wedge \textbf{last}\; empty(E) \;\supset\; \textbf{last}\; Conf(R))$$

for the correctness of the completion strategy KBc.

For proving the correctness of KBc we need a supporting lemma, which is useful also for proofs of other properties.

Lemma 4.3 *For all transition rules $A \in (KB - \{Deduction\}) \cup \{Deduce\}$ the following proposition is valid:* $M_{KB} \models [\textbf{while}\; apply(A)\; \textbf{do}\; A\; \textbf{od}]\,(\textbf{fin} \wedge \textbf{last}\,\neg apply(A))$

Lemma 4.3 implies immediately the following two propositions:

$$M_{KB} \models [rR]\,\textbf{fin} \tag{7}$$
$$M_{KB} \models [rE]\,\textbf{fin} \tag{8}$$

With a little more effort it is possible to prove

Proposition 4.4 $M_{KB} \models [ol]\,\textbf{fin}$.

We cannot have $M_{KB} \models [ol]\,\textbf{last}\,\neg apply(Deduction)$ because this implies $M_{KB} \models [ol]\,\textbf{last}\,(CP(R,R) = E)$, which implies $M_{KB} \models [ol]\,\textbf{f}\,\neg apply(Orient)$ and this is definitely not possible.

Comparing (7), (8), and Proposition 4.4 with ml implies

$$M_{KB} \models [ol]\,\mathbf{last}\,((\neg \bigvee_{A\in KB-\{Deduction\}} apply(A)) \wedge apply(Deduction)) \qquad (9)$$

what indicates that only the *Deduction* rule can be applied after ol. This implies immediately $M_{KB} \models \langle ml\rangle\,\mathbf{inf} \supset \langle ml\rangle\,\mathbf{all}\,\neg lastuse(Deduction)$ or else

$$M_{KB} \models [ml]\,\mathbf{some}\;lastuse(Deduction) \supset [ml]\,\mathbf{fin}$$

The implication $\models [a]\,\mathbf{fin} \supset [a]\,\mathbf{some}\;lastuse(A)$ is trivially satisfied for each atomic program (transition rule) A and program a in each model M, therefore we have

$$M_{KB} \models [ml]\,\mathbf{some}\;lastuse(Deduction) \equiv [ml]\,\mathbf{fin} \qquad (10)$$

Assume that x_{KB} is a *finite* completion path of the strategy KBc and y its suffix. Assume further that z_1 is a suffix of y and $z_2 = next(z_1)$.

$$x_{KB} \;=\; s_1 \ldots \overbrace{s_y \ldots \underbrace{s_{z_1}\, \overbrace{s_{z_2} \ldots s_n}^{z_2}}_{z_1}}^{y}$$

The fact, that each use of the *Deduction* transition rule (at y) is followed by a sequence of *Simplify* rules and then by a sequence of *Delete* rules in the strategy KBc, can be expressed as

$$M_{KB}, y \models use(Deduction) \supset \mathbf{n}\,(use(Simplify)\;\mathbf{suf}\;(apply(Delete) \supset use(Delete))) \qquad (11)$$

From the structure of the transition rule *Delete* we deduce immediately the implication $M_{KB}, z_2 \models use(Delete) \supset \exists e(e \in E \supset e \in Id_{T(\mathcal{F},\mathcal{X})})$. This one implies further

$$M_{KB}, z_2 \models \mathbf{last}\;empty(E) \wedge \mathbf{all}\;use(Delete) \supset (E \subseteq Id_{T(\mathcal{F},\mathcal{X})}) \qquad (12)$$

From the structure of the completion strategy KBc, as well as from (11), follows

$$M_{KB}, z_1 \models lastuse(Simplify) \supset \mathbf{n}\,\mathbf{all}\;use(Delete) \qquad (13)$$

Comparing (12) with (13) and using $\vdash (a \supset b) \supset (a \wedge c \supset b \wedge c)$ gives the implication

$$M_{KB}, z_1 \models \mathbf{last}\;empty(E) \wedge lastuse(Simplify) \supset \mathbf{n}\,(E \subseteq Id_{T(\mathcal{F},\mathcal{X})}) \qquad (14)$$

Applying $\vdash (p \supset q) \wedge (p \supset r) \equiv p \supset (q \wedge r)$ and $\vdash_{PL} \mathbf{n}\,(p \wedge q) \equiv \mathbf{n}\,p \wedge \mathbf{n}\,q$ on (11) and on the implication $M_{KB}, y \models use(Deduction) \supset \mathbf{n}\,(CP(R,R) \subseteq E)$, we get

$$\begin{aligned} M_{KB}, y \;\models\;\; & use(Deduction) \supset \mathbf{n}\,((CP(R,R) \subseteq E)\;\wedge \\ & (use(Simplify)\;\mathbf{suf}\;(apply(Delete) \supset use(Delete)))) \end{aligned} \qquad (15)$$

Comparing (15) and (14) results in the implication

$$M_{KB}, y \models \mathbf{last}\;empty(E) \wedge lastuse(Deduction) \supset \mathbf{n}\,\mathbf{all}\;LConf(R) \qquad (16)$$

By the application of $\vdash_{PL}$ $(a \wedge b \supset c) \supset (\textbf{some}\, a \wedge \textbf{all}\, b \supset \textbf{some}\, c)$ to (16), we can pass from y to x_{KB} and we get

$$M_{KB}, x_{KB} \models \textbf{all last}\ \mathit{empty}(E) \wedge \textbf{some}\ \mathit{lastuse}(\mathit{Deduction}) \supset \textbf{some n all}\ \mathit{LConf}(R) \quad (17)$$

It is clear that $\textbf{all last}\, p$ is equivalent to $\textbf{last}\, p$. We have further $\vdash_{PL} \textbf{n}\, p \supset \textbf{some}\, p$ and $\vdash_{PL} \textbf{some some}\, p \supset \textbf{some}\, p$ which proves

$$M_{KB} \models \textbf{some n all}\ \mathit{LConf}(R) \supset \textbf{some all}\ \mathit{LConf}(R)$$

Therefore (17) implies

$$M_{KB}, x_{KB} \models \textbf{last}\ \mathit{empty}(E) \wedge \textbf{some}\ \mathit{lastuse}(\mathit{Deduction}) \supset \textbf{some all}\ \mathit{LConf}(R) \quad (18)$$

From (10) we imply

$$M_{KB}, x_{KB} \models \textbf{some}\ \mathit{lastuse}(\mathit{Deduction}) \equiv \textbf{fin} \quad (19)$$

Comparing (18) and (19) results in

$$M_{KB}, x_{KB} \models \textbf{fin} \wedge \textbf{last}\ \mathit{empty}(E) \supset \textbf{some all}\ \mathit{LConf}(R)$$

Using $\vdash (a \wedge b \supset c) \equiv (a \wedge b \supset a \wedge c)$ and $\vdash_{PL} \textbf{fin} \wedge \textbf{some all}\, p \supset \textbf{last}\, p$ on the previous implication gives

$$M_{KB}, x_{KB} \models \textbf{fin} \wedge \textbf{last}\ \mathit{empty}(E) \supset \textbf{last}\ \mathit{LConf}(R) \quad (20)$$

The use of a reduction ordering in the transition rule *Orient* subsumes the proposition $[\mathit{norm}(R)]\,\textbf{fin}$ therefore (20) implies

$$M_{KB}, x_{KB} \models \textbf{fin} \wedge \textbf{last}\ \mathit{empty}(E) \supset \textbf{last}\ \mathit{Conf}(R) \quad (21)$$

according to (1). The finiteness of x_{KB} is expressed already in (21), therefore (21) is valid for all completion paths x_{KB} of the strategy *KBc*. Therefore we can generalize (21) to

$$M_{KB} \models [\mathit{KBc}]\,(\textbf{fin} \wedge \textbf{last}\ \mathit{empty}(E) \supset \textbf{last}\ \mathit{Conf}(R))$$

The last implication validates the following theorem.

Theorem 4.5 *The completion strategy KBc is correct.*

4.3 Fairness

Our notion of *fairness* follows, in principle, the ideas of [Fra86, GPSS80, Krö87]. The difference, or additional required property, is the *application determinacy* of strict regular programs. We require that a completion strategy $a \in \Sigma$ must be deterministic with respect to the application of the transition rules Σ_0. The definition of application determinacy is based on the notion of a deterministic program a in dynamic logic which assumes the termination of a.

Definition 4.6 (Application determinacy of strict regular programs)
If $\langle a\rangle(\mathbf{fin} \wedge \mathbf{last}\ apply(A)) \supset [a](\mathbf{fin} \wedge \mathbf{last}\ apply(A))$ *then* a *is* deterministic *with respect to the application of* A. *The program* a *is* deterministic in application *(wrt* Σ_0*) if and only if it is deterministic wrt the application of all* $A \in \Sigma_0$.

If a *and* b *are both deterministic in application, then also* $a; b$ *is* deterministic in application.

If a *is deterministic in application, then also* **while** p **do** a **od** *is* deterministic in application.

If a *and* b *are both deterministic in application, then also* **if** p **then** a **else** b **fi** *is* deterministic in application.

We are ready now to define the *fairness* property in general:

Definition 4.7 *The program* $a \in \Sigma$ *is* **fair** *(wrt* Σ_0*) if it is* deterministic in application *and for all atomic programs* $A \in \Sigma_0$ *the expression* $[a](\mathbf{inf} \supset (\mathbf{all\,some}\ apply(A) \supset \mathbf{all\,some}\ use(A)))$ *is valid.*

This definition expresses exactly the following intuitive property: if there is an atomic program A that can be *applied* infinitely many times during an infinite computation with deterministic application of atomic programs, then the atomic program A is actually *used* infinitely many times during that computation. The definition reflects the general fairness principle expressed by the statement

> *Everything which is enabled infinitely many times within an environment with deterministic application will eventually occur.*

We use the shorthand *fair* to express the fairness property and thus write $[a]\mathit{fair}$ to declare that the strategy a is fair.

The application determinacy of the program a with respect of its fairness is unavoidable. Consider a new completion strategy derived from *KBc*, where the deterministic transition rule *Deduction* is replaced by the nondeterministic one *Deduce*. This new strategy could diverge on the system $R = \{fgfx \to gfx, ggx \to x\}$ if the transition rule *Deduce* never choose the second rule for computing critical pairs. On the other hand, computing the critical pairs of the second rule can leed to a finite canonical system. This is of course in contradiction with the notion of fairness.

For the fairness proof of a completion strategy we need to know the mutual dependence of transition rules with respect to the states where they get enabled or disabled. This is expressed in the following fact by a positive and negative invariant matrices.

Fact 4.8 *The proposition* $M_{KB} \models [A](\mathbf{f}\ apply(B) \supset \mathbf{last}\ apply(B))$ *is valid for the transition rules* A *and* B *according to the following table:*

A	B					
	Delete	*Compose*	*Simplify*	*Orient*	*Collapse*	*Deduction*
Delete		*valid*		*valid*	*valid*	*valid*
Compose	*valid*		*valid*	*valid*	*valid*	*valid*
Simplify		*valid*			*valid*	*valid*
Orient	*valid*	*valid*			*valid*	*valid*
Collapse	*valid*		*valid*	*valid*		*valid*
Deduction	*valid*	*valid*	*valid*	*valid*	*valid*	

The proposition $M_{KB} \models [A]\,(\mathbf{f}\,\neg apply(B) \supset \mathbf{last}\,\neg apply(B))$ *is valid for the transition rules A and B according to the following table:*

A	B: Delete	Compose	Simplify	Orient	Collapse	Deduction
Delete	*valid*	*valid*	*valid*	*valid*	*valid*	
Compose	*valid*	*valid*	*valid*	*valid*	*valid*	*valid*
Simplify		*valid*	*valid*		*valid*	
Orient	*valid*			*valid*		
Collapse		*valid*			*valid*	*valid*
Deduction		*valid*			*valid*	*valid*

The positive and negative invariants on programs $a, b \in \Sigma$ can be extended in a straightforward way on the constructs $a; b$ and **while** p **do** a **od**.

We can prove now the application determinacy of the key parts of the completion strategy *KBc*.

Lemma 4.9 *The programs rR, rE, ol, and therefore also the completion strategy KBc are deterministic in application.*

We continue with the proof of the second part of the fairness condition.

Lemma 4.10 *If* $a \in \Sigma$ *is finite and deterministic in application, and* $[a]\,\mathbf{some}\,use(A)$ *is valid then* **while** q **do** a **od** *is fair wrt* $A \in \Sigma_0$.

Corollary 4.11 *The completion strategy KBc is fair wrt the transition rule Deduction.*

Assume that x_{lb} is a computation path of lb. From the structure of lb follows

$$M_{KB}, x_{lb} \models \mathbf{all}\,\neg use(fail) \supset \mathbf{some}\,use(Orient)$$

which implies $M_{KB} \models [lb]\,(\mathbf{all}\,\neg use(fail) \supset \mathbf{some}\,use(Orient))$ from

$$M_{KB} \models [lb]\,(\mathbf{f}\,apply(Orient) \supset \mathbf{some}\,use(Orient)) \tag{22}$$

The proposition $M_{KB} \models [lb]\,\mathbf{fin}$ follows from (7), (8), Proposition 4.4, and the finiteness of *Orient* and *Deduction*. Using $\vdash_{PL} [a]\,(\mathbf{fin} \wedge p) \supset [\mathbf{while}\ q\ \mathbf{do}\ a\ \mathbf{od}]\,(\mathbf{inf} \supset \mathbf{all}\,p)$ on (22) implies $M_{KB} \models [ml]\,(\mathbf{inf} \supset (\mathbf{all}\,\neg use(fail) \supset \mathbf{all}\,\mathbf{some}\,use(Orient)))$. In general we have $M_{KB}, x \models \mathbf{inf} \supset \mathbf{all}\,\neg(fail)$ from (2), therefore with the use of $\vdash (a \supset b) \wedge (a \supset (b \supset c)) \supset (a \supset c)$ we get $M_{KB} \models [ml]\,(\mathbf{inf} \supset \mathbf{all}\,\mathbf{some}\,use(Orient))$, or else

$$M_{KB} \models [KBc]\,(\mathbf{inf} \supset \mathbf{all}\,\mathbf{some}\,use(Orient))$$

Applying to it $\vdash (a \supset b) \supset (a \supset (c \supset b))$ we get

$$M_{KB} \models [KBc]\,(\mathbf{inf} \supset (\mathbf{all}\,\mathbf{some}\,apply(Orient) \supset \mathbf{all}\,\mathbf{some}\,use(Orient)))$$

Therefore we proved

Lemma 4.12 *The completion strategy KBc is fair wrt the transition rule Orient.*

Corollary 4.11 implies that all critical pairs are generated by the strategy *KBc*:

$$M_{KB} \models [KBc]\,\mathbf{all}\,(e \in CP(R,R) \supset \mathbf{some}\, e \in E) \tag{23}$$

Lemma 4.12 implies that all persistent equations are oriented into rules during an infinite completion by the strategy *KBc*:

$$M_{KB} \models [KBc]\,(\mathbf{inf} \supset \mathbf{all}\,(e \in E \supset \mathbf{some}\, e \notin E)) \tag{24}$$

A statement combined of (23) and (24) was presented as a fairness definition in [Bac87]. Using the statements (23), (24), and the Critical Pair Lemma [BDH86, KB70], we derive by means of proof ordering that for two terms s and t, equal in the equational theory E, an infinite completion by the strategy *KBc* will generate a state $(E_i; R_i)$ such that $s \downarrow_{R_i} t$ [BDH86]: $M_{KB} \models [KBc]\,(\mathbf{inf} \supset (\mathbf{f}\,(s =_E t) \supset \mathbf{some}\,(s \downarrow_R t)))$.

Now we prove that *KBc* is fair wrt the rest of transition rules. We need the following fairness lemma.

Lemma 4.13 *If $a \in \Sigma$ is deterministic in application and $[a]\,(\mathbf{inf} \supset \mathbf{all}\,(apply(A) \supset$ $\mathbf{some}\, use(A)))$ is valid then a is fair wrt $A \in \Sigma_0$.*

Lemma 4.13 in connection with Fact 4.8 is the main tool for proving fairness of a completion strategy wrt the transition rules *Compose*, *Collapse*, *Simplify*, and *Delete*.

Lemma 4.14 *The completion strategy KBc is fair wrt the transition rules Compose, Collapse, Simplify, and Delete.*

We could have proved the fairness of a completion strategy wrt *Compose* and *Collapse* of a completion strategy where the subprogram *rR* would have the form

program *rR* **is**
begin
  while *apply*(*Collapse*) **do** *Collapse* **od**;
  while *apply*(*Compose*) **do** *Compose* **od**
end

The required additional effort would be to prove that each application of *Compose*, disabled by the use of *Collapse*, is replaced by an application of *Simplify*.

To summarize the effort of this section, we state the final theorem

Theorem 4.15 *The completion strategy KBc is fair (wrt KB).*

4.4 Justice

Soundness (a property local to transition rules, and therefore not dealt with here), *success*, *correctness*, and *fairness* are not the only properties to be observed within a completion strategy. We need also the property of *justice*.

Example 4.16 Let us study the completion strategy

```
program ds(E) is
begin
  while apply(Delete) do Delete od;
  while ¬empty(E) do
        while apply(Orient) do Orient; rR; rE od;
        if empty(E) then Deduction else fail fi;
        rE
  od
end
```

The presented strategy is perfectly correct and fair, but it applies the *Deduction* rule only if the set of equations E is empty after *ol* (i.e., all equations were oriented into rules), otherwise it fails. This failure could be premature because a critical pair e could have been produced by *Deduction* and oriented into a rewrite rule r by *Orient*, and this rule r could simplify the previously unorientable equation in E. Therefore the dummy strategy *ds* is *not justified*. It is reasonable to fail only if all critical pairs from already produced rewrite rules were generated and completely simplified (*Deduction* followed by *rE*) and none of the remained equations can be oriented, as it was done in the justified strategy *KBc*.

Now we can define formally the discussed property:

Definition 4.17 *The strategy $b \in \Sigma$ is* **justified** *if and only if b is fair and for all sets of equations E if the strategy b fails on E then every fair strategy $c \in \Sigma$ fails on E, too. Formally:*

$$[b]\,fair \wedge \forall E([b(E)]\,\textbf{some}\ use(fail) \supset \forall c([c]\,fair \supset [c(E)]\,\textbf{some}\ use(fail))) \qquad (25)$$

The justice expression (25) is not an expression in the StSiPL logic any more. For proving it we must use a more subtle variant of process logic than StSiPL. Also the *justice* principle of [MP83] must be modified to cope well with our intentions.

5 Conclusion

Using the process logic, we were able to formulate the *correctness* and *fairness* properties for transition rule based systems in general and proving them for a specific completion strategy *KBc*. During the *fairness* proof we formulated two lemmas suitable for proofs of the *fairness* property of an arbitrary completion strategy based on the transition rules *KB*. Moreover, we showed that the particular formulation of *fairness* for the transition rules *KB*, given by Bachmair and Dershowitz, can be derived from our general one. Finally, we described the necessity of another property, called *justice*, for the analysis of completion strategies.

Acknowledgment

I would like to thank Nachum Dershowitz and Pierre Lescanne for the encouragement to follow this research, and the discussions on the fairness problem in completion.

References

[Bac87] L. Bachmair. *Proof methods for equational theories.* PhD thesis, University of Illinois, Urbana Champaign, Illinois, 1987.

[BDH86] L. Bachmair, N. Dershowitz, and J. Hsiang. Orderings for equational proofs. In *Proceedings 1st IEEE Symposium on Logic in Computer Science, Cambridge, MA*, pages 346–357, June 1986.

[DJ90] N. Dershowitz and J.-P. Jouannaud. Rewrite systems. In J. van Leeuwen, editor, *Handbook of Theoretical Computer Science B: Formal Methods and Semantics*, chapter 6. North-Holland, Amsterdam, 1990.

[FL79] M.J. Fischer and R.E. Ladner. Propositional dynamic logic of regular programs. *Journal of Computer and System Science*, 18:194–211, 1979.

[Fra86] N. Francez. *Fairness.* Springer Verlag, 1986.

[GPSS80] D. Gabbay, A. Pnueli, S. Shelah, and J. Stavi. On the temporal analysis of fairness. In *Proceedings of the 7th ACM Symposium on POPL, Las Vegas*, pages 163–173, January 1980.

[Her90] M. Hermann. On proving properties of completion strategies. Research report 90-R-149, Centre de Recherche en Informatique de Nancy, 1990.

[HKP82] D. Harel, D. Kozen, and R. Parikh. Process logic: Expressiveness, decidability, completeness. *Journal of Computer and System Science*, 25:144–170, 1982.

[HR83] J.Y. Halpern and J.H. Reif. The propositional dynamic logic of deterministic, well-structured programs. *Theoretical Computer Science*, 27:127–165, 1983.

[Hue81] G. Huet. A complete proof of correctness of the Knuth–Bendix completion algorithm. *Journal of Computer and System Science*, 23(1):11–21, August 1981.

[KB70] D.E. Knuth and P.B. Bendix. Simple word problems in universal algebras. In J. Leech, editor, *Computational Problems in Abstract Algebra*, pages 263–297. Pergamon Press, Oxford, 1970.

[Krö87] F. Kröger. *Temporal logic of programs*, volume 8 of *EATCS Monographs on Theoretical Computer Science.* Springer Verlag, 1987.

[MP83] Z. Manna and A. Pnueli. How to cook a temporal proof system for your pet language. In *Proceedings of the 10th ACM POPL Symposium, Austin, TX*, pages 141–154, 1983.

[QS83] J.P. Queille and J. Sifakis. Fairness and related properties in transition systems – A temporal logic to deal with fairness. *Acta Informatica*, 19:195–220, 1983.

On ground AC-completion

Claude Marché
Laboratoire de Recherche en Informatique (UA 410 CNRS)*
Bat 490, Université Paris-Sud
91405 ORSAY CEDEX, FRANCE
E-mail: marche@lri.lri.fr

Abstract

We prove that a canonical set of rules for an equational theory defined by a finite set of ground axioms plus the associativity and commutativity of any number of operators must be finite.

As a corollary, we show that ground AC-completion, when using a total AC-simplification ordering and an appropriate control, must terminate.

Using a recent result of Narendran and Rusinowitch (in this volume), this implies that the word problem for such a theory is decidable.

1 Introduction

The existence of a total AC-simplification ordering when there are at least two AC-operators has been an open problem for a long time. The initial motivation of this work was to show that the existence of such ordering implies the decidability of the word problem in theories defined by a set of ground equations with AC-operators (AC-ground theories).

The existence of such orderings has been proved very recently by Narendran and Rusinowitch (in this volume). As a consequence, we have obtained a complete proof of the decidability of the word problem for AC-ground theories. More precisely, every ground AC-theory has a canonical rewriting system, a much stronger result. Notice that this result has been known for quite some time when there is only one AC-operator [7] (see also [3]).

In this paper, we first show a general result on AC-ground theories, which does not assume a total AC-ordering: a canonical set of rewrite rules (modulo AC) for such a theory must be finite. Then assuming given a total AC-ordering, we prove the existence of a canonical rewriting system modulo AC for the given theory, obtained by a particular completion procedure for AC-ground axioms (general AC-completion has been studied for a long time [5, 6, 11]). To prove termination of this procedure, we use our first result. However, termination is guaranteed only when simplification is applied first: completion may otherwise diverge while generating a finite set of persisting rules. We illustrate this

*This work is partly supported by the "GRECO de programmation du CNRS" and Basic Research Action COMPASS

surprising phenomenon with an exemple of standard completion. To our knowledge, this had not been recognized before.

AC-rewriting is discussed in section 2. Section 3 is devoted to the proof of our result on canonical sets of ground AC-rewrite rules. In section 4 we give our completion algorithm, prove its completeness and termination.

2 Definitions and notations

We assume known the standard definitions and notations from [2].

2.1 AC-ground theories

In this paper, we are interested in ground terms algebras $\mathcal{T}(\mathcal{F})$ where the signature $\mathcal{F}$ has a subset $\mathcal{F}_{AC}$, called the set of AC-operators, which are all binary. We call AC-ground theory an equational theory given by a set of ground axioms plus the following axioms:

$$\text{For all } + \text{ in } \mathcal{F}_{AC}, \begin{cases} (x+y)+z = x+(y+z) \\ x+y = y+x \end{cases}$$

2.2 Total simplification orderings modulo AC

DEFINITION 2.1 *A quasi-ordering* $\succeq$ *on* $\mathcal{T}(\mathcal{F})$ *is a quasi-simplification ordering if*

- $\simeq$ *is a congruence(in particular monotonic).*
- *it is monotonic:* $s \succ t \Longrightarrow w[s] \succ w[t]$
- *it has the subterm property:* $\forall i,\ f(s_1,\ldots,s_n) \succ s_i$

where $\simeq$ *is* $\succeq \cap \preceq$ *and* $\succ$ *is* $\succeq \setminus \simeq$.

LEMMA 2.1 (DERSHOWITZ) *If* $\succeq$ *is a quasi-simplification ordering,* $\succ$ *is nœtherian : there is no infinite sequence* $s_1, s_2, \ldots$ *such that* $s_1 \succ s_2 \succ \ldots$

DEFINITION 2.2 $\succeq$ *is compatible with AC, or is an AC-ordering if* $\simeq$ *is the congruence* $=_{AC}$.

We denote $\geq_{AC}$ an AC-ordering. The *totality* is defined as usual:

$$\forall s,t\ s \geq_{AC} t \text{ or } t \geq_{AC} s$$

Notice that a total AC-ordering must satisfy: $\forall s,t\ s \neq_{AC} t \Longrightarrow \begin{cases} s >_{AC} t \\ \text{or} \\ t >_{AC} s \end{cases}$

LEMMA 2.2 *Let R be a rewrite system and* $\geq_{AC}$ *an AC-simplification ordering. If for all* $l \rightarrow r$ *in R,* $l >_{AC} r$ *then R/AC is terminating.*

2.3 Flattened terms — Basic definitions and properties

We now define the algebra of *flattened terms*. *Flattening* consists in transforming a term so that associativity (but not commutativity) is encoded in the structure of this term. For example, $(a+b)+c$ is written as $a+b+c$.

DEFINITION 2.3 *The algebra $\mathcal{T}_F$ of flattened terms over the signature $\mathcal{F} \cup \mathcal{F}_{AC}$ is inductively defined by:*

- *if $f \in \mathcal{F}$, arity$(f) = n \geq 0$, and $t_1, \ldots, t_n \in \mathcal{T}_F$ then $f(t_1, \ldots, t_n) \in \mathcal{T}_F$.*
- *if $+ \in \mathcal{F}_{AC}$, $n \geq 2$, and $t_1, \ldots, t_n \in \mathcal{T}_F$ then $t_1 + \ldots + t_n \in \mathcal{T}_F$.*

In the following, we usually denote by f a free operator and by $+$ an AC-operator.

DEFINITION 2.4 Flattening *is a map denoted Flat from the free algebra $\mathcal{T}(\mathcal{F} \cup \mathcal{F}_{AC})$ to $\mathcal{T}_F$:*

- *$Flat(f(t_1, \ldots, t_n)) = f(Flat(t_1), \ldots, Flat(t_n))$*
- *if $Flat(s) = s_1 + \ldots + s_n$ and $Flat(t) = t_1 + \ldots + t_m$ ($n \geq 1$ and $m \geq 1$) then $Flat(s+t) = s_1 + \ldots + s_n + t_1 + \ldots + t_m$*

We also write $\bar{s}$ for Flat(s).

The commutativity on free terms corresponds to the *permutation congruence* on flattened terms, which is defined as follows:

DEFINITION 2.5 *For s, t in $\mathcal{T}_F$:*

$$s \equiv t \iff \begin{cases} s = f(s_1, \ldots, s_n) \\ t = f(t_1, \ldots, t_n) \\ \forall i,\ s_i \equiv t_i \end{cases} \quad or \quad \begin{cases} s = s_1 + \ldots + s_n \\ t = t_1 + \ldots + t_m \\ m = n \\ \exists \tau \text{ permutation of } \{1 \ldots n\}\ \forall i,\ s_i \equiv t_{\tau(i)} \end{cases}$$

LEMMA 2.3 *For all s, t in $\mathcal{T}$:*

$$s =_{AC} t \iff \bar{s} \equiv \bar{t}$$

A difficulty with flattened terms is the representation of the positions of its subterms. For example, what is the position of $a+b+c$ in $a+b+c+d$? We use for this purpose the indexes of the beginning and the end of the subsequence equal to $a+b+c$. In this example, we obtain the pair $(1,3)$. Remember that we do not consider commutativity in a flattened term hence a subterm of a flattened term is completely specified by these two indexes ($a+b+c$ is *not* a subterm of $a+b+d+c$)..

DEFINITION 2.6 *Assume that $\mathcal{P}os(t)$ is the set of positions in t defined in the standard way. We define the set of flattened positions $\overline{\mathcal{P}os}(t)$ by:*

- *$\overline{\mathcal{P}os}(f(t_1, \ldots, t_n)) = 1.\overline{\mathcal{P}os}(t_1) \cup \ldots \cup n.\overline{\mathcal{P}os}(t_n) \cup \{\wedge\}$*
- *$\overline{\mathcal{P}os}(t_1 + \ldots + t_n) = 1.\overline{\mathcal{P}os}(t_1) \cup \ldots \cup n.\overline{\mathcal{P}os}(t_n) \cup \mathcal{P}_n$ where $\mathcal{P}_n = \{(i,j) \mid 1 \leq i < j \leq n\}$.*

Positions in $\mathcal{P}_n$ are said to be at the first level *of t. More generally, the* level *of a position is obtained by removing the pair at the end of that position. Formally, it is defined by:*

- $Level(\Lambda) = \Lambda$
- $Level(i.p) = i.Level(p)$
- $Level((i,j)) = \Lambda$

Replacement can be extended to flattened positions as follows:

DEFINITION 2.7

$$(t_1 + \ldots + t_n)[t]_{(i,j)} = Flat(t_1 + \ldots + t_{i-1} + t + t_{j+1} + \ldots + t_n)$$

Notice that flattening is necessary if $\mathcal{H}ead(t)$ is $+$.

Example:

$$(a+b+c+d)[e+f+g]_{(2,3)} = a+e+f+g+d$$

When flattening a free term s, the set of its positions is modified. This transformation defines a map Flat$_s$ from $\mathcal{P}os(s)$ to $\overline{\mathcal{P}os}(s)$. The exact definition of Flat$_s$ is not necessary for our purpose, its existence is sufficient (see [8] for a complete description).

LEMMA 2.4 *If s, t are free terms and $p \in \mathcal{P}os(s)$, $\overline{p} = Flat_s(p)$:*

- $\overline{s|_p} = \overline{s}|_{\overline{p}}$
- $\overline{s[t]_p} = \overline{s}[\overline{t}]_{\overline{p}}$

We give now the definition of the subterm ordering modulo $\equiv$, denoted $\rhd$:

DEFINITION 2.8 *$s \rhd t$ if there exist a term s' and a position $p \in \overline{\mathcal{P}os}(s')$, $p \neq \Lambda$ such that*

$$\begin{cases} s \equiv s' \\ s'|_p = t \end{cases}$$

LEMMA 2.5 *$\rhd$ is terminating.*

PROOF: if $s \rhd t$ then $|s| >_{\mathbb{N}} |t|$ where $|u|$ is the size of u. □

2.4 Rewriting flattened terms

Rewriting flattened terms is defined in a usual way:

DEFINITION 2.9 *If s, t, l, r are flat terms, $p \in \overline{\mathcal{P}os}(s)$, $s \xrightarrow[l \to r]{p} t$ if* $\begin{cases} s|_p = l \\ t = s[r]_p \end{cases}$

Rewriting modulo the permutation congruence is defined as (notice that only the level at which we apply the rule is defined):

DEFINITION 2.10 $s \xrightarrow[l\to r/\equiv]{L} t$ *if* $\begin{cases} \exists s',\ s \equiv s' \\ \exists p \in \overline{Pos}(s'),\ Level(p) = L \\ s'|_p = l \\ t = s'[r]_p \end{cases}$

$\equiv$-extended rewriting is as usual rewriting modulo $\equiv$ but where the application of $\equiv$ is at the position at which we apply the rule (or lower):

DEFINITION 2.11 $s \xrightarrow[\equiv\backslash l\to r]{p} t$ *if* $\begin{cases} s|_p \equiv l \\ t = s[r]_p \end{cases}$

The next lemma shows the equivalence between rewriting modulo AC on free terms and rewriting modulo $\equiv$ on flattened terms:

LEMMA 2.6 *If s, t, l, r are free terms:*

$$s \xrightarrow[l\to r/AC]{} t \text{ iff } \bar{s} \xrightarrow[\bar{l}\to\bar{r}/\equiv]{} \bar{t}$$

We assume known the usual notions of *termination*, *confluence* and *coherence* modulo a set of equations, given in [5]. These definitions are easily extended to rewriting on flattened terms modulo the permutation congruence. A rewriting system R is said to be *convergent* if it is both terminating, coherent and confluent; it is said to be *inter-reduced* if for any rule $l \to r$ in R, r is R-irreducible and l is reducible only by the rule $l \to r$. It is called *canonical* if it is both convergent and inter-reduced.

3 Canonical sets of ground rules modulo AC

In this section, we prove the following result which will allow us to prove that ground AC-completion terminates.

THEOREM 3.1 *Let E be a finite set of ground equations and R a set of ground rules, canonical modulo AC, such that $=_{R\cup AC}$ and $=_{E\cup AC}$ are the same. Then R is finite.*

We first start with two remarks about this theorem in the first subsection. The proof of the theorem is done in the second subsection.

3.1 Remarks

- This theorem is very particular to AC. For instance, it becomes false if we replace AC by A. Here is a counter-example:

  $$E = \{bab = abb, abbc = abc\} \quad R = \{bab \to abb, \forall n \geq 2\ ab^n c \to abc\}$$

  It can be seen easily that E and R have the same equational theory and R is canonical (see [8]). This phenomenon is related to the fact that the word problem in A-ground theories may be undecidable even with a single associative symbol [9, 10, 12].

- This theorem is true modulo the empty theory, i.e. when there are no equation with variables at all. We now give an example which shows the difficulty when there are AC-operators.

The proof is easy if there is no AC-operator: it can be proved that every left member l of a rule in R is an R-successor of a member of an equation of E:

$$\forall l \to r \in R,\ \exists g = d \in E \mid g \overset{*}{\underset{R}{\to}} l$$

The axiom $g = d$ can be found as an axiom which applies at the top of an E-proof of $l = r$. It follows from this remark that there are finitely many left hand-sides, hence R must be finite.

Unfortunately, this result is not true any more when there are AC-operators:

$$E = \begin{cases} a + b = c \\ a + d = e \end{cases} \qquad R = \begin{cases} a + b \to c \\ a + d \to e \\ b + e \to c + d \end{cases}$$

The verification that R is canonical modulo AC is left to the reader (notice that R is the result of applying ground AC-completion to E).

The $E \cup AC$-proof of $b + e = c + d$ is:

$$b + e \overset{2}{\underset{e=a+d}{\longleftarrow}} b + (a + d) \overset{\Lambda}{\underset{AC}{\longleftrightarrow}} (a + b) + d \overset{1}{\underset{a+b=c}{\longrightarrow}} c + d$$

We can see that the only axiom which applies at the top is the associativity of $+$, which is not a ground axiom. Since a term with variables can have an infinite number of instances, we cannot conclude that there are finitely many rules in R.

The idea is then that *each subterm at level one*, in the example b and e, is a R-successor *of a subterm* of a member of an axiom. To deduce that R is finite, we need a lemma on multisets, which is known in the literature as Higman's lemma [4] :

DEFINITION 3.1 *A quasi-ordering $\preceq$ on a set E is a well-quasi-ordering if for all sequence $s_1, s_2, \ldots$ of elements of E there exists i and j such that $i < j$ and $s_i \preceq s_j$.*

THEOREM 3.2 *If F is a finite set and $M_1, M_2, \ldots$ an infinite sequence of multisets of elements of F, then there exist i and j such that $M_i \subseteq M_j$.*

3.2 Proof of the fundamental theorem

We now prove the main theorem. The proof is by contradiction: we assume that $E \cup AC$ and $R \cup AC$ have the same equational theory, and we prove that R cannot be infinite.

For this proof, we assume that the equations in E and the rules in R are flattened, and by hypothesis that R is canonical modulo $\equiv$. We denote by $g = d$ an arbitrary equation of E and by $l \to r$ an arbitrary rule of R.

First we remark that since R is canonical, for any rule $l \to r$, r is $R/\equiv$-irreducible and every strict subterm of l is $R/\equiv$-irreducible.

LEMMA 3.1 *For every left member l of a rule in R, we have:*

- *if $\mathcal{H}ead(l)$ is free, $\exists g = d \in E$ such that $g \xrightarrow[R/\equiv]{*} l$*

- *if $l = l_1 + \ldots + l_n$ then for all i, $\exists g_i = d_i \in E$ such that $g'_i \xrightarrow[R/\equiv]{*} l_i$ where g'_i is a subterm of g_i*

PROOF: For each rule $l \to r$ in R, we consider one of its proofs in E. Two cases are distinguished according to the top function symbol of l.

- $\mathcal{H}ead(l) \notin \mathcal{F}_{AC}$: we first prove by contradiction that there is at least a proof step at the top. Otherwise, l and r would have the same top function symbol, hence $l = f(l_1, \ldots, l_n)$, $r = f(r_1, \ldots, r_n)$ and the proof is made up with proof steps of $l_1 =_{E/\equiv} r_1, \ldots, l_n =_{E/\equiv} r_n$. The whole proof is not empty (since $l \neq_{AC} r$), hence one of these subproofs is not empty, say $l_1 =_{E/\equiv} r_1$. This proof can be done by $R/\equiv$-rewriting so either r_1 or l_1 is R-reducible. But then either r is reducible, or the strict subterm l_1 of l is reducible, a contradiction.

  Therefore there is at least one proof step at the top: the $E/\equiv$-proof of $l = r$ has the form

$$l \xleftrightarrow[E/\equiv]{(\neq\Lambda)*} \xleftrightarrow[g=d]{\Lambda} \xleftrightarrow[E/\equiv]{*} r$$

  hence $l = f(l_1, \ldots, l_n)$ and $g = f(g_1, \ldots, g_n)$ where $l_1 =_{E/\equiv} g_1, \ldots, l_n =_{E/\equiv} g_n$. Since every subterm of l is R-irreducible, we have $g_i \xrightarrow[R/\equiv]{*} l_i$ hence $g \xrightarrow[R/\equiv]{*} l$.

- $\mathcal{H}ead(l) = + \in \mathcal{F}_{AC}$: if there is no proof step at the first level (as defined in definition 2.6) then we have $l = l_1 + \ldots + l_n$ and $r = r_1 + \ldots + r_n$ and the proof is made up with proofs of $l_1 =_{E/\equiv} r_1, \ldots, l_n =_{E/\equiv} r_n$ and we derive a contradiction as in the first case. So there are steps at the first level. The $E/\equiv$-proof of $l = r$ has the form:

$$l \xleftrightarrow{(\not\simeq\Lambda)*} s_1 \xleftrightarrow[g_1=d_1]{\simeq\Lambda} \xleftrightarrow{(\not\simeq\Lambda)*} s_2 \xleftrightarrow[g_2=d_2]{\simeq\Lambda} \ldots \xleftrightarrow{(\not\simeq\Lambda)*} s_m \xleftrightarrow[g_m=d_m]{\simeq\Lambda} \xleftrightarrow{(\not\simeq\Lambda)*} r$$

  Here $\simeq\Lambda$ stands for "at the first level". For an arbitrary l_i, suppose l_i is not involved in the kth step at the first level for any k. Then $l \equiv l_i + l'$ and $r \equiv l_i + r'$, and the previous proof is a proof of $l' =_{E/\equiv} r'$ and $l_i =_{E/\equiv} r_i$. These equalities can be proved in $R/\equiv$ but neither l',r',l_i nor r_i is $R/\equiv$-reducible, a contradiction.

  Therefore l_i is involved in at least one proof step at the first level. Suppose the kth step at level one in the proof is, from the left, the first of such step, that is $g_k \equiv w + g'_k$ where $l_i =_{E/\equiv} g'_k$. Hence $l_i \xrightarrow[R/\equiv]{*} \xleftarrow[R/\equiv]{*} g'_k$ but l_i is irreducible therefore $l_i \xleftarrow[R/\equiv]{*} g'_k$. □

PROOF OF THE THEOREM: We can now prove the fundamental theorem as follows: consider the set S of all the R-successors of the subterms of the members of the equations of E. Since there are finitely many equations, there are finitely many such subterms. On the other hand, every term has a finite number of R-successors since R is terminating (modulo AC). Hence S is finite.

As a consequence of the previous lemma, we now prove that R is finite: by the lemma, for every rule $l \to r$ in R, either $l \in S$ or $l = l_1 + \ldots + l_n$ with $+ \in \mathcal{F}_{AC}$ and $\forall i,\ l_i \in S$. So, there are finitely many rules with a free symbol at the top of the left hand side. If there were infinitely many rules, there would be infinitely many rules with an AC-symbol at the top. Since there are finitely many AC symbols, one of them, say $+$, would be the top function symbol of the left hand side of infinitely many rules. Then these rules would form an infinite sequence $l_1, l_2, \ldots$ of left hand sides of rules such that $l_i = l_{i,1} + \ldots + l_{i,k_i}$ and every $l_{i,j}$ belongs to S. By theorem 3.2, there would exist $i < j$ such that $\{l_{i,1}, \ldots, l_{i,k_i}\} \subseteq \{l_{j,1}, \ldots, l_{j,k_j}\}$ and l_j would be reducible by $l_i \to r_i$, a contradiction.
□

4 Ground AC-completion

In order to prove that the word problem for AC-ground theories is decidable, we describe a completion algorithm which produces a canonical set of rules for the given equational theory.

Usually completion procedures do not terminate, hence they yield semi-decision procedures only. In our case, we are going to prove that our completion procedure always terminates.

We define AC-ground completion as usual by a set of inference rules and a search strategy.

4.1 Overlapping AC-rules

DEFINITION 4.1 *Two rules $l \to r$ and $g \to d$ overlap if there exists a term w such that $l \equiv l' + w$ and $g \equiv g' + w$. The corresponding critical pair is $(l' + d, g' + r)$.*

The equality $l' + d = g' + r$ is an equational consequence of the two rules:

$$\begin{array}{ccc} & l' + w + g' & \\ {\scriptstyle g \to d/\equiv} \swarrow & & \searrow {\scriptstyle l \to r/\equiv} \\ l' + d & & g' + r \end{array}$$

Notice that we can choose w maximal in the sense that l' and g' have no common subterms at level 1. It can be easily seen that critical pairs with a non-maximal w are redundant (they are simplifiable by the corresponding critical pair where w is maximal). Notice also that the standard critical pairs are not needed : if $l \to r$ and $g \to d$ overlap in the usual sense, $l|_p\sigma \equiv g\sigma$ hence $l|_p \equiv g$ because there are no variables, hence $l \to r$ is simplifiable.

We denote by $CP(R)$ the set of critical pairs of the rules in R.

4.2 Inference rules for ground AC-completion

In the following, we assume that $\geq_{AC}$ is a total AC-simplification ordering. The ground AC-completion procedure is described with a set of inference rules. In this system, we consider that every new equation produced is immediately oriented w.r.t $\geq_{AC}$ (or removed if trivial). Since this ordering is total, this is always possible.

DEFINITION 4.2 *Let*

$$\{l \rightleftharpoons r\} = \begin{cases} \phi \text{ if } l =_{AC} r \\ \{l \to r\} \text{ if } l >_{AC} r \\ \{r \to l\} \text{ if } r >_{AC} l \end{cases}$$

The system of inference rules for ground AC-completion is given in figure 1.

SIMPLIFY LEFT

$$R \cup \{l \to r\} \vdash R \cup \{l' \rightleftharpoons r\} \quad \text{if} \begin{cases} g \to d \in R \\ l \xrightarrow[g \to d/\equiv]{} l' \\ l \rhd g \text{ or } l =_{AC} g \text{ and } r >_{AC} d \end{cases}$$

SIMPLIFY RIGHT

$$R \cup \{l \to r\} \vdash R \cup \{l \to r'\} \quad \text{if} \begin{cases} g \to d \in R \\ r \xrightarrow[g \to d/\equiv]{} r' \end{cases}$$

DEDUCE

$$R \vdash R \cup \{l' + d \rightleftharpoons g' + r\} \quad \text{if} \begin{cases} l \to r \in R \\ g \to d \in R \\ l \equiv l' + w \\ g \equiv g' + w \\ w \text{ maximal} \end{cases}$$

Figure 1: Rules of ground AC-completion

The two first rules perform simplification between rules. Notice that in SIMPLIFY RIGHT, the orientation of the generated rule is known but not in SIMPLIFY LEFT.

Notice also that no rule can introduce variables if all starting rules are ground: AC-completion produces only ground rules.

DEFINITION 4.3 *A ground AC-completion procedure is an algorithm which takes for input a total AC-simplification ordering $\geq_{AC}$ and a finite set of ground equations E and produces a (finite or infinite) derivation $R_0 \vdash R_1 \vdash \ldots$ from $R_0 = \bigcup_{l=r \in E} \{l \rightleftharpoons r\}$*

If the sequence is finite, we may assume that it is completed with an infinite number of copies of the last R_n, in order to define the set of persisting rules as:

$$R_\infty = \bigcup_{n=0}^{\infty} \bigcap_{i=n}^{\infty} R_i$$

4.3 Soundness and Completeness

THEOREM 4.1 (SOUNDNESS) *The equational theories of $AC \cup R_\infty$ and $AC \cup E$ are identical*

PROOF: straightforward. □

DEFINITION 4.4 *A derivation $R_0 \vdash R_1 \vdash \ldots$ is* fair *if all possible deductions between rules of R_∞ are done:*

$$CP(R_\infty) \subseteq \bigcup_{i=0}^{+\infty} R_i$$

THEOREM 4.2 (COMPLETENESS) *If the derivation is fair then R_∞ is convergent modulo AC.*

The proof of this theorem use the *normalization proof method* originally introduced in [1]. See [8] for the complete proof.

4.4 Termination of AC-completion

It follows from the completeness of completion that R_∞ is convergent. since all simplifications have been done, it must even be canonical. Therefore, theorem 3.1 implies that R_∞ is finite.

Unfortunately, it does not mean that R_∞ is obtained in finite time ! It is possible that completion keeps generating non persisting rules ! This observation is also valid for standard completion, but has never been made. Before to solve this surprising question, we give an example (for standard completion):

f, g are two unary operators. We start with $E = \{ggx = gx, fgfx = fgx\}$ and we can generate the fair derivation:

	$(1) ggx = gx; (2) fgfx = fgx$
Orient(1) and (2)	$(3) ggx \to gx; (4) fgfx \to fgx$
Deduce(4,4)(overlap $fgfgfx$)	$(5) fggfx = fgfgx$
Simplify(5) with (4), Orient	$(6) fggfx \to fggx$
Deduce(6,4)(overlap $fggfgfx$)	$(7) fgggfx = fggfgx$
Simplify(7) with (6), Orient	$(8) fgggfx \to fgggx$
Simplify(6) with (3)(4), Delete	(6) removed
...	
Deduce(2n,2)(overlap $fg^{n-1}fgfx$)	$(2n+1) fg^n fx = fg^{n-1} fgx$
Simplify(2n+1) with (2n), Orient	$(2n+2) fg^n fx \to fg^n x$
Simplify(2n) with (3)(4), Delete	$(2n)$ removed
...	

We obtain a finite $R_\infty = \{ggx \to gx, fgfx \to fgx\}$.

The problem can indeed be eliminated by an appropriate control: we want to be sure that all possible simplifications are done before deductions. (Simp!Sup)! is described as:

- if R is not inter-reduced, reduce it.

- if R is inter-reduced, compute (fairly) a critical pair.

LEMMA 4.1 *Assume completion does not terminate with the control (Simp!Sup)!. Then R_∞ is infinite.*

PROOF: by contradiction: Assume that R_∞ is finite: $R_\infty = \{l_1 \to r_1, \ldots, l_n \to r_n\}$; and the derivation is infinite:

$$R_0 \overset{*}{\underset{\text{SIMPLIFY}}{\vdash}} R_1 \underset{\text{DEDUCE}}{\vdash} R_1' \overset{*}{\underset{\text{SIMPLIFY}}{\vdash}} R_2 \underset{\text{DEDUCE}}{\vdash} R_2' \overset{*}{\underset{\text{SIMPLIFY}}{\vdash}} \ldots$$

Notice that each SIMPLIFY* step is finite because the simplification rules alone are terminating.

We have

$$R_\infty = \bigcup_{j=0}^{\infty} \bigcap_{i=j}^{\infty} R_i = \{l \to r \mid \exists j, \forall i \geq j \; l \to r \in R_i\}$$

so for all k in $\{1, \ldots, n\}$ $\exists j_k \; \forall i \geq j_k \; l_k \to r_k \in R_i$. Take $N = \max\{j_k \mid 1 \leq k \leq n\}$, then we have $R_\infty = \bigcap_{i=N}^{\infty} R_i$. But then we have for all $i \geq N$ $R_i = R_\infty \cup R_i'$ for some R_i'. Since the theories of R_i and R_∞ are the same, and both R_i and R_∞ are inter-reduced, R_i' must be empty.

Hence $\forall i \geq N, R_i = R_\infty$. It means that after step N, each DEDUCE step compute a critical pair which is immediately removed by the next SIMPLIFY* steps. Since $CP(R_\infty)$ is finite, hence the derivation is finite, a contradiction. □

THEOREM 4.3 *Ground AC-completion terminates when using control (Simp!Sup)!.*

PROOF: By theorem 3.1 and lemma 4.1. □

COROLLARY 4.1 *The word problem for an equational theory defined by a set of ground equations on an algebra $\mathcal{T}(\mathcal{F} \cup \mathcal{F}_{AC})$ with AC operators is decidable.*

5 Conclusion

We have given a completion algorithm devoted to ground equations with any number of AC-operators, by giving a complete set of inference rules.

We have also proved that a canonical set of rules for an equational theory defined by a set of ground axioms plus the associativity and commutativity of any number of operators must be finite.

Thanks to Narendran and Rusinowitch result (this volume) that total AC-reduction orderings do exist, we have obtained as a consequence that the word problem for such a theory is decidable, and more precisely that every AC-ground theory has a canonical rewriting system.

Further work will be devoted to clarify the relationship between ground completion modulo a set of equations, and Buchberger's algorithm for computing a Gröbner basis of a polynomial ideal. Buchberger's algorithm is very close to a ground completion algorithm modulo the theory of polynomial rings, and it has the same property as AC-ground completion: it always terminates.

Acknowledgements I want to express my thanks to Jean-Pierre Jouannaud and Hubert Comon for their help and remarks about this work.

References

[1] L. Bachmair, N. Dershowitz, and J. Hsiang. Orderings for equational proofs. In *Proc. 1st IEEE Symp. Logic in Computer Science, Cambridge, Mass.*, pages 346–357, June 1986.

[2] N. Dershowitz and J.-P. Jouannaud. *Handbook of Theoretical Computer Science*, volume B, chapter Rewrite Systems, pages 243–309. North-Holland, 1990. J. van Leeuwen ed.

[3] P. W. Goldberg. *Ground AC-Completion.* Laboratoire de Recherche en informatique, internal report, 1989.

[4] G. Higman. Ordering by divisibility in abstract algebras. *Proceedings of the London Mathematical Society*, 2(3):326–336, Sept. 1952.

[5] J.-P. Jouannaud and H. Kirchner. Completion of a set of rules modulo a set of equations. *SIAM J. Comput.*, 15(4):1155–1194, 1986.

[6] D. S. Lankford and A. M. Ballantyne. Decision procedures for simple equational theories with commutative-associative axioms: Complete sets of commutative-associative reductions. Research Report Memo ATP-39, Department of Mathematics and Computer Science, University of Texas, Austin, Texas, USA, Aug. 1977.

[7] A. Mal'cev. On homomorphisms of finite groups. *Ivano Gosudarstvenni Pedag. Inst. Ucheneye Zap*, 18:49–90, 1958.

[8] C. Marché. On AC-termination and ground AC-completion. Research Report 598, Laboratoire de Recherche en Informatique, Université de Paris-Sud, Orsay, France, Oct. 1990.

[9] A. A. Markov. On the impossibility of certain algorithms in the theory of associative systems. *Dokl. Akad. Nauk SSSR*, 55:587–590, 1947. In Russian, English translation in C.R. Acad. Sci. URSS, 55, 533-586.

[10] J. V. Matijasevic. Simple examples of undecidable associative calculi. *Soviet Mathematics (Dokladi)*, 8(2):555–557, 1967.

[11] G. E. Peterson and M. E. Stickel. Complete sets of reductions for some equational theories. *J. ACM*, 28(2):233–264, Apr. 1981.

[12] E. L. Post. Recursive unsolvability of a problem of Thue. *Journal of Symbolic Logic*, 12:1–11, 1947.

Any ground associative-commutative theory has a finite canonical system

Paliath Narendran
Inst. of Prog. and Logics
SUNY at Albany
Albany, New York 12222
USA
dran@cs.albany.edu

Michaël Rusinowitch
INRIA & CRIN
BP 239
54506 Vandoeuvre-les-Nancy
France
rusi@loria.fr

Abstract

We show that theories presented by a set of ground equations with several associative-commutative (AC) symbols always admit a finite canonical system. This result is obtained through the construction of a reduction ordering which is AC-compatible and total on the set of congruence classes generated by the associativity and commutativity axioms. As far as we know, this is the first ordering with such properties, when several AC function symbols and free function symbols are allowed. Such an ordering is also a fundamental tool for deriving complete theorem proving strategies with built-in associative commutative unification.

1 Introduction

In this paper, we show that there is an algorithm which, given any finite set E of ground equations which contains associative commutative operation symbols, produces a reduced canonical rewriting system R equivalent to E. The method we use follows the general approach of Knuth and Bendix [KB70] of generating word problem decision algorithms for abstract algebras. It has been known for a long time that this method always succeeds with *purely ground* equations, partly due to the existence of reduction orderings which are total on ground terms. In order to deal with non-orientable axioms, like the commutativity one for instance, the completion technique of Knuth and Bendix has been extended to equivalence class term rewriting systems by [Hue80, LB77, PS81, JK86]. In this framework, Ballantyne and Lankford [BL81] designed a completion algorithm for finitely presented commutative semigroups. This is the case of one binary associative-commutative function and a finite number of constants. Termination of the algorithm was obtained using Dickson's lemma. Our result can be viewed as a generalization of Ballantyne and Lankford's when there are several binary commutative-associative functions and also non-constant function symbols. To prove that completion always succeeds we first need to build a reduction ordering which is total on the set of classes of terms with respect to the congruence generated by the associative commutative axioms. This construction, which generalizes polynomial orderings, is outlined in Section 3. Then, in Section 4, we define a completion procedure and show that it always stops with a finite canonical system. Here, as in the semigroup case, Dickson's lemma is the decisive argument.

2 Preliminaries and notations

Here, we suppose that we are given a signature F which contains a subset F_{AC} of associative commutative operations. Our notations will be adapted to this framework. For a detailed survey on term rewriting the reader may consult [DJ90].

Let $T(F)$ be the set of terms on F (all terms to be considered in the sequel do not have variables). The congruence on $T(F)$ generated by the associative commutative equations satisfied by the symbols in F_{AC} will be written: $=_{F_{AC}}$ or $=_{AC}$ when there is no ambiguity. The extension of this relation to multisets of terms is $==_{AC}$. The root symbol of a term t is denoted by $root(t)$. We write $s[t]_p$ to indicate that a term s contains t as a subterm at position p. Positions are classically coded by sequences of integers. For instance, $t_{|p}$ indicates the subterm of t which occurs at position p.

A *rewrite rule* on $T(F)$ is an ordered pair (l, r) of terms also denoted by $l \rightarrow r$. A *rewrite system* is any set R of rewrite rules. The rewriting relation generated by a rewrite system R is the binary relation $\rightarrow_R$ defined on terms by: $s \rightarrow_R t$ iff $s =_{AC} u[l]_p$ and $t =_{AC} u[r]_p$ for some context u, position p and rule $l \rightarrow r$ in R. The inverse of $\rightarrow_R$ is denoted by $\leftarrow_R$. The transitive reflexive closure of $\rightarrow_R$ (resp. $\leftarrow_R$) is denoted by $\rightarrow_R{}^*$ (resp. $\leftarrow_R{}^*$). A term t is *irreducible* in R if there is no u such that $t \rightarrow_R u$. An irreducible term s is called a *normal form* of t in R if $t \rightarrow_R{}^* s$. The congruence relation generated by R and the AC axioms is denoted by: $\overset{R}{\longleftrightarrow}$. The rewrite system R is *confluent* if $s \leftarrow_R{}^* u \rightarrow_R{}^* t$ implies that there exist v, w such that $s \rightarrow_R{}^* v =_{AC} w \leftarrow_R{}^* t$. The rewrite system R is *locally confluent* if $s \leftarrow_R u \rightarrow_R t$ implies that there exist v, w such that $s \rightarrow_R{}^* v =_{AC} w \leftarrow_R{}^* t$. A rewrite system R *terminates* if $\rightarrow_R$ is noetherian. A set R of rules is *reduced* if for every rule $l \rightarrow r$ in R, l is irreducible under $\rightarrow_{R-\{l \rightarrow r\}}$ and r is irreducible under $\rightarrow_R$. A rewrite system is *canonical* if it is terminating, confluent and reduced. In a canonical system terms have unique normal forms. As a consequence the word problem is decidable in every theory which admits a canonical system.

3 A compatible ordering for AC theories

The design of orderings for proving termination of rewrite systems modulo AC has been considered a hard task. In fact, to our knowledge, very few constructions are available in the literature. Perhaps the best known among them is the *associative path ordering* scheme [BP85], which extends the *recursive path ordering* (see also [BD86]). However, this ordering puts serious limitations on the precedence of AC-symbols. In fact two AC-symbols cannot be compared in the precedence unless they are related by a distributivity law. That explains why there is no hope to get a total ordering from the *associative path ordering*. An extension of Knuth-Bendix ordering has also been proposed by [Ste90]. However, it suffers the same kind of restriction. Recently a very interesting method has been proposed for proving termination of AC rewrite-systems in [KSZ90]. However, it seems difficult to adapt it to derive a total ordering on ground terms.

Another construction uses polynomial interpretations. We are going to elaborate on this method in order to obtain an ordering with the required properties for our purpose. Hence, let us first recall the interpretation technique. Each n-ary function symbol f is interpreted as an integer polynomial P^f with n indeterminates. The interpretation $I(t)$ of a term t is recursively defined by the rule:

$$I(f(t_1, ..., t_n)) = P^f(I(t_1), ..., I(t_n))$$

A set of rules terminates if we can choose polynomial interpretations such that for every rule $l \rightarrow r$ in the set we have $I(l) > I(r)$. Some other properties, to be detailed in this section, are also required.

For proving termination of rewrite systems modulo AC axioms, the ordering should possess the *AC-compatibility* property:

Definition 1 *An ordering $>$ on $T(F)$ is AC-compatible iff whenever we have $s > t$, $s =_{AC} s'$ and $t =_{AC} t'$ we also have $s' > t'$.*

The easiest way to ensure AC-compatibility with polynomial interpretations is to use polynomials which interpret identically the terms of the same AC-class. BenCherifa and Lescanne have pointed out the following necessary and sufficient condition for such a property:

Each polynomial $P^f(x,y)$ interpreting an AC symbol f must be of the form $axy + b(x+y) + c$ where the coefficients satisfy $b^2 = b + ac$.

In order to get an ordering which is also monotonic and total, here, we shall use another construction which may be interesting on its own. Instead of taking numerical values for the coefficients a,b and c, we shall take integer polynomials for them. We now describe the construction precisely.

Our interpretation domain will be the free commutative ring on $\{X_f; f \in F\}$, which is isomorphic to $Z[X_1, ..., X_m]$, where m is the cardinal of F. This algebra will be denoted by ZT. Hence, to each function symbol $f \in F$, is associated an indeterminate X_f.

The subset of elements of ZT whose coefficients are non-negative integers is denoted by NT. Let us define an ordering on NT. We suppose a given total precedence on the indeterminates and we suppose that the indeterminates of every monomial are sorted decreasingly according to this order. Two monomials are compared first by their degree, and second, when these degrees are equal, by lexicographic order.

We define $>_N$ to be the multiset extension of this ordering and we shall use it to compare polynomials, by treating them as multisets of monomials. The main property of $>_N$ is that it is *well-founded.*

The set NT is the target of the interpretation I that we are going to introduce now. For every term $f(t_1, ..., t_n)$ in $T(F)$,

$$I(f(t_1,...,t_n)) = (X_f+1)((X_f^2+2X_f)I(t_1)I(t_2)\cdots I(t_n)+(X_f+1)(I(t_1)+I(t_2)+\cdots+I(t_n))+1)$$

and for any constant a, $I(a) = X_a + 1$.

The following lemma is straightforward:

Lemma 1 *If* $s =_{AC} t$ *then* $I(s) = I(t)$.

Let $root(t)$ denote the root symbol of a term t. When the arity of $root(t)$ is k, the k-tuple $(t_{|1}, ..., t_{|k})$ of immediate subterms of t is denoted by $im(t)$. We also introduce for each AC-symbol f a function ρ_f which maps every term to a multiset of terms and which is defined recursively as:

$$\rho_f(g(s_1,...,s_n)) = \begin{cases} \bigcup_{i=1}^{i=n} \rho_f(s_i) & \text{if } g = f \\ \{g(s_1,...,s_n)\} & \textit{otherwise} \end{cases}$$

The next lemma tells that we can recover the root symbol of s from its interpretation $I(s)$.

Lemma 2 *Given terms* s *and* t, *if* $I(s) = I(t)$ *then* $root(s) = root(t)$.

Proof: Suppose that $s = f(s_1, ..., s_n)$ and $t = g(t_1, ..., t_m)$, where f and g represent two different function symbols. Then

$$\begin{aligned} I(s) &= (X_f+1)((X_f^2+2X_f)(I(s_1)\cdots I(s_n)) + (X_f+1)(I(s_1)+\cdots+I(s_n))+1) \quad (1) \\ I(t) &= (X_g+1)((X_g^2+2X_g)(I(t_1)\cdots I(t_m)) + (X_g+1)(I(t_1)+\cdots+I(t_m))+1) \quad (2) \end{aligned}$$

If $I(s) = I(t)$ then $(X_f + 1)$ divides $I(t)$ in ZT. However, since ZT is a factorial ring and $(X_f + 1)$ does not divide $(X_g + 1)$, we have:

$$(X_f+1) \mid ((X_g^2+2X_g)(I(t_1)\cdots I(t_m)) + (X_g+1)(I(t_1)+\cdots+I(t_m))+1)$$

Let $g_1, ..., g_m$ be the root symbols of $t_1, ..., t_m$ respectively. Applying the substitution $\sigma = \{X_g \leftarrow X_f, X_{g_1} \leftarrow X_f, ..., X_{g_m} \leftarrow X_f\}$ to the previous expression, we get:

$$(X_f+1) \mid ((X_f^2+2X_f)\sigma(I(t_1)\cdots I(t_m)) + (X_f+1)\sigma(I(t_1)+\cdots+I(t_m))+1)$$

This is equivalent to have:

$$(X_f + 1) \mid X_f.\sigma(I(t_1)\cdots I(t_m)) + 1$$

However, since every $I(t_i)$ admits some $X_{g_k} + 1$ as a factor, we can notice that $X_f + 1$ divides $\sigma(I(t_1)\cdots I(t_m))$. This causes a contradiction.

Lemma 3 *Given terms s and t, if $I(s) = I(t)$ then every function symbol has the same number of occurrences in s and t.*

Proof: Notice that the maximal monomial in $I(s)$ is $\prod_{f\in S} X_f^3 \cdot \prod_{f\in C} X_f$ where S (resp. C) is the multiset of symbols in s whose arity is > 0 (resp. 0).

We are now in position to define the relation on $T(F)$ that we want to use as an ordering for proving termination.

Definition 2 *Let $s = f(s_1, ..., s_n)$ and $t = g(t_1, ..., t_m)$. Then $s \succ t$ iff*

- *$I(s) >_N I(t)$ or*
- *$I(s) = I(t)$ and*
  - *if $f \in AC$ then $\rho_f(s) \succ\!\!\succ \rho_f(t)$*
  - *if $f \notin AC$ then $im(s) \succ_{lex} im(t)$*

  where $\succ\!\!\succ$ is defined by: $X = \{a_1, ..., a_n\} \succ\!\!\succ Y = \{b_1, ..., b_m\}$ iff

  - *$X \neq \emptyset$ and $Y = \emptyset$ or*
  - *for some i,j: $a_i =_{AC} b_j$ and $X - \{a_i\} \succ\!\!\succ Y - \{b_j\}$ or*
  - *for some $i,j_1,...,j_k$: $a_i \succ b_{j_1}, ..., b_{j_k}$ $X - \{a_i\} \succ\!\!\succ Y - \{b_{j_1}, ..., b_{j_k}\}$.*
  - *for some $i,j_1,...,j_k$: $a_i \succ b_{j_1}, ..., b_{j_k}$ $X - \{a_i\} ==_{AC} Y - \{b_{j_1}, ..., b_{j_k}\}$.*

  and where $\succ_{lex}$ is defined by: $(a_1, ..., a_n) \succ_{lex} (b_1, ..., b_m)$ iff $\exists i, \forall j < i$ $a_j =_{AC} b_j$ and $a_i \succ b_i$.

The important properties of this relation are quoted in the next proposition:

Proposition 1 *The relation $\succ$ is*

1. *irreflexive and transitive*
2. *well-founded (there is no infinite descending chain $t_1 \succ t_2 \succ ....$)*

3. *monotonic (for any function f and terms s, t, $s_1, ..., s_n$, $s \succ t$ implies $s' = f(s_1, ..., s_{i-1}, s, s_{i+1}, ..., s_n) \succ f(s_1, ..., s_{i-1}, t, s_{i+1}, ..., s_n) = t'$).*

4. *total on the set of AC-classes (for all s,t we have either $s \succ t$ or $t \succ s$ or $s =_{AC} t$)*

5. *AC-compatible.*

Proof of 1: Irreflexivity is obtained easily by induction on the size of terms. Suppose now that $t \succ s$ and $s \succ u$. If $I(t) >_N I(s)$ and $I(s) >_N I(u)$ then $I(t) >_N I(u)$ and therefore $t \succ u$. We can draw the same conclusion if $I(t) = I(s)$ and $I(s) >_N I(u)$ or if $I(t) >_N I(s)$ and $I(s) = I(u)$. Now, if $I(t) = I(s) = I(u)$ then, due to lemma 2, s, t, and u have the same root symbol f. Suppose f is AC then $\rho_f(t) \succ\!\succ \rho_f(s)$ and $\rho_f(s) \succ\!\succ \rho_f(u)$. By induction on the sizes of terms we have $\rho_f(t) \succ\!\succ \rho_f(u)$ and then $t \succ u$. When f is not AC, we apply the same reasoning to $\succ_{lex}$.

Proof of 2: Suppose that there is an infinite antichain: $t_1 \succ t_2 \succ \cdots$. Since $>_N$ is well-founded, there exists k such that for all $j \geq k$, we have $I(t_j) = I(t_k)$. Hence, by lemma 3, the terms t_j for $j \geq k$ have the same number of occurrences of any function symbol. But from a finite multiset of symbols we can build only finitely many ground terms. Therefore, by the pigeon-hole principle, there exist two indices l,m ($l > m$) such that $t_l = t_m$. But, by transitivity, $t_l \succ t_m$. However, this is not compatible with irreflexivity.

Proof of 3: If $I(s) >_N I(t)$ then $I(f(..s..)) >_N I(f(..t..))$ and $f(..s..) \succ f(..t..)$ follows. Suppose now that $I(s) = I(t)$. If f is not an AC symbol, $im(s') \succ_{lex} im(t')$. If f is an AC symbol then we notice that $\rho_f(s') \succ\!\succ \rho_f(t')$.

Proof of 4: We use induction on the size of terms. Suppose that 4 is true for all terms of size $< k$. Then n-uples (resp. multisets) of terms each of size less than k can be ordered with $\succ_{lex}$ (resp. $\succ\!\succ$). Let s, t be such that (at least) one of them has size k. Suppose that neither $s \succ t$ nor $t \succ s$. Then $I(s) = I(t)$. But, if $f = root(s)$ is AC, then by the induction hypothesis, $\rho_f(s) \succ\!\succ \rho_f(t)$ or $\rho_f(t) \succ\!\succ \rho_f(s)$ or $im(s) ==_{AC} im(t)$. The first two cases contradict the hypothesis. Hence there only remains the third from which we derive $s =_{AC} t$. When $root(s)$ is not AC, the proof is similar.

Proof of 5: Suppose that $s' =_{AC} s = f(s_1, ..., s_n) \succ t = g(t_1, ..., t_n) =_{AC} t'$. If $I(s) >_N I(t)$, by lemma 1, we also have $I(s') >_N I(t')$. If $I(s) = I(t)$ and f is not an AC function, the corresponding components of $im(s)$ and $im(s')$ are AC-equal. Since $g = f$, the same is true for $im(t)$ and $im(t')$. Therefore, $im(s') \succ_{lex} im(t')$. If f is AC, notice that $\rho_f(s) ==_{AC} \rho_f(s')$ and $\rho_f(t) ==_{AC} \rho_f(t')$. From $\rho_f(s) ==_{AC} \rho_f(t)$, we derive also $\rho_f(s') ==_{AC} \rho_f(t')$.

4 Completion of ground systems with associative-commutative symbols

Solving the word problem by completion in theories presented by sets of ground equations is known to be always possible, even efficiently [GNP*88, Sny89]. When the signature is built solely from constants and one associative commutative operation, we get finitely presented commutative semigroups. In such structures, the word problem can still be solved by completion [BL81]. We are going to show in the following that this method generalizes to signatures which contain free function symbols and several associative commutative operations.

For convenience, we shall use the flattened representation for terms which contain AC symbols. For instance, if $f \in F_{AC}$, a flattened representation of $f(a, f(b, c))$ is $f(a, b, c)$. When $f(t_1, ..., t_n)$ is a flattened representation of t and $f \in AC$, then for all i the root of t_i differs from f. In that case, we say that the t_i are the *fundamental subterms* of t and we write $fs(t) = \{t_1, ..., t_n\}$. Note also that $\rho_f(t) = fs(t)$ when the root of t is $f \in F_{AC}$.

In the completion approach to word problems, the main operation is the generation of critical pairs, which can be used to test the confluence property of the rewriting relation. Hence, we give the suitable definition of critical pairs for our framework:

Definition 3 *Assume that $l \to r$ and $g \to d$ are two rules in R such that $l =_{AC} f(t_1, ..., t_n)$, $g =_{AC} f(s_1, ..., s_m)$, $f \in F_{AC}$ and there exist i and j such that $t_i =_{AC} s_j$. Let S be a maximal multiset of terms such that there exist $\{k_1, ..., k_v\} \subseteq \{1, ..., n\}$ and $\{l_1, ..., l_u\} \subseteq \{1, ..., n\}$ with:*

$$S \cup \{t_k; k \in \{k_1, ..., k_v\}\} ==_{AC} \{t_k; k = 1, .., n\}$$

and

$$S \cup \{s_l; l \in \{l_1, ..., l_u\}\} ==_{AC} \{s_l; l = 1, .., m\}.$$

The pair of terms $(f(s_{l_1}, ..., s_{l_u}, r), f(t_{k_1}, ..., t_{k_v}, d))$ is called a critical pair.

A critical pair (r, s) of R is *trivial* if there is a term t such that $r \to_R^* t$ and $s \to_R^* t$. The main property of reduced systems of rules is stated in the proposition:

Proposition 2 *Given a reduced system R, the relation $\to_R$ is locally confluent if and only if every critical pair (r, s) of R is trivial.*

Remark: If we compare with the general AC-completion procedure, the critical pairs we need here are between a rule and an extended rule. There are no critical pairs between rules since the left-hand side are kept reduced. Critical pairs between extended rules are covered by the other ones.

We introduce now a completion procedure which, given a ground AC theory, always yields a canonical system. We assume that the equations are oriented according to an ordering

$>$ which is well-founded, monotonic, AC-compatible and total on the set of AC congruence classes. Due to our construction of Section 3, we know that such orderings exist. We extend $>$ to rewrite rules by comparing their left-hand sides and right-hand sides in a lexicographic way. This ordering on rules is noetherian too. Our completion procedure when applied to an initial set L_0 of rules produces a sequence of sets of rules $L_0, R_0, \ldots, L_i, R_i, L_{i+1}, R_{i+1}, \ldots$ which is defined as follows:

- Given L_i, generate a reduced set R_i:
  - While possible do:
    - Choose $(r,s) \in R_i$ and simplify every rule of $R_i - \{(r,s)\}$ which is *not smaller* than (r,s) by (r,s).
    - For any simplified rule (u,v) do
      If $u =_{AC} v$ then delete (u,v) else if $v > u$ then replace (u,v) by (v,u).
  - EndWhile
- Given R_i, compute its set of non-trivial critical pairs CP_i.
- If $CP_i = \emptyset$ then halt else let $L_{i+1} = R_i \cup \{(r,s); (r,s) \text{ or } (s,r) \in CP_i \text{ and } r > s\}$.

Given a set of rules R, and an AC symbol f, we define its *set of generators* $G_f(R)$ to be the set of congruence classes modulo R of the elements of the following set:

$$\bigcup_{(l,r)\in R_f} (\rho_f(l) \cup \rho_f(r))$$

where R_f is the set of rules in R such that at least one of their sides has f as root symbol. We also define $N(R)$ to be the set of sides of rules of R whose root symbol is not AC. For instance, consider the following system R:

$$\begin{aligned} a+b &\rightarrow a*b \\ a*c &\rightarrow g(e) \\ e &\rightarrow f \end{aligned}$$

where $+$ and $*$ are AC. Then $G_+(R) = \{a, b, a*b\}$, $G_*(R) = \{a, b, c, a+b, g(e)\}$ and $N(R) = \{e, f, g(e)\}$. The elements of $G_+(R)$ and $G_*(R)$ have to be considered as *representatives* of their congruence classes modulo R.

The important fact about the set of generators is that it cannot *increase* during the completion procedure. Let us be more specific. The relation $>$ is extended to sets of terms in the following way: $A > B$ if B is obtained either by replacing some element a of A by b such that $a > b$ or by removing some element from A. The relation $>$ is still well-founded on sets. Due to the structure of our completion procedure which simplifies all occurrences of a reducible term in the same step, we notice that $N(R)$ cannot get larger with respect to $>$.

Lemma 4 *For all $i \geq 0$, $G_f(R_i) \subseteq G_f(L_i)$ and $N(R_i) \leq N(L_i)$. If (r,s) is a critical pair of R with $r > s$, then $G_f(R \cup \{(r,s)\}) = G_f(R)$ and $N(R) = N(R \cup \{(r,s)\})$.*

Theorem 1 *When applied to any finite set of rules R, the completion procedure halts with a canonical system.*

Proof: The procedure never fails to orient a new pair of terms since we are provided with an ordering which can always compare two terms which are not congruent modulo AC. Consider now the smallest rule r_1 in $\bigcup_{i \geq 0} R_i$. It belongs to some set R_{i_1}. Notice that r_1 also belongs to any R_j with $j \geq i_1$. Otherwise it would be reducible by some other rule in $\bigcup_{i \geq 0} R_i$. But this possibility is ruled out by the fact that r_1 is minimal and a rule is never reduced by a bigger rule. Consider now the rule r_2 which is minimal in $(\bigcup_{i \geq i_1} R_i) - \{r_{i_1}\}$. It belongs to some set R_{i_2}. Note that r_2 also belongs to any R_j with $j \geq i_2$. The reason is that if it were reducible, that could only be possible by r_1. However, since R_{i_2} is reduced this case is excluded. Hence, by induction, we can build a sequence $r_{i_1}, ..., r_{i_k}, ...$ such that r_{i_k} is minimal in $(\bigcup_{i \geq i_{k-1}} R_i) - \{r_{i_1}, ..., r_{i_{k-1}}\}$. The fact that R_{i_k} is reduced shows that r_{i_k} cannot be reduced later on. Consider now the set $S = \{r_{i_1}, ..., r_{i_k}, ...\}$. We have the following result:

Lemma 5 *The set of rules S is canonical.*

Sketch of proof: It is not difficult to see that S is reduced. Suppose that there is a non-trivial critical pair (u,v) between two rules of S, say r_l and r_m. Consider a minimal proof of (u,v) in $\bigcup_{i \geq 0} R_i$ with respect to the proof ordering defined in [BDH86]. If one step uses a rule which is not in S, this rule will be reduced at some stage of the algorithm. Therefore, we can build a smaller proof. Hence, we can assume that every step in the minimal proof of (u,v) is justified by a rule in S. If this proof is not a rewrite proof, then the computation of a new critical pair allows again to derive a smaller proof.

We prove now that S is finite. Assume that S is infinite. Notice that for any AC symbol f, $G_f(S) \subseteq G_f(R)$. Hence, $G_f(S)$ is finite. Also by lemma 4 we can prove that $N(S)$ is finite. Since S is infinite there are infinitely many rules $(l_i \to r_i)_{i \geq 0}$ whose left-hand sides have the same root symbol f. From the finiteness of $N(S)$ we deduce that f is necessarily AC. Since S is reduced, all the proper subterms of the left-hand sides of S are in normal form. As a consequence, we have:

Lemma 6 *If $s \in fs(l_i)$, $t \in fs(l_j)$ and $s \overset{R}{\longleftrightarrow} t$ then $s =_{AC} t$.*

We introduce the notation K_i for the multiset of congruence classes modulo $\overset{R}{\longleftrightarrow}$ of the elements of $fs(l_i)$. Since $G_f(S)$ is finite (let M be its cardinality) and since every element of K_i is also an element of $G_f(S)$ then we can associate to each K_i a vector of integers $(h_1, h_2, \ldots, h_M)$ where h_g is the number of times that the element of rank g in $G_f(S)$ occurs in K_i. We apply Dickson's lemma [Dic13] to this set of vectors. Let us recall this result:

Lemma 7 (Dickson's Lemma) *Given $n \in N$, every infinite sequence of n-dimensional vectors with nonnegative integer components must contain an infinite subsequence that is nondecreasing with respect to $\leq$ where $W \leq V$ if there exists a vector U with nonnegative components such that $V = W + U$.*

From this lemma we deduce that there exist two rules $l_k \rightarrow r_k$ and $l_j \rightarrow r_j$ such that K_k is a submultiset of K_j. By the previous lemma, this ensures the existence of a submultiset A of $fs(l_j)$ such that $A =\!=_{AC} fs(l_k)$. This implies that l_j can be simplified by $l_k \rightarrow r_k$, contradicting the fact that S is reduced.

We can now achieve the proof of theorem 1. Since S is finite, let m be the smallest index such that $S \subseteq R_m$. One easily sees that $S = R_j$ for all $j \geq m$. In particular $R_{m+1} = R_m$, which means that the completion procedure stops at step m. □

5 Related problems, further works and conclusion

In the previous section, it was shown that in ground associative commutative theories the word problem can always be solved by construction of a canonical system. A natural generalization of the word problem is the unifiability problem. We have investigated this problem by coding it as a reachability problem in Petri nets [PM90]. Then, from the decidability of reachability in Petri nets [May81], the decidability of unifiability in ground AC theories should follow. This idea comes from the observation that the elements of finitely presented commutative semigroups can be represented as vectors of integers. In the general case of several associative commutative symbols, the problem must be coded as a conjunction of reachability problems, each of them being related to one of the AC symbols.

For proving that we can always orient ground AC systems we have developed a generalization of the polynomial ordering. Instead of using polynomials with numerical range we rather use polynomials whose range are polynomial rings. To our knowledge, this is the first occurrence of a reduction ordering which is total on the set of congruence classes modulo AC and which is AC-compatible (for a signature which contains any number of AC symbols). Orderings with such properties are fundamental to derive refutationally complete theorem-proving strategies with built-in associative commutative unification [AH90].

Another interesting issue would be to see what happens when we replace the associativity and commutativity axioms by another well-behaved theory E. Does every ground presentation admit a canonical rewrite system modulo E? An interesting result along these lines is the construction in [KN85] of a Thue system which has a decidable word problem but which has no canonical system. Whether it is possible to extend fast ground completion techniques of [GNP*88, Sny89] to the ground AC case is also an open problem.

Acknowledgements: We thank Eric Domenjoud for reading the manuscript and Uwe Waldmann for correcting a definition.

References

[AH90] S. Anantharaman and J. Hsiang. An automated proof of the moufong identities in alternative rings. *J. Automated Reasoning*, 6:79–109, 1990.

[BD86] L. Bachmair and N. Dershowitz. Commutation, transformation and termination. In J. Siekmann, editor, *Proceedings 8th Conf. on Automated Deduction*, pages 5–20, Springer-Verlag, 1986. Lecture Notes in Computer Science, volume 230.

[BDH86] L. Bachmair, N. Dershowitz, and J. Hsiang. Orderings for equational proofs. In *Proceedings Symp. Logic in Computer Science*, pages 346–357, Boston (Massachusetts USA), 1986.

[BL81] A.M. Ballantyne and D. Lankford. New decision algorithms for finitely presented commutative semigroups. *Comp. & Maths. with Appl.*, 7:159–165, 1981.

[BL87] A. BenCherifa and P. Lescanne. Termination of rewriting systems by polynomial interpretations and its implementation. *Science of Computer Programming*, 9(2):137–160, October 1987.

[BP85] L. Bachmair and D.A. Plaisted. Termination orderings for associative-commutative rewriting systems. *Journal of Symbolic Computation*, 1:329–349, 1985.

[D87] N. Dershowitz. Termination of Rewriting. *Journal of Symbolic Computation*, 1 & 2:69-116, 1987.

[DJ90] N. Dershowitz and J.-P. Jouannaud. Rewrite systems. In Van Leuven, editor, *Handbook of Theoretical Computer Science*, North Holland, 1990.

[Dic13] L. Dickson. Finiteness of the odd perfect and primitive abundant numbers with n distinct prime factors. *Amer. J. Math.*, XXXV:413–422, 1913.

[GNP*88] J. Gallier, P. Narendran, D. Plaisted, S. Raatz, and W. Snyder. Finding canonical rewriting systems equivalent to a finite set of ground equations in polynomial time. In E. Lusk and R. Overbeek, editors, *Proceedings 9th Int. Conf. on Automated Deduction*, pages 182–196, Springer-Verlag, Lecture Notes in Computer Science, 1988.

[HsRus86] J. Hsiang, and M. Rusinowitch. A new method for establishing refutational completeness in theorem proving. Proceedings of the 8th Conference on Automated Deduction, LNCS 230 (1986) 141-152.

[Hue80] G. Huet. Confluent reductions : abstract properties and applications to term rewriting systems. *Journal of the Association for Computing Machinery*, 27(4):797–821, October 1980.

[JK86] J.-P. Jouannaud and H. Kirchner. Completion of a set of rules modulo a set of equations. *SIAM Journal of Computing*, 15(4):1155–1194, 1986.

[KN86b] D. Kapur and P. Narendran. NP-completeness of the associative-commutative unification and related problems. Unpublished Manuscript, Computer Science Branch, General Electric Corporate Research and Development, Schenectady, NY, Dec. 1986.

[KN85] D. Kapur and P. Narendran. A finite Thue system with decidable word problem and without equivalent finite canonical system. *Theoretical Computer Science*, 35:337–344, 1985.

[KSZ90] D. Kapur, G. Sivakumar, and H. Zhang. A New Method for Proving Termination of AC-Rewrite Systems. To be presented at the Conference on the *Foundations of Software Technology and Theoretical Computer Science*, New Delhi, India, December 1990.

[KB70] D.E. Knuth and P.B. Bendix. Simple word problems in universal algebras. In J. Leech, editor, *Computational Problems in Abstract Algebra*, pages 263–297, Pergamon Press, Oxford, 1970.

[Lan79] D.S. Lankford. *On Proving Term Rewriting Systems are Noetherian.* Technical Report, Louisiana Tech. University, Mathematics Dept., Ruston LA, 1979.

[LB77] D.S. Lankford and A. Ballantyne. *Decision procedures for simple equational theories with permutative axioms: complete sets of permutative reductions.* Technical Report, Univ. of Texas at Austin, Dept. of Mathematics and Computer Science, 1977.

[May81] E. W. Mayr. An algorithm for the general petri net reachability problem. In *Proceedings of STOC*, 1981.

[PM90] P. Narendran and M. Rusinowitch. Unifiability in ground AC theories. 1990. In preparation.

[PS81] G. Peterson and M. Stickel. Complete sets of reductions for some equational theories. *Journal of the Association for Computing Machinery*, 28:233–264, 1981.

[Sny89] W. Snyder. Efficient completion: an $O(nlogn)$ algorithm for generating reduced sets of ground rewrite rules equivalent to a set of ground equations E. In N. Dershowitz, editor, *Proceedings 3rd Conf. on Rewriting Techniques and Applications*, pages –, Springer-Verlag, Lecture Notes in Computer Science, 1989.

[Ste90] Steinbach, J. AC-termination of rewrite systems - A modified Knuth-Bendix ordering. Proceedings 2nd International Conference on Algebraic and Logic Programming, Nancy (France), Lecture Notes in Computer Science 463, 372-386, (1990).

A Narrowing-Based Theorem Prover

Ulrich Fraus
Bayerisches Forschungszentrum für Wissensbasierte Systeme,
Universität Passau, Postfach 2540, W-8390 Passau, Germany
fraus@forwiss.uni-passau.de

Heinrich Hußmann
Institut für Informatik, Technische Universität München,
Postfach 20 24 20, W-8000 München 2, Germany
hussmann@lan.informatik.tu-muenchen.dbp.de

Abstract

This work presents a theorem prover for inductive proofs within an equational theory which supports the verification of universally quantified equations. This system, called TIP, is based on a modification of the well-known *narrowing algorithm*. Particulars of the implementation are stated and practical experiences are summarized.

Equational axiomatic specifications are now widely accepted as a promising tool for the early phases of software development. The special style of equational specifications is interesting for the intermediate stage in which the first detailed formal description of the intended product is given. The main advantage of an axiomatic approach to software specification is formal reasoning: Axioms, together with a good calculus, allow computer-assisted verification of propositions about the specified piece of software. Within the software development process, such a tool helps, for instance, to check whether a first operational formulation of a system meets more informal requirements which have been formulated without any consideration for the operationality. It is the aim of the system presented here to contribute to such a verification of software in a very early phase of the development.

In our approach, we try to develop an alternative to the so-called "inductionless induction" method, which has been developed based on early work by Knuth/Bendix, Goguen, Huet/Hullot and others. The aim is here, to keep the proof method rather "natural", i.e. very similar to the method a human being uses when performing a proof with paper and pencil. A successful proof achieved by our system is understandable to everybody who knows the basic mathematical facts of equational reasoning and induction. Moreover, the situation in which a new (usually more general) lemma has to be invented looks more familiar to the user - therefore it is easier to use human intuition. And human intuition will be necessary anyway for mastering a non-trivial proof - regardless of the proof method that is used.

The TIP system (Term Induction Prover) presented in this paper, is restricted to so-called constructor-based specifications, i.e. every ground term has to be equivalent to a constructor ground term (sufficient completeness [Gu 75]). All equations have the form:

$f(c_1, \ldots, c_n) \equiv t$ where f is a function symbol, the c_i are constructor terms (with variables) and t is an arbitrary term.

The basic idea of our variant of term induction is to "misuse" the case analysis which is normally given in the left hand sides of the rules for an inductive proof. This idea allows us to make extensive use of the algorithms that are already available in an implementation of narrowing. Since narrowing enumerates all alternatives for a unification of a subterm of a goal with all patterns of the rules, it automatically generates an appropriate case analysis (when using a constructor-complete specification).

Besides, the TIP system allows the user to work on several theorems in parallel. So one can start to prove a necessary lemma or a generalization whenever one recognizes the need for it. After finishing such a subproof, this lemma can be applied immediately in other proofs.

A short glance at the proof algorithm of the TIP system:

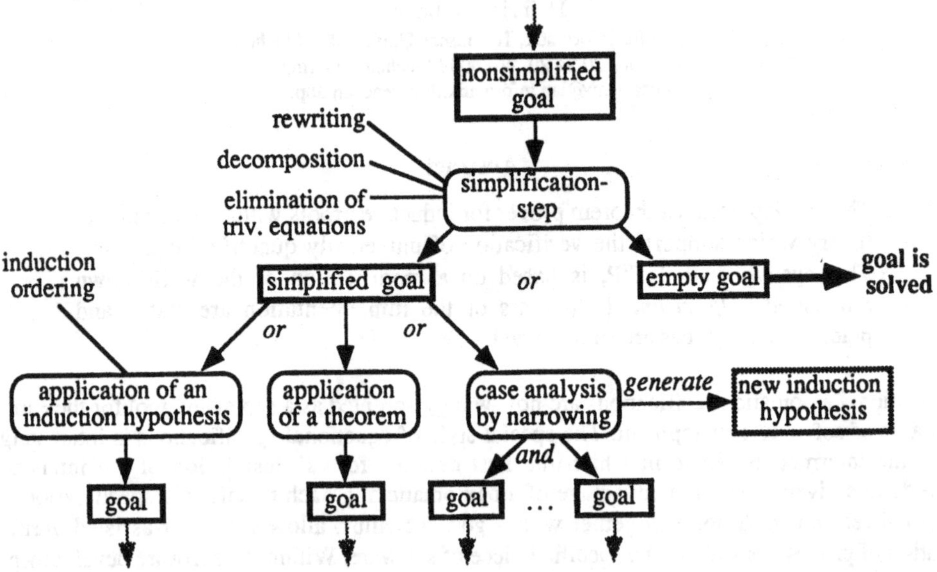

The behavior of the TIP system, in particular the degree of its automation, can be influenced by the user, if some system parameters are adjusted appropriately. Also it is possible to choose one of the four induction orderings, which are implemented in the system.

The theorem prover TIP is quite easy to use, because most parts of the tiresome work are done by the system automatically. So the user can concentrate on choosing the most promising proof steps. Easy proofs like the commutativity of the addition are found without any user interaction, but normally human intuition is needed to find the central idea(s) of a proof. All the standard examples for theorem proving can be done with the TIP system, too.

Up to now TIP can only prove unconditional goals, but we are planning to extend the proof algorithm (with small changes) to conditional goals. Another planned improvement is a comfortable X-Window user interface.

References

[Fr, Hu 90] U. Fraus, H. Hußmann: A Narrowing-Based Theorem Prover (long version). Internal report, University of Passau, 1990

[Gu 75] J. V. Guttag: The specification and application to programming of abstract data types. Ph. D. thesis, University of Toronto, Toronto, 1975.

[Re 90] U. S. Reddy: Term Rewriting Induction. CADE 10, Lecture Notes in Computer Science 249, 1990, 162-177.

[Wi e.a. 83] M. Wirsing, P. Pepper, H. Partsch, W. Dosch, M. Broy: On hierarchies of abstract data types. Acta Informatica 20, 1983, 1-33.

ANIGRAF : An Interactive System for the Animation of Graph Rewriting Systems with Priorities

Michel Billaud

Département Informatique, IUT-A, Université Bordeaux I
F-33405 Talence Cedex - France
Electronic Mail : billaud@geocub.greco-prog.fr

ANIGRAF is a tool for the visualization of Graph Rewriting Systems with Priorities. PGRS have been introduced in [BLMS 89] as a formalism for designing and proving distributed algorithms.

A PGRS is a finite set of rewriting rules, which *do not* modify the underlying graph, but only the *labels* attached to vertices or edges. There is a priority relation between rules, so when two occurrences of rules are overlapping in a graph, the rule with minor priority (if the rules are comparable) can not be applied to the graph. When the occurrences are not overlapping, both rules can be selected ; so checking whether a rule can be applied to a part of the graph does not require global informations.

Various algorithms (computing trees, electing a leader, finding an Hamiltonian path, etc.) are presented under the form of PGRS with their formal proofs in [Bi 89, LM90a, LM90c], and it is shown in [LM 90c] that PGRS can be used to express a wide class of algorithms.

ANIGRAF is written in Turbo-Pascal 4.0 under MS-DOS. It runs on IBM-PC/XT compatible computers with very limited harware : 512 KB of memory, one floppy disk, any graphics card (Hercules, CGA, EGA, ATT ...) supported by TP 4.0, and a Microsoft-compatible mouse. A previous version for X-Windows was written in C on a SUN 3/80 (with 8 MB) by J.L. Lafaye as a student's project [La 90].

Using ANIGRAF

The user must first prepare a *source file* describing the set of rules with their priorities, and one or several sample graphs. Here is a sample source file (building a covering tree by Tremaux algorithm) :

```
COULEURS         A N M F              ! labels for vertices : Active, Not-reached,Marked,Finished
                 n from to            ! for edges : not-used, source, end
REGLES down up                        ! the system has two rules

REGLE  down
PRECEDE up                            ! priority of 'down' over 'up'
SOMMET  father 50 10     A / M        ! the active vertex de-activates itself
SOMMET  neigh 50 90      N / A        ! and activates one of its  unreached neighbours
ARETE   father neigh   n  n / from to ! the edge now belongs to the tree

REGLE  up                             ! if the son  can't activate something
SOMMET  father 50 10     M / A        ! it reactivates its father
SOMMET  son 10 10      A / F          ! and de-activates itself
ARETE   father son     from to        ! the edge stays in the tree

GRAPHE disco                          ! a sample graph with 5 vertices and 8 edges ...
SOMMET x  50 90  N      SOMMET y  90 90  N      SOMMET z  90 50  A      SOMMET t  50 50  N
SOMMET u  10 10  N      ARETE x y n             ARETE y z n             ARETE z t n
ARETE t x n             ARETE x z n             ARETE t u n             ARETE u x n
ARETE z u n
```

ANIGRAF enters a mouse-window dialog. The screen looks like this :

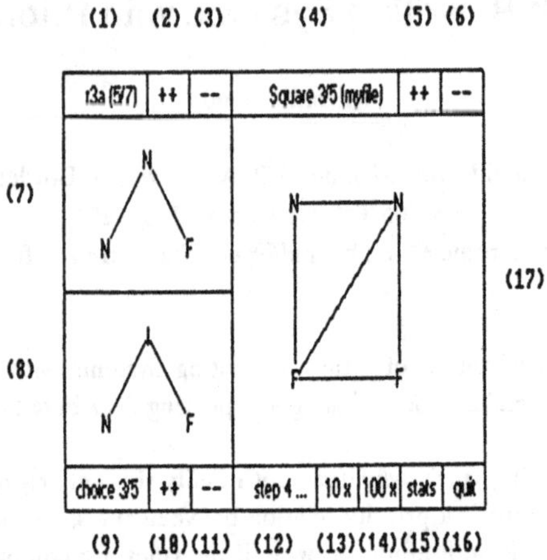

Button (1) displays the name (here 'r3a') of the *current rule* (5th out of the 7 rules). Windows (7,8) show both parts of the current rule. (4) displays the name of the *current graph* 'Square' : the 3rd example out of the 5 from source file 'myfile'. (17) shows the current graph. Buttons (9,12) indicate that there are 5 different ways to apply a rule to the current graph for the 4th step, the 3rd choice being selected now (an animation effect shows *where* the current rule can be applied, it is worth seeing, not explaining !)

Clicking the left button on (1) opens a temporary window with the name of the rules having lower/higher priorities than the current rule. (2, 3) browse the set of rules. (4) resets the current example to its initial labelling. (5, 6) browse the set of examples. (9) shows the animation effect for the current choice. (10) and (11) show the next/previous choices. (12) executes the selected choice, modifies the graph, and computes a new list of choices. (13, 14) start random sequences of 10 (or 100) rewriting steps. (15) opens a new window which tells how many times each rule was used. (16) quits. The right mouse button opens an Helpwindow which tells you what happens when you press the left button at the same location.

References

[Bi 89] M. Billaud, *Some Backtracking Graph Algorithms expressed by Graph Rewriting Systems with Priorities,* Internal Report LABRI n° 8989.

[Bi 90] M. Billaud, *Un Interpréteur pour les Systèmes de Réécriture de Graphes avec Priorités,* Internal Report LABRI n° 9040 (in french).

[BLMS 89] M. Billaud, P. Lafon, Y. Métivier, E. Sopena, *Graph Rewriting Systems with Priorities,* in Graph-theoretic Concepts in Computer Science, 15th Workshop on Graphs'89, LNCS 411, pp. 94-106.

[La 90] J.L. Lafaye, *Mémoire pour l'Obtention du D.U.T.,* Département Informatique, IUT 'A', Université Bordeaux I (1990).

[LM 90a] I. Litovsky, Y. Métivier, *Graph Rewriting Systems as a Tool to design and to prove Graph and Network Algorithms,* Internal Report LABRI n° 9070.

[LM 90b] I. Litovsky, Y. Métivier, *Computing Trees with Graph Rewriting Systems with Priorities,* Internal Report LABRI n° 9085.

[LM 90c] I. Litovsky, Y. Métivier, *Computing with Graph Rewriting Systems with Priorities,* Fourth International Workshop on Graph Grammars and Their Applications to Computer Science, Bremen (1990). Also Report LABRI n° 9087.

EMMY: A REFUTATIONAL THEOREM PROVER FOR FIRST-ORDER LOGIC WITH EQUATIONS*

Aline Deruyver
LIFL UA 369 CNRS, U.F.R. D'I.E.E.A bât. M3
59655 VILLENEUVE D'ASCQ CEDEX FRANCE
deruy@jade.u-strasbg.fr

ABSTRACT: Emmy is an implementation in Quintus Prolog of a refutational theorem prover for first-order logic with equations based on a superposition calculus. This paper describes the structure and the functionalities of this theorem prover.

1- Introduction

Emmy is an implementation in Quintus Prolog (version 2.4.2) of a refutational theorem prover for first-order logic with equations based on a superposition calculus described by Bachmair, Ganzinger (1990). This superposition calculus is refutationally complete and compatible with various simplification and deletion mechanisms, such as term rewriting and tautology deletion.
Emmy runs on Sun 3 machines (system Unix). Size of the source file: 113.872 K. Size of the file after compilation: 75.596 K

2- General overview of the system.

The general structure of the system can be illustrated by the following algorithm:

1- Initialisation.
We have three sets of equational clauses:
Clo = Ø, Clnn = Ø, and Cln is the set of input clauses.

2- Compute the maximal equation for each clause of Cln.

3- Inference engine.
While Cln ≠ Ø and the empty clause is not generated **do**

3.1- Sort the clauses of Cln.
3.2- Select the smallest clause of Cln.
3.3- Apply inference rules to the selected clause. The new clauses are put in the set Clnn and their maximal equation is computed.
3.4- Put the selected clause in Clo.
3.5- Simplify the clauses of Clo, Cln, and Clnn.
3.6- Put the clauses of Clnn in Cln.

4- If the empty clause is generated, then the set of input clauses is unsatisfiable. If it is not generated, the set of input clauses is satisfiable.

Let us explain in more details this algorithm:
i- We have several kinds of set of clauses: The set Clo of marked clauses, the set Cln of non-marked clauses and the set Clnn of new clauses generated at the end of a cycle.

*This work was done while the author was a visitor at SUNY at Stony Brook, N.Y., U.S.A.. Research was supported in part by National Science Fondation under grant CCR-8901322.

ii-The set Cln is sorted in increasing order. This sorting is done, such that, at each cycle we always take the smallest rule. This usually decreases the search space as we try to generate only small clauses.
At the moment the clauses can be sorted with respect to two different criteria:
-according to the number of function symbols and variables which occur in the clauses.
-according to the number of function symbols + 2*(number of variable symbols).

iii- Inference rules of our system are applied to the selected clause C.
We first apply equality resolution and ordered factoring. Next, we apply the strict superposition rules. We superpose the selected clause with each clause of Cln and with each clause of Clo for which this was not done previously.
When we apply the superposition rules, we overlap the maximal equations of each clauses.
Sometimes these equations are maximal if and only if some conditions are satisfied. After the unification, we propagate the substitutions inside the list of conditions and check whether they are still satisfied.
If it is not the case, then the new clause is not generated. Otherwise, it is generated if it is not a tautology and does not already exist.
Each new clause is renamed so that it does not share any variable with the other clauses.

iv-The new generated clauses are put in the set Clnn. The smallest clause C becomes a marked clause.

v- The clauses of each set are simplified in such a way that redundant clauses are deleted.

3- Examples.

About sixty examples have been tested on Emmy. They are splited in different categories (see technical report): Propositional logic, monadic predicate logic, full predicate logic (without identity and functions), full predicate logic with identity (without functions), full predicate logic with identity and arbitrary functions.
These examples are from the paper by Francis Jeffry Pelletier 1985 and from the paper by McCharen, Overbeek and Wos 1976.
From these tests, we could observed that the number of generated critical pairs is low. This number doesn't exceed 208 even in the most difficult cases, which is quite good if we compare with the results obtained by the system of Mc Charen, Overbeek and Wos.

4-References.

Bachmair L. and Dershowitz N. *Equational inference, canonical proofs and proof orderings* (1989).

Bachmair L. and Ganzinger H. *On restrictions of ordered paramodulation with simplification.* Proc. 10th Int. Conf. Aut. Ded. LNCS vol 449, pp. 427-441 (M. Stickel, ed.) (1990).

Dershowitz N. *Termination* Journal of Symbolic Computation (1986).

Deruyver A. *Emmy, a refutational theorem prover for first order logic with equations.* Technical report 90/26 State University of New York at Stony Brook. September 1990.

Huet G. and Oppen Derek C. *Equations and rewrite rules: A survey.* Technical report CSL-111 SRI international January 1980.

Mccharen John D., Overbeek Ross A. and Wos Laurence A. *Problems and experiments for and with automated Theorem-proving programs.* IEEE transactions on computers. Vol C-25 No 8 August 1976.

Pelletier Francis Jeffry *Seventy five problems for testing automatic theorem provers.* Journal of automated reasoning (1986) pp. 191-216.

The Tecton Proof System

Raj Agarwal, David R. Musser
Rensselaer Polytechnic Institute, Troy, NY 12180
Deepak Kapur, Xumin Nie
State University of New York, Albany, NY 12222

The Tecton Proof System is a new automated deduction system designed to support construction of large complex proofs, such as those that typically are necessary in assuring the correctness of computer programs or hardware designs. A main goal of Tecton is to aid the user in finding proofs and in understanding proofs in order to reuse parts of them, for the purpose of building libraries of verified software or hardware components. To achieve this goal, Tecton provides a structured internal representation of proofs and a visual representation using graphics and hypertext links, both of which are designed to help organize and clarify the structure of large proofs. In back of these "front end" facilities is a powerful automated deduction capability obtained from the Rewrite Rule Laboratory, RRL (for a brief overview of the main deductive methods of RRL, which include reduction, cover-set induction, generalization, and case analysis, see [3]).

The main features of Tecton are:

- A proof is displayed as a goal/subgoal tree, allowing easy identification of both the structure of the proof (how goals relate to each other) and the details of how each inference step depends on other information given previously (such as axioms and other theorems). Figure 1 shows a proof tree for a small example, conditional correctness of a procedure for concatenating two queues.

- Tecton makes a natural extension of tree-structured proofs to represent proofs of program correctness. Proof goals may be Hoare formulas as well as ordinary logical formulas, and Hoare-style proof rules for programming language constructs are applied to produce subgoals; this process is repeated until the leaves of the proof tree are ordinary logical formulas (to which inference rules of ordinary first-order logic are then applied). Figure 1 illustrates reduction of Hoare-formulas (three part boxes with preconditions, statements, and postconditions in the left, middle and right parts) with built-in Hoare rules for assignment, loop, composition, and other statement types. In this example, the reduction of ordinary logic formulas (two part boxes, containing assumptions and conclusions) is done with rules that have previously been proved and have separate proof trees (not shown in Figure 1).

- The visual representation of proofs combines a graphical representation of proof trees with *hypertext* links connecting different parts of the visual representation. The visual representation is presented on a series of *pages,* which allows the representation to be displayed on a workstation screen or printed on paper in essentially identical form (no reformatting is required). Several techniques are used to make proof trees relatively compact, including extensive use of references to information stored in tables that are displayed on each proof page; e.g., the Program, Rules, and Parts tables in Figure 1. (In Rules Tables free variables are implicitly universally quantified, but not in Parts Tables.)
- Tecton can explore different paths of inference steps in seaching for a proof, both in its automatic mode and with interactive guidance from the user. All such paths are recorded in the forest of proof trees maintained by the system, so that it is easy to compare different partial or complete proof attempts for length, clarity, or dependence on assumptions. This ability to retain multiple proof attempts (not illustrated in Figure 1) is one difference from an earlier notion of proof forest in [2].

These and other features result in a system that is useful both for educational purposes and for moderate-to-large-scale proof efforts.

Tecton runs on Sun3 and Sun4 Unix workstations, under either Sunview or the X Window System. The implementation is a combination of communicating programs—the front-end command and display interface implemented using KMS [1], a commercially available hypertext system; the program parser and Hoare formula generator implemented in C; and the "theorem proving engine," RRL, implemented in Common Lisp.

1 References

[1] R.M. Akscyn, D.L. McCracken, E.A. Yoder. "KMS: A Distributed Hypermedia System for Managing Knowledge in Organizations," Comm. ACM 31(7): pp. 820-835 (July 1988).

[2] R.W. Erickson and D.R. Musser, "The Affirm Theorem Prover: Proof Forests and Management of Large Proofs," *5th Conference on Automated Deduction, Lecture Notes in Computer Science*, Vol. 87, Springer-Verlag, New York, 1980.

[3] D. Kapur and H. Zhang, "An Overview of RRL (Rewrite Rule Laboratory)," Proc. of *Third International Conf. of Rewriting Techniques and Applications,* Chapel Hill, North Carolina, April 1989.

Proof

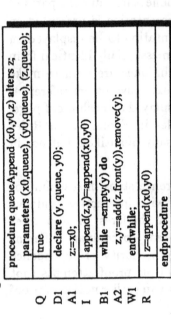

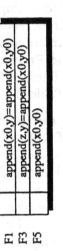

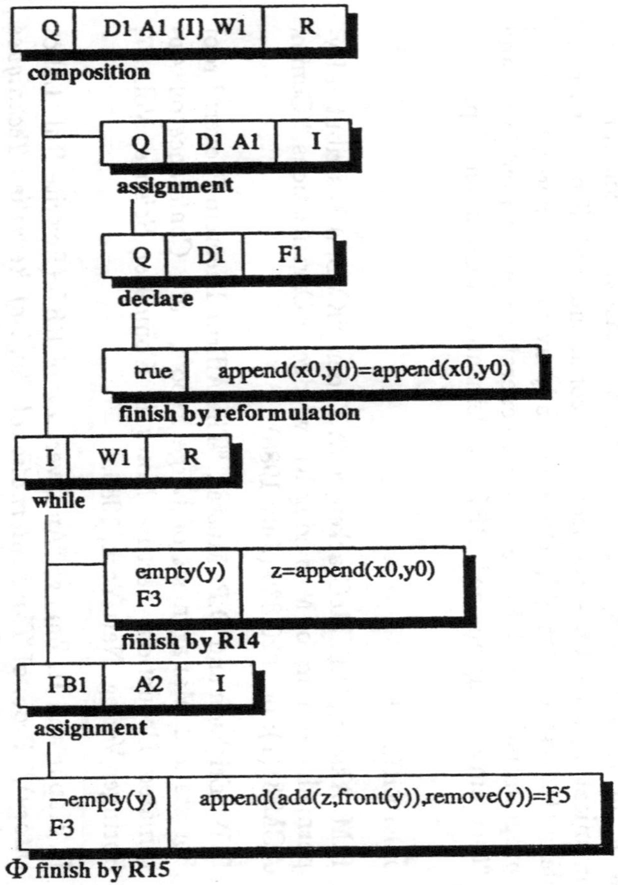

Program

	procedure queueAppend (x0,y0,z) alters z; parameters (x0,queue), (y0,queue), (z,queue);	
Q		true
D1 A1	declare (y, queue, y0); z:=x0;	
I		append(z,y)=append(x0,y0)

B1
A2
W1
	while ¬empty(y) do
	z,y:=add(z,front(y)),remove(y);
	endwhile;	

R		z=append(x0,y0)
	endprocedure	

Rules

R14	empty(y)	append(x,y) = x
R15	¬empty(y)	append(add(x,front(y)),remove(y)) = append(x,y)

Parts

F1		append(x0,y)=append(x0,y0)
F3		append(z,y)=append(x0,y0)
F5		append(x0,y0)

Figure 1: Example of how proof trees and information tables are displayed (actual Tecton Proof System output). The tabular, labeled form of the procedure is generated by Tecton from straight ASCII text entered by the user. The Proof Tree and its accompanying Rules and Parts tables are generated automatically when the proof is found using the deductive methods of RRL.

Open Problems in Rewriting

Nachum Dershowitz*
Department of Computer Science, University of Illinois
1304 West Springfield Avenue, Urbana, IL 61801, U.S.A.
nachum@cs.uiuc.edu

Jean-Pierre Jouannaud
Laboratoire de Recherche en Informatique, Bat. 490
Université de Paris Sud, 91405 Orsay, France
jouannau@lri.lri.fr

Jan Willem Klop†
CWI, Kruislaan 413, 1098 SJ Amsterdam, The Netherlands
Department of Mathematics and Computer Science, Free University
de Boelelaan 1081, 1081 HV Amsterdam, The Netherlands
jwk@cwi.nl

1 Introduction

Interest in the theory and applications of rewriting has been growing rapidly, as evidenced in part by four conference proceedings (including this one) [15, 26, 41, 66]; three workshop proceedings [33, 47, 77]; five special journal issues [5, 88, 24, 40, 67]; more than ten surveys [2, 7, 27, 28, 44, 56, 57, 76, 82, 81]; one edited collection of papers [1]; four monographs [3, 12, 55, 65]; and seven books (four of them still in progress) [8, 9, 35, 54, 60, 75, 84].

To encourage and stimulate continued progress in this area, we have collected (with the help of colleagues) a number of problems that appear to us to be of interest and regarding which we do not know the answer. Questions on rewriting and other equational paradigms have been included; many have not aged sufficiently to be accorded the appellation "open problem". We have limited ourselves to theoretical questions, though there are certainly many additional interesting questions relating to applications and implementations.

Previous lists of questions in this area include one distributed by Leo Marcus and one of us (Dershowitz) at the Sixth International Conference on Automated Deduction (New York, 1982), the questions posed in a set of lecture notes on "Term Rewriting Systems" by one of us (Klop) for a seminar on reduction machines (Ustica, 1985), another list by one of us (Jouannaud) in the *Bulletin of the European Association for Theoretical Computer Science* (Number 31, 1987), and electronic postings to the distribution list (rewriting@crin.crin.fr) maintained by Pierre Lescanne. We use primarily terminology and notation of [27].

*Research supported in part by the National Science Foundation under Grant CCR-9007195.
†Research partially supported by ESPRIT BRA projects 3020: Integration and 3074: Semagraph.

2 Problems

2.1 Rewriting

Problem 1. An important theme that is largely unexplored is definability (or implementability, or interpretability) of rewrite systems in rewrite systems. Which rewrite systems can be directly defined in lambda calculus? Here "directly defined" means that one has to find lambda terms representing the rewrite system operators, such that a rewrite step in the rewrite system translates to a reduction in lambda calculus. For example, Combinatory Logic is directly lambda definable. On the other hand, not every orthogonal rewrite system can be directly defined in lambda calculus. Are there universal rewrite systems, with respect to direct definability? (For alternative notions of definability, see [75].)

Problem 2 (M. Venturini-Zilli [91]). The reduction graph of a term is the set of its reducts structured by the reduction relation. These may be very complicated. The following notion of "spectrum" abstracts away from many inessential details of such graphs: If R is a term-rewriting system and t a term in R, let $Spec(t)$, the "spectrum" of t, be the space of finite and infinite reduction sequences starting with t, modulo the equivalence between reduction sequences generated by the following quasi-order: $t = t_1 \rightarrow_R t_2 \rightarrow_R \cdots \leq t = t'_1 \rightarrow_R t'_2 \rightarrow_R \cdots$ if for all i there is a j such that $t_i \rightarrow^*_R t'_j$. What are the properties of this cpo (complete partial order), in particular for orthogonal (left-linear, non-overlapping) rewrite systems? What influence does the non-erasing property have on the spectrum? (A rewrite system is "non-erasing" if both sides of each rule have exactly the same variables.) The same questions can be asked for the spectrum obtained for orthogonal systems by dividing out the finer notion of "permutation equivalence" due to J.-J. Lévy (see [14, 55, 57]).

Problem 3. A term t is "ground reducible" with respect to a rewrite system R if all its ground (variable-free) instances contain a redex. Ground reducibility is decidable for ordinary rewriting (and finite R) [20, 49, 80]. What is the complexity of this test?

Problem 4. One of the outstanding open problems in typed lambda calculi is the following: Given a term in ordinary untyped lambda calculus, is it decidable whether it can be typed in the second-order $\lambda 2$ calculus? See [7].

Problem 5 (A. Meyer, R. de Vrijer). Do the surjective pairing axioms

$$\begin{aligned} D_1(Dxy) &= x \\ D_2(Dxy) &= y \\ D(D_1x)(D_2x) &= x \end{aligned}$$

conservatively extend $\lambda\beta\eta$-conversion on pure untyped lambda terms? More generally, is surjective pairing *always* conservative, or do there exist lambda theories, or extensions of Combinatory Logic for that matter, for which conservative extension by surjective pairing fails? (Surjective pairing is conservative over the pure $\lambda\beta$-calculus (see [92]). Of course, there are lots of other $\lambda\beta$, indeed $\lambda\beta\eta$, theories where conservative extension holds, simply because the theory consists of the valid equations in some λ model in which surjective pairing functions exist, e.g., D_∞.)

2.2 Normalization

Problem 6 (A. Middeldorp [71]). If R and S are two term-rewriting systems with disjoint vocabularies, such that for each of R and S any two convertible normal forms must be identical, then their union $R \cup S$ also enjoys this property [71]. Accordingly, we say that unicity of normal forms (UN) is a "modular" property of term-rewriting systems. "Unicity of normal forms with respect to reduction" ($UN^{\rightarrow}$) is the weaker property that any two normal forms of the same term must be identical. For non-left-linear systems, this property is not modular. The question remains: Is $UN^{\rightarrow}$ a modular property of left-linear term-rewriting systems?

Problem 7 (H. Comon, M. Dauchet). Is it possible to decide whether the set of ground normal forms with respect to a given (finite) term-rewriting system is a regular tree language? See [34, 62].

Problem 8 (A. Middeldorp). Is the decidability of strong sequentiality for orthogonal term rewriting systems NP-complete? See [39, 58].

Problem 9 (A. Middeldorp). Thatte [87] showed that an orthogonal constructor-based rewrite system is left-sequential if and only if it is strongly sequential. Does this equivalence extend to the whole class of orthogonal term-rewriting systems? If not, is left-sequentiality a decidable property of orthogonal systems? See also [58].

Problem 10 (J. R. Kennaway). Let a term-rewriting system (or more generally, a system with bound variables [57]) have the following properties: it is "finitely generated" (has finitely many function symbols and rules), it is "full" (its terms are all that can be formed from the function symbols), and it is Church-Rosser. Does it follow that it has a recursive, one-step, normalizing reduction strategy? (There are counterexamples if any of the three conditions is dropped.) Kennaway [50] showed that for "weakly" orthogonal systems the answer is yes. So, any counterexample must come from the murky world of non-orthogonal systems.

Problem 11 (A. Middeldorp [72]). A conditional term-rewriting system has rules of the form $p \Rightarrow l \rightarrow r$, which are only applied to instances of l for which the condition p holds. A "standard" (or "join") conditional system is one in which the condition p is a conjunction of conditions $u \downarrow v$, meaning that u and v have a common reduct (are "joinable"). Is unicity of normal forms (UN) a modular property of standard conditional systems?

2.3 Confluence

Problem 12. What is the complexity of the decision problem for the confluence of ground (variable-free) term-rewriting systems? Decidability was shown in [22, 78]; see also [23].

Problem 13 (J.-J. Lévy). By a lemma of G. Huet [38], left-linear term-rewriting systems are confluent if, for every critical pair $t \approx s$ (where $t = u[r\sigma] \leftarrow u[l\sigma] = g\tau \rightarrow d\tau = s$, for some rules $l \rightarrow r$ and $g \rightarrow d$), we have $t \rightarrow\!\!\Vert\; s$ (t reduces in one parallel step to s). (The condition $t \rightarrow\!\!\Vert\; s$ can be relaxed to $t \rightarrow\!\!\Vert\; r \;\Vert\!\!\leftarrow s$ for some r when the critical pair is generated from two rules overlapping at the roots; see [89].) What if $s \rightarrow\!\!\Vert\; t$ for every critical pair $t \approx s$? What if for every $t \approx s$ we have $s \rightarrow^{=} t$? (Here $\rightarrow^{=}$ is the reflexive closure of $\rightarrow$.) What if for every critical pair $t \approx s$, either $s \rightarrow^{=} t$ or $t \rightarrow^{=} s$? In the last case, especially, a confluence proof would be interesting; one would then have confluence after critical-pair completion without regard for termination. If these conditions are insufficient, the counterexamples will have to be (besides left-linear) non-right-linear, non-terminating, and non-orthogonal (have critical pairs). See [57].

Problem 14. Parallel rewriting with orthogonal term-rewriting systems is "subcommutative" (a "strong" form of confluence). Under which interesting syntactic restrictions do conditional rewrite systems enjoy the same property? It is known that orthogonal "normal" conditional rewriting systems (with conditions $u \rightarrow^{!} v$, where v is a ground normal form) are confluent, while "standard" (join) ones are not [13].

Problem 15 (Y. Toyama). Consider the following extension of Combinatory Logic (CL) with constants T (true), F (false), C (conditional):

$$\begin{array}{rrcl} & Ix & \to & x \\ & Kxy & \to & x \\ & Sxyz & \to & (xz)(yz) \\ & CTxy & \to & x \\ & CFxy & \to & y \\ x \leftrightarrow^* y \Rightarrow & Czxy & \to & x \end{array}$$

Is this (non-terminating) "semi-equational" (or "natural", as such are called in [31]) conditional rewrite system confluent? Note that if we take the above system plus the rule $x \leftrightarrow^* y \Rightarrow Czxy \to y$, the resulting conditional rewrite system *is* confluent (cf. [57, 93]).

Problem 16 (Y. Toyama). For a "normal" conditional term-rewriting system $R = \{s \to^! t \Rightarrow l \to r\}$, where t must be a ground normal from of s, we can consider the corresponding semi-equational conditional rewrite system $R' = \{s \leftrightarrow^* t \Rightarrow l \to r\}$. Under what conditions does confluence of R' imply confluence of R? In general, this is not the case, as can be seen from the following non-confluent system R (due to A. Middeldorp):

$$\begin{array}{rl} & a \to b \\ & a \to c \\ b \to^! c \Rightarrow & b \to c \end{array}$$

Problem 17 (R. de Vrijer). Is the following semi-equational conditional term rewriting system (a linearization of Combinatory Logic extended with surjective pairing) confluent:

$$\begin{array}{rrcl} & Ix & \to & x \\ & Kxy & \to & x \\ & Sxyz & \to & (xz)(yz) \\ & D_1(Dxy) & \to & x \\ & D_2(Dxy) & \to & y \\ x \leftrightarrow^* y \Rightarrow & D(D_1x)(D_2y) & \to & x \\ x \leftrightarrow^* y \Rightarrow & D(D_1x)(D_2y) & \to & y \end{array}$$

If yes, does an effective normal form strategy exist for it? See [59, 92].

Problem 18 (J. R. Kennaway, J. W. Klop, M. R. Sleep, F.-J. de Vries). If one wants to consider reductions of transfinite length in the theory of orthogonal term-rewriting systems, one has to be careful. In [51] it is shown that the confluence property "almost" holds for infinite rewriting with orthogonal term-rewriting systems. The only situation in which "infinitary confluence" may fail is when collapsing rules are present. (A rule $t \to s$ is "collapsing" if s is a variable.) Without collapsing rules, or even when only one collapsing rule of the form $f(x) \to x$ is present, infinitary confluence does hold. Now the notion of infinite reduction in [51] is based upon "strong convergence" of infinite sequences of terms in order to define (possibly infinite) limit terms. In related work, Dershowitz, et al. [29] use a more "liberal" notion of convergent sequences (which is referred to in [51] as "Cauchy convergence"). What is unknown (among other questions in this new area) is if this "almost-confluent" result is also valid for the more liberal convergent infinite reduction sequences?

2.4 Termination

Problem 19 (J.-J. Lévy). Can strong normalization (termination) of the typed lambda calculus be proved by a reasonably straightforward mapping from typed terms to a well-founded ordering? Note that the type structure can remain unchanged by β-reduction. The same question arises with polymorphic (second-order) lambda calculus.

Problem 20 (Y. Metivier [70]). What is the best bound on the length of a derivation for a one-rule length-preserving string-rewriting (semi-Thue) system? Is it $O(n^2)$ (n is the size of the initial term) as conjectured in [70], or $O(n^k)$ (k is the size of the rule) as proved there.

Problem 21 (M. Dauchet). Is termination of one linear (left and right) rule decidable? Left linearity alone is not enough for decidability [21].

Problem 22. Devise practical methods for proving termination of (standard) conditional rewriting systems. Part of the difficulty stems from the interdependence of normalization and termination.

Problem 23 (E. A. Cochin [18]). The following system [27], based on the "Battle of Hydra and Hercules" in [52], is terminating, but not provably so in Peano Arithmetic:

$$\begin{aligned} h(z,e(x)) &\rightarrow h(c(z),d(z,x)) \\ d(z,g(0,0)) &\rightarrow e(0) \\ d(z,g(x,y)) &\rightarrow g(e(x),d(z,y)) \\ d(c(z),g(g(x,y),0)) &\rightarrow g(d(c(z),g(x,y)),d(z,g(x,y))) \\ g(e(x),e(y)) &\rightarrow e(g(x,y)) \end{aligned}$$

Transfinite (ϵ_0-) induction is required for a proof of termination. Must any termination *ordering* have the Howard ordinal as its order type, as conjectured in [18]?

Problem 24. The existential fragment of the first-order theory of the "recursive path ordering" (with multiset and lexicographic "status") is decidable when the precedence on function symbols is total [19, 46], but is undecidable for arbitrary formulas. Is the existential fragment decidable for partial precedences?

2.5 Validity

Problem 25 (R. Treinen). Is the theory of multisets (AC) completely axiomatizable? In other words, is it decidable whether a first-order formula containing only equality as predicate symbol is valid in the algebra $\mathcal{T}(\mathcal{F})/AC(F)$? It is known that the Σ_3 fragment is undecidable when there are at least one unary function symbol (besides the AC one) and one constant; the Σ_1 fragment is decidable; the full theory is decidable even when there are no other symbols (besides constants) [90].

Problem 26. Let R be a term-rewriting or combinatory reduction system. Let "decreasing redexes" (DR) be the property that there is a map $\#$ from the set of redexes of R, to some well-founded linear order (or ordinal), satisfying:

- if in rewrite step $t \rightarrow_R t'$ redex r in t and redex r' in t' are such that r' is a descendant (or "residual") of r, then $\#r \geq \#r'$;
- if in rewrite step $t \rightarrow t'$ the redex r in t is reduced and r' in t' is "created" (t' is not the descendant of any redex in t), then $\#r > \#r'$.

Calling $\#r$ the "degree" of redex r, created redexes have a degree strictly less than the degree of the creator redex, while the degree of descendant redexes is not increased. The typical example is reduction in simply typed lambda calculus. In [55] it is proved that for orthogonal term-rewriting systems and combinatory reduction systems, decreasing redexes implies termination (strong normalization). Does this implication also hold for non-orthogonal systems? If not, can some decent subclasses be delineated for which the implication does hold?

Problem 27 (P. Lescanne). In [68] an extension of term embedding, called "well-rewrite orderings", was introduced, leading to an extension of the concept of simplification ordering. Can those ideas be extended to form the basis for some new kind of "recursive path ordering"?

Problem 28 (P. Lescanne). Polynomial and exponential interpretations have been used to prove termination. For the former there are some reasonable methods [11, 63] that can help determine if a particular interpretation decreases with each application of a rule. Are there other implementable methods suitable for exponential interpretations?

Problem 29. Any rewrite relation commutes with the strict-subterm relation; hence, the union of the latter with an arbitrary terminating rewrite relation is terminating, and also "fully invariant" (closed under instantiation). Is subterm the maximal relation with these properties? Is "encompassment" ("containment", the combination of subterm and subsumption) the maximal relation which preserves termination (without full invariance)?

Problem 30 (W. Snyder). What are the complexities of the various term ordering decision problems in the literature (see [25])? Determining if a precedence exists that makes two ground terms comparable in the recursive path ordering is NP-complete [61], but an inequality can be decided in $O(n^2)$, using a dynamic programming algorithm. Snyder [85] has shown that the lexicographic path ordering can be done in $O(n \log n)$ in the ground case with a total precedence, but the technique doesn't extend to non-total precedences or to terms with variables.

Problem 31. Is there a decidable uniform word problem for which there is no variant on the rewriting theme (for example, rewriting modulo a congruence with a decidable matching problem, or ordered rewriting) that can decide it—without adding new symbols to the vocabulary? There are decidable theories that cannot be decided with ordinary rewriting (see, for example, [86]); on the other hand, any theory with decidable word problem can be solved by ordered-rewriting with some ordered system for some conservative extension of the theory (that is, with new symbols) [30], or with a two-phased version of rewriting, wherein normal forms of the first system are inputs to the second [10].

Problem 32. Is there a finite term-rewriting system of some kind for free lattices?

Problem 33. Completion modulo associativity and commutativity (AC) [79] is probably the most important case of "extended completion"; the general case of finite congruence classes is treated in [43]. Adding an axiom (Z) for an identity element, however, gives rise to infinite classes. This case was viewed as conditional completion in [6], and solved completely in [45]. The techniques, however, do not carry over to completion with idempotence (I) added; how to handle ACZI-completion effectively is open.

Problem 34. Ordered rewriting computes a given convergent set of rewrite rules for an equational theory E and an ordering $>$ whenever such a set R exists for $>$, provided $>$ can be made total on ground terms. Unfortunately, this is not always possible, even if $>$ is derivability ($\rightarrow_R^+$) in R. Is there a set of inference rules that will always succeed in computing R whenever R exists for $>$?

2.6 Theorem Proving

Problem 35. Huet's proof [37] of the "completeness" of completion is predicated on the assumption that the ordering supplied to completion does not change during the process. Assume that at step i of completion, the ordering used is able to order the current rewriting relation $\rightarrow_{R_i}$, but not necessarily $\rightarrow_{R_k}$ for $k < i$ (since old rules may have been deleted by completion). Is there an example showing that completion is then incomplete (the persisting rules are not confluent)?

Problem 36 (H. Zhang). Since the work of Hsiang [36], several Boolean-ring based methods have been proposed for resolution-like first-order theorem proving. In [48], superposition rules were defined using multiple overlaps (requiring unifications of products of atoms). It is unknown whether single overlaps (requiring only unifications of atoms) are sufficient in these inference rules. Also, it is not known if unifications of maximal atoms (under a given term ordering) suffice. (The same problem for Hsiang's method was solved positively in [73, 94].) In other respects, too, the set of inference rules in [4, 48] may be larger than necessary and the simplification weaker than possible.

Problem 37 (U. Reddy, F. Bronsard). In [17] a rewriting-like mechanism for clausal reasoning called "contextual deduction" was proposed. It specializes "ordered resolution" by using pattern matching in place of unification, only instantiating clauses to match existing clauses. Does contextual deduction always terminate? (In [17] it was taken to be obvious, but that is not clear; see also [74].) It was shown in [17] that the mechanism is complete for refuting ground clauses using a theory that contains all its "strong-ordered" resolvents. Is there a notion of "complete theory" (like containing all strong-ordered resolvents not provable by contextual refutation) for which contextual deduction is complete for refutation of ground clauses?

2.7 Satisfiability

Problem 38 (J. Siekmann [83]). Is satisfiability of equations in the theory of distributivity (unification modulo a distributivity axiom) decidable?

Problem 39. Rules are given in [42] for computing dag-solved forms of unification problems in equational theories. The *Merge* rule $x \approx s, x \approx t \Rightarrow x \approx s, s \approx t$ given there assumes that s is not a variable and its size is less than or equal to that of t. Can this condition be improved by replacing it with the condition that the rule *Check** does not apply? (In other words, is *Check** complete for finding cycles when *Merge* is modified as above?)

Problem 40. Fages [32] proved that associative-commutative unification terminates when "variable replacement" is made after each step. Boudet, et al. [16] have proven that it terminates when variable replacement is postponed to the end. Does the same (or similar) set of transformation rules terminate with more flexible control?

Problem 41. The complexity of the theory of finite trees when there are finitely many symbols is known to be PSPACE-hard [69]. Is it in PSPACE? The same question applies to infinite trees.

Problem 42 (H. Comon). Given a first-order formula with equality as the only predicate symbol, can negation be effectively eliminated from an arbitrary formula ϕ when ϕ is equivalent to a positive formula? Equivalently, if ϕ has a finite complete set of unifiers, can they be computed? Special cases were solved in [20, 64].

Problem 43. Design a framework for combining constraint solving algorithms.

Problem 44 (H. Comon). "Syntactic" theories enjoy the property that a (semi-) unification algorithm can be derived from the axioms [42, 53]. This algorithm terminates for some particular cases (for instance, if all variable occurrences in the axioms are at depth at most one, and cycles have no solution) but does not in general. For the case of associativity and commutativity (AC), with a seven-axiom syntactic presentation, the derivation tree obtained by the non-deterministic application of the syntactic unification rules (*Decompose, Mutate, Merge, Coalesce, Check*, Delete*) in [42] can be pruned so as to become finite in most cases. The basic idea is that one unification problem (up to renaming) must appear infinitely times on every infinite branch of the tree (since there are finitely many axioms in the syntactic presentation). Hence, it should be possible to prune or freeze every infinite branch from some point on. The problem is to design such pruning rules so as to compute a finite derivation tree (hence, a finite complete set of unifiers) for every finitary unification problem of a syntactic equational theory.

3 Afterword

This list is by no means exhaustive. Please send any contributions by electronic or ordinary mail to the first author. We will periodically publicize new problems and solutions to old ones.

Acknowledgements

We thank all the individuals mentioned above who contributed questions, F.-J. de Vries for helping prepare this list, and Ron Book, for going out of his way to make it possible.

References

[1] H. Ait-Kaci and M. Nivat, eds. *Resolution of Equations in Algebraic Structures.* Vol. 2: Rewriting Techniques, Academic Press, New York, 1989.

[2] J. Avenhaus and K. Madlener. Term rewriting and equational reasoning. In R. B. Banerji, editor, *Formal Techniques in Artificial Intelligence: A Sourcebook*, pp. 1–41, Elsevier, Amsterdam, 1990.

[3] L. Bachmair. *Canonical Equational Proofs.* Birkhäuser, Boston, 1991. To appear.

[4] L. Bachmair and N. Dershowitz. Inference rules for rewrite-based first-order theorem proving. In *Proceedings of the Second IEEE Symposium on Logic in Computer Science*, pp. 331–337, Ithaca, NY, June 1987.

[5] L. Bachmair and J. Hsiang, eds. *Rewrite Techniques in Theorem Proving (Special Issue).* Vol. 11 (1) of *J. Symbolic Computation*, Academic Press, 1991. To appear.

[6] T. Baird, G. Peterson, and R. Wilkerson. Complete sets of reductions modulo Associativity, Commutativity and Identity. In N. Dershowitz, editor, *Proceedings of the Third International Conference on Rewriting Techniques and Applications*, pp. 29–44, Chapel Hill, NC, Apr. 1989. Vol. 355 of *Lecture Notes in Computer Science*, Springer, Berlin.

[7] H. P. Barendregt. Lambda calculi with types. In S. Abramsky, D. M. Gabbay, and T. S. E. Maibaum, eds., *Handbook of Logic in Computer Science*, Oxford University Press, Oxford, 1991. To appear.

[8] H. P. Barendregt. *The Lambda Calculus, its Syntax and Semantics.* North-Holland, Amsterdam, second edition, 1984.

[9] H. P. Barendregt. *The Typed Lambda Calculus, its Syntax and Semantics.* North-Holland, Amsterdam, 1991.

[10] G. Bauer. n-level rewriting systems. *Theoretical Computer Science*, 40:85–99, 1985.

[11] A. Ben Cherifa and P. Lescanne. Termination of rewriting systems by polynomial interpretations and its implementation. *Science of Computer Programming*, 9(2):137–159, Oct. 1987.

[12] B. Benninghofen, S. Kemmerich, and M. M. Richter. *Systems of Reductions.* Vol. 277 of *Lecture Notes in Computer Science*, Springer, Berlin, 1987.

[13] J. A. Bergstra and J. W. Klop. Conditional rewrite rules: Confluency and termination. *J. of Computer and System Sciences*, 32:323–362, 1986.

[14] G. Berry and J. Lévy. Mimimal and optimal computations of recursive programs. *J. of the Association for Computing Machinery*, 26:148–175, 1979.

[15] R. Book, ed. *Proceedings of the Fourth International Conference on Rewriting Techniques and Applications (Como, Italy)*, Springer, Berlin, Apr. 1991.

[16] A. Boudet, E. Contejean, and H. Devie. A new *ac* unification algorithm with an algorithm for solving diophantine equations. In *Proceedings of the Fifth Annual IEEE Symposium on Logic in Computer Science*, pp. 289–299, Philadelphia, PA, June 1990.

[17] F. Bronsard and U. S. Reddy. Conditional rewriting in Focus. In M. Okada, editor, *Proceedings of the Second International Workshop on Conditional and Typed Rewriting Systems*, Springer, Montreal, Canada, 1991. To appear.

[18] E. A. Cichon. *Bounds on Derivation Lengths from Termination Proofs.* Technical Report CSD-TR-622, Department of Computer Science, University of London, Surrey, England, June 1990.

[19] H. Comon. Solving inequations in term algebras (Preliminary version). In *Proceedings of the Fifth Annual IEEE Symposium on Logic in Computer Science*, pp. 62–69, Philadelphia, PA, June 1990.

[20] H. Comon. *Unification et Disunification: Théorie et Applications.* PhD thesis, l'Institut National Polytechnique de Grenoble, 1988.

[21] M. Dauchet. Simulation of Turing machines by a left-linear rewrite rule. In N. Dershowitz, editor, *Proceedings of the Third International Conference on Rewriting Techniques and Applications*, pp. 109–120, Chapel Hill, NC, Apr. 1989. Vol. 355 of *Lecture Notes in Computer Science*, Springer, Berlin.

[22] M. Dauchet, T. Heuillard, P. Lescanne, and S. Tison. Decidability of the confluence of finite ground term rewriting systems and of other related term rewriting systems. *Information and Computation*, 88(2):187–201, October 1990.

[23] M. Dauchet and S. Tison. The theory of ground rewrite systems is decidable. In *Proceedings of the Fifth IEEE Symposium on Logic in Computer Science*, pp. 242–248, Philadelphia, PA, June 1990.

[24] N. Dershowitz, ed. *Rewriting Techniques and Applications III (Special Issue). J. of Symbolic Computation*, Academic Press, 1992. To appear.

[25] N. Dershowitz. Termination of rewriting. *J. of Symbolic Computation*, 3(1&2):69–115, February/April 1987. Corrigendum: *4*, 3 (December 1987), 409–410.

[26] N. Dershowitz, ed. *Third International Conference on Rewriting Techniques and Applications*, Chapel Hill, NC, April 1989. Vol. 355 of *Lecture Notes in Computer Science*, Springer, Berlin.

[27] N. Dershowitz and J. Jouannaud. Rewrite systems. In J. van Leeuwen, editor, *Handbook of Theoretical Computer Science B: Formal Methods and Semantics*, chapter 6, pp. 243–320, North-Holland, Amsterdam, 1990.

[28] N. Dershowitz and J.-P. Jouannaud. Rewrite systems. In S. Shapiro, editor, *Encyclopedia of Artificial Intelligence*, Wiley, 1991. In preparation.

[29] N. Dershowitz, S. Kaplan, and D. A. Plaisted. Rewrite, Rewrite, Rewrite, Rewrite, Rewrite,.... *Theoretical Computer Science*, 1991. In press.

[30] N. Dershowitz, L. Marcus, and A. Tarlecki. *Existence, Uniqueness, and Construction of Rewrite Systems.* Technical Report ATR-85(8354)-7, Computer Science Laboratory, The Aerospace Corporation, El Segundo, CA, December 1985.

[31] N. Dershowitz and M. Okada. A rationale for conditional equational programming. *Theoretical Computer Science*, 75:111–138, 1990.

[32] F. Fages. Associative-commutative unification. *J. Symbolic Computation*, 3(3):257–275, June 1987.

[33] *Proceedings of an NSF Workshop on the Rewrite Rule Laboratory*, Schenectady, NY, Sep. 1983. Report 84GEN008, General Electric Research and Development (April 1984).

[34] R. Gilleron. Decision problems for term rewriting systems and recognizable tree languages. In *Proceddings of STACS*, 1991. To appear.

[35] J. R. Hindley and J. P. Seldin. *Introduction to Combinators and λ-Calculus.* Cambridge University Press, 1986.

[36] J. Hsiang. Refutational theorem proving using term-rewriting systems. *Artificial Intelligence*, 25:255–300, March 1985.

[37] G. Huet. A complete proof of correctness of the Knuth-Bendix completion algorithm. *J. Computer and System Sciences*, 23(1):11–21, 1981.

[38] G. Huet. Confluent reductions: Abstract properties and applications to term rewriting systems. *J. of the Association for Computing Machinery*, 27(4):797–821, Oct. 1980.

[39] G. Huet and J.-J. Lévy. Computations in orthogonal term rewriting systems. In J.-L. Lassez and G. Plotkin, eds., *Computational Logic: Essays in Honour of Alan Robinson*, MIT Press, Cambridge, MA, to appear.

[40] J.-P. Jouannaud, ed. *Rewriting Techniques and Applications.* Academic Press, London, 1987.

[41] J.-P. Jouannaud, ed. *Rewriting Techniques and Applications (Proceedings, Dijon, France)*, Springer, Berlin, May 1985. Vol. 202 of *Lecture Notes in Computer Science.*

[42] J.-P. Jouannaud and C. Kirchner. Solving equations in abstract algebras: A rule-based survey of unification. In J.-L. Lassez and G. Plotkin, eds., *Computational Logic: Essays in Honor of Alan Robinson*, MIT Press, Cambridge, MA, 1991. To appear.

[43] J.-P. Jouannaud and H. Kirchner. Completion of a set of rules modulo a set of equations. *SIAM J. on Computing*, 15:1155–1194, Nov. 1986.

[44] J.-P. Jouannaud and P. Lescanne. Rewriting systems. *Technology and Science of Informatics*, 6(3):181–199, 1987. French version: "La réécriture", *Technique et Science de l'Informatique* (1986), vol. 5, no. 6, pp. 433-452.

[45] J.-P. Jouannaud and C. Marché. Completion modulo associativity, commutativity and identity. In A. Miola, editor, *Proceedings of DISCO*, Capri, Italy, Apr. 1990. Vol. 429 in *Lecture Notes in Computer Science*, Springer, Berlin.

[46] J.-P. Jouannaud and M. Okada. Satisfiability of systems of ordinal notations enjoying the subterm property is decidable. 1991. Submitted.

[47] S. Kaplan and J.-P. Jouannaud, eds. *Conditional Term Rewriting Systems (Proceedings, Orsay, France, July 1987)*, Springer, Berlin, 1988. Vol. 308 of *Lecture Notes in Computer Science.*

[48] D. Kapur and P. Narendran. An equational approach to theorem proving in first-order predicate calculus. In *Proceedings of the Ninth International Joint Conference on Artificial Intelligence*, pp. 1146–1153, Los Angeles, CA, Aug. 1985.

[49] D. Kapur, P. Narendran, and H. Zhang. On sufficient completeness and related properties of term rewriting systems. *Acta Informatica*, 24(4):395–415, Aug. 1987.

[50] J. R. Kennaway. Sequential evaluation strategies for parallel-or and related reduction systems. *Annals of Pure and Applied Logic*, 43:31–56, 1989.

[51] J. R. Kennaway, J. W. Klop, M. R. Sleep, and F. J. de Vries. Transfinite reductions in orthogonal term rewriting systems (Extended abstract). In R. Book, editor, *Proceedings of the Fourth International Conference on Rewriting Techniques and Applications*, Como, Italy, Apr. 1991. In *Lecture Notes in Computer Science*, Springer, Berlin.

[52] L. Kirby and J. Paris. Accessible independence results for Peano arithmetic. *Bulletin London Mathematical Society*, 14:285–293, 1982.

[53] C. Kirchner. Computing unification algorithms. In *Proceedings of the First IEEE Symposium on Logic in Computer Science*, pp. 206–216, Cambridge, Massachussets, June 1986.

[54] C. Kirchner and H. Kirchner. *Rewriting: Theory and Applications.* North-Holland, 1991. In preparation.

[55] J. W. Klop. *Combinatory reduction systems.* Vol. 127 of *Mathematical Centre Tracts*, Mathematisch Centrum, 1980.

[56] J. W. Klop. Term rewriting systems: A tutorial. *Bulletin of the European Association for Theoretical Computer Science*, 32:143–183, June 1987.

[57] J. W. Klop. Term rewriting systems. In S. Abramsky, D. M. Gabbay, and T. S. E. Maibaum, eds., *Handbook of Logic in Computer Science*, Oxford University Press, Oxford, 1991. To appear.

[58] J. W. Klop and A. Middeldorp. *Sequentiality in Orthogonal Term Rewriting Systems.* Report CS-R8932, Centre for Mathematics and Computer Science, Amsterdam, 1989.

[59] J. W. Klop and R. C. de Vrijer. Unique normal forms for lambda calculus with surjective pairing. *Information and Computation*, 80:97–113, 1989.

[60] J. W. Klop and R. C. de Vrijer. *Term Rewriting Systems.* Cambridge University Press, Cambridge, 1991. In preparation.

[61] M. S. Krishnamoorthy and P. Narendran. On recursive path ordering. *Theoretical Computer Science*, 40:323–328, 1985.

[62] G. Kucherov. On relationship between term rewriting systems and regular tree languages. In R. Book, editor, *Proceedings of the Fourth International Conference on Rewriting Techniques and Applications*, Como, Italy, Apr. 1991. In *Lecture Notes in Computer Science*, Springer, Berlin.

[63] D. S. Lankford. *On Proving Term Rewriting Systems are Noetherian.* Memo MTP-3, Mathematics Department, Louisiana Tech. University, Ruston, LA, May 1979. Revised October 1979.

[64] J. Lassez and K. G. Marriott. Explicit representation of terms defined by counter examples. *J. Automated Reasoning*, 3(3):1–17, Sep. 1987.

[65] P. Le Chenadec. *Canonical Forms in Finitely Presented Algebras.* Pitman-Wiley, London, 1985.

[66] P. Lescanne, ed. *Rewriting Techniques and Applications (Proceedings, Bordeaux, France)*, Springer, Berlin, May 1987. Vol. 256 of *Lecture Notes in Computer Science.*

[67] P. Lescanne, ed. *Rewriting Techniques and Applications II (Special Issue).* Vol. 67 (2&3) of *Theoretical Computer Science*, North-Holland, 1989.

[68] P. Lescanne. Well rewrite orderings. In J. Mitchell, editor, *Proceedings of the Fifth IEEE Symposium on Logic in Computer Science*, pp. 239–256, Philadelphia, PA, 1990.

[69] M. J. Maher. Complete axiomatizations of the algebras of the finite, rational and infinite trees. In *Proceedings of the Third IEEE Symposium on Logic in Computer Science*, pp. 348–357, Computer Society Press, Edinburgh, UK, July 1988.

[70] Y. Métivier. Calcul de longueurs de chaînes de réécriture dans le monoïde libre. *Theoretical Computer Science*, 35(1):71–87, Jan. 1985.

[71] A. Middeldorp. Modular aspects of properties of term rewriting systems related to normal forms. In N. Dershowitz, editor, *Proceedings of the Third International Conference on Rewriting Techniques and Applications*, pp. 263–277, Chapel Hill, NC, Apr. 1989. Vol. 355 of *Lecture Notes in Computer Science*, Springer, Berlin.

[72] A. Middeldorp. *Modular Properties of Term Rewriting Systems.* PhD thesis, Vrije Universiteit, Amsterdam, 1990.

[73] J. Müller and R. Socher-Ambrosius. *Topics in completion theorem proving.* SEKI-Report SR-88-13, Fachbereich Informatik, Universität Kaiserslautern, Kaiserslautern, West Germany, 1988. To appear in *J. of Symbolic Computation.*

[74] R. Nieuwenhuis and F. Orejas. Clausal rewriting. In S. Kaplan and M. Okada, eds., *Extended Abstracts of the Second International Workshop on Conditional and Typed Rewriting Systems*, pp. 81–88, Concordia University, Montreal, Canada, June 1990. Revised version to appear in *Lecture Notes in Computer Science*, Springer, Berlin.

[75] M. J. O'Donnell. *Equational Logic as a Programming Language*. MIT Press, Cambridge, MA, 1985.

[76] M. J. O'Donnell. Programming with equations. In D. M. Gabbay, C. J. Hogger, and J. A. Robinson, eds., *Handbook of Logic in Artificial Intelligence and Logic Programming*, Oxford University Press, Oxford, 1991. To appear.

[77] M. Okada, ed. *Proceedings of the Second International Workshop on Conditional and Typed Rewriting Systems (Montreal, Canada)*, 1991. *Lecture Notes in Computer Science*, Springer, Berlin.

[78] M. Oyamaguchi. The Church-Rosser property for ground term rewriting systems is decidable. *Theoretical Computer Science*, 49(1), 1987.

[79] G. E. Peterson and M. E. Stickel. Complete sets of reductions for some equational theories. *J. of the Association for Computing Machinery*, 28(2):233–264, Apr. 1981.

[80] D. A. Plaisted. Semantic confluence tests and completion methods. *Information and Control*, 65(2/3):182–215, May/June 1985.

[81] D. A. Plaisted. Term rewriting systems. In D. M. Gabbay, C. J. Hogger, and J. A. Robinson, eds., *Handbook of Logic in Artificial Intelligence and Logic Programming*, Oxford University Press, Oxford, 1991. To appear.

[82] D. A. Plaisted. Equational reasoning and term rewriting systems. In D. M. Gabbay, C. J. Hogger, and J. A. Robinson, eds., *Handbook of Logic in Artificial Intelligence and Logic Programming*, Oxford University Press, Oxford, 1991. To appear.

[83] J. Siekmann. Universal unification. In R. E. Shostak, editor, *Proceedings of the Seventh International Conference on Automated Deduction*, pp. 1–42, Napa, CA, May 1984. Vol. 170 of *Lecture Notes in Computer Science*, Springer, Berlin.

[84] M. R. Sleep. *An Introduction to Rewriting*. John Wiley, Chichester, England, 1992. In preparation.

[85] W. Snyder. *(Computing the Lexicographic Path Ordering)*. Technical Report, Boston University, Boston, MA, 1990.

[86] C. Squier. Word problems and a homological finiteness condition for monoids. *J. of Pure and Applied Algebra*, 1991. To appear.

[87] S. Thatte. A refinement of strong sequentiality for term rewriting with constructors. *Information and Computation*, 72:46–65, 1987.

[88] M. Thomas, ed. *Term Rewriting (Special Issue)*. Vol. 34 (1) of *Computer J.*, Feb. 1991.

[89] Y. Toyama. Commutativity of term rewriting systems. In K. Fuchi and L. Kott, eds., *Programming of Future Generation Computers II*, pp. 393–407, North-Holland, 1988.

[90] R. Treinen. A new method for undecidability proofs of first order theories. *LNCS* 472, 48–62.

[91] M. Venturini-Zilli. Reduction graphs in the Lambda Calculus. *Theoretical Computer Science*, 29:251–275, 1984.

[92] R. C. de Vrijer. Extending the lambda calculus with surjective pairing is conservative. In *Proceedings of the Fourth IEEE Symposium on Logic in Computer Science*, pp. 204–215, 1989.

[93] R. C. de Vrijer. *Unique normal forms for Combinatory Logic with Parallel Conditional, a case study in conditional rewriting*. Technical Report, Free University, Amsterdam, 1990.

[94] H. Zhang. *A New Strategy for the Boolean Ring Based Approach to First Order Theorem Proving*. Technical Report, Department of Computer Science, The University of Iowa, 1991.

Author Index

'ol. 437: D. Kumar (Ed.), Current Trends in SNePS – Semantic Network ?rocessing System. Proceedings, 1989. VII, 162 pages. 1990. (Subseries LNAI).

'ol. 438: D. H. Norrie, H.-W. Six (Eds.), Computer Assisted Learning – CCAL '90. Proceedings, 1990. VII, 467 pages. 1990.

'ol. 439: P. Gorny, M. Tauber (Eds.), Visualization in Human-Computer nteraction. Proceedings, 1988. VI, 274 pages. 1990.

ol. 440: E. Börger, H. Kleine Büning, M. M. Richter (Eds.), CSL '89. Proeedings, 1989. VI, 437 pages. 1990.

ol. 441: T. Ito, R. H. Halstead, Jr. (Eds.), Parallel Lisp: Languages and Sysms. Proceedings, 1989. XII, 364 pages. 1990.

ol. 442: M. Main, A. Melton, M. Mislove, D. Schmidt (Eds.), Mathematical oundations of Programming Semantics. Proceedings, 1989. VI, 439 ages. 1990.

ol. 443: M. S. Paterson (Ed.), Automata, Languages and Programming. roceedings, 1990. IX, 781 pages. 1990.

ol. 444: S. Ramani, R. Chandrasekar, K. S. R. Anjaneyulu (Eds.), Knowdge Based Computer Systems. Proceedings, 1989. X, 546 pages. 1990. Subseries LNAI).

ol. 445: A. J. M. van Gasteren, On the Shape of Mathematical Arguments. III, 181 pages. 1990.

ol. 446: L. Plümer, Termination Proofs for Logic Programs. VIII, 142 pages. 990. (Subseries LNAI).

ol. 447: J. R. Gilbert, R. Karlsson (Eds.), SWAT 90. 2nd Scandinavian Vorkshop on Algorithm Theory. Proceedings, 1990. VI, 417 pages. 1990.

ol. 448: B. Simons, A. Spector (Eds.), Fault-Tolerant Distributed Comuting. VI, 298 pages. 1990.

ol. 449: M. E. Stickel (Ed.), 10th International Conference on Automated eduction. Proceedings, 1990. XVI, 688 pages. 1990. (Subseries LNAI).

ol. 450: T. Asano, T. Ibaraki, H. Imai, T. Nishizeki (Eds.), Algorithms. Proeedings, 1990. VIII, 479 pages. 1990.

ol. 451: V. Mařík, O. Štěpánková, Z. Zdráhal (Eds.), Artificial Intelligence Higher Education. Proceedings, 1989. IX, 247 pages. 1990. (Subseries NAI).

ol. 452: B. Rovan (Ed.), Mathematical Foundations of Computer Science 990. Proceedings, 1990. VIII, 544 pages. 1990.

ol. 453: J. Seberry, J. Pieprzyk (Eds.), Advances in Cryptology – USCRYPT '90. Proceedings, 1990. IX, 462 pages. 1990.

ol. 454: V. Diekert, Combinatorics on Traces. XII, 165 pages. 1990.

ol. 455: C. A. Floudas, P. M. Pardalos, A Collection of Test Problems for onstrained Global Optimization Algorithms. XIV, 180 pages. 1990.

ol. 456: P. Deransart, J. Maluszyński (Eds.), Programming Language nplementation and Logic Programming. Proceedings, 1990. VIII, 401 ages. 1990.

ol. 457: H. Burkhart (Ed.), CONPAR '90 – VAPP IV. Proceedings, 1990. IV, 900 pages. 1990.

ol. 458: J. C. M. Baeten, J. W. Klop (Eds.), CONCUR '90. Proceedings, 990. VII, 537 pages. 1990.

ol. 459: R. Studer (Ed.), Natural Language and Logic. Proceedings, 989. VII, 252 pages. 1990. (Subseries LNAI).

ol. 460: J. Uhl, H. A. Schmid, A Systematic Catalogue of Reusable bstract Data Types. XII, 344 pages. 1990.

ol. 461: P. Deransart, M. Jourdan (Eds.), Attribute Grammars and their pplications. Proceedings, 1990. VIII, 358 pages. 1990.

ol. 462: G. Gottlob, W. Nejdl (Eds.), Expert Systems in Engineering. Proeedings, 1990. IX, 260 pages. 1990. (Subseries LNAI).

ol. 463: H. Kirchner, W. Wechler (Eds.), Algebraic and Logic Proramming. Proceedings, 1990. VII, 386 pages. 1990.

ol. 464: J. Dassow, J. Kelemen (Eds.), Aspects and Prospects of Theorecal Computer Science. Proceedings, 1990. VI, 298 pages. 1990.

ol. 465: A. Fuhrmann, M. Morreau (Eds.), The Logic of Theory Change. roceedings, 1989. X, 334 pages. 1991. (Subseries LNAI).

ol. 466: A. Blaser (Ed.), Database Systems of the 90s. Proceedings, 990. VIII, 334 pages. 1990.

Vol. 467: F. Long (Ed.), Software Engineering Environments. Proceedings, 1989. VI, 313 pages. 1990.

Vol. 468: S. G. Akl, F. Fiala, W. W. Koczkodaj (Eds.), Advances in Computing and Information – ICCI '90. Proceedings, 1990. VII, 529 pages. 1990.

Vol. 469: I. Guessarian (Ed.), Semantics of Systems of Concurrent Processes. Proceedings, 1990. V, 456 pages. 1990.

Vol. 470: S. Abiteboul, P. C. Kanellakis (Eds.), ICDT '90. Proceedings, 1990. VII, 528 pages. 1990.

Vol. 471: B. C. Ooi, Efficient Query Processing in Geographic Information Systems. VIII, 208 pages. 1990.

Vol. 472: K. V. Nori, C. E. Veni Madhavan (Eds.), Foundations of Software Technology and Theoretical Computer Science. Proceedings, 1990. X, 420 pages. 1990.

Vol. 473: I. B. Damgård (Ed.), Advances in Cryptology – EUROCRYPT '90. Proceedings, 1990. VIII, 500 pages. 1991.

Vol. 474: D. Karagiannis (Ed.), Information Systems and Artificial Intelligence: Integration Aspects. Proceedings, 1990. X, 293 pages. 1991.

Vol. 475: P. Schroeder-Heister (Ed.), Extensions of Logic Programming. Proceedings, 1989. VIII, 364 pages. 1991. (Subseries LNAI).

Vol. 476: M. Filgueiras, L. Damas, N. Moreira, A. P. Tomás (Eds.), Natural Language Processing. Proceedings, 1990. VII, 253 pages. 1991. (Subseries LNAI).

Vol. 477: D. Hammer (Ed.), Compiler Compilers. Proceedings, 1990. VI, 227 pages. 1991.

Vol. 478: J. van Eijck (Ed.), Logics in AI. Proceedings, 1990. IX, 562 pages. 1991. (Subseries LNAI).

Vol. 479: H. Schmidt, Meta-Level Control for Deductive Database Systems. VI, 155 pages. 1991.

Vol. 480: C. Choffrut, M. Jantzen (Eds.), STACS 91. Proceedings, 1991. X, 549 pages. 1991.

Vol. 481: E. Lang, K.-U. Carstensen, G. Simmons, Modelling Spatial Knowledge on a Linguistic Basis. IX, 138 pages. 1991. (Subseries LNAI).

Vol. 482: Y. Kodratoff (Ed.), Machine Learning – EWSL-91. Proceedings, 1991. XI, 537 pages. 1991. (Subseries LNAI).

Vol. 483: G. Rozenberg (Ed.), Advances in Petri Nets 1990. VI, 515 pages. 1991.

Vol. 484: R. H. Möhring (Ed.), Graph-Theoretic Concepts in Computer Science. Proceedings, 1990. IX, 360 pages. 1991.

Vol. 485: K. Furukawa, H. Tanaka, T. Fujisaki (Eds.), Logic Programming '89. Proceedings, 1989. IX, 183 pages. 1991. (Subseries LNAI).

Vol. 487: A. Bode (Ed.), Distributed Memory Computing. Proceedings, 1991. XIII, 506 pages. 1991.

Vol. 488: R. V. Book (Ed.), Rewriting Techniques and Applications. Proceedings, 1991. VII, 458 pages. 1991.